California Na

INVENTORY

OF RARE AND ENDANGERED VASCULAR PLANTS OF CALIFORNIA

Amsinckia grandiflora

THE CALIFORNIA NATIVE PLANT SOCIETY

The California Native Plant Society is an organization of laypersons and professionals united by an interest in California plants. The Society is open to everyone, and exists to conserve and protect California native plants in their natural habitat, and to educate members and the public about the California flora. CNPS achieves this through the publication of educational materials, distribution of scientific information, promotion of legal protection for rare plants and their habitats, local advocacy of plant protection, and fostering of appropriate horticultural use of California native plants.

Inquiries concerning membership, other Society publications, and activities should be directed to CNPS, 1722 J St., Suite 17, Sacramento, CA 95814.

The California Native Plant Society appreciates significant financial assistance from James and Carmen Hickling and the Pacific Gas and Electric Company in support of this project.

CALIFORNIA NATIVE PLANT SOCIETY'S

INVENTORY

OF RARE AND ENDANGERED VASCULAR PLANTS OF CALIFORNIA

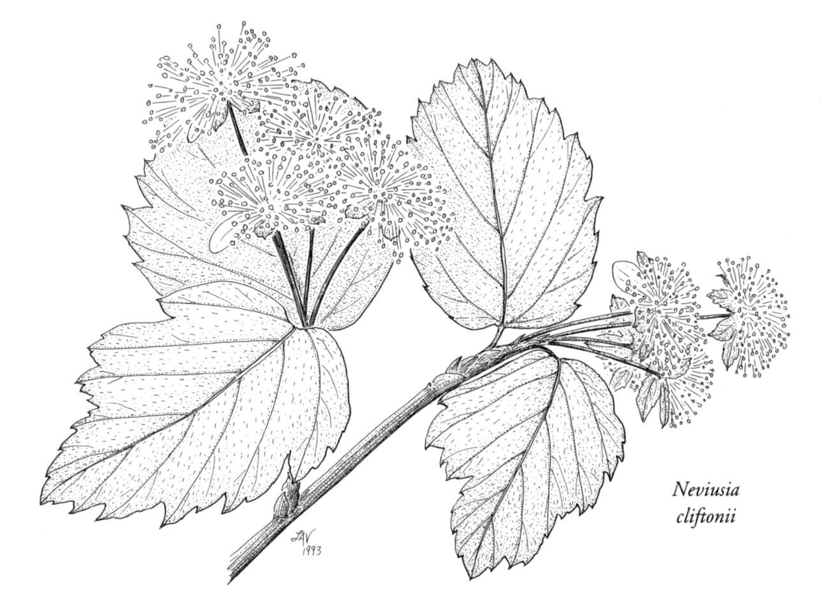

Edited by

MARK W. SKINNER
Botanist
California Native Plant Society

and

BRUCE M. PAVLIK
Associate Professor of Biology
Mills College

Illustrations by

LINDA ANN VOROBIK

Photographs by

MARK W. SKINNER

FEBRUARY 1994 / SPECIAL PUBLICATION NO. 1 / FIFTH EDITION

Inventory of Rare and Endangered Vascular Plants of California
CNPS Special Publication No. 1 (Fifth Edition)

Published February 1994 by The California Native Plant Society

Design by Beth Hansen
Typesetting by Keith Ball
Color separations by Colortec
Printed and bound in the United States by Griffin Printing

Library of Congress Cataloging in Publication Data:

Inventory of Rare and Endangered Vascular Plants of California /
California Native Plant Society Special Publication No. 1 (Fifth Edition) /
edited by Mark W. Skinner, Bruce M. Pavlik
Includes photographic plates, bibliography, and appendices.
1. Rare Plants — California. 2. Endangered Species — California.
3. Plant Conservation — California.
I. Skinner, Mark W., 1955— . II. Pavlik, Bruce. M., 1953— .
III. California Native Plant Society.
IV. Series: Special Publication (California Native Plant Society): No. 1.
ISBN 0-943460-18-2
Library of Congress Catalog Card Number: 93-74796

Introduction

One of California's greatest natural treasures is its plants. Our state's mountain ranges, deserts, and extensive coastline, along with its unusual summer-dry climate, set the stage for the development of a complex and fascinating flora. The numbers themselves are impressive. About 6300 flowering plants, gymnosperms, ferns, and fern allies are native here, more than we find in the entire northeastern United States and adjacent Canada, an area ten times larger than California. Another thousand plants are weedy introductions to the state or escapes from gardens and agricultural fields. Although Texas has more genera and families, California's is the largest state flora in the nation.

One of the California flora's outstanding features is that more than one-third (36%) of its native species, subspecies, and varieties are endemics, plants that are restricted to a particular locality or habitat within the state. If we define the flora in terms of the California Floristic Province— a unit including the Klamath Region of the northwestern part of the state and adjacent southwestern Oregon, along with that portion of California west of the Cascade-Sierra Nevada axis, and a part of northern Baja California— then the degree of endemism (about 48%) is remarkable for a continental flora. Most of these endemics are rare or uncommon plants.

It has been argued that many endemics are plants becoming extinct because of old age, and that we are seeing their demise after a long evolutionary history. Others argued, quite to the contrary, that rare plants are merely evolutionary children that have not had time to spread. We now realize that while some endemics, such as the coast redwood, are indeed ancient relicts, most are of relatively recent origin (see *Rarity in Vascular Plants*, below).

Why does California have so many rare plants, especially so many that are endemic? The primary factor may be our climate, a regime of dry summers and cool, wet winters. Only in the Mediterranean Basin, Chile, and parts of Australia and South Africa do we find similar climates. Each is famous for its array of rare and endemic plants adapted to this unusual climatic regime. A second factor is topography and latitudinal diversity. Within the boundaries of this state are extremes—from low, subtropical desert to habitats above timberline, from northern temperate rain forests in Del Norte County to arid succulent scrub in San Diego County. A third factor is the richness of geological formations and the resulting diversity of soil types. The broken topography of the state, combined with a tremendous diversity of climates and soil types, has resulted in ample opportunities for genetic isolation and speciation. The result has been impressive adaptive radiations within such groups as *Arctostaphylos*, *Astragalus*, *Castilleja*, *Eriogonum*, *Gilia*, *Lupinus*, *Mimulus*, *Phacelia*, and many genera in the Asteraceae, to name only a few. Most of our rare plants, then, are specialists, adapted to a particular combination of climate and substrate. For a more detailed treatment of the California flora, endemism, and rare plants, see Barbour and Major (1988), Elias (1987), Hickman (1993), Lewis (1972), Ornduff (1974), Raven and Axelrod (1978), Smith and Sawyer (1988), Stebbins (1978 a & b, 1980), and Stebbins and Major (1965).

California's flora, especially its rare plants, are increasingly threatened by the spread of urbanization, by our conversion of land to agriculture, by alteration of natural hydrological cycles, by recreational activities and non-native plants and animals, and by pollution. The unique habitats that harbor rare plants are being destroyed. For example, 90-95% of vernal pools in the State are gone, and native grasslands in the Central Valley occupy but 1% of their former extent. About forty plants probably became extinct in the last century. Hundreds more are endangered and could perish if present trends continue. Arguments for conservation of rare plants, animals, and natural communities range from aesthetic, to moral, to economic or ecological. This is not the proper forum for an extended discussion of the value of rare plants and the merits of their preservation, so we refer readers to several excellent reviews that are cited in the *Bibliography*. We strongly believe, however, that failure to conserve the biological diversity of our planet may well prove disastrous.

One of the most effective ways of preserving California's native flora is to assemble, evaluate, and distribute information on our rare and endangered plants. The purpose of this *Inventory* is to summarize this information, and in doing so, to promote thoughtful conservation planning. The material contained herein has been contributed and reviewed by professional and amateur botanists from throughout the state. Without their cooperation, this book would not have been possible.

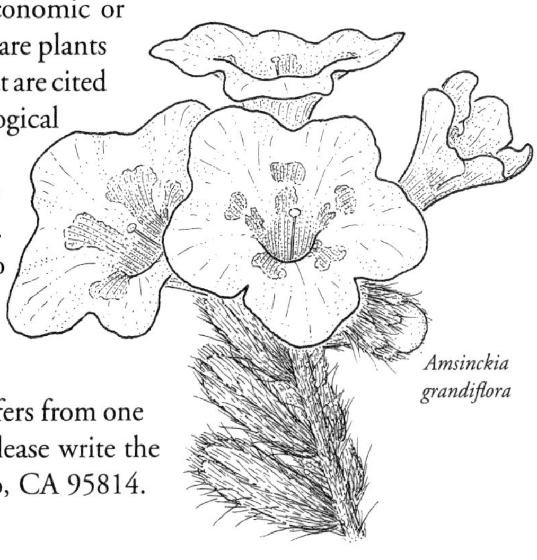

Amsinckia grandiflora

We welcome suggested additions, deletions, nomenclatural changes, and transfers from one list to another. We are especially eager to have errors brought to our attention. Please write the CNPS Botanist, California Native Plant Society, 1722 J St., Suite 17, Sacramento, CA 95814.

Rarity in Vascular Plants

Peggy L. Fiedler

Vascular plant species can be rare for an astounding variety of reasons. Broadly speaking, however, a species is either rare because it lives in a very limited habitat (natural rarity), or because its habitat has been converted by humans to other uses (anthropogenic rarity). Natural rarities are those species that have always been rare during their evolutionary history, or currently are rare by today's standards. Anthropogenic rarities are those species that were formerly widespread, but through negative interactions with human populations, are either greatly fragmented or restricted to a few small, imperiled populations. Each general class of rarity has its own set of consequences that must be considered for a species' conservation and management.

Patterns of Rarity

What does it mean to be rare? Essentially, using the word "rare" makes a statement about the geographic distribution and population abundance of a particular species. Rarity, in fact, describes three very different biological possibilities (Figure 1). A rare taxon can be A) broadly distributed, but never abundant where found (e.g., *Cypripedium californicum*); B) narrowly distributed or clumped, and abundant where found (e.g., *Limnanthes bakeri*); or, C) narrowly distributed or clumped, and not abundant where found (e.g., *Tuctoria mucronata*). Geographic distributions of rarity can also include a temporal dimension, such that a rare species is defined by its abundance, distribution, and persistence through evolutionary time (Fiedler and Ahouse 1992). Patterns of distribution and abundance that define rarity have been generally described for animals (Mayr 1963) and plants (Drury 1974, 1980), but seldom have they been applied to specific problems in biological conservation.

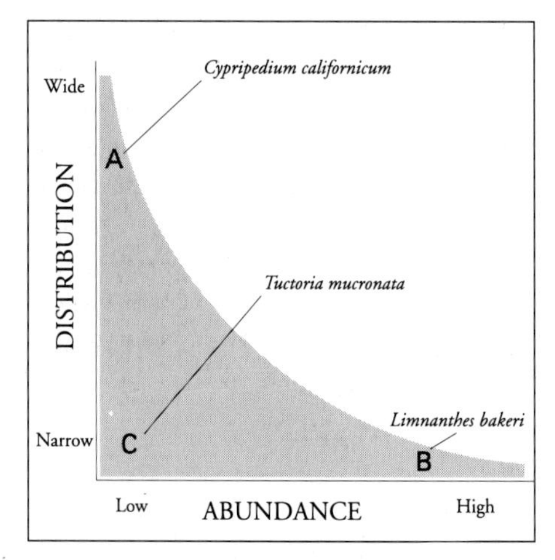

Figure 1. The term "rarity" is used to describe a spectrum of combinations of abundance and distribution patterns. All of the shaded area is often considered "rare" (Adapted from Fiedler and Ahouse 1992).

A particularly persuasive classification of natural rarities was proposed by Rabinowitz (1981), who explicitly tied habitat specificity to geography and abundance (Figure 2). This system identifies seven possible forms of rarity, six of which can be used to classify rarities for floristic regions, as has been done for the British Isles (Rabinowitz, Cairns, and Dillon 1986), but not yet for California. One of the limitations of this scheme, however, is that the causes of rarity (e.g., limited habitats) are not distinguished from the consequences of rarity (e.g., decreased genetic variation), and both may have implications for rare plant management. The goal of conserving biological diversity will be best served if politicians, developers, and conservationists recognize that there are different types of rarity among plants, and that each may require a different form of protection and management.

Causes of Rarity

Although rare plants can be classified by different patterns of distribution and abundance, as is done above, this does not explain the many biological and anthropogenic factors causing these patterns. Fiedler and Ahouse (1992) identified thirteen general categories of factors that probably contribute to rarity, including ten that were related to the biology of rare species and three that were related to human activities. Within each category, they listed up to ten specific causes of rarity. For example, the general category "Human Uses" includes the specific causes of horticultural trade, aboriginal uses, a species' role in ancient and/or modern medicine, and a species' role in past or present industry; the general category of "stochasticity" includes both demographic and environmental causes. There are literally dozens of causes that contribute to plant rarity, but the brief discussion below emphasizes only the major biological explanations.

In California, many vascular plant species are rare because they are either new species (i.e., "neoendemics") or old ones (i.e., "paleoendemics"). Neoendemics are frequently found in geologically youthful habitats, and often their rarity is partly a function of their youth: in some cases these plants have not had time to greatly expand their range from their point of origin to their climatic limits (Stebbins and Major 1965). Such taxa include members of the *Limnanthes floccosa* complex from vernal pools, and *Linanthus arenicola* and *Oenothera californica* ssp. *eurekensis* from the Mojave Desert. *Clarkia lingulata*, of the Sierra Nevada foothills, is another well-known neoendemic. In contrast, some of our most famous rare or restricted species, such as *Abies bracteata*, *Carpenteria californica*, *Lyonothamnus floribundus*, *Pinus radiata*, *P. torreyana*, and *Sequoiadendron giganteum*, are paleoendemics which were once more broadly distributed, but have retreated to their current ranges in response to climatic change.

Stebbins and Major (1965) presented an extensive analysis of California's endemic plants, many of them rarities in the

Geographic Range:	Large		Small	
Habitat Specificity:	**Wide**	**Narrow**	**Wide**	**Narrow**
	Widespread taxa	Predictable taxa	Unlikely endemic taxa	Endemic taxa
Large, Dominant, Local Populations	COMMON PLANTS	*Downingia humilis*	*Allium munzii*	*Calochortus tiburonensis*
Small, Non-dominant, Local Populations	SPARSE PLANTS *Achnatherum lettermanii*	*Torreya californica*	NON-EXISTENT?	*Lathyrus jepsonii* ssp. *jepsonii*

Figure 2. Seven forms of rarity, after Rabinowitz (1981), with examples from the California flora.

flora. Their classification is based upon age, systematic position, and cytology, and includes the paleoendemics, as well as schizoendemics, patroendemics, and apoendemics. Most representatives of the last three categories are neoendemics, though some are of moderate or even relatively old age.

Schizoendemics have more or less simultaneously diverged from a common ancestor, as have many of the rare taxa in *Ceanothus* sect. *Cerastes* such as *C. roderickii* and *C. ophiochilus.* The remaining two categories are especially important because they contain most of California's rare species. Patroendemics are diploid species of limited geographic distribution that are related to, and probably ancestral to, a more recent and widespread species. Patroendemics and their probable descendants include, among many others, the very restricted endemic *Clarkia imbricata* (n = 8) and its derived polyploid *C. purpurea* (n = 26); *Galium clematis* (n = 11) and *G. californicum* (n = 44, 66); and *Tonestus* (formerly *Haplopappus*) *eximius* (n = 9) and *T. peirsonii* (n = 45). In contrast to patroendemics, apoendemics are defined as polyploids of limited geographic distribution that are either sympatric or parapatric to more widespread diploid (or lower degree polyploid) species, from which they are likely descended. Apoendemics pairs include the rare *Dudleya saxosa* ssp. *saxosa* (n = 68, 85) and its probable parent *D. saxosa* ssp. *aloides* (n = 17), which is also a rare plant; *Lomatium repostum* (n = 22) and *L. lucidum* (n = 11); and *Penstemon heterodoxus* var. *shastensis* (n = 16) and the more widespread *P. heterodoxus* var. *heterodoxus* (n = 8). Raven and Axelrod (1978) thoroughly summarize the origin of California endemics, and Kruckeberg and Rabinowitz (1985) discuss the biology of endemic plant species in detail.

Many rare species in California and elsewhere are restricted to specific soil types and are therefore considered "edaphic endemics." The mechanism of plant adaptation and subsequent restriction to unusual soils which generates edaphic endemics is complex, but typically involves physiological tolerance to mineral imbalances or toxic minerals. California's serpentinite flora is well-known worldwide, and contains both common and rare species, including 282 taxa in this *Inventory* (see *Ecological Characteristics of California's Rare Plants*, below). Familiar rarities include *Calochortus tiburonensis, Streptanthus morrisonii*, and *Hesperolinon*, a genus composed almost exclusively of rare serpentinite endemics. Kruckeberg's (1984) monograph on California serpentines provides an excellent starting point for understanding the evolution, distribution, and management of rare serpentinite plant species.

In conclusion, there are many reasons why California has so many rare species, and likely as many ideas about why any vascular plant species might be rare (Fiedler 1986). Only infrequently does a single "cause" by itself truly explain why a species is rare. Indeed, many rare and endangered species in California that began as natural rarities have, through one form or another of human-induced detrimental changes in their populations and/or habitat, become anthropogenic rarities needing immediate protection and recovery.

And although some taxa are rare because of some particular aspect of their biology (such as poor seed dispersal or germination), and still others may be rare because they are old, genetically depauperate, and "on their way out" (Stebbins 1942), it is impossible to generalize about why species are rare. It is important, therefore, to understand the biology of a rare taxon, to know the genesis of its rarity, and to understand current threats to a species and its habitat. Armed with such knowledge, rare plant taxa in California and elsewhere can be appropriately managed and will have much better chances for long-term survival.

Peggy L. Fiedler is an Associate Professor in the Department of Biology, San Francisco State University, San Francisco, CA 94132.

Ecological Characteristics of California's Rare Plants

Bruce M. Pavlik and Mark W. Skinner

The rare plants of California are known to us in very general terms. They have been found, named, described, watched, sometimes utilized, and in a few cases, studied. But overall, many of the specifics regarding these unusual and remarkable organisms remain unknown. How do they live their lives? What are the events that control growth or promote stability in their populations? What other plants, animals, and microbes do they live with or depend on? Are there certain life history traits or habitat preferences that predispose taxa to rarity, endangerment, or extinction? Only when our rare flora is understood ecologically will we have a firm foundation for making the crucial decisions that can secure its future.

This edition of the *Inventory*, along with the new *Electronic Inventory* (see *The CNPS Electronic Inventory...*, below) includes new data and new ways of accessing data that will enhance our ecological understanding of rare plants in the California flora. For example, we have added data fields to describe the duration, growth form, leaf type, mode of nutrition, substrate preference, and blooming period for each taxon. This kind of information can be immediately useful for conducting better searches during environmental impact studies, giving consultants and field biologists a more precise idea of what to look for, when, and how likely it is they will find it at any given time. But beyond its practical applications, this information can also be used to answer questions of concern to ecologists, conservation biologists, and reserve managers. What are ecological characteristics of California's rare plants? Can this knowledge be used proactively for conservation purposes? Will our existing system of nature reserves slow the erosion of plant diversity by protecting the natural communities that contain the most rare plants? Here we present a preliminary analysis.

Duration as an Ecological Characteristic

The data field "duration" describes the longevity of individual plants in the active phase of their life cycle. Annual plants, for which our flora is famous, grow rapidly from seed to reproduce in less than a year. Perennial plants grow steadily and reproduce over two or more years. According to Raunkier (1934) only about 13% of the world's flora are annuals, while 87% are perennial. Raven and Axelrod (1978) determined that 29% of all plants in California are annuals. If we examine *Inventory* List 1A, a relatively high proportion (53%) of plants presumed to be extinct are annuals (Table 1). Although List 2 is deficient in annuals as one might predict (these are mostly plants from the perennial-rich continental flora of other states), the other lists conform to the California norm (26-36% annuals).

This "annual enrichment" on List 1A raises the intriguing question of whether a life cycle with short duration predisposes plants to extinction. Is it trickier to be an annual or perhaps riskier to depend on seeds in the soil for withstanding adversity? Populations of small, short-lived organisms (plants or animals) undergo frequent and sometimes drastic changes in size for a wide variety of reasons. Such changes can exacerbate the effects of competition (annual plants never "hold their ground" for very long), reduced pollen flow (especially in obligate outcrossers), and random but natural events that impact the population (e.g., the chance arrival of hungry herbivores). A growing body of research in conservation biology does suggest that annual plants are more prone to extinction than their perennial relatives, but that the introduction of non-native species, fire suppression, overgrazing, and other byproducts of human activity are usually the most important factors. Most of the extinct annuals in California were subjected to a combination of anthropogenic factors, and dozens of annual plants on List 1B occur in highly modified habitats and are now in serious decline. These plants may swell the ranks of List 1A in future editions of this *Inventory*.

Habitat Preference

Human activity is likely to be the other chief cause of "annual enrichment" for an obvious reason: annual plants tend to be found in lowland habitats that are prone to agricultural and urban development. Lowland habitats are the first to be altered or destroyed as pioneers and settlers establish an intensive pattern of land use. Subsequent migration, population

Table 1. The number and proportion of taxa on each CNPS list that are of annual or perennial duration. The "ann/per" column accounts for taxa that can be both.

CNPS List	Annual Taxa	(%)	Perennial Taxa	(%)	Ann/Per Taxa	(%)	Total Taxa
1A	18	52.9	16	47.1	0	0.0	34
1B	257	30.0	593	69.2	7	0.8	857
2	48	17.7	222	81.6	2	0.7	272
3	17	36.2	29	61.7	1	2.1	47
4	138	25.9	391	73.5	3	0.6	532

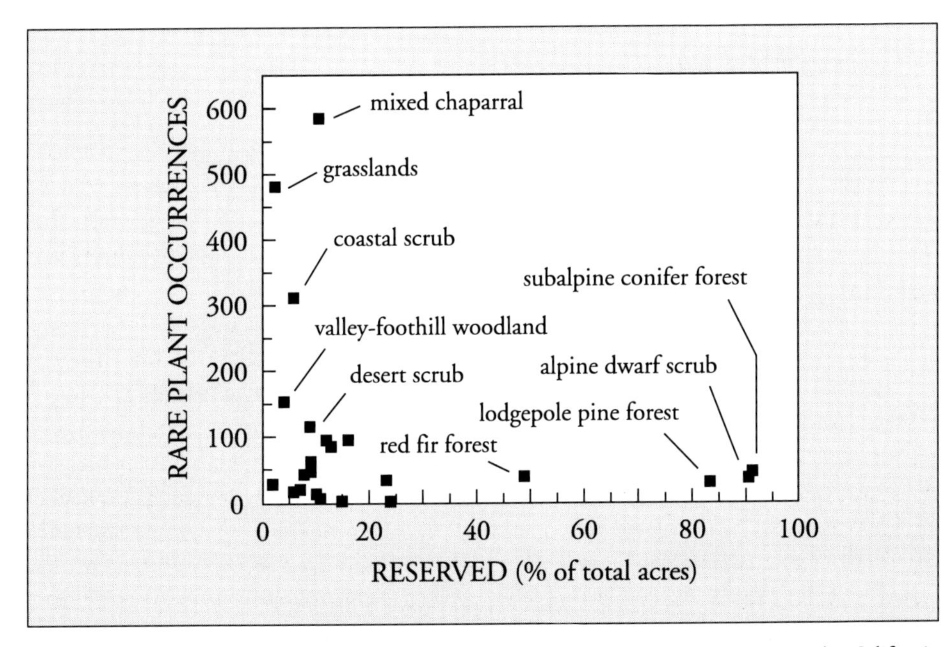

Figure 1. Communities that contribute the largest number of rare plant occurrences to the California Natural Diversity Data Base are those which have a low percentage of their acreage in parks and preserves.

growth, and economic development concentrate along shorelines, navigable rivers, and valley bottoms to ultimately displace the lowland biota. The richness of the lowland biota itself is also a contributing factor to overall losses: temperate grasslands and Mediterranean shrublands rank among the most species-rich communities on the planet (Ashton 1992). A disproportionate number of extinct annuals and the high degree of endangerment in the flora as a whole are the inevitable result of human exploration and exploitation of species-rich, lowland habitats.

This idea is supported if we rank habitats within the *Inventory* according to how many plants each contributes to the CNPS Lists (Table 2). Most extinction (indicated by List 1A) has been associated with low elevation grasslands, coastal scrub, wetlands, and oak woodlands. Endangerment (List 1B) is also associated with grasslands and oak woodlands, but species-rich chaparral rises to the top of the ranking. High elevation habitats that also have fewer native species, such as Great Basin grasslands and alpine scrub, contribute only a handful of rare plants to the *Inventory* despite their considerable acreage. Essentially, rarity in the flora is strongly associated with the lowland, Mediterranean habitats that best characterize the California Floristic Province (see *The Natural Communities Program,* below). These are the habitats of the Spanish missions, the ports, the livestock industry, of rich floodplain soils, of suburbia, and of expensive ocean views where intensive development has taken place (see Barbour *et al.* 1993). Rare plant conservation strategies should, therefore, emphasize the most species-rich, depleted habitats in order to establish an effective system of preserves for the California Floristic Province.

Unfortunately, lowland, species-rich habitats have been relatively ignored by our existing system of preserves and parks. By combining geographic information on rare plant occurrences from the California Natural Diversity Data Base (see *The Natural Diversity Data Base...,* below) with information on land use by habitat (Jensen *et al.* 1990), we see that from the standpoint of rare plants, California's protected areas (wilderness areas, research reserves, national and state parks, wildlife refuges, and recreational sites) are in the wrong places. Areas with the greatest degree of protection from development and high-intensity uses are at high elevations with relatively few rare plant occurrences (Figure 1). For example, approximately 90% of the state's alpine scrub habitat is preserved, but this accounts for fewer than 50 out of over 9,000 rare plant occurrences in the Data Base. On the other hand, the state's coastal scrub contributes 310 occurrences, but

Table 2. The top-ranking habitats of California's rare plants according to CNPS list. The number of taxa in the fifth edition of the *Inventory* is shown for each list and habitat in parentheses.

Rank	List 1A	List 1B	All Lists
1	valley/foothill grasslands (10)	chaparral (299)	chaparral (516)
2	marshes/swamps (9)	cismontane woodland (164)	lower montane conif. forest (359)
3	coastal scrub (8)	valley/foothill grasslands (157)	cismontane woodland (311)
4	meadows (5)	lower montane conif. forest (149)	valley/foothill grasslands (247)
5	chaparral (3), cismontane woodland (3)	coastal scrub (126)	coastal scrub (211)

only about 6% of this habitat is now protected. Although the lack of preserves at low elevations has, in part, produced the current geography of threatened plants, we cannot minimize the need for more lowland protection to benefit plant diversity in California.

Chemical and Physical Substrate

There are other revealing and potentially useful ecological patterns that emerge from the *Inventory.* Serpentinite endemism is a well-known feature of the California flora, and serpentinite substrates (e.g., rocky serpentinite outcrops and their derived soils) contribute the most rare taxa to the CNPS lists (Table 3). Carbonates, such as limestone and dolomite, are also havens for rare and endemic plants, especially when found in the dry, isolated, and ancient mountain ranges of eastern California. Other substrates with unique chemistry are much less likely to support rare plants, including gabbro (16 taxa) and gypsum (2). In terms of physical features, rare plants in the *Inventory* occur most often on sandy and rocky substrates. Many of these taxa are found on coastal and desert dunes, serpentinite rocks, granitic domes, and limestone bajadas. On the other hand, rare plants are seldom found on unstable talus (8 taxa) and scree (2).

Table 3. The top-ranking chemical and physical substrate characteristics of California's rare plants (from all CNPS lists). The number of taxa in the 5th edition of the *Inventory* is shown for each substrate characteristic in parentheses. Chemical and physical categories are not mutually exclusive.

Rank	Chemical Characteristic	Physical Characteristic
1	serpentinite (282)	sandy (166)
2	granitic (103)	rocky (144)
3	carbonates (89)	clay (78)
4	volcanic (77)	gravel (64)
5	alkaline (61)	barrens, openings (35)

Analysis of other data fields in the *Inventory* can provide additional insight into the ecology of California's rare flora. More fields will undoubtedly be added in the future (e.g., climate zone, elevation range) to expand our understanding of the environmental factors associated with these singular organisms. A detailed database, combined with increased accessibility via computer, will also help to focus and strengthen conservation strategy in the future. This brief examination of three fields (duration, habitat, and substrate) suggests that more attention be paid to the problems of annual plants at low elevations, especially those on serpentinite, rocky, or sandy substrates. When combined with new approaches to biodiversity preservation, including multiple listing, habitat-based management plans instead of single species stewardship, development of large preserve networks, and the protection of rare or species-rich natural communities, the ecological view of the *Inventory* will help secure the future of California's unique flora.

Bruce M. Pavlik is an Associate Professor in the Department of Biology, Mills College, Oakland, CA 94613. Mark W. Skinner is the Staff Botanist for the California Native Plant Society, 1722 J Street, Suite 17, Sacramento, CA 95814.

A History of the CNPS Rare Plant Program and Inventory

Rare plant studies have been a focus of the California Native Plant Society since its beginning. In 1968, G. Ledyard Stebbins, who was then President of the Society, started a card file of California plants with a distribution of less than one hundred miles. These cards were sent to botanists to solicit comments. Under the direction of Roman Gankin, Chairman of the newly formed Rare Plant Committee, these comments were evaluated and many more plants were suggested for consideration. Several in-house lists, composed of over 800 plants, were circulated during the next three years. The first widely distributed CNPS list appeared in 1971 and contained 520 plants.

At the same time, W. Robert Powell became Director of the Rare Plant Project. Early in 1973 a major effort was mounted to update and expand our information. Again in-house lists were circulated. The high-point of these activities occurred in July 1974, when a number of professional and amateur botanists from throughout the state met for a mapping session at the University of California at Davis. This provided an excellent opportunity for the participants to share their knowledge, often incomplete, of California's rare and endangered plants. It should be emphasized that from the beginning, CNPS rare plant investigations have been based upon the first-hand knowledge and field experience of a statewide network of botanists who have been generous in their cooperation. We continue to map locations in detail, confirm the existence of plants in the field, and record changes in population size as we refine our database.

In December 1974, the Society published the first edition of its *Inventory of Rare and Endangered Vascular Plants of California*, under the editorship of W. Robert Powell. In its Main List were 704 plants of primary concern. A second list of 554 plants constituted a group of rare, but not currently endangered plants. In addition, 135 plants were listed as having limited distributions, but not as being rare. This brought the total of plants included in the first edition to 1393 species, subspecies, and varieties. At the time of its publication, the CNPS *Inventory* was the most detailed compilation of rare plant data for any state in the nation. It quickly became the most widely used reference on the subject in California, and the prime source of information for botanists, private consultants, conservationists, and the staffs of city, county, state, and federal agencies.

Many of these same individuals, in return, provided new information, refinements, and corrections. CNPS responded by issuing supplementary lists of additions and deletions. With each passing year, it became more obvious that a second edition of the *Inventory* was needed. In November 1979, the United States Fish and Wildlife Service and CNPS co-sponsored a second review conference, once again on the Davis campus. About fifty specialists gave freely of their time and energy to amend and correct provisional lists that CNPS had developed for a proposed second edition.

The second edition appeared in 1980, with James P. Smith, Jr., R. Jane Cole, and John O. Sawyer, Jr. as editors. There were several format changes. This time 1,383 plants were arrayed in four lists, the first one entitled "Presumed Extinct in California." This was done in an effort to bring special attention to 44 plants that were then known from historical collections only. It became a challenge to see if any of them could be rediscovered, and several of them were! The plants on these four lists were adopted by the California Department of Fish and Game as constituting its "species-of-concern" list, now called the Special Plants list. A supplement to the second edition was published in 1981, and another in 1982.

The appearance of the new edition was followed quickly by two other major events in the Society's Rare Plant Program — the hiring of Richard York as a full-time botanist and the signing of a Memorandum-of-Understanding with the California Department of Fish and Game's Natural Diversity Data Base (NDDB). The rare plant data that already existed in our files, together with the accelerated pace of acquiring new information, had outstripped the capacity of our group of volunteers to maintain current manual files and to respond to inquiries in a timely fashion. A full-time employee had become a necessity. Our agreement with the Data Base led to establishing the CNPS botanist's office and files within the facilities of the California Department of Fish and Game, the state agency charged with inventorying rare, threatened, endangered, and sensitive plants and with preparing the documentation needed for state listing of these plants. For a more complete history of the Society's Rare Plant Program and the role of the CNPS Botanist, see Berg and Skinner (1990), Powell (1975, 1978), Powell *et al.* (1981), Smith (1987, 1990), and York *et al.* (1982).

Planning for the third edition of the *Inventory* began in late 1982. Meetings were held the following year to discuss possible format changes and to solicit information on incompletely known plants. In July 1983, through the generosity of many CNPS members, the Rare Plant Program was able to purchase a microcomputer. The work of entering all of our rare plant data into the computer began that summer and was completed in a relatively short time. In February 1984, we once again held a statewide meeting of amateur and professional botanists at the University of California at

Davis. The third edition was published later that same year.

Work on the fourth edition began in March 1987, with the conversion of our computerized data to MS-DOS compatible dBase III+ files. Advances in database management software improved our ability to organize and analyze the data and allowed electronic coordination with Department of Fish and Game botanists at the NDDB. To update our information, a list of proposed status changes was circulated to more than 200 data contributors for review. We also requested that any new status or distribution information be brought to our attention at that time. Meetings were held in Sacramento and Berkeley in December 1987, to discuss final status recommendations and to receive outstanding field survey data. These solicitations resulted in an avalanche of documentation.

In assimilating this information, we systematically reevaluated the rarity and endangerment ratings for more than 700 of our highest priority plants. County and topographic quad data were compared with NDDB computer files. Habitat occurrence data were assembled, categorized, and computerized for more than 1300 plants. This information was reviewed by botanists at the Natural Diversity Data Base, Department of Fish and Game Endangered Plant Project, and The Nature Conservancy. The fourth edition was published in September 1988.

Work on the fifth edition commenced in earnest in Spring 1991, when we conducted primary research for two new fields of information, life form and blooming time. In June 1991 we circulated a request for new information to the data contributors, now numbering about 370. Later that year we began the laborious process of library and herbarium research on the more than 300 plants which are newly added to this addition, and on the more than 35 plants shifted from List 4 to one of the higher priority lists. Also in late 1991, we commenced a detailed comparison with the taxonomic concepts and other information contained in *The Jepson Manual*, the monumental new work on California's flora. This task extended into June 1992, and entailed hundreds of name, status, and distribution changes. In April of 1992, we circulated preliminary review lists of status changes and new plants to the contributors. We received well over 100 responses with thousands of suggested changes and improvements. In August 1992, all proposed changes to date were reviewed in Sacramento by the Rare Plant Scientific Advisory Committee (RPSAC) and several knowledgeable guests, and by July 1993 the first complete draft of the data in this volume appeared for review. RPSAC members and a number of other selected experts reviewed the data one final time in Fall 1993, and hundreds of last minute changes and corrections were made at that time.

Along the way a number of steps were taken to ensure that the fifth edition met the high standards established in the past. Every entry in the *Inventory* has been scrutinized to ensure accuracy of the data and consistency between the geographical information recorded by county, by topographic quad, and by what is implied in the notes. In those few cases where information remains incomplete, we have explicitly requested it in the notes. The notes themselves have been extensively reworked for content, style, and consistency, and greatly expanded in many cases. We compared county and topographic quad data, information on rare plant threats, and habitat with NDDB computer files, and researched all discrepancies. We developed habitat information for the fifty plants which lacked it in the 1988 edition, so that only one plant's habitat is now unknown. Likewise, we researched all the List 1, 2, and 3 plants for which we lacked topographic quad information; but two unknowns remain.

Numerical Analysis of This and Previous Editions

As the following numerical comparison of the plants in the five *Inventory* editions demonstrates, the size of California's rare and endangered flora continues to grow (Table 1). The percentages given below indicate the portion of the total native flora in California represented by the plants on a particular list for the different editions. We estimate that the flora as currently described contains 6300 native species, subspecies, and varieties. This is the number of native taxa that are fully described in *The Jepson Manual* (about 6000), plus the approximate number that receive peripheral mention as minor taxa.

There is a net addition of 182 plants (27% increase) to our highest priority List 1B since the 1988 edition. Almost 80 of these have been upgraded to List 1B from a lower priority list, in some cases because we have learned more about their rarity or endangerment, but usually because conditions have worsened and they are now more seriously endangered than before. This is perhaps the most sobering single fact to emerge during preparation of this edition; nearly 14% of California's native plants are either exceedingly rare or seriously endangered. Our best efforts to date simply have not been sufficient to stem the further deterioration of what is arguably the Nation's most remarkable flora.

During the first fourteen years since publication of the first edition, the *Inventory* grew at a cumulative rate of 0.8% per year. In the last five years we have identified 313 new rare and endangered plants (Table 2), and the *Inventory* has grown 2.5% per year. There are at least four reasons for this increase: rare plants have been overlooked in the past (invoked in 63% of recent cases), more plants are becoming endangered as habitat loss and other threats accelerate (32% of cases), many new plants have been described in California in the last 5 years (15% of cases), and taxonomic changes have "created" new rare plants (7% of cases). (Note that combinations of these factors resulted in some plants being added to the *Inventory*, hence the foregoing percentages total greater than 100%.)

A disproportionate number of the taxa that are new to this *Inventory* have been assigned to List 2 (Table 2). In the 1988 edition, List 2 represented only 11% of the plants, but this list comprises 26% of the newly added plants in the fifth edition. We suspect this is due as much to reexamination of herbarium material during preparation of *The Jepson Manual* as it is to continued botanical exploration on California's fringes, where

Table 1. Numerical comparison of the five CNPS *Inventory* editions. We have reevaluated percent of flora for past editions based on 6300 native plants in California.

List or Appendix	Taxa	%Flora
1974, 1st EDITION		
1. Very Rare & Rare and Endangered	704	11.1%
Appendix I— Rare and Not Endangered	554	8.8%
Appendix II— Mostly of Limited Distribution	135	2.1%
TOTAL	**1393**	**22.0%**
1980, 2nd EDITION		
1. Presumed Extinct in California	44	0.7%
2. Rare and Endangered	656	10.4%
3. Rare, But Not Endangered	446	7.1%
4. Rare in California, But Not Elsewhere	237	3.8%
TOTAL	**1383**	**22.0%**
1984, 3rd EDITION		
1A. Presumed Extinct in California	34	0.5%
1B. Rare or Endangered in California and Elsewhere	604	9.6%
2. R/E in California, More Common Elsewhere	198	3.1%
3. Need More Information	114	1.8%
4. Plants of Limited Distribution	449	7.1%
TOTAL	**1399**	**22.2%**
1988, 4th EDITION		
1A. Presumed Extinct in California	39	0.6%
1B. Rare or Endangered in California and Elsewhere	675	10.7%
2. R/E in California, More Common Elsewhere	177	2.8%
3. Need More Information	149	2.4%
4. Plants of Limited Distribution	508	8.1%
TOTAL	**1548**	**24.6%**
1994, 5th EDITION		
1A. Presumed Extinct in California	34	0.5%
1B. Rare or Endangered in California and Elsewhere	857	13.6%
2. R/E in California, More Common Elsewhere	272	4.3%
3. Need More Information	47	0.8%
4. Plants of Limited Distribution	532	8.4%
TOTAL	**1742**	**27.7%**

most List 2 plants occur. There are undoubtedly many more of these border plants remaining to be discovered in California. Fortunately, we were able to reexamine the plants on List 3 and reclassify or delete many of these, and our thorough research on the new additions meant that we could assign most to Lists 1, 2, or 4. The 47 plants on List 3 in the fifth edition (Table 1) represent just 3% of the plants in this *Inventory*.

California's rare flora is disproportionately rich in subspecies and varieties as compared to the flora as a whole (Table 3). This is unsurprising since subspecies and varieties typically have smaller ranges than species, and are thus biologically rarer to begin with, and consequently more susceptible to disruption. Subspecies and varieties are morphologically, genetically, and geographically distinctive, and much of California's floristic diversity is expressed at this infra-specific level. It is therefore essential that our conservation efforts include these ranks as well as full species if we are to preserve the California flora as the remarkable living evolutionary laboratory that it is.

Table 2. Percentage of fourth edition taxa on each list compared to the number and percentage of new fifth edition taxa on each list.

CNPS List	% 4th Edition	Taxa New to 5th Edition	% New Taxa
1A. Presumed Extinct in California	3%	3	1%
1B. Rare or Endangered in CA and Elsewhere	44%	136	43%
2. R/E in California, More Common Elsewhere	11%	82	26%
3. Need More Information	10%	12	4%
4. Plants of Limited Distribution	33%	80	26%
TOTAL	**100%**	**313**	**100%**

Table 3. Comparison of taxonomic rank of plants in the California flora, in current and previous editions of the *Inventory*, and of the taxa new to this edition.

Taxonomic Rank	CA Native Flora	4th Edition	Taxa New to 5th Edition	5th Edition
No. Full Species	4839	1095	175	1182
No. Subspecies or Varieties	1159	453	138	560
% Full Species	**81%**	**71%**	**56%**	**68%**

Uses of the CNPS *Inventory*

The California Native Plant Society has invested a great deal of time, energy, and money to gather this rare plant information for one purpose: to promote the preservation of rare plants and their habitats. Our responsibilities are best served by complete and accurate information upon which to make sound conservation judgments and recommendations. Although there remain gaps to be filled, errors to be corrected, and refinements to be made, the information is here for all to use. The challenge for us is to use it wisely.

We encourage conservationists, consultants, planners, researchers, and resource managers to use this *Inventory* to educate landowners and public policy makers about rare plant conservation. The information it contains should serve to direct efforts to identify, protect, and manage California's rarest botanical resources. Consultants and planners preparing environmental documents should review this information to determine the potential for resource conflicts and the scope of botanical surveys needed. Conservationists and resource managers can use the same information to review environmental documents and prepare public testimony to influence decision makers, and to guide rare plant protection and preserve acquisition and management. The CNPS policies reproduced as Appendices in this volume should also be consulted for further guidance in these areas.

We are committed to the dissemination of accurate biological information to facilitate plant conservation in California. We offer our assistance to all involved in this arena. Computerization allows us to sort and analyze this information in many useful ways, most of which are now available to the general public through release of the CNPS *Electronic Inventory*. The appendices in this edition are just a few examples of what we can do. For more information, contact the CNPS Botanist.

The Rare, Threatened, and Endangered Plants of California

The heart of the CNPS *Inventory* is the summary of the distribution, rarity, endangerment, and ecology of our state's rare, threatened, and endangered vascular plants. We present these plants in a single alphabetical list that includes all the plants of concern to the CNPS Rare Plant Program. We also include entries for those that were considered but rejected for one or more reasons, as well as other scientific names that have been used in the standard literature or in previous editions of this *Inventory*. Each entry for a plant of concern to CNPS is referred to as a record.

The Basis for Inclusion

Only **vascular** plants, those with a specialized conducting system made up of xylem and phloem tissue, are included in this *Inventory*. This botanical designation refers to the plant groups commonly called ferns, fern allies, gymnosperms (cone-bearing seed plants), and flowering plants. Algae, fungi, lichens, mosses, and liverworts are not vascular plants and are not treated here.

A plant must also be **native** to California to be included. Ornamentals, plants escaped from cultivation, and naturalized plants are excluded. So are the sporadic, sterile hybrids that sometimes occur under natural conditions. The relatively trivial color variants and occasional departures from typical vegetative or floral conditions, referred to by botanists as "forma," are similarly excluded.

A plant must also be **rare** in California. This criterion is, admittedly, more arbitrary. Since no statistically precise definition of rarity has been developed, differences in assessing this critical feature do exist. Most of the plants that were considered but not included in this book were eliminated because we determined them to be too abundant. Because this reference focuses on California, plants that are rare in our state but more common or even widespread elsewhere are included.

A very small number of plants that are still somewhat abundant in California, but which are threatened or declining, are also listed. Most of these are the very attractive or curious ones that have horticultural appeal to the commercial or private collector. The pitcher plant (*Darlingtonia californica*) and Mojave fish-hook cactus (*Sclerocactus polyancistrus*) are examples.

Widespread Plants in Decline

A number of California plants are seriously declining, but remain too common to be appropriately placed on List 4, our "watch" list of limited distribution plants. These plants are fairly widespread and sometimes even common, but each is experiencing systematic decline in all or part of its range. The inclusion on the fourth edition watch list of certain declining plants of somewhat restricted distribution was controversial. Valley oak (*Quercus lobata*) is a notable example of a plant included on the last watch list because it warrants monitoring. It has been drastically reduced in abundance in some areas, it is threatened with extirpation in a portion of its range, and its ability to regenerate is low in many areas. Nevertheless, it cannot be considered rare under any concept.

Consequently, in this edition, valley oak and plants like it have been rejected in favor of stricter adherence to the "rare and endangered" criteria implied by the title of this volume. A few of the other widespread plants in serious decline that were included in the previous edition or considered for inclusion in this one, but ultimately were rejected, are: *Chorizanthe fimbriata* var. *fimbriata*, *C. fimbriata* var. *laciniata*, *Cupressus* (= *Chamaecyparis*) *lawsoniana*, *Erodium macrophyllum*, *Fritillaria biflora*, *Pinus lambertiana*, *Taxus brevifolia*, and *Washingtonia filifera*. There are many others. Overall, we judged that their presence on the lists would compromise the integrity of the endeavor as a whole. However, their exclusion in no way minimizes the genuine threat to these plants, many of which will soon be candidates for inclusion on CNPS priority lists if present trends continue. By raising public concern and focusing research and attention, we hope to keep these plants from declining further.

The Scientific Names, Common Names, and Family Names of Plants

The plants in this *Inventory* are presented alphabetically by their scientific names, the technical names that have been properly published for them according to the *International Code of Botanical Nomenclature*. They are the first line in each of the records.

In its simplest form, a scientific name has three parts. The first is the **genus** or **generic name**. It is always capitalized. The second part is the **specific epithet**, often incorrectly called "the species name." Together, these two components make up the species name. If a scientific name is presented in its most complete form, these two words will be followed by the names of one or more persons, often in an abbreviated form, who first published the specific epithet or subsequently published a taxonomic modification of the plant. These names are the **authorities**. If a portion of an authority occurs within parentheses, then the author in parentheses originally placed the epithet in a different genus or species, or once assigned it to a different taxonomic rank. The name cited outside the parentheses is that of the person who published the combination as it now appears.

Often the scientific name is more complex because botanists have recognized categories below the level of species. The two most useful are the **subspe-**

cies (abbreviated **ssp.**) and the **variety** (abbreviated **var.**) These names are also given according the *Code* and they have their own authorities.

Consider the example *Penstemon newberryi* Gray var. *sonomensis* (Greene) Jeps. *Penstemon* is the genus or generic name; *newberryi* is the specific epithet; Gray, for Asa Gray, is the author of the specific epithet; var. is the abbreviation for variety; *sonomensis* is the varietal epithet; (Greene), for Edward L. Greene, first described the var. *sonomensis* as a full species; and Jeps., for Willis Lynn Jepson, modified its taxonomic position and made it a variety of *P. newberryi.* Note that generic names, along with specific, varietal, and subspecific epithets, appear in italics or are underlined.

We have made every effort to use the currently accepted scientific names that appear in *The Jepson Manual* (Hickman, 1993; see *Coordination with The Jepson Manual,* below). Most of them have been in use for some time and are the scientific names that appear in Munz (1959, 1968, 1974), and Abrams (1923-1960). We have also relied heavily for our nomenclature on recent monographs and revisions, many of which are cited in the notes at the end of each record. Because of this recent systematic work, some names that were used in earlier editions of the *Inventory* have been abandoned; these appear alphabetically with cross-references to the currently preferred name.

Each of the plants also has a common or vernacular name. It is presented on the second line of each record. We include these because it is often easier for many of us to refer to a plant by a more familiar sounding name. Of course, the majority of the plants in this book have no real common names. Most of them were coined by Leroy Abrams for his *Illustrated Flora of the Pacific States.* In other instances, we simply followed his lead by contriving names, usually by translating the Latin or Greek roots into English or by selecting an appropriate geographical reference or person's name. We have attempted to follow Kartesz and Thieret (1991) in matters of capitalization, spelling, and hyphenation of common names.

The second line in each record also includes the technical name of the family to which the plant belongs. Note that all of these names end with the suffix "-aceae." A few plant families have older, alternative names that the *International Code* allows to be used because their widespread acceptance predates formal nomenclature. Gramineae is a perfectly acceptable alternative for Poaceae; Compositae for Asteraceae; Cruciferae for Brassicaceae; Umbelliferae for Apiaceae; Leguminosae for Fabaceae; and Labiatae for Lamiaceae. However, these old names are gradually losing favor, so we have used the standardized, modern names for these families.

Unpublished Names

We have declined to include a small number of putatively rare taxa that have been brought to our attention, because these entities lack effectively published scientific names. As we receive notice that a manuscript has been accepted and scheduled for publication in a botanical journal, we will evaluate each unpublished plant for inclusion in future editions. In a few cases we have included taxa with new names or combinations which are in the process of being published. In these cases we have usually included the journal of imminent publication but not the page numbers.

Coordination with *The Jepson Manual*

Publication of *The Jepson Manual* stimulated a great deal of taxonomic research, much of which changed the way we think about the California flora. Most of these floristic changes needed to be incorporated into this edition of the *Inventory.* Because of the sheer volume of alterations, this process proved to be the single most arduous task we encountered during editing of the fifth edition.

During the simultaneous preparation of the *Inventory* and *The Jepson Manual,* we worked closely with the editors of the *Manual* to achieve correspondence in taxonomy and rarity designations. *The Jepson Manual* editors graciously agreed that entries would include code words indicating placement on the CNPS rarity lists, and this practice has been followed. For our part, we made every attempt to incorporate *Jepson Manual* taxonomic concepts and new names into the fifth edition, but for scientific or logistic reasons this was not always possible. Accordingly, careful readers will find some discrepancies between taxa, rarity designations, and nomenclature in *The Jepson Manual* and those in this edition of the *Inventory.* This is true in part because some rare plants are local forms of highly variable taxa. Authors writing for *The Jepson Manual* sometimes considered them to be subtle variants not warranting taxonomic recognition, but for the several reasons presented below, many are recognized here.

Above all, *The Jepson Manual* was intended to be useful to a wide audience, and explicitly provided instructions to contributing authors to take a "broad brush" approach to their taxonomic groups. This meant that some morphologically, genetically, and geographically distinctive taxa were not included, or were mentioned only peripherally. The "Guide for Contributors," prepared by Editor James Hickman, includes the following instructions (p. 4):

> "Our primary goal is to produce a guide for identification by a wide spectrum of users, most of whom will not be professional botanists. Therefore, insofar as possible, formal recognition should be reserved for taxa that are morphologically distinct, and keys and descriptions should emphasize features visible with little or no magnification. Slight morphological variants and morphologically indistinct taxa that differ only in aspects of chemistry,

cytology, physiology, or ecology will neither be included nor fully treated otherwise, although they will be noted in important cases."

The Jepson Manual clearly benefits from and incorporates a vast amount of research that was unavailable to its predecessors. The new information does not, however, always result in increased taxonomic clarity, thus it is unrealistic to regard *The Jepson Manual* as the final statement in California plant taxonomy. In the "Philosophy" section of the *Manual*, Hickman notes (p. 1): "...intensive study does not necessarily lead to an easier or simpler taxonomic treatment. As taxonomists learn more about the complexities of variation patterns, their best taxonomic assessments automatically change. It may become more difficult to represent what is known in a taxonomic scheme."

As a further complication, because of the increasing gap between the number of taxa and the number of professional monographers, many authors had limited previous knowledge of the groups for which they were preparing treatments. Hickman notes (p. 1.):

> "Some authors approached certain groups as taxonomic authors for the first time; they confronted a different set of challenges. Almost uniformly, they found that time did not allow a study of the intensity they wished and that more was unknown than they believed when they started. Some treatments in *The Jepson Manual* are simultaneously tentative and the best currently possible. In such situations, there was an attempt to point out clearly what problems remain and to suggest specific further work."

The important message here is that, without diminishing its own scientific validity in the least, *The Jepson Manual* itself notes numerous places where there is potential room for different taxonomic conclusions. This could result either from further research, or by an equally valid but alternate scientific interpretation involving both the biological evidence and the rules of nomenclature.

Because taxonomic concepts and the names we apply to them are scientific hypotheses, it is reasonable and expected that here we would adopt a "conservative" approach toward plant taxonomy, an approach we hope will reduce *ex post facto* lamentation over taxa that have been shown to be distinct only after their disappearance. We are concerned about protection of California's plant genetic diversity as it is expressed at all levels of taxonomic variation, not just the level of major variants, about which there is usually widespread scientific agreement. We can partially address this concern through recognition of taxa that are in taxonomic flux or in need of additional clarification.

Therefore, in the event of legitimate disagreement between botanical experts, the *Inventory* will continue to recognize problematic taxa pending further study that may resolve their status. In such cases we have retained existing taxonomic concepts and scientific names, and indicated in the notes section for each entry if and where a plant may be found in *The Jepson Manual*.

Adoption of *The Jepson Manual's* Taxonomic Concepts in the CNPS *Inventory*

Below we indicate explicitly how CNPS will handle the various kinds of taxonomic differences between *The Jepson Manual* and the *Inventory*. The different correspondences between names in these two sources are summarized as follows:

1. **Taxa with full recognition in the *Manual*.** These are obviously not a problem, and are correspondingly treated in the *Inventory*.

   Example: The vast majority of rare and endangered taxa in this edition of the *Inventory*.

2. **Taxa embedded in other entries in the *Manual* with only a very brief treatment.** In general, these will be retained in the *Inventory* fifth edition with the message, "See 'Species X' in *The Jepson Manual*."

   Example: In *The Jepson Manual*, *Castilleja uliginosa* (CNPS List 1A) is considered under *C. miniata* ssp. *miniata* with the note: "Plants from lowland southern outer North Coast Ranges (Pitkin Marsh, Sonoma Co., +/- 240 meters) with yellow inflorescences have been called *C. uliginosa* Eastw., Pitkin Marsh indian paintbrush." The *Inventory* reads: "See *C. miniata* ssp. *miniata* in *The Jepson Manual*."

3. **Taxa synonymized in the *Manual* but retained in the *Inventory*.** In these cases, we have exercised our independent scientific expertise, determined that the evidence is equivocal, and elected to adopt an alternate conclusion supported elsewhere in the taxonomic literature. Rare and endangered plants that are synonymized in *The Jepson Manual* but which merit CNPS's continued recognition will be retained in the *Inventory* fifth edition with the message, "A synonym of 'Species X' in *The Jepson Manual*."

   Example: *Dichanthelium lanuginosum* var. *thermale* (CNPS List 1B) is synonymized under *Panicum acuminatum* var. *acuminatum* in *The Jepson Manual*. The *Inventory* reads: "A synonym of *Panicum acuminatum* var. *acuminatum* in *The Jepson Manual*."

4. **Taxa synonymized in both the *Manual* and the *Inventory*.** Most rare and endangered plants that are synonyms in *The Jepson Manual* will be deleted from the *Inventory* fifth edition with the message, "Considered but rejected: a synonym of 'Species X'; a common taxon."

   Example: *Chorizanthe staticoides* ssp. *chrysacantha* is synonymized under *C. staticoides* in *The Jepson Manual*. The *Inventory* reads:

"Considered but rejected: a synonym of *Chorizanthe staticoides*; a common taxon."

5. **Taxa not mentioned in the *Manual* and retained in the *Inventory*.** *The Jepson Manual* does not include some rare and endangered plants which either are not in Munz (1959, 1968) or else are listed as synonyms in Munz. Because of space constraints these plants are not listed as synonyms in *The Jepson Manual*, so it is difficult to track their synonymy there. If we were unable to identify the accepted synonym in *The Jepson Manual* or elsewhere, but believe the plant to be worthy of recognition, it will be retained with the message, "Not in *The Jepson Manual*."

   Example: *Echinocereus engelmannii* var. *howei* is included in the *Inventory* fifth edition but is not mentioned in *The Jepson Manual* treatment of *Echinocereus*. The *Inventory* reads: "Not in *The Jepson Manual*."

The CNPS Lists

The third line of each record contains three elements of information, the first of which is the CNPS List for that plant. We have created five lists in an effort to categorize degrees of concern. They are described as follows:

List 1A: Plants Presumed Extinct in California

The 34 plants of List 1A are presumed extinct because they have not been seen or collected in the wild in California for many years. Although most of them are restricted to California, a few are found in other states as well. In many cases, repeated attempts have been made to rediscover these plants by visiting known historical locations. Even after such diligent searching, we are constrained against saying that they are extinct, since for most of them rediscovery remains a distinct possibility. Note that care should be taken to distinguish between "extinct" and "extirpated." A plant is extirpated if it has been locally eliminated, but it may be doing quite nicely elsewhere in its range.

We segregate these plants on their own list to highlight their plight and encourage field work to relocate extant populations. Since the publication of the fourth edition, thirteen plants thought to be extinct in California have been rediscovered. These are *Chorizanthe orcuttiana*, *Clarkia mosquinii* ssp. *mosquinii*, *C. mosquinii* ssp. *xerophila*, *Lycium brevipes* var. *hassei*, *Malacothamnus abbotii*, *Malaxis monophyllos* ssp. *brachypoda*, *Monardella douglasii* var. *venosa*, *Phacelia amabilis*, *P. parishii*, *Plagiobothrys diffusus*, *Scheuchzeria palustris* var. *americana*, *Sidalcea keckii*, and *Trifolium amoenum*. These rediscoveries are unfortunately countered by the addition of ten plants to List 1A: *Castilleja uliginosa*, *Erigeron mariposanus*, *Eschscholzia rhombipetala*, *Malacothamnus parishii*, *Montia howelii*, *Plagiobothrys glaber*, *P. hystriculus*, *P. lithocaryus*, *Poliomintha incana*, and *Ranunculus hydrochariodes*.

All of the plants constituting List 1A meet the definitions of Sec. 1901, Chapter 10 (Native Plant Protection Act) or Secs. 2062 and 2067 (California Endangered Species Act) of the California Department of Fish and Game Code, and are eligible for state listing (see *Conserving Plants with Laws and Programs...*, below). Should these taxa be rediscovered, it is mandatory that they be fully considered during preparation of environmental documents relating to the California Environmental Quality Act (CEQA).

List 1B: Plants Rare, Threatened, or Endangered in California and Elsewhere

The 857 plants of List 1B are rare throughout their range. All but a few are endemic to California. All of them are judged to be vulnerable under present circumstances or to have a high potential for becoming so because of their limited or vulnerable habitat, their low numbers of individuals per population (even though they may be wide ranging), or their limited number of populations. Most of the plants of List 1B have declined significantly since the arrival of non-indigenous humanity in California.

All of the plants constituting List 1B meet the definitions of Sec. 1901, Chapter 10 (Native Plant Protection Act) or Secs. 2062 and 2067 (California Endangered Species Act) of the California Department of Fish and Game Code, and are eligible for state listing. It is mandatory that they be fully considered during preparation of environmental documents relating to CEQA.

List 2: Plants Rare, Threatened, or Endangered in California, But More Common Elsewhere

Except for being common beyond the boundaries of California, the 272 plants of List 2 would have appeared on List 1B. From the federal perspective, plants common in other states or countries are not eligible for consideration under the provisions of the Endangered Species Act. Until 1979, a similar policy was followed in California. However, after the passage of the Native Plant Protection Act, plants were considered for protection without regard to their distribution outside the state.

All of the plants constituting List 2 meet the definitions of Sec. 1901, Chapter 10 (Native Plant Protection Act) or Secs. 2062 and 2067 (California Endangered Species Act) of the California Department of Fish and Game Code, and are eligible for state listing. It is mandatory that they be fully considered during preparation of environmental documents relating to CEQA.

List 3: Plants About Which We Need More Information — A Review List

The 47 plants that comprise List 3 are united by one common theme — we lack the necessary information to assign them to one of the other lists or to reject them. For this edition we reviewed the previous 149 plants on List 3, and with the aid of new information offered by our contributors, we shifted most to another list or deleted them.

Nearly all of the plants remaining on List 3 are taxonomically problematic. For each List 3 plant we have provided you with the known information, indicated in the Notes where assistance is needed, and tentatively assigned the taxon to a more definite list. Data regarding distribution, endangerment, ecology, and taxonomic validity will be gratefully received.

Some of the plants constituting List 3 meet the definitions of Sec. 1901, Chapter 10 (Native Plant Protection Act) or Secs. 2062 and 2067 (California Endangered Species Act) of the California Department of Fish and Game Code, and are eligible for state listing. We recommend that List 3 plants be evaluated for consideration during preparation of environmental documents relating to CEQA.

List 4: Plants of Limited Distribution —A Watch List

The 532 plants in this category are of limited distribution or infrequent throughout a broader area in California, and their vulnerability or susceptibility to threat appears low at this time. While we cannot call these plants "rare" from a statewide perspective, they are uncommon enough that their status should be monitored regularly. Should the degree of endangerment or rarity of a List 4 plant change, we will transfer it to a more appropriate list.

Very few of the plants constituting List 4 meet the definitions of Sec. 1901, Chapter 10 (Native Plant Protection Act) or Secs. 2062 and 2067 (California Endangered Species Act) of the California Department of Fish and Game Code, and few, if any, are eligible for state listing. Nevertheless, many of them are significant locally, and we recommend that List 4 plants be evaluated for consideration during preparation of environmental documents relating to CEQA. This may be particularly appropriate for the type locality of a List 4 plant, for populations at the periphery of a species' range or in areas where the taxon is especially uncommon or has sustained heavy losses, or for populations exhibiting unusual morphology or occurring on unusual substrates (see *Ecological Characteristics of California's Rare Plants*, above).

The R-E-D Code

Even before the publication of the first edition, CNPS determined that attempts to categorize plants solely on the degree of threat, as embodied in such terms as rare, threatened, or endangered, were too restrictive. This is so primarily because the question of rarity frequently interferes with the question of endangerment (see *Rarity in Vascular Plants*, above). With few exceptions, endangered plants are also rare. However, some plants of more widespread occurrence are endangered and their numbers have declined because of commercial or private exploitation for horticultural use. Many cacti, lilies, orchids, succulents, and insectivorous plants fall into this category. In other cases, very rare plants occur in stable habitats such as alpine fell fields. Typically these plants cannot realistically be described as endangered, except perhaps through stochastic extinction associated with small population sizes or numbers.

In an attempt to increase the refinement of assigning plants to categories, we use a scheme that combines three complementary elements that are scored independently. These components are: **rarity**, which addresses the extent of the plant, both in terms of numbers of individuals and the nature and extent of distribution; **endangerment**, which embodies the perception of the plant's vulnerability to extinction for any reason; and **distribution**, which focuses on the overall range of the plant.

Together these three elements form the **R-E-D Code**. It is presented on the third line in each paragraph summary. Each element in the code is divided into three classes or degrees of concern, represented by the number 1, 2, or 3. In each case, higher numbers indicate greater concern. The system is summarized as follows:

R (Rarity)

1—Rare, but found in sufficient numbers and distributed widely enough that the potential for extinction is low at this time
2—Distributed in a limited number of occurrences, occasionally more if each occurrence is small
3—Distributed in one to several highly restricted occurrences, or present in such small numbers that it is seldom reported

E (Endangerment)

1—Not endangered
2—Endangered in a portion of its range
3—Endangered throughout its range

D (Distribution)

1—More or less widespread outside California
2—Rare outside California
3—Endemic to California

An R-E-D Code of 3-3-3 indicates that the plant in question is limited to one population or several restricted ones, that it is endangered throughout its range, and that it is endemic to California. A summary of the R-E-D code system appears on the inside back cover for easy reference.

State and Federal Status

The last element on the third line of each record gives the plant's official status under state or federal law (the abbreviations are explained on the inside back cover of the *Inventory*). Our definitions conform to those found in California state law and federal regulations. For a more complete discussion, see *Conserving Plants with Laws and Programs...* and *The Endangered Species Act...*, below, and *The CNPS Lists*, above. For plants that have no current state or federal legal standing we have either printed "CEQA," indicating that CEQA consideration is mandatory (Lists 1 and 2), or "CEQA?," indicating that we recommend evaluation for CEQA consideration (Lists 3 and 4).

Distribution

The fourth line in each record gives the abbreviation of each county or island in California and other states and countries where we know the plant to exist. We record only natural occurrences of rare plants, or occurrences that have been reestablished within the species' historical range as part of an approved recovery plan. For example, although both *Juglans californica* var. *hindsii* and *Pinus radiata* are widely planted within the state, we track only the few natural occurrences of these taxa. When we indicate that a particular plant occurs in a particular county, we are making a positive statement that is based upon specimens, photographs, the literature, or field observations. In no way does this imply that a plant does not occur in other counties in California or in other states. Our understanding of plant distribution constantly improves, and new localities for rare plants are frequently discovered, often in unpredictable circumstances.

County and other geographic abbreviations used here are explained on the inside front cover, along with our usage of the symbols "*" and "?", which, respectively, express extirpation and uncertainty.

Quads

To provide more detailed location information, we have also cited the U.S. Geological Survey (USGS) 7.5 minute quadrangle map for more than 1200 plants on CNPS Lists 1, 2, and 3. This information appears on the fifth line of each record. We employ a modified version of the quadrangle numbering system previously used by the California Department of Water Resources. Please see *Appendix I* to translate this system's quad numbers into USGS topographic map names or vice versa. In those few cases where a quad is listed without a letter following the number, this indicates that our occurrence data are too vague to pinpoint its location on a 7.5 minute quadrangle. As with counties, when we indicate that a plant has been reported from an area on a topographic quad, this is based on hard data. In no way does this imply that a plant does not occur on a topographic quad we have not listed; rather, it may be there but botanists have yet to find it.

Habitat Types

The next line of each record presents one or more habitats in which a rare, threatened, or endangered plant is typically found. This information, compiled from field survey forms, unpublished reports, original descriptions, floras, and herbarium material, is provided for all the plants in the *Inventory* except one, the enigmatic *Arenaria macradenia* var. *kuschei.* Habitat information based upon field observations is given priority whenever possible. Note that for habitats which typically occur within a broader matrix of another habitat, we usually list both. For example, a rare plant from Meadows occurring in a matrix of Lower Montane Coniferous Forest would be coded **LCFrs, Medws.** We are indebted to Robert F. Holland and John O. Sawyer, Jr. for contributing the following brief characterizations, which are presented in taxonomic rather than alphabetical order. Please refer to Holland (1986) for a more complete discussion of the types and their classification. CNPS's upcoming classification of plant communities, due to be published later this year, will not use the system presented here (see *The Natural Communities Program*, below). However, crosswalks will be provided to guide the user from the new system to this one.

CoDns (Coastal dunes)

Herbs or shrubs on coastal sand deposits from Del Norte to San Diego counties. Cover usually low near the beach, increasing with distance from salt spray and blowing sand.

DeDns (Desert dunes)

Sand accumulations east of the Pacific Crest from Modoc to Imperial counties. Vegetation on desert dunes varies considerably. Active dunes usually support only sparse herbs and grasses, but partially stabilized or stabilized dunes often will support shrubs, including mesquite and creosote bush.

InDns (Inland dunes)

Mostly herbs, although shrubs may be locally important. Sand accumulations in and around the Great Valley.

CBScr (Coastal bluff scrub)

Dense shrubs, prostrate to 1-2 m tall. Typically on fairly steep, rocky sites exposed to considerable wind and salt spray because of proximity to the ocean. Many plants succulent, especially to the south. Found from Del Norte to San Diego counties.

CoScr (Coastal scrub)

Dense shrubs 0.5 to 2 m tall with scattered grassy openings. Many plants dormant, even deciduous, during periods of water stress. Most sites have shallow rocky soils, frequently with a southern or western exposure. Many taxa adapted to fire by stump sprouting or high seed production.

SDScr (Sonoran desert scrub)

Widely scattered creosote bushes with the considerable space between them sometimes occupied by ephemeral, colorful shows of annuals following particularly wet winters. Succulents and microphyllous trees conspicuous, especially in rocky environments. The part of Munz's (1959) "Creosote Bush Scrub" found roughly south of the San Bernardino/Riverside county line.

MDScr (Mojavean desert scrub)

Widely scattered creosote bushes with the considerable space between them sometimes occupied by ephemeral, colorful shows of annuals following particularly wet winters. At elevations of 2000' or higher, succulents or microphyllous trees lacking. This habitat type constitutes most of Munz's (1959) "Creosote bush scrub" found north of the San Bernardino/Riverside county line.

GBScr (Great Basin scrub)

Shrubs, ranging in height from very short, <20 cm, on very cold sites or shallow soils to 1 or 2 m tall on warmer sites where soils are deeper. Perennial grasses occupy much of the space between shrubs. Found on the Modoc Plateau, high Cascade Range, Warner Mountains, High Sierra Nevada, and North Coast Ranges.

ChScr (Chenopod scrub)

Usually gray, intricately branched, microphyllous shrubs most commonly on fine-textured, alkaline and/or saline soils in areas of impeded drainage. Diversity usually low to monotonous. Saltbushes and greasewood frequently dominate. This vegetation occurs from Modoc County south to Mexico, including parts of the Great Valley and Inner South Coast Ranges.

Chprl (Chaparral)

Impenetrably dense, evergreen, leathery-leaved shrubs that are active in winter, dormant in summer, and adapted to frequent fires either through resprouting or seed carry-over. There is a characteristic florula of fire-following annuals and short-lived perennials. Mature stands may exceed 3-4m in height. It occurs on diverse substrates, many of which support distinctive suites of edaphic indicators. Chaparral may be successional to conifer forests or oak woodlands, as tree seedlings can be found beneath the shrub canopies.

CoPrr (Coastal prairie)

Dense, fairly tall (<1 m) perennial sod- and tussock-forming grasses and grass-like herbs. They occur in two distinct settings: sandy marine terraces within the zone of coastal fog (usually <350 m elevation, within a matrix of Northern Coastal Scrub), or on fine-textured soils of ridgetops beyond coastal fogs (usually >750 m, within a matrix of Mixed Evergreen or North Coastal Conifer Forests). Intermittent from the Santa Cruz area north to southern Oregon.

GBGrs (Great Basin grassland)

Perennial sod-forming and bunch grasses. Presumed to have once been widespread on the Modoc Plateau and northeastern California. Currently represented as scattered, mostly small, islands in areas where grazing pressure has been low and fire frequencies higher than surrounding scrubs. Both upland and bottom-land forms occur.

VFGrs (Valley and foothill grassland)

Introduced, annual Mediterranean grasses and native herbs. On most sites the native bunch grass species, such as needle grass, have been largely or entirely supplanted by introductions. Stands rich in natives usually found on unusual substrates, such as serpentinite or somewhat alkaline soils.

VnPls (Vernal pools)

Seasonal amphibious environments dominated by annual herbs and grasses adapted to germination and early growth under water. Spring desiccation triggers flowering and fruit set, resulting in colorful concentric bands around the drying pools.

Medws (Meadows and seeps)

More or less dense grasses, sedges, and herbs that thrive, at least seasonally, under moist or saturated conditions. They occur from sea level to treeline and on many different substrates. They may be surrounded by grasslands, forests, or shrublands.

Plyas (Playas)

Non-vascular plants and sparse, gray shrubs on poorly drained soils with usually high salinity and/or alkalinity, due to evaporation of water from closed basins. Found from the Modoc Plateau to Sonoran Desert and in the San Joaquin Valley.

PbPln (Pebble or Pavement plain)

Herb- and grass-dominated openings of low cover, dominated by several cushion-forming plants endemic to dense, clay soils armored by a lag gravel of quartzite pebbles. Many of the dominant taxa are themselves rare plants. Found only in the San Bernardino Mountains.

BgFns (Bogs and fens)

Wetlands, typically occupying sites sub-irrigated by cold, frequently acidic, water. Plant growth dense and low growing, dominated by perennials herbs or low shrubs. Saturated soils frequently allow substantial accumulations of "peat." From the Klamath Ranges to North Coast Ranges, along the North Coast and in the northern Sierra Nevada.

MshSw (Marshes and swamps)

Emergent, suffrutescent herbs adapted to seasonally or permanently saturated soils. These include salt, brackish, alkali, and fresh water marshes, as well as swamps, with their woody dominants and hydrophytic herbs. Found throughout California.

RpFrs (Riparian forest)

Broadleaved, winter deciduous trees, forming closed canopies, associated with low- to mid-elevation perennial and intermittent streams. Most stands even-aged, reflecting their flood-mediated, episodic reproduction. These habitats can be found in every county and climate in California.

RpWld (Riparian woodland)

Broadleaved, winter deciduous trees with open canopies associated with low- to mid-elevation streams. Most stands even-aged, reflecting their flood-controlled, episodic reproduction. This type tends to occupy more intermittent streams, often with cobbly or bouldery bedloads.

RpScr (Riparian scrub)

Streamside thickets dominated by one or more willows, as well as by other fast-growing shrubs and vines. Most plants recolonize following flood disturbance.

CmWld (Cismontane woodland)

Trees deciduous, evergreen, or both, with open canopies. Broadleaved trees, especially oaks, dominate, although conifers may be present in or emergent through the canopy. Understories may be open and herbaceous or closed and shrubby. This type occurs on a variety of sites below the conifer forests in Mediterranean California.

PJWld (Pinyon and juniper woodland)

Open stands of round-topped conifers to 5 m. Understories frequently comprised of shrubs and herbs seen in adjacent stands lacking trees. They often form broad ecotones between higher elevation forests and lower elevation scrublands or grasslands.

JTWld (Joshua tree woodland)

Joshua trees with open canopies are usually the only arborescent species present. Shrubstories typically are diverse mixtures of microphyllous, evergreen shrubs, semi-deciduous shrubs, semi-succulents, and succulents.

STWld (Sonoran thorn woodland)

Succulents, microphyllous herbs and shrubs, especially of rocky environments. Tree-like plants the visual dominant.

BUFrs (Broadleaved upland forest)

Stands of evergreen or deciduous, broadleaved trees 5m or more tall, forming closed canopies. Many, but not all, with very poorly developed understories. Several are seral to montane conifer forests. It includes the "Mixed Evergreen Forest" of the Coast Ranges.

NCFrs (North Coast conifer forest)

Needle-leaved evergreen trees in usually quite dense stands that may attain impressive heights. Usually on well-drained, moist sites within the reach of summer fogs, but not experiencing much winter snow. This type occurs in the wetter parts of the North Coast Ranges.

CCFrs (Closed-cone conifer forest)

Dense, even-aged stands dominated by serotinous-coned conifers. Most stands are even-aged due to fire establishment. Usually associated with sterile, rocky soils, strong and steady winds, and impaired drainage. Many open stands have understories composed of chaparral or coastal scrub species from surrounding areas. Found in most areas, except for the Great Valley or deserts.

LCFrs (Lower montane conifer forest)

Open to dense stands of conifers found at lower and middle elevations in the mountains. Broadleaved trees may be present in the understory. Shrubstories may be dense assemblages of chaparral species, especially in seral stands. The upper limit of lower montane coniferous forests more or less coincides with the elevation of maximum annual precipitation.

UCFrs (Upper montane conifer forest)

Open to dense conifer forests, found at high elevations in the mountains. Trees tend to be somewhat shorter than at lower elevations. Shrubstories tend to be open, drawn from adjacent montane chaparral species, or lacking. Above the elevation of maximum precipitation, with growing seasons curtailed by winter snow accumulations.

SCFrs (Subalpine conifer forest)

Conifer forests and associated clearings of highest elevations of tree establishment. This type occurs in areas where substantial snowpack accumulation and cold temperatures limit the growing season to three months or less.

AlpBR (Alpine boulder and rock field)

Fell-fields, talus slopes, and meadows found above forest line. Favorable sites may develop continuous turf, but in most areas plants are tucked between large nurse rocks that provide protection from harsh winter conditions.

AlpDS (Alpine dwarf scrub)

Compact, woody subshrubs above forest line, adapted to short growing seasons resulting from snow accumulation or harsh winter winds.

Life Form

This field briefly describes plant duration and life form. The information was newly developed for this edition, primarily from published and unpublished literature and from herbarium material. Our simplified classification system is as follows:

Duration:

- **Annuals** grow from seed and reproduce within a single year
- **Perennials** live more than one year
- **Annual/Perennials** are variable depending on environment and conditions

Growth Form:

- **Herbs** are herbaceous and lack aboveground woody tissue
  - **bulbiferous** herbs have fleshy underground storage organs typically derived from scale leaves

(this category includes cormiferous and other similar plants in which storage organs have other origins)

rhizomatous herbs have underground stems (rhizomes), typically bearing shoots which develop into new plants

stoloniferous herbs have above-ground runners (stolons) which typically root and produce new plants

Shrubs are smaller woody perennials that retain most of their above-ground woody tissue and are typically many-stemmed

leaf succulents have thick, fleshy leaves

stem succulents have thick, fleshy stems and reduced or absent leaves

Trees are larger woody perennials that retain all of their above-ground woody tissue and are typically single-stemmed

Vines are twining woody perennials requiring external support for growth

Leaf Condition (shrubs, trees, vines only):

Deciduous plants shed their leaves for part of the year

Evergreen plants retain their leaves for the entire year

Special Habitat:

Aquatic plants are submerged or floating on the water surface

Emergent plants are rooted in water but bear some foliage out of the water

Epiphytic plants are attached above-ground to some other plant

Mode of Nutrition:

Saprophytic plants live on existing organic matter in the soil

Hemiparasitic plants are connected to host plants and derive energy, water, and minerals from them, but also maintain their own functional root systems or photosynthetic surfaces

Parasitic plants are connected to host plants and rely solely on them for energy, water, and nutritional requirements

Carnivorous plants trap insects and other small animals and derive nourishment from them

As in most classifications, some of the above distinctions are somewhat arbitrary, particularly the divisions between growth forms. Furthermore, plant growth form can vary depending on geography and local environmental conditions. Perennials that are often referred to as either suffrutescent herbs or subshrubs present special difficulties. Generally, if these plants die back seasonally to the ground or to a small crown of woody tissue we classified them as herbs, and if they retain much or all of their woody above-ground tissue we called them shrubs.

Blooming Period

This new field of information provides the months when each rare plant is typically in bloom. For ferns and other spore-bearing plants, we have given the months when spores are released and spore-bearing structures such as sori are typically present on the plant. We have not included any comparable information for gymnosperms.

With other ecological information now available in the *Inventory*, biologists can now fine-tune rare plant surveys by specifying plants which occur in particular habitats, their general appearance, and when they are identifiable, that is, when they bloom.

Notes

Many records include additional notes on distribution, endangerment, relationship to names in *The Jepson Manual*, or important literature citations. We have again included information about legal status and endangerment in neighboring states in the notes; official state designations are specifically indicated as such and capitalized, as in "State-listed as Endangered in OR." In this edition we have made a special effort to indicate missing information about distribution, endangerment, or taxonomy for each entry, in the hope that knowledgeable users will fill in the gaps. Abbreviations that are commonly used in the notes are explained on the inside back cover.

Status Report

If a rare plant status summary report is available from the Natural Diversity Data Base, the year of its most recent update is given. Status Reports are available for 475 of the plants in this volume.

The CNPS *Electronic Inventory*—A New Tool for Conserving Rare Plants

The CNPS *Electronic Inventory* is a computer application that provides instant, simplified access to the detailed information contained in this *Inventory*. The application takes advantage of Microsoft *FoxPro*'s state-of-the-art database programming capabilities, and its menu-driven features allow for easy familiarization with the program's sophisticated search and report routines. Purchase of the *FoxPro* run-time license allows CNPS to distribute the *Electronic Inventory* without requiring users to obtain any additional software.

With this new tool, resource planners, conservationists, consultants, and academicians can now view the *Inventory*, search for plants based on hundreds of specific criteria, and report on search results using customized, flexible report formats. The *Electronic Inventory* allows speedy and thorough development of lists of plants that might be affected by development proposals or other human disturbances, and we expect it to rapidly become indispensable for conservation planning activities such as preserve siting and rare plant surveys.

Moreover, the *Electronic Inventory* now replaces this volume as the most up-to-date compilation of rare, threatened, and endangered California plants. We ask users wishing access to the current, official CNPS status to consult the *Electronic Inventory* herewith.

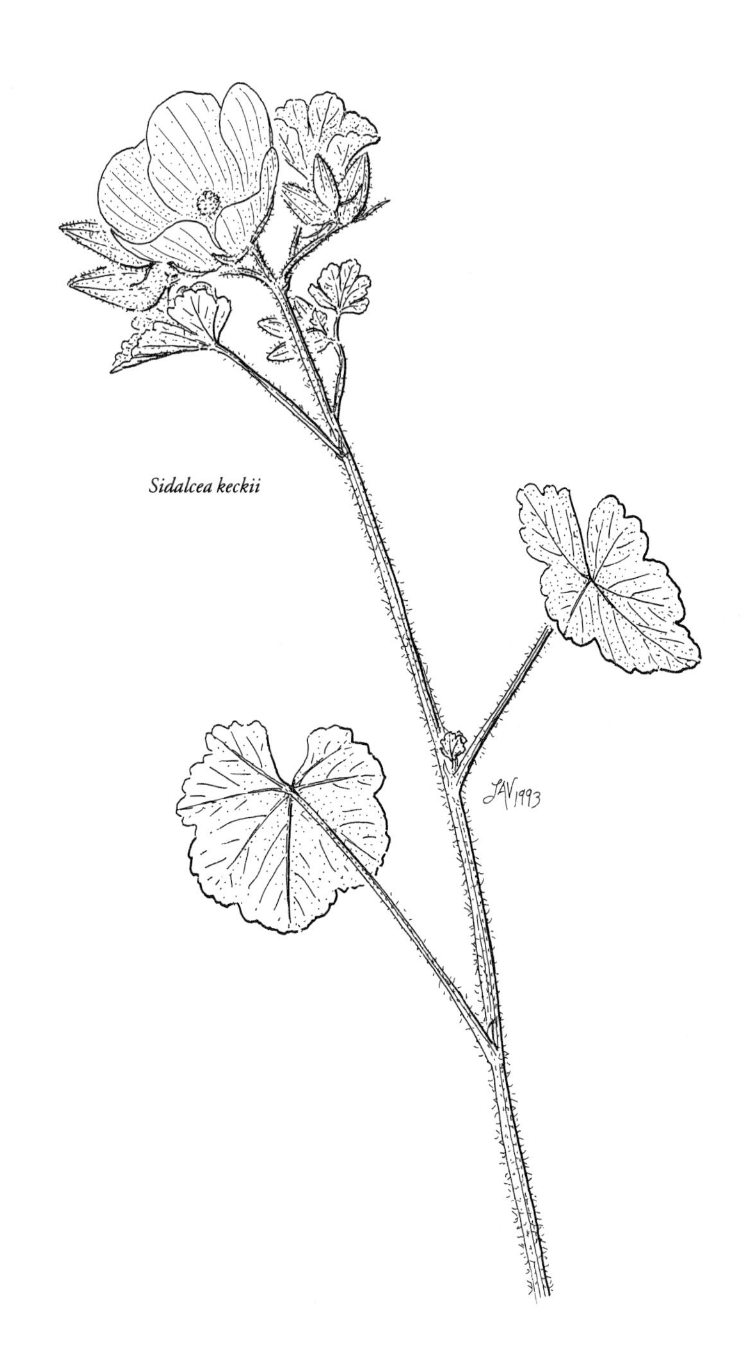

Sidalcea keckii

Conserving Plants with Laws and Programs under the Department of Fish and Game

Sandra Morey and Ken Berg

Californians face the daunting challenge of protecting one of the world's most diverse and unique floras amid the pressures imposed by the Nation's most populous state. Public concern for natural areas and species at risk has resulted in a patchwork of state environmental laws and regulations which agencies and citizens use to protect habitats for native plants and natural plant communities. California's tradition of citizen activism has played a key role in the development and use of these laws. Continuing this history of public support and involvement is critical to our future success.

As the primary trustee agency for wildlife, fish, and plants, the California Department of Fish and Game (DFG) has a central role in identifying, protecting, and managing the rare plants and natural plant communities in California. Within the constraints of its legal authority and limited funding, the Department of Fish and Game works cooperatively with federal, state, and local governments, businesses, conservation organizations, and citizens to conserve plant populations and habitats.

Identification

Identification of rare plants and natural plant communities is the first essential step in the conservation process. The Natural Diversity Data Base (NDDB) is one of the Department of Fish and Game's most visible and successful identification programs. The two articles that follow this one highlight the process and accomplishments of this effort, and the vital roles that the CNPS *Inventory* and membership play.

Another Department identification effort, which is less well-known but equally important, is the Significant Natural Areas Program. This program analyzes NDDB information to identify places that have significant populations or assemblages of rare species, or best examples of common plant communities. The resulting inventory of Significant Natural Areas is used by Department staff, other agencies, and citizens to guide land use planning and public and private land acquisition efforts. For further details on the Significant Natural Areas Program, see Hoshovsky (1992).

Protection

Legal Protection via State Law

The legal framework and authority for the State's program to conserve plants is woven from four pieces of legislation: the California Endangered Species Act (CESA), the Native Plant Protection Act (NPPA), the California Environmental Quality Act (CEQA), and the Natural Communities Conservation Planning Act (NCCPA). Under these laws, the State conserves plants through listing, habitat acquisition and protection, review of local land use planning, multispecies conservation planning, stewardship, recovery, research, and education.

State Listing

Under State law, plant species may be formally designated rare, threatened, or endangered by the California Fish and Game Commission, a five member board appointed by the Governor to establish the policies by which the DFG operates. State listing is a way of formally recognizing the plight of a species and the need to protect its habitat. Once a species is officially listed, it has a greater chance of being protected. Listed plants are generally given greater attention during the land use planning process by local governments, public agencies, and landowners than are plants that have not been listed. State-listed threatened and endangered species and designated candidates are protected from taking except for scientific or management purposes, which requires a permit or agreement from the DFG. In addition, State agencies are required to formally consult with the DFG on State projects which may affect listed threatened or endangered species or candidates.

State listing of plants began in 1977 with the passage of the Native Plant Protection Act. The NPPA directed the DFG to carry out the Legislature's intent to "preserve, protect and enhance endangered plants in this State." The NPPA gave the California Fish and Game Commission the power to designate native plants as endangered or rare, and to require permits for collecting, transporting, or selling such plants.

In 1984 the state legislature enacted the California Endangered Species Act in recognition of the tremendous threats facing California's native plant and animal populations and their habitats. This legislation declares that deserving plants and animals will be given protection by the state because they are of ecological, educational, historical, recreational, aesthetic, economic, and scientific value to the people of the state. CESA established that it is state policy to conserve, protect, restore, and enhance endangered species and their habitats.

The California Endangered Species Act expanded upon the original NPPA and enhanced legal protection for plants. To align with Federal regulations, CESA created the categories of "threatened" and "endangered" species. It grandfathered all "rare" animals into the Act as threatened species, but did not do so for rare plants. Thus, there are *three* listing categories for plants in California: rare, threatened, and endangered.

The following definitions are found within the two acts (Fish and Game Code Sections 1901, 2062, and 2067). A native plant is **endangered** when "its prospects of survival and reproduction are in immediate jeopardy from one or more causes." A native plant is **threatened** when "although not presently threatened with extinction, it is likely to become an endangered species in the foreseeable future in the absence of the special protection and management efforts...". A native plant is **rare** when "although not presently threatened with extinction, it is in such small numbers throughout its range that it may become endangered if its present environment worsens." The CESA also creates a "candidate" category. A **candidate** is a taxon that has been officially noticed by the Commission as being under review by the DFG for addition to the rare, threatened, or endangered species lists.

The CESA establishes a process by which individuals, organizations, or the DFG can submit petitions to the Fish and Game Commission requesting that a species, subspecies, or variety of plant or animal be added to, deleted from, or changed in status on the State lists of rare, threatened, or endangered species. The factors that contribute to determining the need to list a species include the present or threatened modification or destruction of habitat, competition, predation, disease, overexploitation by collectors, or other natural occurrences or human-related activities.

We strongly advise individuals and organizations to consult with the CNPS Botanist and the DFG's Endangered Plant Program staff prior to preparing listing petitions. A standard petition format is available from the Fish and Game Commission Office at 1416 Ninth Street, Sacramento, CA 95814.

Listing Actions, 1988-1993

Table 1 presents a summary of plant listing actions taken by the Fish and Game Commission between 1988-1993. Together with CNPS and concerned individuals, the DFG and the Commission have made great progress in listing and protecting plant species during recent years. However, listings have become increasingly controversial, and more landowners and project proponents have publicly opposed plant listings. With increased controversy surrounding plant listings, the DFG has had to increase its efforts to notify all potentially concerned parties, and to more thoroughly document the need for each listing. Staffing to administer the listing process has not increased, however, and the ability of the Department to adequately address other aspects of plant conservation has been reduced. Recognizing this situation, the CNPS has coordinated with the Department in submitting petitions for only the highest priority plants in the state. Appendix VI contains a complete roster of species listed under State law as of December 1993, as well as candidates for listing.

Table 1. Plant listing actions under the California Endangered Species Act, 1988-1993.

Year	No. of Petitions—Petitioner	No. of Designated Candidates	No. of Plants Listed or Denied
1988	26 - CNPS	25 designated 1 denied	
1989			22 listed 3 denied
1990	1 - individual 1 - DFG	1 designated 1 denied	
1991	8 - CNPS	5 designated	1 listed
1992		2 designated 1 withdrawn*	3 listed
1993			2 listed 1 denied** 1 decision pending
TOTAL	**36 petitions**	**33 designated candidates**	**28 listed (78% of petitions)**

* The petition for *Allium tuolumnense* was withdrawn by CNPS.
** *Arctostaphylos morroensis* is now included in a multispecies regional planning effort, and was not listed.

Habitat Protection

The Department of Fish and Game protects, maintains, and enhances plant and animal populations and natural communities through direct acquisition of habitat, conservation easements on private lands, and management agreements with public and private agencies and organizations. The Department works with conservation partners including The Nature Conservancy, CNPS, local land trusts, U.S. Forest Service, U.S. Bureau of Land Management, and other public agencies and organizations to promote conservation of all wildlife resources. The DFG Lands and Natural Areas Program coordinates Department activities associated with the acquisition, lease, or management of suitable lands, which are approved by the Wildlife Conservation Board.

Currently, the Department administers over 753,000 acres in California, much of which has been designated by the Fish and Game Commission as Wildlife Areas or Ecological Reserves. Over 8,000 acres have been acquired in fee title specifically for the protection of endangered plant populations and their habitats. Examples of DFG plant reserves include Table Bluff in Humboldt County, Pine Hill and Salmon Falls in El Dorado County, Bonny Doon in Santa Cruz County, and Baldwin Lake in San Bernardino County. Many Department lands acquired for other wildlife species also contain populations of rare plants or special natural communities. Funding for acquisitions comes from a variety of sources, including bond acts such as 1988's Proposition 70 that have allowed the purchase and protection of more than 200,000 acres of wildlife habitat.

Local Land Use Planning

California has a strong state law which provides for protection of species during the land use planning process. This law is the California Environmental Quality Act, enacted in 1970. CEQA requires government agencies to consider environmental impacts of projects and to avoid or mitigate them where possible. Under CEQA, Environmental Impact Reports (EIR's) must be prepared to disclose environmental impacts of a project and outline project alternatives. Through the EIR process, the public can review proposed project plans and influence the process through public comment.

DFG regional staff work with project proponents, local governments, and other agencies to guide adequate environmental review and to design appropriate mitigation. CEQA provides protection not only for State-listed species, but for any species which can be shown to meet the criteria for State listing (CEQA Section 15380). The DFG recognizes that Lists 1A, 1B, and 2 of the CNPS *Inventory* consist of plants that, in a majority of cases, would qualify for listing, and the Department recommends they be addressed in EIR's. However, a plant need not be in this *Inventory* to be considered under CEQA if it is a *de facto* rare, threatened, or endangered species. In addition, the DFG recommends, and local governments may require, protection of plants which are regionally significant, such as locally rare species, disjunct populations of more common plants, or plants on the CNPS Lists 3 and 4.

To guide documentation of potential impacts to plants, the DFG has adopted *Guidelines for Assessing Effects of Proposed Developments on Rare and Endangered Plants and Plant Communities,* adapted from those prepared by the CNPS (see *Guidelines...,* below). These guidelines are given out to all project proponents, lead agencies, and the interested public when they request DFG participation or information.

Due to lack of adequate staffing, the DFG is unable to comment on every project that may pose significant environmental impacts. In addition, we have learned that small, piecemeal mitigation efforts, which in the past usually involved the transplantation of endangered plant populations, usually fail. For mitigation of project impacts, the Department now requires the protection of intact habitat and restoration of degraded habitat, rather than transplantation of plant populations.

Multispecies Conservation Planning

The most effective long-term protection for species and habitats comes from protection of larger, ecologically viable sites, including whole watersheds, ecosystems, or "landscapes". Consequently, the DFG is increasingly focusing its efforts on regional multispecies planning throughout the state. In 1991, DFG signed a Memorandum of Understanding with 10 State and Federal agencies to improve statewide coordination of biological diversity conservation. By improving communication and coordination at both state and regional levels, strong conservation partnerships are emerging. Multispecies planning efforts are being carried out throughout California in cooperation with many conservation partners with similar goals, including local, state, and federal agencies, land trusts, conservation organizations, and landowners.

In 1991, under the Natural Communities Conservation Planning Act, a pilot conservation planning program was initiated for the southern California coastal sage scrub plant community. This community is home to an estimated 96 endangered plant and animal species, including the Federally-threatened California gnatcatcher. The NCCPA was enacted to promote long-term protection of species and habitats via regional, multispecies planning, before the special measures provided by CESA become necessary. The NCCPA does not supersede CESA or CEQA, but gives the Department the authority to enter into agreements with any person or entity to protect areas large enough to ensure the continued existence of multiple species and their habitats, while allowing for "reasonable and appropriate" urban

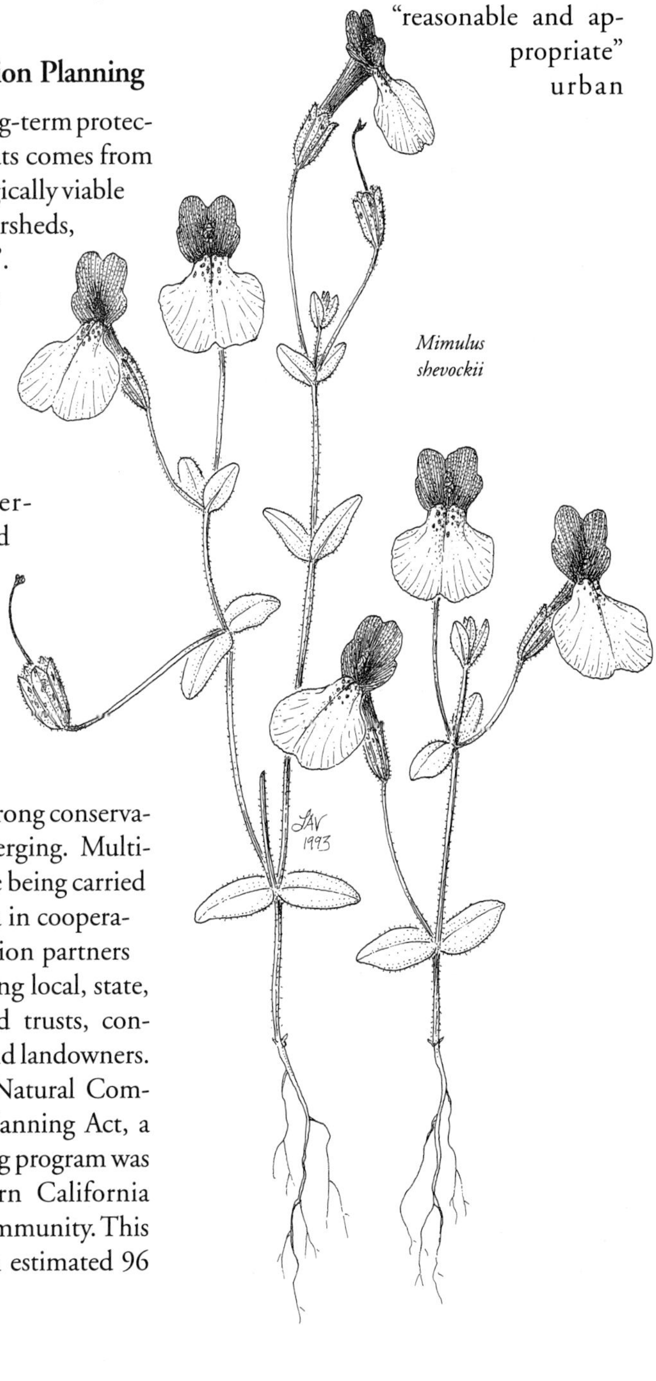

Mimulus shevockii

JAV 1993

growth. Because many rare plants have very narrow distributions, large scale natural community-based conservation efforts must be carefully planned to include their habitat in the preserve areas.

DFG is involved in other multispecies conservation planning efforts for plants throughout California, including western El Dorado County, Sonoma County's Santa Rosa Plains, San Luis Obispo's Morro Bay region, western Riverside County, the Owens Valley, and a large scale effort in the western Mojave Desert.

Management

In 1991, DFG established 15 new positions, including six plant ecologists, which were placed in its five regional offices to work on complex conservation issues and recovery of endangered plants and habitats (see *Appendix X*). The establishment of this program has greatly increased the Department's ability to conserve rare plant species and plant communities throughout the state.

The DFG is conducting interagency recovery workshops to identify management and recovery priorities and to track the implementation and success of on-the-ground efforts. The Department contracts with public agencies, organizations, and individuals to implement these management and recovery activities. Projects can be as simple as fencing a fragile alkaline meadow from livestock, or as complex as developing a regional plan for protection of a rare natural community which spans several counties.

Academic researchers throughout California work with the Department under Memoranda of Understanding to answer questions which may be important to the conservation of listed plant populations. Research may focus on population genetics, reproductive strategies of plants, long-term population trends, habitat characterization, or other pertinent research topics.

A variety of funding sources are available to the DFG to promote recovery of endangered plant populations and to restore degraded habitats. Funding sources include the California Endangered Species Tax Check-Off Fund, U.S. Fish and Wildlife Service support under the Federal Endangered Species Act Section 6 provisions for cooperation with the states, the Tobacco Tax and Health Initiative (Proposition 99), the Department of Parks and Recreation's Off-Highway Vehicle Fund, the Environmental License Plate Fund, the California Emergency Drought Recovery Program, and the California Department of Transportation Environmental Enhancement and Mitigation Grants.

Continued Public Support Is Essential

Every resident and visitor to California has the ability to help conserve our dwindling native plant habitats. Involvement and support from California Native Plant Society members and other concerned citizens is the essential ingredient for effective plant conservation. It affects every aspect of our efforts: the strength of laws, the staffing, funding, and management support available to implement and enforce them, and the political support essential to retain and improve them. Moreover, information and citizen involvement is often what swings the pendulum of decision-making toward a conservation alternative.

The Department of Fish and Game works to improve public awareness of native plants and to increase support for plant conservation in a variety of ways. One recent effort has assisted the Department of Education's development of the Endangered Species Education Project, initiated by state Senator Gary Hart in 1989. Through this program, students in public schools conduct projects to raise public awareness of the plight of local endangered species and help garner support to protect them. Awards are given to the most creative projects. The California Native Plant Society and Audubon Society have helped distribute project information packets to hundreds of teachers and schools.

Other Department public outreach activities include developing interpretive services at Wildlife Areas and Ecological Reserves, publishing the monthly wildlife conservation magazine *Outdoor California*, coordinating Project Wild, a program to train public school teachers, and leading the Endangered Species Campaign to encourage contributions to the Endangered Species Tax Check-Off Fund.

In a democracy, citizen involvement greatly influences the quality of government. To improve State government efforts in plant conservation, citizens must take an active role. California's voiceless plants need us to speak for them. Their futures and ours depend on it.

Sandra Morey is the Coordinator of the Endangered Plant Program, California Department of Fish and Game, 1416 9th St., Sacramento, CA 95814. Ken Berg, former Coordinator of the Endangered Plant Program, is now the National Botanist for the Bureau of Land Management, 2800 Cottage Way, Sacramento, CA 95825.

The Natural Diversity Data Base: California's Natural Heritage Program

Roxanne L. Bittman

The Natural Diversity Data Base (NDDB) is one of California's most valuable tools for rare plant conservation. As part of the Natural Heritage Division of the California Department of Fish and Game, the NDDB inventories the locations of the state's rarest species and natural communities. Its goal is to help conserve California's biotic diversity by providing government agencies and the private sector with information so that wise land-use decisions and resource management can occur. The NDDB is used to identify important natural areas, and also in project planning to avoid conflicts between environmental and development interests.

Using the NDDB to Protect Rare Plant Populations

There are many potential applications for the NDDB records, but those concerning land development and environmental impact review are especially important. Rapid population growth and economic expansion need to be carefully reconciled with protection of California's resource base, and project developers must often consult with DFG's Natural Heritage Division. The developer, usually with a specific geographic location of concern, requests information on the rare plants, animals, and natural communities that may occur at that location. NDDB data, along with data from the CNPS *Inventory*, thus often constitute the first cut for developers or their consultants when devising a "potentials list" of special plants, animals, and communities that may occur in a project area. This is then followed by surveys for rare species and natural communities during the appropriate time of year by qualified biologists and ecologists (see *Guidelines...*, below).

Consider, for example, the results of a search of the Middletown 7.5 minute quadrangle (quad), which hypothetically contains a proposed development project (Figure 1). This search revealed the presence of two rare plants and one rare natural community in the vicinity. The impacts on the biological resources of this proposed project can then be evaluated by the developers' consultants, and issues of project feasibility, cost, and mitigation can be adequately addressed. If such information is obtained early in the planning process, needless conflicts, stalled permits, and the costs of project delay might all be avoided. It should be emphasized that acquisition of NDDB data is not in itself sufficient to satisfy the re-

```
                         California Department of Fish and Game * Natural Diversity Data Base

** Element ID: CTT44131CA ************************************ * ******* List Status ******** ********* Other Lists ********* *
*  NORTHERN BASALT FLOW VERNAL POOL                                  Federal: None                     CDFG:                  *
*                                                                      State: None                  Audubon:                  *
*       NDDB Element Ranks - Global: G1; State: S2.2                                          CNPS List/Code:   /             *
*  Habitat Associations -                                                                                                     *
*     Not available at this time.                                                                                             *
*     Not available at this time.                                                                                             *
** California Department of Fish and Game * * * * * * * * * * * * * * * * * * * * * * * * * * * * * * * * * * * * * * * * * **

   Occurrence #: 11        Last Seen - Element: 1980/XX/XX           Lat/Long: 38d 50m 23s / 122d 31m 57s    Township: 12N
        Quality: Unknown                   Site: 1980/XX/XX               UTM: Zone-10 N4298890 E540576          Range: 06W
           Type: Natural/Native occurrence               Mapping Precision: NON-SPECIFIC (1/5 Mile)        Section: 33 SW Qtr
       Presence: Presumed Extant                               Symbol Type: POINT                         Meridian: M
          Trend: Unknown                                      Group Number: 08649   More Information? N      Acres: 0
    Info Source: BURKE, M. T. 1980 (LIT)                  Map Index Number: 08649    More Map Detail? N  Elevation: 1750  ft
   Quad Summary: Middletown (3812275)
    County(ies): Lake
       Location: STIENHART LAKE. 6 MI NE MIDDLETOWN.
        Threats: DREDGING OF ADJ AREA UNDERWAY TO MAINTAIN & DEVELOP FISH PONDS.
       Comments: Ecological Notes - LARGE VERNAL LAKE. ENTIRE LAKEBED COVERED W/VERNAL POOL SPECIES. DOMINATED BY POGOGYNE
                 DOUGLASII, ERYNGIUM ARISTULATUM & MIMULUS TRICOLOR.

** Element ID: PDCRA0F020 ************************************ * ******* List Status ******** ********* Other Lists ********* *
*  PARVISEDUM LEIOCARPUM                                             Federal: Category 1               CDFG:                  *
*  Lake County Stonecrop                                               State: Endangered            Audubon:                  *
*       NDDB Element Ranks - Global: G1; State: S1.1                                          CNPS List/Code: 1B/3-3-3        *
*  Habitat Associations -                                                                                                     *
*     VALLEY AND FOOTHILL GRASSLAND, VERNAL POOLS                                                                              *
*     LEVEL AREAS THAT ARE SEASONALLY WET AND DRY OUT IN LATE SPRING, SUBSTRATE USU OF VOLCANIC ORIGIN; 1300-2600 FT.         *
** California Department of Fish and Game * * * * * * * * * * * * * * * * * * * * * * * * * * * * * * * * * * * * * * * * * **

   Occurrence #: 5         Last Seen - Element: 1986/04/02           Lat/Long: 38d 51m 29s / 122d 35m 17s    Township: 12N
        Quality: Good                      Site: 1986/04/02               UTM: Zone-10 N4300897 E535743          Range: 07W
           Type: Natural/Native occurrence               Mapping Precision: NON-SPECIFIC (1/5 Mile)        Section: 25 SW Qtr
       Presence: Presumed Extant                               Symbol Type: POINT                         Meridian: M
          Trend: Unknown                                      Group Number: 08477   More Information? Y      Acres: 0
    Info Source: PATTERSON, C. 1986 (LIT)                 Map Index Number: 08477    More Map Detail? Y  Elevation: 1700  ft
   Quad Summary: Middletown (3812275)
    County(ies): Lake
       Location: LITTLE HIGH VALLEY, WEST SIDE, APPROX 5 AIRMI SE OF MANNING FLAT.
        Threats: SOME MINOR GRAZING IN THE AREA.
       Comments: Ecological Notes - SMALL, LEVEL EXPOSED ROCK SURFACES WITH MINIMAL SOIL.  IN ANNUAL GRASSLAND.  ASSOCIATED WITH
                 CRASSULA MUSCOSA, PLANTAGO ERECTA, LASTHENIA CALIFORNICA.  General Notes - 2 POPULATIONS TOTALLING 800 PLANTS
                 IN 1986.  POPULATION DISCOVERED BY JOKERST AND BOWCUTT IN 1985.  Owner/Manager - UNKNOWN

** Element ID: PMPOA4G050 ************************************ * ******* List Status ******** ********* Other Lists ********* *
*  ORCUTTIA TENUIS                                                   Federal: Proposed Threatened      CDFG:                  *
*  Slender Orcutt Grass                                                State: Endangered            Audubon:                  *
*       NDDB Element Ranks - Global: G2; State: S2.2                                          CNPS List/Code: 1B/2-3-3        *
*  Habitat Associations -                                                                                                     *
*     VERNAL POOLS                                                                                                            *
*     Not available at this time.                                                                                             *
** California Department of Fish and Game * * * * * * * * * * * * * * * * * * * * * * * * * * * * * * * * * * * * * * * * * **

   Occurrence #: 36        Last Seen - Element: 1987/06/29           Lat/Long: 38d 50m 45s / 122d 32m 41s    Township: 12N
        Quality: Good                      Site: 1987/06/29               UTM: Zone-10 N4299587 E539488          Range: 06W
           Type: Natural/Native occurrence               Mapping Precision: SPECIFIC     (0   Mile)        Section: 32 NE Qtr
       Presence: Presumed Extant                               Symbol Type: POLYGON                       Meridian: M
          Trend: Unknown                                      Group Number: 08603   More Information? N      Acres: 6.5
    Info Source: MCCARTEN ET AL 1987 (OBS)                Map Index Number: 08603    More Map Detail? N  Elevation: 1840  ft
   Quad Summary: Middletown (3812275)
    County(ies): Lake
       Location: NW LAKE OF STIENHART LAKES, N OF COYOTE VALLEY.
        Threats: NO VISIBLE DISTURBANCES, THOUGH GRAZING MAY OCCUR.
       Comments: Ecological Notes - ARTIFICIAL POND IN VOLCANIC ASH VERNAL POOL.  NAVARRETIA PLIEANTHA IN POOL ALSO. GRATIOLA
                 HETEROSEPALA IN POOLS NEARBY.  General Notes   A FEW HUNDRED PLANTS SEEN.  Owner/Manager - PVT



* S * - Sensitive information; use discretion when disclosing locational details.              Government/Conservation Client
Middletown 7.5' Quadrangle                     Date Information Purchased: 10/25/93        Date of Report: 11/29/93      Page 1
```

Figure 1. Sample NDDB text output from a search of the Middletown 7.5 minute quadrangle.

quirements of the California Environmental Quality Act. Biological surveys, performed by qualified personnel at the appropriate time of year, are essential to document what is actually present on the parcel of concern.

Citizens or conservation groups interested in the protection of biological diversity in California are also encouraged to use the NDDB. Customers can access the NDDB in a number of ways, depending on their needs. The most popular way is by purchasing *RareFind*, a menu-based program in *FoxPro*, which allows access to all of the textual information in the NDDB. Data can be accessed by searching for individual counties, quads, or taxa, and various reports are possible. Another common way to acquire NDDB data is to request the textual and graphic information for a particular area (a quad, for example). These data are presented as a vellum overlay to a specific topographic map along with an accompanying text report on the rare elements appearing on the overlay.

NDDB information is available on a cost reimbursement basis. The most common users are conservationists, state and federal agencies, consultants, and researchers. NDDB data are half-price for conservation organizations and other non-profit groups. Often NDDB reports are offered free to citizens interested in reporting back to us regarding the current condition and quality of populations or stands of rare elements.

In general, NDDB data for plants are made available to CNPS Chapters on a yearly basis. Some CNPS Chapters receive *RareFind*, and others receive text reports, along with copies of vellum overlays at the 1:100,000 scale. You may request information for local areas by contacting us at the address below. Most requests take about two weeks to process. Requests for large amounts of data may require special processing; contact the NDDB for details.

Updating the Data Base: Contributions from the Field

As of November 1993, the NDDB computer files contained approximately 9,210 records on 775 plants (out of 1,751 plants on the Special Plants list), 9,993 records on 337 animals (out of 608 animals on the Special Animals list) and 2,343 records on 87 of our most endangered natural communities (out of the approximately 135 rare natural communities on the list). Biologists throughout the state have contributed to collection of these data which describe the locations, ecology, and status of California's rare, threatened, and endangered plants, animals, and natural communities. The original data set for plants largely stemmed from herbarium records from institutions such as the University of California Berkeley, California Academy of Sciences, U.C. Riverside, Rancho Santa Ana, Humboldt State University and others. Recent, updated plant location data comes largely from surveys done by members of the California Native Plant Society, agency biologists, and consultants. Data contributions are encouraged.

To contribute information on population fluctuations, new locations, and habitat integrity, we encourage you to use California Native Species Field Survey Forms (see *Appendix IX*). We have new and improved versions of the forms for both species and natural communities available from the NDDB to anyone who wants them. Attaching a location map to the form is essential to ensure that we accurately enter the information into the computer.

Roxanne Bittman is the Botanist at the Natural Diversity Data Base, Department of Fish and Game, 1416 Ninth St., Sacramento, CA 95814.

The Natural Communities Program

Todd Keeler-Wolf

The Natural Diversity Data Base (NDDB) Natural Communities program has the responsibility of maintaining up–to–date records of the state's rare natural communities. This is a separate function from that of the NDDB rare plant program but it is in every way analogous. The rationale for inventorying natural communities in conjunction with individual species is grounded in the philosophy of the community as an amalgamation of species interacting within a common physical environment (an ecosystem). As such, the community responds to natural or unnatural environmental changes and can be thought of as an indicator of the overall health of the ecosystem and its component species.

The natural communities inventory is a kind of coarse filter which maintains surveillance on a broader spectrum of biological diversity than can be accomplished through monitoring only the individual rare species. A detailed understanding of the distribution and population dynamics of each species in the entire state would be ideal. However, 1742 taxa of plants are considered rare enough in California to include within the CNPS *Inventory*, and maintaining records on the total native flora (about 6300 taxa) is beyond our current capabilities (a subset of about 775 plant taxa are actively tracked in the NDDB plant program). Hence, the natural communities program is a practical compromise focused on sustaining the biological firmament of the state.

Defining Natural Communities

Although they are complex assemblages of species, communities do share certain similarities with species. Like species, some communities are commonplace, some naturally rare, and some have become rare as a result of anthropogenic influences. California's environmental variation is extreme, from coast to mountains to deserts, and from sand to serpentinite outcrops to flooded alkaline sinks. Like plants, a given natural community typically occurs in only a small portion of this environmental spectrum, and species composition in different communities varies widely and is often non-overlapping. The consequence of this diversity, coupled with a vast range of detrimental impacts, is some communities which contain a wealth of rare species and are indeed rare themselves, and other communities that are rare, but do not contain rare species. Although the NDDB maintains records for only the rare types, the classification system we use considers all of California's naturally occurring communities.

Natural communities are usually most easily defined by their dominant plant species, and many familiar communities are so described. Redwood Forest, Coast Live Oak Woodland, and Valley Needlegrass Grassland are easily classified on this basis. However, some, such as Mixed Evergreen Forest or Northern Mixed Chaparral, are not clearly dominated by a single species and may be co-dominated by several. Other communities, exemplified by Montane Dwarf Scrub or Mojave Mixed Woody and Succulent Scrub, have such distinct structural characteristics that they are most easily described structurally. Yet others have sparse vegetative cover, or overriding physical or chemical properties, and are named solely by their physical characteristics (e.g., Northern Foredune, Fen, Freshwater Seep, Dry Alpine Talus, and Scree Slope).

The NDDB inventory is based upon a classification devised by Cheatham and Haller (1975) to guide selection of sites for inclusion in the University of California Natural Reserve System. Their scheme recognized about 250 community types. California NDDB ecologists have modified this system over the past decade and added several new types. Currently, the Data Base recognizes about 280 individual communities as endpoints within a multi–level hierarchy. About 135 of these are considered rare enough to warrant tracking (Table 1). The habitat clas-

Table 1. Examples of common and rare natural communities identified by the NDDB. Further descriptions can be found in Holland (1986).

Community Name	NDDB Number	Geographic Range
Common Natural Communities:		
Mojave Creosote Scrub	34100	Mojave Desert 2000-4000 ft.
Chamise Chaparral	37200	Cismontane CA 200-5000 ft.
Coast Live Oak Woodland	71160	Coast Ranges S. of Mendocino County
Northern Juniper Woodland	72110	Modoc Plateau, NE. CA
Sierran Mixed Coniferous Forest	84230	Mid-elevation mountains W. of deserts
Rare Natural Communities:		
Relictual Interior Dunes	23200	S. San Joaquin Valley
Ione Chaparral	37000	Central Sierra foothills
Valley Sacaton Grassland	42120	Great Valley
Sycamore Alluvial Woodland	62100S.	Coast Ranges, S. Sierra foothills
Washoe Pine - Fir Forest	85220	NE. CA Mtns. bordering Great Basin

sification system used in this edition of the CNPS *Inventory* is an adaptation of that used by NDDB, equivalent to the mid-levels of the classification hierarchy. These generalized community types are described earlier in this volume (see *Habitat Types,* above).

Over the past few years, the California NDDB inventory has become widely used by various state and federal agencies, conservation groups, and consultants to identify and protect sensitive natural communities. As the CNPS *Inventory* goes to press the NDDB community classification is undergoing major evolution from a descriptive non-technical system to a more finely-honed, data-driven classification. This more technical approach is necessary because of the increasing awareness by conservationists that the ecosystem, and not the individual species, is the level of protection that will afford the greatest ecological security for the long term. Yet, in order to achieve community-level conservation we must have rigorous, defensible, community definitions. The CNPS Plant Communities Program, in conjunction with the NDDB, is currently striving to achieve this long-term goal.

Conservation at the Community Level

California's Natural Community Program provides support to the CNPS Plant Communities Committee, a balanced cross-section of State agencies, academia, private consulting firms, and CNPS members. The State and this Committee are working together to develop a uniform system for describing and recognizing California's plant communities, beginning with the rarest and most imperiled. With diagnostic, unambiguous definitions and accurate distributional information, we hope to be able to present scientifically supportable community descriptions which will eventually be the foundation for legislation to protect the state's rare and threatened ecosystems. Another benefit of this coordinated effort is a forthcoming book describing the state's plant communities, including an inventory of the rare types.

The NDDB currently has about 2300 individual records of rare communities—a small number compared to the almost 10,000 plant records. The natural community database has always encouraged submission of information about rare communities, and has devised its own field survey form to facilitate input of this information (see *Appendix IX*). To further refine our classification, we are now especially interested in receiving detailed records for as many different rare communities as possible. Field survey forms, complete community lists, rarity information, and descriptions of the communities are available from our office.

Todd Keeler-Wolf is the Vegetation Ecologist at the Natural Diversity Data Base, Department of Fish and Game, 1416 9th St., Sacramento, CA 95814.

Guidelines for Assessing Effects of Proposed Developments on Rare Plants and Plant Communities

James R. Nelson

The following recommendations, adopted by the California Native Plant Society and the California Department of Fish and Game, are intended to help those who prepare and review environmental documents determine **when** a botanical survey is needed, **who** should be considered qualified to conduct such surveys, **how** field surveys should be conducted, and **what** information should be contained in the survey report. Additional considerations and techniques for surveying rare plants are found in Nelson (1987).

Survey Guidelines

1. Botanical surveys that are conducted to determine the environmental effects of a proposed development should be directed to all rare, threatened, and endangered plants and rare plant communities. The plants are not necessarily limited to those species which have been "listed" by State and Federal agencies but should include any species that, based on all available data, can be shown to be rare and/or endangered.
2. It is appropriate to conduct a botanical field survey to determine if, or the extent that, rare plants will be affected by a proposed project when:
   a. Based on an initial biological assessment, it appears that the project may damage potential rare plant habitat;
   b. Rare plants have historically been identified on the project site, but adequate information for impact assessment is lacking; or
   c. No initial biological assessment has been conducted and it is unknown whether or not rare plants or their habitat exists on the site.
3. Botanical consultants should be selected on the basis of possession of the following qualifications (in order of importance):
   a. Experience as a botanical field investigator with experience in field sampling design and field methods;
   b. Taxonomic experience and a knowledge of plant ecology;
   c. Familiarity with the plants of the area, including rare species; and
   d. Familiarity with the appropriate State and Federal statutes related to rare plants and plant collecting.
4. Field searches should be conducted in a manner that will locate any rare or endangered species that may be present. Specifically, rare plant surveys should be:
   a. Conducted at the proper time of year when rare or endangered species are both "evident" and identifiable. Field surveys should be scheduled to coincide with known flowering periods, and/or during periods of phenological development that are necessary to identify the plant species of concern.
   b. Floristic in nature. "Predictive surveys" (which predict the occurrence of rare species based on the occurrence of habitat or other physical features rather than actual field inspection) should be reserved for ecological studies, not for impact assessment. Every species noted in the field should be identified to the extent necessary to determine whether it is rare or endangered.
   c. Conducted in a manner that is consistent with conservation ethics. Collections of rare or suspected rare species (voucher specimens) should be made only when such actions would not jeopardize the continued existence of the population and in accordance with applicable State and Federal permit regulations. Voucher specimens should be deposited at recognized public herbaria for future reference. Photography should be used to document plant identification and habitat whenever possible, but especially when the population cannot withstand collection.
   d. Conducted using systematic field techniques in all habitats of the site to ensure a reasonably thorough coverage of potential impact areas.
   e. Well documented. When a rare or endangered plant (or rare plant community) is located, a California Native Species (or Community) Field Survey Form or equivalent written form should be completed and submitted to the Natural Diversity Data Base.
5. Reports of botanical field surveys should be included in or with environmental assessments, negative declarations, EIR's and EIS's, and should contain the following information:
   a. Project description, including a detailed map of the project location and study area.
   b. A written description of biological setting referencing the community nomenclature used, and a vegetation map.
   c. Detailed description of survey methodology.
   d. Dates of field surveys.
   e. Results of survey (including detailed maps).
   f. An assessment of potential impacts.
   g. Discussion of the importance of rare plant populations with consideration of nearby populations and total species distribution.
   h. Recommended mitigation measures to reduce or avoid impacts and a monitoring program to measure the success of the mitigation.
   i. List of all species identified.
   j. Copies of all California Native Species Field Survey Forms or Natural Community Field Survey Forms.
   k. Name of field investigator(s).
   l. References cited, persons contacted, herbaria visited, and disposition of voucher specimens.

James R. Nelson is the Region 1 Natural Heritage Supervisor with the California Department of Fish and Game, 601 Locust St., Redding, CA 96001.

The Endangered Species Act and Rare Plant Protection in California

Jim A. Bartel, Mark W. Skinner, and Jan C. Knight

Designed by Congress to slow or stop anthropogenic extinctions of various species of fish, wildlife, and plants in the United States, the Endangered Species Act also protects foreign species and species threatened by non-human causes. President Richard Nixon signed the Endangered Species Act (Act or ESA) in December 1973, and in so doing repealed much of the Endangered Species Conservation Act of 1969, which had replaced the Endangered Species Preservation Act of 1966. This new law extended Federal protection for the first time to plants and to animals other than vertebrates, mollusks, and crustaceans, and established a category for threatened species to protect plant and animal species before they reach dangerously low numbers. Although amended several times since its passage in 1973, the ESA remains essentially intact. The Act is administered primarily by the U.S. Fish and Wildlife Service (Service) in the Department of the Interior, and to a lesser extent by the National Marine Fisheries Service in the Department of Commerce. For simplicity's sake, the discussion below focuses on Service implementation of the Act.

Terminology under the Endangered Species Act

"Endangered" and "threatened" are words often used loosely or interchangeably in a variety of contexts. Nonetheless, according to the provisions of the Endangered Species Act and its implementing regulations, crucial legal differences exist between "listed" endangered and threatened species, "proposed" endangered and threatened species, and "candidate" species. Discrimination between these terms is essential because the level of protection afforded these three groups varies greatly.

Section 3 of the Act defines an **endangered** species as any species, including subspecies, "in danger of extinction throughout all or a significant portion of its range." The Service considers varieties to be equivalent to subspecies and, thus, these are also covered under the Act. This section further defines **threatened** species as any species "likely to become an endangered species within the foreseeable future throughout all or a significant portion of its range." "Federally-listed" or "listed" indicates that a species has been designated as endangered or threatened through publication of a final rule in the *Federal Register.*

Proposed endangered and threatened species are those species for which a proposed regulation, but not a final rule, has been published in the *Federal Register.* **Candidate** species are taxa the Service is considering for listing as endangered and threatened species which have yet to be the subject of a proposed rule. The Service periodically publishes a notice of review in the *Federal Register* listing the current candidate species. The latest notice was published on September 30, 1993 (*Federal Register* 58:51144-51190).

The Service divides candidate species into two categories. **Category 1** candidates are "taxa for which the Service currently has on file substantial information on biological vulnerability [relating to autecology and distribution] and threat(s) to support the appropriateness of proposing to list the taxa as endangered or threatened species." **Category 2** candidates are "taxa for which information now in the possession of the Service indicates that proposing to list them as endangered or threatened species is possibly appropriate, but for which substantial data on biological vulnerability and threat(s) are not currently known or on file to support the immediate preparation of rules." Thus, the two categories delimit level of information and *not* degree of threat or biological vulnerability.

Category 3 consists of non-candidates, plants previously considered candidates and included on past lists. These former candidate plants have been grouped into three subcategories: extinct (3A), taxonomically invalid or not meeting the Service's definition of a species (3B), or too widespread or not threatened at this time (3C).

California has 626 candidate species, 136 in Category 1 (including 20 that are thought to be extinct) and 490 in Category 2, as well as 240 former candidates in Category 3. There are currently 58 proposed and 48 Federally-listed plants in California (see *Appendix VII* for a complete roster).

Protection under the Endangered Species Act

Candidate Species

Candidate species are afforded no protection under the Endangered Species Act. Although most Federal agencies with resource management responsibilities (i.e., Forest Service, Bureau of Land Management, Fish and Wildlife Service) accord some level of protection or management consideration to candidates, such policies are not mandatory under the Endangered Species Act, nor should they be confused with legal mandates of the Act. However, the Service is required through Section 4(b)(C)(iii) to monitor the status of Category 1 candidate species "to prevent their extinction while awaiting listing" (*Federal Register* 58:51146).

Because many candidate plants may be more imperiled than some listed endangered and threatened species, the Service, upon request, provides "technical assistance" to Federal, state, and local agencies, on the appropriate conservation and management of candidate species. In addition, interagency or conservation agreements between the Service and other Federal agencies can be developed to protect candidates, as illustrated by agreements reached with the Forest Service to protect the Shirley Meadows star-tulip (*Calochortus westonii*) and Rawson's flaming-trumpet (*Collomia rawsoniana*).

Proposed Species

Proposed species are granted limited protection under the Endangered Species Act. Essentially, proposed taxa must be addressed by Federal agencies in biological assessments (a document required by Section 7 of the Act for certain Federal projects or actions). Also, Federal agencies must confer with the Service regarding any action or project "likely to jeopardize the continued existence" of a proposed species. The Service typically reviews project plans and species information to determine the effects of a Federal action on a proposed species. This "conference," like the technical assistance for candidates, is only an advisory process. Any recommendations to modify or abandon the project and/or undertake protective measures for proposed species are not mandatory on the Federal agency conferring with the Service.

Listed Endangered or Threatened Species

Designated endangered and threatened plants receive the full protection of the ESA, but first they must be listed. The listing process is covered in **Section 4** of the Act, and may be initiated either through citizen petitions or internal Service actions. There are no timetables for listings generated internally within the Service, but listings petitioned by individuals or organizations are subject to a schedule imposed in Section 4(b): (1) within 90 days of receipt of a petition, the Service must publish in the Federal Register (FR) whether the petitioned action may be or is not warranted; (2) if the petitioned action may be warranted, the species must be proposed for listing in FR within one year of the petition's receipt; and (3) if proposed, a final rule (regulation) listing the species must be published in FR within one year of the proposal's publication. For proposed species, the one year period before publication of a final listing decision can be extended by the Service for up to six months to allow for receipt of additional information. Opponents of listings occasionally attempt to circumvent the listing process by petitioning for this delay, during which the Service impartially evaluates any new information relevant to a final listing decision.

For internally generated listings, including those covered by CNPS's suit (which is described later in this article), taxa to be protected through listing are generally grouped by the Service into packages based on similar geography, threats, habitat, or taxonomy. Among other advantages, these multi-species packages allow imperiled taxa to be proposed and listed with a minimum of effort, and facilitate the development of recovery plans for coherent clusters of taxa which have similar recovery needs.

Recovery plans address the crux of endangered species protection at the Federal level: The ESA is not a listing law, but on the contrary, its purpose is to recover listed species to the degree that they no longer require the protection of the Act. Section 4(f) requires that the Service develop recovery plans for all endangered or threatened species. These plans direct Service recovery monies and outline actions that can be taken by appropriate public and private agencies in efforts to recover listed species. Recommendations described in recovery plans are binding neither on other public agencies nor on private organizations or individuals. For recovered, delisted taxa, section 4(g) of the Act provides for a system to monitor taxa to ensure their continued survival.

As it now stands, however, the recovery process will require broader and more uniform implementation to be effective for plants. Currently, recovery plans have not been written and approved by the Service for most listed plants in California, nor have most written plans been fully implemented in California or nationwide. Although most listed taxa are plants, almost all recovery monies nationwide are spent on animals: in 1990 the highest ranked plant in terms of recovery expenditures was 57th on the list, and only 24 animals accounted for more than half of all recovery expenditures. Overall, very few taxa and only one plant nationwide have been delisted because of recovery. However, this record is expected to improve if Congress acknowledges that the Service and other Federal resource managers, including the U.S. Forest Service, Bureau of Land Management, and Bureau of Reclamation, have been unable to fully recover listed species because of insufficient staffing and funding.

Section 5 of the Act governs land acquisition, and allows Federal agencies within the Department of Interior (e.g., Bureau of Land Management, Park Service, and Fish and Wildlife Service) and Forest Service to implement a program to conserve listed plants via "the land acquisition and other authority under the Fish and Wildlife Act of 1956, as amended, the Fish and Wildlife Coordination Act, as amended, and the Migratory Bird Conservation Act, as appropriate" to acquire endangered or threatened plant habitat by purchase, donation, or otherwise. More importantly, this section allows these Federal agencies to use funds from the Land and Water Conservation Fund Act of 1965, as amended, for habitat acquisition.

Section 6 of the Act enables the Service to enter into management and conservation agreements with state agencies. Under management agreements, states can administer or manage any area established for the conservation of a listed species. Conservation agreements (also called cooperative agreements) enable the Service to financially assist state agencies such as the California Department of Fish and Game (DFG)) to develop programs for endangered and threatened species protection. Actions typically funded by the Service include research, species management, monitoring candidate and listed species, and other recovery activities described in recovery plans. The Service and DFG have undertaken a number of such projects for plants under California's cooperative agreement. Recent creative uses of Section 6 cooperative funds include promotion of multispecies conservation planning for the ensemble of eight gabbro endemic plants in western El Dorado County, habitat suitability studies for *Cordylanthus palmatus*, and reintroduction efforts for *Amsinckia grandiflora.*

Section 7, which mandates interagency cooperation to protect listed species, provides plants with the most significant protection conferred by any of the 18 sections of the Act. Section 7(a)(1) directs Federal agencies, in consultation with the Service, to use their respective resources in furtherance of the purposes of the Act by carrying out programs for the conservation of listed species. Section 7(a)(2) requires Federal agencies, in consultation with the Service, to ensure "that any action authorized, funded, or carried out by such agency... is not likely to jeopardize the continued existence of any listed species or result in the destruction or adverse modification of [critical] habitat."

If a proposed activity affects a listed species and is contingent on one or more Federal actions, "formal consultation" with the Service is initiated by the affected Federal agencies, and the Service is requested to issue a biological opinion. Formal consultation may result in a required modification or rarely abandonment of a proposed Federal action or project, if the Service determines that such activity is likely to jeopardize the species or adversely modify its critical habitat. Although a Federal agency or state governor may seek an exemption, typically the "reasonable and prudent alternatives" resulting from a Service "jeopardy opinion" must be adopted by the Federal agency to eliminate jeopardy to listed species. However, most agency consultations (89%) are resolved informally, and contrary to public opinion, most formal consultation (91%) does not result in issuance of jeopardy opinions (GAO 1992). Moreover, fewer than 1% of formal consultations result in cancelling of projects; in the five years 1987-1991, only 18 projects were cancelled nationwide as a result of Section 7 formal consultations (Barry *et al.* 1992). The viewpoint espoused by some critics of the ESA, that Section 7 consultations under the Act result in cancelling urgently needed projects throughout the country, is simply not true. Instead, in the vast majority of cases, economic development and endangered species protection go hand in hand.

Formal consultations for plants are increasing dramatically, in part because the number of listed plants is increasing, but more significantly, because many recently listed plants either occur on Federal land or there is common Federal "nexus" with otherwise private actions that may affect their habitat. For example, private development can be affected by Section 7 if Federal authorization or Federal funds are involved, as they are in the U.S. Army Corps of Engineers jurisdiction over placing of fill in vernal or other wetlands. Development of vernal pools, which are isolated wetlands according to the provisions of the Clean Water Act, requires a Section 404 permit from the Corps if a listed species is affected. This Federal action triggers the consultation process, and provides the Service with an opportunity to encourage project redesign or other mitigation measures to avoid jeopardy to listed species. (For further information on Section 7, refer to the final procedural regulations published on June 3, 1986 [*Federal Register* 51:19925-19963].)

Section 7(e) establishes the Endangered Species Committee, popularly referred to as the "God Squad." This seven member Committee is composed of cabinet level and other presidential appointees. If a Federal agency receives a jeopardy opinion during formal consultation, it may apply to the Committee for exemption from the requirement of Section 7(a)(2) that no Federal agency action jeopardize an endangered or threatened species. An exemption has never been sought for an action affecting a Federally-listed plant, but the process recently received national attention over the exemption requested (and partially granted) for agency timber sales affecting the northern spotted owl.

Section 9 of the Endangered Species Act prohibits the removal and reduction to possession (i.e., collection) of endangered and threatened plants from lands under Federal jurisdiction. It further prohibits the removal, cutting, digging, damage, or destruction of listed plants on any other area in knowing violation of a state law or regulation. Plants, however, do *not* enjoy the full protection accorded animals under this section against "take" (i.e., harass, harm [which includes significant habitat modification or degradation], pursue, hunt, shoot, wound, kill, trap, capture, or attempt to engage in any such conduct). Nevertheless, as with animals, Section 9 makes illegal the international and interstate transport, import, export, and sale or offer for sale of endangered and threatened plants.

Habitat Conservation Plans (HCP's) are developed to grant the incidental take of listed species on non-Federal land, and are authorized in **Section 10(a)** of the Act. Individuals, corporations, and state or local agencies may apply to the Service for a Section 10(a) permit to initiate the planning process. Issuance of incidental take permits allows the sacrifice of individuals or habitat in one area in exchange for protection and enhancement of populations or habitat in other areas. This flexibility typically promotes concentration of impacts and development, and enables the protection of essential habitat in the large blocks required for survival of most taxa. Although the scope of Habitat Conservation Planning varies from parcels to ecosystems, larger scale efforts are typically more worthwhile. One of the most successful HCP's established the 13,000 acre Coachella Valley Preserve to protect the Coachella Valley fringe-toed lizard (*Uma inornata*) from development in the Palm Springs area. HCP's are not written or implemented for plants; since listed plants are not protected from take or destruction on private land. However, if they occur within the boundaries of an HCP, listed and candidate plants are typically incorporated into the planning process. Furthermore, ecosystem planning processes similar to HCP's have been initiated for plants on private lands, but only when there is sufficient interest from local government, and a Section 7 nexus such as the Clean Water Act Section 404 permitting process which governs fill of wetlands. Such ecosystem planning is currently underway for three vernal pool plants of Sonoma County's Santa Rosa Plain, *Lasthenia burkei*, *Limnanthes vinculans*, and *Blennosperma bakeri*, and their associated sensitive animals.

Recent Legal Settlements Will Speed Listings

The numbers of listed and proposed plant species throughout the nation will increase dramatically as the result of recently settled lawsuits brought against the Service to eliminate the existing listing backlog for both plants and animals. Nationally, the number of Category 1 plants that comprise the listing backlog is nearly an order of magnitude greater than that for all classes of animals combined. In California, the 1991 out-of-court settlement between the Service and the California Native Plant Society requires the Service to propose 159 plant taxa for endangered or threatened status by early 1996. The Service is making steady progress toward this goal; in addition to 52 taxa listed and proposed under the settlement agreement, over 60 plant species are contained in draft proposed rules undergoing internal Service review. CNPS is participating in this process in many ways. Local experts and the Rare Plant Program provide the Service with distributional and status information, and the CNPS Botanist has used the *Inventory* database to help Service biologists establish a relevant priority order for listings which will ensure that the most threatened taxa are protected first. In the near future, numbers and categories of candidate and listed plant species will change relatively rapidly as status reviews are completed and qualified species achieve endangered or threatened status. We ask that you consult the Service for the latest information.

Reauthorization of the ESA

Congressional action on the scheduled 1992 Endangered Species Act reauthorization was delayed by elections until 1993, and still the battle rages on. Early hearings on the issue indicate strong support to substantially modify key provisions of virtually all critical sections of the Act. These proposed modifications emphasize incorporation of economic analyses and assessment of "takings" of private property into the Section 4 listing process, establishing mandatory time schedules for preparation of recovery plans in Section 4, weakening the requirements for interagency cooperation under Section 7, weakening the prohibitions against "take" of listed species in Section 9, changing habitat conservation planning and the permit granting process for "take" in Section 10, and adding requirements to compensate private individuals for economic losses resulting from listed species protection. The next reauthorization clearly promises to be a debate over the fundamental concepts, processes, and protections that comprise endangered and threatened species conservation as we now know it under the Act.

The Future

Although few taxa have been formally recovered and delisted since the Act's initial passage in 1973, the Endangered Species Act has been instrumental in protecting habitat and slowing the population decline of hundreds of species across the United States, including many of California's rarest plants. The Service's responsibilities to protect and recover California's rare plants are expanding geometrically as more taxa are listed and the pace of development and population growth in California accelerates. It is essential that the U.S. Congress reauthorize a strong ESA, and appropriate sufficient funds in the future for the Service to add essential staff for Section 7 consultations, listing activities, and preparation and implementation of multi-species recovery plans. Given the anticipated success of Interior Secretary Bruce Babbitt's fledgling National Biological Survey, with its emphasis on ecosystem protection to minimize the need for future listings, and the continued cooperation of agencies and organizations like the DFG and CNPS, California's rare flora will stand a fighting chance.

Jim A. Bartel is the Chief of Listing, Region 1, U.S. Fish and Wildlife Service, 911 N.E. 11th Ave., Portland, OR 97232; Mark W. Skinner is the CNPS Botanist, 1722 J Street, Sacramento, CA 95814; and Jan C. Knight is the lead Botanist, U.S. Fish and Wildlife Service, 2800 Cottage Way, Room E-1823, Sacramento, CA 95825.

Sensitive Plant Management on the National Forests and Grasslands in California

Ronald E. Stewart

California has a total land area of 101.5 million acres. Of this total area, 20 million acres are National Forest System (NFS) lands. These lands are divided into 18 national forests and one national grassland managed by the Pacific Southwest Region (Region 5) of the Forest Service, U.S. Department of Agriculture. Portions of the Toiyabe National Forest managed by the Intermountain Region (Region 4), and the Siskiyou and Rogue River National Forests managed by the Pacific Northwest Region (Region 6), also are within California.

The Forest Service is mandated by Federal law to manage these lands for multiple uses. Multiple-use resource management provides a sustainable supply of water, forage, wildlife, wood, recreation, and other renewable resources to benefit the American people, while ensuring the productivity of the land and protecting the quality of the environment. The Forest Service is committed to practicing the highest standards of land and resource stewardship. Management of national forests in California is intended to promote integrity of ecosystems, biological diversity, fish and wildlife habitat, and forest and rangeland health, as well as provide a sustainable supply of renewable resources. Managing for endangered, threatened, and sensitive species, the rarest of the resources under our care, is essential to meeting these objectives.

What Is a Sensitive Plant Species?

Sensitive is a term used by the Forest Service to designate plant species known or highly suspected to occur on NFS lands that are considered valid candidates for Federal threatened or endangered classification under the Endangered Species Act of 1973 (as amended, 1988; see *The Endangered Species Act...*, above). The term "sensitive" is used to distinguish potential candidates for listing from plants officially listed as "rare," "threatened," or "endangered," terms that have legal meanings under Federal and state laws.

Of all the National Forest Regions, Region 5 contains the largest assemblage of sensitive plant species in comparison to its land base. In fact, of the more than 7,000 vascular plants occurring in California, well over half are known to occur on NFS lands. This is due to the diversity of topography, geography, geology and soils, climate, and vegetation that occur on NFS lands in California, the same factors that account for the exceptionally high endemic flora of the State.

Nearly 300 vascular plants known to occur on NFS lands have been identified as Forest Service sensitive and need further evaluation or require special management to ensure long-term species viability. Of these, 100 are endemic to NFS lands. The Region has the sole responsibility for the viability and long-term conservation of these species. At present, seven plants that occur on NFS lands in Region 5 are Federally-listed as threatened or endangered, and five are proposed for listing. Additional listings under the Federal ESA are anticipated.

Early in 1975, Region 5 issued plant policy direction concerning endangered, threatened, and sensitive species which developed into the current Sensitive Plant Program. Activities of the Sensitive Plant Program include field verification of known or reported locations of sensitive plants, preparation of individual population records, field reconnaissance of projects such as timber harvests and input to environmental documents, identifying basic research needs, monitoring key populations, and preparing individual species management guides.

How Is the Region 5 Sensitive Plant List Developed?

Region 5 developed its list of sensitive plants from a variety of sources, including plant species published in the *Federal Register* as under review for Federal listing by the U.S. Fish & Wildlife Service. The list is periodically validated by botanists affiliated with major scientific institutions, the California Native Plant Society (CNPS), and the Endangered Plant Program and Natural Diversity Data Base (NDDB) of the California Department of Fish and Game. The Region 5 sensitive plant list also takes into account the taxa listed under State law as endangered, threatened, or rare. Any State-listed species which may need special management on NFS lands is considered for the list. In addition, the Region 5 sensitive plant list incorporates the professional field knowledge of Forest Service botanists and ecologists.

The Region 5 sensitive plant list is subject to additions and deletions as new data are obtained, taxonomic problems clarified, or as revisions to the source documents are made. Generally, the plant list is revised every other year. The Region 5 sensitive plant list is available upon request from the Forest Service Regional Office in San Francisco.

What Is the Forest Service Policy for Sensitive Plants?

Implementation of the Sensitive Plant Program is outlined in Forest Service Manual (FSM) Section 2670, "Threatened, Endangered, and Sensitive Plants and Animals." The FSM and the Forest Land Management Plans provide the working policies and framework for implementing all Forest Service activities and evaluating possible effects on endangered, threatened, proposed, or sensitive species.

Additionally, a Region 5 "Threatened and Endangered Plants Program Handbook" (R-5 FSH 2609.25), developed in 1988, provides direction for day-to-day management of the sensitive plant program and technical procedures to implement the program at the project level. Both the FSM direction and Region 5 handbook are available for public review at any Forest Service office.

Key parts of FSM 2670.22 concerning sensitive taxa are:

1. Develop and implement management practices to ensure that species do not become threatened or endangered because of Forest Service actions.
2. Maintain viable populations of all native and desired non-native wildlife, fish, and plant species in habitats distributed throughout their geographic range on NFS lands.
3. Develop and implement management objectives for populations and/or habitat of sensitive species.

In addition, FSM 2670.32 states that the Forest Service will:

1. Assist states in achieving their goals for the conservation of endemic species.
2. Review programs and activities, through a biological evaluation, to determine their potential effect on sensitive species.
3. Avoid or minimize impacts to species whose viability has been identified as a concern.
4. If impacts cannot be avoided, analyze the significance of potential adverse effects on the population or its habitat within the area of concern and on the species as a whole.
5. Establish management objectives, in cooperation with the states, when projects on National Forest System lands may have a significant effect on sensitive species' population numbers or distributions.

Each national forest has a Threatened & Endangered Species Program Manager responsible for the implementation of the Sensitive Plant Program within that forest. Region 5 national forests with high concentrations of sensitive plants have a professional botanist on staff and/or other highly trained biologists working on sensitive plant management issues. The Region currently has 28 full-time botanists.

How Are Sensitive Plants Actually Managed?

Distribution patterns, habitats, and ecological parameters differ for each of the 300 sensitive plants in Region 5. We have learned first-hand that protecting and conserving these taxa does not necessarily entail segregating sensitive plants from other forest development or management activities.

Past management activities and practices often provide important insights for assessing ecological requirements, and management opportunities and constraints for species. Some species require frequent burning, others are early successional taxa, while others prefer a specific microenvironment for optimal population size and vigor. Timing, intensity, and frequency of a proposed action are the key factors in biological evaluations of proposed forest activities.

For example, a timber harvest could have no effect, adverse effect, or beneficial effect on a sensitive plant occurrence depending on whether or not the proposed action is evaluated and planned in terms of the species' ecological needs. This is well illustrated by the Shirley Meadows star-tulip (*Calochortus westonii*), which is endemic to the Sequoia National Forest. The ecological requirements of this species suggested that selective thinnings of dense conifers could be accomplished when the plants were dormant in the fall, thus creating a more open, park-like environment for this species. Numbers of Shirley Meadows star-tulips have increased markedly under this prescription. Historically, tree thinning was probably accomplished by low-intensity ground fires, but with the successful fire suppression efforts of this century, a shift towards denser white fir and incense-cedar has occurred in what would otherwise be an open mixed conifer-black oak forest.

Experience in maintaining viable populations of sensitive species in dynamic forest, woodland, chaparral, and grassland ecosystems throughout their range on NFS lands has taught us that "fence them and leave them" is not always the best prescription. The key objective for long-term sensitive species management is not how much forest management the species can *take*, but rather, what kind of forest management does the species *need* to assure long-term conservation. Some sensitive plants require prescribed management treatments, while others, including some sensitive plants occurring on restrictive or unique habitats such as the Pebble Plains on the San Bernardino National Forest or species of serpentinite substrates, simply need to be "left alone." In these situations, we manage by reducing ground disturbing activities. Sensitive plant stewardship in Region 5 is indeed a dynamic and challenging part of multiple-use resource conservation.

As the national forest inventories for sensitive plants are completed, long-range species and/or habitat management guides are prepared and incorporated into forest plans. These guides are not intended to be exhaustive, but are designed as "work plans" providing site-specific objectives, activities, and time tables for implementation. The guides specify monitoring and periodic review to ensure that the guide is working to benefit the species. As new data become available, they are incorporated into species management guides. Effective implementation of these guides should ensure the long-term viability of sensitive species, thereby preventing the need to list the species under Federal law.

Forest Supervisors and District Rangers can best manage sensitive plants on the NFS lands they administer if they have the most current information. Therefore, forests have developed specific methodologies for collecting and maintaining sensitive plant data. In addition, Region 5 has signed a Memorandum of Understanding with CNPS and NDDB. All of the sensitive plant data generated by Region 5 are forwarded annually to NDDB and shared freely with CNPS. This provides all interested parties with the current distribution, population trends, condition, and vigor for each sensitive species that is inventoried and managed on NFS lands in California.

What Are Region 5 Forest Service Accomplishments Regarding Botanical Conservation?

Region 5 developed the first Sensitive Plant Program in the Forest Service, and has a long history of conserving rare and unique plants and plant communities. Throughout NFS lands in California there exists a great wealth of places with unusual scenic, historic, prehistoric, and biological values that merit special attention and management. Botanical Areas are one of the categories of "Special Interest Areas" (SIA's) that are identified in FSM section 2462. Like other SIA's, botanical areas are established to protect sensitive resources, and where appropriate, to foster public education and enjoyment. As of January 1, 1992, 35 botanical areas had been formally established by the Regional Forester pursuant to 36 CFR 294.1a and incorporated into final Forest Land Management Plans. Several potential botanical areas, many containing sensitive plant species, are being evaluated with CNPS's assistance. Qualifying botanical areas will be established through amendments to Forest Plans with full public review under the National Environmental Policy Act.

Other sensitive plant occurrences are within the 26 Research Natural Areas (RNA's, FSM Section 4060) established for non-manipulative research and study. Many proposed RNA's await establishment. RNA's are recommended jointly by the Station Director (of the research branch of the Forest Service) and Regional Forester and are established by the Chief of the Forest Service in Washington, DC.

Botanical values and biological diversity are also provided by the 3.9 million acres of Congressionally designated Wilderness, and the nearly 320 acres per mile of protected riparian area along the 959 miles of Congressionally-designated Wild and Scenic Rivers in California's NFS lands. For more detailed information on native plant diversity and special area designations on the national forests in California, see Shevock (1988).

Conclusion

Sensitive plants on the national forests and grasslands are a unique and scientifically valuable resource. I personally encourage CNPS to continue to assist the Forest Service in our sensitive plant inventory and conservation strategy efforts. CNPS members concerned with sensitive plant management need to continue to coordinate conservation efforts with Forest personnel at the local level, as they have during CNPS's participation in the recent development of Forest Land Management Plans (Meyer 1990). This involvement has been beneficial for sensitive resources on California's National Forests, and educational for both CNPS members and Forest Service planners and biologists. Working together, we can continue to conserve and manage for viable populations of sensitive plant species occurring on the national forests and grasslands in California.

Ronald E. Stewart is the Regional Forester, Pacific Southwest Region, USDA-Forest Service, 630 Sansome St., San Francisco, CA 94111.

Rare Plant Conservation on Bureau of Land Management Lands

Edward L. Hastey

The Bureau of Land Management (BLM) manages about 17.1 million acres of public lands in California, including some in almost every one of California's 58 counties. These public lands are habitat for many of the rare plants included in this *Inventory*. Currently, 205 special status plants are known to occur on BLM lands in California. An additional 146 special status plants are suspected to occur on BLM lands, but their presence has not yet been documented.

Special Status Plants

BLM uses the term "special status plants" to include all of the following: 1) Federally-listed and proposed species; 2) Federal candidate species; 3) State-listed species; and 4) sensitive species. Sensitive species are those species that do not meet any of the first three criteria, but which are designated by the State Director for special management consideration. Plants on List 1B (Plants Rare, Threatened, or Endangered in California and Elsewhere) of the CNPS *Inventory* that do not meet any of the first three criteria are considered sensitive by BLM in California. Sensitive plants receive the same level of protection as Federal candidate species.

Nationwide Bureau policy on the management of special status species (both plants and animals) is given in BLM Manual 6840. California's policy on the management of candidate and sensitive plants predates the national policy and is generally more restrictive. California's policy was first developed in 1977 in recognition of the fact that many candidate and sensitive plants require the same level of conservation as Federally-listed plants. The policy, largely unchanged since 1977, reads as follows:

"Pending formal listing all candidate and sensitive plant species are afforded the full protection of the Endangered Species Act unless the BLM State Director judges on a case-by-case basis that the evidence against listing a particular plant species is sufficient to allow a specific action."

The BLM Program in California

In California the BLM is continuing to pursue the eight-point program outlined in the previous edition of this *Inventory*. This program consists of the following elements: 1) inventory; 2) designation of Areas of Critical Environmental Concern (ACEC's); 3) monitoring; 4) research; 5) protection; 6) public education; 7) land acquisition; and 8) volunteer assistance. Significant advances have been made in all of these areas. Examples include extensive inventories in the San Joaquin Valley and adjacent foothills to find and document new locations of several rare plant species (including *Acanthomintha obovata* ssp. *cordata*, *Antirrhinum ovatum*, *Caulanthus californicus*, *Eremalche kernensis*, *Eriastrum hooveri*, and *Lembertia congdonii*), acquisition of more than 100,000 acres of valuable rare plant and animal habitat in the Carrizo Plain, and designation of 26 ACEC's covering more than 300,000 acres specifically to protect rare plants (six more are currently proposed for this purpose).

In California, BLM currently employs botanists in its State Office, in all four of its District Offices, and in eight of its 15 Resource Area offices, for a total of 13 botanists. The target is to have at least 22 botanists in place by the year 2000. In addition to botanists, other BLM personnel spend considerable time

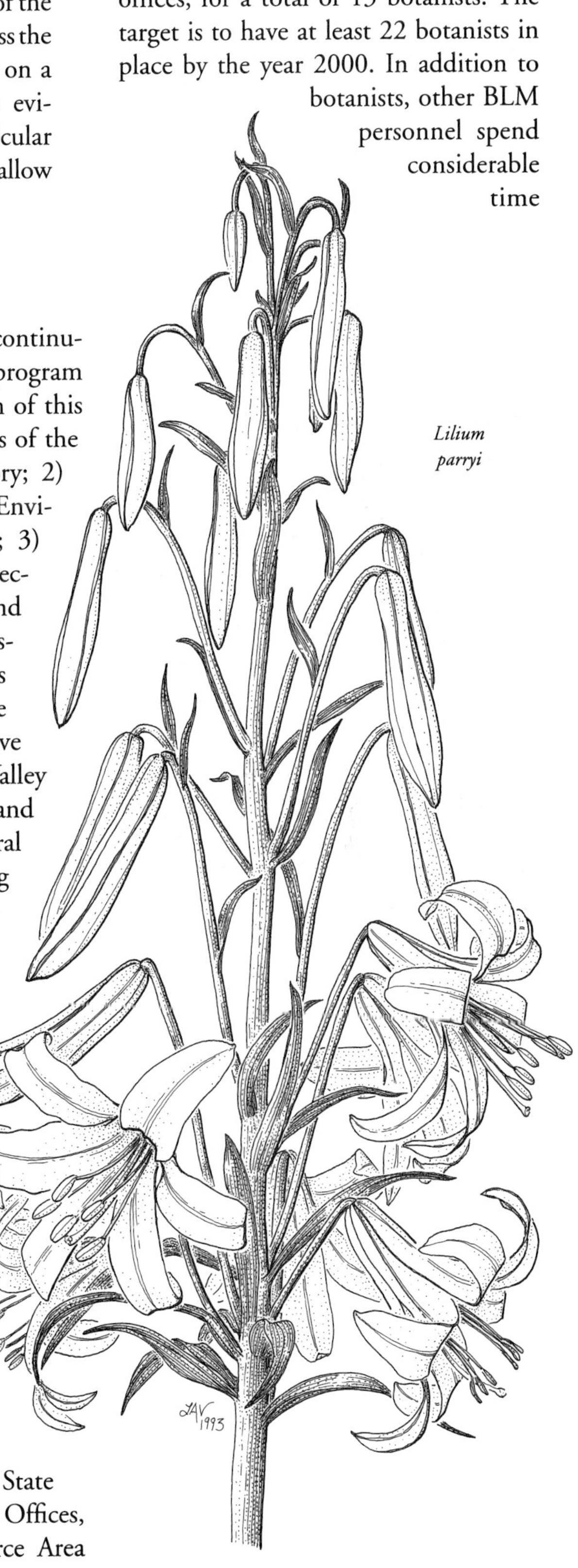

Lilium parryi

and effort on rare plant management.

The Bureau has recently prepared a nationwide plan entitled *Rare Plants and Natural Plant Communities: A Strategy for the Future* (BLM 1992). This plan will guide the Bureau's management of rare plants through the year 2000. It sets forth the objectives of the plan and identifies the funding and personnel required to accomplish them. California's objectives are to inventory more than six million acres of public lands for rare plants, to monitor all of the rare plants on public lands, to conduct 65 studies on the biology and management of rare plants, to complete 95 rare plant management plans, to construct habitat improvement projects benefitting 40 rare plants, and to acquire more than 6,000 acres of significant rare plant habitat.

A very exciting chapter in biological conservation began recently with the signing, on September 19, 1991, of the Memorandum of Understanding (MOU) on California's Coordinated Regional Strategy to Conserve Biological Diversity (Strategy). Ten State and Federal agencies are party to the MOU: The Resources Agency, California Department of Fish and Game, California Department of Forestry and Fire Protection, California Department of Parks and Recreation, State Lands Commission, University of California, U.S. Forest Service, U.S. Fish and Wildlife Service, National Park Service, and BLM. County governments and public interest groups, including the California Native Plant Society, will play important roles in implementing the Strategy.

The Strategy is an important step toward conserving biological diversity in California, and implementation of the strategy's concepts will do much to preserve California's rare flora. The emphasis will be on the conservation of communities and ecosystems, including those that harbor rare plants and plant communities. Planning will cross all land ownerships, both public and private, and focus on *biological* boundaries rather than political ones.

The most important tool in executing the Strategy on the ground will be Coordinated Resource Management and Planning (CRMP). CRMP is a resource planning, problem solving, and management process that allows for direct participation of everyone in a given planning area concerned with natural resource management. The grass roots involvement that CRMP fosters is vital to developing effective and workable conservation plans.

The CRMP process has a long and successful history in California. This cooperative approach has been used successfully in the Coachella Valley and the Carrizo Plain to preserve the rare biological resources of those areas, while at the same time allowing needed development to continue. In the Western Mojave Desert a CRMP is underway that will result in a plan to conserve the biological diversity of eight million acres of public and private lands, while providing for continued socio-economic growth. Many other efforts are either already underway or will be shortly.

All those interested in conserving California's rare flora should make an effort to be involved in these local planning efforts. Only by working together can we preserve the biological diversity for which California is famous.

Edward L. Hastey is the California State Director of the Bureau of Land Management, 2800 Cottage Way, Room E-2845, Sacramento, CA, 95825.

Rare Plant Conservation — A Global Perspective

Tim Messick

Native plant societies are now fairly common throughout the United States, but few conservation organizations outside this country are devoted entirely to native plant conservation. Many conservation organizations both inside and outside the United States work to save threatened native plants and plant communities through their efforts to protect threatened sites or wildlife species or through specialized working groups for botanists. The number and variety of these organizations and their activities is too great to summarize here, but a useful perspective can be gained from looking at the coordination and facilitation provided by one organization, The World Conservation Union.

The Coordinating Role of IUCN

The World Conservation Union, or IUCN (the acronym is a holdover from its former name, the International Union for Conservation of Nature and Natural Resources), is an umbrella organization of conservation organizations, scientific institutions, resource management agencies, and national governments joined together to promote the protection and sustainable use of living resources. IUCN counts more than 50 nations, 90 government agencies, and 450 nongovernmental organizations among its members, with over 100 countries represented by one institution or another. More than 75 conservation, scientific, and educational organizations in the United States are members of IUCN; CNPS has been a member since 1985.

In addition to its organizational and governmental members, over 3000 individual scientists and other conservation specialists are members of six commissions: the Commission on Ecology, Commission on Education and Communication, Commission on National Parks and Protected Areas, Commission on Environmental Strategy and Planning, Commission on Environmental Law, and Species Survival Commission. Working groups within each commission provide advice and carry out projects on specific sites, resources, problems, needs, or groups of organisms. CNPS members have participated in projects of the North American Plants Specialist Group of the Species Survival Commission by helping to identify California's most critically threatened plants, and by contributing sections to a book on centers of plant diversity.

IUCN provides training and organizational support for hundreds of conservation projects each year, many of which benefit threatened plants as well as threatened ecosystems and impoverished people. The World Wide Fund for Nature, or WWF (formerly World Wildlife Fund), works closely with IUCN (whose international headquarters are in the same building) to provide economic support for many plant conservation projects in developing countries.

Many of IUCN's plant conservation efforts have been organized through their Plant Conservation Office at Kew Gardens, and the World Conservation Monitoring Centre in Cambridge, England. Projects have included the development of a worldwide database on threatened plants (similar to those developed for California by CNPS and the Natural Diversity Data Base), development of computer software to help botanic gardens acquire and manage collections of threatened plants, development of the Botanic Gardens Conservation Strategy, consultation on projects saving plants from extinction, and publication of books. Among the plant conservation books produced or sponsored by IUCN are *The IUCN Plant Red Data Book* (Lucas and Synge 1978), *The Biological Aspects of Rare Plant Conservation* (Synge 1981), *Plants in Danger: What Do We Know?* (Davis *et al.* 1986), *Green Inheritance: The World Wildlife Fund Book of Plants* (Huxley 1985), and the forthcoming *Centres of Plant Diversity: A Guide and Strategy for their Conservation* (Heywood, in prep.).

IUCN and its member organizations have aggressively promoted the establishment, implementation, and strengthening of international treaties that benefit plants and other resources, such as the Convention on International Trade in Endangered Species (CITES) and the Convention on Wetlands of International Importance Especially as Waterfowl Habitat (Ramsar Convention).

CNPS's involvement with IUCN has included participation in the 1984 and 1988 IUCN General Assemblies, providing information to IUCN and others on CNPS's Rare Plant Program, sending complimentary copies of the *Inventory* and the proceedings of CNPS's 1986 rare plant conference to selected botanical institutions in 15 lesser developed countries, and contributing chapters on important botanical sites in California to the *Centres of Plant Diversity* book.

"Think Globally, Act Locally"

The World Conservation Union recognizes CNPS as a strong leader in plant conservation in the western United States. But one of the challenges facing CNPS in the coming years will be to "think globally" more often while continuing to "act locally." Understanding conservation goals and principles in a global context can lead to better action on conservation priorities and methods at home. Thinking globally can help focus the momentum of global conservation priorities on site-specific and species-specific native plant conservation issues in California. Membership in the global conservation community gives CNPS opportunities to encourage developing plant conservation movements abroad and to learn from the problems, experiments, and creativity of other groups with goals similar to ours.

The World Conservation Union has, appropriately, provided leadership in "global thinking" in recent years. In 1980, IUCN published the *World Conservation Strategy* (IUCN 1980). This strategy for "living resource conservation for sustainable development" emphasized that humanity exists as a part of nature and that the future of humanity is dependent on the conservation of nature and natural resources. It asserted that conservation cannot be achieved without development that alleviates poverty. The World Conservation Strategy stressed three objectives:

- maintenance of essential ecological processes and life-support systems
- preservation of genetic diversity
- sustainable use of species and ecosystems.

Throughout the 1980's the World Conservation Strategy was tested through development of over 50 national and sub-national conservation strategies. A decade of experience and debate clarified the complexity of environmental problems and increased the urgency for action.

In October 1991, The World Conservation Union, the United Nations Environment Program, and the World Wide Fund for Nature released *Caring for the Earth: A Strategy for Sustainable Living* (IUCN, UNEP, and WWF 1991), as a successor to the World Conservation Strategy. *Caring for the Earth* is a strategy of mutually reinforcing actions at individual, local, national, and international levels based on an ethic of caring for nature and for people. It strengthens the original objectives of the 1980 document by identifying specific actions to promote a set of conservation principles which include: conserving the Earth's vitality and diversity, minimizing depletion of nonrenewable resources, enabling communities to care for their own environments, and integrating development and conservation.

CNPS has already earned international recognition for its contribution to local plant conservation in a region of global botanical importance. Let us use the wisdom and knowledge provided by worldly perspectives and global collaboration to increase CNPS's recognition and effectiveness in local conservation action in the years to come.

[Note: *Caring for the Earth* is available in the United States for $19.95 plus tax and shipping from Island Press, P. O. Box 7, Covelo, CA 95428 (800/828-1302). IUCN's mailing address is Rue Mauverney 28, 1196 Gland, Switzerland.]

Tim Messick, the World Conservation Union representative for CNPS, is a botanist with the environmental consulting firm of Jones & Stokes Associates, 2600 V St., Suite 100, Sacramento, CA 95818.

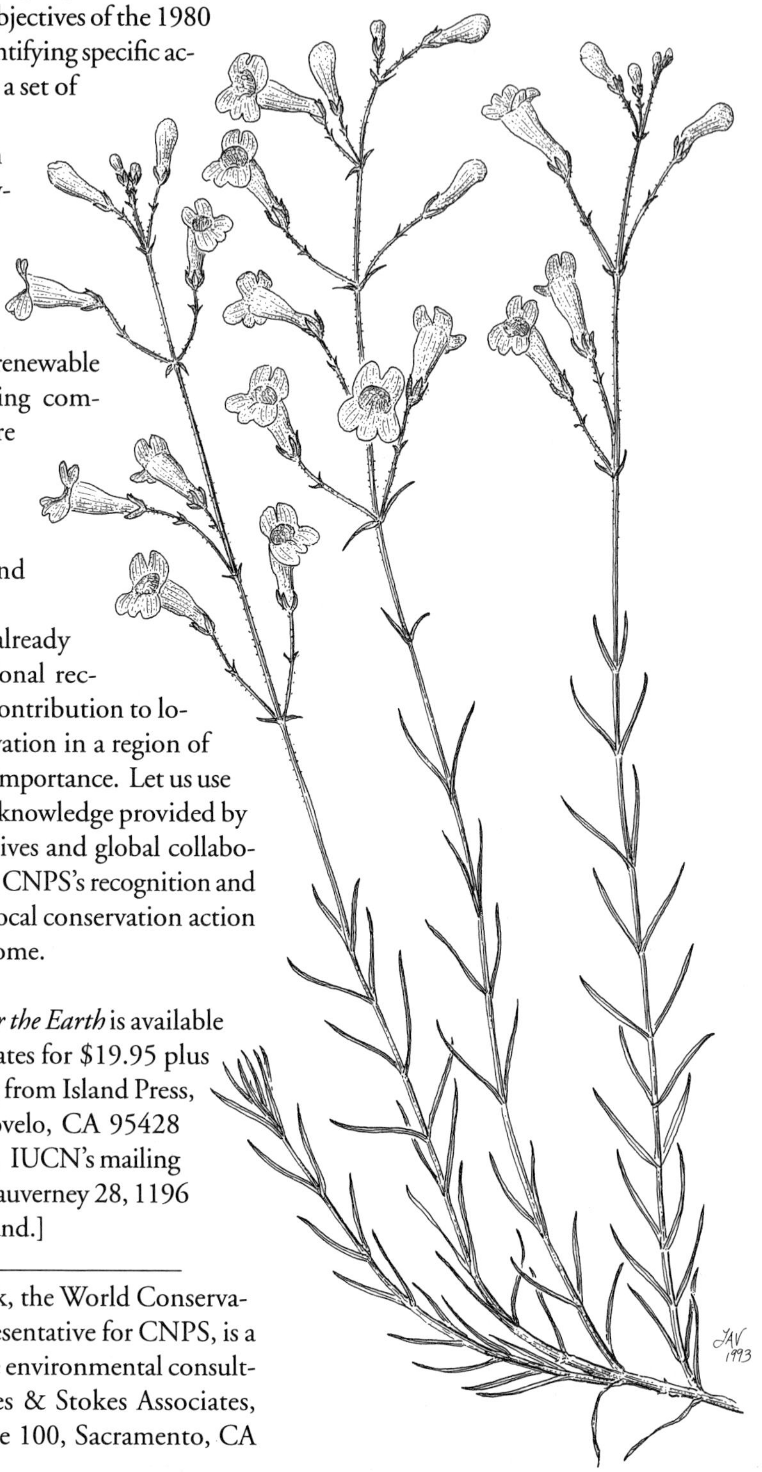

JAV 1993

Penstemon filiformis

Literature Cited in the Introductory Sections

Introduction

Barbour, M., and J. Major, eds. 1988. *Terrestrial Vegetation of California.* New expanded edition. Special Publication No. 9. California Native Plant Society. Sacramento, CA. 1002 pp. + Supplement.

Elias, T.S., ed. 1987. *Conservation and Management of Rare and Endangered Plants.* Proceedings from a conference of the California Native Plant Society. California Native Plant Society. Sacramento, CA. 630 pp.

Hickman, J.C., ed. 1993. *The Jepson Manual: Higher Plants of California.* University of California Press. Berkeley, CA. 1400 pp.

Lewis, H. 1972. The origin of endemics in the California flora. Pages 179-188 *in*: D.H. Valentine, ed. *Taxonomy, phytogeography and evolution.* Academic Press. New York, NY.

Ornduff, R. 1974. *Introduction to California plant life.* University of California Press. Berkeley, CA. 152 pp.

Raven, P.H. and D.I. Axelrod. 1978. Origin and relationships of the California flora. *University of California Publications in Botany* 72:1-134.

Smith, J.P., Jr., and J.O. Sawyer, Jr. 1988. Endemic vascular plants of northwestern California and southwestern Oregon. *Madroño* 35(1):54-69.

Stebbins, G.L. 1978a. Why are there so many rare plants in California? I. Environmental factors. *Fremontia* 5(4):6-10.

Stebbins, G.L. 1978b. Why are there so many rare plants in California? II. Youth and age of species. *Fremontia* 6(1):17-20.

Stebbins, G.L. 1980. Rarity of plant species: a synthetic viewpoint. *Rhodora* 82:77-86.

Stebbins, G.L., and J. Major. 1965. Endemism and speciation in the California flora. *Ecological Monographs* 35:1-35.

Rarity in Vascular Plants

Drury, W.H. 1974. Rare species. *Biological Conservation* 6:162-169.

Drury, W.H. 1980. Rare species of plants. *Rhodora* 82:3-48.

Fiedler, P.L. 1986. Concepts of rarity in vascular plant species, with special reference to the genus *Calochortus* Pursh (Liliaceae). *Taxon* 35:502-518.

Fiedler, P.L., and J.J. Ahouse. 1992. Hierarchies of cause: Toward an understanding of rarity in vascular plant species. Pages 23-47 *in*: P.L. Fiedler and S.K. Jain, eds. *Conservation Biology: The Theory and Practice of Nature Conservation, Preservation and Management.* Chapman and Hall. New York, NY.

Kruckeberg, A.R. 1984. California serpentines: Flora, vegetation, geology, soils, and management problems. *University of California Publications in Botany* 78:1-180.

Kruckeberg, A.R., and D. Rabinowitz. 1985. Biological aspects of endemism in higher plants. *Annual Review of Ecology and Systematics* 16:447-479.

Mayr, E. 1963. *Animal Species and Evolution.* The Belknap Press of Harvard University Press. Cambridge, MA.

Rabinowitz, D. 1981. Seven forms of rarity. Pages 205-217 *in*: H. Synge, ed. *The Biological Aspects of Rare Plant Conservation.* John Wiley & Sons. New York, NY.

Rabinowitz, D., S. Cairns, and T. Dillon. 1986. Seven forms of rarity and their frequency in the flora of the British Isles. Pages 182-204 *in*: M. Soule, ed. *Conservation Biology: The Science of Scarcity and Diversity.* Sinauer Associates. Sunderland, MA.

Raven, P. H., and D. I. Axelrod. 1978. Origin and relationships of the California flora. *University of California Publications in Botany* 72:1-134.

Stebbins, G.L., 1942. The genetic approach to problems of rare and endemic species. *Madroño* 6:241-272.

Stebbins, G.L., and J. Major. 1965. Endemism and speciation in the California flora. *Ecological Monographs* 35:1-35.

Ecological Characteristics of California's Rare Plants

Ashton, P.S. 1992. Species richness in plant communities. Pages 3-22 *in*: P.L. Fiedler and S.K. Jain, eds. *Conservation Biology: The Theory and Practice of Nature Conservation, Preservation and Management.* Chapman and Hall. New York, NY.

Barbour, M., B. Pavlik, F. Drysdale, and S. Lindstrom. 1993. *California's Changing Landscapes: Diversity and Conservation of California Vegetation.* California Native Plant Society. Sacramento, CA. 244 pp.

Jensen, D.B., M. Torn, and J. Harte. 1990. *In Our Own Hands: A Strategy for Conserving Biological Diversity in California.* California Policy Seminar Research Report, University of California. Berkeley, CA. 184 pp.

Raven, P.H., and D.I. Axelrod. 1978. Origin and relationships of the California flora. *University of California Publications in Botany* 72:1-134.

Raunkier, C. 1934. *The Life Forms of Plants and Statistical Plant Geography.* Clarendon Press. Oxford. 632 pp.

A History of the CNPS Rare Plant Program and Inventory

Berg, K., and M. Skinner. 1990. The CNPS botanist is a reflection of the Society. *Fremontia* 18(4):13-15.

Powell, W.R., ed. 1974. *Inventory of Rare and Endangered Vascular Plants of California.* Special Publication No. 1 (first edition). California Native Plant Society. Berkeley, CA. iii + 56 pp.

Powell, W.R. 1975. The CNPS Rare Plant Project. *Fremontia* 2(4):14-19.

Powell, W.R. 1978. The CNPS *Inventory* — a progress report. *Fremontia* 5(4):28-29.

Powell, W.R., T. Duncan, and A.Q. Howard. 1981. The California Native Plant Society Rare Plant Project. Pages 193-198 *in*: L.E. Morse and M.S. Henifin, eds. *Rare plant conservation: geographical data organization.* New York Botanical Garden. Bronx, NY

Smith, J.P., Jr., ed. 1981. *First Supplement, Inventory of Rare and Endangered Vascular Plants of California.* Special Publication No. 1 (second edition). California Native Plant Society. Berkeley, CA. 28 pp.

Smith, J. P., Jr. 1987. California: leader in endangered plant protection. *Fremontia* 15(1):3-7.

Smith, J. P., Jr. 1990. A history of the rare plant program and *Inventory. Fremontia* 18(4):9-12.

Smith, J.P., Jr., and K. Berg, eds. 1988. *Inventory of Rare and Endangered Vascular Plants of California.* Special Publication No. 1 (fourth edition). California Native Plant Society. Sacramento, CA. xviii + 168 pp.

Smith, J.P., Jr., and R. York, eds. 1982. *Second Supplement, Inventory of Rare and Endangered Vascular Plants of California.* Special Publication No. 1 (second edition). California Native Plant Society. Berkeley, CA. 28 pp.

Smith, J.P., Jr., and R. York, eds. 1984. *Inventory of Rare and Endangered Vascular Plants of California.* Special Publication No. 1 (third edition). California Native Plant Society. Berkeley, CA. xviii + 174 pp.

Smith, J.P., Jr., R.J. Cole, and J.O. Sawyer, Jr., eds. 1980. *Inventory of Rare and Endangered Vascular Plants of California.* Special Publication No. 1 (second edition). California Native Plant Society. Berkeley, CA. vii + 115 pp.

York, R., J.P. Smith, Jr., and S. Cochrane. 1982. New developments in the rare plant program. *Fremontia* 9(4):11-13.

The Rare, Threatened, and Endangered Plants of California

Abrams, L.R. 1923-1960. *An Illustrated Flora of the Pacific States, Washington, Oregon and California.* Vol. 4 by R. Ferris. Stanford University Press. Stanford, CA. 4 vols.

Hickman, J.C., ed. 1993. *The Jepson Manual: Higher Plants of California.* University of California Press. Berkeley, CA. 1400 pp.

Holland, R.F. 1986. *Preliminary Descriptions of the Terrestrial Natural Communities of California.* Nongame-Heritage Program, California Department of Fish and Game. Sacramento, CA. 156 pp.

Kartesz, J.T., and J.W. Thieret. 1991. Common names for vascular plants: Guidelines for use and application. *Sida* 14(3):421-434.

Munz, P.A. 1959. *A California Flora.* In collaboration with D.D. Keck. University of California Press. Berkeley, CA. 1681 pp.

Munz, P.A. 1968. *Supplement to a California Flora.* University of California Press. Berkeley, CA. 224 pp.

Munz, P.A. 1974. *A Flora of Southern California.* University of California Press. Berkeley, CA. 1086 pp.

Conserving Plants with Laws and Programs under the Department of Fish and Game

Hoshovsky, M. 1992. Developing partnerships in conserving California's biological diversity. *Fremontia* 20(1):19-23.

The Natural Communities Program

Holland, R.F., 1986. *Preliminary Descriptions of the Terrestrial Natural Communities of California.* Nongame-Heritage Program, California Department of Fish and Game. Sacramento, CA. 156 pp.

Cheatham, N.D., and J. R. Haller. 1975. *An Annotated List of California Habitat Types.* University of California Natural Land and Water Reserves System. Berkeley, CA. 77 pp.

Guidelines for Assessing Effects of Proposed Developments on Rare Plants and Plant Communities

Nelson, J.R. 1987. Rare plant surveys: Techniques for impact assessment. Pages 159-166 *in*: T.S. Elias, ed. *Conservation and Management of Rare and Endangered Plants.* California Native Plant Society. Sacramento, CA.

The Endangered Species Act and Rare Plant Protection in California

Barry, D., L. Haroun, and C. Halvorson. 1992. *For Conserving Listed Species, Talk is Cheaper Than We Think: The Consultation Process Under the Endangered Species Act.* World Wide Fund for Nature. Washington, DC. iii + 30 pp.

General Accounting Office. 1992. *Endangered Species Act: Types and Number of Implementing Actions.* Report No. RCED-92-131BR. Gaithersburg, MD. 40 pp.

Sensitive Plant Management on the National Forests and Grasslands in California

Meyer, M. 1990. Monitoring Forest Service Planning. *Fremontia* 18(4): 16-22.

Shevock, J.R. 1988. Native plant diversity and special area designations on the national forests in California. *Fremontia* 16(2):21-27.

Rare Plant Conservation on Bureau of Land Management Lands

Bureau of Land Management. 1992. *Rare Plants and Natural Communities: A Strategy for the Future.* U.S. Government Printing Office. 60 pp.

Rare Plant Conservation — A Global Perspective

Davis, S., S. Droop, P. Gregerson, L. Henson, C. Leon, J. Villa-Lobos, H. Synge, and J. Zantovska. 1986. *Plants in Danger: What Do We Know?* International Union for Conservation of Nature and Natural Resources. Gland, Switzerland. 461 pp.

Heywood, V., ed. 1994 (in prep.). *Centres of Plant Diversity: A Guide and Strategy for Their Conservation.* World Conservation Union. Gland, Switzerland.

Huxley, A. 1985. *Green Inheritance: The World Wildlife Fund Book of Plants.* Anchor/Doubleday. Garden City, NY. 193 pp.

IUCN. 1980. *World Conservation Strategy: Living Resource Conservation for Sustainable Development.* International Union for Conservation of Nature and Natural Resources. Gland, Switzerland.

IUCN, UNEP, and WWF. 1991. *Caring for the Earth: A Strategy for Sustainable Living.* World Conservation Union, United Nations Environment Programme, and World Wide Fund for Nature. Gland, Switzerland. 227 pp.

Lucas, G., and H. Synge, eds. 1978. *The IUCN Plant Red Data Book.* International Union for Conservation of Nature and Natural Resources. Morges, Switzerland. 540 pp.

Synge, H., ed. 1981. *The Biological Aspects of Rare Plant Conservation.* John Wiley. New York, NY. 558 pp.

Caulanthus californicus

Amsinckia furcata

Lasthenia conjugens

Opuntia basilaris ssp. *treleasei*

Limnanthes floccosa ssp. *californica*

Mimulus pictus

Rare plants of grasslands, oak woodland, and vernal pools. *Amsinckia furcata, Caulanthus californicus,* and *Opuntia basilaris* ssp. *treleasei* all occur in the grasslands of the southern San Joaquin Valley and surrounding hillsides. Conversion to agriculture, urbanization, and energy development have extirpated many occurrences of the latter two species. *Mimulus pictus,* from the oak woodland belt of the southern Sierra, is one of the prettiest of the rare monkey-flowers. *Lasthenia conjugens* occurs in vernal pools to the north and east of San Francisco Bay, and *Limnanthes floccosa* ssp. *californica* is known from vernal pools and swales near Chico in BUT Co.; both are severely endangered by urbanization and conversion to agriculture.

Suaeda californica

Chorizanthe howellii

Ferocactus viridescens

Penstemon albomarginatus

Phacelia nashiana

Brodiaea filifolia

Linanthus killipii

Rare plants of Southern California, the Central Coast, and dunes and deserts. *Ferocactus viridescens* occurs in maritime succulent scrub in coastal San Diego County, a habitat much favored by land developers. *Suaeda californica* has been eliminated from San Francisco Bay by destruction of its coastal salt marsh habitat, and is now found only at Morro Bay in San Luis Obispo Co. *Chorizanthe howellii* is a dune-loving plant with only five occurrences in MEN Co., where it is being obliterated by the ice plant (*Carpobrotus edulis*) in the photo. *Phacelia nashiana* and *Penstemon albomarginatus* are rare plants that occur only in specialized microhabitats within the vast Mojave Desert. *Clarkia speciosa* ssp. *immaculata*, *Eriogonum crocatum*, and *Calochortus obispoensis* occur along the Central Coast, the last two on special soil types: *Eriogonum* occurs only in volcanic soils, and, like so many California rare plants, the *Calochortus* occurs in serpentinite grasslands. *Linanthus killipii* mostly occurs near rapidly developing Big Bear Lake. *Brodiaea filifolia* has been much reduced by urbanization in coastal southern California.

Eriogonum crocatum

◀ *Calochortus obispoensis*

Clarkia speciosa ssp. *immaculata*

Fritillaria biflora var. *ineziana*

Calochortus tiburonensis

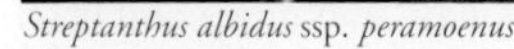

Streptanthus albidus ssp. *peramoenus*

Arctostaphylos imbricata

Clarkia imbricata

Rare plants of the San Francisco Bay area, the Coast Ranges, and Channel Islands. Although recognized by Jepson many years ago, *Fritillaria biflora* var. *ineziana* is a new addition to the fifth edition. Other plants of serpentinite grasslands of the Bay Area, all threatened primarily by urbanization, include *Calochortus tiburonensis*, *Streptanthus albidus* ssp. *peramoenus*, and *Arctostaphylos imbricata*. *Clarkia imbricata* is known from only one natural occurrence in the world, in the front yard of a private residence. *Erysimum teretifolium* is threatened by sand mining, and by urbanization of its curious habitat, ponderosa pine forest on inland marine sands in SCR Co. *Lotus argophyllus* var. *niveus*, *Malacothamnus fasciculatus* var. *nesioticus*, and *Dudleya nesiotica* are endemic to SCZ Isl., where they have been reduced by non-native plants, and overgrazing and browsing by livestock and pigs. Although controlled or eliminated on most of the Channel Islands, feral animals continue to brutalize the vegetation on SCT, SCZ, and SRO islands.

Lotus argophyllus var. *niveus*

Erysimum teretifolium

Malacothamnus fasciculatus var. *nesioticus*

Dudleya nesiotica

Pale swallowtail butterflies pollinate *L. kelloggii*

L. occidentale

A white-lined sphinx moth visits *L. parryi*

Pollination diversity in rare true lilies. Most plants require pollinators for reproduction, and insect and other herbivores are dependent on plants for food. Rare plants are embedded in complex and mysterious interactions with other members of their ecosystems, and mutual dependence is often the rule. Therefore, the ecosystem consequences of altered plant diversity are significant, but are unpredictable and often unnoticed by human observers. The lack of appropriate plant and animal associates is one reason that most transplantations of rare plants fail.

Western *Lilium* feed a wide variety of pollinating animals, and illustrate just part of the broad range of animal associates of rare plants. *L. kelloggii* is visited here by a pale swallowtail butterfly (*Papilio eurymedon*), *L. maritimum* by a *Bombus* bumblebee, *L. parryi* by a white-lined sphinx moth (*Hyles lineata*), and *L. bolanderi* by an immature/female Allen's hummingbird (*Selasphorus sasin*). *L. rubescens*, shown without any visitors, is pollinated by a combination of bees, butterflies, and probably moths, and *L. occidentale* by Allen and rufous hummingbirds. Rare lilies support non-pollinating associates too, such as this black and yellow syrphid fly feeding on pollen at an anther of *L. parryi*, and this camel cricket resting on a flower of *L. kelloggii*, probably after stealing nectar from the flower.

L. rubescens ➧

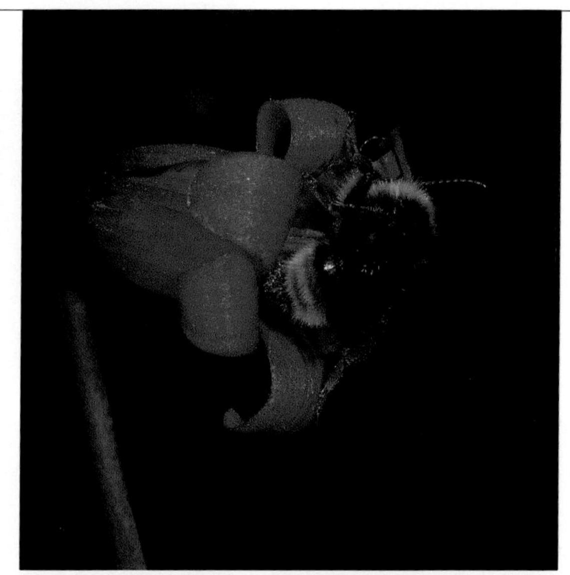

A bumblebee pollinates *L. maritimum*

An Allen's hummingbird pollinates *L. bolanderi*

Syrphid flies often eat *L. parryi* pollen

A camel cricket resting on *L. kelloggii*

Galium californicum ssp. *sierrae*

Allium tuolumnense

Fremontodendron decumbens

Carpenteria californica

Rare endemic shrubs, and plants of the Sierra foothills. *Fremontodendron decumbens* and *Galium californicum* ssp. *sierrae* are part of a rare plant ensemble that occurs primarily in the gabbro chaparral of ELD Co., where urbanization is fragmenting this unique habitat. *Allium tuolumnense* and *Verbena californica* are serpentinite endemics from the unusual outcrop called the Red Hills in TUO Co., where they are partially protected on BLM land. *Carpenteria californica* is in a monospecific genus from the Sierra foothills, and an example of a paleoendemic which once had a much broader range. Although it occurs by roadsides, *Neviusia cliftonii* was only just discovered in 1992 on SHA Co. limestones, illustrating in dramatic fashion how much more there is to know about California's rare flora.

Neviusia cliftonii

◀ *Verbena californica*

Abies amabilis (Dougl.) Forbes

"Pacific silver fir" Pinaceae
CNPS List: 2 **R-E-D Code:** 2-1-1 **State/Fed. Status:** CEQA
Distribution: SIS, OR, WA+
Quads: 702B, 736A
Habitat: UCFrs
Life Form: Tree (evergreen)
Blooming: N/A
Notes: Known in CA only from isolated groves. See *Fremontia* 16(3):19-20 (1988) for information on 736A occurrence.

Abies bracteata (D. Don) Nutt.

"bristlecone fir" Pinaceae
CNPS List: 4 **R-E-D Code:** 1-1-3 **State/Fed. Status:** CEQA?
Distribution: MNT, SLO
Habitat: LCFrs
Life Form: Tree (evergreen)
Blooming: N/A
Notes: Endemic to the Santa Lucia Mtns.

Abies lasiocarpa

See *Abies lasiocarpa* var. *lasiocarpa*

Abies lasiocarpa (Hook.) Nutt. var. *lasiocarpa*

"subalpine fir" Pinaceae
CNPS List: 2 **R-E-D Code:** 2-1-1 **State/Fed. Status:** CEQA
Distribution: SIS, AZ, ID, NV, OR, WA+
Quads: 684A, 701C, 702A, 719C, 719D
Habitat: Medws, SCFrs, UCFrs
Life Form: Tree (evergreen)
Blooming: N/A
Notes: See *Madroño* 20(8):413-415 (1971) for distributional information.

Abronia alpina Bdg.

"Ramshaw Meadows abronia" Nyctaginaceae
CNPS List: 1B **R-E-D Code:** 3-3-3 **State/Fed. Status:** /C1
Distribution: TUL
Quads: 306B*, 329C, 330D
Habitat: Medws (gravelly margins)
Life Form: Perennial herb
Blooming: July-August
Notes: Known from only two extant occurrences at Ramshaw Meadows and Templeton Meadows. Threatened by trampling from cattle and trail construction. See *Botanical Gazette* 27:444-457 (1899) for original description, and *Aliso* 7(2):201-205 (1970) for discussion of rediscovery.
Status Report: 1989

Abronia maritima Wats.

"red sand-verbena" Nyctaginaceae
CNPS List: 4 **R-E-D Code:** 1-2-2 **State/Fed. Status:** CEQA?
Distribution: ANA, LAX, ORA, SBA, SCM, SCT, SCZ, SDG, SLO, SMI, SNI, VEN, BA
Habitat: CoDns
Life Form: Perennial herb
Blooming: February-November
Notes: Nearly extirpated in southern California; more common to north? Hybridizes with *A. latifolia* and *A. umbellata.* See *Botany of California* 2:4 (1880) for original description.

Abronia nana Wats. ssp. *covillei* (Heimerl) Munz

"Coville's dwarf abronia" Nyctaginaceae
CNPS List: 4 **R-E-D Code:** 1-2-1 **State/Fed. Status:** CEQA?
Distribution: INY, MNO, SBD, NV
Habitat: GBScr, JTWld, PJWld, UCFrs / carbonate, sandy
Life Form: Perennial herb
Blooming: May-August
Notes: Threatened by carbonate mining and vehicles in the San Bernardino Mtns., and by grazing in the White and Inyo Mtns. See *Smithsonian Miscellaneous Collections* 52:197 (1908) for original description, and *Manual of Southern California Botany,* pp. 150, 598 (1935) by P. Munz for taxonomic treatment.

Abronia umbellata Lam. ssp. *breviflora* (Standl.) Munz

"pink sand-verbena" Nyctaginaceae
CNPS List: 1B **R-E-D Code:** 2-2-2 **State/Fed. Status:** /C2
Distribution: DNT, HUM, MEN, SON, OR
Quads: 503D, 537B, 569A, 569D, 585A, 585D, 654B, 655A, 672A, 672B, 672C, 689D, 706A, 706D, 723B, 723D, 740C
Habitat: CoDns
Life Form: Perennial herb
Blooming: July-September
Notes: Most occurrences have few plants. Threatened by vehicles and foot traffic. State-listed as Endangered in OR.

Acacia minuta ssp. *minuta*

Considered but rejected: Not in CA; name misapplied to *A. farnesiana* var. *farnesiana*; a non-native plant

Acacia smallii

Considered but rejected: Not native; misapplied to *A. farnesiana* var. *farnesiana*

Acalypha californica

Considered but rejected: Too common

Acanthomintha duttonii (Abrams) Jokerst

"San Mateo thorn-mint" Lamiaceae
CNPS List: 1B **R-E-D Code:** 3-3-3 **State/Fed. Status:** CE/FE
Distribution: SMT
Quads: 428B*, 429A, 448D*
Habitat: Chprl, VFGrs / serpentinite
Life Form: Annual herb
Blooming: April-June
Notes: Known from only two extant natural occurrences; most historical occurrences have been extirpated. Seriously threatened by development. State-listed as *A. obovata* ssp. *duttonii*; USFWS also uses this name. See *Illustrated Flora of the Pacific States* 3:635 (1951) by L. Abrams for original description, and *Madroño* 38(4):278-286 (1991) for revised nomenclature.
Status Report: 1986

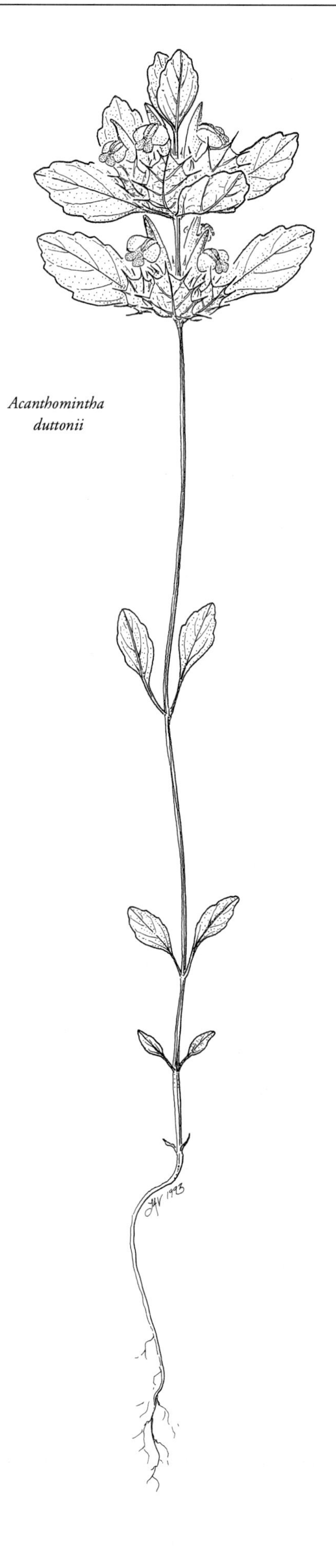

Acanthomintha duttonii

Acanthomintha ilicifolia (Gray) Gray

"San Diego thorn-mint" Lamiaceae
CNPS List: 1B **R-E-D Code:** 2-3-2 **State/Fed. Status:** CE/C1
Distribution: SDG, BA
Quads: 10A, 10B*, 10C, 11A*, 20C, 21B, 21D, 22A, 22B, 22C*, 22D, 34C, 35B, 35C, 35D, 36D
Habitat: Chprl, CoScr, VFGrs, VnPls / clay
Life Form: Annual herb
Blooming: April-June
Notes: Approximately one-third of the historical occurrences have been extirpated; those remaining threatened by urbanization, road construction, vehicles, and grazing. Several occurrences introduced as mitigation for development, but few have survived. See *Proceedings of the American Academy of Arts and Sciences* 8:368 (1872) for original description.
Status Report: 1993

Acanthomintha lanceolata Curran

"Santa Clara thorn-mint" Lamiaceae
CNPS List: 4 **R-E-D Code:** 1-2-3 **State/Fed. Status:** CEQA?
Distribution: ALA, FRE, MER, MNT, SBT, SCL, STA
Habitat: Chprl (serpentinite), CoScr
Life Form: Annual herb
Blooming: March-June
Notes: Possibly threatened by grazing, and by proposed reservoir in MER Co.

Acanthomintha obovata Jeps. ssp. *cordata* Jokerst

"heart-leaved thorn-mint" Lamiaceae
CNPS List: 4 **R-E-D Code:** 1-2-3 **State/Fed. Status:** CEQA?
Distribution: LAX, SBA, SLO, VEN
Habitat: Chprl (openings), CmWld, PJWld, VFGrs / clay
Life Form: Annual herb
Blooming: April-July
Notes: Possibly threatened by vehicles and grazing. See *Madroño* 38(4):278-286 (1991) for original description.

Acanthomintha obovata ssp. *duttonii*

See *Acanthomintha duttonii*

Acanthomintha obovata Jeps. ssp. *obovata*

"San Benito thorn-mint" Lamiaceae
CNPS List: 4 **R-E-D Code:** 1-2-3 **State/Fed. Status:** /C2
Distribution: FRE, MNT, SBT, SLO
Habitat: Chprl, VFGrs / heavy clay, alkaline, serpentinite
Life Form: Annual herb
Blooming: April-June
Notes: Threatened by grazing.

Achillea millefolium var. *gigantea*

Considered but rejected: A synonym of *A. millefolium*; a common taxon

Achnatherum aridum (M.E. Jones) Barkworth

"Mormon needle grass" Poaceae
CNPS List: 2 **R-E-D Code:** 2-1-1 **State/Fed. Status:** CEQA
Distribution: INY, MNO, SBD, AZ, NV+
Quads: 200A, 248B, 248C, 249C, 249D, 369C, 390D, 431A
Habitat: JTWld, PJWld / carbonate
Life Form: Perennial herb
Blooming: May-July
Notes: See *Phytologia* 74(1):1-25 (1993) for revised nomenclature.

Achnatherum diegoense (Swallen) Barkworth

"San Diego County needle grass" Poaceae
CNPS List: 4 **R-E-D Code:** 1-2-1 **State/Fed. Status:** CEQA?
Distribution: ANA, SCZ, SDG, SMI, SNI, SRO, BA
Habitat: Chprl, CoScr / often mesic
Life Form: Perennial herb
Blooming: May-June
Notes: Previously known only from the mainland; now also from several of the Channel Islands. See *Phytologia* 74(1):1-25 (1993) for revised nomenclature.

Achnatherum lemmonii (Vasey) Barkworth var. *pubescens* (Crampton) Barkworth

"pubescent needle grass" Poaceae
CNPS List: 3 **R-E-D Code:** 3-2-3 **State/Fed. Status:** /C3b
Distribution: LAK, TEH
Quads: 565D, 596A, 596B
Habitat: Chprl, LCFrs / serpentinite
Life Form: Perennial herb
Blooming: May-July
Notes: Move to List 1B? Taxonomic questions remain, since plant co-occurs with nominate variety. A synonym of *A. lemmonii* in *The Jepson Manual*, but author now believes it is distinct. Threatened by grazing. See *Leaflets of Western Botany* 7(9):220 (1955) for original description, *Madroño* 31(1):48-56 (1984) for alternate taxonomic treatment, and *Phytologia* 74(1):1-25 (1993) for revised nomenclature.
Status Report: 1977

Acleisanthes longiflora Gray

"angel trumpets" Nyctaginaceae
CNPS List: 2 **R-E-D Code:** 3-1-1 **State/Fed. Status:** CEQA
Distribution: RIV, AZ, NM, TX+
Quads: 58A
Habitat: SDScr (carbonate)
Life Form: Perennial herb
Blooming: May

Adolphia californica Wats.

"California adolphia" Rhamnaceae
CNPS List: 2 **R-E-D Code:** 1-2-1 **State/Fed. Status:** CEQA
Distribution: SDG, BA
Quads: 10B, 10C, 11A, 11D, 22A, 22B, 22D, 35B, 35C, 35D, 36A, 36D
Habitat: Chprl, CoScr, VFGrs / clay
Life Form: Shrub (deciduous)
Blooming: December-April
Notes: Threatened by urbanization, road construction, and grazing.

Agastache parvifolia

Considered but rejected: Too common

Agave shawii (Engelm.) Gentry

"Shaw's agave" Liliaceae
CNPS List: 2 **R-E-D Code:** 3-3-1 **State/Fed. Status:** /C2
Distribution: SDG, BA
Quads: 11B, 11D, 22B
Habitat: CBScr, CoScr
Life Form: Shrub (leaf succulent)
Blooming: September-May
Notes: Known in CA from fewer than five native occurrences. Occurrence at Border Field SP is very small. Torrey Pines SP and Cabrillo NM occurrences are probably planted.

Agave utahensis (Engelm.) Gentry

"Utah agave" Liliaceae
CNPS List: 4 **R-E-D Code:** 1-2-2 **State/Fed. Status:** /C3c
Distribution: INY, SBD, AZ, NV, UT
Habitat: JTWld, MDScr, PJWld / carbonate or volcanic
Life Form: Shrub (leaf succulent)
Blooming: May-July
Notes: Includes *A. utahensis* vars. *eborispina* and *nevadensis*. See *Four Seasons* 3(3):6 (1970) and *Intermountain Flora* 6:538 (1977) for distributional information.

Agave utahensis var. *eborispina*

See *Agave utahensis*

Agave utahensis var. *nevadensis*

See *Agave utahensis*

Ageratina shastensis (D.W. Taylor & Stebb.) R.M. King & H. Robinson

"Shasta ageratina" Asteraceae
CNPS List: 4 **R-E-D Code:** 1-1-3 **State/Fed. Status:** /C3c
Distribution: SHA
Habitat: Chprl, LCFrs / carbonate
Life Form: Perennial herb
Blooming: June-September
Notes: See *Madroño* 25(4):218-220 (1978) for original description.

Agoseris elata

Considered but rejected: Too common

Agropyron scribneri

See *Elymus scribneri*

Agrostis aristiglumis

Considered but rejected: A synonym of *A. microphylla*; a common taxon

Agrostis blasdalei Hitchc.

"Blasdale's bent grass" Poaceae
CNPS List: 1B **R-E-D Code:** 3-2-3 **State/Fed. Status:** /C2
Distribution: MEN, MRN, SCR, SON
Quads: 409D, 485B, 485C, 502C, 503A, 503D, 520B, 569A, 569D
Habitat: CBScr, CoDns, CoPrr
Life Form: Perennial herb (rhizomatous)
Blooming: May-July
Notes: Known from fewer than fifteen occurrences. Historical occurrences need field surveys. Threatened by agriculture and recreation. Includes *A. blasdalei* var. *marinensis*, which is state-listed Rare.
Status Report: 1988

Agrostis blasdalei var. *blasdalei*

See *Agrostis blasdalei*

Agrostis blasdalei var. *marinensis*

See *Agrostis blasdalei*

Agrostis clivicola var. *clivicola*

Considered but rejected: A synonym of *A. densiflora*; a common taxon

Agrostis clivicola var. *punta-reyesensis*

Considered but rejected: A synonym of *A. densiflora*; a common taxon

Agrostis hendersonii Hitchc.

"Henderson's bent grass" Poaceae
CNPS List: 3 **R-E-D Code:** 3-2-2 **State/Fed. Status:** /C2
Distribution: BUT?, CAL, MER, SHA, OR*
Quads: 421B, 459A, 477B*, 575A?, 646A
Habitat: VFGrs (mesic), VnPls
Life Form: Annual herb
Blooming: April-May
Notes: Move to List 1B? Field surveys needed. Identity of BUT Co. plants remains questionable. One occurrence extirpated by construction of Comanche Reservoir; possibly threatened by development. Candidate for state listing in OR, but probably extinct there. Very similar to *A. microphylla*; study needed. USFWS uses the name *A. microphylla* var. *hendersonii*. See *Journal of the Washington Academy of Sciences* 20(15):381 (1930) for original description, *Madroño* 3:230 (1936) for first CA record, and *Leaflets of Western Botany* 3(12):258 (1943) for second CA record.
Status Report: 1979

Agrostis hooveri Swall.

"Hoover's bent grass" Poaceae
CNPS List: 4 **R-E-D Code:** 1-2-3 **State/Fed. Status:** CEQA?
Distribution: SBA, SLO
Habitat: Chprl, CmWld, VFGrs / sandy
Life Form: Perennial herb (stoloniferous)
Blooming: June

Agrostis humilis Vasey

"mountain bent grass" Poaceae
CNPS List: 2 **R-E-D Code:** 3-1-1 **State/Fed. Status:** CEQA
Distribution: ALP, MPA, TUO, NV, OR, WA, ++
Quads: 454A, 454C, 489C, 506D
Habitat: AlpBR, Medws, SCFrs
Life Form: Perennial herb
Blooming: July-September
Notes: To be expected elsewhere in CA; need information. On review list in OR. Intergrades with *A. thurberiana.* See *Madroño* 25(4):232 (1978) for first CA record.

Agrostis microphylla var. *hendersonii*

See *Agrostis hendersonii*

Allium anserinum

Considered but rejected: Known only from the type; see *A. parishii*, of which it is probably a synonym

Allium atrorubens Wats. var. *atrorubens*

"Great Basin onion" Liliaceae
CNPS List: 2 **R-E-D Code:** 2-1-1 **State/Fed. Status:** CEQA
Distribution: LAS, MNO, AZ, NV, OR+
Quads: 470A, 470B, 470C, 620D
Habitat: GBScr, PJWld
Life Form: Perennial herb (bulbiferous)
Blooming: May-June
Notes: See *Botany of the King Exploration*, p. 352 (1871) for original description, and *Madroño* 39(2):83-89 (1992) for distribution in CA.

Allium cratericola

Considered but rejected: Too common

Allium davisiae

Considered but rejected: Too common; a synonym of *A. lacunosum* var. *davisiae*

Allium fimbriatum var. *munzii*

See *Allium munzii*

Allium fimbriatum Wats. var. *purdyi* (Eastw.) Ownbey & Aase

"Purdy's onion" Liliaceae
CNPS List: 4 **R-E-D Code:** 1-1-3 **State/Fed. Status:** CEQA?
Distribution: COL, LAK
Habitat: Chprl, CmWld / serpentinite
Life Form: Perennial herb (bulbiferous)
Blooming: May

Allium fimbriatum var. *sharsmithiae*

See *Allium sharsmithae*

Allium hickmanii Eastw.

"Hickman's onion" Liliaceae
CNPS List: 1B **R-E-D Code:** 2-2-3 **State/Fed. Status:** /C1
Distribution: MNT, SLO
Quads: 271B, 272A, 295B, 365B, 366C, 366D
Habitat: CCFrs, Chprl, CoPrr, CoScr, VFGrs
Life Form: Perennial herb (bulbiferous)
Blooming: April-May
Notes: Known from fewer than twenty occurrences. Threatened by urbanization, grazing, road construction, and military activities. See *Bulletin of the Torrey Botanical Club* 30:483-502 (1903) for original description.
Status Report: 1977

Allium hoffmanii Traub

"Beegum onion" Liliaceae
CNPS List: 4 **R-E-D Code:** 1-1-3 **State/Fed. Status:** /C3c
Distribution: HUM, SHA, TEH, TRI
Habitat: LCFrs (serpentinite)
Life Form: Perennial herb (bulbiferous)
Blooming: June-July
Notes: See *Plant Life* 28:63 (1972) for original description.
Status Report: 1979

Allium howellii Eastw. var. *clokeyi* Traub

"Mt. Pinos onion" Liliaceae
CNPS List: 4 **R-E-D Code:** 1-1-3 **State/Fed. Status:** CEQA?
Distribution: VEN
Habitat: GBScr
Life Form: Perennial herb (bulbiferous)
Blooming: April-June

Allium jepsonii (Traub) Denison & McNeal

"Jepson's onion" Liliaceae
CNPS List: 1B **R-E-D Code:** 3-2-3 **State/Fed. Status:** /C2
Distribution: BUT, TUO
Quads: 457B, 458B, 591C, 592D
Habitat: CmWld, LCFrs / serpentinite or volcanic
Life Form: Perennial herb (bulbiferous)
Blooming: May-June
Notes: Known from only two more or less extended occurrences. See *Plant Life* 28:63 (1972) for original description, and *Madroño* 36(2):122-130 (1989) for revised nomenclature.

Allium membranaceum

Considered but rejected: Too common

Allium monticola

Considered but rejected: Too common

Allium monticola var. *keckii*

Considered but rejected: A synonym of *A. monticola*; a common taxon

Allium munzii (Traub) McNeal

"Munz's onion" Liliaceae
CNPS List: 1B **R-E-D Code:** 3-3-3 **State/Fed. Status:** CT/C1
Distribution: RIV
Quads: 68A, 68B, 68D, 69A, 69B, 69D, 86C, 86D, 87D
Habitat: Chprl, CmWld, CoScr, PJWld, VFGrs / clay
Life Form: Perennial herb (bulbiferous)
Blooming: March-May
Notes: Known from fewer than ten occurrences. Threatened by urbanization, mining, agriculture, and non-native plants. Cleveland NF has adopted species management guidelines. State-listed as *A. fimbriatum* var. *munzii*.
Status Report: 1989

Allium nevadense Wats.

"Nevada onion" Liliaceae
CNPS List: 2 **R-E-D Code:** 3-1-1 **State/Fed. Status:** CEQA
Distribution: SBD, AZ, NV, OR+
Quads: 176A, 224C
Habitat: PJWld
Life Form: Perennial herb (bulbiferous)
Blooming: April-May
Notes: Known in CA from about five occurrences. See *Botany of the King Exploration*, p. 351 (1871) for original description, and *Madroño* 39(2):83-89 (1992) for distribution in CA.

Allium parishii Wats.

"Parish's onion" Liliaceae
CNPS List: 4 **R-E-D Code:** 1-1-2 **State/Fed. Status:** CEQA?
Distribution: IMP, RIV, SBD, AZ
Habitat: JTWld
Life Form: Perennial herb (bulbiferous)
Blooming: April-May

Allium sanbornii Wood var. *congdonii* Jeps.

"Congdon's onion" Liliaceae
CNPS List: 4 **R-E-D Code:** 1-1-3 **State/Fed. Status:** CEQA?
Distribution: ELD, MPA, NEV, PLA, TUO
Habitat: Chprl, CmWld / serpentinite or volcanic
Life Form: Perennial herb (bulbiferous)
Blooming: April-July
Notes: See *Flora of California* 1(6):275 (1922) by W.L. Jepson for original description, and *Madroño* 36(2):122-130 (1989) for taxonomic treatment.

Allium sanbornii Wood var. *sanbornii*

"Sanborn's onion" Liliaceae
CNPS List: 4 **R-E-D Code:** 1-2-2 **State/Fed. Status:** CEQA?
Distribution: BUT, CAL, ELD, NEV, PLA, SHA*, TEH, YUB, OR
Habitat: Chprl, CmWld, LCFrs / serpentinite, gravelly
Life Form: Perennial herb (bulbiferous)
Blooming: May-September
Notes: Consists of small, widely scattered populations. On review list in OR. See *Proceedings of the Academy of Natural Sciences of Philadelphia* 20:171 (1868) for original description, and *Madroño* 36(2):122-130 (1989) for taxonomic treatment.

Allium sanbornii var. *tuolumnense*

See *Allium tuolumnense*

Allium sharsmithae (Traub) McNeal

"Sharsmith's onion" Liliaceae
CNPS List: 1B **R-E-D Code:** 2-1-3 **State/Fed. Status:** CEQA
Distribution: ALA, SCL, STA
Quads: 425B, 445D
Habitat: CmWld (serpentinite)
Life Form: Perennial herb (bulbiferous)
Blooming: March-May
Notes: Known only from the Mt. Hamilton Range.

Allium shevockii McNeal

"Spanish Needle onion" Liliaceae
CNPS List: 1B **R-E-D Code:** 3-1-3 **State/Fed. Status:** /C2
Distribution: KRN
Quads: 282C, 283D
Habitat: UCFrs (rocky)
Life Form: Perennial herb (bulbiferous)
Blooming: June
Notes: Known from only two occurrences at Spanish Needle. See *Madroño* 34(2):150-154 (1987) for original description.

Allium siskiyouense Traub

"Siskiyou onion" Liliaceae
CNPS List: 4 **R-E-D Code:** 1-1-1 **State/Fed. Status:** CEQA?
Distribution: DNT, HUM, SIS, TRI, OR
Habitat: LCFrs, UCFrs / sometimes serpentinite
Life Form: Perennial herb (bulbiferous)
Blooming: June-July

Allium tribracteatum Torr.

"three-bracted onion" Liliaceae
CNPS List: 1B **R-E-D Code:** 3-2-3 **State/Fed. Status:** /C2
Distribution: CAL, TUO
Quads: 472C, 473B, 473D, 474A, 474B, 474C, 474D, 475A, 475D, 491A, 491D
Habitat: Chprl, LCFrs, UCFrs / volcanic
Life Form: Perennial herb (bulbiferous)
Blooming: April-July
Notes: Recent field work has located several new populations. Threatened by vehicles. See *Report of the Pacific Railroad Expedition* 4:148 (1857) for original description, and *Aliso* 11(1):27-35 (1985) for revised treatment.

Allium tuolumnense (Traub) Denison & McNeal

"Rawhide Hill onion" Liliaceae
CNPS List: 1B **R-E-D Code:** 2-2-3 **State/Fed. Status:** /C1
Distribution: TUO
Quads: 458B, 458C, 458D
Habitat: CmWld (serpentinite)
Life Form: Perennial herb (bulbiferous)
Blooming: May
Notes: Known from approximately twenty occurrences. Threatened by grazing and mining. See *Plant Life* 28:63 (1972) for original description, and *Madroño* 36(2):122-130 (1989) for revised nomenclature.

Allium yosemitense Eastw.

"Yosemite onion" Liliaceae
CNPS List: 1B **R-E-D Code:** 2-1-3 **State/Fed. Status:** CR/C3c
Distribution: MPA, TUO
Quads: 437B, 437C, 438A, 438D, 456C
Habitat: BUFrs, Chprl, CmWld, LCFrs / rocky, metamorphic
Life Form: Perennial herb (bulbiferous)
Blooming: May-July
Notes: Known from fewer than twenty occurrences, most of which are inaccessible. See *Leaflets of Western Botany* 1:132-133 (1934) for original description, *Fremontia* 8(1):15-18 (1980) for information on rediscovery, and *Aliso* 11(1):27-35 (1985) for taxonomic treatment.
Status Report: 1990

Alopecurus aequalis Sobol. var. *sonomensis* Rubtzoff

"Sonoma alopecurus" Poaceae
CNPS List: 1B **R-E-D Code:** 3-3-3 **State/Fed. Status:** /C2
Distribution: MRN, SON
Quads: 467B, 485C, 485D, 501A, 502A, 502B, 502D, 503A
Habitat: MshSw (freshwater), RpScr
Life Form: Perennial herb
Blooming: May-July
Notes: Known from fewer than five native occurrences. Two occurrences introduced in 1987 (485C, 485D), but as of 1993 both appear to have failed. Historical localities need field surveys. Threatened by cattle trampling and wetland habitat loss. See *A. aequalis* in *The Jepson Manual*.

Ambrosia chenopodiifolia (Benth.) Payne

"San Diego bur-sage" Asteraceae
CNPS List: 2 **R-E-D Code:** 2-2-1 **State/Fed. Status:** CEQA
Distribution: SDG, BA
Quads: 10C, 11A, 11D
Habitat: CoScr
Life Form: Shrub
Blooming: April-June
Notes: Known in CA from approximately ten occurrences. Threatened by development.

Ambrosia pumila (Nutt.) Gray

"San Diego ambrosia" Asteraceae
CNPS List: 1B **R-E-D Code:** 3-3-2 **State/Fed. Status:** /C2
Distribution: RIV, SDG, BA
Quads: 9B, 10B, 11A, 11D, 21C, 22C, 22D, 35D, 36A, 50C*, 68C, 68D
Habitat: Chprl, CoScr, VFGrs, VnPls / often in disturbed areas
Life Form: Perennial herb (rhizomatous)
Blooming: June-September
Notes: Historical occurrences need field surveys. Many occurrences have been extirpated in SDG Co. Threatened by development. See *Madroño* 39(2):157 (1992) for first RIV Co. record.

Ammobroma sonorae

See *Pholisma sonorae*

Ammoselinum giganteum Coult. & Rose

"desert sand-parsley" Apiaceae
CNPS List: 2 **R-E-D Code:** 3-1-1 **State/Fed. Status:** CEQA
Distribution: RIV, AZ+
Quads: 62A, 62B
Habitat: SDScr
Life Form: Annual herb
Blooming: March-April
Notes: Known in CA only from Hayfields Dry Lake. Field surveys needed.
Status Report: 1989

Amsinckia furcata

See *Amsinckia vernicosa* var. *furcata*

Amsinckia grandiflora (Gray) Greene

"large-flowered fiddleneck" Boraginaceae
CNPS List: 1B **R-E-D Code:** 3-3-3 **State/Fed. Status:** CE/FE
Distribution: ALA, CCA, SJQ
Quads: 444B, 444C, 445A, 464A, 464D
Habitat: CmWld, VFGrs
Life Form: Annual herb
Blooming: April-May
Notes: Known from only three natural occurrences. Reduced by agriculture, development, and grazing; currently threatened by non-native plants and possibly by altered fire frequency. Recent reintroductions have occurred (444B, 444C, 464A, and 464D) as part of State and Federal recovery work. See *Botany of California* 1:525 (1876) for original description, and *Conservation Biology* 7(3):510-526 (1993) for population biology.
Status Report: 1985

Amsinckia lunaris Macbr.

"bent-flowered fiddleneck" Boraginaceae
CNPS List: 4 **R-E-D Code:** 1-1-3 **State/Fed. Status:** CEQA?
Distribution: ALA, CCA, LAK, MRN, SCR, SHA, SIS
Habitat: CmWld, VFGrs
Life Form: Annual herb
Blooming: March-June
Notes: Most Bay Area records are old; current status unknown.

Amsinckia spectabilis var. *macrocarpa*

Considered but rejected: Too common

Amsinckia vernicosa H. & A. var. *furcata* (Suksd.) Hoov.

"forked fiddleneck" Boraginaceae
CNPS List: 4 **R-E-D Code:** 1-2-3 **State/Fed. Status:** /C2
Distribution: FRE, KNG, KRN, SBT, SLO
Habitat: CmWld, VFGrs
Life Form: Annual herb
Blooming: March-May
Notes: Many new populations found in 1991 and 1992; more common than previously thought. Most new populations are relatively unthreatened; others threatened by mining and grazing. See *American Journal of Botany* 44:529-536 (1957) for discussion of taxonomy.

Amsinckia vernicosa var. *vernicosa*

Considered but rejected: Too common

Androsace elongata L. ssp. *acuta* (Greene) Robbins

"California androsace" Primulaceae
CNPS List: 4 **R-E-D Code:** 1-2-2 **State/Fed. Status:** CEQA?
Distribution: ALA, CCA, FRE, KRN, LAX*, SBD, SDG, SIS, SJQ, SLO, TEH?, OR, BA
Habitat: Chprl, CmWld, CoScr
Life Form: Annual herb
Blooming: March-June
Notes: Highly localized and often overlooked; many occurrences extirpated. Does plant occur in TEH Co.? Very rare in Southern California. See *Manual of the Botany of the Region of San Francisco Bay*, p. 238 (1894) by E. Greene for original description, and *American Midland Naturalist* 32:132-163 (1944) for taxonomic treatment.

Androsace filiformis Retz.

"slender-stemmed androsace" Primulaceae
CNPS List: 2 **R-E-D Code:** 3-1-1 **State/Fed. Status:** CEQA
Distribution: SIS, OR, WA, ++
Quads: 732C, 733D
Habitat: Medws, UCFrs
Life Form: Annual herb
Blooming: June-August

Androstephium breviflorum Wats.

"small-flowered androstephium" Liliaceae
CNPS List: 2 **R-E-D Code:** 3-1-1 **State/Fed. Status:** CEQA
Distribution: INY, SBD, AZ, NV+
Quads: 158C, 204C, 204D
Habitat: MDScr (bajadas)
Life Form: Perennial herb (bulbiferous)
Blooming: March-April
Notes: Need quads for INY Co. See *Madroño* 31(3):192 (1984) for distributional information.

Angelica arguta

Considered but rejected: Too common

Angelica callii Math. & Const.

"Call's angelica" Apiaceae
CNPS List: 4 **R-E-D Code:** 1-1-3 **State/Fed. Status:** /C3c
Distribution: FRE, KRN, TUL
Habitat: CmWld, LCFrs / mesic
Life Form: Perennial herb
Blooming: June-July
Notes: See *Madroño* 24(2):78-83 (1977) for original description.

Antennaria flagellaris (Gray) Gray

"stoloniferous pussytoes" Asteraceae
CNPS List: 4 **R-E-D Code:** 1-2-1 **State/Fed. Status:** CEQA?
Distribution: LAS, MOD, ID, NV, OR, WA+
Habitat: GBScr
Life Form: Perennial herb (stoloniferous)
Blooming: April-June
Notes: Previously known in CA from fewer than ten occurrences, but approximately 50 new populations discovered in last four years. Somewhat threatened by mining. See *Madroño* 30(2):129 (1983) for first CA record.

Antennaria marginata Greene

"white-margined everlasting" Asteraceae
CNPS List: 2 **R-E-D Code:** 3-1-1 **State/Fed. Status:** CEQA
Distribution: SBD, AZ, NM+
Quads: 105A
Habitat: LCFrs, UCFrs
Life Form: Perennial herb (stoloniferous)
Blooming: May-August
Notes: Known in CA only from the South Fork Santa Ana River area. See *Pittonia* 3:290 (1898) for original description.

Antennaria pulchella Greene

"beautiful pussy-toes" Asteraceae
CNPS List: 4 **R-E-D Code:** 1-1-2 **State/Fed. Status:** CEQA?
Distribution: ALP, ELD, FRE, INY, MNO, TUL, TUO, NV
Habitat: AlpBR (stream margins), Medws
Life Form: Perennial herb (stoloniferous)
Blooming: June-September
Notes: Mostly unthreatened due to remoteness of habitat. See *Leaflets of Botanical Observation and Criticism* 2:149 (1911) for original description, and *Madroño* 37(3):171-183 (1990) for taxonomic treatment.

Antennaria racemosa

Considered but rejected: Too common

Antennaria suffrutescens Greene

"evergreen everlasting" Asteraceae
CNPS List: 4 **R-E-D Code:** 1-1-2 **State/Fed. Status:** /C3c
Distribution: DNT, HUM, OR
Habitat: LCFrs (serpentinite)
Life Form: Perennial herb (stoloniferous)
Blooming: January-July
Notes: See *Pittonia* 3:277 (1898) for original description.
Status Report: 1979

Antirrhinum cyathiferum Benth.

"Deep Canyon snapdragon" Scrophulariaceae
CNPS List: 2 **R-E-D Code:** 3-1-1 **State/Fed. Status:** CEQA
Distribution: RIV, AZ, BA, SO
Quads: 65A, 65B
Habitat: SDScr (rocky)
Life Form: Annual herb
Blooming: February-April
Notes: Known in CA only from the Deep Cyn. area.

Antirrhinum filipes

Considered but rejected: Too common

Antirrhinum ovatum Eastw.

"oval-leaved snapdragon" Scrophulariaceae
CNPS List: 4 **R-E-D Code:** 1-2-3 **State/Fed. Status:** /C3c
Distribution: KRN, MNT, SBA, SBT, SLO, VEN
Habitat: Chprl, CmWld, PJWld, VFGrs / clay or gypsum, often alkaline
Life Form: Annual herb
Blooming: May-November
Notes: Relatively abundant in 1991, but appears only in favorable years. Threatened by grazing and vehicles. See *Bulletin of the Torrey Botanical Club* 32:213 (1905) for original description.

Antirrhinum subcordatum Gray

"dimorphic snapdragon" Scrophulariaceae
CNPS List: 1B **R-E-D Code:** 2-2-3 **State/Fed. Status:** /C3c
Distribution: COL, GLE, LAK, TEH
Quads: 533A, 533C, 534B, 563C, 564A, 564B, 564C, 564D, 580A, 596A, 596C, 596D, 630D
Habitat: Chprl, LCFrs / sometimes serpentinite
Life Form: Annual herb
Blooming: April-July
Notes: Threatened by road maintenance and grazing. See *Proceedings of the American Academy of Arts and Sciences* 20:306 (1884) for original description.
Status Report: 1978

Antirrhinum virga Gray

"tall snapdragon" Scrophulariaceae
CNPS List: 4 **R-E-D Code:** 1-1-3 **State/Fed. Status:** CEQA?
Distribution: LAK, MEN, NAP, SON
Habitat: Chprl (rocky openings, often serpentinite)
Life Form: Perennial herb
Blooming: June-July
Notes: See *Proceedings of the American Academy of Arts and Sciences* 7:373 (1868) for original description, and *Systematic Botany Monographs* 22:53-57 (1988) for taxonomic treatment.

Aphanisma blitoides Moq.

"aphanisma" Chenopodiaceae
CNPS List: 1B **R-E-D Code:** 2-2-2 **State/Fed. Status:** /C2
Distribution: ANA, LAX, ORA, SBA, SBR, SCM, SCT, SCZ, SDG, SNI, SRO, VEN, BA
Quads: 11B, 22B, 71B, 71D, 73A, 90C*, 141D, ANAC, SBRA, SCMC, SCMN, SCMS, SCTE, SCTW, SCZA, SNIC, SROS
Habitat: CBScr, CoScr / sandy
Life Form: Annual herb
Blooming: April-May
Notes: In steep decline on mainland, also declining on islands. Known from only three occurrences on SNI Isl. Last seen on SCZ and SRO islands in 1932. Threatened by urbanization, recreational development, and foot traffic, and by feral herbivores on SCT, SCZ, and SRO islands.

Arabis aculeolata Greene

"Waldo rock cress" Brassicaceae
CNPS List: 2 **R-E-D Code:** 3-2-1 **State/Fed. Status:** /C3c
Distribution: DNT, SIS, OR
Quads: 720B, 736B, 736D, 738B, 738D, 739A, 739B
Habitat: BUFrs, LCFrs, UCFrs / serpentinite
Life Form: Perennial herb
Blooming: April-June
Notes: Known in CA from fewer than ten occurrences. Threatened by logging and mining. On watch list in OR. See *Rhodora* 43(511):352-353 (1941) for taxonomic treatment, and *Contributions from the Gray Herbarium* 204:151 (1973) for taxonomic information.

Arabis blepharophylla H. & A.

"coast rock cress" Brassicaceae
CNPS List: 4 **R-E-D Code:** 1-1-3 **State/Fed. Status:** /C3c
Distribution: CCA, MRN, SCR, SFO, SMT, SON
Habitat: BUFrs, CoPrr, CoScr
Life Form: Perennial herb
Blooming: February-April
Notes: See *Rhodora* 43(511):348-349 (1941) for taxonomic treatment, and *Contributions from the Gray Herbarium* 204:149-154 (1973) for taxonomic information.

Arabis bodiensis Roll.

"Bodie Hills rock cress" Brassicaceae
CNPS List: 1B **R-E-D Code:** 2-1-2 **State/Fed. Status:** /C2
Distribution: FRE, INY, MNO, TUL, NV
Quads: 330A, 391C, 393C, 470A, 470B, 487A, 487B, 487C
Habitat: GBScr, PJWld, SCFrs?
Life Form: Perennial herb
Blooming: June-August
Notes: On watch list in NV. See *Contributions from the Gray Herbarium* 212:113 (1982) for original description.

Arabis breweri var. *austinae*

Considered but rejected: Too common

Arabis breweri Wats. var. *pecuniaria* Roll.

"San Bernardino rock cress" Brassicaceae
CNPS List: 1B **R-E-D Code:** 3-2-3 **State/Fed. Status:** /C2
Distribution: SBD
Quads: 105D
Habitat: SCFrs (rocky)
Life Form: Perennial herb
Blooming: March-August
Notes: Known from only two occurrences in the San Gorgonio Wilderness Area. See *Rhodora* 43(511):409 (1941) for taxonomic treatment.
Status Report: 1979

Arabis cobrensis M.E. Jones

"Masonic rock cress" Brassicaceae
CNPS List: 2 **R-E-D Code:** 3-1-1 **State/Fed. Status:** CEQA
Distribution: MNO, MOD, NV, OR+
Quads: 451C, 470A, 486C, 487C, 690C
Habitat: GBScr, PJWld / sandy
Life Form: Perennial herb
Blooming: June-July
Notes: See *Rhodora* 43(511):455-457 (1941) for taxonomic treatment.

Arabis constancei Roll.

"Constance's rock cress" Brassicaceae
CNPS List: 1B **R-E-D Code:** 1-2-3 **State/Fed. Status:** /C3c
Distribution: PLU, SIE
Quads: 573B, 589A, 589B, 589C, 590A, 590B, 605B, 606A, 606C
Habitat: Chprl, LCFrs / serpentinite
Life Form: Perennial herb
Blooming: May-July
Notes: Threatened by road widening, logging, mining, and urbanization. See *Contributions from the Gray Herbarium* 201:4-6 (1971) for original description.
Status Report: 1979

Arabis dispar M.E. Jones

"pinyon rock cress" Brassicaceae
CNPS List: 2 **R-E-D Code:** 2-1-1 **State/Fed. Status:** CEQA
Distribution: INY, MNO, SBD, TUL, NV
Quads: 103D, 131D, 283B, 302A, 302C, 304A, 304B, 372A, 412C, 434C
Habitat: JTWLD, MDScr, PJWld / granitic, gravelly
Life Form: Perennial herb
Blooming: March-June
Notes: Rare in SBD Co. See *Contributions to Western Botany* 8:41 (1898) for original description.

Arabis fernaldiana Roll. var. *stylosa* (Wats.) Roll.

"stylose rock cress" Brassicaceae
CNPS List: 1B **R-E-D Code:** 3-1-2 **State/Fed. Status:** CEQA
Distribution: MNO, NV
Quads: 433B
Habitat: GBScr (carbonate)
Life Form: Perennial herb
Blooming: June-July
Notes: Known only from three occurrences. See *Rhodora* 43(511):430-432 (1941) for taxonomic treatment.

Arabis hoffmannii (Munz) Roll.

"Hoffmann's rock cress" Brassicaceae
CNPS List: 1B **R-E-D Code:** 3-3-3 **State/Fed. Status:** /C1
Distribution: ANA*, SCZ
Quads: ANAC*, SCZB
Habitat: CBScr (volcanic cliff ledges)
Life Form: Perennial herb
Blooming: February-April
Notes: Rediscovered independently in 1985 by T. Hesseldenz and S. Junak on SCZ Isl.; now known from only three extant occurrences. Reduced by feral herbivores, but now relatively unthreatened due to reduction of feral sheep population and inaccessibility of one occurrence. See *Rhodora* 43(511):407 (1941) for taxonomic treatment.

Arabis johnstonii Munz

"Johnston's rock cress" Brassicaceae
CNPS List: 1B **R-E-D Code:** 3-2-3 **State/Fed. Status:** /C1
Distribution: RIV
Quads: 66A, 66B, 66C, 66D
Habitat: Chprl, LCFrs / often on eroded clay
Life Form: Perennial herb
Blooming: February-June
Notes: Known from fewer than ten occurrences in the San Jacinto Mtns. Threatened by residential development in Garner Valley and fuelbreak maintenance on Desert Divide. See *Bulletin of the Southern California Academy of Sciences* 31:62-63 (1932) for original description, and *Rhodora* 43(511):467-468 (1941) for taxonomic treatment.

Arabis koehleri Howell var. *stipitata* Roll.

"Koehler's stipitate rock cress" Brassicaceae
CNPS List: 1B **R-E-D Code:** 3-1-2 **State/Fed. Status:** /C3c
Distribution: DNT, SIS, OR
Quads: 722A, 738B, 738D, 739B, 739C, 740D
Habitat: Chprl, LCFrs / serpentinite
Life Form: Perennial herb
Blooming: March-July
Notes: On watch list in OR. See *Rhodora* 43(511):426 (1941) for taxonomic treatment.

Arabis lignifera

Considered but rejected: Unable to verify the two reported CA locations in INY and MNO counties.

Arabis macdonaldiana Eastw.

"McDonald's rock cress" Brassicaceae
CNPS List: 1B **R-E-D Code:** 2-3-2 **State/Fed. Status:** CE/FE
Distribution: DNT, MEN, TRI, OR
Quads: 600B, 686B, 739B, 739C, 740A
Habitat: LCFrs, UCFrs / serpentinite
Life Form: Perennial herb
Blooming: May-June
Notes: Threatened by mining. Endangered in OR. Protected in part at Red Mtn. ACEC (BLM), MEN Co. See *Rhodora* 43(511):350 (1941) for taxonomic treatment, and *Contributions from the Gray Herbarium* 204:149-154 (1973) for taxonomic information.
Status Report: 1988

Arabis microphylla Nutt. var. *microphylla*

"small-leaved rock cress" Brassicaceae
CNPS List: 4 **R-E-D Code:** 1-1-1 **State/Fed. Status:** CEQA?
Distribution: MNO, MOD, PLU, NV, OR, WA+
Habitat: PJWld (volcanic or granitic crevices)
Life Form: Perennial herb
Blooming: July
Notes: See *Rhodora* 43(511):426-428 (1941) for taxonomic treatment.

Arabis modesta Roll.

"modest rock cress" Brassicaceae
CNPS List: 4 **R-E-D Code:** 1-1-2 **State/Fed. Status:** /C3c
Distribution: NAP, SIS, TRI, OR
Habitat: Chprl, LCFrs
Life Form: Perennial herb
Blooming: March-July
Notes: Intergrades with *A. oregana* in SIS Co.; may be a variety of that plant. Endangered in OR. See *Rhodora* 43(511):350-352 (1941) for taxonomic treatment, and *Contributions from the Gray Herbarium* 204:149-154 (1973) for taxonomic information.
Status Report: 1979

Arabis oregana Roll.

"Oregon rock cress" Brassicaceae
CNPS List: 4 **R-E-D Code:** 1-1-1 **State/Fed. Status:** /C3c
Distribution: LAK, MOD, NAP, SIS, TRI, OR
Habitat: Chprl, LCFrs / serpentinite
Life Form: Perennial herb
Blooming: May
Notes: Many collections in Klamath region. Exact taxonomic status of NAP Co. occurrence needs to be determined. See *Rhodora* 43(511):349-350 (1941) for taxonomic treatment, and *Contributions from the Gray Herbarium* 204:149-154 (1973) for taxonomic information.

Arabis parishii Wats.

"Parish's rock cress" Brassicaceae
CNPS List: 1B **R-E-D Code:** 2-2-3 **State/Fed. Status:** /C2
Distribution: SBD
Quads: 104B, 105A, 105B, 126D, 131C, 131D, 132D
Habitat: PbPln, PJWld, UCFrs / quartzite on clay, or sometimes carbonate
Life Form: Perennial herb
Blooming: April-May
Notes: Endemic to the San Bernardino Mtns. Threatened by vehicles, carbonate mining, development, grazing, and road construction. See *Rhodora* 43(511):468-469 (1941) for taxonomic treatment.

Arabis pinzlae Roll.

"Pinzl's rock cress" Brassicaceae
CNPS List: 1B **R-E-D Code:** 3-1-2 **State/Fed. Status:** /C2
Distribution: MNO, NV
Quads: 450D
Habitat: AlpBR, SCFrs (scree or sandy)
Life Form: Perennial herb
Blooming: July
Notes: Known in CA from only one occurrence in the White Mtns. On watch list in NV. See *Contributions from the Gray Herbarium* 212:110 (1982) for original description, and *Madroño* 37(1):64 (1990) for first CA record.

Arabis pulchra M.E. Jones var. *munciensis* M.E. Jones

"Darwin rock cress" Brassicaceae
CNPS List: 2 **R-E-D Code:** 3-1-1 **State/Fed. Status:** CEQA
Distribution: INY, SBD, NV+
Quads: 104D, 327, 349B, 349C, 370A, 370B, 372A, 390C, 412D
Habitat: ChScr, MDScr / carbonate
Life Form: Perennial herb
Blooming: April
Notes: Need precise locality information for quad 327. See *Rhodora* 43(511):459-460 (1941) for taxonomic treatment.

Arabis pygmaea Roll.

"Tulare County rock cress" Brassicaceae
CNPS List: 4 **R-E-D Code:** 1-1-3 **State/Fed. Status:** /C3c
Distribution: INY, TUL
Habitat: Medws (edges), SCFrs / volcanic or granitic, gravelly
Life Form: Perennial herb
Blooming: June-July
Notes: See *Rhodora* 43(511):476 (1941) for taxonomic treatment.

Arabis rigidissima

See *Arabis rigidissima* var. *rigidissima*

Arabis rigidissima Roll. var. *demota* Roll.

"Carson Range rock cress" Brassicaceae
CNPS List: 1B **R-E-D Code:** 3-2-2 **State/Fed. Status:** /C1
Distribution: PLA, NV
Quads: 554D
Habitat: BUFrs, UCFrs / rocky
Life Form: Perennial herb
Blooming: August
Notes: Known in CA from only two occurrences near Martis Pk., and in NV from eleven occurrences in the Carson Range. Threatened by logging. Proposed for state listing as Critically Endangered in NV. Not in *The Jepson Manual*. See *Journal of the Arnold Arboretum* 64:498 (1983) for original description.

Arabis rigidissima Roll. var. *rigidissima*

"Trinity Mtns. rock cress" Brassicaceae
CNPS List: 4 **R-E-D Code:** 1-1-3 **State/Fed. Status:** CEQA?
Distribution: HUM, SIS, TRI
Habitat: UCFrs (gravelly or rocky)
Life Form: Perennial herb
Blooming: July-August
Notes: See *Rhodora* 43(511):380-381 (1941) for original description.

Arabis serpentinicola Roll.

"Preston Peak rock cress" Brassicaceae
CNPS List: 1B **R-E-D Code:** 3-1-3 **State/Fed. Status:** /C2
Distribution: DNT, SIS, OR*
Quads: 738A, 738D
Habitat: LCFrs, NCFrs, UCFrs / serpentinite
Life Form: Perennial herb
Blooming: June-July
Notes: Known from fewer than five occurrences in the Preston Pk. area. On review list in OR. See *A. macdonaldiana* in *The Jepson Manual*; *A. serpentinicola* is probably a variety of that plant. See *Contributions from the Gray Herbarium* 204:150 (1973) for original description.
Status Report: 1979

Arabis shockleyi Munz

"Shockley's rock cress" Brassicaceae
CNPS List: 2 **R-E-D Code:** 3-2-1 **State/Fed. Status:** /C3c
Distribution: INY, SBD, NV+
Quads: 104B, 130C, 131C, 131D, 326A, 390D, 410C
Habitat: PJWld (carbonate or quartzite)
Life Form: Perennial herb
Blooming: May-June
Notes: See *Bulletin of the Southern California Academy of Sciences* 31(2):61 (1932) for original description, and *Rhodora* 43(511):457 (1941) for taxonomic treatment.
Status Report: 1979

Arabis tiehmii Roll.

"Tiehm's rock cress" Brassicaceae
CNPS List: 1B **R-E-D Code:** 3-1-2 **State/Fed. Status:** /C2
Distribution: MNO, NV
Quads: 453B, 454A, 457B
Habitat: AlpBR (granitic)
Life Form: Perennial herb
Blooming: August
Notes: Known in CA from fewer than five occurrences near Tioga Crest. On watch list in NV. See *Journal of the Arnold Arboretum* 64:496 (1983) for original description.

Arctomecon merriamii Cov.

"white bear poppy" Papaveraceae
CNPS List: 1B **R-E-D Code:** 2-2-2 **State/Fed. Status:** /C2
Distribution: INY, SBD, NV
Quads: 248B, 248C, 249A, 249D, 275D, 323B, 323C, 325D, 346B, 349A, 368C, 368D, 369B, 370A, 370B, 370D, 389C
Habitat: ChScr, MDScr
Life Form: Perennial herb
Blooming: April-May
Notes: Historical occurrences need field surveys. Threatened by mining and vehicles. On watch list in NV. See *Proceedings of the Biological Society of Washington* 7:65-80 (1892) for original description, and *Mentzelia* 3:2-5 (1977) for species account.
Status Report: 1977

Arctostaphylos acutifolia

Considered but rejected: A synonym of *A. patula*; a common taxon

Arctostaphylos andersonii Gray

"Santa Cruz manzanita" Ericaceae
CNPS List: 1B **R-E-D Code:** 2-2-3 **State/Fed. Status:** /C2
Distribution: SCL, SCR, SMT
Quads: 406C, 407D, 408B, 408C, 408D, 428A, 428D, 429A, 448C, 448D
Habitat: BUFrs, Chprl, NCFrs / openings, edges
Life Form: Shrub (evergreen)
Blooming: November-April
Notes: Known only from the Santa Cruz Mtns.; has been confused with other species merged with it as varieties. Threatened by development. See *Proceedings of the American Academy of Arts and Sciences* 11:83 (1876) for original description, and *North American Flora* 29:98 (1914) for additional information.

Arctostaphylos auriculata Eastw.

"Mt. Diablo manzanita" Ericaceae
CNPS List: 1B **R-E-D Code:** 3-1-3 **State/Fed. Status:** /C3c
Distribution: CCA
Quads: 464A, 464B, 464C, 464D
Habitat: Chprl (sandstone)
Life Form: Shrub (evergreen)
Blooming: January-March
Notes: Known from fewer than twenty occurrences. See *Bulletin of the Torrey Botanical Club* 32:202 (1905) for original description, and *American Midland Naturalist* 23:622 (1940) for taxonomic treatment.
Status Report: 1977

Arctostaphylos bakeri

See *Arctostaphylos bakeri* ssp. *bakeri*

Arctostaphylos bakeri Eastw. ssp. *bakeri*

"Baker's manzanita" Ericaceae
CNPS List: 1B **R-E-D Code:** 3-3-3 **State/Fed. Status:** CR/C3c
Distribution: SON
Quads: 500C, 502B, 503A, 518A, 518C
Habitat: BUFrs, Chprl / often serpentinite
Life Form: Shrub (evergreen)
Blooming: February-April
Notes: Known from fewer than ten occurrences. Threatened by road construction and widening, non-native plants, and dumping, and potentially by development. See *Leaflets of Western Botany* 1:115 (1934) for original description.
Status Report: 1988

Arctostaphylos bakeri Eastw. ssp. *sublaevis* Wells

"The Cedars manzanita" Ericaceae
CNPS List: 1B **R-E-D Code:** 3-2-3 **State/Fed. Status:** CEQA
Distribution: SON
Quads: 518C, 519A, 519B, 519C, 519D
Habitat: CCFrs, Chprl / serpentinite seeps
Life Form: Shrub (evergreen)
Blooming: April-May
Notes: See *Four Seasons* 8(2):58-68 (1988) for original description.

Arctostaphylos canescens ssp. *malloryi*

See *Arctostaphylos malloryi*

Arctostaphylos canescens Eastw. ssp. *sonomensis* (Eastw.) Wells

"Sonoma manzanita" Ericaceae
CNPS List: 1B **R-E-D Code:** 2-2-3 **State/Fed. Status:** CEQA
Distribution: HUM, LAK, MEN, SON, TEH
Quads: 501B, 550D, 597C, 597D, 670C, 670D
Habitat: Chprl, LCFrs
Life Form: Shrub (evergreen)
Blooming: January-March
Notes: Much of Rincon Ridge (SON Co.) threatened by development. See *Four Seasons* 7(3):42-46 (1985) for status update, and *Madroño* 35(4):330-341 (1988) for revised nomenclature.

Arctostaphylos canescens var. *sonomensis*

See *Arctostaphylos canescens* ssp. *sonomensis*

Arctostaphylos catalinae Wells

"Santa Catalina Island manzanita" Ericaceae
CNPS List: 1B **R-E-D Code:** 2-2-3 **State/Fed. Status:** /C2
Distribution: SCT
Quads: SCTN
Habitat: Chprl (volcanic)
Life Form: Shrub (evergreen)
Blooming: February-May
Notes: Recovering vigorously after removal of most feral goats. See *Madroño* 19(6):193 (1968) for original description.

Arctostaphylos cinerea

Considered but rejected: A hybrid between *A. canescens* and *A. viscida*

Arctostaphylos confertiflora Eastw.

"Santa Rosa Island manzanita" Ericaceae
CNPS List: 1B **R-E-D Code:** 3-2-3 **State/Fed. Status:** /C1
Distribution: SRO
Quads: SROE, SRON
Habitat: BUFrs, CCFrs, Chprl / sandstone
Life Form: Shrub (evergreen)
Blooming: February-April
Notes: Threatened by cattle grazing and feral herbivores.

Arctostaphylos cruzensis Roof

"La Cruz manzanita" Ericaceae
CNPS List: 1B **R-E-D Code:** 2-2-3 **State/Fed. Status:** /C2
Distribution: MNT, SLO
Quads: 247D, 271B, 271D, 272A, 296B
Habitat: BUFrs, CBScr, CCFrs, Chprl, CoScr, VFGrs / sandy
Life Form: Shrub (evergreen)
Blooming: December-March
Notes: Known from fewer than twenty occurrences.

Arctostaphylos densiflora M.S. Baker

"Vine Hill manzanita" Ericaceae
CNPS List: 1B **R-E-D Code:** 3-3-3 **State/Fed. Status:** CE/C2
Distribution: SON
Quads: 502A
Habitat: Chprl (acid marine sand)
Life Form: Shrub (evergreen)
Blooming: February-March
Notes: Known from only one extant occurrence on the Sonoma Barren near Forestville. Threatened by fungal infection. See *Leaflets of Western Botany* 1(4):31-32 (1932) for original description.
Status Report: 1987

Arctostaphylos edmundsii J.T. Howell

"Little Sur manzanita" Ericaceae
CNPS List: 1B **R-E-D Code:** 3-2-3 **State/Fed. Status:** /C2
Distribution: MNT
Quads: 320E, 344B, 344C, 344D
Habitat: CBScr, Chprl / sandy
Life Form: Shrub (evergreen)
Blooming: November-April
Notes: Known from fewer than ten occurrences. Threatened by foot traffic and non-native plants. Includes *A. edmundsii* var. *parvifolia*, which is state-listed Rare. See *Leaflets of Western Botany* 9(12):188-196 (1961) for information.
Status Report: 1977

Arctostaphylos edmundsii var. *edmundsii*

See *Arctostaphylos edmundsii*

Arctostaphylos edmundsii var. *parvifolia*

See *Arctostaphylos edmundsii*

Arctostaphylos elegans

Considered but rejected: A synonym of *A. manzanita* ssp. *elegans*; a common taxon

Arctostaphylos franciscana

See *Arctostaphylos hookeri* ssp. *franciscana*

Arctostaphylos gabrielensis Wells

"San Gabriel manzanita" Ericaceae
CNPS List: 1B **R-E-D Code:** 3-2-3 **State/Fed. Status:** /C2
Distribution: LAX
Quads: 136A
Habitat: Chprl (rocky)
Life Form: Shrub (evergreen)
Blooming: March
Notes: Endemic to Mill Creek Summit divide in the San Gabriel Mtns. See *Four Seasons* 9(2):46 (1992) for original description.

Arctostaphylos glandulosa Eastw. ssp. *crassifolia* (Jeps.) Wells

"Del Mar manzanita" Ericaceae
CNPS List: 1B **R-E-D Code:** 3-3-2 **State/Fed. Status:** /PE
Distribution: SDG, BA
Quads: 22A, 22B, 35B, 35C, 36A, 36D
Habitat: Chprl (maritime, sandy mesas and bluffs)
Life Form: Shrub (evergreen)
Blooming: December-April
Notes: Threatened by agricultural conversion and development on coastal bluffs. See *Madroño* 1(4):86 (1922) for original description, and *Four Seasons* 7(4):5-27 (1987) for taxonomic treatment.

Arctostaphylos glutinosa Schreib.

"Schreiber's manzanita" Ericaceae
CNPS List: 1B **R-E-D Code:** 3-2-3 **State/Fed. Status:** /C2
Distribution: SCR
Quads: 408B, 408C, 409D
Habitat: CCFrs, Chprl / diatomaceous shale
Life Form: Shrub (evergreen)
Blooming: November-April
Notes: Known from fewer than ten occurrences. Threatened by road construction. See *American Midland Naturalist* 23:620-621 (1940) for original description.
Status Report: 1977

Arctostaphylos hearstiorum

See *Arctostaphylos hookeri* ssp. *hearstiorum*

Arctostaphylos hispidula Howell

"Howell's manzanita" Ericaceae
CNPS List: 4 **R-E-D Code:** 1-2-2 **State/Fed. Status:** /C3c
Distribution: DNT, HUM, SON, OR
Habitat: Chprl (serpentinite or sandstone)
Life Form: Shrub (evergreen)
Blooming: March-April
Notes: Threatened by mining. Endangered in OR.

Arctostaphylos hookeri D. Don ssp. *franciscana* (Eastw.) Munz

"Franciscan manzanita" Ericaceae
CNPS List: 1A **Last Seen:** 1942 **State/Fed. Status:** /C2
Distribution: SFO*
Quads: 448B*, 466C*
Habitat: CoScr (serpentinite)
Life Form: Shrub (evergreen)
Blooming: February-April
Notes: Plant now occurs only in cultivation. See *Bulletin of the Torrey Botanical Club* 32:201-202 (1905) for original description.
Status Report: 1977

Arctostaphylos hookeri D. Don ssp. *hearstiorum* (Hoov. & Roof) Wells

"Hearst's manzanita" Ericaceae
CNPS List: 1B **R-E-D Code:** 3-2-3 **State/Fed. Status:** CE/C2
Distribution: SLO
Quads: 271B, 272A
Habitat: Chprl (maritime), CoPrr, CoScr, VFGrs / sandy
Life Form: Shrub (evergreen)
Blooming: February-April
Notes: Known from fewer than five occurrences in the Arroyo de la Cruz area. Threatened by grazing and rangeland conversion. See *Four Seasons* 2(1):2-4 (1966) for original description.
Status Report: 1987

Arctostaphylos hookeri D. Don ssp. *hookeri*

"Hooker's manzanita" Ericaceae
CNPS List: 1B **R-E-D Code:** 2-2-3 **State/Fed. Status:** CEQA
Distribution: MNT, SCR
Quads: 366C, 387A
Habitat: CCFrs, Chprl, CoScr / sandy
Life Form: Shrub (evergreen)
Blooming: February-May
Notes: Threatened by agriculture, development, fire suppression, and competition with introduced *Eucalyptus*.

Arctostaphylos hookeri D. Don ssp. *montana* (Eastw.) Wells

"Mt. Tamalpais manzanita" Ericaceae
CNPS List: 1B **R-E-D Code:** 3-1-3 **State/Fed. Status:** /C2
Distribution: MRN
Quads: 467A, 467B
Habitat: Chprl, VFGrs / serpentinite
Life Form: Shrub (evergreen)
Blooming: February-April
Notes: Known from fewer than twenty occurrences in the Mt. Tamalpais area. See *Madroño* 19(6):193-210 (1968) for revised nomenclature.

Arctostaphylos hookeri D. Don ssp. *ravenii* Wells

"Presidio manzanita" Ericaceae
CNPS List: 1B **R-E-D Code:** 3-3-3 **State/Fed. Status:** CE/FE
Distribution: SFO
Quads: 448B*, 466C
Habitat: Chprl, CoPrr, CoScr / serpentinite
Life Form: Shrub (evergreen)
Blooming: February-March
Notes: Known from only one extant native occurrence at the Presidio in San Francisco; plants there apparently belong to a single clone. Five of six historical occurrences extirpated by urbanization; currently threatened by non-native plants, and possibly by foot traffic and vandalism. Plantings made at the Presidio in 1987. USFWS uses the name *A. pungens* var. *ravenii*. See *Madroño* 19(6):200-201 (1968) for original description.
Status Report: 1988

Arctostaphylos hooveri Wells

"Hoover's manzanita" Ericaceae
CNPS List: 4 **R-E-D Code:** 1-1-3 **State/Fed. Status:** CEQA?
Distribution: MNT, SLO
Habitat: Chprl (rocky)
Life Form: Shrub (evergreen)
Blooming: April-June

Arctostaphylos imbricata Eastw.

"San Bruno Mtn. manzanita" Ericaceae
CNPS List: 1B **R-E-D Code:** 3-3-3 **State/Fed. Status:** CE/C1
Distribution: SMT
Quads: 448B
Habitat: Chprl
Life Form: Shrub (evergreen)
Blooming: February-May
Notes: Known from fewer than five occurrences near San Bruno Mtn. Threatened by urbanization and alteration of fire regimes. See *Proceedings of the American Academy of Arts and Sciences* 20:149-150 (1931) for original description.
Status Report: 1988

Arctostaphylos insularis

Considered but rejected: Too common

Arctostaphylos intricata var. *intricata*

Considered but rejected: Too common; a synonym of *A. glandulosa* ssp. *glandulosa*

Arctostaphylos intricata var. *oblongifolia*

Considered but rejected: Too common; a synonym of *A. glandulosa* ssp. *glandulosa*

Arctostaphylos klamathensis Edwards, Keeler-Wolf & Knight

"Klamath manzanita" Ericaceae
CNPS List: 1B **R-E-D Code:** 3-2-3 **State/Fed. Status:** /C2
Distribution: SHA, SIS, TRI
Quads: 682B, 682C, 683A, 683D, 700C, 700D
Habitat: Chprl (montane), SCFrs, UCFrs / sometimes serpentinite
Life Form: Shrub (evergreen)
Blooming: May-June
Notes: Possibly threatened by road maintenance. See *Four Seasons* 6(4):17-21 (1983) for original description.

Arctostaphylos luciana Wells

"Santa Lucia manzanita" Ericaceae
CNPS List: 1B **R-E-D Code:** 2-2-3 **State/Fed. Status:** /C2
Distribution: SLO
Quads: 221A, 246B, 246C, 246D, 270C
Habitat: Chprl (shale)
Life Form: Shrub (evergreen)
Blooming: February-March

Arctostaphylos malloryi (Knight & Gankin) Wells

"Mallory's manzanita" Ericaceae
CNPS List: 4 **R-E-D Code:** 1-1-3 **State/Fed. Status:** CEQA?
Distribution: COL, SHA, TRI
Habitat: Chprl, LCFrs / volcanic
Life Form: Shrub (evergreen)
Blooming: April-July
Notes: To be expected in GLE and TEH counties. See *Four Seasons* 6(4):23 (1983) for original description and 9(2):54-59 (1992) for revised nomenclature.

Arctostaphylos manzanita ssp. *elegans*

Considered but rejected: Too common

Arctostaphylos manzanita Parry ssp. *laevigata* (Eastw.) Munz

"Contra Costa manzanita" Ericaceae
CNPS List: 1B **R-E-D Code:** 3-2-3 **State/Fed. Status:** CEQA
Distribution: CCA
Quads: 464A, 464B, 464C
Habitat: Chprl (rocky)
Life Form: Shrub (evergreen)
Blooming: January-February

Arctostaphylos manzanita ssp. *roofii*

Considered but rejected: Too common

Arctostaphylos mendocinoensis Wells

"pygmy manzanita" Ericaceae
CNPS List: 1B **R-E-D Code:** 3-2-3 **State/Fed. Status:** CEQA
Distribution: MEN
Quads: 569D
Habitat: CCFrs (acidic sandy clay)
Life Form: Shrub (evergreen)
Blooming: January
Notes: Narrowly endemic to the Mendocino Plains. See *Four Seasons* 8(3):25-35 (1988) for original description.

Arctostaphylos mewukka Merriam ssp. *truei* (Knight) Wells

"True's manzanita" Ericaceae
CNPS List: 4 **R-E-D Code:** 1-2-3 **State/Fed. Status:** CEQA?
Distribution: BUT, PLU, YUB
Habitat: Chprl, LCFrs
Life Form: Shrub (evergreen)
Blooming: March-May
Notes: See *Four Seasons* 3(1):19 (1969) for original description and 9(2):60-63 (1992) for revised nomenclature.

Arctostaphylos montana

See *Arctostaphylos hookeri* ssp. *montana*

Arctostaphylos montaraensis Roof

"Montara manzanita" Ericaceae
CNPS List: 1B **R-E-D Code:** 3-2-3 **State/Fed. Status:** /C2
Distribution: SMT
Quads: 448B, 448C
Habitat: Chprl (maritime), CoScr
Life Form: Shrub (evergreen)
Blooming: January-March
Notes: Known from approximately ten occurrences. Threatened by development and vehicles.

Arctostaphylos montereyensis Hoov.

"Monterey manzanita" Ericaceae
CNPS List: 1B **R-E-D Code:** 3-2-3 **State/Fed. Status:** /C2
Distribution: MNT
Quads: 343A, 365B, 365C, 366D
Habitat: Chprl, CmWld, CoScr / sandy
Life Form: Shrub (evergreen)
Blooming: February-March
Notes: Known from fewer than ten occurrences.

Arctostaphylos morroensis Wies. & Schreib.

"Morro manzanita" Ericaceae
CNPS List: 1B **R-E-D Code:** 2-3-3 **State/Fed. Status:** /PE
Distribution: SLO
Quads: 246C, 247D
Habitat: Chprl, CmWld, CoDns (pre-Flandrian), CoScr
Life Form: Shrub (evergreen)
Blooming: January-March
Notes: Known from fewer than twenty occurrences in the Morro Bay area. Threatened by urbanization. See *Madroño* 5:38-47 (1939) for original description.

Arctostaphylos morroensis

Arctostaphylos myrtifolia Parry

"Ione manzanita" Ericaceae
CNPS List: 1B **R-E-D Code:** 2-2-3 **State/Fed. Status:** /C1
Distribution: AMA, CAL
Quads: 476A, 476B, 477A, 477B, 493C, 494B, 494C, 494D, 495A
Habitat: Chprl, CmWld / acidic Ione clay or sandy
Life Form: Shrub (evergreen)
Blooming: November-February
Notes: Threatened by mining and fungal infection. Protected in part at ACEC (BLM), AMA Co. See *Pittonia* 1:34 (1887) for original description, *Ecology* 45(4):792-808 (1964) for ecological discussion, and *Changing Seasons* 1(4):4-19 (1982) for species account.
Status Report: 1977

Arctostaphylos nissenana Merriam

"Nissenan manzanita" Ericaceae
CNPS List: 1B **R-E-D Code:** 3-2-3 **State/Fed. Status:** /C2
Distribution: ELD, TUO
Quads: 458B, 509B, 510A, 525C, 526A, 526D
Habitat: CCFrs, Chprl
Life Form: Shrub (evergreen)
Blooming: February-March
Notes: Known from approximately ten occurrences. Threatened by development. See *Proceedings of the Biological Society of Washington* 31:102 (1918) for original description, and *Four Seasons* 1(4):7-15 (1966) for species account and 6(4):12-16 (1983) for range extension information.

Arctostaphylos nortensis Wells

"Del Norte manzanita" Ericaceae
CNPS List: 4 **R-E-D Code:** 1-1-3 **State/Fed. Status:** /C2
Distribution: DNT, OR?
Habitat: Chprl, LCFrs (open) / often serpentinite
Life Form: Shrub (evergreen)
Blooming: February
Notes: Plant may occur in adjacent OR. See *Four Seasons* 8(1):50 (1988) for original description and 9(2):54-59 (1992) for revised nomenclature.

Arctostaphylos obispoensis Eastw.

"Bishop manzanita" Ericaceae
CNPS List: 4 **R-E-D Code:** 1-1-3 **State/Fed. Status:** CEQA?
Distribution: MNT, SLO
Habitat: CCFrs (serpentinite)
Life Form: Shrub (evergreen)
Blooming: February-March

Arctostaphylos osoensis Wells

"Oso manzanita" Ericaceae
CNPS List: 1B **R-E-D Code:** 3-3-3 **State/Fed. Status:** /C2
Distribution: SLO
Quads: 247D
Habitat: Chprl, CmWld / dacite porphyry buttes
Life Form: Shrub (evergreen)
Blooming: February-March
Notes: Narrowly endemic to the mountains north of Los Osos Valley. Eventual urbanization a threat. Occurs with *A. tomentosa* ssp. *daciticola.* See *Four Seasons* 9(2):45 (1992) for original description.

Arctostaphylos otayensis Wies. & Schreib.

"Otay manzanita" Ericaceae
CNPS List: 1B **R-E-D Code:** 3-2-3 **State/Fed. Status:** /C2
Distribution: SDG
Quads: 10A, 10B, 10D, 20D, 33D, 48D
Habitat: Chprl, CmWld / volcanic
Life Form: Shrub (evergreen)
Blooming: January-March
Notes: Historical occurrences need field surveys. See *Madroño* 5:43 (1939) for original description.

Arctostaphylos pacifica

Considered but rejected: A hybrid

Arctostaphylos pajaroensis Adams

"Pajaro manzanita" Ericaceae
CNPS List: 1B **R-E-D Code:** 2-3-3 **State/Fed. Status:** /C2
Distribution: MNT, SCR*
Quads: 386B, 386C, 386D, 387A, 408B*
Habitat: Chprl (sandy)
Life Form: Shrub (evergreen)
Blooming: December-March
Notes: Never abundant, and now severely threatened by development in Pajaro Hills area, where habitat loss has occurred.

Arctostaphylos pallida Eastw.

"pallid manzanita" Ericaceae
CNPS List: 1B **R-E-D Code:** 3-3-3 **State/Fed. Status:** CE/C1
Distribution: ALA, CCA
Quads: 465B, 465C, 466A
Habitat: BUFrs, Chprl, CmWld / siliceous shale
Life Form: Shrub (evergreen)
Blooming: December-March
Notes: Known from fewer than twenty occurrences. Threatened by urbanization, alteration of fire regimes, non-native plants, and fungal infection. See *Leaflets of Western Botany* 1:76-77 (1933) for original description, and *Four Seasons* 7(4):28-46 (1987) for ecological assessment.
Status Report: 1987

Arctostaphylos parvifolia

Considered but rejected: A hybrid between *A. glandulosa* and *A. nevadensis*

Arctostaphylos pechoensis Dudl.

"Pecho manzanita" Ericaceae
CNPS List: 1B **R-E-D Code:** 2-3-3 **State/Fed. Status:** /C2
Distribution: SLO
Quads: 221A, 221B, 222A, 246B, 246C, 247B, 247D, 270C
Habitat: CCFrs, Chprl, CoScr / siliceous shale
Life Form: Shrub (evergreen)
Blooming: November-March
Notes: Narrowly endemic to siliceous shale in coastal mountains of SLO Co. Threatened by urbanization.

Arctostaphylos peninsularis Wells ssp. *peninsularis*

"Peninsular manzanita" Ericaceae
CNPS List: 2 **R-E-D Code:** 3-1-1 **State/Fed. Status:** CEQA
Distribution: RIV, BA
Quads: 65C
Habitat: Chprl
Life Form: Shrub (evergreen)
Blooming: April-May
Notes: Plants from northwestern SDG Co. and adjacent RIV Co. formerly considered *A. peninsularis* ssp. *peninsularis* are *A. rainbowensis*. Not in *The Jepson Manual.* See *Madroño* 21(5):268-273 (1972) for original description and 39(4):285-287 (1992) for taxonomic treatment.

Arctostaphylos pilosula Jeps. & Wies.

"Santa Margarita manzanita" Ericaceae
CNPS List: 1B **R-E-D Code:** 3-2-3 **State/Fed. Status:** /C2
Distribution: MNT, SLO
Quads: 220B, 245C, 245D, 246B, 270C, 271A, 296A
Habitat: CCFrs, Chprl / shale
Life Form: Shrub (evergreen)
Blooming: December-March
Notes: Threatened by road maintenance and vehicles.

Arctostaphylos pilosula ssp. *pilosula*

See *Arctostaphylos pilosula*

Arctostaphylos pilosula ssp. *pismoensis*

See *Arctostaphylos wellsii*

Arctostaphylos pseudopungens

Considered but rejected: A synonym of *A. pungens*; a common taxon

Arctostaphylos pumila Nutt.

"sandmat manzanita" Ericaceae
CNPS List: 1B **R-E-D Code:** 3-2-3 **State/Fed. Status:** /C2
Distribution: MNT
Quads: 366A, 366C, 366D
Habitat: CCFrs, Chprl, CoDns, CoScr / sandy
Life Form: Shrub (evergreen)
Blooming: February-May
Notes: Known from fewer than twenty occurrences. Threatened by urbanization and military activities at Fort Ord.
Status Report: 1977

Arctostaphylos pungens ssp. *chaloneorum*

Considered but rejected: A synonym of *A. pungens*; a common taxon

Arctostaphylos pungens ssp. *laevigata*

See *Arctostaphylos manzanita* ssp. *laevigata*

Arctostaphylos purissima Wells

"La Purisima manzanita" Ericaceae
CNPS List: 1B **R-E-D Code:** 2-3-3 **State/Fed. Status:** CEQA
Distribution: SBA
Quads: 145A, 145B, 170A, 170B, 170C, 170D, 171A, 195D, 196B, 196D
Habitat: Chprl (sandy)
Life Form: Shrub (evergreen)
Blooming: November-May
Notes: Threatened by urbanization.

Arctostaphylos rainbowensis J. Keeley & Massihi

"Rainbow manzanita" Ericaceae
CNPS List: 1B **R-E-D Code:** 3-3-3 **State/Fed. Status:** CEQA
Distribution: RIV, SDG
Quads: 50A, 50B, 51A, 51B?, 69D
Habitat: Chprl
Life Form: Shrub (evergreen)
Blooming: January-February
Notes: Previously called *A. peninsularis* or considered to be a hybrid between *A. glandulosa* and *A. glauca*. Threatened by conversion to avocado orchards. Not in *The Jepson Manual*. See *Madroño* 41(1):x-x (1994, in press) for original description.

Arctostaphylos refugioensis Gankin

"Refugio manzanita" Ericaceae
CNPS List: 1B **R-E-D Code:** 2-2-3 **State/Fed. Status:** /C2
Distribution: SBA
Quads: 143A, 145B, 169A, 169D, 170C
Habitat: Chprl (sandstone)
Life Form: Shrub (evergreen)
Blooming: December-March
Notes: See *Four Seasons* 2(2):13-14 (1967) for original description.
Status Report: 1979

Arctostaphylos regismontana Eastw.

"Kings Mtn. manzanita" Ericaceae
CNPS List: 4 **R-E-D Code:** 1-1-3 **State/Fed. Status:** CEQA?
Distribution: SCR, SMT
Habitat: BUFrs, Chprl, NCFrs / granitic or sandstone
Life Form: Shrub (evergreen)
Blooming: January-April
Notes: See *Leaflets of Western Botany* 1:77 (1933) for original description.

Arctostaphylos roofii

Considered but rejected: A synonym of *A. manzanita* ssp. *roofii*; a common taxon

Arctostaphylos rudis Jeps. & Wies.

"sand mesa manzanita" Ericaceae
CNPS List: 1B **R-E-D Code:** 2-2-3 **State/Fed. Status:** /C1
Distribution: SBA, SLO
Quads: 145B, 170B, 170C, 171A, 195B, 196A, 196B, 196D, 220C, 221D
Habitat: Chprl, CoScr / sandy
Life Form: Shrub (evergreen)
Blooming: November-February
Notes: Severely reduced on Nipomo Mesa; more widespread on Burton Mesa.

Arctostaphylos silvicola Jeps. & Wies.

"Bonny Doon manzanita" Ericaceae
CNPS List: 1B **R-E-D Code:** 2-2-3 **State/Fed. Status:** /C2
Distribution: SCR
Quads: 408A, 408B, 408C, 408D
Habitat: Chprl, CCFrs, LCFrs / inland marine sands
Life Form: Shrub (evergreen)
Blooming: February-March
Notes: Known from fewer than twenty occurrences. Threatened by sand mining and urbanization.

Arctostaphylos stanfordiana Parry ssp. *decumbens* Wells

"Rincon manzanita" Ericaceae
CNPS List: 1B **R-E-D Code:** 3-3-3 **State/Fed. Status:** CEQA
Distribution: SON
Quads: 501A, 501B, 518B
Habitat: Chprl (rhyolitic)
Life Form: Shrub (evergreen)
Blooming: February-April
Notes: Known from fewer than ten occurrences. Seriously threatened by development, road construction, vehicles, and viticulture. See *Four Seasons* 4(2):16-17 (1972) for original description and 9(2):60-63 (1992) for revised nomenclature.

Arctostaphylos stanfordiana ssp. *hispidula*

See *Arctostaphylos hispidula*

Arctostaphylos stanfordiana Parry ssp. *raichei* Knight

"Raiche's manzanita" Ericaceae
CNPS List: 1B **R-E-D Code:** 2-3-3 **State/Fed. Status:** /C2
Distribution: LAK, MEN
Quads: 534A?, 535B, 550A, 550D
Habitat: Chprl (often serpentinite)
Life Form: Shrub (evergreen)
Blooming: February-April
Notes: Threatened by urbanization. See *Four Seasons* 7(3):16-20 (1985) for original description.

Arctostaphylos stanfordiana var. *repens*

See *Arctostaphylos stanfordiana* ssp. *decumbens*

Arctostaphylos tomentosa (Pursh) Lindl. ssp. *daciticola* Wells

"dacite manzanita" Ericaceae
CNPS List: 1B **R-E-D Code:** 3-3-3 **State/Fed. Status:** /C2
Distribution: SLO
Quads: 247D
Habitat: Chprl, CmWld / dacite porphyry buttes
Life Form: Shrub (evergreen)
Blooming: March
Notes: Narrowly endemic to the mountains north of western Los Osos Valley. Eventual urbanization a threat. Occurs with *A. osoensis*. See *Four Seasons* 9(2):60 (1992) for original description.

Arctostaphylos tomentosa (Pursh) Lindl. ssp. *eastwoodiana* Wells

"Eastwood's manzanita" Ericaceae
CNPS List: 1B **R-E-D Code:** 2-3-3 **State/Fed. Status:** CEQA
Distribution: SBA
Quads: 170A, 170B, 171A, 196B
Habitat: Chprl (sandy)
Life Form: Shrub (evergreen)
Blooming: March
Notes: Known only from the La Purisima Ridge, Burton Mesa, and Point Sal areas. Threatened by urbanization. See *Madroño* 19(6):197 (1968) for original description.

Arctostaphylos tomentosa (Pursh) Lindl. ssp. *insulicola* Wells

"island manzanita" Ericaceae
CNPS List: 4 **R-E-D Code:** 1-2-3 **State/Fed. Status:** CEQA?
Distribution: SCZ, SRO
Habitat: CCFrs, Chprl
Life Form: Shrub (evergreen)
Blooming: December-February
Notes: Possibly threatened by feral herbivores. See *Madroño* 19(6):197 (1968) for original description.

Arctostaphylos tomentosa (Pursh) Lindl. ssp. *subcordata* (Eastw.) Wells

"Santa Cruz Island manzanita" Ericaceae
CNPS List: 4 **R-E-D Code:** 1-2-3 **State/Fed. Status:** CEQA?
Distribution: SCZ, SRO
Habitat: CCFrs, Chprl
Life Form: Shrub (evergreen)
Blooming: March-April
Notes: Possibly threatened by feral herbivores. See *Madroño* 19(6):198 (1968) for original description.

Arctostaphylos truei

See *Arctostaphylos mewukka* ssp. *truei*

Arctostaphylos uva-ursi var. *leobreweri*

Considered but rejected: A synonym of *A. uva-ursi*; a common taxon

Arctostaphylos uva-ursi var. *marinensis*

Considered but rejected: A synonym of *A. uva-ursi*; a common taxon

Arctostaphylos uva-ursi ssp. *monoensis*

Considered but rejected: A synonym of *A. uva-ursi*; a common taxon

Arctostaphylos uva-ursi var. *suborbiculata*

Considered but rejected: A synonym of *A. uva-ursi*; a common taxon

Arctostaphylos virgata Eastw.

"Marin manzanita" Ericaceae
CNPS List: 1B **R-E-D Code:** 2-2-3 **State/Fed. Status:** /C3c
Distribution: MRN
Quads: 467A, 467B, 467E, 484C, 485C, 485D
Habitat: BUFrs, CCFrs, Chprl, NCFrs / sandstone or granitic
Life Form: Shrub (evergreen)
Blooming: January-March
Notes: Known from fewer than twenty occurrences. Threatened by fire suppression.

Arctostaphylos viridissima (Eastw.) McMinn

"white-haired manzanita" Ericaceae
CNPS List: 4 **R-E-D Code:** 1-2-3 **State/Fed. Status:** CEQA?
Distribution: SCZ
Habitat: CCFrs, Chprl / shale
Life Form: Shrub (evergreen)
Blooming: January-March
Notes: Possibly threatened by feral herbivores.

Arctostaphylos wellsii Knight

"Wells's manzanita" Ericaceae
CNPS List: 1B **R-E-D Code:** 2-3-3 **State/Fed. Status:** CEQA
Distribution: SLO
Quads: 220C, 221A, 221B, 246B, 247D?, 270C
Habitat: CCFrs, Chprl / sandstone
Life Form: Shrub (evergreen)
Blooming: December-April
Notes: Threatened by urbanization. See *Madroño* 19(6):192-210 (1968) for original description, and *Four Seasons* 8(3):12-16 (1989) for revised nomenclature.

Arenaria howellii

See *Minuartia howellii*

Arenaria kingii ssp. *compacta*

Considered but rejected: Too common

Arenaria macradenia Wats. var. *kuschei* (Eastw.) Maguire

"Forest Camp sandwort" Caryophyllaceae
CNPS List: 3 **R-E-D Code:** ?-?-3 **State/Fed. Status:** /C2
Distribution: SBD
Quads: unknown
Habitat: unknown
Life Form: Perennial herb
Blooming: June-July
Notes: Move to List 1B? Known from only a single collection made in 1929 at "Forest Camp, SBD Co., 4000 feet"; need quad. See *Proceedings of the California Academy of Sciences* IV 20:140 (1931) for original description.

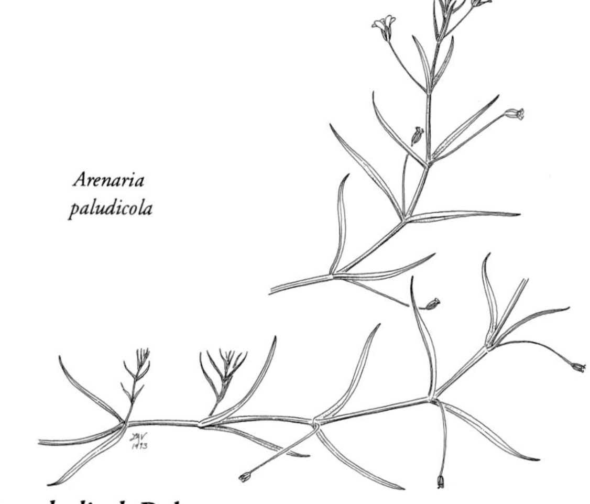

Arenaria paludicola

Arenaria paludicola Rob.

"marsh sandwort" Caryophyllaceae
CNPS List: 1B **R-E-D Code:** 3-3-2 **State/Fed. Status:** CE/FE
Distribution: LAX*, SBD*, SCR*, SFO*, SLO, WA*
Quads: 107D*, 221D, 408D*, 466C*
Habitat: MshSw (freshwater)
Life Form: Perennial herb (stoloniferous)
Blooming: May-August
Notes: Nearly extinct: fewer than twenty plants found in 1993. Only known extant occurrence near Black Lake on Nipomo Mesa is threatened by development and encroaching *Eucalyptus* groves, which may be altering the hydrology of the area. See *Proceedings of the California Academy of Natural Sciences* 3:61 (1863) for original description.
Status Report: 1990

Arenaria rosei
See *Minuartia rosei*

Arenaria ursina Rob.
"Big Bear Valley sandwort" Caryophyllaceae
CNPS List: 1B **R-E-D Code:** 2-2-3 **State/Fed. Status:** /C1
Distribution: SBD
Quads: 104B, 105A, 105B, 131C, 131D
Habitat: PbPln, PJWld / mesic, rocky
Life Form: Perennial herb
Blooming: May-August
Notes: Threatened by development, grazing, foot traffic, trampling, and vehicles. See *Proceedings of the American Academy of Arts and Sciences* 29:294 (1894) for original description.
Status Report: 1979

Argemone munita ssp. *robusta*
Considered but rejected: A synonym of *A. munita*; a common taxon

Argyrochosma limitanea (Maxon) Windham var. *limitanea*
"cloak fern" Pteridaceae
CNPS List: 2 **R-E-D Code:** 3-1-1 **State/Fed. Status:** CEQA
Distribution: SBD, AZ, BA, ++
Quads: 225D
Habitat: PJWld (carbonate)
Life Form: Perennial herb (rhizomatous)
Fertile: April-October
Notes: Known in CA only from the New York Mtns. See *Madroño* 25:55 (1978) for first CA collection, *Phytologia* 41(6):431-437 (1979) for nomenclature, and *American Fern Journal* 77:37-41 (1987) for revised nomenclature.

Aristocapsa insignis (Curran) Reveal & Hardham
"Indian Valley spineflower" Polygonaceae
CNPS List: 4 **R-E-D Code:** 1-2-3 **State/Fed. Status:** CEQA?
Distribution: MNT, SBT?, SLO
Habitat: CmWld
Life Form: Annual herb
Blooming: May-September
Notes: Does plant occur in SBT Co.? Threatened by development. See *Bulletin of the California Academy of Sciences* 1:275 (1885) for original description, *Great Basin Naturalist Memoirs* 2:169-190 (1978) for taxonomic treatment, and *Phytologia* 66(2):83-88 (1989) for revised nomenclature and taxonomic treatment.

Arnica cernua Howell
"serpentine arnica" Asteraceae
CNPS List: 4 **R-E-D Code:** 1-1-2 **State/Fed. Status:** CEQA?
Distribution: DNT, HUM, SIS, OR
Habitat: LCFrs (serpentinite)
Life Form: Perennial herb (rhizomatous)
Blooming: April-July
Notes: Possibly threatened by logging.

Arnica fulgens Pursh.
"hillside arnica" Asteraceae
CNPS List: 2 **R-E-D Code:** 3-1-1 **State/Fed. Status:** CEQA
Distribution: LAS, MOD, PLU, NV, OR, WA, ++
Quads: 590D, 674D, 691B?, 693C
Habitat: GBScr, LCFrs, Medws / mesic
Life Form: Perennial herb (rhizomatous)
Blooming: May-August
Notes: Is plant more widespread in CA? Similar to *A. sororia.* See *Rhodora* 90:245-275 (1988) for taxonomic treatment.

Arnica sororia Greene
"twin arnica" Asteraceae
CNPS List: 2 **R-E-D Code:** 2-1-1 **State/Fed. Status:** CEQA
Distribution: LAS, MNO, MOD, NV, OR, WA, ++
Quads: 451C, 487C, 660C, 674C, 708A, 724A
Habitat: GBScr, PJWld
Life Form: Perennial herb (rhizomatous)
Blooming: May-July
Notes: Similar to *A. fulgens.* See *Ottawa Naturalist* 23:213 (1910) for original description, and *Rhodora* 90:245-275 (1988) for taxonomic treatment.

Arnica spathulata Greene
"Klamath arnica" Asteraceae
CNPS List: 4 **R-E-D Code:** 1-1-2 **State/Fed. Status:** CEQA?
Distribution: DNT, HUM, SIS, TRI, OR
Habitat: LCFrs (serpentinite)
Life Form: Perennial herb (rhizomatous)
Blooming: May-July

Arnica spathulata ssp. *spathulata*
See *Arnica spathulata*

Arnica tomentella
Considered but rejected: Taxonomic problem, but needs reevaluation

Arnica venosa Hall
"Shasta County arnica" Asteraceae
CNPS List: 4 **R-E-D Code:** 1-2-3 **State/Fed. Status:** /C3c
Distribution: SHA, TRI
Habitat: CmWld, LCFrs / often in disturbed areas
Life Form: Perennial herb (rhizomatous)
Blooming: May-June
Notes: Threatened by logging. Some plants seem intermediate to *A. discoidea.*

Arnica viscosa Gray
"Mt. Shasta arnica" Asteraceae
CNPS List: 4 **R-E-D Code:** 1-1-2 **State/Fed. Status:** /C3c
Distribution: SIS, TRI, OR
Habitat: SCFrs
Life Form: Perennial herb (rhizomatous)
Blooming: August
Notes: Endangered in OR. See *Proceedings of the American Academy of Arts and Sciences* 13:374 (1878) for original description.

Artemisia cana ssp. *bolanderi*
Considered but rejected: Too common

Artemisia palmeri Gray

"San Diego sagewort" Asteraceae
CNPS List: 2 **R-E-D Code:** 2-2-1 **State/Fed. Status:** CEQA
Distribution: SDG, BA
Quads: 9B, 10A, 11D, 21C, 21D, 22A, 22B, 22C, 35C, 35D
Habitat: Chprl, CoScr, RpFrs, RpScr / sandy
Life Form: Shrub (deciduous)
Blooming: July-September
Notes: Known in CA from fewer than twenty occurrences. Threatened by development.

Asarum marmoratum Piper

"marbled wild-ginger" Aristolochiaceae
CNPS List: 2 **R-E-D Code:** 3-1-1 **State/Fed. Status:** CEQA
Distribution: DNT, SIS, OR
Quads: 738A, 739A
Habitat: LCFrs
Life Form: Perennial herb (rhizomatous)
Blooming: April-July
Notes: See *Proceedings of the Biological Society of Washington* 29:99 (1916) for original description, and *Brittonia* 42(1):33-37 (1990) for taxonomic treatment.

Asclepias cryptoceras ssp. *cryptoceras*

Considered but rejected: Not in CA

Asclepias solanoana Woodson

"serpentine milkweed" Asclepiadaceae
CNPS List: 4 **R-E-D Code:** 1-2-3 **State/Fed. Status:** CEQA?
Distribution: COL, GLE, LAK, MEN, NAP, SHA, SON, TEH, TRI, YOL
Habitat: Chprl, CmWld, LCFrs / serpentinite
Life Form: Perennial herb
Blooming: May-August
Notes: Not common where it occurs. Some occurrences threatened by grazing, vehicles, logging, mining, or geothermal development. See *Proceedings of the American Academy of Arts and Sciences* 10:76 (1874) for original description, and *Madroño* 24(3):159-177 (1977) for information on floral ecology.

Aspidotis carlotta-halliae (Wagn. & Gilb.) Lellinger

"Carlotta Hall's lace fern" Pteridaceae
CNPS List: 4 **R-E-D Code:** 1-2-3 **State/Fed. Status:** CEQA?
Distribution: ALA, MNT, MRN, SBT, SLO
Habitat: Chprl, CmWld / generally serpentinite
Life Form: Perennial herb (rhizomatous)
Fertile: January-December
Notes: Fertile hybrid between *A. californica* and *A. densa*; sometimes backcrosses. See *American Journal of Botany* 44:738-743 (1957) for original description, and *American Fern Journal* 58:141 (1968) for revised nomenclature.

Asplenium septentrionale (L.) Hoffm.

"northern spleenwort" Aspleniaceae
CNPS List: 2 **R-E-D Code:** 3-1-1 **State/Fed. Status:** CEQA
Distribution: SHA, TEH, TUL, OR, ++
Quads: 331A, 626A, 643C, 644D
Habitat: Chprl, LCFrs, SCFrs, UCFrs / rocky
Life Form: Perennial herb (rhizomatous)
Fertile: July-August
Notes: Known in CA from fewer than ten occurrences. Endangered in OR. See *Madroño* 25(4):232 (1978) for distributional information.

Asplenium trichomanes

See *Asplenium trichomanes* ssp. *trichomanes*

Asplenium trichomanes L. ssp. *trichomanes*

"maidenhair spleenwort" Aspleniaceae
CNPS List: 2 **R-E-D Code:** 3-1-1 **State/Fed. Status:** CEQA
Distribution: DNT, ID*, OR, ++
Quads: 740C
Habitat: LCFrs (rocky)
Life Form: Perennial herb (rhizomatous)
Fertile: May-July
Notes: Extirpated in ID. See *Madroño* 29(1):57 (1982) for first CA record.

Asplenium trichomanes-ramosum L.

"green spleenwort" Aspleniaceae
CNPS List: 2 **R-E-D Code:** 3-1-1 **State/Fed. Status:** CEQA
Distribution: SIE, ID, NV, OR, WA, ++
Quads: 572C
Habitat: SCFrs (carbonate in granitic cliffs)
Life Form: Perennial herb (rhizomatous)
Fertile: June-August
Notes: Known in CA only from the Sierra Buttes. Endangered in ID and OR. See *Madroño* 12:128 (1953) for first CA record.
Status Report: 1980

Asplenium vespertinum

Considered but rejected: Too common

Asplenium viride

See *Asplenium trichomanes-ramosum*

Aster brickellioides var. *brickellioides*

Considered but rejected: Too common; a synonym of *A. brickellioides*

Aster chilensis var. *lentus*

See *Aster lentus*

Aster greatae Parish

"Greata's aster" Asteraceae
CNPS List: 4 **R-E-D Code:** 1-1-3 **State/Fed. Status:** /C3c
Distribution: LAX
Habitat: Chprl (mesic canyons)
Life Form: Perennial herb (rhizomatous)
Blooming: August-October
Notes: To be expected in SBD Co. See *Phytologia* 71(3):167-170 (1991) for nomenclatural correction.

Aster lentus Greene
"Suisun Marsh aster" Asteraceae
CNPS List: 1B **R-E-D Code:** 2-2-3 **State/Fed. Status:** /C2
Distribution: CCA, NAP, SAC, SJQ, SOL
Quads: 463A, 479C, 480B, 480C, 480D, 481C, 481D, 482A, 482C, 483A, 483D, 498D
Habitat: MshSw (brackish and fresh water)
Life Form: Perennial herb (rhizomatous)
Blooming: August-November
Notes: Seriously threatened by marsh habitat alteration and loss. Grades into *A. chilensis.* USFWS uses the name *A. chilensis* var. *lentus.* See *Manual of the Botany of the Region of San Francisco Bay,* p. 180 (1894) by E. Greene for original description.
Status Report: 1977

Aster paludicola
Considered but rejected: A synonym of *A. occidentalis* var. *yosemitanus*; a common taxon

Aster peirsonii
Considered but rejected: Too common

Astragalus agnicidus Barneby
"Humboldt milk-vetch" Fabaceae
CNPS List: 1B **R-E-D Code:** 3-3-3 **State/Fed. Status:** CE/C1
Distribution: HUM
Quads: 617A
Habitat: BUFrs (disturbed openings)
Life Form: Perennial herb
Blooming: June-August
Notes: Rediscovered in 1987 by R. Sutherland, R. Bittman, and K. Berg near Miranda. Only known occurrence is voluntarily protected by landowner; previously grazed by deer, but now fenced. See *Madroño* 14:37-40 (1957) for original description, *Fremontia* 16(1):13-14 (1988) for discussion of rediscovery, and *Fremontia* 20(4):21-22 (1992) for management update.
Status Report: 1987

Astragalus albens Greene
"Cushenbury milk-vetch" Fabaceae
CNPS List: 1B **R-E-D Code:** 3-3-3 **State/Fed. Status:** /PE
Distribution: SBD
Quads: 130C, 131B, 131C, 131D
Habitat: JTWld, PJWld / carbonate or granitic
Life Form: Perennial herb
Blooming: March-May
Notes: Known from fewer than twenty occurrences. Threatened by carbonate mining and vehicles. See *Bulletin of the California Academy of Sciences* 1:156 (1885) for original description, and *Fremontia* 16(1):20-21 (1988) for discussion of mining threats.
Status Report: 1988

Astragalus allochrous Gray var. *playanus* (M.E. Jones) Isely
"playa milk-vetch" Fabaceae
CNPS List: 2 **R-E-D Code:** 2-2-1 **State/Fed. Status:** CEQA
Distribution: SBD, AZ, NM, TX, UT+
Quads: 173B, 174A
Habitat: MDScr (sandy)
Life Form: Perennial herb
Blooming: April
Notes: Known in CA only from near Goffs. Includes *A. wootoni* var. *wootoni.* See *Contributions to Western Botany* 8:6 (1898) for original description.

Astragalus anxius Meinke & Kaye
"troubled milk-vetch" Fabaceae
CNPS List: 1B **R-E-D Code:** 3-1-3 **State/Fed. Status:** CEQA
Distribution: LAS
Quads: 675C
Habitat: GBScr, PJWld, UCFrs / volcanic
Life Form: Perennial herb
Blooming: May-July
Notes: Known from only one large occurrence in the Ash Valley RNA (BLM) and on adjacent private lands. See *Fremontia* 16(1):15-17 (1988) for brief species account (as *A. tegetarioides*) and BLM management plans, and *Madroño* 39(3):193-204 (1992) for original description.

Astragalus argophyllus T. & G. var. *argophyllus*
"silverleaf milk-vetch" Fabaceae
CNPS List: 2 **R-E-D Code:** 3-2-1 **State/Fed. Status:** CEQA
Distribution: INY, LAS, MNO, AZ, ID, NV, UT+
Quads: 413B, 432C, 638D
Habitat: Medws, Plyas / alkaline
Life Form: Perennial herb
Blooming: May-July
Notes: Threatened by grazing, but BLM will fence both LAS Co. occurrences. See *Memoirs of the New York Botanical Garden* 13:626-628 (1964) for revised nomenclature.

Astragalus atratus Wats. var. *mensanus* M.E. Jones
"Darwin Mesa milk-vetch" Fabaceae
CNPS List: 1B **R-E-D Code:** 3-1-3 **State/Fed. Status:** CEQA
Distribution: INY
Quads: 304C, 304D, 349D
Habitat: GBScr, JTWld, PJWld / volcanic clay, gravelly
Life Form: Perennial herb
Blooming: April-June
Notes: Known from only five collections near Darwin. Historical occurrences need field surveys. See *Memoirs of the New York Botanical Garden* 13:473 (1964) for taxonomic treatment.

Astragalus austinae
Considered but rejected: Too common

Astragalus bicristatus Gray

"crested milk-vetch" Fabaceae
CNPS List: 4 **R-E-D Code:** 1-1-3 **State/Fed. Status:** CEQA?
Distribution: RIV, SBD
Habitat: LCFrs, UCFrs / rocky
Life Form: Perennial herb
Blooming: May-August

Astragalus brauntonii Parish

"Braunton's milk-vetch" Fabaceae
CNPS List: 1B **R-E-D Code:** 3-2-3 **State/Fed. Status:** /PE
Distribution: LAX, ORA, VEN
Quads: 87C, 87D, 109B, 110A, 111B, 111C*, 111D*, 112B, 112C*, 112D, 113A, 138D
Habitat: CCFrs, Chprl, CoScr, VFGrs / recent burns or disturbed areas, carbonate
Life Form: Perennial herb
Blooming: March-July
Notes: Known from fewer than ten extant occurrences. Threatened by development and alteration of local fire regimes. See *Bulletin of the Southern California Academy of Sciences* 2:26-27 (1903) for original description, and *Fremontia* 19(3):6-7 (1991) for species account.
Status Report: 1980

Astragalus breweri Gray

"Brewer's milk-vetch" Fabaceae
CNPS List: 4 **R-E-D Code:** 1-2-3 **State/Fed. Status:** CEQA?
Distribution: COL, LAK, MEN, MRN, NAP, SON, YOL
Habitat: Chprl, CmWld, Medws, VFGrs / often serpentinite, volcanic
Life Form: Annual herb
Blooming: April-June
Notes: Populations have been lost to development and road construction. See *Systematic Botany* 17(3):367-379 (1992) for distributional information.

Astragalus californicus

Considered but rejected: Too common

Astragalus cimae M.E. Jones var. *cimae*

"Cima milk-vetch" Fabaceae
CNPS List: 1B **R-E-D Code:** 3-2-2 **State/Fed. Status:** /C3c
Distribution: SBD, NV
Quads: 200A, 200B, 200C, 201A, 201B, 201D, 224C, 225D, 226A, 226B, 226D
Habitat: GBScr, JTWld, PJWld / clay
Life Form: Perennial herb
Blooming: April-May
Notes: Known in CA from fewer than twenty occurrences. Threatened by grazing. On watch list in NV. See *Memoirs of the New York Botanical Garden* 13:505-508 (1964) for taxonomic treatment.

Astragalus cimae var. *sufflatus*

Considered but rejected: Too common

Astragalus clarianus Jeps.

"Clara Hunt's milk-vetch" Fabaceae
CNPS List: 1B **R-E-D Code:** 3-3-3 **State/Fed. Status:** CT/C1
Distribution: NAP, SON
Quads: 500B, 516C, 517C, 517D
Habitat: CmWld, VFGrs / serpentinite, volcanic clay
Life Form: Annual herb
Blooming: March-April
Notes: Known from only five occurrences. Threatened by recreational development and non-native plants; largest known population covered by dredge material in 1990. See *Manual of the Flowering Plants of California*, p. 578 (1925) by W.L. Jepson for original description, and *Systematic Botany* 17(3):367-379 (1992) for distributional information.
Status Report: 1989

Astragalus clevelandii Greene

"Cleveland's milk-vetch" Fabaceae
CNPS List: 4 **R-E-D Code:** 1-1-3 **State/Fed. Status:** CEQA?
Distribution: COL, LAK, NAP, SBT, YOL
Habitat: Chprl, CmWld / serpentinite seeps
Life Form: Perennial herb
Blooming: June-September

Astragalus congdonii

Considered but rejected: Too common

Astragalus crotalariae (Benth.) Gray

"Salton milk-vetch" Fabaceae
CNPS List: 4 **R-E-D Code:** 1-1-2 **State/Fed. Status:** CEQA?
Distribution: IMP, RIV, SDG, AZ, BA
Habitat: SDScr (sandy or gravelly)
Life Form: Perennial herb
Blooming: January-April

Astragalus deanei (Rydb.) Barneby

"Dean's milk-vetch" Fabaceae
CNPS List: 1B **R-E-D Code:** 3-3-3 **State/Fed. Status:** /C2
Distribution: SDG
Quads: 9B, 9C, 10B, 11A, 20B, 21C, 21D
Habitat: Chprl, CoScr, RpFrs
Life Form: Perennial herb
Blooming: March-May
Notes: Known from approximately ten occurrences. Seriously threatened by development in foothills. See *North American Flora* 24:355 (1929) for original description, and *Memoirs of the New York Botanical Garden* 13:845-846 (1964) for taxonomic treatment.
Status Report: 1979

Astragalus douglasii (T. & G.) Gray var. *perstrictus* (Rydb.) Munz

"Jacumba milk-vetch" Fabaceae
CNPS List: 1B **R-E-D Code:** 2-2-2 **State/Fed. Status:** /C2
Distribution: IMP, SDG, BA
Quads: 7A, 7B, 8A, 8B, 8D, 19D
Habitat: Chprl, CmWld, VFGrs / rocky
Life Form: Perennial herb
Blooming: April-June
Notes: Many occurrences are small.

Astragalus ertterae Barneby & Shevock

"Walker Pass milk-vetch" Fabaceae
CNPS List: 1B **R-E-D Code:** 3-1-3 **State/Fed. Status:** /C2
Distribution: KRN
Quads: 259A
Habitat: PJWld (sandy, granitic)
Life Form: Perennial herb
Blooming: April-May
Notes: Known in CA from only three occurrences near Walker Pass. See *Aliso* 11(4):585-588 (1987) for original description.

Astragalus ertterae

Astragalus funereus M.E. Jones

"black milk-vetch" Fabaceae
CNPS List: 1B **R-E-D Code:** 3-2-2 **State/Fed. Status:** /C2
Distribution: INY, NV
Quads: 326B, 368A, 368D, 369C
Habitat: MDScr (sometimes carbonate)
Life Form: Perennial herb
Blooming: March-May
Notes: Known in CA from only five occurrences. On watch list in NV. See *Contributions to Western Botany* 12:11-12 (1908) for original description, and *Memoirs of the New York Botanical Garden* 13:686-688 (1964) for taxonomic treatment.
Status Report: 1979

Astragalus geyeri Gray var. *geyeri*

"Geyer's milk-vetch" Fabaceae
CNPS List: 2 **R-E-D Code:** 3-2-1 **State/Fed. Status:** CEQA
Distribution: INY, MNO, NV, OR, WA+
Quads: 372B, 412D
Habitat: ChScr, GBScr / sandy
Life Form: Annual herb
Blooming: May-August
Notes: State-listed as Sensitive in WA. See *Memoirs of the New York Botanical Garden* 13:894-895 (1964) for taxonomic treatment.

Astragalus gilmanii Tides.

"Gilman's milk-vetch" Fabaceae
CNPS List: 4 **R-E-D Code:** 1-2-2 **State/Fed. Status:** /C2
Distribution: INY, NV
Habitat: GBScr, PJWld / gravelly
Life Form: Perennial herb
Blooming: May-July
Notes: On watch list in NV.

Astragalus insularis Kell. var. *harwoodii* Munz & McBurn.

"Harwood's milk-vetch" Fabaceae
CNPS List: 2 **R-E-D Code:** 2-2-1 **State/Fed. Status:** CEQA
Distribution: IMP, RIV, SDG, AZ, SO
Quads: 6B, 13D, 18B, 19A, 30A, 61A, 61D
Habitat: DeDns
Life Form: Annual herb
Blooming: January-May

Astragalus inversus M.E. Jones

"Susanville milk-vetch" Fabaceae
CNPS List: 4 **R-E-D Code:** 1-1-3 **State/Fed. Status:** CEQA?
Distribution: LAS, MOD, SHA, SIS
Habitat: GBScr, LCFrs
Life Form: Perennial herb
Blooming: May-September

Astragalus inyoensis

Considered but rejected: Too common

Astragalus jaegerianus Munz

"Lane Mtn. milk-vetch" Fabaceae
CNPS List: 1B **R-E-D Code:** 3-3-3 **State/Fed. Status:** /PE
Distribution: SBD
Quads: 206A, 206B, 206C
Habitat: JTWld, MDScr / granitic, sandy or gravelly
Life Form: Perennial herb
Blooming: April-June
Notes: Threatened by grazing and vehicles, and by Army activities at Ft. Irwin. See *Leaflets of Western Botany* 3:49-80 (1941) for original description.
Status Report: 1977

Astragalus johannis-howellii Barneby

"Long Valley milk-vetch" Fabaceae
CNPS List: 1B **R-E-D Code:** 2-2-2 **State/Fed. Status:** CR/C3c
Distribution: MNO, NV
Quads: 433B, 433C, 433D, 434A, 486C
Habitat: GBScr (sandy loam)
Life Form: Perennial herb
Blooming: June-August
Notes: Known in CA from approximately twelve occurrences. Threatened by grazing, vehicles, and mining. Usually found in swales in vicinity of former or present hot springs activity. See *Leaflets of Western Botany* 8(5):124-125 (1957) for original description, and *Memoirs of the New York Botanical Garden* 13:467-468 (1964) for taxonomic treatment.
Status Report: 1986

Astragalus kentrophyta Gray var. *danaus* (Barneby) Barneby

"Sweetwater Mtns. milk-vetch" Fabaceae
CNPS List: 4 **R-E-D Code:** 1-1-3 **State/Fed. Status:** CEQA?
Distribution: FRE, INY, MNO, TUO
Habitat: AlpBR, SCFrs (rocky)
Life Form: Perennial herb
Blooming: July-September

Astragalus kentrophyta Gray var. *elatus* Wats.

"spiny-leaved milk-vetch" Fabaceae
CNPS List: 2 **R-E-D Code:** 2-2-1 **State/Fed. Status:** CEQA
Distribution: INY, NV
Quads: 372A, 412B
Habitat: SCFrs (sometimes carbonate)
Life Form: Perennial herb
Blooming: June-September
Notes: Threatened by parking lot construction at Schulman Grove in the Ancient Bristlecone Forest area. See *Memoirs of the New York Botanical Garden* 13:367 (1964) for taxonomic treatment.

Astragalus lentiformis Gray

"lens-pod milk-vetch" Fabaceae
CNPS List: 1B **R-E-D Code:** 3-2-3 **State/Fed. Status:** /C2
Distribution: PLU
Quads: 586B, 586C, 587A, 587B, 587C, 588A, 588D, 603C
Habitat: GBScr, LCFrs / volcanic
Life Form: Perennial herb
Blooming: May-June
Notes: Threatened by road maintenance.

Astragalus lentiginosus Hook. var. *antonius* Barneby

"San Antonio milk-vetch" Fabaceae
CNPS List: 1B **R-E-D Code:** 3-1-3 **State/Fed. Status:** /C2
Distribution: LAX, SBD
Quads: 131C, 134C, 134D, 135A
Habitat: LCFrs, UCFrs
Life Form: Perennial herb
Blooming: April-July
Notes: See *Leaflets of Western Botany* 4(5):100 (1945) for original description, and *Memoirs of the New York Botanical Garden* 13:926-927 (1964) for taxonomic treatment.

Astragalus lentiginosus Hook. var. *borreganus* M.E. Jones

"Borrego milk-vetch" Fabaceae
CNPS List: 4 **R-E-D Code:** 1-1-1 **State/Fed. Status:** CEQA?
Distribution: IMP, RIV, SBD, SDG, AZ, BA, SO
Habitat: MDScr, SDScr / sandy
Life Form: Annual herb
Blooming: February-May
Notes: See *Memoirs of the New York Botanical Garden* 13:953-954 (1964) for revised nomenclature.

Astragalus lentiginosus var. *chartaceus*

Considered but rejected: Not in CA

Astragalus lentiginosus Hook. var. *coachellae* F. Shreve & Wiggins

"Coachella Valley milk-vetch" Fabaceae
CNPS List: 1B **R-E-D Code:** 2-2-3 **State/Fed. Status:** /PE
Distribution: RIV
Quads: 82B, 82C, 82D, 83A, 83B
Habitat: SDScr (sandy)
Life Form: Annual herb
Blooming: February-May
Notes: Known from fewer than twenty occurrences in the Coachella Valley. Threatened by vehicles, wind energy development, and urbanization. See *Memoirs of the New York Botanical Garden* 13:951-952 (1964) for taxonomic treatment.

Astragalus lentiginosus Hook. var. *kernensis* (Jeps.) Barneby

"Kern Plateau milk-vetch" Fabaceae
CNPS List: 2 **R-E-D Code:** 2-2-1 **State/Fed. Status:** CEQA
Distribution: INY, TUL, NV
Quads: 306B, 306C, 307D, 329C, 330D, 394A
Habitat: Medws (xeric), SCFrs / sandy
Life Form: Perennial herb
Blooming: June-July
Notes: See *Memoirs of the New York Botanical Garden* 13:929 (1964) for taxonomic treatment.

Astragalus lentiginosus Hook. var. *micans* Barneby

"shining milk-vetch" Fabaceae
CNPS List: 1B **R-E-D Code:** 3-2-3 **State/Fed. Status:** /PT
Distribution: INY
Quads: 390C, 391D
Habitat: DeDns
Life Form: Perennial herb
Blooming: April-June
Notes: Recovery plan has been completed by BLM for the Eureka Dunes. Threatened by vehicle trespass and non-native Russian thistle. See *Leaflets of Western Botany* 8(1):22 (1956) for original description, *Memoirs of the New York Botanical Garden* 13:952-953 (1964) for taxonomic treatment, and *Biological Conservation* 46:217-242 (1988) for population biology.

Astragalus lentiginosus Hook. var. *piscinensis* Barneby

"Fish Slough milk-vetch" Fabaceae
CNPS List: 1B **R-E-D Code:** 3-3-3 **State/Fed. Status:** /PE
Distribution: INY, MNO
Quads: 413B, 413C, 432A, 432C
Habitat: Plyas (alkaline)
Life Form: Perennial herb
Blooming: June-July
Notes: Known from eight occurrences near Fish Slough. Threatened by fisheries development, agriculture, grazing, and vehicles. Protected in part at Fish Slough ACEC (BLM). See *Brittonia* 29:376-381 (1977) for original description.

Astragalus lentiginosus var. *semotus*

Considered but rejected: Too common

Astragalus lentiginosus Hook. var. *sesquimetralis* (Rydb.) Barneby

"Sodaville milk-vetch" Fabaceae
CNPS List: 1B **R-E-D Code:** 3-3-2 **State/Fed. Status:** CE/PT
Distribution: INY, NV
Quads: 390A
Habitat: Medws (alkaline)
Life Form: Perennial herb
Blooming: June-July
Notes: Known in CA from only one occurrence at Big Sand Spring; only two occurrences known in NV. Threatened by trampling by cattle and feral burros. Protected at Big/Little Sand Springs ACEC (BLM). State-listed as Critically Endangered in NV. See *Leaflets of Western Botany* 4(5):116 (1945) for original description, and *Brittonia* 29:376-381 (1977) for discussion of CA occurrence.
Status Report: 1987

Astragalus lentiginosus Hook. var. *sierrae* M.E. Jones

"Big Bear Valley milk-vetch" Fabaceae
CNPS List: 1B **R-E-D Code:** 2-2-3 **State/Fed. Status:** /C2
Distribution: SBD
Quads: 105A, 105B, 131C, 131D
Habitat: MDScr, Medws, PJWld, UCFrs / gravelly
Life Form: Perennial herb
Blooming: April-August
Notes: Endemic to Big Bear Valley and Baldwin Lake region, where it is threatened by urbanization. See *Revision of North American Species of Astragalus*, p. 124 (1923) by M. E. Jones for original description, and *Memoirs of the New York Botanical Garden* 13:925 (1964) for taxonomic treatment.

Astragalus leucolobus M.E. Jones

"Big Bear Valley woollypod" Fabaceae
CNPS List: 1B **R-E-D Code:** 2-2-3 **State/Fed. Status:** /C2
Distribution: LAX, RIV, SBD, SDG?
Quads: 65C, 104B, 105A, 130C, 131C, 131D, 134B, 134C, 134D
Habitat: LCFrs, PbPln, PJWld, UCFrs / rocky
Life Form: Perennial herb
Blooming: May-July
Notes: Does plant occur in SDG Co.? Threatened by development and vehicles.

Astragalus macrodon (H. & A.) Gray

"Salinas milk-vetch" Fabaceae
CNPS List: 4 **R-E-D Code:** 1-1-3 **State/Fed. Status:** CEQA?
Distribution: KRN, MNT, SBT, SLO
Habitat: Chprl (openings), CmWld, VFGrs / sandstone, shale, or serpentinite
Life Form: Perennial herb
Blooming: April-June

Astragalus magdalenae Greene var. *peirsonii* (Munz & McBurn.) Barneby

"Peirson's milk-vetch" Fabaceae
CNPS List: 1B **R-E-D Code:** 2-2-2 **State/Fed. Status:** CE/PE
Distribution: IMP, SDG(*?), AZ, BA
Quads: 2A, 2B, 13C, 14A, 14B, 14D, 27C, 28A, 28D, 30A, 30D, 47D(*?)
Habitat: DeDns
Life Form: Perennial herb
Blooming: December-April
Notes: Known in CA from fewer than twenty occurrences. Is plant extant in SDG Co.? Threatened by vehicles in the Algodones Dunes. See *Bulletin of the Southern California Academy of Sciences* 31:67 (1932) for original description, and *Memoirs of the New York Botanical Garden* 13:928 (1964) for taxonomic treatment.
Status Report: 1987

Astragalus miguelensis Greene

"San Miguel Island milk-vetch" Fabaceae
CNPS List: 4 **R-E-D Code:** 1-1-3 **State/Fed. Status:** CEQA?
Distribution: ANA, SCM, SCZ, SMI, SRO
Habitat: CBScr, CoDns
Life Form: Perennial herb
Blooming: March-July
Notes: Blooms intermittently throughout the year.

Astragalus mojavensis Wats. var. *hemigyrus* (Clokey) Barneby

"curved-pod milk-vetch" Fabaceae
CNPS List: 1A **Last Seen:** 1941 **State/Fed. Status:** /C2
Distribution: INY*, NV
Quads: 327A*
Habitat: JTWld, MDScr / carbonate
Life Form: Annual herb
Blooming: April-June
Notes: Only CA record is from Darwin Mesa in 1941. Field work needed. Proposed for state listing as Critically Endangered in NV. See *Madroño* 6:220 (1942) for original description, and *Memoirs of the New York Botanical Garden* 13:1025 (1964) for taxonomic treatment.
Status Report: 1980

Astragalus monoensis

See *Astragalus monoensis* var. *monoensis*

Astragalus monoensis Barneby var. *monoensis*

"Mono milk-vetch" Fabaceae
CNPS List: 1B **R-E-D Code:** 2-2-3 **State/Fed. Status:** CR/C2
Distribution: MNO
Quads: 433C, 434A, 434B, 451C, 452A, 452B, 452C, 453D
Habitat: GBScr, UCFrs / pumice flats
Life Form: Perennial herb
Blooming: June-August
Notes: Known from fewer than twenty occurrences. Threatened by road maintenance, vehicles, and sheep grazing. See *Leaflets of Western Botany* 4:55 (1944) for original description, and *Systematic Botany* 8(4):423 (1983) for revised nomenclature.
Status Report: 1986

Astragalus monoensis Barneby var. *ravenii* (Barneby) Isley

"Raven's milk-vetch" Fabaceae
CNPS List: 1B **R-E-D Code:** 3-1-3 **State/Fed. Status:** /C2
Distribution: FRE, INY, MNO
Quads: 373A, 373B, 414B, 451C
Habitat: AlpBR, UCFrs / gravelly
Life Form: Perennial herb
Blooming: July-September
Notes: Known from approximately five occurrences. See *Aliso* 4:131 (1958) for original description, *Memoirs of the New York Botanical Garden* 13:967-968 (1964) for taxonomic treatment, and *Systematic Botany* 8(4):423 (1983) for revised nomenclature.
Status Report: 1979

Astragalus nevinii Gray

"San Clemente Island milk-vetch" Fabaceae
CNPS List: 1B **R-E-D Code:** 2-2-3 **State/Fed. Status:** /C1
Distribution: SCM
Quads: SCMC, SCMN
Habitat: CoDns, CoScr, VFGrs
Life Form: Perennial herb
Blooming: February-July
Notes: Known from fewer than twenty occurrences.

Astragalus nutans M.E. Jones

"Providence Mtns. milk-vetch" Fabaceae
CNPS List: 4 **R-E-D Code:** 1-1-3 **State/Fed. Status:** CEQA?
Distribution: IMP, RIV, SBD
Habitat: JTWld, MDScr, PJWld, SDScr / sandy or gravelly
Life Form: Annual herb
Blooming: March-June

Astragalus oocarpus Gray

"San Diego milk-vetch" Fabaceae
CNPS List: 1B **R-E-D Code:** 3-2-3 **State/Fed. Status:** /C2
Distribution: SDG
Quads: 20A, 20B, 20D, 33A, 33B, 33C, 33D, 34A, 49D, 66B
Habitat: Chprl (openings), CmWld
Life Form: Perennial herb
Blooming: May-August
Notes: Known from approximately twenty occurrences. Threatened by development, road maintenance, and recreation.

Astragalus oophorus Wats. var. *lavinii* Barneby

"Lavin's milk-vetch" Fabaceae
CNPS List: 1B **R-E-D Code:** 3-2-2 **State/Fed. Status:** /C2
Distribution: MNO, NV
Quads: 487C, 487D
Habitat: GBScr
Life Form: Perennial herb
Blooming: June
Notes: Known in CA only from the Bodie Hills. On watch list in NV. See *Brittonia* 36(2):167-173 (1984) for original description.

Astragalus pachypus Greene var. *jaegeri* Munz & McBurney

"Jaeger's milk-vetch" Fabaceae
CNPS List: 1B **R-E-D Code:** 3-3-3 **State/Fed. Status:** /C2
Distribution: RIV
Quads: 48A, 49A, 49B, 50B, 84A, 84B, 84C, 85D
Habitat: Chprl, CmWld, CoScr, VFGrs / sandy or rocky
Life Form: Shrub
Blooming: December-June
Notes: Known from about six occurrences; to be looked for in northern SDG Co., west of Warner Springs. Some habitat probably lost during creation of Vail Lake; also threatened by urbanization. See *Bulletin of the Southern Academy of Sciences* 31:67 (1932) for original description, and *Memoirs of the New York Botanical Garden* 13:489 (1964) for taxonomic treatment.

Astragalus panamintensis

Considered but rejected: Too common

Astragalus pauperculus Greene

"depauperate milk-vetch" Fabaceae
CNPS List: 4 **R-E-D Code:** 1-1-3 **State/Fed. Status:** /C3c
Distribution: BUT, PLA, SHA, TEH, YUB
Habitat: CmWld, VFGrs / vernally mesic, volcanic
Life Form: Annual herb
Blooming: March-May
Notes: See *Systematic Botany* 17(3):367-379 (1992) for distributional information.

Astragalus platytropis Gray

"little big-pod milk-vetch" Fabaceae
CNPS List: 2 **R-E-D Code:** 2-2-1 **State/Fed. Status:** CEQA
Distribution: INY, MNO, NV, ++
Quads: 369B, 369C, 372A, 410C, 412B, 413A, 488A, 488B, 488D, 489C,
Habitat: AlpBR, PJWld, SCFrs / usually carbonate, granitic, or pumice-sand
Life Form: Perennial herb
Blooming: June-September
Notes: Threatened by grazing, trampling, and vehicles. See *Proceedings of the American Academy of Arts and Sciences* 6:526 (1865) for original description, and *Memoirs of the New York Botanical Garden* 13:979-981 (1964) for taxonomic treatment.

Astragalus preussii Gray var. *laxiflorus* Gray

"Lancaster milk-vetch" Fabaceae
CNPS List: 1B **R-E-D Code:** 3-3-2 **State/Fed. Status:** CEQA
Distribution: LAX, AZ, NV
Quads: 161A, 161B
Habitat: ChScr
Life Form: Perennial herb
Blooming: March-May
Notes: Known in CA only from near Lancaster, where extremely rare; only reported once in recent years. More field work needed. See *Proceedings of the American Academy of Arts and Sciences* 13:369 (1878) for original description, and *Memoirs of the New York Botanical Garden* 13:576 (1964) for taxonomic treatment.

Astragalus preussii Gray var. *preussii*

"Preuss's milk-vetch" Fabaceae
CNPS List: 2 **R-E-D Code:** 3-1-1 **State/Fed. Status:** CEQA
Distribution: INY, SBD, AZ, NV, UT
Quads: 273C, 273D
Habitat: ChScr, MDScr / clay
Life Form: Perennial herb
Blooming: May-June
Notes: Known in CA from only one or two occurrences. More field work needed. See *Proceedings of the American Academy of Arts and Sciences* 6:222 (1864) for original description, and *Memoirs of the New York Botanical Garden* 13:575-576 (1964) for taxonomic treatment.

Astragalus pseudiodanthus Barneby

"Tonopah milk-vetch" Fabaceae
CNPS List: 1B **R-E-D Code:** 3-2-2 **State/Fed. Status:** /C3c
Distribution: MNO, NV
Quads: 469B, 469D
Habitat: GBScr (stabilized dunes)
Life Form: Perennial herb
Blooming: May-June
Notes: Known in CA from fewer than ten occurrences. Threatened by grazing. See *Leaflets of Western Botany* 3:99 (1942) for original description, and *Systematic Botany* 8:422 (1983) for alternate taxonomic treatment as *A. iodanthus* var. *pseudiodanthus.*

Astragalus pulsiferae Gray var. *pulsiferae*

"Pulsifer's milk-vetch" Fabaceae
CNPS List: 1B **R-E-D Code:** 2-1-2 **State/Fed. Status:** CEQA
Distribution: LAS, MOD, PLU, SIE, NV
Quads: 586A, 586C, 586D, 604C, 621A, 624C, 639D, 656C, 676B, 692A
Habitat: GBScr, LCFrs, PJWld / volcanic, sandy or rocky
Life Form: Perennial herb
Blooming: May-August
Notes: Need quads for SIE Co. Plant is fairly widespread, but populations are small. See *Proceedings of the American Academy of Arts and Sciences* 10:69 (1874) for original description, and *Memoirs of the New York Botanical Garden* 13:970-972 (1964) for taxonomic treatment.

Astragalus pulsiferae Gray var. *suksdorfii* (Howell) Barneby

"Suksdorf's milk-vetch" Fabaceae
CNPS List: 1B **R-E-D Code:** 3-1-2 **State/Fed. Status:** /C2
Distribution: LAS, PLU, SHA, WA
Quads: 606B, 624C, 625D, 676B
Habitat: GBScr, LCFrs / volcanic, often rocky
Life Form: Perennial herb
Blooming: May-August
Notes: Endangerment information needed. Known from fewer than ten herbarium collections; rarer than var. *pulsiferae.* Need quads for SHA Co. State-listed as Threatened in WA. See *Erythea* 1:111 (1893) for original description, and *Memoirs of the New York Botanical Garden* 13:968-971 (1964) for taxonomic treatment.

Astragalus pycnostachyus Gray var. *lanosissimus* (Rydb.) Munz & McBurn.

"Ventura marsh milk-vetch" Fabaceae
CNPS List: 1A **Last Seen:** 1967 **State/Fed. Status:** /C2*
Distribution: LAX*, ORA*, VEN*
Quads: 72A*, 90B*, 111D*, 112D*, 114B*, 141D*
Habitat: MshSw (coastal salt)
Life Form: Perennial herb
Blooming: July-October
Notes: Recent attempts to rediscover this plant have been unsuccessful. Habitat lost to urbanization. See *North American Flora* 24:357-358 (1929) for original description, and *Memoirs of the New York Botanical Garden* 13:813 (1964) for taxonomic treatment.
Status Report: 1978

Astragalus rattanii Gray var. *jepsonianus* Barneby

"Jepson's milk-vetch" Fabaceae
CNPS List: 1B **R-E-D Code:** 2-2-3 **State/Fed. Status:** CEQA
Distribution: COL, GLE, LAK, NAP, TEH, YOL
Quads: 516A, 516B, 517A, 532D, 533C, 533D, 547C, 564A, 612B
Habitat: CmWld, VFGrs / often serpentinite
Life Form: Annual herb
Blooming: April-June
Notes: Most occurrences are small and widely scattered. See *Aliso* 4:137 (1958) for original description, *Memoirs of the New York Botanical Garden* 13:1052 (1964) for taxonomic treatment, and *Systematic Botany* 17(3):367-379 (1992) for distributional information.

Astragalus rattanii Gray var. *rattanii*

"Rattan's milk-vetch" Fabaceae
CNPS List: 4 **R-E-D Code:** 1-1-3 **State/Fed. Status:** CEQA?
Distribution: COL, GLE, HUM, LAK, MEN, SON, TEH, TRI
Habitat: CmWld, LCFrs / gravelly streambanks
Life Form: Perennial herb
Blooming: April-July
Notes: See *Systematic Botany* 17(3):367-379 (1992) for distributional information.

Astragalus ravenii

See *Astragalus monoensis* var. *ravenii*

Astragalus sepultipes

Considered but rejected: Too common

Astragalus serenoi var. *serenoi*

Considered but rejected: Not in CA

Astragalus serenoi (Kuntze) Sheld. var. *shockleyi* (M.E. Jones) Barneby

"naked milk-vetch" Fabaceae
CNPS List: 2 **R-E-D Code:** 2-2-1 **State/Fed. Status:** CEQA
Distribution: INY, MNO, NV
Quads: 350B, 391B, 392B, 392C, 411C, 411D, 412B, 412C, 431A
Habitat: ChScr, GBScr, PJWld / alkaline, granitic alluvium
Life Form: Perennial herb
Blooming: May-July

Astragalus shevockii Barneby

"Shevock's milk-vetch" Fabaceae
CNPS List: 1B **R-E-D Code:** 3-1-3 **State/Fed. Status:** /C3c
Distribution: TUL
Quads: 283D, 307B, 308A
Habitat: UCFrs (granitic sand)
Life Form: Perennial herb
Blooming: June-July
Notes: See *Brittonia* 29:376-381 (1977) for original description.

Astragalus subvestitus (Jeps.) Barneby

"Kern County milk-vetch" Fabaceae
CNPS List: 4 **R-E-D Code:** 1-1-3 **State/Fed. Status:** /C3c
Distribution: KRN, TUL
Habitat: GBScr, Medws (xeric), PJWld / gravelly or sandy
Life Form: Perennial herb
Blooming: June-July
Notes: See *Flora of California* 2(1):361 (1936) by W.L. Jepson for original description, and *Memoirs of the New York Botanical Garden* 13:685-686 (1964) for revised nomenclature.

Astragalus tegetarioides

See *Astragalus anxius*

Astragalus tener Gray var. *ferrisiae* Liston

"Ferris's milk-vetch" Fabaceae
CNPS List: 1B **R-E-D Code:** 3-3-3 **State/Fed. Status:** /C2
Distribution: BUT, COL*, SOL*, SUT*, YOL*
Quads: 497B*, 497C*, 498D*, 530B*, 544A*, 546A*, 546D*, 560B, 561A, 561B, 593C*
Habitat: Medws (vernally mesic), VFGrs (subalkaline flats)
Life Form: Annual herb
Blooming: April-May
Notes: Rediscovered in 1989 by V. Oswald in Butte Sink WA (DFG); known only from two extant occurrences. Most historical habitat destroyed by agriculture. See *Brittonia* 42(2):100-104 (1990) for original description, and *Systematic Botany* 17(3):367-379 (1992) for distributional information.

Astragalus tener Gray var. *tener*

"alkali milk-vetch" Fabaceae
CNPS List: 1B **R-E-D Code:** 3-2-3 **State/Fed. Status:** CEQA
Distribution: ALA*, CCA*, MER, MNT*, NAP*, SBT*, SCL*, SFO*, SJQ*, SOL, SON*, STA*, YOL
Quads: 365B*, 385B*, 385C*, 403A, 423B*, 423C*, 423D, 424A*, 427B*, 428A*, 446A*, 447A*, 447B*, 447D*, 462A*, 463C*, 465A*, 465C*, 466A*, 466C*, 466D*, 481B, 484B*, 497B*, 498A*, 498C, 498D, 499D*, 500D*, 513B, 513C*, 514A*, 514D*
Habitat: Plyas, VFGrs (adobe clay), VnPls (alkaline)
Life Form: Annual herb
Blooming: March-June
Notes: Last Bay Area collection in 1959. Threatened by habitat destruction, especially agricultural conversion, and protected only at Jepson Prairie Preserve (TNC), SOL Co. See *Proceedings of the American Academy of Arts and Sciences* 6:206 (1864) for original description, and *Systematic Botany* 17(3):367-379 (1992) for distributional information.

Astragalus tener Gray var. *titi* (Eastw.) Barneby

"coastal dunes milk-vetch" Fabaceae
CNPS List: 1B **R-E-D Code:** 3-3-3 **State/Fed. Status:** CE/C1
Distribution: LAX*, MNT, SDG
Quads: 11B*, 22B, 36B, 90A*, 112D*, 366C
Habitat: CBScr (sandy), CoDns
Life Form: Annual herb
Blooming: March-May
Notes: Known from only three extant occurrences; SDG Co. occurrences have not been documented since the 1970's. Threatened by urbanization, recreational activities, and non-native plants. See *Bulletin of the Torrey Botanical Club* 32:195-196 (1905) for original description, and *Memoirs of the New York Botanical Garden* 13:1048 (1964) for taxonomic treatment.
Status Report: 1987

Astragalus traskiae Eastw.

"Trask's milk-vetch" Fabaceae
CNPS List: 1B **R-E-D Code:** 3-2-3 **State/Fed. Status:** CR/C2
Distribution: SBR, SNI
Quads: SBRA, SNIC
Habitat: CBScr, CoDns
Life Form: Perennial herb
Blooming: March-July
Notes: Populations on SBR Isl. declined severely (65%) between 1986 and 1989. Possibly threatened by Navy activities on SNI Isl. See *Proceedings of the California Academy of Sciences* III 1:102 (1898) for original description, and *Memoirs of the New York Botanical Garden* 13:461-463 (1964) for taxonomic treatment.
Status Report: 1990

Astragalus tricarinatus Gray

"triple-ribbed milk-vetch" Fabaceae
CNPS List: 1B **R-E-D Code:** 3-1-3 **State/Fed. Status:** /PE
Distribution: RIV, SBD
Quads: 65D, 83A, 83B, 104D
Habitat: JTWld, SDScr / sandy or gravelly
Life Form: Perennial herb
Blooming: February-May
Notes: Known from approximately four occurrences near Whitewater and Morongo Valley. Threatened by vehicles.

Astragalus trichopodus var. *trichopodus*

Considered but rejected: Too common

Astragalus umbraticus Sheld.

"Bald Mtn. milk-vetch" Fabaceae
CNPS List: 4 **R-E-D Code:** 1-1-2 **State/Fed. Status:** CEQA?
Distribution: HUM, OR
Habitat: CmWld
Life Form: Perennial herb
Blooming: May-July
Notes: Possibly threatened by logging. Endangered in OR.

Astragalus webberi Gray

"Webber's milk-vetch" Fabaceae
CNPS List: 1B **R-E-D Code:** 3-2-3 **State/Fed. Status:** /C2
Distribution: PLU, SIE
Quads: 605C, 605D, 606D
Habitat: LCFrs
Life Form: Perennial herb
Blooming: May-July
Notes: Known from approximately ten occurrences. Undocumented in SIE Co.; need quads. See *Botany of California* 1:154 (1876) for original description.

Astrolepis cochisensis (Goodd.) Benham & Windham

"scaly cloak fern" Pteridaceae
CNPS List: 2 **R-E-D Code:** 2-1-1 **State/Fed. Status:** CEQA
Distribution: SBD, AZ, BA+
Quads: 249D
Habitat: JTWld, PJWld / carbonate
Life Form: Perennial herb (rhizomatous)
Fertile: April-October
Notes: Need quad for occurrence in the Providence Mtns. See *Madroño* 25:57 (1978) for distributional information, *Phytologia* 41(6):431-437 (1979) for nomenclature, *American Journal of Botany* 75:138 (1988) for taxonomic discussion, and *American Fern Journal* 82(2):57 (1992) for revised nomenclature.

Atriplex cordulata Jeps.

"heartscale" Chenopodiaceae
CNPS List: 1B **R-E-D Code:** 2-2-3 **State/Fed. Status:** /C2
Distribution: ALA, BUT, CCA*, FRE, GLE, KNG, KRN, MAD, MER, SJQ*, SOL, STA*, TUL
Quads: 215B, 216A, 216C, 287B, 287C, 288A, 288B, 288C, 288D, 359A*, 360A, 381A*, 381D, 401B*, 401D*, 403A, 403B*, 403C*, 403D, 423B, 423C, 424A*, 445B, 463A*, 463D, 481C, 481D*, 561D, 562B
Habitat: ChScr, VFGrs (sandy) / saline or alkaline
Life Form: Annual herb
Blooming: May-October
Notes: Need quads for KNG and SJQ counties.

Atriplex coronata Wats. var. *coronata*

"crownscale" Chenopodiaceae
CNPS List: 4 **R-E-D Code:** 1-2-3 **State/Fed. Status:** CEQA?
Distribution: ALA, CCA, FRE, KNG, KRN, MER, MNT, SJQ?, SLO, STA
Habitat: ChScr?, VFGrs, VnPls / alkaline
Life Form: Annual herb
Blooming: April-October
Notes: Does plant occur in SJQ Co.? Similar to *A. cordulata* and *A. vallicola*. See *Proceedings of the American Academy of Arts and Sciences* 9:114 (1874) for original description.

Atriplex coronata Wats. var. *notatior* Jeps.

"San Jacinto Valley crownscale" Chenopodiaceae
CNPS List: 1B **R-E-D Code:** 3-3-3 **State/Fed. Status:** /C1
Distribution: RIV
Quads: 68A, 84C*, 85A, 85C, 85D
Habitat: Plyas, VnPls / alkaline
Life Form: Annual herb
Blooming: April-August
Notes: Known from one extended but fragmented population in the San Jacinto Valley. Threatened by flood control, agriculture, urbanization, grazing, and vehicles. See *Manual of the Flowering Plants of California*, p. 325 (1925) by W.L. Jepson for original description.
Status Report: 1988

Atriplex coulteri (Moq.) D. Dietr.

"Coulter's saltbush" Chenopodiaceae
CNPS List: 1B **R-E-D Code:** 2-2-2 **State/Fed. Status:** CEQA
Distribution: ANA, LAX, ORA, RIV, SBA, SBD, SCM, SCT, SCZ, SDG, SMI, SRO, BA
Quads: 10C, 11A, 11B, 11D, 22C, 34A, 49D, 52B, 70C, 71B, 71D, 87B, 113D, 142A, 142B, 143A, 145B, 170C, ANAC, SCMC, SCTN, SCTS, SCTW, SCZA, SCZB, SCZC, SMIE, SROE, SRON
Habitat: CBScr, CoDns, CoScr, VFGrs / alkaline or clay
Life Form: Perennial herb
Blooming: March-October
Notes: Few recent sightings. Need quads for RIV Co. Threatened by development, and probably by feral herbivores.

Atriplex depressa Jeps.

"brittlescale" Chenopodiaceae
CNPS List: 1B **R-E-D Code:** 2-2-3 **State/Fed. Status:** CEQA
Distribution: ALA, CCA, COL, FRE, GLE, KRN, MAD, MER, SOL, STA*, TUL, YOL
Quads: 265A, 288D, 334B*, 334D*, 335B*, 358C*, 359A, 359B, 359D*, 360A, 381A*, 401B*, 401D*, 402A, 403A, 423B*, 424A*, 443B*, 445B, 463C, 481B, 498D, 513B, 514A*, 546A, 547A, 562B, 562D, 578C*
Habitat: ChScr, Plyas, VFGrs / alkaline or clay
Life Form: Annual herb
Blooming: May-October
Notes: Closely related to *A. minuscula* and *A. parishii*; a synonym of the latter in *A California Flora* (1959) by P. Munz. See *Pittonia* 2:304 (1892) for original description.

Atriplex joaquiniana A. Nels.

"San Joaquin spearscale" Chenopodiaceae
CNPS List: 1B **R-E-D Code:** 2-2-3 **State/Fed. Status:** /C2
Distribution: ALA, CCA, COL, GLE, MER, NAP, SAC, SBT, SCL*, SJQ*, SOL, TUL*, YOL
Quads: 340D, 385B*, 385C*, 403A, 403B, 406D*, 423C, 427B*, 445B, 462A*, 463C, 463D, 464A, 464B*, 464C*, 465D*, 480B*, 480C*, 481D*, 483A, 498C*, 498D, 499D*, 500D, 511A, 513B*, 530B*, 547A, 547C, 562B, 564A, 578C*
Habitat: ChScr, Medws, VFGrs / alkaline
Life Form: Annual herb
Blooming: April-September
Notes: Need historical quads for TUL Co. Threatened by grazing, agriculture, and development.

Atriplex minuscula Standl.

"lesser saltscale" Chenopodiaceae
CNPS List: 1B **R-E-D Code:** 3-3-3 **State/Fed. Status:** CEQA
Distribution: FRE*, KRN, MAD, MER*, TUL*
Quads: 241B, 287C*, 288D*, 334C*, 334D*, 359B*, 359D*, 381A, 381C, 401B*, 401C, 401D*, 402A*
Habitat: ChScr, Plyas, VFGrs / alkaline
Life Form: Annual herb
Blooming: May-October
Notes: Known from fewer than five extant occurrences. Historical occurrences extirpated by agriculture. Closely related to *A. depressa* and *A. parishii*; a synonym of the latter in *A California Flora* (1959) by P. Munz. See *North American Flora* 21:51 (1916) for original description.

Atriplex pacifica Nels.

"South Coast saltscale" Chenopodiaceae
CNPS List: 1B **R-E-D Code:** 3-2-2 **State/Fed. Status:** /C2
Distribution: ANA, LAX, ORA*, RIV, SCM, SCT, SCZ, SDG, SNI, SRO, VEN*, BA
Quads: 10C(*?), 11B, 22A, 22B, 22C*, 52A, 52D, 68A, 71B*, 71D*, 73A, 85C, 85D, 90C*, 141D*, ANAC, SCTE, SCTN, SCTS, SCZC, SNIC
Habitat: CBScr, CoScr, Plyas
Life Form: Annual herb
Blooming: March-October
Notes: Many known occurrences extirpated; need information. Need quads for SCM and SRO islands. Greatly reduced by urbanization on mainland. See *Proceedings of the Biological Society of Washington* 17:99 (1904) for original description.

Atriplex parishii Wats.

"Parish's brittlescale" Chenopodiaceae
CNPS List: 1B **R-E-D Code:** 3-3-2 **State/Fed. Status:** /C2
Distribution: LAX*, ORA*, RIV, SBD*, SDG*, BA
Quads: 66D*, 68A, 71B?*, 71D*, 72A?*, 83D*, 85C, 89D*, 90C*, 111A*, 111C*, 131D*
Habitat: ChScr, Plyas, VnPls
Life Form: Annual herb
Blooming: June-October
Notes: Plant collected only once (1993) in CA since 1974; probably still extant in BA. Threatened by development, agricultural conversion, and grazing. Taxonomic reevaluation indicates plant is only from southern California, but is closely related to more northern *A. depressa* and *A. minuscula.* See *Proceedings of the American Academy of Arts and Sciences* 17:377 (1882) for original description.

Atriplex patula ssp. *spicata*

See *Atriplex joaquiniana*

Atriplex serenana Nels. var. *davidsonii* (Standl.) Munz

"Davidson's saltscale" Chenopodiaceae
CNPS List: 1B **R-E-D Code:** 3-2-2 **State/Fed. Status:** CEQA
Distribution: LAX(*?), ORA, RIV, SBA, SDG, SRO, VEN, BA
Quads: 71B, 71D, 73A(*?), 85D, 88A, 140B
Habitat: CBScr, CoScr / alkaline
Life Form: Annual herb
Blooming: April-October
Notes: Is plant extirpated from LAX Co.? Need quads for SBA and SDG counties and SRO Isl. See *North American Flora* 21:57 (1916) for original description.

Atriplex tularensis Cov.

"Bakersfield smallscale" Chenopodiaceae
CNPS List: 1B **R-E-D Code:** 3-3-3 **State/Fed. Status:** CE/C2*
Distribution: KRN(*?)
Quads: 214B?, 215A*, 239C*
Habitat: ChScr
Life Form: Annual herb
Blooming: June-October
Notes: Possibly extinct. Three historical occurrences extirpated by agriculture; only remaining occurrence at Kern Lake Preserve (TNC) is probably an undescribed form of *A. serenana,* not *A. tularensis* as thought. Immediate taxonomic study warranted. Threatened by lowering of water table. See *Contributions from the U.S. National Herbarium* 4:182 (1893) for original description, and *Fremontia* 19(2):15-18 (1991) for species account and discussion of management.
Status Report: 1993

Atriplex vallicola Hoov.

"Lost Hills crownscale" Chenopodiaceae
CNPS List: 1B **R-E-D Code:** 2-2-3 **State/Fed. Status:** /C2
Distribution: FRE, KNG, KRN, MER, SLO
Quads: 218A, 218B, 242A, 242D, 243C, 265B*, 265D, 266A, 289C, 290D, 359B, 381C, 383B
Habitat: ChScr, VFGrs, VnPls / alkaline
Life Form: Annual herb
Blooming: May-August
Notes: Threatened by grazing and agricultural conversion. Plants from SLO Co. are probably an unnamed new taxon. See *Leaflets of Western Botany* 2(8):130-131 (1938) for original description.
Status Report: 1988

Ayenia compacta Rose

"ayenia" Sterculiaceae
CNPS List: 2 **R-E-D Code:** 2-1-1 **State/Fed. Status:** CEQA
Distribution: RIV, SBD, SDG, BA
Quads: 7C, 19A, 19B, 20D, 32A, 32B, 32C, 32D, 33A, 47D, 61B, 62B, 65C, 65D, 66A, 79D, 83D, 176A
Habitat: MDScr, SDScr / washes
Life Form: Perennial herb
Blooming: March-April

Azolla mexicana C. Presl

"Mexican mosquito fern" Azollaceae
CNPS List: 4 **R-E-D Code:** 1-2-1 **State/Fed. Status:** CEQA?
Distribution: BUT, KRN, LAK, MOD, NEV, PLU, SCL, SDG, TUL, AZ, BA, GU, NV, OR, ++
Habitat: MshSw (ponds, slow water)
Life Form: Annual/Perennial herb
Fertile: August
Notes: Too common? Difficult to distinguish from *A. filiculoides*, which is common. See *American Fern Journal* 34(3):69-84 (1944) for a review of New World *Azolla.*

Baccharis malibuensis

Considered but rejected: Not yet published

Baccharis plummerae

See *Baccharis plummerae* ssp. *plummerae*

Baccharis plummerae Gray ssp. *glabrata* Hoov.

"San Simeon baccharis" Asteraceae
CNPS List: 1B **R-E-D Code:** 2-2-3 **State/Fed. Status:** CEQA
Distribution: SLO
Quads: 271A, 271B, 272A
Habitat: CoScr
Life Form: Shrub (deciduous)
Blooming: June
Notes: Known only from San Simeon and Arroyo de la Cruz. Probably threatened by grazing. See *Vascular Plants of San Luis Obispo County*, p. 302 (1970) by R. Hoover for original description.

Baccharis plummerae Gray ssp. *plummerae*

"Plummer's baccharis" Asteraceae
CNPS List: 4 **R-E-D Code:** 1-1-3 **State/Fed. Status:** CEQA?
Distribution: ANA, LAX, SBA, SCZ, VEN
Habitat: BUFrs, Chprl, CmWld, CoScr / rocky
Life Form: Shrub (deciduous)
Blooming: August-October

Baccharis vanessae Beauch.

"Encinitas baccharis" Asteraceae
CNPS List: 1B **R-E-D Code:** 2-3-3 **State/Fed. Status:** CE/PE
Distribution: SDG
Quads: 22A, 34C, 35C, 35D, 36D, 51B
Habitat: Chprl (sandstone)
Life Form: Shrub (deciduous)
Blooming: August-November
Notes: Known from fewer than twenty occurrences. Almost extirpated from Encinitas area. Threatened by development and recreation. See *Phytologia* 46(4):216-222 (1980) for original description, and *Madroño* 40(2):133 (1993) for range extension information.
Status Report: 1987

Bacopa nobsiana

Considered but rejected: A synonym of *B. rotundifolia*; a common, non-native taxon

Balsamorhiza hookeri Nutt. var. *lanata* Sharp

"woolly balsamroot" Asteraceae
CNPS List: 1B **R-E-D Code:** 3-3-3 **State/Fed. Status:** CEQA
Distribution: SIS
Quads: 716C, 717B
Habitat: CmWld
Life Form: Perennial herb
Blooming: April-June
Notes: Known only from the Shasta Valley area. Probably reduced by grazing. See *Annals of the Missouri Botanical Garden* 22:130 (1935) for original description.

Balsamorhiza macrolepis Sharp var. *macrolepis*

"big-scale balsamroot" Asteraceae
CNPS List: 1B **R-E-D Code:** 2-2-3 **State/Fed. Status:** CEQA
Distribution: ALA, BUT, MPA, NAP, PLA, SCL, TEH
Quads: 406D, 438D, 446A, 447A, 465C, 482B, 528A, 528D, 575A, 630D
Habitat: CmWld, VFGrs / sometimes serpentinite
Life Form: Perennial herb
Blooming: March-June
Notes: See *Annals of the Missouri Botanical Garden* 22:132 (1935) for original description.

Balsamorhiza sericea W.A. Weber

"silky balsamroot" Asteraceae
CNPS List: 4 **R-E-D Code:** 2-1-2 **State/Fed. Status:** /C2
Distribution: SIS, TRI, OR
Habitat: LCFrs (serpentinite)
Life Form: Perennial herb
Blooming: April-May
Notes: On watch list in OR. Perhaps not distinct from *B. macrolepis* var. *platylepis.* See *Phytologia* 50(5):357-359 (1982) for original description.

Benitoa occidentalis

See *Lessingia occidentalis*

Bensoniella oregona (Abrams & Bacig.) Morton

"bensoniella" Saxifragaceae
CNPS List: 1B **R-E-D Code:** 3-3-2 **State/Fed. Status:** CR/C2
Distribution: HUM, OR
Quads: 653A, 671D
Habitat: BgFns, LCFrs (openings), Medws / mesic
Life Form: Perennial herb
Blooming: July
Notes: Known in CA from fewer than ten occurrences. Threatened by logging and grazing. Candidate for state listing in OR. See *Contributions from the Dudley Herbarium* 1:95 (1929) for original description, and *Leaflets of Western Botany* 10:181 (1965) for revised nomenclature.
Status Report: 1987

Berberis fremontii Torr.

"Fremont barberry" Berberidaceae
CNPS List: 3 **R-E-D Code:** ?-?-1 **State/Fed. Status:** CEQA?
Distribution: SBD, SDG, AZ, BA, NV, SO, ++
Quads: 7B, 7C, 8A*, 9B, 131D*, 200A, 200B, 224C, 225D
Habitat: Chprl, JTWld, PJWld / rocky
Life Form: Shrub (evergreen)
Blooming: April-June
Notes: Taxonomy extremely complex; as treated here includes *B. haematocarpa* and *B. higginsiae*, but these may be distinct taxa. Definitive study needed. See *Report on the U.S. and Mexican Boundary Survey*, p. 30 (1859) by W. Emory for original description, and *Phytologia* 52:221-226 (1982) for relationship between *Berberis* and *Mahonia*.

Berberis higginsiae

See *Berberis fremontii*

***Berberis nevinii* Gray**

"Nevin's barberry" Berberidaceae
CNPS List: 1B **R-E-D Code:** 3-3-3 **State/Fed. Status:** CE/C1
Distribution: LAX, RIV, SBD, SDG
Quads: 33A, 49A, 49B, 49C, 50A, 67C, 106C, 108B, 110B, 111A, 137C*, 137D*, 138A, 163D
Habitat: Chprl, CmWld, CoScr, RpScr / sandy or gravelly
Life Form: Shrub (evergreen)
Blooming: March-April
Notes: Many historical occurrences have been extirpated. Threatened by development and road maintenance. Angeles NF has adopted species management guidelines. State-listed as *Mahonia nevinii*. See *Synoptical Flora of North America* 1(1):69 (1895) for original description, and *Journal of the Linnean Society of London* 57(369):354 (1961) for revised nomenclature.
Status Report: 1987

***Berberis pinnata* Lag. ssp. *insularis* Munz**

"island barberry" Berberidaceae
CNPS List: 1B **R-E-D Code:** 3-2-3 **State/Fed. Status:** CE/C1
Distribution: ANA*, SCZ, SRO
Quads: ANAC*, SCZA, SCZB, SROW
Habitat: CCFrs, CmWld, CoScr
Life Form: Shrub (evergreen)
Blooming: March
Notes: Known from fewer than ten occurrences. Last seen on ANA Isl. in 1978. Threatened by feral herbivores. State-listed as *Mahonia pinnata* ssp. *insularis*. See *Changing Seasons* 1(3):14 (1981) for revised nomenclature.
Status Report: 1987

Berberis sonnei

Considered but rejected: A synonym of *B. aquifolium* var. *repens*; a common taxon

***Bergerocactus emoryi* (Engelm.) Britt. & Rose**

"golden-spined cereus" Cactaceae
CNPS List: 2 **R-E-D Code:** 2-2-1 **State/Fed. Status:** CEQA
Distribution: ORA*, SCM, SCT, SDG, BA
Quads: 10C, 11A, 11B, 11D, 22B, SCMC, SCMN, SCMS, SCTN, SCTS, SCTW
Habitat: CCFrs, Chprl, CoScr / sandy
Life Form: Shrub (stem succulent)
Blooming: May-June
Notes: Need historical quads for ORA Co. Threatened by development and horticultural collecting.

***Blennosperma bakeri* Heiser**

"Sonoma sunshine" Asteraceae
CNPS List: 1B **R-E-D Code:** 2-3-3 **State/Fed. Status:** CE/FE
Distribution: SON
Quads: 483B, 500C, 501B, 501C, 501D, 502A
Habitat: VFGrs (mesic), VnPls
Life Form: Annual herb
Blooming: March-April
Notes: Threatened by urbanization, grazing, and agriculture. See *Madroño* 9:103-104 (1947) for original description.
Status Report: 1977

***Blennosperma nanum* (Hook.) Blake var. *robustum* J.T. Howell**

"Point Reyes blennosperma" Asteraceae
CNPS List: 1B **R-E-D Code:** 3-2-3 **State/Fed. Status:** CR/C2
Distribution: MEN, MRN
Quads: 485B, 485C, 569A
Habitat: CoPrr, CoScr
Life Form: Annual herb
Blooming: March
Notes: Known from fewer than fifteen occurrences. Possibly threatened by grazing. Plants found near Fort Bragg in 1988 appear closest to this taxon. Some Pt. Reyes populations intermediate to var. *nanum*. See *Leaflets of Western Botany* 5:105-108 (1948) for original description.
Status Report: 1988

***Blepharidachne kingii* (Wats.) Hack.**

"King's eyelash grass" Poaceae
CNPS List: 2 **R-E-D Code:** 2-1-1 **State/Fed. Status:** CEQA
Distribution: INY, MNO, NV+
Quads: 302A, 327C, 347C, 390C, 412D, 431A
Habitat: PJWld (carbonate)
Life Form: Perennial herb
Blooming: May

Blepharizonia plumosa (Kell.) Greene ssp. *plumosa*

"big tarplant" Asteraceae
CNPS List: 1B **R-E-D Code:** 3-3-3 **State/Fed. Status:** CEQA
Distribution: ALA, CCA(*?), SJQ*, STA*, SOL*
Quads: 443C*, 444B*, 445B, 446A, 462C*, 465A(*?), 481C(*?), 481D(*?)
Habitat: VFGrs
Life Form: Annual herb
Blooming: July-October
Notes: Need historical quads for SOL Co. Very limited; possibly extant only on private property in the hills near Livermore, ALA Co. Historical occurrences probably extirpated by agriculture and non-native plants. See *Proceedings of the California Academy of Sciences* I 5:49 (1873) for original description.

Bloomeria humilis Hoov.

"dwarf goldenstar" Liliaceae
CNPS List: 1B **R-E-D Code:** 3-2-3 **State/Fed. Status:** CR/C1
Distribution: MNT?, SLO
Quads: 272A
Habitat: CBScr, Chprl, VFGrs
Life Form: Perennial herb (bulbiferous)
Blooming: June
Notes: Known in SLO Co. from only two occurrences in the Arroyo de la Cruz area. Does plant occur in MNT Co.? See *Plant Life* 11:20-22 (1955) for original description.
Status Report: 1988

Boerhavia annulata

Considered but rejected: Too common; a synonym of *Anulocaulis annulatus*

Bolandra californica Gray

"Sierra bolandra" Saxifragaceae
CNPS List: 4 **R-E-D Code:** 1-1-3 **State/Fed. Status:** CEQA?
Distribution: ALP, AMA, CAL, ELD, MPA, STA, TUO
Habitat: LCFrs, UCFrs / mesic, rocky
Life Form: Perennial herb
Blooming: June-July

Boschniakia hookeri Walp.

"small groundcone" Orobanchaceae
CNPS List: 2 **R-E-D Code:** 3-1-1 **State/Fed. Status:** CEQA
Distribution: DNT, HUM, MEN, MRN, OR, WA+
Quads: 467A, 467B, 550D
Habitat: NCFrs
Life Form: Perennial herb (rhizomatous, parasitic)
Blooming: April-August
Notes: How rare is plant outside CA? Need quads for DNT and HUM counties. Parasitic on *Gaultheria shallon* and *Vaccinium* spp. See *Report of the Pacific Railroad Expedition* 3:479 (1844-45) for original description.

Botrychium ascendens W. Wagner

"upswept moonwort" Ophioglossaceae
CNPS List: 2 **R-E-D Code:** 3-1-1 **State/Fed. Status:** /C2
Distribution: BUT, ELD, TEH, NV, OR, ++
Quads: 607B, 607C
Habitat: LCFrs (mesic)
Life Form: Perennial herb (rhizomatous)
Fertile: July-August
Notes: Known in CA from only two occurrences: near Jonesville on the BUT and TEH Co. border, and south of Fallen Leaf Lake, ELD Co. Candidate for state listing in OR. See *American Fern Journal* 76(2):33-47 (1986) for original description, and *Madroño* 36(2):131-136 (1989) for discussion of Jonesville records.

Botrychium crenulatum W. Wagner

"scalloped moonwort" Ophioglossaceae
CNPS List: 1B **R-E-D Code:** 2-1-2 **State/Fed. Status:** /C2
Distribution: BUT, LAX, MOD?, SBD, TEH, TUL, ID, OR, UT+
Quads: 105A, 105B, 135D, 306C, 607B, 607C
Habitat: BgFns, LCFrs, Medws, MshSw (freshwater)
Life Form: Perennial herb (rhizomatous)
Fertile: June-July
Notes: Scattered but not common anywhere in CA. Does plant occur in MOD Co.? On review list in ID, and a candidate for state listing in OR. See *American Fern Journal* 71(1):20-30 (1981) for original description, and *Madroño* 36(2):131-136 (1989) for distributional information.

Botrychium lunaria (L.) Sw.

"moonwort" Ophioglossaceae
CNPS List: 2 **R-E-D Code:** 3-1-1 **State/Fed. Status:** CEQA
Distribution: MNO, MOD, TUO, NV, OR, WA, ++
Quads: 690B
Habitat: Medws, SCFrs, UCFrs
Life Form: Perennial herb (rhizomatous)
Fertile: August
Notes: Need quads for MNO and TUO counties. Endangered in OR, and state-listed as Sensitive in WA.

Botrychium minganense Victorin

"Mingan moonwort" Ophioglossaceae
CNPS List: 2 **R-E-D Code:** 3-1-1 **State/Fed. Status:** CEQA
Distribution: BUT, FRE, TEH, AZ, ID, OR, WA, ++
Quads: 394D, 607B, 607C
Habitat: LCFrs (mesic)
Life Form: Perennial herb (rhizomatous)
Fertile: July-August
Notes: Threatened in ID, endangered in OR, and state-listed as Sensitive in WA. See *Bulletin of the Torrey Botanical Club* 83(4):261-280 (1956) for comparison with *B. lunaria,* and *Madroño* 36(2):131-136 (1989) for CA records.

Botrychium montanum W. Wagner

"western goblin" Ophioglossaceae
CNPS List: 2 **R-E-D Code:** 3-1-1 **State/Fed. Status:** CEQA
Distribution: BUT, TEH, OR, WA, ++
Quads: 607B, 607C
Habitat: LCFrs (mesic)
Life Form: Perennial herb (rhizomatous)
Fertile: July-August
Notes: Endangered in OR, and state-listed as Sensitive in WA. See *American Fern Journal* 71(1):20-30 (1981) for original description, and *Madroño* 36(2):131-136 (1989) for first CA records.

Botrychium pinnatum St. John

"northwestern moonwort" Ophioglossaceae
CNPS List: 2 **R-E-D Code:** 3-1-1 **State/Fed. Status:** CEQA
Distribution: SIS, ID, NV, OR, WA, ++
Quads: 701D
Habitat: LCFrs?, Medws
Life Form: Perennial herb (rhizomatous)
Fertile: September-October
Notes: Known in CA only from near Etna. Threatened in ID, endangered in OR, and state-listed as Sensitive in WA. See *American Fern Journal* 19:11-15 (1929) for original description.

Botrychium pumicola

Considered but rejected: Not in CA; only known CA occurrence on Mt. Shasta a misidentification

Bouteloua simplex

Considered but rejected: Not in CA

Bouteloua trifida Thurb.

"red grama" Poaceae
CNPS List: 2 **R-E-D Code:** 3-1-1 **State/Fed. Status:** CEQA
Distribution: INY, SBD, AZ, NV, SO, ++
Quads: 176A, 201A, 201C, 225D, 249A, 274C, 298D, 323B
Habitat: MDScr (carbonate, crevices)
Life Form: Perennial herb
Blooming: May-September
Notes: See *Proceedings of the American Academy of Arts and Sciences* 18:177 (1883) for original description, and *Annals of the Missouri Botanical Garden* 66(3):399-400 (1979) for taxonomic treatment.
Status Report: 1987

Boykinia rotundifolia Parry

"round-leaved boykinia" Saxifragaceae
CNPS List: 4 **R-E-D Code:** 1-1-3 **State/Fed. Status:** CEQA?
Distribution: LAX, ORA, RIV, SBA, SBD, SDG, VEN
Habitat: Chprl (mesic), RpWld
Life Form: Perennial herb
Blooming: June-July

Brickellia frutescens

Considered but rejected: Too common

Brickellia knappiana

Considered but rejected: A hybrid

Brickellia nevinii

Considered but rejected: Too common

Brodiaea coronaria (Salisb.) Engler ssp. *rosea* (Greene) Niehaus

"Indian Valley brodiaea" Liliaceae
CNPS List: 1B **R-E-D Code:** 3-3-3 **State/Fed. Status:** CE/C2
Distribution: COL, GLE, LAK, TEH
Quads: 547B, 547C, 548A, 548C, 564A, 596B, 596C
Habitat: CCFrs, Chprl, VFGrs / serpentinite
Life Form: Perennial herb (bulbiferous)
Blooming: May-June
Notes: Known from fewer than twenty occurrences. Threatened by vehicles and dumping. Protected in part at ACEC (BLM), LAK Co. See *Bulletin of the California Academy of Sciences* 2:137 (1886) for original description, and *American Midland Naturalist* 22:560-561 (1939) for revised nomenclature.
Status Report: 1987

Brodiaea filifolia Wats.

"thread-leaved brodiaea" Liliaceae
CNPS List: 1B **R-E-D Code:** 3-3-3 **State/Fed. Status:** CE/C1
Distribution: LAX, ORA, RIV, SBD, SDG
Quads: 35B, 35C, 36A, 36D, 65C, 68A, 68B, 69A, 69C, 69D, 70D, 85C, 85D, 107A, 109A
Habitat: CoScr, CmWld, VFGrs, VnPls / clay
Life Form: Perennial herb (bulbiferous)
Blooming: March-June
Notes: Seriously threatened by residential development, agriculture, and vehicles. May hybridize with *B. orcuttii* and *B. terrestris* ssp. *kernensis*. See *Proceedings of the American Academy of Arts and Sciences* 17:381 (1882) for original description.
Status Report: 1987

Brodiaea insignis (Jeps.) Niehaus

"Kaweah brodiaea" Liliaceae
CNPS List: 1B **R-E-D Code:** 2-2-3 **State/Fed. Status:** CE/C2
Distribution: TUL
Quads: 308A, 308B, 308C, 309D, 332A, 332B, 332C, 332D, 333A
Habitat: CmWld, VFGrs / granitic or clay
Life Form: Perennial herb (bulbiferous)
Blooming: April-June
Notes: Known only from the Tule and Kaweah River drainages. Threatened by residential development, road maintenance, grazing, and non-native plants. See *Flora of California* 1(6):288 (1922) by W.L. Jepson for original description, and *University of California Publications in Botany* 60:52 (1971) for revised nomenclature.
Status Report: 1993

Brodiaea kinkiensis Niehaus

"San Clemente Island brodiaea" Liliaceae
CNPS List: 1B **R-E-D Code:** 3-2-3 **State/Fed. Status:** /C2
Distribution: SCM
Quads: SCMC, SCMS
Habitat: VFGrs (clay)
Life Form: Perennial herb (bulbiferous)
Blooming: May-June
Notes: Known from fewer than twenty occurrences. Feral herbivores removed from SCM Isl., and vegetation recovering. See *Madroño* 19:223-225 (1966) for original description.

Brodiaea leptandra
Considered but rejected: Too common; a synonym of *B. californica* var. *leptandra*

Brodiaea orcuttii (Greene) Baker
"Orcutt's brodiaea" Liliaceae
CNPS List: 1B **R-E-D Code:** 1-3-2 **State/Fed. Status:** /C2
Distribution: ORA, RIV, SDG, BA
Quads: 9B, 10A, 10C, 20A, 20C, 21B, 21C, 22A, 22B, 22C, 22D, 33C, 33D, 34A, 34D, 35A, 35B, 35C, 49D, 50B, 51A, 65C, 68C, 69C, 69D
Habitat: CCFrs, Chprl, CmWld, Medws, VFGrs, VnPls / clay, sometimes serpentinite
Life Form: Perennial herb (bulbiferous)
Blooming: May-July
Notes: Historical occurrences need field surveys. Santa Rosa Plateau (RIV Co.) populations are rare and declining. Threatened by development, vehicles, road construction, and dumping. May hybridize with *B. filifolia.* See *Bulletin of the California Academy of Sciences* 2:138 (1886) for original description, and *University of California Publications in Botany* 60:59-61 (1971) for taxonomic treatment.
Status Report: 1977

Brodiaea pallida Hoov.
"Chinese Camp brodiaea" Liliaceae
CNPS List: 1B **R-E-D Code:** 3-3-3 **State/Fed. Status:** CE/C1
Distribution: TUO
Quads: 458B, 458C
Habitat: VFGrs (vernal streambeds, serpentinite)
Life Form: Perennial herb (bulbiferous)
Blooming: May-June
Notes: Known from only one extended occurrence near Chinese Camp. Threatened by residential development. Hybridizes with *B. elegans.* See *Leaflets of Western Botany* 2:129-130 (1938) for original description.
Status Report: 1977

Bromus polyanthus
Considered but rejected: A synonym of *B. carinatus* var. *carinatus*; a common taxon

Buddleja utahensis
Considered but rejected: Too common

Bursera hindsiana
Considered but rejected: Unable to verify presumed CA occurrence in southern SDG Co.

Bursera microphylla Gray
"elephant tree" Burseraceae
CNPS List: 2 **R-E-D Code:** 3-2-1 **State/Fed. Status:** CEQA
Distribution: IMP, SDG, AZ, BA
Quads: 7A, 18B, 18C, 19A, 31C, 31D, 46B
Habitat: SDScr (rocky)
Life Form: Tree (deciduous)
Blooming: June-July
Notes: Known in CA from fewer than twenty occurrences. Threatened by horticultural collecting. State-listed as Highly Safeguarded in AZ.

Cakile edentula ssp. *californica*
Considered but rejected: Not native; a synonym of *C. edentula*

Calamagrostis bolanderi Thurb.
"Bolander's reed grass" Poaceae
CNPS List: 4 **R-E-D Code:** 1-1-3 **State/Fed. Status:** CEQA?
Distribution: HUM, MEN, SON
Habitat: BgFns, CCFrs, CoScr, Medws?, MshSw (freshwater)
Life Form: Perennial herb (rhizomatous)
Blooming: June-August

Calamagrostis crassiglumis Thurb.
"Thurber's reed grass" Poaceae
CNPS List: 2 **R-E-D Code:** 3-3-1 **State/Fed. Status:** /C2
Distribution: DNT, HUM?, MEN, MRN, SON, WA, ++
Quads: 485C, 502A, 569D, 585D, 740C
Habitat: CoScr (mesic), MshSw (freshwater)
Life Form: Perennial herb (rhizomatous)
Blooming: June-July
Notes: Known in CA from fewer than ten occurrences. Undocumented in HUM Co.; need quads. Threatened by grazing at Pt. Reyes NS. State-listed as Threatened in WA. See *C. stricta* ssp. *inexpansa* in *The Jepson Manual.* See *Botany of California* 2:281 (1880) for original description.

Calamagrostis densa
Considered but rejected: A synonym of *C. koelerioides*; a common taxon

Calamagrostis foliosa Kearn.
"leafy reed grass" Poaceae
CNPS List: 4 **R-E-D Code:** 1-2-3 **State/Fed. Status:** CR/C3c
Distribution: DNT, HUM, MEN
Habitat: CBScr, NCFrs / rocky
Life Form: Perennial herb
Blooming: May-August
Notes: Many occurrences located in the King Range of HUM Co. Threatened by grazing. See *Contributions from the U.S. National Herbarium* 3:83 (1892) for original description.
Status Report: 1987

Calamagrostis ophitidis (J.T. Howell) Nygren
"serpentine reed grass" Poaceae
CNPS List: 4 **R-E-D Code:** 1-1-3 **State/Fed. Status:** CEQA?
Distribution: LAK, MRN, NAP, SON
Habitat: Chprl, Medws, VFGrs / serpentinite
Life Form: Perennial herb
Blooming: April-June

Calamintha chandleri
See *Satureja chandleri*

Calandrinia breweri Wats.

"Brewer's calandrinia" Portulacaceae
CNPS List: 4 **R-E-D Code:** 1-2-2 **State/Fed. Status:** CEQA?
Distribution: CCA, LAX, MEN, MNT, MPA, MRN, NAP, SBA, SBD, SCL, SCR, SCZ, SDG, SLO, SMT, SON, VEN, BA
Habitat: Chprl, CoScr / disturbed sites, burns
Life Form: Annual herb
Blooming: March-June
Notes: How common is this plant? Plant appears to be widely scattered but uncommon everywhere, and most collections are old. Field surveys needed. See *Proceedings of the American Academy of Arts and Sciences* 11:124 (1876) for original description.

Calandrinia maritima Nutt.

"seaside calandrinia" Portulacaceae
CNPS List: 4 **R-E-D Code:** 1-2-1 **State/Fed. Status:** CEQA?
Distribution: ANA, LAX, ORA, SBA, SBR, SCM, SCT, SCZ, SDG, SRO, VEN, BA
Habitat: CBScr, VFGrs / sandy
Life Form: Annual herb
Blooming: March-May

Calliandra eriophylla Benth.

"fairyduster" Fabaceae
CNPS List: 2 **R-E-D Code:** 3-1-1 **State/Fed. Status:** CEQA
Distribution: IMP, SDG, AZ, BA+
Quads: 12B, 13A, 13B, 13C, 13D, 26A, 26C, 26D, 32C, 32D
Habitat: SDScr (sandy)
Life Form: Shrub (deciduous)
Blooming: January-March

Calochortus catalinae Wats.

"Catalina mariposa lily" Liliaceae
CNPS List: 4 **R-E-D Code:** 1-2-3 **State/Fed. Status:** CEQA?
Distribution: LAX, ORA, SBA, SCT, SCZ, SDG, SLO, SRO, VEN
Habitat: Chprl, CmWld, CoScr, VFGrs
Life Form: Perennial herb (bulbiferous)
Blooming: February-May
Notes: Threatened by development.

Calochortus clavatus Wats. var. *avius* Jeps.

"Pleasant Valley mariposa lily" Liliaceae
CNPS List: 1B **R-E-D Code:** 2-2-3 **State/Fed. Status:** /C1
Distribution: AMA, ELD, MPA*
Quads: 420D*, 492A, 508B, 508C, 508D, 509A, 509B*, 524C
Habitat: LCFrs (Josephine silt loam and volcanic)
Life Form: Perennial herb (bulbiferous)
Blooming: June-July
Notes: Threatened by development and logging.

Calochortus clavatus Wats. var. *clavatus*

"club-haired mariposa lily" Liliaceae
CNPS List: 4 **R-E-D Code:** 1-1-3 **State/Fed. Status:** CEQA?
Distribution: LAX, SBA, SBT, SLO
Habitat: Chprl, CmWld, VFGrs / generally serpentinite
Life Form: Perennial herb (bulbiferous)
Blooming: June

Calochortus clavatus Wats. var. *gracilis* Ownbey

"slender mariposa lily" Liliaceae
CNPS List: 1B **R-E-D Code:** 3-2-3 **State/Fed. Status:** /C2
Distribution: LAX
Quads: 108B, 109A, 109B, 135D, 137B, 138A,
Habitat: Chprl
Life Form: Perennial herb (bulbiferous)
Blooming: March
Notes: Threatened by development. See *Annals of the Missouri Botanical Garden* 27(4):371-561 (1940) for original description.

Calochortus clavatus Wats. ssp. *recurvifolius* (Hoov.) Munz

"Arroyo de la Cruz mariposa lily" Liliaceae
CNPS List: 1B **R-E-D Code:** 3-2-3 **State/Fed. Status:** /C2
Distribution: SLO
Quads: 272A
Habitat: CBScr, Chprl (maritime), CoPrr, LCFrs
Life Form: Perennial herb (bulbiferous)
Blooming: June-July
Notes: Known from fewer than ten occurrences. Threatened by grazing. See *Leaflets of Western Botany* 10:121-128 (1964) for original description.
Status Report: 1977

Calochortus coeruleus var. *nanus*

Considered but rejected: A synonym of *C. elegans*; a common taxon

Calochortus coeruleus var. *westonii*

See *Calochortus westonii*

Calochortus concolor

Considered but rejected: Too common

Calochortus dunnii Purdy

"Dunn's mariposa lily" Liliaceae
CNPS List: 1B **R-E-D Code:** 2-2-2 **State/Fed. Status:** CR/C2
Distribution: SDG, BA
Quads: 9C, 10A, 10B, 10D, 19C, 20A, 20D, 21A, 33D, 49C
Habitat: CCFrs, Chprl / gabbroic
Life Form: Perennial herb (bulbiferous)
Blooming: May-June
Notes: See *Proceedings of the California Academy of Sciences* III 2:147 (1901) for original description.
Status Report: 1987

Calochortus elegans var. *oreophyllus*

Considered but rejected: A synonym of *C. elegans*; a common taxon

Calochortus excavatus Greene

"Inyo County star-tulip" Liliaceae
CNPS List: 1B **R-E-D Code:** 2-3-3 **State/Fed. Status:** /C2
Distribution: INY, MNO
Quads: 351B, 351C, 351D, 372B, 372C, 393A, 393B, 393C, 413A, 413B, 413C, 414A, 432C, 432D, 451A, 451D
Habitat: ChScr, Medws (alkaline)
Life Form: Perennial herb (bulbiferous)
Blooming: April-July
Notes: Most occurrences are remnants of former populations and have few plants. Threatened by groundwater pumping, development, road maintenance, and grazing.

Calochortus greenei Wats.

"Greene's mariposa lily" Liliaceae
CNPS List: 1B **R-E-D Code:** 3-2-2 **State/Fed. Status:** /C1
Distribution: MOD, SIS, OR
Quads: 682B, 695D, 716A, 732B, 732C, 733B, 733C, 733D, 734D
Habitat: PJWld, UCFrs / volcanic
Life Form: Perennial herb (bulbiferous)
Blooming: June-August
Notes: Known in CA from fewer than twenty occurrences. Threatened by development, horticultural collecting, grazing, and vehicles. Candidate for state listing in OR. See *Proceedings of the American Academy of Arts and Sciences* 14:264 (1879) for original description.

Calochortus longebarbatus

See *Calochortus longebarbatus* var. *longebarbatus*

Calochortus longebarbatus Wats. var. *longebarbatus*

"long-haired star-tulip" Liliaceae
CNPS List: 1B **R-E-D Code:** 1-2-2 **State/Fed. Status:** /C2
Distribution: MOD, SHA, SIS, OR, WA
Quads: 662B, 662C, 677B, 679B, 680A, 693B, 693C, 694A, 694B, 694C, 694D, 695C, 695D, 696C, 696D, 710D, 711D
Habitat: LCFrs (openings and drainages), Medws / clay
Life Form: Perennial herb (bulbiferous)
Blooming: June-July
Notes: Many occurrences have few plants. Threatened by grazing, and possibly by logging and vehicles. Endangered in OR, and state-listed as Threatened in WA. See *Proceedings of the American Academy of Arts and Sciences* 17:381 (1882) for original description.
Status Report: 1980

Calochortus monanthus Ownbey

"single-flowered mariposa lily" Liliaceae
CNPS List: 1A **Last Seen:** 1876 **State/Fed. Status:** /C2*
Distribution: SIS*
Quads: 734D*
Habitat: Medws
Life Form: Perennial herb (bulbiferous)
Blooming: June
Notes: Known only from the type locality along the Shasta River, now mostly converted to agriculture. Rediscovery attempt in 1990 was unsuccessful. See *Annals of the Missouri Botanical Garden* 27:371-560 (1940) for original description.
Status Report: 1977

Calochortus nudus var. *shastensis*

Considered but rejected: A hybrid

Calochortus obispoensis Lemmon

"San Luis mariposa lily" Liliaceae
CNPS List: 1B **R-E-D Code:** 2-2-3 **State/Fed. Status:** /C3c
Distribution: SLO
Quads: 221A, 221B, 221D*, 246B, 246C, 246D, 247D, 271A
Habitat: Chprl, CoScr, VFGrs / often serpentinite
Life Form: Perennial herb (bulbiferous)
Blooming: May-July
Notes: Threatened by grazing, development, road construction, and recreation, and potentially by mining. See *Journal of Ecology* 75:977-995 (1987) for population biology.

Calochortus palmeri Wats. var. *munzii* Ownbey

"Munz's mariposa lily" Liliaceae
CNPS List: 1B **R-E-D Code:** 3-2-3 **State/Fed. Status:** /C2
Distribution: RIV
Quads: 66A, 66B, 67A, 84D
Habitat: Chprl, LCFrs
Life Form: Perennial herb (bulbiferous)
Blooming: June-July
Notes: Known from only a few occurrences in the San Jacinto Mtns. See *Aliso* 4:88 (1958) for original description.

Calochortus palmeri Wats. var. *palmeri*

"Palmer's mariposa lily" Liliaceae
CNPS List: 1B **R-E-D Code:** 2-2-3 **State/Fed. Status:** /C2
Distribution: KRN, LAX, RIV, SBD, SLO
Quads: 65C, 66D, 104A, 104B, 105A, 105B, 131C, 132C, 133C, 133D, 135C, 212B, 212C, 237A, 238A, 245D
Habitat: Chprl, LCFrs, Medws / mesic
Life Form: Perennial herb (bulbiferous)
Blooming: May-July
Notes: Declining rapidly; occurs in wet meadows which are heavily grazed. See *Proceedings of the American Academy of Arts and Sciences* 14:266 (1879) for original description, and *Annals of the Missouri Botanical Garden* 27:459 (1940) for taxonomic treatment.

Calochortus panamintensis Ownbey (Reveal)

"Panamint mariposa lily" Liliaceae
CNPS List: 4 **R-E-D Code:** 1-1-3 **State/Fed. Status:** CEQA?
Distribution: INY
Habitat: PJWld
Life Form: Perennial herb (bulbiferous)
Blooming: June-July
Notes: See *Annals of the Missouri Botanical Garden* 27(4):371-561 (1940) for original description.

Calochortus persistens Ownbey

"Siskiyou mariposa lily" Liliaceae
CNPS List: 1B **R-E-D Code:** 3-3-3 **State/Fed. Status:** CR/C2
Distribution: SIS
Quads: 717B, 718A, 735D
Habitat: LCFrs, NCFrs / rocky
Life Form: Perennial herb (bulbiferous)
Blooming: June-July
Notes: Threatened by disturbance and non-native plants. Klamath NF has adopted species management guidelines. See *Annals of the Missouri Botanical Garden* 27:448 (1940) for original description.
Status Report: 1987

Calochortus plummerae Greene

"Plummer's mariposa lily" Liliaceae
CNPS List: 1B **R-E-D Code:** 2-2-3 **State/Fed. Status:** /C2
Distribution: LAX, RIV, SBD, VEN
Quads: 66B, 67A, 69A, 84A*, 84B, 85B, 86B, 105C, 105D, 106B, 106C, 106D, 107B, 107C, 108A, 108C*, 109A, 109D?, 110A, 110B*, 110C*, 111D*, 112C, 112D*, 113A, 113C, 113D, 114A, 131D, 133C, 134D*, 135D, 136C*, 136D, 137D, 138A, 138C*, 138D, 140B
Habitat: Chprl, CmWld, CoScr, LCFrs, VFGrs / granitic
Life Form: Perennial herb (bulbiferous)
Blooming: May-July
Notes: Significantly reduced by development, and continues to decline. Less common at higher elevations. Hybridizes with *C. weedii* var. *intermedius* in the San Jose Hills and Puente Hills. See *Pittonia* 2:70 (1890) for original description, and *Annals of the Missouri Botanical Garden* 27:515 (1940) for taxonomic treatment.

Calochortus pulchellus Benth.

"Mt. Diablo fairy-lantern" Liliaceae
CNPS List: 1B **R-E-D Code:** 2-2-3 **State/Fed. Status:** CEQA
Distribution: CCA, SOL?
Quads: 464A, 464B, 464C, 464D, 465A, 465B, 482C, 499A?
Habitat: Chprl, CmWld, VFGrs
Life Form: Perennial herb (bulbiferous)
Blooming: April-June
Notes: Does plant occur in SOL Co.? Threatened by grazing and urbanization. See *Journal of Ecology* 75:977-995 (1987) for population biology.

Calochortus raichei Farwig & Girard

"The Cedars fairy-lantern" Liliaceae
CNPS List: 1B **R-E-D Code:** 3-2-3 **State/Fed. Status:** /C2
Distribution: SON
Quads: 519A, 519B, 519C, 519D
Habitat: CCFrs, Chprl / serpentinite
Life Form: Perennial herb (bulbiferous)
Blooming: May-August
Notes: Endemic to The Cedars near Guerneville. Potentially threatened by mining and road construction. See *Herbertia* 43(1):2-9 (1987) for original description, and *Fremontia* 15(2):18 (1987) for species account.
Status Report: 1990

Calochortus simulans (Hoov.) Munz

"San Luis Obispo mariposa lily" Liliaceae
CNPS List: 4 **R-E-D Code:** 1-1-3 **State/Fed. Status:** /C3c
Distribution: SLO
Habitat: Chprl, CmWld, VFGrs / decomposed granitic
Life Form: Perennial herb (bulbiferous)
Blooming: April-May

Calochortus striatus Parish

"alkali mariposa lily" Liliaceae
CNPS List: 1B **R-E-D Code:** 2-2-2 **State/Fed. Status:** /C2
Distribution: KRN, LAX, SBD, NV
Quads: 102A, 131A, 131B, 131D, 135C, 161A, 161B, 185C, 185D, 186C, 186D, 206C, 215B, 235B?, 235C, 236B, 260A, 260B, 260C, 260D, 283C, 284C
Habitat: Chprl, ChScr, Medws (alkaline)
Life Form: Perennial herb (bulbiferous)
Blooming: April-June
Notes: Threatened by grazing, urbanization, and road construction. On watch list in NV.
Status Report: 1980

Calochortus raichei

Calochortus tiburonensis Hill

"Tiburon mariposa lily" Liliaceae
CNPS List: 1B **R-E-D Code:** 3-3-3 **State/Fed. Status:** CT/PT
Distribution: MRN
Quads: 466B
Habitat: VFGrs (serpentinite)
Life Form: Perennial herb (bulbiferous)
Blooming: March-June
Notes: Known from only one occurrence at Ring Mtn. Preserve (TNC) on the Tiburon Peninsula. See *Madroño* 22(2):100-104 (1973) for original description, and *Journal of Ecology* 75:977-995 (1987) for population biology.
Status Report: 1988

Calochortus umbellatus Wood

"Oakland star-tulip" Liliaceae
CNPS List: 4 **R-E-D Code:** 1-2-3 **State/Fed. Status:** CEQA?
Distribution: ALA, CCA, MRN, SCL, SCR*, SMT
Habitat: BUFrs, Chprl, LCFrs, VFGrs / often serpentinite
Life Form: Perennial herb (bulbiferous)
Blooming: March-May
Notes: Protected at Ring Mtn. Preserve (TNC) on Tiburon Peninsula, MRN Co.

Calochortus uniflorus

Considered but rejected: Too common

Calochortus venustus var. *sanguineus*

Considered but rejected: A synonym of *C. venustus*; a common taxon

Calochortus weedii Wood var. *intermedius* Ownbey

"intermediate mariposa lily" Liliaceae
CNPS List: 1B **R-E-D Code:** 2-2-3 **State/Fed. Status:** /C2
Distribution: LAX, ORA, RIV
Quads: 49B, 52B, 68A, 68D, 69B, 70A, 70B, 70C, 70D, 71A, 71D, 87C, 88A, 88B*, 109D
Habitat: Chprl, CoScr, VFGrs / rocky
Life Form: Perennial herb (bulbiferous)
Blooming: May-July
Notes: Threatened by development, road construction, and fuel modification. Hybridizes with *C. plummerae* in San Jose Hills and Puente Hills. See *Annals of the Missouri Botanical Garden* 27(4):519 (1940) for original description.

Calochortus weedii Wood var. *vestus* Purdy

"late-flowered mariposa lily" Liliaceae
CNPS List: 1B **R-E-D Code:** 2-2-3 **State/Fed. Status:** /C2
Distribution: MNT, SBA, SLO, VEN
Quads: 140B, 141A, 141B, 142A, 142B, 143A, 164A, 166D, 168D, 169D, 190C, 296D
Habitat: Chprl, CmWld
Life Form: Perennial herb (bulbiferous)
Blooming: June-August
Notes: Field studies needed, especially in the Santa Lucia Range. Threatened by grazing and development. See *Proceedings of the California Academy of Sciences* III 2:133 (1901) for original description.

Calochortus westonii Eastw.

"Shirley Meadows star-tulip" Liliaceae
CNPS List: 1B **R-E-D Code:** 3-2-3 **State/Fed. Status:** /C1
Distribution: KRN, TUL
Quads: 258D, 259D, 261A, 285A, 285D, 332A
Habitat: BUFrs, LCFrs, Medws / granitic
Life Form: Perennial herb (bulbiferous)
Blooming: May-June
Notes: Many new occurrences found in 1991. Threatened by logging and road construction. Sequoia NF has adopted species management guidelines. See *Proceedings of the California Academy of Sciences* IV 20:136 (1931) for original description.
Status Report: 1977

Calycadenia fremontii

Considered but rejected: Too common

Calycadenia hooveri Carr

"Hoover's calycadenia" Asteraceae
CNPS List: 1B **R-E-D Code:** 2-1-3 **State/Fed. Status:** /C2
Distribution: CAL, MAD, MER, MPA, STA
Quads: 400A, 420B, 420C, 421A, 440B, 441A, 441B, 459D, 477B
Habitat: CmWld, VFGrs / rocky
Life Form: Annual herb
Blooming: July-September
Notes: Most of occurrences are on private land. See *Brittonia* 27:136 (1975) for original description.

Calycadenia oppositifolia (Greene) Greene

"Butte County calycadenia" Asteraceae
CNPS List: 4 **R-E-D Code:** 1-1-3 **State/Fed. Status:** CEQA?
Distribution: BUT
Habitat: Chprl, CmWld, VFGrs / volcanic or serpentinite
Life Form: Annual herb
Blooming: April-July

Calycadenia tenella

Considered but rejected: Too common

Calycadenia villosa DC.

"dwarf calycadenia" Asteraceae
CNPS List: 1B **R-E-D Code:** 3-3-3 **State/Fed. Status:** CEQA
Distribution: KRN*, MNT, SLO
Quads: 213D*, 244B*, 245A, 245D*, 246C*, 269D*, 270A*, 271B*, 292B*, 294C*, 295B, 318C*
Habitat: Chprl, CmWld, Medws, VFGrs / rocky
Life Form: Annual herb
Blooming: May-October
Notes: Known from only two extant occurrences. Probably consists of northern and southern unrecognized subspecies. Habitat lost to construction of San Antonio Reservoir; also threatened by urbanization, vehicles, and non-native plants.

Calyptridium parryi Gray var. *hesseae* Thomas

"Santa Cruz Mtns. pussypaws" Portulacaceae
CNPS List: 3 **R-E-D Code:** ?-?-3 **State/Fed. Status:** CEQA?
Distribution: MNT, SBT, SCL, SCR*
Quads: 407D*, 408C*, 408D*
Habitat: Chprl, CmWld
Life Form: Annual herb
Blooming: June-July
Notes: Move to List 1B? Location, rarity, and endangerment information needed, especially quads for MNT and SBT counties.

Calyptridium pulchellum (Eastw.) Hoov.

"Mariposa pussypaws" Portulacaceae
CNPS List: 1B **R-E-D Code:** 3-3-3 **State/Fed. Status:** /C2
Distribution: FRE, MAD, MPA
Quads: 397C, 398B, 418C, 418D, 419B
Habitat: CmWld (sandy)
Life Form: Annual herb
Blooming: April-May
Notes: Known from fewer than twenty occurrences. Threatened by development and grazing. USFWS uses the name *Cistanthe pulchella*. See *Bulletin of the Torrey Botanical Club* 29:79 (1902) for original description, *Leaflets of Western Botany* 2(13):222-225 (1940) for revised nomenclature, *Brittonia* 27:197-208 (1975) for taxonomic treatment, and *Madroño* 28:188 (1981) for distributional information.
Status Report: 1977

Calyptridium quadripetalum Wats.

"four-petaled pussypaws" Portulacaceae
CNPS List: 4 **R-E-D Code:** 1-1-3 **State/Fed. Status:** CEQA?
Distribution: GLE, LAK, NAP, SON, TEH, TRI
Habitat: Chprl (sandy or gravelly, usually serpentinite)
Life Form: Annual herb
Blooming: April-June
Notes: See *Proceedings of the American Academy of Arts and Sciences* 20:356 (1885) for original description.

Calystegia atriplicifolia Hallier f. ssp. *buttensis* Brummitt

"Butte County morning-glory" Convolvulaceae
CNPS List: 3 **R-E-D Code:** ?-?-3 **State/Fed. Status:** /C2
Distribution: BUT
Quads: 592D
Habitat: LCFrs
Life Form: Perennial herb (rhizomatous)
Blooming: May-July
Notes: Move to List 1B? Location, rarity, and endangerment information needed. Variable. See *Kew Bulletin* 35(2):327 (1980) for original description.

Calystegia collina (Greene) Brummitt ssp. *oxyphylla* Brummitt

"Mt. Saint Helena morning-glory" Convolvulaceae
CNPS List: 4 **R-E-D Code:** 1-2-3 **State/Fed. Status:** /C2
Distribution: LAK, NAP, SON
Habitat: Chprl (serpentinite)
Life Form: Perennial herb (rhizomatous)
Blooming: May-June
Notes: See *Kew Bulletin* 35(2):328 (1980) for original description.

Calystegia collina (Greene) Brummitt ssp. *venusta* Brummitt

"South Coast Range morning-glory" Convolvulaceae
CNPS List: 4 **R-E-D Code:** 1-1-3 **State/Fed. Status:** /C2
Distribution: FRE, MNT, SBA, SBT
Habitat: Chprl, CmWld, VFGrs / serpentinite or sedimentary
Life Form: Perennial herb (rhizomatous)
Blooming: May-June
Notes: Many new occurrences located in 1991-92. Relatively abundant and tolerant of disturbance. See *Kew Bulletin* 35(2):328 (1980) for original description.

Calystegia macrostegia (Greene) Brummitt ssp. *amplissima* Brummitt

"island morning-glory" Convolvulaceae
CNPS List: 4 **R-E-D Code:** 1-1-3 **State/Fed. Status:** /C2
Distribution: SBR, SCM, SNI
Habitat: CBScr, VFGrs / rocky
Life Form: Perennial herb
Blooming: February-May
Notes: Doing well on SBR and SNI islands. See *Kew Bulletin* 35(2):327 (1980) for original description.

Calystegia malacophylla (Greene) Munz var. *berryi* (Eastw.) Brummitt

"Berry's morning-glory" Convolvulaceae
CNPS List: 4 **R-E-D Code:** 1-1-3 **State/Fed. Status:** CEQA?
Distribution: FRE, TUL
Habitat: Chprl
Life Form: Perennial herb (rhizomatous)
Blooming: July-August
Notes: May hybridize with *C. occidentalis* ssp. *fulcrata.* See *C. malacophylla* var. *malacophylla* in *The Jepson Manual.* See *Kew Bulletin* 29:502 (1974) for taxonomic revision.

Calystegia peirsonii (Abrams) Brummitt

"Peirson's morning-glory" Convolvulaceae
CNPS List: 4 **R-E-D Code:** 1-2-3 **State/Fed. Status:** /C2
Distribution: LAX
Habitat: Chprl, ChScr, CmWld, CoScr, LCFrs
Life Form: Perennial herb (rhizomatous)
Blooming: May-June
Notes: Threatened by grazing. Intergrades with *C. longipes, C. occidentalis* ssp. *occidentalis,* mainland coastal *C. macrostegia* sspp., and perhaps *C. sepium* ssp. *limnophila.* See *Illustrated Flora of the Pacific States* 3:387 (1951) by L. Abrams for original description, and *Annals of the Missouri Botanical Garden* 52(2):214-216 (1965) for revised nomenclature.
Status Report: 1979

***Calystegia sepium* (L.) R. Br. ssp. *binghamiae* (Greene) Brummitt**

"Santa Barbara morning-glory" Convolvulaceae
CNPS List: 1B **R-E-D Code:** 3-3-3 **State/Fed. Status:** CEQA
Distribution: LAX*, ORA, SBA
Quads: 71A, 71B, 72A, 111D*, 142B
Habitat: MshSw (coastal)
Life Form: Perennial herb (rhizomatous)
Blooming: April-May
Notes: Move to List 1A? Need information on current status of known occurrences. Probably reduced by wetland modification and urbanization. See *Bulletin of the California Academy of Sciences* 2:417 (1887) for original description.

Calystegia soldanella

Considered but rejected: Too common

***Calystegia stebbinsii* Brummitt**

"Stebbins's morning-glory" Convolvulaceae
CNPS List: 1B **R-E-D Code:** 3-3-3 **State/Fed. Status:** CE/C2
Distribution: ELD, NEV
Quads: 510B, 526C, 527D, 542A, 542D
Habitat: Chprl (openings), CmWld / serpentinite or gabbroic
Life Form: Perennial herb (rhizomatous)
Blooming: May-June
Notes: Known from fewer than fifteen occurrences. Threatened by development, vehicles, and road maintenance. See *Kew Bulletin* 29:499 (1975) for original description.
Status Report: 1993

***Calystegia subacaulis* H. & A. ssp. *episcopalis* Brummitt**

"Cambria morning-glory" Convolvulaceae
CNPS List: 1B **R-E-D Code:** 3-2-3 **State/Fed. Status:** /C2
Distribution: SLO
Quads: 271D
Habitat: Chprl, CmWld
Life Form: Perennial herb (rhizomatous)
Blooming: April
Notes: Intergrades with ssp. *subacaulis.* See *Kew Bulletin* 35(2):327 (1980) for original description.

***Camissonia benitensis* Raven**

"San Benito evening-primrose" Onagraceae
CNPS List: 1B **R-E-D Code:** 3-3-3 **State/Fed. Status:** /FT
Distribution: FRE, SBT
Quads: 339B, 339C, 339D, 340B, 340D, 363D
Habitat: Chprl, CmWld / serpentinite alluvium
Life Form: Annual herb
Blooming: May-June
Notes: Known only from the New Idria area. Threatened by vehicles. Protected in part at ACEC (BLM). See *Contributions from the U.S. National Herbarium* 37(5):332-333 (1969) for original description.
Status Report: 1988

***Camissonia boothii* (Dougl.) Raven ssp. *alyssoides* (H. & A.) Raven**

"Pine Creek evening-primrose" Onagraceae
CNPS List: 4 **R-E-D Code:** 1-1-1 **State/Fed. Status:** CEQA?
Distribution: LAS, NV
Habitat: GBScr (sandy)
Life Form: Annual herb
Blooming: April-July
Notes: Usually common during favorable springs, as in LAS Co. in 1993. See *Contributions from the U.S. National Herbarium* 37(5):362-364 (1969) for taxonomic treatment.

Camissonia boothii* (Dougl.) Raven ssp. *boothii

"Booth's evening-primrose" Onagraceae
CNPS List: 4 **R-E-D Code:** 1-1-1 **State/Fed. Status:** CEQA?
Distribution: INY, MNO, SBD, AZ, NV
Habitat: JTWld, PJWld
Life Form: Annual herb
Blooming: April-May
Notes: See *Contributions from the U.S. National Herbarium* 37(5):365-366 (1969) for taxonomic treatment.

Camissonia boothii* ssp. *inyoensis

Considered but rejected: Too common

Camissonia graciliflora

Considered but rejected: Too common

***Camissonia guadalupensis* (Wats.) Raven ssp. *clementina* (Raven) Raven**

"San Clemente Island evening-primrose" Onagraceae
CNPS List: 1B **R-E-D Code:** 3-2-3 **State/Fed. Status:** /C2
Distribution: SCM
Quads: SCMC, SCMN
Habitat: CoDns
Life Form: Annual herb
Blooming: April-June
Notes: Known from fewer than ten occurrences. See *Contributions from the U.S. National Herbarium* 37(5):275 (1969) for taxonomic treatment.

***Camissonia hardhamiae* Raven**

"Hardham's evening-primrose" Onagraceae
CNPS List: 1B **R-E-D Code:** 3-2-3 **State/Fed. Status:** /C2
Distribution: MNT, SLO
Quads: 245A, 245B, 246A, 293B
Habitat: Chprl, CmWld / decomposed carbonate
Life Form: Annual herb
Blooming: May
Notes: Known from approximately ten occurrences. Threatened by proposed road construction in SLO Co. See *Contributions from the U.S. National Herbarium* 37(5):301-302 (1969) for original description.

Camissonia heterochroma

Considered but rejected: Too common

Camissonia integrifolia Raven

"Kern River evening-primrose" Onagraceae
CNPS List: 4 **R-E-D Code:** 1-1-3 **State/Fed. Status:** CEQA?
Distribution: KRN
Habitat: Chprl
Life Form: Annual herb
Blooming: May
Notes: See *Contributions from the U.S. National Herbarium* 37(5):344-345 (1969) for original description.

Camissonia kernensis (Munz) Raven ssp. *kernensis*

"Kern County evening-primrose" Onagraceae
CNPS List: 4 **R-E-D Code:** 1-1-3 **State/Fed. Status:** CEQA?
Distribution: KRN
Habitat: JTWld (sandy, granitic)
Life Form: Annual herb
Blooming: May
Notes: See *Contributions from the U.S. National Herbarium* 37(5):309-310 (1969) for taxonomic treatment.

Camissonia lewisii Raven

"Lewis's evening-primrose" Onagraceae
CNPS List: 3 **R-E-D Code:** ?-?-2 **State/Fed. Status:** CEQA?
Distribution: LAX, ORA*, SDG, BA
Quads: 10B, 10C, 11A, 11D, 22C, 34D, 36A, 51A, 70B*, 73A, 90A, 90B, 111A, 111C, 111D, 113D
Habitat: CBScr, CmWld, CoDns, CoScr, VFGrs / sandy or clay
Life Form: Annual herb
Blooming: March-June
Notes: Move to List 4? Location, rarity, and endangerment information needed. Dried material difficult to identify; apparently other taxa are often misidentified as *C. lewisii.* See *Contributions from the U.S. National Herbarium* 37:161-396 (1969) for original description.

Camissonia minor (A. Nels.) Raven

"Nelson's evening-primrose" Onagraceae
CNPS List: 4 **R-E-D Code:** 1-1-1 **State/Fed. Status:** CEQA?
Distribution: LAS, MOD, NV, ++
Habitat: GBScr (sandy)
Life Form: Annual herb
Blooming: May-June
Notes: See *Bulletin of the Torrey Botanical Club* 26:130 (1899) for original description, and *Contributions from the U.S. National Herbarium* 37(5):371-373 (1969) for taxonomic treatment.

Camissonia sierrae Raven ssp. *alticola* Raven

"Mono Hot Springs evening-primrose" Onagraceae
CNPS List: 1B **R-E-D Code:** 3-2-3 **State/Fed. Status:** /C2
Distribution: FRE, MPA
Quads: 395B, 414B, 415B, 415C, 416D, 436B
Habitat: UCFrs (gravel and sand pans)
Life Form: Annual herb
Blooming: June-August
Notes: Known from fewer than ten occurrences. MPA Co. record (436B) needs verification. See *Contributions from the U.S. National Herbarium* 37(5):329 (1969) for original description.

Camissonia tanacetifolia (T. & G.) Raven ssp. *quadriperforata* Raven

"Sierra Valley evening-primrose" Onagraceae
CNPS List: 4 **R-E-D Code:** 1-1-3 **State/Fed. Status:** /C3c
Distribution: LAS, PLU, SIE
Habitat: LCFrs (clay, sandy)
Life Form: Perennial herb
Blooming: May-July
Notes: Difficult to identify. See *Contributions from the U.S. National Herbarium* 37(5):248-249 (1969) for taxonomic treatment.

Campanula californica (Kell.) Heller

"swamp harebell" Campanulaceae
CNPS List: 1B **R-E-D Code:** 1-2-3 **State/Fed. Status:** /C2
Distribution: MEN, MRN, SCR*, SON
Quads: 408D*, 485B, 485C, 485D, 502A*, 503A*, 503D*, 520A, 520B, 520D, 537B, 537C, 537D, 552A, 553A, 568B, 569A, 569D, 585D
Habitat: BgFns, CCFrs, CoPrr, Medws, MshSw (freshwater), NCFrs / mesic
Life Form: Perennial herb (rhizomatous)
Blooming: June-September
Notes: Many occurrences have few plants. Threatened by grazing, development, and marsh habitat loss. See *Proceedings of the California Academy of Sciences* I 2:158 (1861) for original description.
Status Report: 1977

Campanula exigua Rattan

"chaparral harebell" Campanulaceae
CNPS List: 4 **R-E-D Code:** 1-1-3 **State/Fed. Status:** CEQA?
Distribution: ALA, CCA, SBT, SCL, STA
Habitat: Chprl (rocky, usually serpentinite)
Life Form: Annual herb
Blooming: May-June
Notes: See *Botanical Gazette* 11:339 (1886) for original description, and *Madroño* 27(4):149-163 (1980) for taxonomic treatment.

Campanula griffinii

Considered but rejected: Too common

Campanula scabrella Engelm.

"rough harebell" Campanulaceae
CNPS List: 4 **R-E-D Code:** 1-1-2 **State/Fed. Status:** CEQA?
Distribution: MOD?, SHA, SIS, TRI, ID, OR, WA+
Habitat: AlpBR (serpentinite or volcanic)
Life Form: Perennial herb (rhizomatous)
Blooming: August-September
Notes: Does plant occur in MOD Co.? On review list in OR. See *Madroño* 27(4):179-180 (1980) for distributional information.

Campanula sharsmithiae Morin

"Sharsmith's harebell" Campanulaceae
CNPS List: 1B **R-E-D Code:** 3-3-3 **State/Fed. Status:** /C2
Distribution: SCL, STA
Quads: 425B, 426A
Habitat: Chprl (serpentinite)
Life Form: Annual herb
Blooming: May-June
Notes: Known from only three occurrences. See *Madroño* 27(4):149-163 (1980) for original description.

Campanula shetleri Heckard

"Castle Crags harebell" Campanulaceae
CNPS List: 1B **R-E-D Code:** 3-1-3 **State/Fed. Status:** /C3c
Distribution: SHA, SIS
Quads: 682A, 682B
Habitat: LCFrs (rocky)
Life Form: Perennial herb (rhizomatous)
Blooming: June-September
Notes: Known from fewer than ten occurrences. See *Madroño* 20(4):231-235 (1969) for original description.

Campanula wilkinsiana Greene

"Wilkin's harebell" Campanulaceae
CNPS List: 1B **R-E-D Code:** 2-2-3 **State/Fed. Status:** /C2
Distribution: SIS, TEH, TRI
Quads: 625C, 667A, 667B, 668A, 684C, 698A, 698B, 698C
Habitat: Medws, SCFrs, UCFrs
Life Form: Perennial herb (rhizomatous)
Blooming: July-September
Notes: Known from fewer than twenty occurrences. Threatened by grazing and recreation.

Canbya candida Parry

"pygmy poppy" Papaveraceae
CNPS List: 1B **R-E-D Code:** 2-2-3 **State/Fed. Status:** CEQA
Distribution: KRN, LAX, SBD
Quads: 131B, 133A, 133C, 133D, 134B, 161A, 161B, 183B, 184A, 185C, 208B, 211C, 259A, 259C, 260A, 260D
Habitat: JTWld, MDScr
Life Form: Annual herb
Blooming: March-June
Notes: See *Proceedings of the American Academy of Arts and Sciences* 12:51 (1876) for original description.

Canotia holacantha

Considered but rejected: Not in CA

Cardamine gambellii

See *Rorippa gambellii*

Cardamine gemmata

See *Cardamine nuttallii* var. *gemmata*

Cardamine nuttallii Greene var. *gemmata* (Greene) Roll.

"yellow-tubered toothwort" Brassicaceae
CNPS List: 1B **R-E-D Code:** 3-1-2 **State/Fed. Status:** /C2
Distribution: DNT, OR
Quads: 739B, 739C, 740A
Habitat: LCFrs, NCFrs / serpentinite
Life Form: Perennial herb (rhizomatous)
Blooming: April-May
Notes: Known from fewer than ten occurrences. Possibly threatened by mining. Candidate for state listing in OR. USFWS uses the name *C. gemmata.*

Cardamine pachystigma (Wats.) Roll. var. *dissectifolia* (Detl.) Roll.

"dissected-leaf toothwort" Brassicaceae
CNPS List: 3 **R-E-D Code:** ?-?-3 **State/Fed. Status:** CEQA?
Distribution: BUT, MEN, PLA, SON
Quads: 534D, 540D, 569D, 581C, 592A, 592B, 597B, 600B
Habitat: Chprl? (serpentinite)
Life Form: Perennial herb (rhizomatous)
Blooming: March-April
Notes: Move to List 1B? Location, rarity, and endangerment information needed.

Carex albida Bailey

"white sedge" Cyperaceae
CNPS List: 1B **R-E-D Code:** 3-3-3 **State/Fed. Status:** CE/C1
Distribution: SON
Quads: 502A, 502B*
Habitat: BgFns, MshSw (freshwater)
Life Form: Perennial herb (rhizomatous)
Blooming: May-July
Notes: Known from only one confirmed extant occurrence at Pitkin Marsh; three historical occurrences extirpated by wetland drainage and spraying of chemical effluents. Also threatened by competition with other plants. See *Memoirs of the Torrey Botanical Club* 1:9 (1889) for original description, and *Leaflets of Western Botany* 8(7):178-180 (1957) for taxonomic discussion.
Status Report: 1993

Carex californica Bailey

"California sedge" Cyperaceae
CNPS List: 2 **R-E-D Code:** 3-1-1 **State/Fed. Status:** CEQA
Distribution: MEN, SON, ID, OR, WA+
Quads: 502A, 503A, 537B, 569A, 569D
Habitat: BgFns, CCFrs, CoPrr, Medws, MshSw (margins)
Life Form: Perennial herb (rhizomatous)
Blooming: May-August
Notes: Sensitive in ID. See *Memoirs of the Torrey Botanical Club* 1:9 (1889) for original description.

Carex comosa Boott

"bristly sedge" Cyperaceae
CNPS List: 2 **R-E-D Code:** 3-3-1 **State/Fed. Status:** CEQA
Distribution: LAK, SBD(*?), SCR*, SFO*, SHA, SJQ, SON, ID, OR, WA, ++
Quads: 407C*, 408D?*, 462B, 503D, 518C, 550A, 678C
Habitat: MshSw (lake margins)
Life Form: Perennial herb (rhizomatous)
Blooming: May-September
Notes: Location, rarity, and endangerment information needed; need quads for SBD and SFO counties. Fairly widely distributed, but apparently rarely collected. Threatened by marsh drainage. Endangered in ID, on review list in OR, and state-listed as Sensitive in WA.

Carex congdonii Bailey

"Congdon's sedge" Cyperaceae
CNPS List: 4 **R-E-D Code:** 1-1-3 **State/Fed. Status:** CEQA?
Distribution: FRE, INY, MAD, MNO, MPA, TUL, TUO
Habitat: AlpBR, SCFrs (rocky)
Life Form: Perennial herb (rhizomatous)
Blooming: July-August

Carex davyi Mkze.

"Davy's sedge" Cyperaceae
CNPS List: 4 **R-E-D Code:** 1-1-3 **State/Fed. Status:** CEQA?
Distribution: ALP, AMA, CAL, ELD, NEV, PLA, TUO
Habitat: SCFrs, UCFrs
Life Form: Perennial herb
Blooming: May-June
Notes: Similar to *C. petasata.*

Carex eleocharis Bailey

"spikerush sedge" Cyperaceae
CNPS List: 2 **R-E-D Code:** 2-1-1 **State/Fed. Status:** CEQA
Distribution: INY, MNO, NV, ++
Quads: 431C, 432A, 450D
Habitat: GBScr, SCFrs
Life Form: Perennial herb (rhizomatous)
Blooming: July-August
Notes: On review list in OR.

Carex geyeri Boott

"Geyer's sedge" Cyperaceae
CNPS List: 4 **R-E-D Code:** 1-2-1 **State/Fed. Status:** CEQA?
Distribution: BUT, HUM, PLU, SIE, SIS, OR, ++
Habitat: GBScr, LCFrs (volcanic)
Life Form: Perennial herb (rhizomatous)
Blooming: May-June
Notes: Threatened by logging.

Carex gigas (Holm) Mkze.

"Siskiyou sedge" Cyperaceae
CNPS List: 4 **R-E-D Code:** 1-1-2 **State/Fed. Status:** CEQA?
Distribution: DNT, PLU, SIS, TRI, OR
Habitat: LCFrs, Medws, UCFrs / mesic, sometimes serpentinite seeps
Life Form: Perennial herb (rhizomatous)
Blooming: May-July
Notes: Endangered in OR.

Carex halliana Bailey

"Hall's sedge" Cyperaceae
CNPS List: 2 **R-E-D Code:** 3-1-1 **State/Fed. Status:** CEQA
Distribution: SIS, OR, WA+
Quads: 703B, 713D, 720C
Habitat: Medws, PJWld / often pumice
Life Form: Perennial herb (rhizomatous)
Blooming: July-August

Carex hystricina Muhl.

"bottlebrush sedge" Cyperaceae
CNPS List: 2 **R-E-D Code:** 3-3-1 **State/Fed. Status:** CEQA
Distribution: TRI, AZ, OR, WA, ++
Quads: 649A, 667D
Habitat: MshSw (streambanks)
Life Form: Perennial herb (rhizomatous)
Blooming: June
Notes: Known in CA only from Rush Creek. Endangered in OR, and state-listed as Sensitive in WA.

Carex incurviformis Mkze. var. *danaensis* (Stacey) F.J. Herm.

"Dana's sedge" Cyperaceae
CNPS List: 4 **R-E-D Code:** 1-1-1 **State/Fed. Status:** CEQA?
Distribution: INY, MNO, TUL, TUO, +
Habitat: AlpBR
Life Form: Perennial herb (rhizomatous)
Blooming: July-August
Notes: See *Leaflets of Western Botany* 2:166 (1939) for original description and 7(12):288 (1955) for revised nomenclature, and *Madroño* 37(1):64 (1990) for information on MNO Co. collection.

Carex incurviformis var. *incurviformis*

Considered but rejected: Not in CA

Carex jepsonii

Considered but rejected: A synonym of *C. whitneyi*; a common taxon

Carex lasiocarpa Ehrh.

"slender sedge" Cyperaceae
CNPS List: 2 **R-E-D Code:** 3-1-1 **State/Fed. Status:** CEQA
Distribution: LAS, PLU, ID, OR, WA, ++
Quads: 572B, 623B, 625A, 625B
Habitat: BgFns, MshSw (freshwater, lake margins)
Life Form: Perennial herb (rhizomatous)
Blooming: June-July
Notes: Known in CA from only four occurrences.

Carex leptalea Wahl.

"flaccid sedge" Cyperaceae
CNPS List: 2 **R-E-D Code:** 3-2-1 **State/Fed. Status:** CEQA
Distribution: DNT, HUM, MRN*, TRI, ID, OR, ++
Quads: 485C*, 654B, 689A, 689C, 722B
Habitat: BgFns, Medws, MshSw
Life Form: Perennial herb (rhizomatous)
Blooming: May-July
Notes: Need quads for TRI Co. Apparently extirpated in MRN Co. by wetland conversion. Sensitive in ID.

Carex limosa L.

"shore sedge" Cyperaceae
CNPS List: 2 **R-E-D Code:** 2-2-1 **State/Fed. Status:** CEQA
Distribution: ELD, PLU, NV, ++
Quads: 522C, 590D, 625A, 625B
Habitat: BgFns, LCFrs, UCFrs
Life Form: Perennial herb (rhizomatous)
Blooming: June-August
Notes: Possibly more widespread in the Sierra Nevada.

Carex livida (Wahl.) Willd.

"livid sedge" Cyperaceae
CNPS List: 1A **Last Seen:** 1866 **State/Fed. Status:** CEQA
Distribution: MEN*, ID, OR, WA, ++
Quads: 569D*
Habitat: BgFns, MshSw
Life Form: Perennial herb (rhizomatous)
Blooming: June
Notes: Extirpated from CA; sensitive in ID, and endangered in OR.

Carex norvegica Retz.

"Scandinavian sedge" Cyperaceae
CNPS List: 2 **R-E-D Code:** 3-1-1 **State/Fed. Status:** CEQA
Distribution: MNO, ID, NM, OR, UT, WA, ++
Quads: 432A, 450D
Habitat: AlpBR, Medws (mesic)
Life Form: Perennial herb (rhizomatous)
Blooming: August
Notes: Endangered in OR, and state-listed as Sensitive in WA. See *Madroño* 37(1):64 (1990) for first CA reports (1986) of this circumboreal species.

Carex obispoensis Stacey

"San Luis Obispo sedge" Cyperaceae
CNPS List: 1B **R-E-D Code:** 2-2-3 **State/Fed. Status:** /C3c
Distribution: SLO
Quads: 245C, 246B, 246C, 247D, 271A, 271B, 272A, 295C
Habitat: CCFrs, Chprl, CoPrr, CoScr, VFGrs / often serpentinite seeps
Life Form: Perennial herb (rhizomatous)
Blooming: April-June
Notes: Threatened by grazing.

Carex parryana Dewey var. *hallii* (Olney) Kukenth.

"Hall's sedge" Cyperaceae
CNPS List: 2 **R-E-D Code:** 2-1-1 **State/Fed. Status:** CEQA
Distribution: MNO, NV, ++
Quads: 431D
Habitat: Medws, SCFrs
Life Form: Perennial herb (rhizomatous)
Blooming: July
Notes: Known in CA from only one collection near Station Pk. in the White Mtns. See *Madroño* 35(2):164 (1988) for first two CA reports (second erroneous; actually *C. norvegica*) and 37(1):65 (1990) for explanation of error.

Carex paucifructus

Considered but rejected: A synonym of *C. mariposana*; a common taxon

Carex petasata Dewey

"Liddon's sedge" Cyperaceae
CNPS List: 2 **R-E-D Code:** 2-1-1 **State/Fed. Status:** CEQA
Distribution: ALP, LAS, MNO, MOD, OR, ++
Quads: 432A, 642B, 690C
Habitat: LCFrs, Medws
Life Form: Perennial herb
Blooming: June-July
Notes: Need quads for ALP Co.

Carex praticola Rydb.

"meadow sedge" Cyperaceae
CNPS List: 2 **R-E-D Code:** 2-2-1 **State/Fed. Status:** CEQA
Distribution: DNT, HUM, MAD, MNO, TUO, OR, ++
Quads: 454A, 472C, 670B, 671A, 672C, 706D, 740C
Habitat: Medws
Life Form: Perennial herb
Blooming: May-July
Notes: Need quads for MAD Co.

Carex scoparia Schk.

"pointed broom sedge" Cyperaceae
CNPS List: 2 **R-E-D Code:** 3-2-1 **State/Fed. Status:** CEQA
Distribution: PLU, SHA, OR, ++
Quads: 629A
Habitat: GBScr (mesic)
Life Form: Perennial herb
Blooming: May
Notes: Need quads for PLU Co.

Carex sheldonii Mkze.

"Sheldon's sedge" Cyperaceae
CNPS List: 2 **R-E-D Code:** 2-1-1 **State/Fed. Status:** CEQA
Distribution: MOD, PLA, PLU, ID, OR, UT+
Quads: 587B, 588A, 604D, 605C
Habitat: LCFrs (mesic), RpScr
Life Form: Perennial herb (rhizomatous)
Blooming: August
Notes: Need quads for MOD and PLA counties. On review list in ID, and watch list in OR.

Carex tiogana D.W. Taylor & J. Mastrogiuseppe

"Tioga Pass sedge" Cyperaceae
CNPS List: 1B **R-E-D Code:** 3-1-3 **State/Fed. Status:** CEQA
Distribution: MNO
Quads: 453C, 471D
Habitat: Medws (mesic, lake margins)
Life Form: Perennial herb
Blooming: July-August
Notes: Known from only three small occurrences. See *Novon* (1994, in press) for original description.

Carex tompkinsii J.T. Howell

"Tompkins's sedge" Cyperaceae
CNPS List: 1B **R-E-D Code:** 2-2-3 **State/Fed. Status:** CR/C3c
Distribution: FRE, MPA
Quads: 374A, 374B, 374C, 374D, 375A, 375C, 375D, 437B, 438A
Habitat: Chprl, CmWld, LCFrs, UCFrs / sometimes granitic
Life Form: Perennial herb (rhizomatous)
Blooming: May-July
Notes: Threatened by development and trail maintenance and use. See *Leaflets of Western Botany* 9(12):185-187 (1961) for original description.
Status Report: 1990

Carex vulpinoidea Michx.

"fox sedge" Cyperaceae
CNPS List: 2 **R-E-D Code:** 2-2-1 **State/Fed. Status:** CEQA
Distribution: BUT, SHA, SIS, TRI, AZ, OR, ++
Quads: 560A, 577B, 594D, 629A
Habitat: MshSw, RpWld
Life Form: Perennial herb
Blooming: June
Notes: Need quads for SIS and TRI counties.

Carex whitneyi

Considered but rejected: Too common

Carlowrightia arizonica Gray

"Arizona carlowrightia" Acanthaceae
CNPS List: 2 **R-E-D Code:** 3-2-1 **State/Fed. Status:** CEQA
Distribution: SDG, AZ, BA, ++
Quads: 32B, 47C
Habitat: SDScr (sandy, granitic alluvium)
Life Form: Shrub (deciduous)
Blooming: March-May
Notes: Known in CA from only one extended population at Anza Borrego SP and on adjacent private land. Possibly threatened by recreational activities. See *Madroño* 35(3):279 (1988) for distribution in CA.

Carnegiea gigantea (Engelm.) Britt. & Rose

"saguaro" Cactaceae
CNPS List: 2 **R-E-D Code:** 3-2-1 **State/Fed. Status:** CEQA
Distribution: IMP, RIV, SBD, AZ, SO
Quads: 11E, 26A, 26B, 41D, 75, 94B, 120C, 121C, 121D
Habitat: SDScr (rocky)
Life Form: Shrub (stem succulent)
Blooming: May-June
Notes: Need precise locality information for quad 75. State flower of AZ, where state-listed as Highly Safeguarded.

Carpenteria californica Torr.

"tree-anemone" Hydrangeaceae
CNPS List: 1B **R-E-D Code:** 3-2-3 **State/Fed. Status:** CT/C1
Distribution: FRE
Quads: 377A, 396C, 397C, 397D, 398A
Habitat: Chprl, CmWld / granitic
Life Form: Shrub (evergreen)
Blooming: May-July
Notes: Known from fewer than ten occurrences. There are historical reports of introduced plants in MAD Co. Threatened by proposed road construction, vehicles, logging, and development. Special management is necessary since reproduction is fire dependent. See *Fremontia* 10(4):21-22 (1983) for discussion of propagation.
Status Report: 1989

Cassia covesii

See *Senna covesii*

Castela emoryi (Gray) Moran & Felger

"crucifixion thorn" Simaroubaceae
CNPS List: 2 **R-E-D Code:** 2-1-1 **State/Fed. Status:** CEQA
Distribution: IMP, RIV, SBD, AZ, SO
Quads: 6A, 6B, 26B, 30D, 41B, 62A, 62B, 77D, 100A, 122A, 122B, 147C, 148B, 151C, 152D, 154A, 154B?, 174A, 181C, 204C
Habitat: MDScr, SDScr
Life Form: Shrub (deciduous)
Blooming: June-July
Notes: Protected in part at Crucifixion Thorn Natural Area (BLM) along Highway 98.

Castilleja affinis H. & A. ssp. *neglecta* (Zeile) Chuang & Heckard

"Tiburon Indian paintbrush" Scrophulariaceae
CNPS List: 1B **R-E-D Code:** 3-2-3 **State/Fed. Status:** CT/PE
Distribution: MRN, NAP, SCL
Quads: 406B, 466B, 482B, 484C
Habitat: VFGrs / serpentinite
Life Form: Perennial herb (hemiparasitic)
Blooming: April-June
Notes: Known from six occurrences. Protected in part at Ring Mtn. Preserve (TNC), MRN Co. Threatened by development, gravel mining, and grazing. State-listed as *C. neglecta*; USFWS also uses this name.
Status Report: 1989

Castilleja ambigua H. & A. ssp. *humboldtiensis* (Keck) Chuang & Heckard

"Humboldt Bay owl's-clover" Scrophulariaceae
CNPS List: 1B **R-E-D Code:** 2-2-3 **State/Fed. Status:** /C2
Distribution: HUM, MRN
Quads: 485D, 655A, 672B, 672C, 672D, 689A
Habitat: MshSw (coastal salt)
Life Form: Annual herb (hemiparasitic)
Blooming: May-August
Notes: Threatened by coastal development. See *Proceedings of the American Academy of Arts and Sciences* IV 16:536 (1927) for original description, and *Systematic Botany* 16(4):644-666 (1991) for revised nomenclature.
Status Report: 1977

Castilleja arachnoidea ssp. *arachnoidea*

Considered but rejected: Too common; a synonym of *C. arachnoidea*

Castilleja arachnoidea ssp. *schizotricha*

See *Castilleja schizotricha*

Castilleja arachnoidea ssp. *shastensis*

Considered but rejected: A synonym of *C. arachnoidea*; a common taxon

Castilleja brevilobata

See *Castilleja hispida* ssp. *brevilobata*

Castilleja campestris (Benth.) Chuang & Heckard ssp. *succulenta* (Hoov.) Chuang & Heckard

"succulent owl's-clover" Scrophulariaceae
CNPS List: 1B **R-E-D Code:** 2-2-3 **State/Fed. Status:** CE/PT
Distribution: FRE, MAD, MER, MPA, STA
Quads: 377B, 378B, 378D, 379A, 379D*, 398C, 398D, 399C, 420C, 421A, 421B, 421D, 422B, 441A
Habitat: VnPls
Life Form: Annual herb (hemiparasitic)
Blooming: April-May
Notes: Threatened by urbanization, agriculture, grazing, flood control, and trampling. State-listed as *Orthocarpus campestris* var. *succulentus.* See *Leaflets of Western Botany* 1(19):228-229 (1936) for original description, and *Systematic Botany* 16(4):644-666 (1991) for revised nomenclature.
Status Report: 1986

Castilleja cinerea Gray

"ash-gray Indian paintbrush" Scrophulariaceae
CNPS List: 1B **R-E-D Code:** 2-2-3 **State/Fed. Status:** /C2
Distribution: SBD
Quads: 104B, 105A, 105B, 106A, 126D, 131C, 131D
Habitat: MDScr, Medws, PbPln, PJWld, UCFrs (clay openings)
Life Form: Perennial herb (hemiparasitic)
Blooming: June-July
Notes: Threatened by grazing, development, non-native plants, and vehicles. See *Proceedings of the American Academy of Arts and Sciences* 19:93 (1883) for original description.
Status Report: 1980

Castilleja culbertsonii

Considered but rejected: A synonym of *C. lemmonii*; a common taxon

Castilleja disticha

Considered but rejected: Too common; a synonym of *C. applegatei* ssp. *disticha*

Castilleja elata

See *Castilleja miniata* ssp. *elata*

Castilleja ewanii

See *Castilleja montigena*

Castilleja franciscana

Considered but rejected: Too common; a synonym of *C. subinclusa* ssp. *franciscana*

Castilleja gleasonii Elmer

"Mt. Gleason Indian paintbrush" Scrophulariaceae
CNPS List: 1B **R-E-D Code:** 3-2-3 **State/Fed. Status:** CR/C2
Distribution: LAX
Quads: 136B, 136C, 136D
Habitat: LCFrs (granitic)
Life Form: Perennial herb (hemiparasitic)
Blooming: May-June
Notes: Known from fewer than ten occurrences. Threatened by proximity to campgrounds, fuelwood cutting, and vehicles. See *C. pruinosa* in *The Jepson Manual.* Angeles NF has adopted species management guidelines. See *Botanical Gazette* 39:51 (1905) for original description.
Status Report: 1990

Castilleja grisea Dunkle

"San Clemente Island Indian paintbrush" Scrophulariaceae
CNPS List: 1B **R-E-D Code:** 1-2-3 **State/Fed. Status:** CE/FE
Distribution: SCM
Quads: SCMC, SCMN, SCMS
Habitat: CBScr, CoScr / rocky
Life Form: Perennial herb (hemiparasitic)
Blooming: February-August
Notes: Feral herbivores removed from SCM Isl., and vegetation recovering. See *Bulletin of the Southern California Academy of Sciences* 42:31 (1943) for original description.
Status Report: 1987

Castilleja hispida Benth. ssp. *brevilobata* (Piper) Chuang & Heckard

"short-lobed Indian paintbrush" Scrophulariaceae
CNPS List: 4 **R-E-D Code:** 1-2-1 **State/Fed. Status:** /C3c
Distribution: DNT, SIS, OR
Habitat: LCFrs (serpentinite, edges and openings)
Life Form: Perennial herb (hemiparasitic)
Blooming: April-July
Notes: On review list in OR. See *Proceedings of the Biological Society of Washington* 33:164 (1920) for original description, and *Memoirs of the New York Botanical Garden* 21:33-35 (1971) for revised nomenclature.
Status Report: 1979

Castilleja hololeuca

See *Castilleja lanata* ssp. *hololeuca*

Castilleja lanata Gray ssp. *hololeuca* (Greene) Chuang & Heckard

"white-felted Indian paintbrush" Scrophulariaceae
CNPS List: 1B **R-E-D Code:** 2-2-3 **State/Fed. Status:** /C3c
Distribution: ANA, SCZ, SMI, SRO
Quads: ANAC, SCZA, SCZB, SCZC, SCZD, SMIE, SMIW, SRON
Habitat: Chprl, CoScr
Life Form: Perennial herb (hemiparasitic)
Blooming: March-August
Notes: Threatened by feral herbivores on SCZ and SRO islands; also by cattle grazing on SRO Isl.

Castilleja lasiorhyncha (Gray) Chuang & Heckard

"San Bernardino Mtns. owl's-clover" Scrophulariaceae
CNPS List: 1B **R-E-D Code:** 1-2-3 **State/Fed. Status:** /C2
Distribution: RIV?, SBD, SDG
Quads: 20A*, 33D, 66B?, 83C?, 105A, 105B, 105C, 106A, 106B, 107A, 131C, 131D, 132C, 133D
Habitat: Chprl, Medws, PbPln, UCFrs / mesic
Life Form: Annual herb (hemiparasitic)
Blooming: June-August
Notes: RIV Co. occurrences need confirmation. Threatened by vehicles and recreation. See *Proceedings of the American Academy of Arts and Sciences* 12:82 (1876) for original description, and *Systematic Botany* 16(4):644-666 (1991) for revised nomenclature.
Status Report: 1980

Castilleja lassenensis

Considered but rejected: A synonym of *C. lemmonii*; a common taxon

Castilleja latifolia H. & A.

"Monterey Indian paintbrush" Scrophulariaceae
CNPS List: 4 **R-E-D Code:** 1-1-3 **State/Fed. Status:** CEQA?
Distribution: MNT, SCR
Habitat: CoDns, CoScr
Life Form: Perennial herb (hemiparasitic)
Blooming: February-September

Castilleja latifolia ssp. *latifolia*

See *Castilleja latifolia*

Castilleja latifolia ssp. *mendocinensis*

See *Castilleja mendocinensis*

Castilleja leschkeana

Considered but rejected: Not in CA; name misapplied to *C. chrymactis*, native to Alaska, extinct waif in CA

Castilleja longispica

Considered but rejected: Too common

Castilleja martinii var. *clokeyi*

Considered but rejected: A synonym of *C. applegatei* ssp. *martinii*; a common taxon

Castilleja mendocinensis (Eastw.) Penn.

"Mendocino coast Indian paintbrush" Scrophulariaceae
CNPS List: 1B **R-E-D Code:** 2-2-3 **State/Fed. Status:** /C2
Distribution: HUM, MEN
Quads: 537C, 537D, 552B, 552C, 553A, 569A, 569D, 585A, 585D, 601D, 689C
Habitat: CBScr, CCFrs, CoPrr, CoScr
Life Form: Perennial herb (hemiparasitic)
Blooming: April-August
Notes: Threatened by coastal development, recreation, and non-native plants. Related to *C. affinis* ssp. *litoralis.*

Castilleja miniata Hook. ssp. *elata* (Piper) Munz

"Siskiyou Indian paintbrush" Scrophulariaceae
CNPS List: 2 **R-E-D Code:** 1-2-1 **State/Fed. Status:** /C3c
Distribution: DNT, SIS?, OR
Quads: 698C?, 738A, 739A, 739C
Habitat: BgFns, LCFrs (seeps) / often serpentinite
Life Form: Perennial herb (hemiparasitic)
Blooming: May-August
Notes: SIS Co. record may be erroneous; some specimens confused with *C. miniata* ssp. *miniata.*

Castilleja mollis Penn.

"soft-leaved Indian paintbrush" Scrophulariaceae
CNPS List: 1B **R-E-D Code:** 3-3-3 **State/Fed. Status:** /C1
Distribution: SMI?, SRO
Quads: SMIW?, SRON
Habitat: CBScr, CoDns
Life Form: Perennial herb (hemiparasitic)
Blooming: April-August
Notes: Known from Carrington Pt. Peninsula and one occurrence west of Jaw Gulch (both SRO Isl.), and one unconfirmed location at Point Bennett (SMI Isl.). Mainland plants previously included in *C. mollis* are *C. affinis* ssp. *affinis.* Possibly threatened by cattle grazing and feral herbivores. See *Madroño* 38(2):141-142 (1991) for taxonomic reassessment.

Castilleja montigena Heckard

"Heckard's Indian paintbrush" Scrophulariaceae
CNPS List: 4 **R-E-D Code:** 1-1-3 **State/Fed. Status:** CEQA?
Distribution: SBD
Habitat: LCFrs, PJWld, UCFrs
Life Form: Perennial herb (hemiparasitic)
Blooming: May-August
Notes: Presumed to be a stabilized species of hybrid origin, with *C. applegatei* ssp. *martinii* and *C. angustifolia* as parents; see the former in *The Jepson Manual.* See *Systematic Botany* 5(1):71-85 (1980) for discussion of origin and taxonomy.

Castilleja neglecta

See *Castilleja affinis* ssp. *neglecta*

Castilleja payneae

Considered but rejected: A synonym of *C. arachnoidea*; a common taxon

Castilleja peirsonii

Considered but rejected: Too common

Castilleja pilosa ssp. *jusselii*

Considered but rejected: Too common; a synonym of *C. pilosa*

Castilleja plagiotoma Gray

"Mojave Indian paintbrush" Scrophulariaceae
CNPS List: 4 **R-E-D Code:** 1-1-3 **State/Fed. Status:** CEQA?
Distribution: KRN, LAX, SBD, SLO
Habitat: GBScr (alluvial), PJWld
Life Form: Perennial herb (hemiparasitic)
Blooming: April-June

Castilleja praeterita

Considered but rejected: Too common

Castilleja psittacina

Considered but rejected: A synonym of *C. pilosa*; a common taxon

Castilleja schizotricha Greenm.

"split-hair Indian paintbrush" Scrophulariaceae
CNPS List: 4 **R-E-D Code:** 1-1-2 **State/Fed. Status:** CEQA?
Distribution: SIS, OR
Habitat: UCFrs (decomposed granitic or marble)
Life Form: Perennial herb (hemiparasitic)
Blooming: July-August
Notes: Endangered in OR.

Castilleja uliginosa Eastw.

"Pitkin Marsh Indian paintbrush" Scrophulariaceae
CNPS List: 1A **Last Seen:** 1987 **State/Fed. Status:** CE/C2
Distribution: SON*
Quads: 502A*
Habitat: MshSw (freshwater)
Life Form: Perennial herb (hemiparasitic)
Blooming: June-July
Notes: Last known remaining plant died in 1987; field surveys needed. See *C. miniata* ssp. *miniata* in *The Jepson Manual*. See *Leaflets of Western Botany* 3:166-117 (1942) for original description.
Status Report: 1987

Caulanthus amplexicaulis Wats. var. *barbarae* (J.T. Howell) Munz

"Santa Barbara jewelflower" Brassicaceae
CNPS List: 1B **R-E-D Code:** 3-1-3 **State/Fed. Status:** /C1
Distribution: SBA
Quads: 168B, 193C, 194D
Habitat: CCFrs, CmWld / serpentinite
Life Form: Annual herb
Blooming: May-July
Notes: Known from fewer than five occurrences in the San Rafael Mtns. See *Leaflets of Western Botany* 9:223 (1952) for original description.

Caulanthus californicus (Wats.) Pays.

"California jewelflower" Brassicaceae
CNPS List: 1B **R-E-D Code:** 3-3-3 **State/Fed. Status:** CE/FE
Distribution: FRE, KNG*, KRN, SBA, SLO, TUL*
Quads: 191B*, 191C*, 192A, 192B*, 192D, 214A*, 216B*, 217C, 217D, 218A*, 219D*, 239B*, 239A*, 239D*, 240A*, 240B*, 242B*, 244C, 262D*, 264D*, 265A, 265C*, 266A*, 266B*, 267A*, 287B*, 287C*, 287D*, 288D*, 291B*, 291C*, 291D*, 310C*, 311A*, 314A*, 314C*, 315A*, 315C*, 315D*
Habitat: ChScr, PJWld, VFGrs
Life Form: Annual herb
Blooming: February-May
Notes: Known from approximately twenty extant occurrences; over 35 historical occurrences extirpated. Threatened by agriculture, urbanization, energy development, and grazing. See *Fremontia* 16(1):18-19 (1988) for species account.
Status Report: 1993

Caulanthus glaucus

Considered but rejected: Too common

Caulanthus heterophyllus var. *pseudosimulans*

Considered but rejected: Not yet published

Caulanthus simulans Pays.

"Payson's jewelflower" Brassicaceae
CNPS List: 4 **R-E-D Code:** 1-2-3 **State/Fed. Status:** /C2
Distribution: RIV, SDG
Habitat: Chprl, CoScr / sandy, granitic
Life Form: Annual herb
Blooming: March-June
Notes: Confused with *C. heterophyllus* var. *pseudosimulans* (unpublished), which is more coastal and, unlike *C. simulans*, appears after fires. Some populations threatened by proposed reservoir construction, but many populations occur on public lands (Anza Borrego SP, BLM, and USFS). Also threatened by urbanization, grazing, and road construction.

Caulanthus stenocarpus

Considered but rejected: A synonym of *C. heterophyllus* var. *heterophyllus*; a common taxon

Caulostramina jaegeri (Roll.) Roll.

"Jaeger's caulostramina" Brassicaceae
CNPS List: 1B **R-E-D Code:** 3-2-3 **State/Fed. Status:** /C2
Distribution: INY
Quads: 327B, 349C, 350D, 392D
Habitat: PJWld, SCFrs / carbonate
Life Form: Perennial herb
Blooming: May-June
Notes: Known from fewer than five occurrences in the Inyo Mtns. See *Contributions from the Dudley Herbarium* 3:174 (1941) for original description, and *Contributions from the Gray Herbarium* 204:155-157 (1973) for revised nomenclature.

Ceanothus arboreus

Considered but rejected: Too common

Ceanothus confusus J.T. Howell

"Rincon Ridge ceanothus" Rhamnaceae
CNPS List: 1B **R-E-D Code:** 3-3-3 **State/Fed. Status:** /C2
Distribution: LAK, MEN, NAP, SON
Quads: 500B, 501A, 501B, 502A*, 517A, 517B, 517C, 517D?*, 518B, 533C, 535A, 535D, 549A
Habitat: CCFrs, Chprl, CmWld / volcanic or serpentinite
Life Form: Shrub (evergreen)
Blooming: February-April
Notes: Threatened by development. Closely related to *C. prostratus*. See *Leaflets of Western Botany* 2:160-162 (1939) for original description.
Status Report: 1979

Ceanothus cuneatus var. *decumbens*

Considered but rejected: Not published

Ceanothus cuneatus (Hook.) Nutt. var. *rigidus* (Nutt.) Hoov.

"Monterey ceanothus" Rhamnaceae
CNPS List: 4 **R-E-D Code:** 1-2-3 **State/Fed. Status:** /C2
Distribution: MNT, SLO, SCR*
Habitat: CCFrs, CoScr
Life Form: Shrub (evergreen)
Blooming: February-March
Notes: Threatened by development. Intergrades with var. *fascicularis* in SLO Co.

Ceanothus cyaneus Eastw.

"Lakeside ceanothus" Rhamnaceae
CNPS List: 1B **R-E-D Code:** 3-2-2 **State/Fed. Status:** /C2
Distribution: RIV?, SDG, BA
Quads: 10A, 10C, 20A, 20D, 21A, 21B, 21C, 21D, 22B, 49B?, 50A?
Habitat: CCFrs, Chprl
Life Form: Shrub (evergreen)
Blooming: April-June
Notes: Need verification of RIV Co. specimen. Threatened by development.

Ceanothus divergens Parry

"Calistoga ceanothus" Rhamnaceae
CNPS List: 1B **R-E-D Code:** 3-2-3 **State/Fed. Status:** /C2
Distribution: LAK, NAP, SON
Quads: 500B, 501A, 501B, 516C, 517B, 517C, 517D, 533C
Habitat: Chprl (serpentinite or volcanic)
Life Form: Shrub (evergreen)
Blooming: February-March
Notes: Threatened by development in The Geysers geothermal area. Closely related to *C. purpureus.*
Status Report: 1979

Ceanothus ferrisae McMinn

"Coyote ceanothus" Rhamnaceae
CNPS List: 1B **R-E-D Code:** 3-3-3 **State/Fed. Status:** /PE
Distribution: SCL
Quads: 406A, 406B
Habitat: Chprl, CoScr, VFGrs / serpentinite
Life Form: Shrub (evergreen)
Blooming: January-March
Notes: Known from fewer than five occurrences in the Mt. Hamilton Range. Threatened by expansion of Anderson Reservoir spillway, development, and alteration of fire regimes. See *Madroño* 2:89-90 (1933) for original description.
Status Report: 1977

Ceanothus foliosus Parry var. *vineatus* McMinn

"Vine Hill ceanothus" Rhamnaceae
CNPS List: 1B **R-E-D Code:** 3-3-3 **State/Fed. Status:** /C2
Distribution: MEN*, SON
Quads: 502A
Habitat: Chprl
Life Form: Shrub (evergreen)
Blooming: March-May
Notes: Nearly extirpated in SON Co.; now confirmed from only one native population. Known from one historical occurrence in MEN Co., but unable to relocate; need quads.

Ceanothus fresnensis Abrams

"Fresno ceanothus" Rhamnaceae
CNPS List: 4 **R-E-D Code:** 1-1-3 **State/Fed. Status:** CEQA?
Distribution: CAL, FRE, MAD, MPA, TUL, TUO
Habitat: CmWld, LCFrs
Life Form: Shrub (evergreen)
Blooming: May-June

Ceanothus gloriosus J.T. Howell var. *gloriosus*

"Point Reyes ceanothus" Rhamnaceae
CNPS List: 4 **R-E-D Code:** 1-1-3 **State/Fed. Status:** CEQA?
Distribution: MEN, MRN, SON
Habitat: CBScr, CCFrs, CoDns, CoScr / sandy
Life Form: Shrub (evergreen)
Blooming: March-May

Ceanothus gloriosus J.T. Howell var. *porrectus* J.T. Howell

"Mt. Vision ceanothus" Rhamnaceae
CNPS List: 1B **R-E-D Code:** 3-1-3 **State/Fed. Status:** /C2
Distribution: MRN
Quads: 485B, 485C, 485D
Habitat: CCFrs, CoPrr, CoScr, VFGrs
Life Form: Shrub (evergreen)
Blooming: March-May
Notes: Known from fewer than fifteen occurrences in the Mt. Vision area near Pt. Reyes. Some plants destroyed by quarrying; occurs in areas grazed by cattle.

Ceanothus hearstiorum Hoov. & Roof

"Hearst's ceanothus" Rhamnaceae
CNPS List: 1B **R-E-D Code:** 3-2-3 **State/Fed. Status:** CR/C2
Distribution: SLO
Quads: 271B, 272A
Habitat: Chprl (maritime), CoPrr, CoScr
Life Form: Shrub (evergreen)
Blooming: March-April
Notes: Known from fewer than ten occurrences. Possibly threatened by grazing. See *Four Seasons* 2:1-5 (1966) for original description.
Status Report: 1988

Ceanothus xhumboldtiensis

Considered but rejected: A hybrid

Ceanothus impressus var. *impressus*

Considered but rejected: Too common

Ceanothus impressus var. *nipomensis*

Considered but rejected: A synonym of *C. impressus*; a common taxon

Ceanothus insularis

See *Ceanothus megacarpus* var. *insularis*

Ceanothus jepsonii var. *jepsonii*

Considered but rejected: Too common

Ceanothus maritimus Hoov.

"maritime ceanothus" Rhamnaceae
CNPS List: 1B **R-E-D Code:** 3-2-3 **State/Fed. Status:** CR/C2
Distribution: SLO
Quads: 271B, 272A
Habitat: Chprl (maritime), VFGrs
Life Form: Shrub (evergreen)
Blooming: January-March
Notes: Known from fewer than ten occurrences. Possibly threatened by grazing. See *Leaflets of Western Botany* 7(4):111-112 (1953) for original description.
Status Report: 1988

Ceanothus masonii McMinn

"Mason's ceanothus" Rhamnaceae
CNPS List: 1B **R-E-D Code:** 3-2-3 **State/Fed. Status:** CR/C2
Distribution: MRN
Quads: 467B, 485A
Habitat: Chprl (serpentinite)
Life Form: Shrub (evergreen)
Blooming: March-April
Notes: Known from approximately five occurrences. May be a variety of *C. gloriosus*. See *Madroño* 6:171-173 (1942) for original description.
Status Report: 1988

Ceanothus megacarpus Nutt. var. *insularis* (Eastw.) Munz

"island ceanothus" Rhamnaceae
CNPS List: 4 **R-E-D Code:** 1-1-3 **State/Fed. Status:** CEQA?
Distribution: SCZ, SRO
Habitat: Chprl (sandy)
Life Form: Shrub (evergreen)
Blooming: January-March

Ceanothus ophiochilus Boyd, Ross & Arnseth

"Vail Lake ceanothus" Rhamnaceae
CNPS List: 1B **R-E-D Code:** 3-3-3 **State/Fed. Status:** CE/C2
Distribution: RIV
Quads: 49B, 50A
Habitat: Chprl (gabbroic or pyroxenite-rich outcrops)
Life Form: Shrub (evergreen)
Blooming: February-March
Notes: Known from only three occurrences near Vail Lake; first discovered in 1989. Threatened by proposed development and long-term hybridization with *C. crassifolius*. See *Phytologia* 70(1):28-41 (1991) for original description.

Ceanothus papillosus var. *roweanus*

Considered but rejected: Too common

Ceanothus pinetorum

Considered but rejected: Too common

Ceanothus prostratus var. *laxus*

Considered but rejected: A synonym of *C. prostratus*; a common taxon

Ceanothus purpureus Jeps.

"holly-leaf ceanothus" Rhamnaceae
CNPS List: 4 **R-E-D Code:** 1-1-3 **State/Fed. Status:** CEQA?
Distribution: NAP, SOL
Habitat: Chprl (volcanic)
Life Form: Shrub (evergreen)
Blooming: February-April
Notes: See *Fremontia* 15(4):25-26 (1988) for species account.

Ceanothus rigidus

See *Ceanothus cuneatus* var. *rigidus*

Ceanothus roderickii Knight

"Pine Hill ceanothus" Rhamnaceae
CNPS List: 1B **R-E-D Code:** 3-2-3 **State/Fed. Status:** CR/C1
Distribution: ELD
Quads: 510B, 511A, 527D
Habitat: Chprl, CmWld / often serpentinite or gabbroic
Life Form: Shrub (evergreen)
Blooming: May-June
Notes: Known from approximately ten occurrences. Threatened by residential development in the Sierra Nevada foothills. See *Four Seasons* 2(4):23-24 (1968) for original description.
Status Report: 1993

Ceanothus sonomensis J.T. Howell

"Sonoma ceanothus" Rhamnaceae
CNPS List: 1B **R-E-D Code:** 3-2-3 **State/Fed. Status:** /C2
Distribution: NAP, SON
Quads: 500B, 500C, 501A
Habitat: Chprl (sandy, serpentinite or volcanic)
Life Form: Shrub (evergreen)
Blooming: February-April
Notes: Known from approximately ten occurrences; only one occurrence known from NAP Co. Seriously threatened by development. Closely related to *C. cuneatus*.
Status Report: 1991

Ceanothus verrucosus Nutt.

"white coast ceanothus" Rhamnaceae
CNPS List: 2 **R-E-D Code:** 1-2-1 **State/Fed. Status:** /C2
Distribution: SDG, BA
Quads: 22B, 22C, 22D, 35C, 35D, 36A, 36D
Habitat: Chprl
Life Form: Shrub (evergreen)
Blooming: January-April
Notes: Threatened by development.

Celtis reticulata

Considered but rejected: Too common

Centaurium exaltatum

Considered but rejected: Too common

Centaurium namophilum var. *namophilum*

Considered but rejected: Not in CA; misidentification of *C. exaltatum*; a common taxon

Centaurium namophilum var. *nevadensis*

Considered but rejected: A synonym of *C. exaltatum*; a common taxon

Centrostegia insignis
See *Aristocapsa insignis*

Centrostegia leptoceras
See *Dodecahema leptoceras*

Centrostegia vortriedei
See *Systenotheca vortriedei*

Cerastium beeringianum var. *capillare*
Considered but rejected: Too common

Cercocarpus betuloides ssp. *blancheae*
See *Cercocarpus betuloides* var. *blancheae*

Cercocarpus betuloides T. & G. var. *blancheae* (C. Schneider) Little
"island mountain-mahogany" Rosaceae
CNPS List: 4 **R-E-D Code:** 1-1-3 **State/Fed. Status:** CEQA?
Distribution: LAX, SCT, SCZ, SRO, VEN
Habitat: Chprl
Life Form: Shrub (evergreen)
Blooming: March-April
Notes: Does plant occur on other Channel Islands?

Cercocarpus minutiflorus
Considered but rejected: Too common

Cercocarpus traskiae Eastw.
"Catalina Island mountain-mahogany" Rosaceae
CNPS List: 1B **R-E-D Code:** 3-3-3 **State/Fed. Status:** CE/C1
Distribution: SCT
Quads: SCTS
Habitat: Chprl (rocky)
Life Form: Shrub (evergreen)
Blooming: March-May
Notes: May be the rarest tree in CA; only seven adult plants surviving in 1990. Threatened by feral herbivores. Recovery work in progress. See *Proceedings of the California Academy of Sciences* III 1:136 (1898) for original description, and *Conservation Biology* 3(1):52-58 (1989) for discussion of hybridization.
Status Report: 1987

Cereus giganteus
See *Carnegiea gigantea*

Chaenactis carphoclinia Gray var. *peirsonii* (Jeps.) Munz
"Peirson's pincushion" Asteraceae
CNPS List: 1B **R-E-D Code:** 2-1-3 **State/Fed. Status:** CEQA
Distribution: IMP, RIV?, SDG
Quads: 31D, 32A, 46D
Habitat: SDScr
Life Form: Annual herb
Blooming: March-April
Notes: Endemic to the eastern Santa Rosa Mtns, where known from few collections. Does plant occur in RIV Co.? See *Madroño* 1(17):259 (1929) for original description, and *Manual of Southern California Botany*, p. 567 (1935) by P. Munz for revised nomenclature.

Chaenactis douglasii (Hook.) H. & A. var. *alpina* Gray
"alpine dusty maidens" Asteraceae
CNPS List: 2 **R-E-D Code:** 2-1-1 **State/Fed. Status:** CEQA
Distribution: ALP, ELD, INY, SIS, TUO, NV, OR, ++
Quads: 302A, 506B?, 507A?, 522C, 523B, 698B, 698C
Habitat: AlpBR (granitic)
Life Form: Perennial herb
Blooming: July-September
Notes: Intergrades downslope with var. *douglasii*. See *Synoptical Flora of North America* 1(2):341 (1884) for original description, and *Proceedings of the California Academy of Sciences* II 5:699 (1895) for additional information.

Chaenactis parishii Gray
"Parish's chaenactis" Asteraceae
CNPS List: 4 **R-E-D Code:** 1-1-2 **State/Fed. Status:** /C3c
Distribution: RIV, SDG, BA
Habitat: Chprl
Life Form: Perennial herb
Blooming: May-July
Notes: See *Proceedings of the American Academy of Arts and Sciences* 20:299 (1885) for original description.
Status Report: 1979

Chaenactis suffrutescens Gray
"Shasta chaenactis" Asteraceae
CNPS List: 1B **R-E-D Code:** 2-1-3 **State/Fed. Status:** CEQA
Distribution: SIS, TRI
Quads: 668B, 682A, 682B, 683B, 683C, 684C, 685B, 686D, 698B, 698C, 699B, 699C, 699D, 700A, 701B, 716A, 717D?, 718C, 718D, 733A
Habitat: LCFrs, UCFrs / sandy, serpentinite
Life Form: Perennial herb
Blooming: May-September
Notes: Rare on the Shasta-Trinity NF.

Chamaebatia australis (Bdg.) Abrams
"southern mountain misery" Rosaceae
CNPS List: 4 **R-E-D Code:** 1-2-1 **State/Fed. Status:** CEQA?
Distribution: LAX, SDG, BA
Habitat: Chprl
Life Form: Shrub (evergreen)
Blooming: November-May
Notes: Threatened by agriculture. See *Madroño* 27(4):111 (1980) for range extension information.

Chamaecyparis lawsoniana
Considered but rejected: A synonym of *Cupressus lawsoniana*; a common taxon

Chamaecyparis nootkatensis
See *Cupressus nootkatensis*

Chamaesyce arizonica (Engelm.) Arthur

"Arizona spurge" Euphorbiaceae
CNPS List: 2 **R-E-D Code:** 2-1-1 **State/Fed. Status:** CEQA
Distribution: IMP?, RIV, SDG, AZ, BA+
Quads: 47C, 83D
Habitat: SDScr (sandy)
Life Form: Perennial herb
Blooming: March-April
Notes: Undocumented in IMP Co.; need quads. See *Madroño* 32(3):187-189 (1985) for revision of *Chamaesyce* nomenclature.

Chamaesyce hooveri (Wheeler) Koutnik

"Hoover's spurge" Euphorbiaceae
CNPS List: 1B **R-E-D Code:** 3-2-3 **State/Fed. Status:** /PT
Distribution: BUT, GLE, STA, TEH, TUL
Quads: 333B, 334A*, 441A, 441C, 441D, 562B, 576B, 593B, 593C, 594A
Habitat: VnPls
Life Form: Annual herb
Blooming: July
Notes: Known from approximately twenty occurrences. Threatened by grazing, agriculture, and non-native plants. See *Proceedings of the Biological Society of Washington* 53:9 (1940) for original description, and *Madroño* 32(3):187-189 (1985) for revised nomenclature.

Chamaesyce ocellata (Dur. & Hilg.) Millsp. ssp. *rattanii* (Wats.) Koutnik

"Stony Creek spurge" Euphorbiaceae
CNPS List: 4 **R-E-D Code:** 1-1-3 **State/Fed. Status:** /C3c
Distribution: COL, GLE, TEH
Habitat: VFGrs (sandy or rocky)
Life Form: Annual herb
Blooming: May-October
Notes: See *Proceedings of the American Academy of Arts and Sciences* 20:372 (1885) for original description, and *Madroño* 32(3):187-189 (1985) for revised nomenclature.

Chamaesyce ocellata var. *rattanii*

See *Chamaesyce ocellata* ssp. *rattanii*

Chamaesyce parishii

Considered but rejected: Too common

Chamaesyce pediculifera

Considered but rejected: Too common

Chamaesyce platysperma (Engelm.) Shinners

"flat-seeded spurge" Euphorbiaceae
CNPS List: 1B **R-E-D Code:** 3-2-2 **State/Fed. Status:** /C2
Distribution: IMP, SBD?, SDG, RIV, AZ, SO
Quads: 30D, 32C, 82C, 130B?
Habitat: SDScr (sandy), DeDns
Life Form: Annual herb
Blooming: February-September
Notes: Known in CA from only four herbarium collections (need confirmation) and a 1987 collection from IMP Co. See *Desert Plants* 2(2):104-105 (1980) for species account.

Cheilanthes clevelandii

Considered but rejected: Too common

Cheilanthes clokeyi

Considered but rejected: Not published

Cheilanthes cochisensis

See *Astrolepis cochisensis*

Cheilanthes fibrillosa

Considered but rejected: A sterile hybrid

Cheilanthes limitanea var. *limitanea*

See *Argyrochosma limitanea* var. *limitanea*

Cheilanthes newberryi

Considered but rejected: Too common

Cheilanthes viscida

Considered but rejected: Too common

Cheilanthes wootonii Maxon

"Wooton's lace fern" Pteridaceae
CNPS List: 2 **R-E-D Code:** 2-1-1 **State/Fed. Status:** CEQA
Distribution: INY?, SBD, AZ, BA, SO+
Quads: 200A, 225D
Habitat: JTWld, PJWld / rocky
Life Form: Perennial herb (rhizomatous)
Fertile: May-October
Notes: Reported from the Panamint Mtns. in *A Flora of Southern California* (1974) by P. Munz; does plant occur in INY Co.? See *Madroño* 25:56 (1978) for distributional information, and *Phytologia* 41(6):431-437 (1979) for nomenclature.

Chenopodium gigantospermum

See *Chenopodium simplex*

Chenopodium nevadense

Considered but rejected: Too common

Chenopodium simplex (Torr.) Raf.

"large-seeded goosefoot" Chenopodiaceae
CNPS List: 4 **R-E-D Code:** 1-1-1 **State/Fed. Status:** CEQA?
Distribution: INY, MOD, PLU, ++
Habitat: LCFrs (openings, disturbed areas) , PJWld (carbonate)
Life Form: Annual herb
Blooming: June-October

Chlorogalum grandiflorum Hoov.

"Red Hills soaproot" Liliaceae
CNPS List: 1B **R-E-D Code:** 2-2-3 **State/Fed. Status:** /C2
Distribution: ELD, PLA, TUO
Quads: 458B, 458C, 459A, 459D, 510B, 511A, 525B, 527D
Habitat: Chprl, CmWld / serpentinite or gabbroic
Life Form: Perennial herb (bulbiferous)
Blooming: May-June
Notes: Threatened by development, mining, and vehicles. Protected in part at Red Hills ACEC (BLM), TUO Co. See *Leaflets of Western Botany* 2(8):128 (1938) for original description.
Status Report: 1977

Chlorogalum pomeridianum (DC.) Kunth var. *minus* Hoov.

"dwarf soaproot" Liliaceae
CNPS List: 1B **R-E-D Code:** 2-2-3 **State/Fed. Status:** CEQA
Distribution: COL, LAK, SLO, SON, TEH
Quads: 246C?, 247D, 519D, 548A, 596A?, 596B
Habitat: Chprl (serpentinite)
Life Form: Perennial herb (bulbiferous)
Blooming: May-August
Notes: See *Madroño* 5(5):144 (1940) for original description.

Chlorogalum purpureum Bdg. var. *purpureum*

"purple amole" Liliaceae
CNPS List: 1B **R-E-D Code:** 3-3-3 **State/Fed. Status:** /C1
Distribution: MNT
Quads: 295B, 295C, 318C*
Habitat: CmWld, VFGrs
Life Form: Perennial herb (bulbiferous)
Blooming: May-June
Notes: Known from only five occurrences near Jolon on Ft. Hunter Liggett. Threatened by foot traffic, grazing, and vehicles.
Status Report: 1977

Chlorogalum purpureum Bdg. var. *reductum* Hoov.

"Camatta Canyon amole" Liliaceae
CNPS List: 1B **R-E-D Code:** 3-3-3 **State/Fed. Status:** CR/C1
Distribution: SLO
Quads: 245A
Habitat: CmWld (serpentinite)
Life Form: Perennial herb (bulbiferous)
Blooming: April-May
Notes: Known from only one occurrence near La Panza, on the Los Padres NF. Threatened by vehicles. A portion of the occurrence has been fenced by CNPS and USFS. See *Leaflets of Western Botany* 10:123 (1964) for original description.
Status Report: 1988

Chorizanthe angustifolia

Considered but rejected: Too common

Chorizanthe biloba Goodm. var. *immemora* Reveal & Hardham

"San Benito spineflower" Polygonaceae
CNPS List: 1B **R-E-D Code:** 2-2-3 **State/Fed. Status:** /C2
Distribution: FRE, MNT, SBT
Quads: 316B, 340A?, 340B, 340D, 385C
Habitat: Chprl, CmWld
Life Form: Annual herb
Blooming: May-September
Notes: See *Phytologia* 66(2):137-139 (1989) for original description.

Chorizanthe blakleyi Hardham

"Blakley's spineflower" Polygonaceae
CNPS List: 4 **R-E-D Code:** 1-1-3 **State/Fed. Status:** /C3c
Distribution: SBA, SLO
Habitat: Chprl
Life Form: Annual herb
Blooming: April-June
Notes: Closely related to *C. palmeri.* See *Leaflets of Western Botany* 10:95 (1964) for original description, and *Phytologia* 66(2):141-142 (1989) for taxonomic treatment.

Chorizanthe breweri Wats.

"Brewer's spineflower" Polygonaceae
CNPS List: 1B **R-E-D Code:** 3-1-3 **State/Fed. Status:** /C3c
Distribution: SLO
Quads: 221A, 221B, 246B, 246C, 246D
Habitat: CCFrs, Chprl, CmWld, CoScr / serpentinite
Life Form: Annual herb
Blooming: May-June
Notes: Known from approximately twenty occurrences. Possibly threatened by road construction. Closely related to *C. staticoides.* See *Proceedings of the American Academy of Arts and Sciences* 12:270 (1877) for original description, and *Phytologia* 66(2):163-164 (1989) for taxonomic treatment.

Chorizanthe californica var. *suksdorfii*

See *Mucronea californica*

Chorizanthe cuspidata Wats. var. *cuspidata*

"San Francisco Bay spineflower" Polygonaceae
CNPS List: 1B **R-E-D Code:** 2-2-3 **State/Fed. Status:** /C2
Distribution: ALA*, MRN, SCL?, SFO, SMT, SON
Quads: 448B, 448C, 448D, 466C, 466D*, 467A, 503D
Habitat: CBScr, CoDns, CoPrr, CoScr / sandy
Life Form: Annual herb
Blooming: April-July
Notes: Plant may occur in SCL Co.; need more information. Closely related to *C. pungens.* Some plants from Point Reyes (MRN Co.) probably intermediate to var. *villosa.* See *C. cuspidata* in *The Jepson Manual.* See *Proceedings of the Davenport Academy of Natural Sciences* 4:60 (1884) for original description, and *Phytologia* 66(2):127-129 (1989) for taxonomic treatment.

Chorizanthe cuspidata Wats. var. *villosa* (Eastw.) Munz

"woolly-headed spineflower" Polygonaceae
CNPS List: 1B **R-E-D Code:** 2-2-3 **State/Fed. Status:** CEQA
Distribution: MRN, SON
Quads: 485B, 485C, 502C, 503D
Habitat: CoDns, CoPrr, CoScr / sandy
Life Form: Annual herb
Blooming: May-August
Notes: Endemic to coastline from Bodega Bay to Point Reyes. See *C. cuspidata* in *The Jepson Manual.* See *Bulletin of the Torrey Botanical Club* 30:485 (1903) for original description, and *Phytologia* 66(2):127-130 (1989) for taxonomic treatment.

Chorizanthe douglasii Benth.

"Douglas's spineflower" Polygonaceae
CNPS List: 4 **R-E-D Code:** 1-1-3 **State/Fed. Status:** CEQA?
Distribution: MNT, SBT, SLO
Habitat: Chprl, CmWld, LCFrs
Life Form: Annual herb
Blooming: April-July
Notes: See *Phytologia* 66(2):118-120 (1989) for taxonomic treatment.

Chorizanthe fimbriata var. *fimbriata*

Considered but rejected: Too common

Chorizanthe fimbriata var. *laciniata*

Considered but rejected: Too common

Chorizanthe howellii Goodm.

"Howell's spineflower" Polygonaceae
CNPS List: 1B **R-E-D Code:** 3-2-3 **State/Fed. Status:** CT/FE
Distribution: MEN
Quads: 569A, 585D
Habitat: CoDns, CoPrr (sandy areas near dunes)
Life Form: Annual herb
Blooming: May-July
Notes: Known from only five occurrences near Ten Mile River dunes north of Fort Bragg. Threatened by recreational activities and non-native plants. Closely related to *C. pungens.* See *Annals of the Missouri Botanical Garden* 21:44 (1934) for original description, and *Phytologia* 66(2):131-132 (1989) for taxonomic treatment.
Status Report: 1985

Chorizanthe insignis

See *Aristocapsa insignis*

Chorizanthe leptoceras

See *Dodecahema leptoceras*

Chorizanthe leptotheca Goodm.

"Peninsular spineflower" Polygonaceae
CNPS List: 4 **R-E-D Code:** 1-2-2 **State/Fed. Status:** CEQA?
Distribution: RIV, SBD, SDG, BA
Habitat: Chprl, CoScr, LCFrs / alluvial fan, granitic
Life Form: Annual herb
Blooming: May-August
Notes: Much habitat already lost to development; also threatened by non-native grasses. Closely related to and difficult to distinguish from *C. staticoides.* See *Annals of the Missouri Botanical Garden* 21:61 (1934) for original description, and *Phytologia* 66(2):159-160 (1989) for taxonomic treatment.

Chorizanthe orcuttiana Parry

"Orcutt's spineflower" Polygonaceae
CNPS List: 1B **R-E-D Code:** 3-3-3 **State/Fed. Status:** CE/PE
Distribution: SDG
Quads: 11B*, 22B, 22C*, 35C*, 36D
Habitat: Chprl, CCFrs, CoScr
Life Form: Annual herb
Blooming: March-April
Notes: Only known occurrence confirmed in 1991 by C. Reiser. Most historical habitat urbanized. Threatened by foot traffic. See *Phytologia* 66(2):183 (1989) for taxonomic treatment.
Status Report: 1987

Chorizanthe palmeri Wats.

"Palmer's spineflower" Polygonaceae
CNPS List: 4 **R-E-D Code:** 1-2-3 **State/Fed. Status:** CEQA?
Distribution: MNT, SBA, SBT?, SLO
Habitat: Chprl, CmWld, VFGrs / serpentinite
Life Form: Annual herb
Blooming: May-August
Notes: Too common? Does plant occur in SBT Co.? Isolated populations show local differences. Taxonomic revision in *Phytologia* 66(4):295-441 (1989) indicates species occurs mainly in the Santa Lucia Mtns. of MNT and SLO counties. See *Proceedings of the American Academy of Arts and Sciences* 12:271 (1877) for original description, and *Phytologia* 66(2):135-137 (1989) for taxonomic treatment.

Chorizanthe parryi Wats. var. *fernandina* (Wats.) Jeps.

"San Fernando Valley spineflower" Polygonaceae
CNPS List: 1A **Last Seen:** 1940 **State/Fed. Status:** /C2*
Distribution: LAX*, ORA*, SDG*
Quads: 22B*, 22C*, 70B*, 87C*, 88D*, 110A*, 112B*, 137C*, 137D*, 138A*, 138B*, 138D*, 162B*
Habitat: CoScr (sandy)
Life Form: Annual herb
Blooming: April-June
Notes: Most historical habitat is now heavily urbanized. Most likely to be rediscovered in northwestern LAX Co. See *Botany of California* 2:481 (1880) for original description, and *Phytologia* 66(2):147-149 (1989) for taxonomic treatment.
Status Report: 1979

Chorizanthe parryi Wats. var. *parryi*

"Parry's spineflower" Polygonaceae
CNPS List: 3 **R-E-D Code:** ?-2-3 **State/Fed. Status:** /C2
Distribution: LAX(*?), RIV, SBD
Quads: 49B, 68A, 68B, 68C, 69A, 84A, 84B, 86C, 86D, 106C, 106D, 107A, 107B, 107D, 108B, 108C, 110A(*?), 110B(*?), 113D(*?), 133C?
Habitat: Chprl, CoScr / sandy openings
Life Form: Annual herb
Blooming: April-June
Notes: Move to List 1B? Location and rarity information needed. Known from about twenty occurrences in RIV Co. Habitat dwindling rapidly due to urbanization; may be extirpated from LAX Co. Previously confused with *C. procumbens*; often misidentified as this plant. See *Proceedings of the American Academy of Arts and Sciences* 12:271 (1877) for original description, and *Phytologia* 66(2):147-149 (1989) for taxonomic treatment.

Chorizanthe polygonoides T. & G. var. *longispina* (Goodm.) Munz

"long-spined spineflower" Polygonaceae
CNPS List: 1B **R-E-D Code:** 2-2-2 **State/Fed. Status:** /C2
Distribution: RIV, SDG, BA
Quads: 11B, 20A, 21C?, 22B, 33D?, 50B, 51B, 66B, 68C, 68D, 69C, 69D, 86C, 86D
Habitat: Chprl, CoScr, Medws, VFGrs / often clay
Life Form: Annual herb
Blooming: April-July
Notes: Much habitat already lost to development; also threatened by non-native grasses. See *Leaflets of Western Botany* 7(10):236 (1955) for original description, and *Phytologia* 66(2):176-179 (1989) for taxonomic treatment.

Chorizanthe procumbens Nutt.

"prostrate spineflower" Polygonaceae
CNPS List: 4 **R-E-D Code:** 1-2-2 **State/Fed. Status:** CEQA?
Distribution: LAX, ORA, RIV, SBD, SDG, VEN, BA
Habitat: Chprl, CoScr, PJWld, VFGrs / gabbroic clay, granitic
Life Form: Annual herb
Blooming: April-June
Notes: Too common? Much habitat already lost to development; also threatened by non-native grasses. Includes *C. procumbens* var. *albiflora.* Subsection is taxonomically difficult; see *Phytologia* 66(2):151-154 (1989) for discussion.

Chorizanthe procumbens var. *albiflora*

See *Chorizanthe procumbens*

Chorizanthe pungens Benth. var. *hartwegiana* Reveal & Hardham

"Ben Lomond spineflower" Polygonaceae
CNPS List: 1B **R-E-D Code:** 2-3-3 **State/Fed. Status:** /PE
Distribution: SCR
Quads: 387B, 407C, 408A?, 408B, 408D
Habitat: LCFrs (maritime ponderosa pine sandhills)
Life Form: Annual herb
Blooming: April-July
Notes: Endemic to sandhill parkland communities in the Santa Cruz Mtns. Threatened by sand mining and residential development. See *C. pungens* in *The Jepson Manual.* See *Annals of the Missouri Botanical Garden* 21:37 (1934) for original description, and *Phytologia* 66(2):123-126 (1989) for taxonomic treatment.

Chorizanthe pungens Benth. var. *pungens*

"Monterey spineflower" Polygonaceae
CNPS List: 1B **R-E-D Code:** 2-2-3 **State/Fed. Status:** /PE
Distribution: MNT, SCR
Quads: 342A, 366A, 366C, 366D, 386B, 386C, 387A, 387D
Habitat: CoDns, CoScr
Life Form: Annual herb
Blooming: April-June
Notes: See *C. pungens* in *The Jepson Manual.* See *Phytologia* 66(2):123-125 (1989) for taxonomic treatment.

Chorizanthe rectispina Goodm.

"straight-awned spineflower" Polygonaceae
CNPS List: 1B **R-E-D Code:** 3-1-3 **State/Fed. Status:** /C2
Distribution: MNT, SBA, SLO
Quads: 196D, 245A, 245B, 246A, 246B, 294B, 295A, 295B
Habitat: Chprl, CmWld, CoScr
Life Form: Annual herb
Blooming: June-July
Notes: Known from approximately twenty occurrences. See *Annals of the Missouri Botanical Garden* 21:72 (1934) for original description, and *Phytologia* 66(2):143 (1989) for taxonomic treatment.

Chorizanthe robusta

See *Chorizanthe robusta* var. *robusta*

Chorizanthe robusta Parry var. *hartwegii* (Benth.) Reveal & Morgan

"Scott's Valley spineflower" Polygonaceae
CNPS List: 1B **R-E-D Code:** 3-3-3 **State/Fed. Status:** /PE
Distribution: SCR
Quads: 407C, 408D
Habitat: Medws (xeric, sandy), VFGrs (mudstone and Purisima outcrops)
Life Form: Annual herb
Blooming: April-July
Notes: Known from only one extended population in Scotts Valley. Threatened by development. See *C. robusta* in *The Jepson Manual.* See *Phytologia* 67(5):357-360 (1989) for revised nomenclature.

Chorizanthe robusta Parry var. *robusta*

"robust spineflower" Polygonaceae
CNPS List: 1B **R-E-D Code:** 3-3-3 **State/Fed. Status:** /PE
Distribution: ALA*, MNT, SCL*, SCR, SMT*
Quads: 318D*, 342A*, 366D*, 386B(*?), 387A, 387B, 387D, 407B*, 407C*, 408D(*?), 427C*, 448B*, 465C*
Habitat: CmWld (openings), CoDns, CoScr
Life Form: Annual herb
Blooming: May-September
Notes: Most populations extirpated and now known from only four extended occurrences. Threatened by development, recreation, mining, and dune stabilization. See *C. robusta* in *The Jepson Manual.* See *Phytologia* 66(2):130-131 (1989) for taxonomic treatment.

Chorizanthe spinosa Wats.

"Mojave spineflower" Polygonaceae
CNPS List: 4 **R-E-D Code:** 1-2-3 **State/Fed. Status:** /C3c
Distribution: KRN, LAX, SBD
Habitat: ChScr, MDScr
Life Form: Annual herb
Blooming: April-July
Notes: Threatened by vehicles and energy development. See *Flora of California* 2:481 (1880) by S. Watson for original description, and *Phytologia* 66(2):109-110 (1989) for taxonomic treatment.
Status Report: 1977

Chorizanthe staticoides ssp. *chrysacantha*

Considered but rejected: A synonym of *C. staticoides*; a common taxon

Chorizanthe staticoides var. *compacta*

Considered but rejected: A synonym of *C. staticoides*; a common taxon

Chorizanthe stellulata

Considered but rejected: Too common

Chorizanthe valida Wats.

"Sonoma spineflower" Polygonaceae
CNPS List: 1B **R-E-D Code:** 3-3-3 **State/Fed. Status:** CE/FE
Distribution: MRN, SON*
Quads: 467D*, 485C, 502A*, 519C*
Habitat: CoPrr (sandy)
Life Form: Annual herb
Blooming: June-August
Notes: Thought extinct for 77 years; only known extant occurrence was rediscovered in 1980 at Point Reyes NS. Closely related to *C. pungens.* See *Proceedings of the American Academy of Arts and Sciences* 12:271 (1877) for original description, *Phytologia* 66(2):132-134 (1989) for taxonomic treatment, *Fremontia* 18(1):17-18 (1990) for species account, and *Madroño* 39(4):271-280 (1992) for ecological study.
Status Report: 1990

Chorizanthe vortriedei

See *Systenotheca vortriedei*

Chorizanthe wheeleri Wats.

"Wheeler's spineflower" Polygonaceae
CNPS List: 4 **R-E-D Code:** 1-1-3 **State/Fed. Status:** CEQA?
Distribution: SCZ, SRO
Habitat: Chprl
Life Form: Annual herb
Blooming: April-July
Notes: Mainland occurrences are *C. staticoides*; a closely-related widespread taxon. See *Proceedings of the American Academy of Arts and Sciences* 12:272 (1877) for original description, and *Phytologia* 66(2):164-166 (1989) for taxonomic treatment.

Chorizanthe xanti Wats. var. *leucotheca* Goodm.

"white-bracted spineflower" Polygonaceae
CNPS List: 4 **R-E-D Code:** 1-2-3 **State/Fed. Status:** CEQA?
Distribution: RIV, SBD
Habitat: MDScr, PJWld
Life Form: Annual herb
Blooming: April-June
Notes: See *Annals of the Missouri Botanical Garden* 21:60 (1934) for original description, and *Phytologia* 66(2):160-163 (1989) for taxonomic treatment.

Chrysothamnus axillaris

Considered but rejected: Too common; a synonym of *C. viscidiflorus* ssp. *axillaris*

Chrysothamnus gramineus Hall

"Panamint rock-goldenrod" Asteraceae
CNPS List: 4 **R-E-D Code:** 1-1-1 **State/Fed. Status:** CEQA?
Distribution: INY, SBD?, NV
Habitat: PJWld, SCFrs / carbonate, rocky
Life Form: Perennial herb
Blooming: June-July
Notes: Does plant occur in SBD Co.?

Chrysothamnus parryi ssp. *bolanderi*

Considered but rejected: A hybrid between *C. nauseosus* ssp. *albicaulis* and *Ericameria discoidea*

Cicuta bolanderi

Considered but rejected: A synonym of *C. maculata* var. *bolanderi*; a common taxon

Cicuta maculata var. *bolanderi*

Considered but rejected: Too common

Cirsium andrewsii (Gray) Jeps.

"Franciscan thistle" Asteraceae
CNPS List: 4 **R-E-D Code:** 1-1-3 **State/Fed. Status:** CEQA?
Distribution: MRN, SFO, SMT, SON
Habitat: BUFrs, CBScr / sometimes serpentinite
Life Form: Perennial herb
Blooming: June-July

Cirsium campylon

See *Cirsium fontinale* var. *campylon*

Cirsium ciliolatum (Henders.) J.T. Howell

"Ashland thistle" Asteraceae
CNPS List: 2 **R-E-D Code:** 3-3-1 **State/Fed. Status:** CE/C3b
Distribution: SIS, OR
Quads: 733C, 733D
Habitat: CmWld, VFGrs
Life Form: Perennial herb
Blooming: June-August
Notes: Known in CA from fewer than ten occurrences. Threatened by agriculture and road maintenance. See *Bulletin of the Torrey Botanical Club* 27:348 (1900) for original description.
Status Report: 1987

Cirsium crassicaule (Greene) Jeps.

"slough thistle" Asteraceae
CNPS List: 1B **R-E-D Code:** 3-2-3 **State/Fed. Status:** /C2
Distribution: KNG, KRN, SJQ
Quads: 241D, 264C, 265A, 265B, 265D, 289C, 444A, 462D
Habitat: ChScr, MshSw (sloughs), RpScr
Life Form: Annual/Perennial herb
Blooming: May-August
Notes: Threatened by agriculture. Population sizes fluctuate widely from year to year.
Status Report: 1979

Cirsium fontinale Greene var. *campylon* (H.K. Sharsm.) Keil & C. Turner

"Mt. Hamilton thistle" Asteraceae
CNPS List: 1B **R-E-D Code:** 2-2-3 **State/Fed. Status:** /C2
Distribution: ALA, SCL, STA
Quads: 406A*, 406B, 407A, 407B, 425B, 426A, 426C, 426D, 427D, 445D
Habitat: Chprl, CmWld, VFGrs / serpentinite seeps
Life Form: Perennial herb
Blooming: April-October
Notes: Threatened by urbanization and cattle trampling and grazing. See *Phytologia* 73(4):312-317 (1992) for revised nomenclature.
Status Report: 1979

Cirsium fontinale (Greene) Jeps. var. *fontinale*

"fountain thistle" Asteraceae
CNPS List: 1B **R-E-D Code:** 3-3-3 **State/Fed. Status:** CE/PE
Distribution: SMT
Quads: 429A, 448D
Habitat: Chprl (openings), VFGrs / serpentinite seeps
Life Form: Perennial herb
Blooming: June-October
Notes: Known from only four occurrences in the vicinity of Crystal Springs Reservoir. Seriously threatened by urbanization, dumping, road maintenance, and non-native plants. See *Bulletin of the California Academy of Sciences* 2:151-152 (1886) for original description.
Status Report: 1988

Cirsium fontinale (Greene) Jeps. var. *obispoense* J.T. Howell

"Chorro Creek bog thistle" Asteraceae
CNPS List: 1B **R-E-D Code:** 3-2-3 **State/Fed. Status:** CE/PE
Distribution: SLO
Quads: 221B, 246C, 271A
Habitat: Chprl, CmWld / serpentinite seeps
Life Form: Perennial herb
Blooming: February-July
Notes: Known from fewer than ten occurrences. Threatened by grazing, development, and proposed water diversions. See *Leaflets of Western Botany* 2:71 (1938) for original description.
Status Report: 1977

Cirsium hydrophilum (Greene) Jeps. var. *hydrophilum*

"Suisun thistle" Asteraceae
CNPS List: 1B **R-E-D Code:** 3-3-3 **State/Fed. Status:** /C1
Distribution: SOL
Quads: 482A
Habitat: MshSw (salt)
Life Form: Perennial herb
Blooming: July-September
Notes: Only known occurrence rediscovered in 1989 by N. Havlik on Grizzly Isl. in the Suisun Marsh. Possibly threatened by altered hydrology and competition from native and non-native plants. Protected in part at Grizzly Island WA (DFG) and Peytonia Slough ER (DFG). See *Proceedings of the Academy of Natural Sciences of Philadelphia* 44:358 (1892) for original description.
Status Report: 1977

Cirsium hydrophilum (Greene) Jeps. var. *vaseyi* (Gray) J.T. Howell

"Mt. Tamalpais thistle" Asteraceae
CNPS List: 1B **R-E-D Code:** 3-2-3 **State/Fed. Status:** /C2
Distribution: MRN
Quads: 467A, 467B
Habitat: BUFrs, Chprl / serpentinite seeps
Life Form: Perennial herb
Blooming: May-July
Notes: Known from fewer than ten occurrences on Mt. Tamalpais. Threatened by road construction and non-native plants. See *Synoptical Flora of North America* 1(2):403-404 (1884) for original description.

Cirsium loncholepis Petrak

"La Graciosa thistle" Asteraceae
CNPS List: 1B **R-E-D Code:** 3-2-3 **State/Fed. Status:** CT/C1
Distribution: SBA, SLO
Quads: 171A*, 195C*, 195D, 196A, 196B, 221B*, 221D
Habitat: CoDns (mesic), MshSw (brackish)
Life Form: Perennial herb
Blooming: June-August
Notes: Known from approximately twenty occurrences. Threatened by coastal development, vehicles, and grazing.
Status Report: 1989

Cirsium nidulum

Considered but rejected: Too common

Cirsium occidentale (Nutt.) Jeps. var. *compactum* Hoov.

"compact cobwebby thistle" Asteraceae
CNPS List: 1B **R-E-D Code:** 2-2-3 **State/Fed. Status:** /C2
Distribution: MNT?, SFO*, SLO
Quads: 271B, 271C, 271D, 272A, 344C?, 448B*
Habitat: Chprl, CoDns, CoPrr, CoScr
Life Form: Perennial herb
Blooming: April-June
Notes: Known from fewer than twenty occurrences. Threatened by grazing, and potentially by road construction and development. Some inland plants weakly separated from var. *occidentale.* Compact, low-growing plants from MNT Co. (344C) are probably not var. *compactum.*

Cirsium rhothophilum Blake

"surf thistle" Asteraceae
CNPS List: 1B **R-E-D Code:** 2-2-3 **State/Fed. Status:** CT/C1
Distribution: SBA, SLO
Quads: 145B, 171A, 171C, 196B, 196D, 221D, 247D
Habitat: CBScr, CoDns
Life Form: Perennial herb
Blooming: April-June
Notes: Threatened by vehicles, foot traffic, and non-native plants. See *Botanical Gazette* 39:45 (1905) for original description.
Status Report: 1990

Cirsium walkerianum

Considered but rejected: A synonym of *C. quercetorum*; a common taxon

Clarkia amoena (Lehm.) Nels. & Macbr. ssp. *whitneyi* (Gray) Lewis & Lewis

"Whitney's farewell-to-spring" Onagraceae
CNPS List: 4 **R-E-D Code:** 1-1-3 **State/Fed. Status:** /C3c
Distribution: HUM, MEN
Habitat: CBScr, CoScr
Life Form: Annual herb
Blooming: June-August

Clarkia australis E. Small

"Small's southern clarkia" Onagraceae
CNPS List: 1B **R-E-D Code:** 2-2-3 **State/Fed. Status:** /C3c
Distribution: MAD, MPA, TUO
Quads: 418B, 438B, 439A, 455C, 456A, 456C, 456D, 457D, 474C
Habitat: CmWld, LCFrs
Life Form: Annual herb
Blooming: May-August
Notes: Threatened by logging and reforestation with herbicides. See *Canadian Journal of Botany* 49:1211-1217 (1971) for taxonomic treatment.

Clarkia biloba (Durand) Nels. & Macbr. ssp. *australis* Lewis & Lewis

"Mariposa clarkia" Onagraceae
CNPS List: 1B **R-E-D Code:** 3-2-3 **State/Fed. Status:** /C3c
Distribution: MPA, TUO?
Quads: 419B, 438A, 438B, 438C, 439B, 439C, 439D
Habitat: Chprl, CmWld
Life Form: Annual herb
Blooming: May-July
Notes: Known from fewer than twenty occurrences. Does plant occur in TUO Co.? Threatened by road maintenance and biocide spraying.

Clarkia borealis E. Small ssp. *arida* E. Small

"Shasta clarkia" Onagraceae
CNPS List: 1B **R-E-D Code:** 3-3-3 **State/Fed. Status:** /C2
Distribution: SHA
Quads: 646D
Habitat: CmWld
Life Form: Annual herb
Blooming: June
Notes: Known from only one occurrence near Redding. Field surveys needed. See *Canadian Journal of Botany* 49:1211-1217 (1971) for original description.
Status Report: 1977

Clarkia borealis E. Small ssp. *borealis*

"northern clarkia" Onagraceae
CNPS List: 4 **R-E-D Code:** 1-1-3 **State/Fed. Status:** CEQA?
Distribution: SHA, TRI
Habitat: Chprl, CmWld, LCFrs
Life Form: Annual herb
Blooming: June-September
Notes: See *Canadian Journal of Botany* 49:1211-1217 (1971) for taxonomic revision.

Clarkia bottae

See *Clarkia lewisii*, a name previously misapplied to this taxon

Clarkia breweri (Gray) Greene

"Brewer's clarkia" Onagraceae
CNPS List: 4 **R-E-D Code:** 1-2-3 **State/Fed. Status:** CEQA?
Distribution: ALA, FRE, MER, MNT, SBT, SCL, STA
Habitat: Chprl, CmWld, CoScr / often serpentinite
Life Form: Annual herb
Blooming: April-May
Notes: Threatened by cattle grazing, and potentially by reservoir construction.

Clarkia calientensis

See *Clarkia tembloriensis* ssp. *calientensis*

Clarkia concinna Fisch. & Mey. (Greene) ssp. *automixa* Bowman

"Santa Clara red ribbons" Onagraceae
CNPS List: 1B **R-E-D Code:** 2-2-3 **State/Fed. Status:** /C2
Distribution: ALA, SCL
Quads: 406A, 407B, 408A, 426B, 426C, 427A, 427D, 428B, 428C, 428D, 445D, 446C
Habitat: CmWld
Life Form: Annual herb
Blooming: April-July
Notes: See *Madroño* 34(1):41-47 (1987) for original description.

Clarkia concinna Fisch. & Mey. (Greene) ssp. *raichei* G. Allen, V. Ford & L. Gottlieb

"Raiche's red ribbons" Onagraceae
CNPS List: 1B **R-E-D Code:** 3-1-3 **State/Fed. Status:** /C2
Distribution: MRN
Quads: 485B
Habitat: CBScr
Life Form: Annual herb
Blooming: April-May
Notes: Known from only one occurrence near Tomales. See *Madroño* 37(4):305-310 (1990) for original description.

Clarkia delicata (Abrams) Nels. & Macbr.

"delicate clarkia" Onagraceae
CNPS List: 2 **R-E-D Code:** 1-2-1 **State/Fed. Status:** CEQA
Distribution: SDG, BA
Quads: 21D, 34C, 34D
Habitat: Chprl, CmWld
Life Form: Annual herb
Blooming: May-June
Notes: Threatened by development and road improvement.

Clarkia exilis Lewis & Vasek

"slender clarkia" Onagraceae
CNPS List: 4 **R-E-D Code:** 1-1-3 **State/Fed. Status:** CEQA?
Distribution: KRN, TUL
Habitat: CmWld
Life Form: Annual herb
Blooming: April-May
Notes: See *Evolution* 18:26-42 (1964) for occurrence information.

Clarkia franciscana Lewis & Raven

"Presidio clarkia" Onagraceae
CNPS List: 1B **R-E-D Code:** 3-3-3 **State/Fed. Status:** CE/PE
Distribution: ALA, SFO
Quads: 465C, 466C
Habitat: CoScr, VFGrs (serpentinite)
Life Form: Annual herb
Blooming: May-July
Notes: Known from fewer than five occurrences. Threatened by Army activities, vehicles, urbanization and non-native plants. See *Brittonia* 10:7-13 (1958) for original description, and *Madroño* 39(1):1-7 (1992) for information on ALA Co. occurrence.
Status Report: 1988

Clarkia gracilis (Piper) Nels. & Macbr. ssp. *albicaulis* (Jeps.) Lewis & Lewis

"white-stemmed clarkia" Onagraceae
CNPS List: 1B **R-E-D Code:** 3-2-3 **State/Fed. Status:** CEQA
Distribution: BUT
Quads: 575B, 575D, 576A, 591C, 592B, 592C, 592D, 593A, 593D
Habitat: Chprl, CmWld / sometimes serpentinite
Life Form: Annual herb
Blooming: May-July
Notes: Known from fewer than twenty occurrences. Threatened in foothills by urbanization. See *University of California Publications in Botany* 2:239 (1907) for original description and 20:241-392 (1955) for taxonomic treatment.

Clarkia imbricata Lewis & Lewis

"Vine Hill clarkia" Onagraceae
CNPS List: 1B **R-E-D Code:** 3-3-3 **State/Fed. Status:** CE/C1
Distribution: SON
Quads: 502A
Habitat: Chprl, Medws, VFGrs
Life Form: Annual herb
Blooming: June-July
Notes: Known from only two occurrences, one of which is transplanted and doing poorly; a third, natural occurrence has been extirpated. Threatened by development and road maintenance. See *Madroño* 12:33-39 (1953) for original description.
Status Report: 1988

Clarkia jolonensis Parnell

"Jolon clarkia" Onagraceae
CNPS List: 4 **R-E-D Code:** 1-1-3 **State/Fed. Status:** CEQA?
Distribution: MNT
Habitat: CmWld
Life Form: Annual herb
Blooming: June
Notes: See *Madroño* 20(6):322 (1970) for original description.

Clarkia lewisii Raven & Parnell

"Lewis's clarkia" Onagraceae
CNPS List: 4 **R-E-D Code:** 1-1-3 **State/Fed. Status:** CEQA?
Distribution: MNT, SBT
Habitat: Chprl, CmWld, CoScr
Life Form: Annual herb
Blooming: May-July
Notes: See *Annals of the Missouri Botanical Garden* 64:642 (1977) for revised taxonomy.

Clarkia lingulata Lewis & Lewis

"Merced clarkia" Onagraceae
CNPS List: 1B **R-E-D Code:** 3-3-3 **State/Fed. Status:** CE/C1
Distribution: MPA
Quads: 438B
Habitat: CCFrs, Chprl, CmWld
Life Form: Annual herb
Blooming: May-June
Notes: Known from only two occurrences along the Merced River. Largest occurrence damaged by herbicide spraying in 1984; also threatened by road widening. See *Madroño* 12:33-39 (1953) for original description.
Status Report: 1985

Clarkia mildrediae (Heller) Lewis & Lewis

"Mildred's clarkia" Onagraceae
CNPS List: 4 **R-E-D Code:** 1-1-3 **State/Fed. Status:** CEQA?
Distribution: BUT, PLU, YUB
Habitat: LCFrs
Life Form: Annual herb
Blooming: June-July

Clarkia mosquinii E. Small ssp. *mosquinii*

"Mosquin's clarkia" Onagraceae
CNPS List: 1B **R-E-D Code:** 3-3-3 **State/Fed. Status:** /C2
Distribution: BUT
Quads: 575B
Habitat: CmWld
Life Form: Annual herb
Blooming: May-July
Notes: Rediscovered in 1991 by L. Janeway. Identification difficult. See *Canadian Journal of Botany* 49:1211-1217 (1971) for original description.
Status Report: 1977

Clarkia mosquinii E. Small ssp. *xerophila* E. Small

"Enterprise clarkia" Onagraceae
CNPS List: 1B **R-E-D Code:** 3-3-3 **State/Fed. Status:** /C2
Distribution: BUT
Quads: 575A, 575B, 575D
Habitat: CmWld, LCFrs
Life Form: Annual herb
Blooming: May-July
Notes: Rediscovered from herbarium collections made by R. Schlising in 1981 and L. Ahart in 1983, but these occurrences need field study to confirm their current existence. Type locality covered by Oroville Reservoir in 1968. See *Canadian Journal of Botany* 49:1211-1217 (1971) for original description.

Clarkia rostrata W.S. Davis

"beaked clarkia" Onagraceae
CNPS List: 1B **R-E-D Code:** 2-1-3 **State/Fed. Status:** /C2
Distribution: MER, MPA, STA
Quads: 419C, 420C, 421A, 439C, 440C, 441A, 442A, 459C
Habitat: CmWld, VFGrs
Life Form: Annual herb
Blooming: April-May
Notes: See *Brittonia* 22:270-284 (1970) for original description.

Clarkia rubicunda ssp. *blasdalei*

Considered but rejected: A synonym of *C. rubicunda*; a common taxon

Clarkia rubicunda ssp. *rubicunda*

Considered but rejected: Too common and taxonomic problem

Clarkia speciosa Lewis & Lewis ssp. *immaculata* Lewis & Lewis

"Pismo clarkia" Onagraceae
CNPS List: 1B **R-E-D Code:** 3-3-3 **State/Fed. Status:** CR/PE
Distribution: SLO
Quads: 221A, 221D
Habitat: Chprl (margins, openings), CmWld, VFGrs
Life Form: Annual herb
Blooming: May-June
Notes: Known from only four extant occurrences. Threatened by development and road maintenance, and possibly by grazing. USFWS uses the name *C. speciosa* var. *immaculata*. See *University of California Publications in Botany* 20:291 (1955) for original description.
Status Report: 1987

Clarkia springvillensis Vasek

"Springville clarkia" Onagraceae
CNPS List: 1B **R-E-D Code:** 3-2-3 **State/Fed. Status:** CE/C1
Distribution: TUL
Quads: 308A, 308B, 309A, 332A
Habitat: Chprl, CmWld, VFGrs
Life Form: Annual herb
Blooming: May-July
Notes: Known from fewer than ten occurrences. Threatened by grazing, vehicles, road maintenance, logging, and residential development. Sequoia NF has adopted species management guidelines. See *Madroño* 17:220 (1964) for original description.
Status Report: 1993

Clarkia tembloriensis Vasek ssp. *calientensis* (Vasek) Holsinger

"Vasek's clarkia" Onagraceae
CNPS List: 1B **R-E-D Code:** 3-3-3 **State/Fed. Status:** /C1
Distribution: KRN
Quads: 238C, 239D
Habitat: VFGrs
Life Form: Annual herb
Blooming: April
Notes: Known from only three occurrences near Caliente Creek. Threatened by grazing and non-native plants. Perhaps best treated as *C. calientensis*. See *Systematic Botany* 2:252-255 (1977) for original description and 10(2):155-165 (1985) for taxonomic treatment.

Clarkia tembloriensis ssp. *tembloriensis*

Considered but rejected: Too common

Clarkia virgata Greene

"Sierra clarkia" Onagraceae
CNPS List: 4 **R-E-D Code:** 1-1-3 **State/Fed. Status:** CEQA?
Distribution: AMA, CAL, ELD, MPA, TUO
Habitat: CmWld, LCFrs
Life Form: Annual herb
Blooming: May-July
Notes: May form sterile hybrids with *C. australis*.

Clarkia xantiana Gray ssp. *parviflora* (Eastw.) Lewis & Raven

"Kern Canyon clarkia" Onagraceae
CNPS List: 1B **R-E-D Code:** 3-2-3 **State/Fed. Status:** CEQA
Distribution: KRN
Quads: 284B, 284C
Habitat: CmWld
Life Form: Annual herb
Blooming: May-June
Notes: Known only from the Kern River drainage. Threatened by road construction. See *Bulletin of the Torrey Botanical Club* 30:492 (1903) for original description, and *Madroño* 39(3):163-169 (1992) for revised nomenclature.

Claytonia bellidifolia

See *Claytonia megarhiza*

Claytonia lanceolata Pursh var. *peirsonii* Munz & Jtn.

"Peirson's spring beauty" Portulacaceae
CNPS List: 1B **R-E-D Code:** 3-3-3 **State/Fed. Status:** /C1
Distribution: SBD
Quads: 108A, 134D
Habitat: SCFrs, UCFrs (scree)
Life Form: Perennial herb
Blooming: May-June
Notes: Threatened by trampling and proposed ski area expansion. A synonym of *C. lanceolata* in *The Jepson Manual*. Angeles NF has adopted species management guidelines. See *Bulletin of the Torrey Botanical Club* 49:352 (1922) for original description.
Status Report: 1980

Claytonia megarhiza (Gray) Wats.

"fell-fields claytonia" Portulacaceae
CNPS List: 2 **R-E-D Code:** 2-1-1 **State/Fed. Status:** /C3c
Distribution: ALP, MNO, MOD, MPA, NEV, TUO, OR+
Quads: 435B, 454C, 454D, 472D, 506D, 523B, 555A, 690C
Habitat: AlpBR, SCFrs (rocky)
Life Form: Perennial herb
Blooming: July-August

Claytonia megarhiza var. *bellidifolia*

See *Claytonia megarhiza*

Claytonia palustris Swanson & Kelley

"marsh claytonia" Portulacaceae
CNPS List: 4 **R-E-D Code:** 1-1-3 **State/Fed. Status:** CEQA?
Distribution: BUT, FRE, PLU, SIS, TEH, TUL
Habitat: Medws (mesic), MshSw (montane)
Life Form: Perennial herb
Blooming: June-August
Notes: See *Madroño* 34(2):155-161 (1987) for original description.

Claytonia saxosa
Considered but rejected: Too common

Claytonia spathulata var. *rosulata*
Considered but rejected: Too common; a synonym of *C. exigua* ssp. *exigua*

Claytonia umbellata Wats.
"Great Basin claytonia" Portulacaceae
CNPS List: 1B **R-E-D Code:** 3-1-2 **State/Fed. Status:** CEQA
Distribution: ALP, LAS, MNO, MOD, SIS, NV, OR
Quads: 488A, 505B, 603A, 699C
Habitat: SCFrs (talus)
Life Form: Perennial herb
Blooming: June-August
Notes: Known from fewer than twenty occurrences throughout its range. Need quads for MOD Co. On watch list in OR. See *Botany of the King Exploration*, p. 43 (1871) for original description.

Cneoridium dumosum
Considered but rejected: Too common

Cochlearia groenlandica
See *Cochlearia officinalis* var. *arctica*

Cochlearia officinalis L. var. *arctica* (DC.) Gelert
"Arctic spoonwort" Brassicaceae
CNPS List: 2 **R-E-D Code:** 3-1-1 **State/Fed. Status:** CEQA
Distribution: DNT, OR+
Quads: 740C
Habitat: CBScr (on basaltic sea stack)
Life Form: Annual herb
Blooming: May-July
Notes: Known in CA only from a sea stack off Crescent City. Endangered in OR. See *Madroño* 28(3):86 (1981) for CA record.

Collinsia antonina
Considered but rejected: A synonym of *C. parryi*; a common taxon

Collinsia bartsiifolia var. *bartsiifolia*
Considered but rejected: Too common

Collinsia concolor
Considered but rejected: Too common

Collinsia corymbosa Herder
"round-headed chinese houses" Scrophulariaceae
CNPS List: 1B **R-E-D Code:** 2-2-3 **State/Fed. Status:** CEQA
Distribution: HUM, MEN, MRN?, SFO*, SON
Quads: 466C*, 569A, 585D
Habitat: CoDns
Life Form: Annual herb
Blooming: April-June
Notes: Scattered distribution. Does plant occur in MRN Co.? Need quads for HUM Co. and for "Russian colony" (SON Co.). May intergrade with *C. bartsiifolia* var. *bartsiifolia*.

Collinsia franciscana
See *Collinsia multicolor*

Collinsia greenei
Considered but rejected: Too common

Collinsia linearis
Considered but rejected: Too common

Collinsia multicolor Lindl. & Paxton
"San Francisco collinsia" Scrophulariaceae
CNPS List: 4 **R-E-D Code:** 1-1-3 **State/Fed. Status:** CEQA?
Distribution: MNT, SCR, SFO, SMT
Habitat: CCFrs, CoScr
Life Form: Annual herb
Blooming: March-May
Notes: Known in SCR Co. from only a few populations.

Collinsia parryi
Considered but rejected: Too common

Collinsia sparsiflora
Considered but rejected: Too common

Collomia debilis var. *larsenii*
See *Collomia larsenii*

Collomia diversifolia Greene
"serpentine collomia" Polemoniaceae
CNPS List: 4 **R-E-D Code:** 1-1-3 **State/Fed. Status:** CEQA?
Distribution: CCA, COL, GLE, LAK, MEN, NAP, YOL, SHA, STA
Habitat: Chprl, CmWld / serpentinite
Life Form: Annual herb
Blooming: May-June

Collomia larsenii (Gray) Payson
"talus collomia" Polemoniaceae
CNPS List: 2 **R-E-D Code:** 3-2-1 **State/Fed. Status:** /C3c
Distribution: SHA, SIS, OR, WA
Quads: 626A, 644A, 644D, 713C
Habitat: AlpBR, CCFrs, UCFrs / volcanic talus
Life Form: Perennial herb (rhizomatous)
Blooming: July-October
Notes: Known in CA from fewer than five occurrences. Threatened by foot traffic. On watch list in OR.

Collomia rawsoniana Greene
"flaming trumpet" Polemoniaceae
CNPS List: 1B **R-E-D Code:** 2-2-3 **State/Fed. Status:** /C1
Distribution: MAD, MPA
Quads: 397A, 397B, 417C, 417D, 418A, 418C, 418D
Habitat: LCFrs, RpFrs
Life Form: Perennial herb (rhizomatous)
Blooming: July-August
Notes: Need quads for MPA Co. Potentially threatened by hydroelectric development and logging. Interagency agreement has been established between USFWS and USFS, and Sierra NF has adopted species management guidelines. See *American Midland Naturalist* 31:216-231 (1944) for original description.

Collomia tracyi Mason

"Tracy's collomia" Polemoniaceae
CNPS List: 4 **R-E-D Code:** 1-1-3 **State/Fed. Status:** CEQA?
Distribution: DNT, HUM, LAS, SIS, TEH, TRI
Habitat: LCFrs
Life Form: Annual herb
Blooming: June-July
Notes: Similar to *C. tinctoria*, but does not intergrade.

Colubrina californica Jtn.

"Las Animas colubrina" Rhamnaceae
CNPS List: 4 **R-E-D Code:** 1-1-2 **State/Fed. Status:** /C3c
Distribution: IMP, RIV, SDG, AZ, BA, SO
Habitat: MDScr
Life Form: Shrub (evergreen)
Blooming: April-May
Notes: See *Proceedings of the California Academy of Sciences* IV 12:1085 (1924) for original description, and *Brittonia* 23:36 (1971) for distributional information.

Comarostaphylis diversifolia (Parry) Greene ssp. *diversifolia*

"summer holly" Ericaceae
CNPS List: 1B **R-E-D Code:** 2-2-2 **State/Fed. Status:** /C2
Distribution: ORA, RIV, SDG, BA
Quads: 10B, 10C, 10D, 22B, 22C, 22D, 35B, 35C, 35D, 36D, 47C?, 51B, 69C, 70C
Habitat: Chprl
Life Form: Shrub (evergreen)
Blooming: April-June
Notes: Threatened by development and gravel mining.

Condalia globosa Jtn. var. *pubescens* Jtn.

"spiny abrojo" Rhamnaceae
CNPS List: 4 **R-E-D Code:** 1-2-1 **State/Fed. Status:** CEQA?
Distribution: IMP, RIV, AZ, BA, SO+
Habitat: SDScr
Life Form: Shrub (deciduous)
Blooming: March-May
Notes: See *Proceedings of the California Academy of Sciences* IV 12:1087 (1924) for original description, and *Brittonia* 14:332-368 (1972) for taxonomic treatment.

Conioselinum chinense

Considered but rejected: Too common; a synonym of *C. pacificum*

Convolvulus simulans Perry

"small-flowered morning-glory" Convolvulaceae
CNPS List: 4 **R-E-D Code:** 1-2-2 **State/Fed. Status:** CEQA?
Distribution: CCA, KRN, LAX, RIV, SBA, SBT, SCM, SCT, SCZ, SDG, SJQ, SLO, STA, BA
Habitat: CoScr, VFGrs / clay, serpentinite seeps
Life Form: Annual herb
Blooming: March-June
Notes: Rare in southern CA. See *Rhodora* 33:76 (1931) for original description.

Corallorhiza trifida Chatel.

"northern coralroot" Orchidaceae
CNPS List: 2 **R-E-D Code:** 3-3-1 **State/Fed. Status:** CEQA
Distribution: PLU, NV, OR, ++
Quads: 590B
Habitat: LCFrs, Medws (edges) / mesic
Life Form: Perennial herb (rhizomatous, saprophytic)
Blooming: June-July
Notes: Known in CA from only one occurrence near Buck's Lake. See *The Wasmann Journal of Biology* 36:199-200 (1978) for information on CA occurrence, and *Fremontia* 19(1):22-23 (1991) for account of recent discovery (1990).

Cordylanthus bernardinus

See *Cordylanthus eremicus* ssp. *eremicus*

Cordylanthus brunneus ssp. *capillaris*

See *Cordylanthus tenuis* ssp. *capillaris*

Cordylanthus capitatus Benth.

"Yakima bird's-beak" Scrophulariaceae
CNPS List: 2 **R-E-D Code:** 1-2-1 **State/Fed. Status:** CEQA
Distribution: LAS, MOD, ID, NV, OR, WA
Quads: 622C, 707C, 724A, 724C
Habitat: LCFrs, PJWld
Life Form: Annual herb (hemiparasitic)
Blooming: July-September
Notes: See *Systematic Botany Monographs* 10:69-73 (1986) for taxonomic treatment.
Status Report: 1980

Cordylanthus eremicus (Cov. & Mort.) Munz ssp. *eremicus*

"desert bird's-beak" Scrophulariaceae
CNPS List: 4 **R-E-D Code:** 1-1-3 **State/Fed. Status:** /C3c
Distribution: INY, SBD
Habitat: JTWld, MDScr / rocky
Life Form: Annual herb (hemiparasitic)
Blooming: August-October
Notes: Includes *C. bernardinus.* See *Systematic Botany Monographs* 10:89-92 (1986) for revised taxonomic treatment.
Status Report: 1977

Cordylanthus eremicus (Cov. & Mort.) Munz ssp. *kernensis* Chuang & Heckard

"Kern Plateau bird's-beak" Scrophulariaceae
CNPS List: 4 **R-E-D Code:** 1-1-3 **State/Fed. Status:** CEQA?
Distribution: INY, KRN, TUL
Habitat: UCFrs
Life Form: Annual herb (hemiparasitic)
Blooming: July-September
Notes: Endemic to the Kern Plateau region. See *Systematic Botany Monographs* 10:89-92 (1986) for original description.

Cordylanthus ferrisianus

Considered but rejected: A synonym of *C. rigidus* ssp. *rigidus*; a common taxon

Cordylanthus helleri

Considered but rejected: Too common; a synonym of *C. kingii* ssp. *helleri*

Cordylanthus littoralis ssp. *littoralis*

See *Cordylanthus rigidus* ssp. *littoralis*

Cordylanthus maritimus Benth. ssp. *maritimus*

"salt marsh bird's-beak" Scrophulariaceae
CNPS List: 1B **R-E-D Code:** 2-2-2 **State/Fed. Status:** CE/FE
Distribution: LAX, ORA, SBA, SDG, SLO, VEN, BA
Quads: 11A, 11B*, 11D, 71B, 72A, 89C, 89D*, 112D*, 114B, 114D, 142A, 247D
Habitat: CoDns, MshSw (coastal salt)
Life Form: Annual herb (hemiparasitic)
Blooming: May-October
Notes: Threatened by vehicles, road construction, foot traffic, and loss of salt marsh habitat. See *Madroño* 31(3):185-190 (1984) for information on parasitism.
Status Report: 1988

Cordylanthus maritimus Benth. ssp. *palustris* (Behr) Chuang & Heckard

"Point Reyes bird's-beak" Scrophulariaceae
CNPS List: 1B **R-E-D Code:** 2-2-2 **State/Fed. Status:** /C2
Distribution: ALA*, HUM, MRN, SCL*, SMT*, SON, OR
Quads: 427B*, 428A*, 447B*, 447C*, 448D*, 466B*, 466D*, 467A, 467B, 484D, 485B, 485C, 485D, 502C, 503D, 654B, 655A, 672B, 672C, 672D
Habitat: MshSw (coastal salt)
Life Form: Annual herb (hemiparasitic)
Blooming: June-October
Notes: Habitat much reduced due to development; also threatened by foot traffic, non-native plants, altered hydrology, and cattle grazing and trampling. Candidate for state listing in OR. See *Brittonia* 25:135-158 (1973) for original description.
Status Report: 1990

Cordylanthus mollis Gray ssp. *hispidus* (Penn.) Chuang & Heckard

"hispid bird's-beak" Scrophulariaceae
CNPS List: 1B **R-E-D Code:** 2-3-3 **State/Fed. Status:** /C2
Distribution: ALA, KRN, MER, PLA, SOL
Quads: 215D, 382B, 402C, 403A, 403B, 403C, 403D, 423C, 445B, 481B, 528D
Habitat: Medws (alkaline), Plyas
Life Form: Annual herb (hemiparasitic)
Blooming: June-September
Notes: Extirpated from much of the lower San Joaquin Valley? Threatened by agricultural conversion, development, and grazing. See *Brittonia* 25:135-158 (1973) for revised nomenclature.

Cordylanthus mollis Gray ssp. *mollis*

"soft bird's-beak" Scrophulariaceae
CNPS List: 1B **R-E-D Code:** 3-2-3 **State/Fed. Status:** CR/C1
Distribution: CCA, MRN*, NAP, SOL, SON*
Quads: 481B, 481C, 482A, 482C, 482D, 483A, 483D, 484A*
Habitat: MshSw (coastal salt)
Life Form: Annual herb (hemiparasitic)
Blooming: July-September
Notes: Known from fewer than ten occurrences; some populations declining. Threatened by erosion and marsh drainage. See *Proceedings of the American Academy of Arts and Sciences* 7:327-402 (1867) for original description, and *Madroño* 25:107 (1978) for rediscovery in NAP Co.
Status Report: 1988

Cordylanthus nevinii

Considered but rejected: Too common

Cordylanthus nidularius J.T. Howell

"Mt. Diablo bird's-beak" Scrophulariaceae
CNPS List: 1B **R-E-D Code:** 3-3-3 **State/Fed. Status:** CR/C1
Distribution: CCA
Quads: 464B
Habitat: Chprl (serpentinite)
Life Form: Annual herb (hemiparasitic)
Blooming: July-August
Notes: Our most narrowly restricted bird's-beak; known from only one occurrence on Mt. Diablo. Possibly threatened by recreational activities. See *Leaflets of Western Botany* 3(9):207 (1943) for original description, and *Systematic Botany Monographs* 10:48-50 (1986) for taxonomic treatment.
Status Report: 1988

Cordylanthus orcuttianus Gray

"Orcutt's bird's-beak" Scrophulariaceae
CNPS List: 2 **R-E-D Code:** 3-3-1 **State/Fed. Status:** /C2
Distribution: SDG, BA
Quads: 10B, 10C, 11D
Habitat: CoScr
Life Form: Annual herb (hemiparasitic)
Blooming: March-July
Notes: Seriously threatened by urbanization.

Cordylanthus pallescens

See *Cordylanthus tenuis* ssp. *pallescens*

Cordylanthus palmatus (Ferris) Macbr.

"palmate-bracted bird's-beak" Scrophulariaceae
CNPS List: 1B **R-E-D Code:** 3-3-3 **State/Fed. Status:** CE/FE
Distribution: ALA, COL, FRE, MAD*, SJQ*, YOL
Quads: 359A*, 360A, 381A*, 445B, 446A, 462A*, 513B, 545C*, 546D, 562D
Habitat: ChScr, VFGrs (alkaline)
Life Form: Annual herb (hemiparasitic)
Blooming: May-October
Notes: Known from approximately six extant occurrences. Threatened by agriculture, urbanization, vehicles, and altered hydrology. See *Bulletin of the Torrey Botanical Club* 45:399-423 (1918) for original description, *Fremontia* 17(1):20-23 (1989) for information on Springtown Alkali Sink occurrence (ALA Co.), and *Environmental Management* 17:115-127 (1993) for population biology.
Status Report: 1986

Cordylanthus parviflorus (Ferris) Wiggins

"purple bird's-beak" Scrophulariaceae
CNPS List: 2 **R-E-D Code:** 3-1-1 **State/Fed. Status:** CEQA
Distribution: SBD, AZ, ID, NV, UT
Quads: 224C, 225D
Habitat: JTWld, MDScr, PJWld
Life Form: Annual herb (hemiparasitic)
Blooming: August-October
Notes: See *Systematic Botany Monographs* 10:75-77 (1986) for taxonomic treatment.

Cordylanthus pringlei

Considered but rejected: Too common

Cordylanthus rigidus (Benth.) Jeps. ssp. *littoralis* (Ferris) Chuang & Heckard

"seaside bird's-beak" Scrophulariaceae
CNPS List: 1B **R-E-D Code:** 2-3-3 **State/Fed. Status:** CE/C1
Distribution: MNT, SBA
Quads: 170B, 170D, 195C, 196D, 365B, 365C, 366A*, 366C*, 366D, 387D
Habitat: CCFrs, Chprl, CmWld, CoDns, CoScr / sandy
Life Form: Annual herb (hemiparasitic)
Blooming: May-September
Notes: Known from fewer than twenty occurrences. Threatened by coastal development. See *Bulletin of the Torrey Botanical Club* 45:399-423 (1918) for original description, and *Systematic Botany Monographs* 10:35-48 (1986) for taxonomic treatment.
Status Report: 1987

Cordylanthus tecopensis Munz & Roos

"Tecopa bird's-beak" Scrophulariaceae
CNPS List: 1B **R-E-D Code:** 3-2-2 **State/Fed. Status:** /C2
Distribution: INY, SBD, NV
Quads: 252B, 275B, 275C, 276A, 322D*
Habitat: MDScr, Medws / mesic alkaline
Life Form: Annual herb (hemiparasitic)
Blooming: July-October
Notes: Known in CA from approximately five occurrences. Threatened in NV. See *Aliso* 3(2):111-129 (1955) for original description.
Status Report: 1988

Cordylanthus tenuis Gray ssp. *barbatus* Chuang & Heckard

"Fresno County bird's-beak" Scrophulariaceae
CNPS List: 4 **R-E-D Code:** 1-1-3 **State/Fed. Status:** /C2
Distribution: FRE
Habitat: LCFrs
Life Form: Annual herb (hemiparasitic)
Blooming: July-August
Notes: See *Systematic Botany Monographs* 10:50-62 (1986) for original description.

Cordylanthus tenuis Gray ssp. *brunneus* (Jeps.) Munz

"serpentine bird's-beak" Scrophulariaceae
CNPS List: 4 **R-E-D Code:** 1-1-3 **State/Fed. Status:** CEQA?
Distribution: LAK, NAP, SON
Habitat: CCFrs, Chprl, CmWld / serpentinite
Life Form: Annual herb (hemiparasitic)
Blooming: July-August
Notes: See *Systematic Botany Monographs* 10:50-62 (1986) for taxonomic treatment.

Cordylanthus tenuis Gray ssp. *capillaris* (Penn.) Chuang & Heckard

"Pennell's bird's-beak" Scrophulariaceae
CNPS List: 1B **R-E-D Code:** 3-2-3 **State/Fed. Status:** CR/PE
Distribution: SON
Quads: 502B
Habitat: CCFrs, Chprl / serpentinite
Life Form: Annual herb (hemiparasitic)
Blooming: June-July
Notes: Known from fewer than five occurrences near Occidental. Threatened by dumping, vehicles, and road maintenance, and potentially by development. See *Systematic Botany Monographs* 10:50-62 (1986) for taxonomic treatment.
Status Report: 1988

Cordylanthus tenuis Gray ssp. *pallescens* (Penn.) Chuang & Heckard

"pallid bird's-beak" Scrophulariaceae
CNPS List: 1B **R-E-D Code:** 3-2-3 **State/Fed. Status:** /C2
Distribution: SIS
Quads: 682A, 682B, 699B, 699C, 699D, 700A
Habitat: LCFrs (gravelly, volcanic alluvium)
Life Form: Annual herb (hemiparasitic)
Blooming: July-August
Notes: Known from only one extended occurrence near Black Butte. Threatened by development and road maintenance. Intergrades with ssp. *viscidus*; how distinct is it? Further research needed. See *Systematic Botany Monographs* 10:50-62 (1986) for taxonomic treatment.
Status Report: 1977

Coreopsis gigantea

Considered but rejected: Too common

Coreopsis hamiltonii (Elmer) H.K. Sharsm.

"Mt. Hamilton coreopsis" Asteraceae
CNPS List: 1B **R-E-D Code:** 3-2-3 **State/Fed. Status:** /C2
Distribution: SCL, STA
Quads: 425C, 426C, 426D, 444D
Habitat: CmWld (rocky)
Life Form: Annual herb
Blooming: March-May
Notes: Known from fewer than ten occurrences in the Mt. Hamilton Range. See *Botanical Gazette* 41:323-324 (1906) for original description, and *Madroño* 4:214-215 (1938) for revised nomenclature.
Status Report: 1979

Coreopsis maritima (Nutt.) Hook. f.

"sea dahlia" Asteraceae
CNPS List: 2 **R-E-D Code:** 2-2-1 **State/Fed. Status:** CEQA
Distribution: SDG, BA
Quads: 11B, 22B, 35C, 36D, 51C
Habitat: CBScr, CoScr
Life Form: Perennial herb
Blooming: March-May

Corethrogyne filaginifolia (H. & A.) Nutt. var. *incana* (Nutt.) Canby

"San Diego sand aster" Asteraceae
CNPS List: 1B **R-E-D Code:** 2-2-2 **State/Fed. Status:** CEQA
Distribution: SDG, BA
Quads: 11B, 11D, 22B
Habitat: CoScr
Life Form: Perennial herb
Blooming: June-August
Notes: Known in CA from only three occurrences. Threatened by development. A synonym of *Lessingia filaginifolia* var. *filaginifolia* in *The Jepson Manual*; taxonomic study needed.

Corethrogyne filaginifolia (H. & A.) Nutt. var. *linifolia* Hall

"Del Mar Mesa sand aster" Asteraceae
CNPS List: 1B **R-E-D Code:** 3-2-3 **State/Fed. Status:** /PT
Distribution: SDG
Quads: 22B, 22C, 35C, 36D
Habitat: Chprl, CoScr
Life Form: Perennial herb
Blooming: July-September
Notes: Threatened by development. A synonym of *Lessingia filaginifolia* var. *filaginifolia* in *The Jepson Manual*; requires additional taxonomic work.

Corethrogyne leucophylla Jeps.

"branching beach aster" Asteraceae
CNPS List: 4 **R-E-D Code:** 1-1-3 **State/Fed. Status:** CEQA?
Distribution: MNT, SCR, SLO
Habitat: CCFrs, CoDns
Life Form: Perennial herb
Blooming: July-October
Notes: Needs taxonomic study; a synonym of *Lessingia filaginifolia* var. *filaginifolia* in *The Jepson Manual*.

Corydalis caseana Gray ssp. *caseana*

"Sierra corydalis" Papaveraceae
CNPS List: 4 **R-E-D Code:** 1-1-3 **State/Fed. Status:** /C3c
Distribution: BUT, LAS, PLA, PLU, SHA, SIE, TEH, TUL
Habitat: Medws, UCFrs / mesic
Life Form: Perennial herb
Blooming: June-September
Notes: See *Proceedings of the American Academy of Arts and Sciences* 10:69 (1874) for original description.

Coryphantha vivipara var. *alversonii*

See *Escobaria vivipara* var. *alversonii*

Coryphantha vivipara var. *rosea*

See *Escobaria vivipara* var. *rosea*

Crepis nana ssp. *ramosa*

Considered but rejected: Too common

Crepis runcinata T. & G. ssp. *hallii* Babc. & Steb.

"Hall's meadow hawksbeard" Asteraceae
CNPS List: 2 **R-E-D Code:** 3-3-1 **State/Fed. Status:** CEQA
Distribution: INY, MNO, NV
Quads: 276A, 324C, 325D, 411C, 413B, 413C, 432C, 434A, 450C, 451D, 469C, 487C
Habitat: MDScr, PJWld / mesic, alkaline
Life Form: Perennial herb
Blooming: May-July
Notes: Threatened by grazing and groundwater drawdown. More common in NV. See *Carnegie Institution of Washington Publication* 504:104 (1938) for original description.

Crossosoma bigelovii

Considered but rejected: Too common

Crossosoma californicum Nutt.

"Catalina crossosoma" Crossosomataceae
CNPS List: 4 **R-E-D Code:** 1-2-2 **State/Fed. Status:** /C3c
Distribution: LAX, SCM, SCT, GU
Habitat: CoScr (rocky)
Life Form: Shrub (deciduous)
Blooming: February-May
Notes: Mainland occurrences threatened by development.

Croton wigginsii Wheeler

"Wiggins's croton" Euphorbiaceae
CNPS List: 2 **R-E-D Code:** 2-2-1 **State/Fed. Status:** CR/C3c
Distribution: IMP, AZ, BA, SO
Quads: 2A, 2B, 13B, 13C, 13D, 14A, 14B, 14D, 15D, 27C, 27D, 28A, 28D
Habitat: DeDns, SDScr
Life Form: Shrub
Blooming: March-May
Notes: Known in CA from one large extended population. See *Contributions from the Gray Herbarium* 124:37 (1939) for original description.
Status Report: 1987

Cryptantha clevelandii Greene var. *dissita* (Jtn.) Jeps. & Hoov.

"serpentine cryptantha" Boraginaceae
CNPS List: 1B **R-E-D Code:** 2-2-3 **State/Fed. Status:** CEQA
Distribution: LAK, NAP
Quads: 499B, 534B, 534D, 549C
Habitat: Chprl (serpentinite)
Life Form: Annual herb
Blooming: April-June
Notes: See *C. clevelandii* in *The Jepson Manual.* See *Flora of California* 3(2):348 (1943) by W.L. Jepson for original description.

Cryptantha clokeyi Jtn.

"Clokey's cryptantha" Boraginaceae
CNPS List: 1B **R-E-D Code:** 3-3-3 **State/Fed. Status:** CEQA
Distribution: SBD
Quads: 182A
Habitat: MDScr
Life Form: Annual herb
Blooming: April
Notes: Known only from near Barstow. See *Journal of the Arnold Arboretum* 20:387 (1939) for original description.

Cryptantha costata Bdg.

"ribbed cryptantha" Boraginaceae
CNPS List: 4 **R-E-D Code:** 1-1-2 **State/Fed. Status:** CEQA?
Distribution: IMP, INY, RIV, SBD, SDG, AZ, BA
Habitat: MDScr, SDScr / sandy
Life Form: Annual herb
Blooming: February-May

Cryptantha crinita Greene

"silky cryptantha" Boraginaceae
CNPS List: 1B **R-E-D Code:** 3-2-3 **State/Fed. Status:** /C2
Distribution: SHA, TEH
Quads: 610B, 611A, 611C, 628B, 628C, 629A, 629D, 645D, 646C, 647A, 647D
Habitat: CmWld, LCFrs, RpFrs, RpWld, VFGrs / gravelly streambeds
Life Form: Annual herb
Blooming: April-May
Notes: Threatened by vehicles and gravel mining. See *Erythea* 3:66 (1895) for original description.
Status Report: 1979

Cryptantha crymophila Jtn.

"subalpine cryptantha" Boraginaceae
CNPS List: 4 **R-E-D Code:** 1-1-3 **State/Fed. Status:** /C3c
Distribution: ALP, TUO
Habitat: SCFrs (volcanic talus)
Life Form: Perennial herb
Blooming: July-August

Cryptantha excavata Bdg.

"deep-scarred cryptantha" Boraginaceae
CNPS List: 4 **R-E-D Code:** 1-1-3 **State/Fed. Status:** CEQA?
Distribution: COL, LAK, YOL
Habitat: CmWld (sandy or gravelly)
Life Form: Annual herb
Blooming: April-May

Cryptantha ganderi Jtn.

"Gander's cryptantha" Boraginaceae
CNPS List: 1B **R-E-D Code:** 3-3-2 **State/Fed. Status:** /C2
Distribution: SDG, BA, SO
Quads: 31B, 47C, 47D
Habitat: DeDns, SDScr (sandy)
Life Form: Annual herb
Blooming: February-May
Notes: Known in CA from fewer than five occurrences. Seriously threatened by development and vehicles. How rare is this plant in Mexico?

Cryptantha hispidula

Considered but rejected: Too common

Cryptantha hoffmannii

Considered but rejected: A synonym of *C. virginensis*; a common taxon

Cryptantha holoptera (Gray) Macbr.

"winged cryptantha" Boraginaceae
CNPS List: 4 **R-E-D Code:** 1-1-2 **State/Fed. Status:** CEQA?
Distribution: IMP, INY, RIV, SBD, SDG, AZ, NV
Habitat: MDScr, SDScr
Life Form: Annual herb
Blooming: March-April

Cryptantha hooveri Jtn.

"Hoover's cryptantha" Boraginaceae
CNPS List: 4 **R-E-D Code:** 1-2-3 **State/Fed. Status:** CEQA?
Distribution: ALA, CCA, MAD, MER, SJQ, STA
Habitat: VFGrs (sandy)
Life Form: Annual herb
Blooming: April-May

Cryptantha incana

Considered but rejected: Too common

Cryptantha mariposae Jtn.

"Mariposa cryptantha" Boraginaceae
CNPS List: 4 **R-E-D Code:** 1-1-3 **State/Fed. Status:** CEQA?
Distribution: CAL, MPA, TUO
Habitat: Chprl (serpentinite)
Life Form: Annual herb
Blooming: April-May

Cryptantha nubigena

Considered but rejected: Too common

Cryptantha rattanii Greene

"Rattan's cryptantha" Boraginaceae
CNPS List: 4 **R-E-D Code:** 1-1-3 **State/Fed. Status:** CEQA?
Distribution: FRE, MER, MNT, SBT
Habitat: CmWld, VFGrs
Life Form: Annual herb
Blooming: April-July
Notes: See *C. decipiens* in *The Jepson Manual.*

Cryptantha roosiorum Munz

"bristlecone cryptantha" Boraginaceae
CNPS List: 1B **R-E-D Code:** 3-2-3 **State/Fed. Status:** CR/C2
Distribution: INY
Quads: 372A
Habitat: SCFrs (carbonate)
Life Form: Perennial herb
Blooming: June-July
Notes: Known from only two occurrences in the Inyo Mtns. Threatened by vehicles. See *Aliso* 3:124-125 (1955) for original description.
Status Report: 1987

Cryptantha scoparia A. Nels.

"gray cryptantha" Boraginaceae
CNPS List: 4 **R-E-D Code:** 1-1-1 **State/Fed. Status:** CEQA?
Distribution: INY, NV, OR, WA+
Habitat: GBScr, PJWld
Life Form: Annual herb
Blooming: June-July
Notes: See *C. nevadensis* in *The Jepson Manual.*

Cryptantha subretusa

Considered but rejected: A synonym of *C. sobolifera*; a common taxon

Cryptantha traskiae Jtn.

"Trask's cryptantha" Boraginaceae
CNPS List: 1B **R-E-D Code:** 3-2-3 **State/Fed. Status:** /C1
Distribution: SCM, SNI
Quads: SCMC, SCMN, SNIC
Habitat: CBScr (rocky)
Life Form: Annual herb
Blooming: April-June
Notes: Possibly threatened by non-native plants and Navy activities.

Cryptantha tumulosa (Pays.) Pays.

"New York Mtns. cryptantha" Boraginaceae
CNPS List: 4 **R-E-D Code:** 1-1-2 **State/Fed. Status:** /C3c
Distribution: INY, SBD, NV
Habitat: MDScr, PJWld / often carbonate
Life Form: Perennial herb
Blooming: April-June
Notes: On watch list in NV.

Cupressus abramsiana C.B. Wolf

"Santa Cruz cypress" Cupressaceae
CNPS List: 1B **R-E-D Code:** 3-2-3 **State/Fed. Status:** CE/FE
Distribution: SCR, SMT
Quads: 408B, 408C, 408D, 409A
Habitat: CCFrs (sandstone or granitic)
Life Form: Tree (evergreen)
Blooming: N/A
Notes: Known from fewer than ten occurrences. Threatened by development, agriculture, and alteration of fire regimes, and possibly by introgression from planted *C. macrocarpa.* Largest known specimen was cut down in 1983. See *Aliso* 1:215-222 (1948) for original description, and *Madroño* 2(4):189-194 (1952) for distributional information.
Status Report: 1985

Cupressus arizonica Greene ssp. *nevadensis* (Abrams) E. Murray

"Piute cypress" Cupressaceae
CNPS List: 1B **R-E-D Code:** 2-2-3 **State/Fed. Status:** /C3c
Distribution: KRN, TUL
Quads: 237D, 260B, 260C, 260D, 261A, 261D, 284C
Habitat: CCFrs, Chprl, CmWld, PJWld
Life Form: Tree (evergreen)
Blooming: N/A
Notes: Threatened by grazing and mining. BA and RNA established by USFS and BLM for this plant. See *Torreya* 19:92 (1919) for original description.

Cupressus bakeri Jeps.

"Baker's cypress" Cupressaceae
CNPS List: 4 **R-E-D Code:** 1-2-2 **State/Fed. Status:** CEQA?
Distribution: MOD, PLU, SHA, SIS, OR
Habitat: Chprl, LCFrs / serpentinite or volcanic
Life Form: Tree (evergreen)
Blooming: N/A
Notes: Threatened by fire suppression. Endangered in OR. Includes *C. bakeri* ssp. *matthewsii.* See *Fremontia* 16(3):17-18 (1988) for species account, and *Madroño* 39(1):79 (1992) for distributional information.

Cupressus bakeri ssp. *bakeri*

See *Cupressus bakeri*

Cupressus bakeri ssp. *matthewsii*

See *Cupressus bakeri*

Cupressus forbesii Jeps.

"Tecate cypress" Cupressaceae
CNPS List: 1B **R-E-D Code:** 3-2-2 **State/Fed. Status:** /C2
Distribution: ORA, SDG, BA
Quads: 9C, 10A, 10B, 10C, 10D, 20D, 70B, 87C, 87D
Habitat: CCFrs, Chprl
Life Form: Tree (evergreen)
Blooming: N/A
Notes: Now known from fewer than five occurrences. Threatened by too frequent wildfires, and development in ORA Co. RIV Co. trees planted. Protected in part at Otay Mtn. ACEC (BLM), SDG Co. See *Aliso* 9:189-196 (1978) for information, and *Fremontia* 13(3):3-10 (1985) for species account.
Status Report: 1990

Cupressus goveniana

See *Cupressus goveniana* ssp. *goveniana*

Cupressus goveniana Gord. ssp. *goveniana*

"Gowen cypress" Cupressaceae
CNPS List: 1B **R-E-D Code:** 3-2-3 **State/Fed. Status:** /C2
Distribution: MNT
Quads: 366C
Habitat: CCFrs
Life Form: Tree (evergreen)
Blooming: N/A
Notes: Known from only two occurrences in the Monterey area. Threatened by development.

Cupressus goveniana Gord. ssp. *pigmaea* (Lemmon) Bartel

"pygmy cypress" Cupressaceae
CNPS List: 1B **R-E-D Code:** 1-2-3 **State/Fed. Status:** /C2
Distribution: MEN, SON
Quads: 520D, 537B, 537D, 552B, 568C, 568D, 569A, 569D
Habitat: CCFrs (podzol-like soil)
Life Form: Tree (evergreen)
Blooming: N/A
Notes: Threatened by development and vehicles. See *Phytologia* 70(4):229-230 (1990) for revised nomenclature.

Cupressus guadalupensis ssp. *forbesii*

See *Cupressus forbesii*

Cupressus lawsoniana

Considered but rejected: Too common

Cupressus macrocarpa Gord.

"Monterey cypress" Cupressaceae
CNPS List: 1B **R-E-D Code:** 3-2-3 **State/Fed. Status:** /C2
Distribution: MNT
Quads: 366C
Habitat: CCFrs
Life Form: Tree (evergreen)
Blooming: N/A
Notes: Known from only two native occurrences in the Monterey area; widely planted and naturalized elsewhere.

Cupressus nevadensis

See *Cupressus arizonica* ssp. *nevadensis*

Cupressus nootkatensis D. Don

"Alaska cedar" Cupressaceae
CNPS List: 4 **R-E-D Code:** 1-1-1 **State/Fed. Status:** CEQA?
Distribution: DNT, SIS, OR, ++
Habitat: UCFrs
Life Form: Tree (evergreen)
Blooming: N/A

Cupressus pygmaea

See *Cupressus goveniana* ssp. *pigmaea*

Cupressus stephensonii C.B. Wolf

"Cuyamaca cypress" Cupressaceae
CNPS List: 1B **R-E-D Code:** 3-3-3 **State/Fed. Status:** /C1
Distribution: SDG
Quads: 20A, 20B
Habitat: CCFrs, Chprl, RpFrs / gabbroic rock
Life Form: Tree (evergreen)
Blooming: N/A
Notes: Known from only two occurrences near Cuyamaca Pk. Threatened by frequent wildfire; the 1950 Conejo fire extirpated plant over part of its range. A synonym of *C. arizonica* ssp. *arizonica* in *The Jepson Manual*, but genetic evidence does not support this interpretation.

Cuscuta howelliana

Considered but rejected: Too common

Cuscuta jepsonii

Considered but rejected: Too common; a synonym of *C. indecora* var. *indecora*

Cusickiella quadricostata (Roll.) Roll.

"Bodie Hills cusickiella" Brassicaceae
CNPS List: 1B **R-E-D Code:** 2-2-2 **State/Fed. Status:** /C2
Distribution: MNO, NV
Quads: 470A, 470B, 470D, 486C, 487C, 487D, 504C, 504D
Habitat: GBScr, PJWld / clay or rocky
Life Form: Perennial herb
Blooming: May-July
Notes: Not as common as previously thought. Threatened by mining, grazing, and vehicles. On watch list in NV. See *Contributions from the Dudley Herbarium* 3:366 (1946) for original description.
Status Report: 1979

Cymopterus cinerarius

Considered but rejected: Too common

Cymopterus deserticola Bdg.

"desert cymopterus" Apiaceae
CNPS List: 1B **R-E-D Code:** 3-2-3 **State/Fed. Status:** /C1
Distribution: KRN, LAX, SBD
Quads: 132B*, 157C*, 183A, 183B, 184A, 184B, 184C, 185A*, 185C, 185D, 208D, 210D*, 233C
Habitat: JTWld, MDScr / sandy
Life Form: Perennial herb
Blooming: March-May
Notes: Known from fewer than twenty occurrences, apparently all located on Edwards AFB. Threatened by sheep grazing, vehicles, and urbanization. See *University of California Publications in Botany* 6:168 (1915) for original description.
Status Report: 1978

Cymopterus gilmanii Mort.

"Gilman's cymopterus" Apiaceae
CNPS List: 2 **R-E-D Code:** 2-1-1 **State/Fed. Status:** CEQA
Distribution: INY, SBD, NV
Quads: 248B, 248C, 249A, 249D, 274D, 323B, 326A, 345C, 368B, 368D, 369B, 369C, 390B, 390C, 390D, 391D, 411D
Habitat: MDScr (often carbonate)
Life Form: Perennial herb
Blooming: April-May
Notes: Occurrences are usually very small.

Cymopterus ripleyi Barneby

"Ripley's cymopterus" Apiaceae
CNPS List: 2 **R-E-D Code:** 3-2-1 **State/Fed. Status:** CEQA
Distribution: INY, NV
Quads: 305B, 323B, 327A, 327B, 328C, 349C, 349D
Habitat: JTWld, MDScr / sandy, carbonate
Life Form: Perennial herb
Blooming: April-June
Notes: Known in CA from fewer than ten occurrences. Threatened by cattle grazing on BLM land at Lee Flat. See *Leaflets of Western Botany* 3:81-83 (1941) for original description.

Cynanchum utahense (Engelm.) Woodson

"Utah vine milkweed" Asclepiadaceae
CNPS List: 4 **R-E-D Code:** 1-1-1 **State/Fed. Status:** CEQA?
Distribution: IMP?, RIV, SBD, SDG, AZ, NV+
Habitat: MDScr, SDScr / sandy, gravelly
Life Form: Perennial herb
Blooming: April-June
Notes: Does plant occur in IMP Co.?

Cypripedium californicum Gray

"California lady's-slipper" Orchidaceae
CNPS List: 4 **R-E-D Code:** 1-2-2 **State/Fed. Status:** /C3c
Distribution: BUT, DNT, HUM, MRN*, PLU, SHA, SIS, SON, TRI, OR
Habitat: BgFns, LCFrs / serpentinite seeps and streambanks
Life Form: Perennial herb (rhizomatous)
Blooming: April-July
Notes: Threatened by horticultural collecting and logging. Many protected populations on USFS land not reproducing. On watch list in OR. See *Fremontia* 17(2):17-19 (1989) for species account.

Cypripedium fasciculatum Wats.

"clustered lady's-slipper" Orchidaceae
CNPS List: 4 **R-E-D Code:** 1-2-2 **State/Fed. Status:** /C2
Distribution: BUT, DNT, HUM, NEV, PLU, SCL, SCR*, SHA, SIE, SIS, SMT, TEH, TRI, YUB, ID, OR, UT, WA+
Habitat: LCFrs, NCFrs / usually serpentinite seeps and streambanks
Life Form: Perennial herb (rhizomatous)
Blooming: March-July
Notes: Many occurrences but most contain few plants. Threatened by logging and horticultural collecting. Monitoring needed for protected populations on USFS lands to assess reproduction, which may be inadequate. Threatened in ID, a candidate for state listing in OR, and state-listed as Threatened in WA. See *Proceedings of the American Academy of Arts and Sciences* 17:380 (1882) for original description, *Lindleyana* 2(1):553-57 (1987) for distributional information, and *Fremontia* 17(2):17-19 (1989) for species account.

Cypripedium montanum Lindl.

"mountain lady's-slipper" Orchidaceae
CNPS List: 4 **R-E-D Code:** 1-1-2 **State/Fed. Status:** /C3c
Distribution: DNT, HUM, MAD, MEN, MOD, MPA, PLU, SIE, SIS, SMT, SON, TEH, TRI, TUO, OR, WA, ++
Habitat: BUFrs, LCFrs
Life Form: Perennial herb (rhizomatous)
Blooming: March-July
Notes: Many protected populations on USFS land not reproducing. Possibly threatened by logging. On watch list in OR. See *Fremontia* 17(2):17-19 (1989) for species account.

Dalea arborescens

See *Psorothamnus arborescens* var. *arborescens*

Dalea californica

Considered but rejected: A synonym of *Psorothamnus aborescens* var. *simplicifolius*; a common taxon

Dalea ornata (Hook.) Eat. & Wright

"ornate dalea" Fabaceae
CNPS List: 2 **R-E-D Code:** 3-3-1 **State/Fed. Status:** CEQA
Distribution: LAS, ID, NV, OR, WA
Quads: 621B
Habitat: PJWld (clay)
Life Form: Perennial herb
Blooming: June
Notes: Known in CA from only one occurrence on tablelands west of Shaffer Mtn. See *Madroño* 32(2):123 (1985) for CA record.

Darlingtonia californica Torr.

"California pitcherplant" Sarraceniaceae
CNPS List: 4 **R-E-D Code:** 1-2-1 **State/Fed. Status:** /C3c
Distribution: DNT, NEV, PLU, SHA, SIE, SIS, TRI, YUB, OR
Habitat: BgFns, Medws / mesic, generally serpentinite seeps
Life Form: Perennial herb (carnivorous)
Blooming: April-June
Notes: Threatened by horticultural collecting and mining. On watch list in OR. See *Fremontia* 14(2):18 (1986) for habitat information.

Dedeckera eurekensis Reveal & J.T. Howell

"July gold" Polygonaceae
CNPS List: 1B **R-E-D Code:** 2-1-3 **State/Fed. Status:** CR/C2
Distribution: INY, MNO
Quads: 350A, 350B, 370A, 370B, 390C, 390D, 413A, 413C, 413D
Habitat: MDScr (carbonate)
Life Form: Shrub (deciduous)
Blooming: June-August
Notes: Known from approximately twenty occurrences. Reproductive capabilities extremely limited, and no juvenile plants or seedlings currently known. See *Brittonia* 28:245-251 (1976) for original description, *Madroño* 28:86-87 (1981) and 32(2):122-123 (1985) for range extension information, and *Phytologia* 66(3):238-241 (1989) for taxonomic treatment.
Status Report: 1987

Delphinium bakeri Ewan

"Baker's larkspur" Ranunculaceae
CNPS List: 1B **R-E-D Code:** 3-3-3 **State/Fed. Status:** CR/C1
Distribution: MRN, SON*
Quads: 484B, 485A, 485B, 502B*
Habitat: CoScr
Life Form: Perennial herb
Blooming: March-May
Notes: Known from fewer than five occurrences. Threatened by agriculture and grazing, and potentially by road maintenance. See *Bulletin of the Torrey Botanical Club* 69:144 (1942) for original description.
Status Report: 1988

Delphinium californicum T. & G. ssp. *interius* (Eastw.) Ewan

"Hospital Canyon larkspur" Ranunculaceae
CNPS List: 1B **R-E-D Code:** 3-2-3 **State/Fed. Status:** /C2
Distribution: ALA, CCA, SCL, SJQ, SLO
Quads: 243D, 426A, 426D, 444C, 444D, 445D, 464B
Habitat: CmWld (mesic)
Life Form: Perennial herb
Blooming: April-June
Notes: See *Leaflets of Western Botany* 2:137 (1938) for original description.

Delphinium gypsophilum Ewan ssp. *gypsophilum*

"gypsum-loving larkspur" Ranunculaceae
CNPS List: 4 **R-E-D Code:** 1-1-3 **State/Fed. Status:** CEQA?
Distribution: FRE, KNG, KRN, MAD, MER, MNT, SJQ, SLO, STA
Habitat: ChScr, CmWld, VFGrs
Life Form: Perennial herb
Blooming: April-May

Delphinium gypsophilum Ewan ssp. *parviflorum* Lewis & Epl.

"small-flowered gypsum-loving larkspur" Ranunculaceae
CNPS List: 4 **R-E-D Code:** 1-1-3 **State/Fed. Status:** CEQA?
Distribution: MNT, SLO
Habitat: CmWld
Life Form: Perennial herb
Blooming: March-June
Notes: See *Brittonia* 8:5 (1954) for original description.

Delphinium hansenii (Greene) Greene ssp. *ewanianum* Warnock

"Ewan's larkspur" Ranunculaceae
CNPS List: 4 **R-E-D Code:** 1-2-3 **State/Fed. Status:** CEQA?
Distribution: CAL, KRN, MAD
Habitat: CmWld, VFGrs / rocky
Life Form: Perennial herb
Blooming: March-May
Notes: To be expected in other foothill areas of the Sierra Nevada; need information. Populations are very local; many in developing areas. See *Phytologia* 68(1):1-6 (1990) for original description.

Delphinium hesperium Gray ssp. *cuyamacae* (Abrams) Lewis & Epl.

"Cuyamaca larkspur" Ranunculaceae
CNPS List: 1B **R-E-D Code:** 2-2-3 **State/Fed. Status:** CR/C2
Distribution: RIV, SDG
Quads: 19B, 20A, 33C, 49C, 49D, 66B, 67D
Habitat: LCFrs, Medws / mesic
Life Form: Perennial herb
Blooming: June-July
Notes: Threatened by development. See *Bulletin of the Torrey Botanical Club* 32:538 (1905) for original description.
Status Report: 1980

Delphinium hutchinsoniae Ewan

"Hutchinson's larkspur" Ranunculaceae
CNPS List: 1B **R-E-D Code:** 3-2-3 **State/Fed. Status:** /C2
Distribution: MNT
Quads: 320B, 320D, 320F, 344B, 344C, 365B, 365C, 366C
Habitat: BUFrs, Chprl, CoPrr, CoScr
Life Form: Perennial herb
Blooming: March-June
Notes: Known from fewer than ten occurrences.

Delphinium inopinum (Jeps.) Lewis & Epl.

"unexpected larkspur" Ranunculaceae
CNPS List: 1B **R-E-D Code:** 2-2-3 **State/Fed. Status:** /C3c
Distribution: FRE, INY, KRN, TUL, VEN
Quads: 165A, 165B, 165C, 165D, 166D, 189C, 190D, 237A, 237B, 260C, 260D, 283D, 285D, 308B, 308D, 330B, 331D, 352C, 374A, 375A, 375D
Habitat: UCFrs (rocky)
Life Form: Perennial herb
Blooming: May-July
Notes: Protected in part at Slate Mtn. BA (USFS), TUL Co.

Delphinium kinkiense

See *Delphinium variegatum* ssp. *kinkiense*

Delphinium luteum Heller

"yellow larkspur" Ranunculaceae
CNPS List: 1B **R-E-D Code:** 3-3-3 **State/Fed. Status:** CR/C1
Distribution: SON
Quads: 502A*, 502C*, 503D
Habitat: Chprl, CoPrr, CoScr
Life Form: Perennial herb
Blooming: March-May
Notes: Known from only two occurrences near Bodega Bay. Plants from MRN Co. apparently hybrids with *D. decorum* ssp. *decorum*; also hybridizes with *D. nudicaule.* Threatened by development and grazing. See *Bulletin of the Southern California Academy of Sciences* 2:68-69 (1903) for original description.
Status Report: 1988

Delphinium parishii ssp. *purpureum*

See *Delphinium parryi* ssp. *purpureum*

Delphinium parishii Gray ssp. *subglobosum* (Wiggins) Lewis & Epl.

"intermediate larkspur" Ranunculaceae
CNPS List: 4 **R-E-D Code:** 1-1-2 **State/Fed. Status:** CEQA?
Distribution: IMP, SDG, BA
Habitat: Chprl, PJWld, SDScr
Life Form: Perennial herb
Blooming: March-May

Delphinium parryi Gray ssp. *blochmaniae* (Greene) Lewis & Epl.

"dune larkspur" Ranunculaceae
CNPS List: 1B **R-E-D Code:** 3-2-3 **State/Fed. Status:** /C2
Distribution: SBA, SLO, VEN
Quads: 113A, 114A, 196D, 221D, 246A, 271A
Habitat: Chprl (maritime), CoDns
Life Form: Perennial herb
Blooming: April-May
Notes: Field work needed. Threatened by development.

Delphinium parryi Gray ssp. *purpureum* (Lewis & Epl.) Warnock

"Mt. Pinos larkspur" Ranunculaceae
CNPS List: 4 **R-E-D Code:** 1-1-3 **State/Fed. Status:** /C3c
Distribution: KRN, VEN
Habitat: Chprl, MDScr, PJWld
Life Form: Perennial herb
Blooming: May-June
Notes: Possibly threatened by development. See *Brittonia* 8:15 (1954) for original description, and *Phytologia* 68(1):1-3 (1990) for revised nomenclature.

Delphinium pratense

Considered but rejected: A synonym of *D. gracilentum*; a common taxon

Delphinium purpusii Bdg.

"Kern County larkspur" Ranunculaceae
CNPS List: 4 **R-E-D Code:** 1-1-3 **State/Fed. Status:** CEQA?
Distribution: KRN, TUL
Habitat: Chprl, CmWld, PJWld / rocky, often carbonate
Life Form: Perennial herb
Blooming: April-May
Notes: Precise location and endangerment information needed. Historical occurrences need field surveys.

Delphinium recurvatum Greene

"recurved larkspur" Ranunculaceae
CNPS List: 1B **R-E-D Code:** 1-2-3 **State/Fed. Status:** /C2
Distribution: ALA, CCA, COL, FRE, KNG, KRN, MER, SLO, SOL, TUL
Quads: 218A, 240C, 240D, 241C, 241D, 264A, 264D, 265A, 265B, 265D, 267A, 287B, 288A, 288D, 290C, 291B, 291D, 312B, 315C, 338D, 359B, 359D, 383B, 463C, 463D, 498B, 547D
Habitat: ChScr, CmWld, VFGrs / alkaline
Life Form: Perennial herb
Blooming: March-May
Notes: Many historical occurrences need field surveys. Much habitat converted to agriculture; also threatened by grazing.

Delphinium stachydeum (Gray) Tides.

"spiked larkspur" Ranunculaceae
CNPS List: 4 **R-E-D Code:** 1-1-1 **State/Fed. Status:** CEQA?
Distribution: MOD, ID, OR, UT, WA+
Habitat: GBScr, UCFrs (edges)
Life Form: Perennial herb
Blooming: July-August
Notes: Known in CA only from the Warner Mtns.

Delphinium trolliifolium

Considered but rejected: Too common

Delphinium uliginosum Curran

"swamp larkspur" Ranunculaceae
CNPS List: 4 **R-E-D Code:** 1-2-3 **State/Fed. Status:** CEQA?
Distribution: COL, LAK, NAP, SIS
Habitat: Chprl, VFGrs / serpentinite seeps
Life Form: Perennial herb
Blooming: May-June
Notes: Highly localized. Hybridizes with *D. hesperium* ssp. *pallescens.* See *Bulletin of the California Academy of Sciences* 1:151 (1885) for original description.

Delphinium umbraculorum Lewis & Epl.

"umbrella larkspur" Ranunculaceae
CNPS List: 4 **R-E-D Code:** 1-1-3 **State/Fed. Status:** CEQA?
Distribution: MNT, SLO
Habitat: CmWld (mesic)
Life Form: Perennial herb
Blooming: May-June
Notes: Hybridizes with *D. parryi* ssp. *parryi.* See *Brittonia* 8:19 (1954) for original description.

Delphinium variegatum T. & G. ssp. *kinkiense* (Munz) Warnock

"San Clemente Island larkspur" Ranunculaceae
CNPS List: 1B **R-E-D Code:** 3-3-3 **State/Fed. Status:** CE/FE
Distribution: SCM
Quads: SCMC, SCMN, SCMS
Habitat: VFGrs (coastal)
Life Form: Perennial herb
Blooming: March-April
Notes: Feral herbivores removed from SCM Isl., and vegetation recovering. Dubiously distinct from *D. variegatum* in *Phytologia* 68(1):1-3 (1990); may be difficult to distinguish from ssp. *thornei.* State-listed as *D. kinkiense*; USFWS uses the name *D. kinkiense* ssp. *kinkiense.* See *Aliso* 7:69 (1969) for original description.
Status Report: 1987

Delphinium variegatum T. & G. ssp. *thornei* Munz

"Thorne's royal larkspur" Ranunculaceae
CNPS List: 1B **R-E-D Code:** 3-3-3 **State/Fed. Status:** /C1
Distribution: SCM
Quads: SCMC, SCMS
Habitat: CmWld, VFGrs (coastal)
Life Form: Perennial herb
Blooming: March-May
Notes: Known from fewer than ten occurrences. Feral herbivores removed from SCM Isl., and vegetation recovering. May be difficult to distinguish from ssp. *kinkiense.*

Dendromecon harfordii Kell. var. *harfordii*

"Channel Island tree poppy" Papaveraceae
CNPS List: 4 **R-E-D Code:** 1-2-3 **State/Fed. Status:** /C2
Distribution: SCZ, SRO
Habitat: Chprl
Life Form: Shrub (evergreen)
Blooming: April-June
Notes: Threatened by feral herbivores. See *D. harfordii* in *The Jepson Manual.* USFWS uses the name *D. rigida* ssp. *harfordii.*

Dendromecon harfordii Kell. var. *rhamnoides* (Greene) Munz

"island tree poppy" Papaveraceae
CNPS List: 1B **R-E-D Code:** 3-3-3 **State/Fed. Status:** /C2
Distribution: SCM*, SCT
Quads: SCMN*, SCMS*, SCTE, SCTN, SCTW
Habitat: Chprl, CmWld, CoScr
Life Form: Shrub (evergreen)
Blooming: April-June
Notes: Known from approximately five occurrences. Last seen on SCM Isl. in 1966. Threatened by feral herbivores. Most of the year a few plants are in flower. See *D. harfordii* in *The Jepson Manual.* USFWS uses the name *D. rigida* ssp. *rhamnoides.*

Dendromecon rigida ssp. *harfordii*

See *Dendromecon harfordii* var. *harfordii*

Dendromecon rigida ssp. *rhamnoides*

See *Dendromecon harfordii* var. *rhamnoides*

Dentaria gemmata

See *Cardamine nuttallii* var. *gemmata*

Dentaria pachystigma var. *dissectifolia*

See *Cardamine pachystigma* var. *dissectifolia*

Deschampsia atropurpurea (Wahl.) Scheele

"mountain hair grass" Poaceae
CNPS List: 4 **R-E-D Code:** 1-1-1 **State/Fed. Status:** CEQA?
Distribution: SIS, TRI, OR, ++
Habitat: Medws, SCFrs, UCFrs
Life Form: Perennial herb
Blooming: July-August

Dicentra chrysantha

Considered but rejected: Too common

Dicentra formosa (Haw.) Walp. ssp. *oregana* (Eastw.) Munz

"Oregon bleeding heart" Papaveraceae
CNPS List: 4 **R-E-D Code:** 1-2-2 **State/Fed. Status:** /C3c
Distribution: DNT, HUM, SIS, OR
Habitat: LCFrs (serpentinite)
Life Form: Perennial herb
Blooming: April-May
Notes: Commercially exploited in OR; on watch list. See *D. formosa* in *The Jepson Manual.* See *Proceedings of the California Academy of Sciences* IV 20:144 (1931) for original description, and *Brittonia* 13:1-57 (1961) for taxonomic treatment.
Status Report: 1977

Dicentra nevadensis Eastw.

"Tulare County bleeding heart" Papaveraceae
CNPS List: 4 **R-E-D Code:** 1-1-3 **State/Fed. Status:** CEQA?
Distribution: FRE, TUL
Habitat: SCFrs (gravelly openings)
Life Form: Perennial herb (rhizomatous)
Blooming: June-August
Notes: Can be confused with *D. formosa.* See *Proceedings of the California Academy of Sciences* IV 20:143 (1931) for original description, and *Brittonia* 13:1-57 (1961) for taxonomic treatment.

Dicentra ochroleuca

Considered but rejected: Too common

Dicentra pauciflora

Considered but rejected: Too common

Dichanthelium lanuginosum (Ell.) Gould var. *thermale* (Boland.) Spellenberg

"Geysers's dichanthelium" Poaceae
CNPS List: 1B **R-E-D Code:** 3-3-3 **State/Fed. Status:** CE/C2
Distribution: SON
Quads: 533C, 534D
Habitat: CCFrs, RpFrs / hydrothermally-altered soil
Life Form: Perennial herb
Blooming: June-August
Notes: Endemic to The Geysers geothermal area. Threatened by energy development. A synonym of *Panicum acuminatum* var. *acuminatum* in *The Jepson Manual.* See *Madroño* 23(3):151 (1975) for taxonomic treatment.
Status Report: 1988

Dichelostemma lacuna-vernalis
Considered but rejected: A synonym of *D. capitatum* ssp. *capitatum*; a common taxon

Dichelostemma venustum
Considered but rejected: A sporadic hybrid

Dichondra donnelliana
Considered but rejected: Too common

Dichondra occidentalis House
"western dichondra" Convolvulaceae
CNPS List: 4 **R-E-D Code:** 1-2-1 **State/Fed. Status:** /C3c
Distribution: LAX?, MRN?, ORA, SBA, SCT, SCZ, SDG, SMI, SRO, VEN, BA
Habitat: Chprl, CmWld, CoScr, VFGrs
Life Form: Perennial herb (rhizomatous)
Blooming: March-May
Notes: Records for MRN Co. are questionable; report from LAX Co. needs confirmation. See *Muhlenbergia* 1:130-131 (1906) for original description.
Status Report: 1977

Dimeresia howellii Gray
"doublet" Asteraceae
CNPS List: 4 **R-E-D Code:** 1-1-1 **State/Fed. Status:** CEQA?
Distribution: LAS, MOD, SIS, ID, NV, OR
Habitat: PJWld (volcanic)
Life Form: Annual herb
Blooming: May-July
Notes: Threatened in ID. Protected in part at Ash Valley RNA (BLM), LAS Co. See *Fremontia* 16(1):15-17 (1988) for brief species account and BLM management plans.

Diplacus aridus
See *Mimulus aridus*

Diplacus clevelandii
See *Mimulus clevelandii*

Diplacus grandiflorus
Considered but rejected: A synonym of *Mimulus bifidus* ssp. *bifidus*; a common taxon

Diplacus parviflorus
See *Mimulus flemingii*

Dirca occidentalis Gray
"western leatherwood" Thymelaeaceae
CNPS List: 1B **R-E-D Code:** 2-2-3 **State/Fed. Status:** CEQA
Distribution: ALA, CCA, MRN, SCL, SMT, SON
Quads: 428A, 428B, 428C, 429A, 429D, 448C, 465B, 465C, 465D, 466A, 467B, 482C, 484C, 485D, 503D
Habitat: BUFrs, CCFrs, Chprl, CmWld, NCFrs, RpFrs, RpWld / mesic
Life Form: Shrub (deciduous)
Blooming: January-April

Dissanthelium californicum (Nutt.) Benth.
"California dissanthelium" Poaceae
CNPS List: 1A **Last Seen:** 1912 **State/Fed. Status:** /C2*
Distribution: SCM*, SCT*, GU*
Quads: SCMC*, SCTN*
Habitat: CoScr
Life Form: Annual herb
Blooming: unknown
Notes: Feral goats may have caused extinction; population now removed from SCM Isl. and controlled on SCT Isl.
Status Report: 1977

Ditaxis adenophora
See *Ditaxis clariana*

Ditaxis californica (Bdg.) Pax & K. Hoffm.
"California ditaxis" Euphorbiaceae
CNPS List: 1B **R-E-D Code:** 3-2-3 **State/Fed. Status:** /C2
Distribution: RIV, SDG
Quads: 47D, 61A, 61B, 62A, 62B, 63A, 63C, 65A, 78C
Habitat: SDScr
Life Form: Perennial herb
Blooming: March-December
Notes: Known from fewer than twenty occurrences, most with few plants. Threatened by vehicles.
Status Report: 1980

Ditaxis clariana (Jeps.) Webster
"glandular ditaxis" Euphorbiaceae
CNPS List: 2 **R-E-D Code:** 3-2-1 **State/Fed. Status:** CEQA
Distribution: IMP, RIV, SBD, AZ, SO
Quads: 25C, 26D, 64B, 65A, 65B, 94B
Habitat: MDScr, SDScr / sandy
Life Form: Perennial herb
Blooming: December-March

Dithyrea maritima Davids.
"beach spectaclepod" Brassicaceae
CNPS List: 1B **R-E-D Code:** 3-3-2 **State/Fed. Status:** CT/C1
Distribution: LAX, SBA, SCT*, SLO, SMI*, SNI, BA
Quads: 90B*, 90C, 112D, 171A, 196A, 196B, 196D, 221B, 221D, 247D, SMIE*, SMIW*, SNIC
Habitat: CoDns, CoScr (sandy)
Life Form: Perennial herb (rhizomatous)
Blooming: April-May
Notes: Known in CA from fewer than twenty occurrences; extirpated from half of its historical range. Need historical quads for SCT Isl. Last seen on SMI Isl. in 1932. Threatened by trampling, vehicles, and non-native plants. See *Erythea* 2:179 (1894) for original description.
Status Report: 1990

Dodecahema leptoceras (Gray) Rev. & Hardham

"slender-horned spineflower" Polygonaceae
CNPS List: 1B **R-E-D Code:** 3-3-3 **State/Fed. Status:** CE/FE
Distribution: LAX, RIV, SBD
Quads: 49B, 50A, 67A, 67B*, 69A*, 69B, 70B, 84C*, 104C, 106C, 106D*, 107A*, 107B, 107D*, 110A*, 133C*, 137A, 137B*, 137C*, 137D, 138A
Habitat: Chprl, CoScr (alluvial fan)
Life Form: Annual herb
Blooming: April-June
Notes: Many historical occurrences lost to urbanization; currently threatened by development, flood control, vehicles, and proposed reservoir. State-listed as *Centrostegia leptoceras.* See *Proceedings of the American Academy of Arts and Sciences* 12:269 (1877) for original description, *Great Basin Naturalist Memoirs* 2:169-190 (1978) for taxonomic treatment, and *Phytologia* 66(2):83-88 (1989) for revised nomenclature.
Status Report: 1987

Downingia concolor Greene var. *brevior* McVaugh

"Cuyamaca Lake downingia" Campanulaceae
CNPS List: 1B **R-E-D Code:** 3-3-3 **State/Fed. Status:** CE/C1
Distribution: SDG
Quads: 20A, 33D
Habitat: Medws (mesic), VnPls
Life Form: Annual herb
Blooming: May-July
Notes: Known from seven occurrences in the Cuyamaca Lake area. Threatened by development, altered hydrology, grazing, and recreation. See *Memoirs of the Torrey Botanical Club* 19(4):1-57 (1941) for original description.
Status Report: 1987

Downingia humilis

See *Downingia pusilla*

Downingia pusilla (D. Don) Torr.

"dwarf downingia" Campanulaceae
CNPS List: 2 **R-E-D Code:** 1-2-1 **State/Fed. Status:** /C3c
Distribution: MER, MPA, NAP, PLA, SAC, SOL, SON, STA, TEH, SA
Quads: 420C, 440B, 440C, 441A, 459C, 481B, 483A, 483B, 496A, 496B, 496D, 497C, 498C, 498D, 499B, 499C, 500A, 500D, 501A, 501D, 502A, 512B, 518D, 528A, 528B, 528C, 528D, 594B, 595A, 610B, 610C, 611D
Habitat: VFGrs (mesic), VnPls
Life Form: Annual herb
Blooming: March-May
Notes: Threatened by urbanization, agriculture, grazing, and vehicles.

Draba asterophora Pays. var. *asterophora*

"Tahoe draba" Brassicaceae
CNPS List: 1B **R-E-D Code:** 3-1-2 **State/Fed. Status:** /C3c
Distribution: ALP, ELD, MNO, TUO, NV
Quads: 453B, 522C, 522D, 523D
Habitat: AlpBR, SCFrs
Life Form: Perennial herb
Blooming: July-August
Notes: Known in CA from fewer than ten occurrences. On watch list in NV. See *American Journal of Botany* 4:263 (1917) for original description.
Status Report: 1979

Draba asterophora Pays. var. *macrocarpa* C.L. Hitchc.

"Cup Lake draba" Brassicaceae
CNPS List: 1B **R-E-D Code:** 3-1-3 **State/Fed. Status:** /C2
Distribution: ELD
Quads: 523D
Habitat: SCFrs (rocky)
Life Form: Perennial herb
Blooming: July
Notes: Known from only two occurrences near Cup Lake below Ralston Pk. See *Proceedings of the Biological Society of Washington* 11:64 (1941) for original description.
Status Report: 1979

Draba aureola Wats.

"golden draba" Brassicaceae
CNPS List: 1B **R-E-D Code:** 3-1-2 **State/Fed. Status:** CEQA
Distribution: SHA, SIS, TRI, OR, WA
Quads: 625B, 626A, 699C, 700D
Habitat: AlpBR, SCFrs / serpentinite or volcanic
Life Form: Perennial herb
Blooming: July-August
Notes: Known in CA from eight occurrences near Mt. Eddy and Mt. Lassen. On watch list in OR, and state-listed as Sensitive in WA.

Draba californica (Jeps.) Roll. & Price

"California draba" Brassicaceae
CNPS List: 4 **R-E-D Code:** 1-2-3 **State/Fed. Status:** CEQA?
Distribution: INY, MNO
Habitat: AlpBR, Medws
Life Form: Perennial herb
Blooming: July-August
Notes: Can be locally common. Threatened by grazing and road construction. See *Manual of the Flowering Plants of California,* p. 443 (1925) by W.L. Jepson for original description, and *Aliso* 12(1):17-27 (1988) for revised nomenclature.

Draba cana Rydb.

"hoary draba" Brassicaceae
CNPS List: 2 **R-E-D Code:** 3-1-1 **State/Fed. Status:** CEQA
Distribution: MNO, WA, ++
Quads: 434C, 488A
Habitat: AlpBR, Medws, SCFrs / carbonate
Life Form: Perennial herb
Blooming: July
Notes: Known in CA from only two occurrences near Lake Genevieve and Wheeler Pk. State-listed as Sensitive in WA.

Draba carnosula O.E. Schulz

"Mt. Eddy draba" Brassicaceae
CNPS List: 1B **R-E-D Code:** 3-1-3 **State/Fed. Status:** /C2
Distribution: DNT, SIS, TRI
Quads: 668B, 684C, 698B, 699C, 699D, 700D, 738C, 738D
Habitat: SCFrs, UCFrs / serpentinite
Life Form: Perennial herb
Blooming: July-August
Notes: Known from fewer than twenty occurrences.
Status Report: 1980

Draba crassifolia var. *nevadensis*

Considered but rejected: Not in CA; name misapplied to *D. albertina*; a common taxon

Draba cruciata Pays.

"Mineral King draba" Brassicaceae
CNPS List: 4 **R-E-D Code:** 1-1-3 **State/Fed. Status:** /C3c
Distribution: MNO, TUL
Habitat: SCFrs (gravelly)
Life Form: Perennial herb
Blooming: July-August
Notes: See *American Journal of Botany* 4:265 (1917) for original description.
Status Report: 1980

Draba cruciata var. *cruciata*

See *Draba cruciata*

Draba cruciata var. *integrifolia*

See *Draba sharsmithii*

Draba douglasii var. *crockeri*

Considered but rejected: A synonym of *Cusickiella douglasii*; a common taxon

Draba douglasii var. *douglasii*

Considered but rejected: A synonym of *Cusickiella douglasii*; a common taxon

Draba howellii Wats.

"Howell's draba" Brassicaceae
CNPS List: 4 **R-E-D Code:** 1-1-2 **State/Fed. Status:** CEQA?
Distribution: HUM, SHA, SIS, TRI, OR
Habitat: SCFrs (rocky)
Life Form: Perennial herb
Blooming: June-July
Notes: Endangered in OR.

Draba howellii var. *carnosula*

See *Draba carnosula*

Draba howellii var. *howellii*

See *Draba howellii*

Draba incrassata (Roll.) Roll.

"Sweetwater Mtns. draba" Brassicaceae
CNPS List: 4 **R-E-D Code:** 1-2-3 **State/Fed. Status:** /C3c
Distribution: MNO
Habitat: AlpBR (rhyolite talus)
Life Form: Perennial herb (stoloniferous)
Blooming: July-August

Draba lemmonii var. *incrassata*

See *Draba incrassata*

Draba monoensis Roll. & Price

"White Mtns. draba" Brassicaceae
CNPS List: 1B **R-E-D Code:** 2-2-3 **State/Fed. Status:** CEQA
Distribution: MNO
Quads: 431C, 450D
Habitat: AlpBR, Medws
Life Form: Perennial herb
Blooming: August
Notes: Narrowly endemic to the White Mtns. of MNO Co. Possibly threatened by grazing. See *Aliso* 12(1):17-27 (1988) for original description.

Draba nivalis var. *elongata*

Considered but rejected: Too common, but needs reevaluation; a synonym of *D. lonchocarpa* var. *lonchocarpa*

Draba pterosperma Pays.

"winged-seed draba" Brassicaceae
CNPS List: 4 **R-E-D Code:** 1-1-3 **State/Fed. Status:** CEQA?
Distribution: SIS
Habitat: UCFrs (rocky, often carbonate)
Life Form: Perennial herb
Blooming: June-July
Notes: Known only from the Marble and Salmon Mtns.

Draba quadricostata

See *Cusickiella quadricostata*

Draba sharsmithii Roll. & Price

"Mt. Whitney draba" Brassicaceae
CNPS List: 1B **R-E-D Code:** 3-1-3 **State/Fed. Status:** /C3c
Distribution: FRE, INY, TUL
Quads: 308A, 352A, 352D, 373D, 416D
Habitat: AlpBR
Life Form: Perennial herb
Blooming: July-August
Notes: Known from fewer than ten occurrences. Need quads for FRE Co. See *Madroño* 5:151 (1940) for original description.
Status Report: 1980

Draba sierrae C.W. Sharsm.

"Sierra draba" Brassicaceae
CNPS List: 4 **R-E-D Code:** 1-1-3 **State/Fed. Status:** CEQA?
Distribution: FRE, INY
Habitat: AlpBR (granitic or carbonate)
Life Form: Perennial herb
Blooming: June-July

Draba stenoloba var. *ramosa*

Considered but rejected: A synonym of *D. albertina*; a common taxon

Draba subumbellata Roll. & Price

"mound draba" Brassicaceae
CNPS List: 4 **R-E-D Code:** 1-1-3 **State/Fed. Status:** CEQA?
Distribution: INY, MNO
Habitat: AlpBR, AlpDS? / carbonate
Life Form: Perennial herb
Blooming: July
Notes: See *Aliso* 12(1):17-27 (1988) for original description.

Drosera anglica Huds.

"English sundew" Droseraceae
CNPS List: 2 **R-E-D Code:** 2-1-1 **State/Fed. Status:** CEQA
Distribution: LAS, NEV, PLU, SIE, SIS, ID, OR, WA, ++
Quads: 554B, 555A, 606B, 607C, 624D, 625A, 662B, 673B, 674A, 719D
Habitat: BgFns, Medws
Life Form: Perennial herb (carnivorous)
Blooming: July-August
Status Report: 1980

Drosera rotundifolia

Considered but rejected: Too common

Dryopteris filix-mas (L.) Schott

"male fern" Dryopteridaceae
CNPS List: 2 **R-E-D Code:** 3-1-1 **State/Fed. Status:** CEQA
Distribution: INY, MNO, SBD*, AZ, OR, ++
Quads: 131C*, 412A, 431C, 431D
Habitat: PJWld (granitic)
Life Form: Perennial herb (rhizomatous)
Fertile: July-September
Notes: Known in CA from only five occurrences in the White Mtns. Endangered in OR.

Dudleya abramsii Rose ssp. *affinis* K. Nakai

"San Bernardino Mtns. dudleya" Crassulaceae
CNPS List: 1B **R-E-D Code:** 2-2-3 **State/Fed. Status:** /C2
Distribution: SBD
Quads: 105A, 131D, 132C, 132D
Habitat: PbPln, PJWld, UCFrs / granitic, quartzite, or carbonate
Life Form: Perennial herb
Blooming: April-June
Notes: Endemic to the San Bernardino Mtns. Threatened by development and limestone mining. See *Madroño* 34(4):334-353 (1987) for original description.

Dudleya abramsii Rose ssp. *bettinae* (Hoov.) Bartel

"San Luis Obispo serpentine dudleya" Crassulaceae
CNPS List: 1B **R-E-D Code:** 3-1-3 **State/Fed. Status:** /C2
Distribution: SLO
Quads: 246C, 247A, 247B, 247D
Habitat: Chprl, CoScr, VFGrs / serpentinite
Life Form: Perennial herb
Blooming: May-July
Notes: Known from fewer than ten occurrences. See *Leaflets of Western Botany* 10(11):186 (1965) for original description, and *Phytologia* 70(4):229-230 for revised nomenclature.
Status Report: 1977

Dudleya abramsii Rose ssp. *murina* (Eastw.) Moran

"San Luis Obispo dudleya" Crassulaceae
CNPS List: 4 **R-E-D Code:** 1-1-3 **State/Fed. Status:** /C3c
Distribution: SLO
Habitat: Chprl, CmWld / serpentinite
Life Form: Perennial herb
Blooming: May-June

Dudleya abramsii Rose ssp. *parva* (Rose & Davids.) J. Bartel

"Conejo dudleya" Crassulaceae
CNPS List: 1B **R-E-D Code:** 3-2-3 **State/Fed. Status:** /PT
Distribution: VEN
Quads: 113A, 113B
Habitat: CoScr, VFGrs / rocky, clay
Life Form: Perennial herb
Blooming: May-June
Notes: Known from approximately ten occurrences. Threatened by horticultural collecting, recreation, vehicles, and urbanization. See *Bulletin of the Southern California Academy of Sciences* 22:5 (1923) for original description, *Madroño* 30(3):191 (1983) for distributional information, and *Phytologia* 70(4):229-230 (1991) for revised nomenclature.

Dudleya alainae Reiser

"Banner dudleya" Crassulaceae
CNPS List: 3 **R-E-D Code:** 3-2-3 **State/Fed. Status:** CEQA?
Distribution: SDG
Quads: 33D
Habitat: LCFrs (rocky)
Life Form: Perennial herb
Blooming: July
Notes: Move to List 1B? A synonym of *D. saxosa* ssp. *aloides* in *The Jepson Manual.* See *Cactus and Succulent Journal* 56:147-148 (1984) for original description. Taxonomic questions remain unanswered by treatment in *Cactus and Succulent Journal* 58:111-115 (1986); critical evaluation needed.

Dudleya attenuata (Wats.) Moran ssp. *orcuttii* (Rose) Moran

"Orcutt's dudleya" Crassulaceae
CNPS List: 2 **R-E-D Code:** 3-3-1 **State/Fed. Status:** /C2
Distribution: SDG, BA
Quads: 11D
Habitat: CBScr, Chprl, CoScr
Life Form: Perennial herb
Blooming: May-July
Notes: Known in CA only from Border Field SP. Possibly threatened by trampling. See *Bulletin of the New York Botanical Garden* 3:36 (1903) for original description.

Dudleya bettinae

See *Dudleya abramsii* ssp. *bettinae*

Dudleya blochmaniae (Eastw.) Moran ssp. *blochmaniae*

"Blochman's dudleya" Crassulaceae
CNPS List: 1B **R-E-D Code:** 2-2-2 **State/Fed. Status:** /C2
Distribution: LAX, ORA, SBA, SDG, SLO, VEN, BA
Quads: 36B, 51C, 52A, 52B, 112C, 113B, 113D, 114A, 196B, 221B, 246C, 247B, 247D
Habitat: CBScr, CoScr, VFGrs / rocky, often clay or serpentinite
Life Form: Perennial herb
Blooming: April-June
Notes: Known from fewer than twenty occurrences in CA, and fewer than five in BA. Threatened by development.

Dudleya blochmaniae (Eastw.) Moran ssp. *brevifolia* Moran

"short-leaved dudleya" Crassulaceae
CNPS List: 1B **R-E-D Code:** 3-3-3 **State/Fed. Status:** CE/PE
Distribution: SDG
Quads: 22B, 22C
Habitat: Chprl, CoScr / Torrey sandstone
Life Form: Perennial herb
Blooming: April
Notes: Known from fewer than five occurrences in the Del Mar and La Jolla areas. Seriously threatened by urbanization and vehicles. Probably a distinct species. State-listed as *D. brevifolia.* See *Baileya* 19:146 (1975) for revised nomenclature, and *Fremontia* 13(1):21 (1985) for species account.
Status Report: 1987

Dudleya blochmaniae (Eastw.) Moran ssp. *insularis* Moran

"Santa Rosa Island dudleya" Crassulaceae
CNPS List: 1B **R-E-D Code:** 3-3-3 **State/Fed. Status:** /C1
Distribution: SRO
Quads: SROE
Habitat: CBScr
Life Form: Perennial herb
Blooming: March-April
Notes: Known from only one occurrence near Old Ranch Pt. Threatened by cattle trampling and grazing by feral herbivores.

Dudleya brevifolia

See *Dudleya blochmaniae* ssp. *brevifolia*

Dudleya calcicola Bartel & Shevock

"limestone dudleya" Crassulaceae
CNPS List: 4 **R-E-D Code:** 1-1-3 **State/Fed. Status:** CEQA?
Distribution: INY, KRN, TUL
Habitat: Chprl, PJWld / carbonate
Life Form: Perennial herb
Blooming: April-August
Notes: See *Madroño* 30(4):210-216 (1983) for original description and 34(4):334-353 (1987) for alternate treatment.

Dudleya candelabrum Rose

"candleholder dudleya" Crassulaceae
CNPS List: 1B **R-E-D Code:** 2-2-3 **State/Fed. Status:** /C2
Distribution: SCZ, SMI, SRO
Quads: SCZA, SCZB, SCZC, SMIE, SROE, SROW
Habitat: CoScr (rocky)
Life Form: Perennial herb
Blooming: April-July
Notes: Threatened by feral herbivores. See *Bulletin of the New York Botanical Garden* 3(9):1-45 (1903) for original description.
Status Report: 1979

Dudleya cymosa ssp. *agourensis*

See *Dudleya cymosa* ssp. *ovatifolia*

Dudleya cymosa (Lem.) Britt. & Rose ssp. *costafolia* Bartel & Shevock

"Pierpoint Springs dudleya" Crassulaceae
CNPS List: 1B **R-E-D Code:** 3-2-3 **State/Fed. Status:** /C1
Distribution: TUL
Quads: 308B, 331A
Habitat: Chprl, CmWld / carbonate
Life Form: Perennial herb
Blooming: May-July
Notes: Endemic to the Middle Fork of the Tule River. Threatened by potential limestone mining and horticultural collecting. See *Aliso* 12(4):701-704 (1990) for original description.

Dudleya cymosa (Lem.) Britt. & Rose ssp. *crebrifolia* K. Nakai & Verity

"San Gabriel River dudleya" Crassulaceae
CNPS List: 1B **R-E-D Code:** 3-1-3 **State/Fed. Status:** /C2
Distribution: LAX
Quads: 109B
Habitat: Chprl (granitic)
Life Form: Perennial herb
Blooming: April-July
Notes: Known only from the type locality along the San Gabriel River. See *Madroño* 34(4):334-353 (1987) for original description.

Dudleya cymosa (Lem.) Britt. & Rose ssp. *marcescens* Moran

"marcescent dudleya" Crassulaceae
CNPS List: 1B **R-E-D Code:** 3-2-3 **State/Fed. Status:** CR/PT
Distribution: LAX, VEN
Quads: 112C, 113B, 113C, 113D
Habitat: Chprl (volcanic)
Life Form: Perennial herb
Blooming: May-June
Notes: Known from only eight occurrences. Threatened by development and foot traffic. See *Madroño* 14:106-108 (1957) for original description.
Status Report: 1986

Dudleya cymosa (Lem.) Britt. & Rose ssp. *ovatifolia* (Britt.) Moran

"Santa Monica Mtns. dudleya" Crassulaceae
CNPS List: 1B **R-E-D Code:** 2-2-3 **State/Fed. Status:** /PT
Distribution: LAX, ORA, VEN
Quads: 70A, 112B, 112C, 112D, 113A, 113C, 113D
Habitat: Chprl, CoScr / volcanic
Life Form: Perennial herb
Blooming: March-June
Notes: Known from fewer than ten occurrences; some threatened by development and recreation. Includes *D. cymosa* ssp. *agourensis.*

Dudleya densiflora (Rose) Moran

"San Gabriel Mtns. dudleya" Crassulaceae
CNPS List: 1B **R-E-D Code:** 3-3-3 **State/Fed. Status:** /C1
Distribution: LAX
Quads: 109A, 109B
Habitat: Chprl, CoScr, LCFrs / granitic, cliffs and canyon walls
Life Form: Perennial herb
Blooming: March-July
Notes: Known from approximately five extant occurrences. Threatened by mining and development.
Status Report: 1979

Dudleya greenei Rose

"Greene's dudleya" Crassulaceae
CNPS List: 4 **R-E-D Code:** 1-2-3 **State/Fed. Status:** CEQA?
Distribution: SCT, SCZ, SMI, SRO
Habitat: CBScr, Chprl, CmWld, CoScr / volcanic cliffs
Life Form: Perennial herb
Blooming: May-July
Notes: Threatened by cattle trampling and feral herbivores on SRO Isl.

Dudleya hassei

Considered but rejected: Too common

Dudleya multicaulis (Rose) Moran

"many-stemmed dudleya" Crassulaceae
CNPS List: 1B **R-E-D Code:** 1-2-3 **State/Fed. Status:** /C2
Distribution: LAX, ORA, RIV, SBD, SDG
Quads: 51B, 52A, 52D, 69A, 69B, 69C, 70B, 70C, 70D, 71A, 71B*, 71D, 86C, 87B, 87C, 87D, 88B, 88D, 107A, 108B, 109A, 109D, 111D, 112B, 112D
Habitat: Chprl, CoScr, VFGrs / often clay
Life Form: Perennial herb
Blooming: May-July
Notes: Threatened by development, road construction, grazing, and recreation. See *Bulletin of the New York Botanical Garden* 3:38 (1903) for original description.
Status Report: 1977

Dudleya nesiotica (Moran) Moran

"Santa Cruz Island dudleya" Crassulaceae
CNPS List: 1B **R-E-D Code:** 3-3-3 **State/Fed. Status:** CR/C2
Distribution: SCZ
Quads: SCZA
Habitat: CBScr
Life Form: Perennial herb
Blooming: May-June
Notes: Known from only two occurrences. Possibly threatened by feral herbivores. See *Desert Plant Life* 22:99 (1951) for original description.
Status Report: 1979

Dudleya parva

See *Dudleya abramsii* ssp. *parva*

Dudleya saxosa (M.E. Jones) Britt. & Rose ssp. *saxosa*

"Panamint dudleya" Crassulaceae
CNPS List: 4 **R-E-D Code:** 1-2-3 **State/Fed. Status:** /C2
Distribution: INY, RIV, SBD
Habitat: MDScr, PJWld / granitic or carbonate
Life Form: Perennial herb
Blooming: April-September

Dudleya setchellii (Jeps.) Britt. & Rose

"Santa Clara Valley dudleya" Crassulaceae
CNPS List: 1B **R-E-D Code:** 3-3-3 **State/Fed. Status:** /PE
Distribution: SCL
Quads: 406A, 406B, 406C, 406D, 407A, 407B, 427D
Habitat: VFGrs (serpentinite)
Life Form: Perennial herb
Blooming: May-June
Notes: Known from fewer than fifteen occurrences in the Santa Clara Valley. Threatened by urbanization, vehicles, and grazing.

Dudleya stolonifera Moran

"Laguna Beach dudleya" Crassulaceae
CNPS List: 1B **R-E-D Code:** 3-3-3 **State/Fed. Status:** CT/C1
Distribution: ORA
Quads: 70C, 71B, 71D
Habitat: Chprl, CmWld, CoScr, VFGrs / rocky
Life Form: Perennial herb (stoloniferous)
Blooming: May-July
Notes: Known from fewer than ten occurrences. Threatened by development, and possibly by horticultural collecting. See *Bulletin of the Southern California Academy of Sciences* 48:105-114 (1950) for original description.
Status Report: 1987

Dudleya traskiae (Rose) Moran

"Santa Barbara Island dudleya" Crassulaceae
CNPS List: 1B **R-E-D Code:** 3-3-3 **State/Fed. Status:** CE/FE
Distribution: SBR
Quads: SBRA
Habitat: CBScr, CoScr
Life Form: Perennial herb
Blooming: April-May
Notes: Known from eleven occurrences. Reduced by feral herbivores, but these have now been removed. See *Bulletin of the New York Botanical Garden* 3:34 (1903) for original description, and *Fremontia* 5(4):37-38 (1978) for discussion of rediscovery and 14(4):3-6 (1987) for information on recovery work.
Status Report: 1987

Dudleya variegata (Wats.) Moran

"variegated dudleya" Crassulaceae
CNPS List: 1B **R-E-D Code:** 2-2-2 **State/Fed. Status:** /C2
Distribution: SDG, BA
Quads: 10A, 10B, 10C, 10D, 11A, 11D, 21B, 21C, 21D, 22A, 22B, 22C, 22D, 35D, 36A
Habitat: Chprl, CmWld, CoScr, VFGrs, VnPls
Life Form: Perennial herb
Blooming: May-June
Notes: Threatened by development and grazing.

Dudleya verityi K. Nakai

"Verity's dudleya" Crassulaceae
CNPS List: 1B **R-E-D Code:** 3-2-3 **State/Fed. Status:** /PT
Distribution: VEN
Quads: 113B, 114A
Habitat: Chprl, CmWld, CoScr / volcanic
Life Form: Perennial herb
Blooming: May-June
Notes: Known from only three occurrences near Conejo Mtn. Threatened by rock mining, flood control activities, and development. See *Cactus and Succulent Journal* 55:196-200 (1983) for original description.

Dudleya virens (Rose) Moran

"bright green dudleya" Crassulaceae
CNPS List: 1B **R-E-D Code:** 2-2-2 **State/Fed. Status:** /C2
Distribution: LAX, SCM, SCT, SNI, GU
Quads: 73A, 90C, SCMC, SCMS, SCTE, SCTN, SNIC
Habitat: CBScr, Chprl, CoScr
Life Form: Perennial herb
Blooming: April-June
Notes: Threatened by development on mainland. See *Bulletin of the New York Botanical Garden* 3:34 (1903) for original description, and *Desert Plant Life* 14:191 (1943) for revised nomenclature.

Dudleya viscida (Wats.) Moran

"sticky dudleya" Crassulaceae
CNPS List: 1B **R-E-D Code:** 3-2-3 **State/Fed. Status:** /C1
Distribution: ORA, RIV, SDG
Quads: 22C, 35C, 36A, 36B, 50B, 51A, 51B, 51C, 69C, 70D
Habitat: CBScr, Chprl, CoScr / rocky
Life Form: Perennial herb
Blooming: May-June
Notes: Known from fewer than twenty occurrences. Threatened by development and road construction. See *Proceedings of the American Academy of Arts and Sciences* 17:372 (1882) for original description.
Status Report: 1979

Eatonella congdonii

See *Lembertia congdonii*

Echinocactus polycephalus var. *polycephalus*

Considered but rejected: Too common

Echinocereus engelmannii (Parry) Ruempler var. *howei* L. Benson

"Howe's hedgehog cactus" Cactaceae
CNPS List: 1B **R-E-D Code:** 3-3-3 **State/Fed. Status:** /C2
Distribution: SBD
Quads: 173A, 174A
Habitat: MDScr
Life Form: Shrub (stem succulent)
Blooming: April-May
Notes: Probably threatened by horticultural collecting. Not in *The Jepson Manual.* See *Cactus and Succulent Journal* 46:80 (1974) for original description.
Status Report: 1977

Echinocereus engelmannii var. *munzii*

Considered but rejected: A synonym of *E. engelmannii*; a common taxon

Eleocharis decumbens

Considered but rejected: Too common; a synonym of *E. montevidensis*

Eleocharis parvula (R. & S.) Link

"small spikerush" Cyperaceae
CNPS List: 4 **R-E-D Code:** 1-1-1 **State/Fed. Status:** CEQA?
Distribution: CCA, HUM, NAP, ORA, SLO, SON, ++
Habitat: MshSw (coastal salt)
Life Form: Perennial herb
Blooming: June-September
Notes: On review list in OR. See *Wasmann Journal of Biology* 33(1-2):98 (1975) for discussion of CA distribution.

Eleocharis quadrangulata (Michx.) R. & S.

"four-angled spikerush" Cyperaceae
CNPS List: 2 **R-E-D Code:** 3-2-1 **State/Fed. Status:** CEQA
Distribution: BUT, MER, TEH, ++
Quads: 403B, 559C, 628D
Habitat: MshSw (freshwater)
Life Form: Perennial herb
Blooming: July-September

Elodea brandegeae

Considered but rejected: A synonym of *E. canadensis*; a common taxon

Elymus californicus (Bol.) Gould

"California bottle-brush grass" Poaceae
CNPS List: 4 **R-E-D Code:** 1-1-3 **State/Fed. Status:** /C3c
Distribution: MNT, MRN, SCR, SMT, SON
Habitat: NCFrs
Life Form: Perennial herb
Blooming: June-August

Elymus scribneri M.E. Jones

"Scribner's wheat grass" Poaceae
CNPS List: 2 **R-E-D Code:** 2-1-1 **State/Fed. Status:** CEQA
Distribution: MNO, NV, ++
Quads: 432A, 434D
Habitat: AlpBR
Life Form: Perennial herb
Blooming: July-August
Notes: Known in CA only from the White Mtns.

Empetrum hermaphroditum

See *Empetrum nigrum* ssp. *hermaphroditum*

Empetrum nigrum (L.) ssp. *hermaphroditum* (Lange) Bocher

"black crowberry" Empetraceae
CNPS List: 2 **R-E-D Code:** 3-2-1 **State/Fed. Status:** CEQA
Distribution: DNT, HUM, OR, ++
Quads: 689C, 723B, 740C
Habitat: CBScr, CoPrr
Life Form: Shrub (evergreen)
Blooming: April-June
Notes: Threatened by trampling and cattle grazing. See *E. nigrum* in *The Jepson Manual*. See *Four Seasons* 4(4):19-20 (1974) for species account, and *Madroño* 23:299 (1975) for discussion of HUM Co. discovery.

Enceliopsis covillei (Nels.) Blake

"Panamint daisy" Asteraceae
CNPS List: 1B **R-E-D Code:** 3-2-3 **State/Fed. Status:** /C2
Distribution: INY
Quads: 302A, 302B, 302C, 325B, 325C
Habitat: MDScr (subalkaline)
Life Form: Perennial herb
Blooming: March-June
Notes: Threatened by horticultural collecting and mining. See *Proceedings of the California Academy of Sciences* II 5:702 (1895) for original description.
Status Report: 1979

Enceliopsis nudicaulis (Gray) A. Nels.

"naked-stemmed daisy" Asteraceae
CNPS List: 4 **R-E-D Code:** 1-1-1 **State/Fed. Status:** CEQA?
Distribution: INY, SBD, AZ, ID, NV, UT+
Habitat: GBScr, MDScr / volcanic or carbonate
Life Form: Perennial herb
Blooming: May

Enceliopsis nudicaulis var. *nudicaulis*

See *Enceliopsis nudicaulis*

Enneapogon desvauxii Beauv.

"nine-awned pappus grass" Poaceae
CNPS List: 2 **R-E-D Code:** 3-1-1 **State/Fed. Status:** CEQA
Distribution: SBD, ++
Quads: 176A, 200A, 226A, 249D
Habitat: PJWld (rocky, carbonate)
Life Form: Perennial herb
Blooming: August-September

Ephedra funerea

Considered but rejected: Too common

Epilobium canum ssp. *septentrionale*

See *Epilobium septentrionale*

Epilobium howellii Hoch

"subalpine fireweed" Onagraceae
CNPS List: 1B **R-E-D Code:** 3-1-3 **State/Fed. Status:** CEQA
Distribution: FRE, MNO, SIE
Quads: 395D, 396A, 396B, 571C
Habitat: Medws, SCFrs / mesic
Life Form: Perennial herb (stoloniferous)
Blooming: July-August
Notes: Rarity and endangerment information needed. Need quads for MNO Co. See *Phytologia* 73(6):460-462 (1992) for original description.

Epilobium latifolium

Considered but rejected: Too common

Epilobium luteum Pursh

"yellow willowherb" Onagraceae
CNPS List: 2 **R-E-D Code:** 3-1-1 **State/Fed. Status:** CEQA
Distribution: PLU, SIS, OR, ++
Quads: 589D, 738A
Habitat: LCFrs (alongs streams and seeps)
Life Form: Perennial herb (stoloniferous)
Blooming: July-September
Notes: On review list in OR. See *Systematic Botany* 18(2):218-228 (1993) for information on hybridization.

Epilobium nivium Bdg.

"Snow Mtn. willowherb" Onagraceae
CNPS List: 1B **R-E-D Code:** 2-2-3 **State/Fed. Status:** /C3c
Distribution: COL, GLE, LAK, MEN, TRI
Quads: 564B, 564C, 565A, 565D, 581C, 598A, 598B
Habitat: Chprl, UCFrs / rocky
Life Form: Perennial herb
Blooming: July-October
Notes: Threatened by recreation and foot traffic. See *Zoe* 3:242-243 (1892) for original description.
Status Report: 1977

Epilobium oreganum Greene

"Oregon fireweed" Onagraceae
CNPS List: 1B **R-E-D Code:** 2-2-2 **State/Fed. Status:** /C2
Distribution: DNT, ELD, HUM, MEN, SHA, SIS, TEH, TRI, OR
Quads: 523D, 598A, 632D, 633D, 634D, 652A, 652B, 653B, 667A, 667B, 669C, 670B, 670C, 682A, 686B, 699C, 699D*, 703C, 720D, 721C, 721D, 739A
Habitat: BgFns, LCFrs (mesic)
Life Form: Perennial herb
Blooming: June-August
Notes: Unable to confirm many historical occurrences on Shasta-Trinity NF. Is it more common elsewhere? Threatened by logging. Known from fewer than 1000 plants at about twenty localities in OR, where a candidate for state listing. See *Pittonia* 1:255 (1888) for original description.

Epilobium rigidum Hausskn.

"Siskiyou Mtns. willowherb" Onagraceae
CNPS List: 4 **R-E-D Code:** 1-1-2 **State/Fed. Status:** CEQA?
Distribution: DNT, SIS, OR
Habitat: LCFrs (serpentinite)
Life Form: Perennial herb
Blooming: July-August
Notes: On review list in OR.

Epilobium septentrionale (Keck) Raven

"Humboldt County fuchsia" Onagraceae
CNPS List: 4 **R-E-D Code:** 1-1-3 **State/Fed. Status:** CEQA?
Distribution: HUM, MEN, TRI
Habitat: BUFrs, NCFrs / sandy or rocky
Life Form: Perennial herb
Blooming: August-September
Notes: See *Carnegie Institution of Washington Publication* 520:219 (1940) for original description, and *Annals of the Missouri Botanical Garden* 63:335 (1976) for revised treatment.

Epilobium siskiyouense (Munz) Hoch & Raven

"Siskiyou fireweed" Onagraceae
CNPS List: 1B **R-E-D Code:** 3-1-2 **State/Fed. Status:** /C3c
Distribution: SIS, TRI, OR
Quads: 667A, 667B, 682B, 684A, 684C, 699C, 700C, 700D, 736A, 736B, 738D
Habitat: SCFrs, UCFrs / rocky, serpentinite
Life Form: Perennial herb
Blooming: July-September
Notes: Known in CA from fewer than twenty occurrences. Candidate for state listing in OR. See *North American Flora* II 5:205 (1965) for original description, and *Madroño* 27(3):146 (1980) for revised nomenclature.

Equisetum palustre L.

"marsh horsetail" Equisetaceae
CNPS List: 3 **R-E-D Code:** 3-?-1 **State/Fed. Status:** CEQA?
Distribution: LAK, SFO, SMT, OR, ++
Quads: 466C
Habitat: MshSw
Life Form: Perennial herb (rhizomatous)
Fertile: unknown
Notes: Move to List 2? Location, rarity, and endangerment information needed; need quads. Scarcity poorly understood.

Eremalche kernensis C.B. Wolf

"Kern mallow" Malvaceae
CNPS List: 1B **R-E-D Code:** 3-3-3 **State/Fed. Status:** /FE
Distribution: KRN
Quads: 216C, 217A, 241A, 241C, 242A, 242B, 242C, 242D, 265B*
Habitat: ChScr, VFGrs
Life Form: Annual herb
Blooming: March-May
Notes: White-flowered plants of KRN Co. (*E. kernensis* sensu stricto) known from fewer than twenty occurrences. Seriously threatened by agriculture, grazing, and oil development. See *E. parryi* ssp. *kernensis* in *The Jepson Manual*. See *Phytologia* 72(1):48-54 (1992) for alternate taxonomic treatment which includes lavender-flowered plants and broadens range to west.
Status Report: 1990

Eriastrum brandegeae Mason

"Brandegee's eriastrum" Polemoniaceae
CNPS List: 1B **R-E-D Code:** 2-2-3 **State/Fed. Status:** /C2
Distribution: COL, GLE, LAK, SCL, TEH, TRI
Quads: 426D, 533B*, 534A, 534D, 564A, 564C, 564D, 565A, 596A, 596D, 612D, 650C
Habitat: Chprl, CmWld / volcanic
Life Form: Annual herb
Blooming: May-August
Notes: Threatened by grazing and vehicles. Includes *E. tracyi*, which is state-listed Rare. See *Madroño* 8:88-89 (1945) for original description.
Status Report: 1979

Eriastrum densifolium (Benth.) Mason ssp. _sanctorum_ (Mlkn.) Mason

"Santa Ana River woollystar" Polemoniaceae
CNPS List: 1B **R-E-D Code:** 3-3-3 **State/Fed. Status:** CE/FE
Distribution: ORA*, SBD
Quads: 87C*, 106C, 106D, 107A, 107B*, 107D
Habitat: Chprl, CoScr (alluvial fan)
Life Form: Perennial herb
Blooming: June-August
Notes: Known from one extended but fragmented population. Threatened by development, sand and gravel mining, grazing, flood control projects, and non-native plants. See *Crossosoma* 10(5):1-8 (1984) and *Fremontia* 13(3):19-20 (1985) for species accounts, and *Fremontia* 17(3):20-21 (1989) for discussion of ecology.
Status Report: 1987

Eriastrum hooveri (Jeps.) Mason

"Hoover's eriastrum" Polemoniaceae
CNPS List: 4 **R-E-D Code:** 1-2-3 **State/Fed. Status:** /FT
Distribution: FRE, KNG, KRN, SBA, SBT, SLO, TUL
Habitat: ChScr, VFGrs
Life Form: Annual herb
Blooming: April-July
Notes: Threatened by agriculture, grazing, urbanization, and energy development.
Status Report: 1988

Eriastrum luteum (Benth.) Mason

"yellow-flowered eriastrum" Polemoniaceae
CNPS List: 4 **R-E-D Code:** 1-1-3 **State/Fed. Status:** CEQA?
Distribution: MNT, SLO
Habitat: BUFrs, Chprl, CmWld
Life Form: Annual herb
Blooming: May-June

Eriastrum pluriflorum ssp. _sherman-hoytae_

Considered but rejected: A synonym of *E. pluriflorum*; a common taxon

Eriastrum tracyi

See *Eriastrum brandegeae*

Eriastrum virgatum (Benth.) Mason

"virgate eriastrum" Polemoniaceae
CNPS List: 4 **R-E-D Code:** 1-1-3 **State/Fed. Status:** CEQA?
Distribution: MNT, SBT
Habitat: Chprl (sandy), CoDns
Life Form: Annual herb
Blooming: May-July

Ericameria cuneata (Gray) McClatchie var. _macrocephala_ Urbatsch

"Laguna Mtns. goldenbush" Asteraceae
CNPS List: 1B **R-E-D Code:** 2-1-3 **State/Fed. Status:** CEQA
Distribution: SDG
Quads: 19B, 19C, 19D
Habitat: Chprl (granitic)
Life Form: Shrub
Blooming: September-December
Notes: Endemic to the Laguna Mtns. See *Madroño* 23(6):338-345 (1976) for original description.

Ericameria fasciculata (Eastw.) Macbr.

"Eastwood's goldenbush" Asteraceae
CNPS List: 1B **R-E-D Code:** 3-3-3 **State/Fed. Status:** /C2
Distribution: MNT
Quads: 365C, 366A, 366C, 366D, 386C
Habitat: CCFrs, Chprl (maritime), CoDns, CoScr
Life Form: Shrub (evergreen)
Blooming: July-October
Notes: Known from fewer than twenty occurrences. Threatened by development. See *Bulletin of the Torrey Botanical Club* 32:215 (1905) for original description.
Status Report: 1977

Ericameria gilmanii (Blake) Nesom

"Gilman's goldenbush" Asteraceae
CNPS List: 1B **R-E-D Code:** 3-1-3 **State/Fed. Status:** CEQA
Distribution: INY, KRN
Quads: 258B, 283D, 392D
Habitat: SCFrs, UCFrs / carbonate or granitic
Life Form: Shrub
Blooming: August-September
Notes: See *Madroño* 37:63 (1990) for first KRN Co. record, and *Phytologia* 68(2):144-155 (1990) for revised nomenclature.

Ericameria ophitidis (J.T. Howell) Nesom

"serpentine goldenbush" Asteraceae
CNPS List: 4 **R-E-D Code:** 1-1-3 **State/Fed. Status:** /C3c
Distribution: SHA, TEH, TRI
Habitat: LCFrs (serpentinite)
Life Form: Shrub
Blooming: June-August
Notes: See *Leaflets of Western Botany* 6:85 (1950) for original description, and *Phytologia* 68(2):144-155 (1990) for revised nomenclature.
Status Report: 1979

Ericameria palmeri

See *Ericameria palmeri* ssp. *palmeri*

Ericameria palmeri (Gray) Hall ssp. *palmeri*

"Palmer's goldenbush" Asteraceae
CNPS List: 2 **R-E-D Code:** 2-2-1 **State/Fed. Status:** /C2
Distribution: SDG, BA
Quads: 9B, 10B, 11B*, 21C, 22D
Habitat: CoScr
Life Form: Shrub (evergreen)
Blooming: September-November
Notes: Known in CA from only three occurrences. Threatened by development. USFWS uses the name *Haplopappus palmeri* ssp. *palmeri.*

Erigeron acris var. *debilis*

See *Trimorpha acris* var. *debilis*

Erigeron aequifolius Hall

"Hall's daisy" Asteraceae
CNPS List: 1B **R-E-D Code:** 3-1-3 **State/Fed. Status:** /C3c
Distribution: FRE, KRN, TUL
Quads: 258B, 307B, 307C, 330C, 374C, 374D, 375C, 375D
Habitat: BUFrs, LCFrs, PJWld, UCFrs / rocky
Life Form: Perennial herb (rhizomatous)
Blooming: July-August
Notes: Known from fewer than twenty occurrences. Relatively unthreatened due to inaccessibility. See *University of California Publications in Botany* 6:174 (1915) for original description, and *Phytologia* 72(3):157-208 (1992) for taxonomic treatment.

Erigeron angustatus Greene

"narrow-leaved daisy" Asteraceae
CNPS List: 1B **R-E-D Code:** 2-2-3 **State/Fed. Status:** CEQA
Distribution: LAK, NAP, SON
Quads: 500A, 500B, 502B, 516C, 517A, 517B, 519D, 533C, 548B
Habitat: Chprl (serpentinite)
Life Form: Perennial herb
Blooming: May-September
Notes: See *Bulletin of the Southern California Academy of Sciences* 1:88 (1885) for original description, and *Phytologia* 72(3):157-208 (1992) for taxonomic treatment.

Erigeron biolettii Greene

"streamside daisy" Asteraceae
CNPS List: 3 **R-E-D Code:** ?-?-3 **State/Fed. Status:** CEQA?
Distribution: HUM, MEN, MRN, NAP, SOL, SON
Quads: 482B, 500C, 501A, 502B, 503A, 516C, 517C, 517D, 551C, 617D
Habitat: BUFrs, CmWld, NCFrs / rocky, mesic
Life Form: Perennial herb
Blooming: June-September
Notes: Move to List 1B? Location, rarity, and endangerment information needed, especially quads for MRN Co. Most collections are very old. Intergrades with *E. inornatus.* See *Manual of the Botany of the Region of San Francisco Bay*, p. 181 (1894) by E. Greene for original description, and *Phytologia* 72(2):157-208 (1992) for taxonomic treatment.

Erigeron blochmaniae Greene

"Blochman's leafy daisy" Asteraceae
CNPS List: 1B **R-E-D Code:** 1-2-3 **State/Fed. Status:** /C3c
Distribution: SBA, SLO
Quads: 145B, 171A, 171C, 195B, 196B, 196D, 221B, 221D, 247D
Habitat: CoDns
Life Form: Perennial herb (rhizomatous)
Blooming: July-August
Notes: Field work needed. Threatened by coastal development. See *Pittonia* 3:27-28 (1896) for original description, and *Phytologia* 72(3):157-208 (1992) for taxonomic treatment.
Status Report: 1977

Erigeron bloomeri Gray var. *nudatus* (Gray) Cronq.

"Waldo daisy" Asteraceae
CNPS List: 2 **R-E-D Code:** 2-1-1 **State/Fed. Status:** /C3c
Distribution: DNT, SIS, OR
Quads: 701A, 737B, 738A, 738B, 738C, 739A
Habitat: LCFrs, UCFrs / serpentinite
Life Form: Perennial herb
Blooming: June-July
Status Report: 1980

Erigeron breweri Gray var. *bisanctus* Nesom

"pious daisy" Asteraceae
CNPS List: 1B **R-E-D Code:** 2-2-3 **State/Fed. Status:** CEQA
Distribution: LAX, SBD
Quads: 108A, 108B, 108C, 109A
Habitat: Chprl, LCFrs
Life Form: Perennial herb
Blooming: May-September
Notes: See *Phytologia* 72(3):157-208 (1992) for original description.

Erigeron breweri Gray var. *jacinteus* (Hall) Cronq.

"San Jacinto Mtns. daisy" Asteraceae
CNPS List: 4 **R-E-D Code:** 1-1-3 **State/Fed. Status:** CEQA?
Distribution: LAX, RIV, SBD
Habitat: SCFrs, UCFrs / rocky
Life Form: Perennial herb (rhizomatous)
Blooming: June-September
Notes: See *University of California Publications in Botany* 1:127 (1902) for original description, and *Phytologia* 72(3):157-208 (1992) for taxonomic treatment.

Erigeron calvus Cov.

"bald daisy" Asteraceae
CNPS List: 1B **R-E-D Code:** 3-3-3 **State/Fed. Status:** /C3b
Distribution: INY
Quads: 350D
Habitat: GBScr
Life Form: Perennial herb
Blooming: May
Notes: Known only from near Keeler; field surveys needed. Closely related to *E. divergens*, and also confused with *E. aphanactis* var. *aphanactis*; detailed study needed. See *Proceedings of the Biological Society of Washington* 7:69 (1892) for original description, and *Brittonia* 6:121-302 (1947) for taxonomic treatment.

Erigeron cervinus Greene

"Siskiyou daisy" Asteraceae
CNPS List: 4 **R-E-D Code:** 1-1-2 **State/Fed. Status:** CEQA?
Distribution: DNT, SIS, TRI, OR
Habitat: LCFrs, Medws
Life Form: Perennial herb (rhizomatous)
Blooming: June-August
Notes: Includes *E. delicatus.* Endangered in OR.

Erigeron decumbens Nutt. var. *robustior* (Cronq.) Cronq.

"robust daisy" Asteraceae
CNPS List: 4 **R-E-D Code:** 1-1-3 **State/Fed. Status:** CEQA?
Distribution: HUM, TRI
Habitat: Medws (sometimes serpentinite)
Life Form: Perennial herb
Blooming: June-July
Notes: See *Madroño* 35(2):81-82 (1988) for taxonomic treatment.

Erigeron delicatus

See *Erigeron cervinus*

Erigeron elegantulus Greene

"volcanic daisy" Asteraceae
CNPS List: 4 **R-E-D Code:** 1-1-1 **State/Fed. Status:** CEQA?
Distribution: LAS, MOD, SHA, SIS, OR
Habitat: AlpBR, GBScr, PJWld, SCFrs / volcanic
Life Form: Perennial herb
Blooming: March-August
Notes: Protected in part at Ash Valley RNA (BLM), LAS Co. See *Fremontia* 16(1):15-17 (1988) for brief species account and BLM management plans.

Erigeron flexuosus

Considered but rejected: Too common; a synonym of *E. lassenianus* var. *lassenianus*

Erigeron foliosus var. *blochmaniae*

See *Erigeron blochmaniae*

Erigeron inornatus Gray var. *calidipetris* Nesom

"hot rock daisy" Asteraceae
CNPS List: 4 **R-E-D Code:** 1-1-3 **State/Fed. Status:** CEQA?
Distribution: BUT, MOD, PLU, SHA, SIS
Habitat: LCFrs (sandy, volcanic)
Life Form: Perennial herb
Blooming: June-September
Notes: See *Phytologia* 72(3):157-208 (1992) for original description.

Erigeron inornatus Gray var. *keilii* Nesom

"Keil's daisy" Asteraceae
CNPS List: 1B **R-E-D Code:** 2-2-3 **State/Fed. Status:** CEQA
Distribution: FRE, TUL
Quads: 308B, 354B, 374D, 375A, 375C, 375D
Habitat: LCFrs, Medws
Life Form: Perennial herb
Blooming: June-September
Notes: Potentially threatened by logging. See *Phytologia* 72(3):157-208 (1992) for original description.

Erigeron inornatus var. *reductus*

Considered but rejected: A synonym of *E. reductus* var. *reductus*; a common taxon

Erigeron lobatus

Considered but rejected: Not in CA; misidentification of *E. divergens*; a common taxon

Erigeron mariposanus Congd.

"Mariposa daisy" Asteraceae
CNPS List: 1A **Last Seen:** 1900 **State/Fed. Status:** CEQA
Distribution: MPA*
Quads: 419B*
Habitat: CmWld
Life Form: Perennial herb
Blooming: June-August
Notes: Probably extinct. Suspected to occur in specialized habitat. See *Erythea* 7:185 (1900) for original description, and *Phytologia* 72(3):157-208 (1992) for taxonomic treatment.

Erigeron miser Gray

"starved daisy" Asteraceae
CNPS List: 1B **R-E-D Code:** 2-1-3 **State/Fed. Status:** CEQA
Distribution: NEV, PLA
Quads: 555C, 555D
Habitat: UCFrs (rocky)
Life Form: Perennial herb
Blooming: July-October
Notes: See *Proceedings of the American Academy of Arts and Sciences* 13:272 (1878) for original description, and *Phytologia* 72(3):157-208 (1992) for taxonomic treatment.

Erigeron multiceps Greene

"Kern River daisy" Asteraceae
CNPS List: 1B **R-E-D Code:** 3-2-3 **State/Fed. Status:** /C2
Distribution: TUL
Quads: 306B, 306C, 307A, 307B, 330C
Habitat: JTWld, Medws, UCFrs (openings)
Life Form: Perennial herb
Blooming: June-September
Notes: Known from fewer than twenty occurrences on the Kern Plateau. Possibly threatened by grazing and vehicles. Similar to *E. divergens.* See *Pittonia* 2:167 (1891) for original description, and *Phytologia* 73(3):186-202 (1992) for taxonomic information.
Status Report: 1979

Erigeron parishii Gray

"Parish's daisy" Asteraceae
CNPS List: 1B **R-E-D Code:** 2-3-3 **State/Fed. Status:** /PE
Distribution: RIV, SBD
Quads: 103B, 103C, 104B, 130C, 131C, 131D, 132D
Habitat: MDScr, PJWld / often carbonate
Life Form: Perennial herb
Blooming: May-June
Notes: Threatened by carbonate mining, vehicles, and residential development. See *Synoptical Flora of North America* 1(2):221 (1884) for original description, and *Fremontia* 16(1):20-21 (1988) for discussion of mining threats.
Status Report: 1988

Erigeron petrophilus var. *petrophilus*
Considered but rejected: Too common

Erigeron petrophilus Greene var. *sierrensis* Nesom
"northern Sierra daisy" Asteraceae
CNPS List: 4 **R-E-D Code:** 1-1-3 **State/Fed. Status:** CEQA?
Distribution: BUT, ELD, NEV, PLU, SIE, YUB
Habitat: CmWld, LCFrs, UCFrs / sometimes serpentinite
Life Form: Perennial herb (rhizomatous)
Blooming: June-September
Notes: See *Phytologia* 72(3):157-208 (1992) for original description.

Erigeron petrophilus Greene var. *viscidulus* (Gray) Nesom
"Klamath daisy" Asteraceae
CNPS List: 4 **R-E-D Code:** 1-1-2 **State/Fed. Status:** CEQA?
Distribution: HUM, SHA, SIS, TRI, OR
Habitat: Chprl, LCFrs, Medws, UCFrs / sometimes serpentinite
Life Form: Perennial herb (rhizomatous)
Blooming: July-September
Notes: Intergrades with *E. reductus. E. petrophilus* endangered in OR. See *Synoptical Flora of North America* 1(2):215 (1884) for original description, and *Phytologia* 72(3):157-208 (1992) for taxonomic treatment.

Erigeron sanctarum Wats.
"Saint's daisy" Asteraceae
CNPS List: 4 **R-E-D Code:** 1-2-3 **State/Fed. Status:** CEQA?
Distribution: SBA, SCZ, SLO, SRO
Habitat: Chprl, CmWld, CoScr
Life Form: Perennial herb (rhizomatous)
Blooming: March-June
Notes: Threatened by development.

Erigeron serpentinus Nesom
"serpentine daisy" Asteraceae
CNPS List: 1B **R-E-D Code:** 3-1-3 **State/Fed. Status:** CEQA
Distribution: SON
Quads: 519C
Habitat: Chprl (serpentinite)
Life Form: Perennial herb
Blooming: May-August
Notes: Endemic to The Cedars. Similar to *E. angustatus.* See *Phytologia* 72(3):157-208 (1992) for original description.

Erigeron supplex Gray
"supple daisy" Asteraceae
CNPS List: 1B **R-E-D Code:** 3-2-3 **State/Fed. Status:** /C2
Distribution: MEN, MRN*, SON
Quads: 485C*, 520B, 520D, 537A, 537B, 537C, 537D, 569D
Habitat: CBScr, CoPrr
Life Form: Perennial herb
Blooming: May-June
Notes: Threatened by coastal development. See *Proceedings of the American Academy of Arts and Sciences* 24:83 (1889) for original description, and *Madroño* 33(4): 308-309 (1986) for distributional information.

Erigeron uncialis ssp. *uncialis*
See *Erigeron uncialis* var. *uncialis*

Erigeron uncialis S.F. Blake var. *uncialis*
"limestone daisy" Asteraceae
CNPS List: 2 **R-E-D Code:** 3-2-1 **State/Fed. Status:** CEQA
Distribution: INY, SBD, NV
Quads: 131D, 249D, 369B, 372A
Habitat: GBScr, SCFrs / carbonate
Life Form: Perennial herb
Blooming: June-July

Eriochloa aristata
Considered but rejected: Not native; introduced from AZ

Eriodictyon altissimum Wells
"Indian Knob mountainbalm" Hydrophyllaceae
CNPS List: 1B **R-E-D Code:** 3-3-3 **State/Fed. Status:** CE/PE
Distribution: SLO
Quads: 221B, 247D
Habitat: Chprl (maritime), CmWld
Life Form: Shrub (evergreen)
Blooming: March-June
Notes: Known from six occurrences in the Irish Hills. Threatened by urbanization, energy development, and vehicles. Taxonomic relationship to *E. angustifolium* needs clarification. See *Madroño* 16:184-186 (1962) for original description.
Status Report: 1987

Eriodictyon angustifolium Nutt.
"narrow-leaved yerba santa" Hydrophyllaceae
CNPS List: 4 **R-E-D Code:** 1-1-1 **State/Fed. Status:** CEQA?
Distribution: SBD, AZ, BA, NV+
Habitat: PJWld
Life Form: Shrub (evergreen)
Blooming: June-July

Eriodictyon capitatum Eastw.
"Lompoc yerba santa" Hydrophyllaceae
CNPS List: 1B **R-E-D Code:** 3-2-3 **State/Fed. Status:** CR/C1
Distribution: SBA
Quads: 145A, 145B, 170B, 171A, 195C
Habitat: CCFrs, Chprl / sandy
Life Form: Shrub (evergreen)
Blooming: May-August
Notes: Known from fewer than ten occurrences. Reproduction under study. See *Leaflets of Western Botany* 1:40-41 (1933) for original description.
Status Report: 1987

Eriodictyon tomentosum
Considered but rejected: Too common

Eriodictyon traskiae ssp. *traskiae*
Considered but rejected: A synonym of *E. traskiae*, a common taxon

Eriogonum alpinum Engelm.

"Trinity buckwheat" Polygonaceae
CNPS List: 1B **R-E-D Code:** 3-1-3 **State/Fed. Status:** CE/C2
Distribution: SIS, TRI
Quads: 683A, 699C, 700D
Habitat: AlpBR, SCFrs, UCFrs / serpentinite
Life Form: Perennial herb (rhizomatous)
Blooming: June-September
Notes: Known from fewer than ten occurrences in the Mt. Eddy and Cory Pk. areas. See *Botanical Gazette* 7:6 (1882) for original description, and *Phytologia* 66(4):356-357 (1989) for taxonomic treatment.
Status Report: 1987

Eriogonum ampullaceum

Considered but rejected: Too common

Eriogonum apricum J.T. Howell var. *apricum*

"Ione buckwheat" Polygonaceae
CNPS List: 1B **R-E-D Code:** 3-3-3 **State/Fed. Status:** CE/C1
Distribution: AMA
Quads: 494C, 495A
Habitat: Chprl (Ione soil)
Life Form: Perennial herb
Blooming: July-October
Notes: Known from fewer than ten occurrences near Ione. Threatened by vehicles and clay mining. Protected in part at Apricum Hill ER (DFG). See *Leaflets of Western Botany* 7(10):237-238 (1955) for original description, and *Phytologia* 66(4):320-321 (1989) for taxonomic treatment.
Status Report: 1993

Eriogonum apricum J.T. Howell var. *prostratum* Myatt

"Irish Hill buckwheat" Polygonaceae
CNPS List: 1B **R-E-D Code:** 3-3-3 **State/Fed. Status:** CE/C1
Distribution: AMA
Quads: 494B, 495A
Habitat: Chprl (Ione soil)
Life Form: Perennial herb
Blooming: June-July
Notes: Known from only two occurrences near Irish Hill and Carbondale Mesa on the Ione Formation. Threatened by vehicles, clay mining, and erosion. See *Madroño* 20:320 (1970) for original description, and *Phytologia* 66(4):320-321 (1989) for taxonomic treatment.
Status Report: 1993

Eriogonum arborescens

Considered but rejected: Too common

Eriogonum argillosum J.T. Howell

"clay-loving buckwheat" Polygonaceae
CNPS List: 4 **R-E-D Code:** 1-1-3 **State/Fed. Status:** CEQA?
Distribution: MNT, SBT, SCL
Habitat: CmWld (serpentinite or clay)
Life Form: Annual herb
Blooming: March-June
Notes: See *Phytologia* 66(4):376 (1989) for taxonomic treatment.

Eriogonum beatleyae Reveal

"Beatley's buckwheat" Polygonaceae
CNPS List: 2 **R-E-D Code:** 3-3-1 **State/Fed. Status:** /C3c
Distribution: MNO, NV
Quads: 470B
Habitat: GBScr (volcanic)
Life Form: Perennial herb
Blooming: May-August
Notes: Known in CA from only one occurrence in the Bodie Hills. Threatened by sheep grazing, and potentially by mining. Similar to *E. ochrocephalum* and *E. rosense*; study needed. See *Aliso* 7(4):415-419 (1972) for original description, and *Phytologia* 66(4):322-323 (1989) for taxonomic treatment.

Eriogonum bifurcatum Reveal

"forked buckwheat" Polygonaceae
CNPS List: 1B **R-E-D Code:** 3-2-2 **State/Fed. Status:** /C2
Distribution: INY, SBD, NV
Quads: 273D, 274A, 298A
Habitat: ChScr (sandy)
Life Form: Annual herb
Blooming: April-June
Notes: Known in CA from six occurrences. Threatened in NV. See *Aliso* 7(3):357-360 (1971) for original description, and *Phytologia* 66(4):363 (1989) for taxonomic treatment.

Eriogonum breedlovei (J.T. Howell) Reveal var. *breedlovei*

"Breedlove's buckwheat" Polygonaceae
CNPS List: 1B **R-E-D Code:** 3-2-3 **State/Fed. Status:** /C1
Distribution: KRN
Quads: 237A, 237B, 260C, 260D
Habitat: PJWld, UCFrs / often carbonate
Life Form: Perennial herb
Blooming: June-August
Notes: Known from approximately ten occurrences. See *Mentzelia* 1:19-21 (1975) for original description, and *Phytologia* 66(4):323 (1989) for taxonomic treatment.
Status Report: 1977

Eriogonum breedlovei (J.T. Howell) Reveal var. *shevockii* J.T. Howell

"The Needles buckwheat" Polygonaceae
CNPS List: 4 **R-E-D Code:** 1-1-3 **State/Fed. Status:** /C3c
Distribution: KRN, TUL
Habitat: PJWld, UCFrs / granitic
Life Form: Perennial herb
Blooming: July-September
Notes: More common than previously thought, and relatively unthreatened. See *Mentzelia* 1:19-21 (1975) for original description, and *Phytologia* 66(4):323 (1989) for taxonomic treatment.
Status Report: 1979

Eriogonum butterworthianum J.T. Howell

"Butterworth's buckwheat" Polygonaceae
CNPS List: 1B **R-E-D Code:** 3-1-3 **State/Fed. Status:** CR/C2
Distribution: MNT
Quads: 319B, 319C
Habitat: Chprl (sandstone)
Life Form: Perennial herb
Blooming: June-July
Notes: Known from only four occurrences near Arroyo Seco in the Santa Lucia Mtns. See *Leaflets of Western Botany* 9(9-10):153-154 (1961) for original description, and *Phytologia* 66(4):328 (1989) for taxonomic treatment.
Status Report: 1988

Eriogonum caninum

See *Eriogonum luteolum* var. *caninum*

Eriogonum cithariforme

Considered but rejected: Too common

Eriogonum collinum

Considered but rejected: Not in CA

Eriogonum congdonii (S. Stokes) Reveal

"Congdon's buckwheat" Polygonaceae
CNPS List: 4 **R-E-D Code:** 1-1-3 **State/Fed. Status:** /C3c
Distribution: SHA, SIS, TRI
Habitat: LCFrs (serpentinite)
Life Form: Shrub (deciduous)
Blooming: June-August
Notes: See *Aliso* 7(2):220 (1970) for original description, and *Phytologia* 66(4):349 (1989) for taxonomic treatment.

Eriogonum contiguum (Reveal) Reveal

"Reveal's buckwheat" Polygonaceae
CNPS List: 2 **R-E-D Code:** 2-1-1 **State/Fed. Status:** /C3c
Distribution: INY, SBD, NV
Quads: 298A, 298C, 324D, 346D, 367C
Habitat: MDScr
Life Form: Annual herb
Blooming: February-June
Notes: Need quads for SBD Co. See *Phytologia* 23:175 (1972) for original description and 66(4):359 (1989) for taxonomic treatment.

Eriogonum crocatum A. Davids.

"Conejo buckwheat" Polygonaceae
CNPS List: 1B **R-E-D Code:** 2-2-3 **State/Fed. Status:** CR/C2
Distribution: VEN
Quads: 113A, 113B, 113C, 114A
Habitat: Chprl, CoScr, VFGrs / Conejo volcanic outcrops
Life Form: Perennial herb
Blooming: April-July
Notes: Known from fewer than twenty occurrences. Threatened by potential development. See *Bulletin of the Southern California Academy of Sciences* 23:17 (1924) for original description, and *Phytologia* 66(4):338-339 (1989) for taxonomic treatment.
Status Report: 1986

Eriogonum dasyanthemum

Considered but rejected: Too common

Eriogonum deserticola

Considered but rejected: Too common

Eriogonum diclinum Reveal

"Jaynes Canyon buckwheat" Polygonaceae
CNPS List: 4 **R-E-D Code:** 1-1-2 **State/Fed. Status:** /C3c
Distribution: SIS, TRI, OR
Habitat: UCFrs (often serpentinite)
Life Form: Perennial herb
Blooming: June-September
Notes: Endangered in OR. See *Aliso* 7(2):218 (1970) for original description, and *Phytologia* 66(4):354 (1989) for taxonomic treatment.
Status Report: 1979

Eriogonum eastwoodianum J.T. Howell

"Eastwood's buckwheat" Polygonaceae
CNPS List: 4 **R-E-D Code:** 1-1-3 **State/Fed. Status:** /C3c
Distribution: FRE, MNT
Habitat: CmWld (sandy or clay)
Life Form: Annual herb
Blooming: June-July
Notes: Marginally distinct from *E. temblorense* and *E. vestitum.* See *Phytologia* 66(4):374-375 (1989) for taxonomic treatment.

Eriogonum elegans

Considered but rejected: Too common

Eriogonum eremicola J.T. Howell & Reveal

"Wildrose Canyon buckwheat" Polygonaceae
CNPS List: 1B **R-E-D Code:** 3-1-3 **State/Fed. Status:** /C2
Distribution: INY
Quads: 302A, 350B, 350D
Habitat: PJWld, UCFrs / sandy or gravelly
Life Form: Annual herb
Blooming: June-September
Notes: Known from only five occurrences. See *Phytologia* 66(4):365 (1989) for taxonomic treatment.

Eriogonum ericifolium T. & G. var. *thornei* Reveal & Hendrickson

"Thorne's buckwheat" Polygonaceae
CNPS List: 1B **R-E-D Code:** 3-2-3 **State/Fed. Status:** CE/C1
Distribution: SBD
Quads: 225D
Habitat: PJWld
Life Form: Perennial herb
Blooming: July-August
Notes: Known from only two occurrences in the New York Mtns. See *Madroño* 23(4):205-209 (1975) for original description, and *Phytologia* 66(4):313 (1989) for taxonomic treatment.
Status Report: 1986

Eriogonum esmeraldense var. *esmeraldense*

Considered but rejected: Too common

Eriogonum foliosum Wats.

"leafy buckwheat" Polygonaceae
CNPS List: 1B **R-E-D Code:** 2-2-2 **State/Fed. Status:** CEQA
Distribution: RIV, SBD, SDG, BA
Quads: 20D, 48C, 66B, 105A, 105B, 131C, 131D
Habitat: Chprl, LCFrs, PJWld / sandy
Life Form: Annual herb
Blooming: July-October
Notes: Known in SDG Co. from only two collections. Easily confused with *E. davidsonii*, so possibly overlooked. See *Phytologia* 66(4):382 (1989) for taxonomic treatment.

Eriogonum giganteum Wats. var. *compactum* Dunkle

"Santa Barbara Island buckwheat" Polygonaceae
CNPS List: 1B **R-E-D Code:** 3-1-3 **State/Fed. Status:** CR/C2
Distribution: SBR
Quads: SBRA
Habitat: CBScr (rocky)
Life Form: Shrub (deciduous)
Blooming: May-August
Notes: Known from fewer than fifteen occurrences, but population numbers are increasing. See *Bulletin of the Southern California Academy of Sciences* 41:130 (1943) for original description, and *Phytologia* 66(4):318-319 (1989) for taxonomic treatment.
Status Report: 1986

Eriogonum giganteum Wats. var. *formosum* K. Bdg.

"San Clemente Island buckwheat" Polygonaceae
CNPS List: 1B **R-E-D Code:** 3-2-3 **State/Fed. Status:** /C2
Distribution: SCM
Quads: SCMC, SCMN, SCMS
Habitat: CBScr (rocky)
Life Form: Shrub (deciduous)
Blooming: March-October
Notes: Possibly threatened by Navy activities. Feral herbivores removed from SCM Isl., and vegetation recovering. See *Phytologia* 66(4):318-319 (1989) for taxonomic treatment.

Eriogonum giganteum var. *giganteum*

Considered but rejected: Too common

Eriogonum gilmanii S. Stokes

"Gilman's buckwheat" Polygonaceae
CNPS List: 4 **R-E-D Code:** 1-1-3 **State/Fed. Status:** /C3c
Distribution: INY
Habitat: MDScr (gravelly)
Life Form: Perennial herb
Blooming: May-August
Notes: See *Leaflets of Western Botany* 3(1):16 (1941) for original description, and *Phytologia* 66(4):335 (1989) for taxonomic treatment.
Status Report: 1977

Eriogonum gossypinum Curran

"cottony buckwheat" Polygonaceae
CNPS List: 4 **R-E-D Code:** 1-2-3 **State/Fed. Status:** /C3c
Distribution: FRE, KNG, KRN, SLO
Habitat: ChScr, VFGrs / clay
Life Form: Annual herb
Blooming: April-September
Notes: See *Phytologia* 66(4):372 (1989) for taxonomic treatment.

Eriogonum gracilipes

Considered but rejected: Too common

Eriogonum grande var. *dunklei*

See *Eriogonum grande* var. *rubescens*

Eriogonum grande Greene var. *grande*

"island buckwheat" Polygonaceae
CNPS List: 4 **R-E-D Code:** 1-2-3 **State/Fed. Status:** CEQA?
Distribution: ANA, SCM, SCT, SCZ
Habitat: CBScr
Life Form: Perennial herb
Blooming: June-October
Notes: See *Phytologia* 66(4):333-334 (1989) for taxonomic treatment.

Eriogonum grande Greene var. *rubescens* (Greene) Munz

"red-flowered buckwheat" Polygonaceae
CNPS List: 4 **R-E-D Code:** 1-2-3 **State/Fed. Status:** /C2
Distribution: ANA?, SCZ, SMI, SRO
Habitat: CBScr, CoScr
Life Form: Perennial herb
Blooming: June-October
Notes: Does plant occur on ANA Isl.? Threatened by feral herbivores, and by cattle grazing on SRO Isl. Includes *E. grande* var. *dunklei*. See *Phytologia* 66(4):333-334 (1989) for taxonomic treatment.

Eriogonum grande Greene var. *timorum* Reveal

"San Nicolas Island buckwheat" Polygonaceae
CNPS List: 1B **R-E-D Code:** 2-3-3 **State/Fed. Status:** CE/C2
Distribution: SNI
Quads: SNIC
Habitat: CBScr
Life Form: Perennial herb
Blooming: June-October
Notes: Possibly threatened by Navy activities, erosion, and non-native plants. See *Aliso* 7(2):229 (1970) for original description, and *Phytologia* 66(4):333-334 (1989) for taxonomic treatment.
Status Report: 1987

Eriogonum heermannii Dur. & Hilg. var. *floccosum* Munz

"Clark Mtn. buckwheat" Polygonaceae
CNPS List: 4 **R-E-D Code:** 1-1-3 **State/Fed. Status:** /C3c
Distribution: SBD
Habitat: PJWld (carbonate)
Life Form: Shrub (deciduous)
Blooming: August-October
Notes: See *Phytologia* 66(4):314-316 (1989) for taxonomic treatment.

Eriogonum heermannii Dur. & Hilg. var. *occidentale* S. Stokes

"western Heermann's buckwheat" Polygonaceae
CNPS List: 4 **R-E-D Code:** 1-2-3 **State/Fed. Status:** CEQA?
Distribution: FRE, MNT, SBT
Habitat: CmWld (clay or shale)
Life Form: Shrub (deciduous)
Blooming: July-October
Notes: See *Leaflets of Western Botany* 1(4):30 (1932) for original description, and *Phytologia* 66(4):314-316 (1989) for taxonomic treatment.

Eriogonum hirtellum J.T. Howell & Bacig.

"Klamath Mtn. buckwheat" Polygonaceae
CNPS List: 1B **R-E-D Code:** 2-1-3 **State/Fed. Status:** /C3c
Distribution: DNT, SIS
Quads: 719A, 736D, 737B, 737C, 738A, 738D
Habitat: LCFrs, UCFrs / serpentinite
Life Form: Perennial herb (rhizomatous)
Blooming: July-September
Notes: Known from fewer than twenty occurrences. Possibly threatened by logging and grazing. Klamath NF has adopted species management guidelines. See *Leaflets of Western Botany* 9:174-176 (1961) for original description, and *Phytologia* 66(4):355 (1989) for taxonomic treatment.
Status Report: 1977

Eriogonum hoffmannii S. Stokes var. *hoffmannii*

"Hoffmann's buckwheat" Polygonaceae
CNPS List: 4 **R-E-D Code:** 1-1-3 **State/Fed. Status:** /C3c
Distribution: INY
Habitat: MDScr (washes, roadsides)
Life Form: Annual herb
Blooming: June-September
Notes: See *Phytologia* 66(4):365 (1989) for taxonomic treatment.

Eriogonum hoffmannii S. Stokes var. *robustius* S. Stokes

"robust Hoffmann's buckwheat" Polygonaceae
CNPS List: 4 **R-E-D Code:** 1-1-3 **State/Fed. Status:** /C3c
Distribution: INY
Habitat: MDScr, PJWld / washes, roadsides
Life Form: Annual herb
Blooming: August-November
Notes: See *Phytologia* 66(4):365 (1989) for taxonomic treatment.

Eriogonum hookeri

Considered but rejected: Too common

Eriogonum intrafractum Cov. & Mort.

"jointed buckwheat" Polygonaceae
CNPS List: 4 **R-E-D Code:** 1-1-3 **State/Fed. Status:** /C2
Distribution: INY
Habitat: MDScr (carbonate)
Life Form: Perennial herb
Blooming: May-October
Notes: See *Journal of the Washington Academy of Sciences* 26:305 (1936) for original description, and *Phytologia* 66(4):359 (1989) for taxonomic treatment.
Status Report: 1977

Eriogonum kearneyi var. *kearneyi*

Considered but rejected: A synonym of *E. nummulare*; a common taxon

Eriogonum kelloggii Gray

"Kellogg's buckwheat" Polygonaceae
CNPS List: 1B **R-E-D Code:** 3-3-3 **State/Fed. Status:** CE/C1
Distribution: MEN
Quads: 600B, 600C
Habitat: LCFrs (serpentinite)
Life Form: Perennial herb
Blooming: May-August
Notes: Known from only five occurrences in the Red Mtn. area near Leggett. Threatened by mining. Protected in part at Red Mtn. ACEC (BLM). See *Proceedings of the American Academy of Arts and Sciences* 8:293 (1870) for original description, and *Phytologia* 66(4):351 (1989) for taxonomic treatment.
Status Report: 1987

Eriogonum kennedyi Wats. var. *alpigenum* M. & J.

"southern alpine buckwheat" Polygonaceae
CNPS List: 1B **R-E-D Code:** 2-1-3 **State/Fed. Status:** CEQA
Distribution: LAX, SBD, VEN
Quads: 105C, 105D, 134D, 135D, 190C
Habitat: AlpBR, SCFrs / granitic, gravelly
Life Form: Perennial herb
Blooming: July-September
Notes: Rarity information needed. See *Bulletin of the Torrey Botanical Club* 51:296 (1924) for original description, and *Phytologia* 66(4):326-327 (1989) for taxonomic treatment.

Eriogonum kennedyi Wats. var. *austromontanum* M. & J.

"southern mountain buckwheat" Polygonaceae
CNPS List: 1B **R-E-D Code:** 2-2-3 **State/Fed. Status:** /C2
Distribution: SBD
Quads: 105A, 105B, 126D, 130C, 131C, 131D, 132D
Habitat: LCFrs (gravelly), PbPln
Life Form: Perennial herb
Blooming: July-August
Notes: Threatened by development, grazing, vehicles, and non-native plants. See *Bulletin of the Torrey Botanical Club* 51:295 (1924) for original description, and *Phytologia* 66(4):326-327 (1989) for taxonomic treatment.
Status Report: 1979

Eriogonum kennedyi Wats. var. *pinicola* Reveal

"Kern buckwheat" Polygonaceae
CNPS List: 1B **R-E-D Code:** 3-3-3 **State/Fed. Status:** /C2
Distribution: KRN
Quads: 211B, 212A, 212D
Habitat: Chprl, PJWld
Life Form: Perennial herb
Blooming: May-June
Notes: Known from only three occurrences in the vicinity of Sweet Ridge. Field work needed. Threatened by wind energy development. See *Phytologia* 66(4):326-328 (1989) for taxonomic treatment.
Status Report: 1977

Eriogonum latens

Considered but rejected: Too common

Eriogonum libertini Reveal

"Dubakella Mtn. buckwheat" Polygonaceae
CNPS List: 4 **R-E-D Code:** 1-2-3 **State/Fed. Status:** /C3c
Distribution: SHA, TEH, TRI
Habitat: Chprl (serpentinite)
Life Form: Perennial herb
Blooming: June-August
Notes: See *Madroño* 28(3):163-166 (1981) for original description, and *Phytologia* 66(4):349 (1989) for taxonomic treatment.

Eriogonum luteolum Greene var. *caninum* (Greene) Reveal

"Tiburon buckwheat" Polygonaceae
CNPS List: 3 **R-E-D Code:** ?-2-3 **State/Fed. Status:** /C3c
Distribution: ALA, CCA?, COL, LAK, MRN, NAP, SCL, SMT, SON*
Quads: 407A, 448C, 465B?, 465C, 465D?, 466B, 466C, 467A, 467B, 467D, 484B, 484C, 499B, 499C, 502B*, 516A, 517A, 517B(*?), 549D
Habitat: Chprl, CoPrr, VFGrs / serpentinite
Life Form: Annual herb
Blooming: June-September
Notes: Move to List 1B? Location information needed, especially quads for COL Co. Does plant occur in CCA Co.? Easily confused with var. *luteolum*. Threatened by development and non-native plants. Protected in part at Ring Mtn. Preserve (TNC), MRN Co. See *Flora Franciscana*, pp. 150-151 (1891) by E. Greene for original description, and *Phytologia* 66(4):378-379 (1989) for alternative treatment which restricts var. *caninum* to ALA and MRN counties.
Status Report: 1977

Eriogonum microthecum Nutt. var. *corymbosoides* Reveal

"San Bernardino buckwheat" Polygonaceae
CNPS List: 4 **R-E-D Code:** 1-1-3 **State/Fed. Status:** CEQA?
Distribution: SBD
Habitat: PJWld (granitic)
Life Form: Shrub (deciduous)
Blooming: July-September
Notes: See *Phytologia* 66(4):312-313 (1989) for taxonomic treatment.

Eriogonum microthecum Nutt. var. *johnstonii* Reveal

"Johnston's buckwheat" Polygonaceae
CNPS List: 1B **R-E-D Code:** 3-1-3 **State/Fed. Status:** /C2
Distribution: LAX, SBD
Quads: 108A, 134B, 134C
Habitat: SCFrs, UCFrs / rocky
Life Form: Shrub (deciduous)
Blooming: July-September
Notes: Known from fewer than ten occurrences, most of which are in designated wilderness areas. Possibly threatened by foot traffic in a portion of its range. See *Phytologia* 66(4):313 (1989) for taxonomic treatment.
Status Report: 1979

Eriogonum microthecum Nutt. var. *lapidicola* Reveal

"Inyo Mtns. buckwheat" Polygonaceae
CNPS List: 4 **R-E-D Code:** 1-1-2 **State/Fed. Status:** CEQA?
Distribution: INY, NV
Habitat: PJWld (carbonate)
Life Form: Shrub (deciduous)
Blooming: July-September
Notes: See *Phytologia* 66(4):313 (1989) for taxonomic treatment.

Eriogonum microthecum Nutt. var. *panamintense* S. Stokes

"Panamint Mtns. buckwheat" Polygonaceae
CNPS List: 1B **R-E-D Code:** 3-1-3 **State/Fed. Status:** /C2
Distribution: INY
Quads: 302A, 348C, 350B, 350D
Habitat: PJWld (rocky)
Life Form: Shrub (deciduous)
Blooming: July-October
Notes: Known from fewer than ten occurrences. See *Phytologia* 66(4):313 (1989) for taxonomic treatment.

Eriogonum nervulosum (S. Stokes) Reveal

"Snow Mtn. buckwheat" Polygonaceae
CNPS List: 1B **R-E-D Code:** 3-2-3 **State/Fed. Status:** /C2
Distribution: COL, GLE?, LAK, NAP, SON, YOL
Quads: 517B, 519C, 532C, 532D, 533C, 534D, 548A, 548C, 564D, 565A
Habitat: Chprl (serpentinite)
Life Form: Perennial herb (rhizomatous)
Blooming: June-September
Notes: Known from approximately twenty occurrences. Undocumented in GLE Co.; need quads. See *Phytologia* 40:467 (1978) for revised nomenclature and 66(4):350 (1989) for taxonomic treatment.
Status Report: 1979

Eriogonum nortonii Greene

"Pinnacles buckwheat" Polygonaceae
CNPS List: 1B **R-E-D Code:** 2-1-3 **State/Fed. Status:** /C3c
Distribution: MNT, SBT
Quads: 341A, 341B, 343A, 343D, 363C, 364B, 364D, 385C, 386D
Habitat: Chprl, VFGrs / sandy, often on recent burns
Life Form: Annual herb
Blooming: May-June
Notes: Known from approximately twenty occurrences. See *Phytologia* 66(4):376 (1989) for taxonomic treatment.

Eriogonum nudum Benth. var. *decurrens* (S. Stokes) Bowerman

"Ben Lomond buckwheat" Polygonaceae
CNPS List: 1B **R-E-D Code:** 3-3-3 **State/Fed. Status:** CEQA
Distribution: CCA, SCR
Quads: 408D
Habitat: Chprl, CmWld, LCFrs (maritime ponderosa pine sandhills)
Life Form: Perennial herb
Blooming: June-October
Notes: Need quads for Mt. Diablo area (CCA Co.). Threatened by development and sand mining. See *Phytologia* 66(4):329-333 (1989) for taxonomic treatment.

Eriogonum nudum Benth. var. *indictum* (Jeps.) Reveal

"protruding buckwheat" Polygonaceae
CNPS List: 4 **R-E-D Code:** 1-2-3 **State/Fed. Status:** CEQA?
Distribution: FRE, KRN, MER, MNT, SBT, SLO
Habitat: Chprl, ChScr, CmWld / clay, serpentinite
Life Form: Perennial herb
Blooming: May-October
Notes: Not always distinct from var. *auriculatum* in FRE and SBT counties. See *Flora of California* 1(4):421 (1914) by W.L. Jepson for original description, and *Phytologia* 66(4):329-332 (1989) for taxonomic treatment.

Eriogonum nudum Benth. var. *murinum* Reveal

"mouse buckwheat" Polygonaceae
CNPS List: 1B **R-E-D Code:** 2-2-3 **State/Fed. Status:** /C2
Distribution: FRE, TUL
Quads: 332A, 332B, 332D, 354C, 354D, 376D
Habitat: Chprl, CmWld, VFGrs / sandy
Life Form: Perennial herb
Blooming: June-November
Notes: Known from fewer than twenty occurrences; some degraded by road improvements along General's Highway. Sequoia-Kings Canyon NP has adopted species management guidelines. See *Aliso* 7(2):228 (1976) for original description, and *Phytologia* 66(4):329-332 (1989) for taxonomic treatment.
Status Report: 1977

Eriogonum nudum Benth. var. *paralinum* Reveal

"Del Norte buckwheat" Polygonaceae
CNPS List: 2 **R-E-D Code:** 2-2-1 **State/Fed. Status:** CEQA
Distribution: DNT, OR
Quads: 723B, 723D
Habitat: CBScr, CoPrr
Life Form: Perennial herb
Blooming: June-September
Notes: This variety is the coastal expression of *E. nudum.* See *Phytologia* 66(3):258 (1989) for original description and 66(4):329-331 (1989) for taxonomic treatment.

Eriogonum nudum Benth. var. *regivirum* Reveal & J. Stebbins

"Kings River buckwheat" Polygonaceae
CNPS List: 1B **R-E-D Code:** 3-2-3 **State/Fed. Status:** CEQA
Distribution: FRE
Quads: 355A, 376A, 376D, 377D
Habitat: CmWld (carbonate)
Life Form: Perennial herb
Blooming: August-November
Notes: Known only from the Kings River Cyn. See *Phytologia* 66(3):246-248 (1989) for original description and 66(4):329-332 (1989) for taxonomic treatment.

Eriogonum nutans T. & G.

"nodding buckwheat" Polygonaceae
CNPS List: 2 **R-E-D Code:** 2-1-1 **State/Fed. Status:** CEQA
Distribution: LAS, MNO, NV, OR, UT
Quads: 470A, 620A
Habitat: GBScr (sandy or gravelly)
Life Form: Annual herb
Blooming: May-October
Notes: Grayish plants dubiously called var. *glabrum* appear in the high Sierra as roadside waifs from NV. On review list in OR. See *Madroño* 27(3):142 (1980) for first CA occurrence, and *Phytologia* 66(4):366 (1989) for taxonomic treatment.

Eriogonum nutans var. *glabrum*

Considered but rejected: Not native and taxonomic problem; occurs in CA only as a roadside waif from NV, and is dubiously distinct from *E. nutans*

Eriogonum nutans var. *nutans*

See *Eriogonum nutans*

Eriogonum ochrocephalum Wats. var. *alexanderae* Reveal

"Alexander's buckwheat" Polygonaceae
CNPS List: 2 **R-E-D Code:** 3-2-1 **State/Fed. Status:** CEQA
Distribution: MNO, NV
Quads: 470A
Habitat: GBScr, PJWld / shale, gravelly
Life Form: Perennial herb
Blooming: May-July
Notes: See *Phytologia* 66(4):321 (1989) for taxonomic treatment.

Eriogonum ovalifolium Nutt. var. *eximium* (Tides.) J.T. Howell

"brown-margined buckwheat" Polygonaceae
CNPS List: 4 **R-E-D Code:** 1-1-1 **State/Fed. Status:** CEQA?
Distribution: ALP, ELD, NV
Habitat: AlpBR, SCFrs / granitic sand
Life Form: Perennial herb
Blooming: June-August
Notes: Intergrades with var. *nivale.* See Proceedings of the Biological Society of Washington 36:181 (1923) for original description, *Mentzelia* 1:19 (1976) for revised nomenclature, and *Phytologia* 66(4):335-336 (1989) for taxonomic treatment.

Eriogonum ovalifolium Nutt. var. *vineum* (Small) Jeps.

"Cushenbury buckwheat" Polygonaceae
CNPS List: 1B **R-E-D Code:** 3-3-3 **State/Fed. Status:** /PE
Distribution: SBD
Quads: 104B, 105A, 130C, 131C, 131D, 132D
Habitat: MDScr, PJWld / carbonate
Life Form: Perennial herb
Blooming: May-June
Notes: Threatened by carbonate mining and vehicles. See *Bulletin of the Torrey Botanical Club* 25:40-53 (1898) for original description, *Fremontia* 16(1):20-21 (1988) for discussion of mining threats, and *Phytologia* 66(4):335-337 (1989) for taxonomic treatment.
Status Report: 1979

Eriogonum panamintense var. *mensicola*

Considered but rejected: A synonym of *E. panamintense*; a common taxon

Eriogonum parvifolium var. *lucidum*

Considered but rejected: A synonym of *E. parvifolium*; a common taxon

Eriogonum parvifolium var. *paynei*

Considered but rejected: A synonym of *E. parvifolium*; a common taxon

Eriogonum pendulum Wats.

"Waldo buckwheat" Polygonaceae
CNPS List: 2 **R-E-D Code:** 3-2-1 **State/Fed. Status:** /C3c
Distribution: DNT, OR
Quads: 739B, 739C, 740A
Habitat: LCFrs, UCFrs / serpentinite
Life Form: Perennial herb
Blooming: August-September
Notes: Known in CA from fewer than ten occurrences. Threatened by mining. On watch list in OR. See *Phytologia* 66(4):314 (1989) for taxonomic treatment.

Eriogonum polypodum Small

"Tulare County buckwheat" Polygonaceae
CNPS List: 4 **R-E-D Code:** 1-1-3 **State/Fed. Status:** CEQA?
Distribution: FRE, TUL
Habitat: SCFrs (granitic sand)
Life Form: Perennial herb
Blooming: July-August
Notes: See *Phytologia* 66(4):354-355 (1989) for taxonomic treatment.

Eriogonum prattenianum Durand var. *avium* Reveal & Shevock

"Kettle Dome buckwheat" Polygonaceae
CNPS List: 4 **R-E-D Code:** 1-2-3 **State/Fed. Status:** CEQA?
Distribution: FRE, MAD
Habitat: UCFrs (granitic)
Life Form: Perennial herb
Blooming: June-August
Notes: See *Phytologia* 66(3):249-250 (1989) for original description and 66(4):350 (1989) for taxonomic treatment.

Eriogonum prociduum Reveal

"prostrate buckwheat" Polygonaceae
CNPS List: 1B **R-E-D Code:** 2-2-2 **State/Fed. Status:** /C2
Distribution: LAS, MOD, NV, OR
Quads: 658C, 659A, 675C, 690B, 690C, 691A, 692A, 692B, 707C, 708C, 725D
Habitat: GBScr, PJWld, UCFrs / volcanic
Life Form: Perennial herb
Blooming: May-July
Notes: Occasionally threatened by cattle trampling. On watch list in NV, and a candidate for state listing in OR. Protected in part at Ash Valley RNA (BLM), LAS Co. See *Aliso* 7(4):415-419 (1972) for original description, *Fremontia* 16(1):15-17 (1988) for brief species account and BLM management plans, and *Phytologia* 66(4):321 (1989) for taxonomic treatment.
Status Report: 1979

Eriogonum puberulum Wats.

"downy buckwheat" Polygonaceae
CNPS List: 2 **R-E-D Code:** 3-1-1 **State/Fed. Status:** CEQA
Distribution: INY, NV+
Quads: 369B
Habitat: PJWld (sandy or gravelly, carbonate)
Life Form: Annual herb
Blooming: May-September
Notes: Known in CA from only one occurrence in the Cottonwood Mtns. See *Phytologia* 66(4):383 (1989) for taxonomic treatment.

Eriogonum rixfordii

Considered but rejected: Too common

Eriogonum rosense

Considered but rejected: Too common

Eriogonum rupinum

Considered but rejected: A synonym of *E. panamintense*; a common taxon

Eriogonum shockleyi Wats. var. *shockleyi*

"Shockley's buckwheat" Polygonaceae
CNPS List: 4 **R-E-D Code:** 1-1-1 **State/Fed. Status:** CEQA?
Distribution: INY, AZ, ID, NV+
Habitat: PJWld (gravelly, carbonate)
Life Form: Perennial herb
Blooming: May-July
Notes: Sensitive in ID. See *Phytologia* 66(4):323-324 (1989) for taxonomic treatment.

Eriogonum siskiyouense Small

"Siskiyou buckwheat" Polygonaceae
CNPS List: 4 **R-E-D Code:** 1-1-3 **State/Fed. Status:** /C3c
Distribution: SIS, TRI
Habitat: LCFrs (often serpentinite)
Life Form: Perennial herb
Blooming: August-September
Notes: See *Bulletin of the Torrey Botanical Club* 25:44 (1898) for original description, and *Phytologia* 66(4):348 (1989) for taxonomic treatment.
Status Report: 1979

Eriogonum strictum Benth. var. *greenei* (Gray) Reveal

"Greene's buckwheat" Polygonaceae
CNPS List: 4 **R-E-D Code:** 1-1-3 **State/Fed. Status:** CEQA?
Distribution: COL, MEN, SIS, TEH, TRI
Habitat: LCFrs (serpentinite)
Life Form: Perennial herb
Blooming: July-September
Notes: See *Proceedings of the American Academy of Arts and Sciences* 12:83 (1870) for original description, and *Phytologia* 40:467 (1978) for revised nomenclature and 66(4):337-338 (1989) for taxonomic treatment.

Eriogonum temblorense J.T. Howell & Twisselmann

"Temblor buckwheat" Polygonaceae
CNPS List: 4 **R-E-D Code:** 1-1-3 **State/Fed. Status:** /C2
Distribution: KRN, MNT, SLO
Habitat: VFGrs (clay or sandstone)
Life Form: Annual herb
Blooming: May-September
Notes: Marginally distinct from *E. eastwoodianum* and *E. vestitum.* See *Phytologia* 66(4):375 (1989) for taxonomic treatment.

Eriogonum ternatum Howell

"ternate buckwheat" Polygonaceae
CNPS List: 4 **R-E-D Code:** 1-1-2 **State/Fed. Status:** CEQA?
Distribution: DNT, SIS, SON, TEH, OR
Habitat: LCFrs (serpentinite)
Life Form: Perennial herb
Blooming: June-August
Notes: On watch list in OR. See *Phytologia* 66(4):348-349 (1989) for taxonomic treatment.

Eriogonum tripodum Greene

"tripod buckwheat" Polygonaceae
CNPS List: 4 **R-E-D Code:** 1-2-3 **State/Fed. Status:** CEQA?
Distribution: AMA, COL, ELD, LAK, MPA, NAP, PLA, TEH, TUO
Habitat: Chprl, CmWld / often serpentinite
Life Form: Shrub (deciduous)
Blooming: May-July
Notes: Some occurrences threatened by mining. See *Pittonia* 1:39 (1887) for original description, and *Phytologia* 66(4):350-351 (1989) for taxonomic treatment.

Eriogonum truncatum T. & G.

"Mt. Diablo buckwheat" Polygonaceae
CNPS List: 1A **Last Seen:** 1940 **State/Fed. Status:** /C3a
Distribution: ALA*, CCA*, SOL*
Quads: 445A*, 445B*, 464A*, 464B*, 464C*, 481D*, 482A*
Habitat: Chprl, CoScr, VFGrs / sandy
Life Form: Annual herb
Blooming: April-September
Notes: Recent attempts to rediscover this plant have been unsuccessful. See *Proceedings of the American Academy of Arts and Sciences* 8:173 (1870) for original description, and *Phytologia* 66(4):375-376 (1989) for taxonomic treatment.
Status Report: 1988

Eriogonum twisselmannii (J.T. Howell) Reveal

"Twisselmann's buckwheat" Polygonaceae
CNPS List: 1B **R-E-D Code:** 2-2-3 **State/Fed. Status:** CR/C2
Distribution: TUL
Quads: 308A, 308D
Habitat: UCFrs (granitic)
Life Form: Perennial herb
Blooming: July-September
Notes: Endemic to Sequoia NF; protected in part at Slate Mtn. BA, which contains the largest of approximately ten known occurrences. See *Leaflets of Western Botany* 10(1):13 (1963) for original description, and *Phytologia* 66(4):352 (1989) for taxonomic treatment.
Status Report: 1979

Eriogonum umbellatum var. *aureum*

See *Eriogonum umbellatum* var. *glaberrimum*

Eriogonum umbellatum Torr. var. *glaberrimum* (Gand.) Reveal

"green buckwheat" Polygonaceae
CNPS List: 2 **R-E-D Code:** 3-1-1 **State/Fed. Status:** CEQA
Distribution: MOD, OR
Quads: 724B
Habitat: LCFrs, UCFrs / sandy or gravelly
Life Form: Perennial herb
Blooming: June-September
Notes: Known in CA only from the Warner Mtns., where it is either rare or undercollected. Status in OR unknown. See *Taxon* 17:531-532 (1968) for revised nomenclature, and *Phytologia* 66(4):341-347 (1989) for taxonomic treatment.

Eriogonum umbellatum var. *hausknechtii*

Considered but rejected: Not in CA

Eriogonum umbellatum Torr. var. *humistratum* Reveal

"Mt. Eddy buckwheat" Polygonaceae
CNPS List: 4 **R-E-D Code:** 1-1-3 **State/Fed. Status:** /C3c
Distribution: SIS, TRI
Habitat: AlpBR, Chprl, Medws, SCFrs, UCFrs / rocky, usually serpentinite
Life Form: Perennial herb
Blooming: May-September
Notes: Intergrades with var. *polyanthum.* See *Phytologia* 66(3):260 (1989) for original description and 66(4):341-345 (1989) for taxonomic treatment.

Eriogonum umbellatum Torr. var. *juniporinum* Reveal

"juniper buckwheat" Polygonaceae
CNPS List: 2 **R-E-D Code:** 3-1-1 **State/Fed. Status:** CEQA
Distribution: SBD, NV
Quads: 176A, 200B, 225D, 249C, 250B
Habitat: MDScr, PJWld
Life Form: Perennial herb
Blooming: July-October
Notes: Similar to var. *subaridum.* See *Great Basin Naturalist* 45:279 (1985) for original description, and *Phytologia* 66(4):341-347 (1989) for taxonomic treatment.

Eriogonum umbellatum Torr. var. *minus* Jtn.

"alpine sulfur-flowered buckwheat" Polygonaceae
CNPS List: 4 **R-E-D Code:** 1-2-3 **State/Fed. Status:** /C3c
Distribution: LAX, SBD
Habitat: SCFrs, UCFrs / gravelly
Life Form: Perennial herb
Blooming: July-September
Notes: See *Bulletin of the California Academy of Sciences* 17:64 (1918) for original description, and *Phytologia* 66(4):341-344 (1989) for taxonomic treatment.
Status Report: 1979

Eriogonum umbellatum Torr. var. *torreyanum* (Gray) M.E. Jones

"Donner Pass buckwheat" Polygonaceae
CNPS List: 1B **R-E-D Code:** 3-2-3 **State/Fed. Status:** /C2
Distribution: NEV, PLA, SIE
Quads: 538B, 539A, 555A, 555B, 555D
Habitat: Medws, UCFrs / volcanic
Life Form: Perennial herb
Blooming: July-September
Notes: Known from fewer than ten occurrences. Similar to var. *glaberrimum.* Draft interim management guide prepared by USFS in 1993. See *Fremontia* 1(3):20 (1973) for discussion of rediscovery at type locality, and *Phytologia* 66(4):341-347 (1989) for taxonomic treatment.
Status Report: 1977

Eriogonum vestitum J.T. Howell

"Idria buckwheat" Polygonaceae
CNPS List: 4 **R-E-D Code:** 1-1-3 **State/Fed. Status:** /C3c
Distribution: FRE, MER, SBT
Habitat: VFGrs (semi-siliceous diatomaceous shale)
Life Form: Annual herb
Blooming: May-June
Notes: Marginally distinct from *E. eastwoodianum* or *E. temblorense.* See *Phytologia* 66(4):375 (1989) for taxonomic treatment.

Eriogonum wrightii Benth. var. *olanchense* (J.T. Howell) Reveal

"Olancha Peak buckwheat" Polygonaceae
CNPS List: 1B **R-E-D Code:** 3-1-3 **State/Fed. Status:** /C3c
Distribution: TUL
Quads: 329D
Habitat: AlpBR, SCFrs (gravelly or rocky)
Life Form: Perennial herb
Blooming: July-September
Notes: Known from only two occurrences on the uppermost slopes of Olancha Pk. See *Leaflets of Western Botany* 6:151 (1951) for original description, and *Phytologia* 66(4):324-326 (1989) for taxonomic treatment.
Status Report: 1979

Erioneuron pilosum (Buckl.) Nash

"hairy erioneuron" Poaceae
CNPS List: 2 **R-E-D Code:** 2-1-1 **State/Fed. Status:** CEQA
Distribution: INY, SBD, NV, ++
Quads: 226A, 249D, 369C
Habitat: PJWld (rocky, sometimes carbonate)
Life Form: Perennial herb
Blooming: May-June
Notes: See *American Journal of Botany* 48:565-573 (1961) for taxonomic treatment.

Eriophyllum congdonii Bdg.

"Congdon's woolly sunflower" Asteraceae
CNPS List: 1B **R-E-D Code:** 2-1-3 **State/Fed. Status:** CR/C3c
Distribution: MPA
Quads: 437C, 438A, 438B, 438D
Habitat: Chprl, CmWld, LCFrs
Life Form: Annual herb
Blooming: May-June
Notes: See *Botanical Gazette* 27:449-450 (1899) for original description.
Status Report: 1990

Eriophyllum jepsonii Greene

"Jepson's woolly sunflower" Asteraceae
CNPS List: 4 **R-E-D Code:** 1-1-3 **State/Fed. Status:** CEQA?
Distribution: ALA, CCA, KRN, SBT, SCL, STA, VEN
Habitat: Chprl, CmWld, CoScr / sometimes serpentinite
Life Form: Perennial herb
Blooming: April-June

Eriophyllum lanatum (Pursh) Forbes var. *hallii* Const.

"Fort Tejon woolly sunflower" Asteraceae
CNPS List: 1B **R-E-D Code:** 3-3-3 **State/Fed. Status:** /C1
Distribution: KRN, SBA
Quads: 189C, 192C
Habitat: Chprl, CmWld
Life Form: Perennial herb
Blooming: May-July
Notes: Seriously threatened by grazing. See *University of California Publications in Botany* 18:94-96 (1937) for revised nomenclature.
Status Report: 1977

Eriophyllum lanatum (Pursh) Forbes var. *obovatum* (Greene) Hall

"southern Sierra woolly sunflower" Asteraceae
CNPS List: 4 **R-E-D Code:** 1-1-3 **State/Fed. Status:** CEQA?
Distribution: FRE, KRN, SBD, TUL
Habitat: LCFrs, UCFrs
Life Form: Perennial herb
Blooming: June-July
Notes: Similar to var. *lanceolatum.*

Eriophyllum latilobum Rydb.

"San Mateo woolly sunflower" Asteraceae
CNPS List: 1B **R-E-D Code:** 3-3-3 **State/Fed. Status:** CE/PE
Distribution: SMT
Quads: 429D*, 448D
Habitat: CmWld (serpentinite, often on roadcuts)
Life Form: Perennial herb
Blooming: May-June
Notes: Known from only one extant occurrence, which was rediscovered in 1981 by K. Culligan and M. Showers. Threatened by development, erosion, and road maintenance.

Eriophyllum mohavense (Jtn.) Jeps.

"Barstow woolly sunflower" Asteraceae
CNPS List: 1B **R-E-D Code:** 2-2-3 **State/Fed. Status:** /C2
Distribution: SBD
Quads: 181B, 182A, 183A, 183B, 184A, 206C, 208B, 208C, 208D, 209A, 209D, 233C
Habitat: ChScr, MDScr, Plyas
Life Form: Annual herb
Blooming: April-May
Notes: Threatened by energy development, road improvements, vehicles, and grazing. Protected in part at an ACEC (BLM).
Status Report: 1977

Eriophyllum nevinii Gray

"Nevin's woolly sunflower" Asteraceae
CNPS List: 1B **R-E-D Code:** 2-1-3 **State/Fed. Status:** /C2
Distribution: SBR, SCM, SCT
Quads: SBRA, SCMC, SCMN, SCMS, SCTE
Habitat: CBScr, CoScr
Life Form: Shrub (evergreen)
Blooming: April-August
Notes: Feral herbivores removed from SCM Isl., and vegetation recovering. Populations appear to be recovering on SBR Isl.

Eriophyllum nubigenum Greene

"Yosemite woolly sunflower" Asteraceae
CNPS List: 1B **R-E-D Code:** 2-1-3 **State/Fed. Status:** /C2
Distribution: MAD, MPA, TUO
Quads: 437A, 437D, 454C, 456C
Habitat: Chprl, LCFrs, UCFrs / gravelly
Life Form: Annual herb
Blooming: May-August
Notes: Known from fewer than twenty occurrences, but several large occurrences have been found since 1980. See *Madroño* 29(2):123 (1982) for information on rediscovery.
Status Report: 1977

Eriophyllum nubigenum var. *congdonii*

See *Eriophyllum congdonii*

Erodium macrophyllum

Considered but rejected: Too common

Eryngium aristulatum Jeps. var. *hooveri* Sheikh

"Hoover's button-celery" Apiaceae
CNPS List: 4 **R-E-D Code:** 1-1-3 **State/Fed. Status:** /C1
Distribution: SBT, SCL, SLO
Habitat: VnPls
Life Form: Annual/Perennial herb
Blooming: July
Notes: See *Madroño* 30(2):93-101 (1983) for original description.

Eryngium aristulatum Jeps. var. *parishii* (Coult. & Rose) Math. & Const.

"San Diego button-celery" Apiaceae
CNPS List: 1B **R-E-D Code:** 2-3-2 **State/Fed. Status:** CE/FE
Distribution: RIV, SDG, BA
Quads: 7A, 10C, 11A, 11D, 22A, 22B, 22C, 22D, 35B, 36A, 36B, 36D, 51C, 68C, 69D
Habitat: CoScr, VFGrs, VnPls / mesic
Life Form: Annual/Perennial herb
Blooming: April-June
Notes: Threatened by agriculture, urbanization, road maintenance, vehicles, and foot traffic. See *Contributions from the U.S. National Herbarium* 7:57 (1900) for original description, and *American Midland Naturalist* 25(2):361-387 (1941) for revised nomenclature.
Status Report: 1987

Eryngium constancei Sheikh

"Loch Lomond button-celery" Apiaceae
CNPS List: 1B **R-E-D Code:** 3-3-3 **State/Fed. Status:** CE/FE
Distribution: LAK
Quads: 533C
Habitat: VnPls
Life Form: Annual/Perennial herb
Blooming: April-June
Notes: Only known occurrence protected at Loch Lomond ER (DFG), but entire watershed not protected. Previously damaged by dredging of vernal lake. See *Madroño* 30(2):93-101 (1983) for original description.
Status Report: 1985

Eryngium mathiasiae

Considered but rejected: Too common

Eryngium pinnatisectum Jeps.

"Tuolumne button-celery" Apiaceae
CNPS List: 4 **R-E-D Code:** 1-2-3 **State/Fed. Status:** /C2
Distribution: AMA, CAL, SAC, TUO
Habitat: CmWld, LCFrs, VnPls / mesic
Life Form: Annual/Perennial herb
Blooming: June

Eryngium racemosum Jeps.

"delta button-celery" Apiaceae
CNPS List: 1B **R-E-D Code:** 2-3-3 **State/Fed. Status:** CE/C2
Distribution: CAL, MER, SJQ*, STA*
Quads: 402A, 402B, 402D, 403A, 423A, 423C, 423D, 424A*, 441D, 443B*, 443C*, 444A*, 461A*, 462D*, 477D
Habitat: RpScr (vernally mesic clay depressions)
Life Form: Annual/Perennial herb
Blooming: June-August
Notes: Threatened by agriculture and flood control activities. See *Flora of California* 2(1):659 (1936) by W.L. Jepson for original description.
Status Report: 1986

Eryngium spinosepalum Math.

"spiny-sepaled button-celery" Apiaceae
CNPS List: 1B **R-E-D Code:** 3-2-3 **State/Fed. Status:** /C2
Distribution: FRE, MAD, STA, TUL
Quads: 309A*, 332B, 333A, 333B, 333D, 355B, 356A, 356D, 377B, 378D*, 398C, 399D, 424D
Habitat: VFGrs, VnPls
Life Form: Annual/Perennial herb
Blooming: April-May
Notes: Known from approximately twenty occurrences. Threatened by development, grazing, and agriculture. Apparently intergrades with *E. castrense,* and possibly *E. vaseyi.*

Erysimum ammophilum Heller

"coast wallflower" Brassicaceae
CNPS List: 1B **R-E-D Code:** 2-2-3 **State/Fed. Status:** /C2
Distribution: MNT, SCR, SMT, SRO
Quads: 366A, 366D, 387A, 387D, 409D
Habitat: CoDns
Life Form: Perennial herb
Blooming: February-June
Notes: Need quads for SRO Isl. Occurrences from SDG Co. previously included in this species are *E. capitatum* ssp. *capitatum.* Threatened by coastal development.

Erysimum capitatum (Douglas) Greene ssp. *angustatum* (Greene) R. Price

"Contra Costa wallflower" Brassicaceae
CNPS List: 1B **R-E-D Code:** 3-3-3 **State/Fed. Status:** CE/FE
Distribution: CCA
Quads: 481D
Habitat: InDns
Life Form: Perennial herb
Blooming: March-July
Notes: Known from only two occurrences at the Antioch Dunes. Seriously threatened by mining, agricultural conversion, non-native plants, and industrial development. Recovery work in progress. State and federally listed as *E. capitatum* var. *angustatum.* See *Pittonia* 3:132 (1896) for original description, and *Biological Conservation* 65:257-278 (1993) for population biology.
Status Report: 1988

Erysimum capitatum var. *angustatum*

See *Erysimum capitatum* ssp. *angustatum*

Erysimum capitatum (Douglas) Greene ssp. *lompocense* (Rossbach) R. Price

"San Luis Obispo wallflower" Brassicaceae
CNPS List: 4 **R-E-D Code:** 1-2-3 **State/Fed. Status:** CEQA?
Distribution: SBA, SLO
Habitat: Chprl, CoScr / sandy
Life Form: Perennial herb
Blooming: February-May
Notes: Intergrades locally with ssp. *capitatum* and *E. insulare* ssp. *suffrutescens.*

Erysimum franciscanum G. Rossb.

"San Francisco wallflower" Brassicaceae
CNPS List: 4 **R-E-D Code:** 1-2-3 **State/Fed. Status:** /C2
Distribution: MRN, SCL, SCR, SFO, SMT, SON
Habitat: CoDns, CoScr, VFGrs (often serpentinite or granitic)
Life Form: Perennial herb
Blooming: March-June
Notes: Rare and declining in SCR Co. Includes *E. franciscanum* var. *crassifolium.* Inland plants approach *E. capitatum.* See *Aliso* 4(1):118-121 (1958) for original description.
Status Report: 1977

Erysimum franciscanum var. *crassifolium*

See *Erysimum franciscanum*

Erysimum franciscanum var. *franciscanum*

See *Erysimum franciscanum*

Erysimum insulare

See *Erysimum insulare* ssp. *insulare*

Erysimum insulare Greene ssp. *insulare*

"island wallflower" Brassicaceae
CNPS List: 1B **R-E-D Code:** 2-1-3 **State/Fed. Status:** /C2
Distribution: ANA, SCZ*, SMI, SRO
Quads: ANAC, SMIE, SMIW, SRON
Habitat: CBScr
Life Form: Perennial herb
Blooming: March-May
Notes: Last seen on SCZ Isl. in 1880's on offshore rocks; need quad.

Erysimum insulare Greene ssp. *suffrutescens* (Abrams) R. Price

"suffrutescent wallflower" Brassicaceae
CNPS List: 4 **R-E-D Code:** 1-2-3 **State/Fed. Status:** CEQA?
Distribution: LAX, SBA, SLO, VEN
Habitat: CBScr, CoDns, CoScr
Life Form: Perennial herb
Blooming: January-June
Notes: Threatened by coastal development and vehicles. Includes *E. suffrutescens* var. *grandifolium.* Hybridizes locally with *E. capitatum.*

Erysimum menziesii

See *Erysimum menziesii* sspp. *eurekense, menziesii,* and *yadonii*

Erysimum menziesii (Hook.) Wettst. ssp. *eurekense* R. Price

"Humboldt Bay wallflower" Brassicaceae
CNPS List: 1B **R-E-D Code:** 3-3-3 **State/Fed. Status:** CE/FE
Distribution: HUM
Quads: 655A, 672B, 672C
Habitat: CoDns
Life Form: Perennial herb
Blooming: March-June
Notes: Known only from the Humboldt Bay area. Threatened by development, vehicles, and non-native plants. Formerly included in state-listed Endangered *E. menziesii.* Protected in part at Manila Dunes ACEC (BLM).

Erysimum menziesii (Hook.) Wettst. ssp. *menziesii*

"Menzies's wallflower" Brassicaceae
CNPS List: 1B **R-E-D Code:** 3-3-3 **State/Fed. Status:** CE/FE
Distribution: MEN, MNT
Quads: 366C, 569A, 585D
Habitat: CoDns
Life Form: Perennial herb
Blooming: March-June
Notes: Nearly extirpated on the Monterey Peninsula. Seriously threatened by development, vehicles, and non-native plants.
Status Report: 1987

Erysimum menziesii (Hook.) Wettst. ssp. *yadonii* R. Price

"Yadon's wallflower" Brassicaceae
CNPS List: 1B **R-E-D Code:** 3-2-3 **State/Fed. Status:** CE/FE
Distribution: MNT
Quads: 366A
Habitat: CoDns
Life Form: Perennial herb
Blooming: March-May
Notes: Known only from the Monterey Bay near Marina. Threatened by development and sand mining. Formerly included in state-listed Endangered *E. menziesii.*

Erysimum suffrutescens var. *grandifolium*

See *Erysimum insulare* ssp. *suffrutescens*

Erysimum suffrutescens var. *lompocense*

See *Erysimum capitatum* ssp. *lompocense*

Erysimum suffrutescens var. *suffrutescens*

See *Erysimum insulare* ssp. *suffrutescens*

Erysimum teretifolium Eastw.

"Santa Cruz wallflower" Brassicaceae
CNPS List: 1B **R-E-D Code:** 2-3-3 **State/Fed. Status:** CE/PE
Distribution: SCR
Quads: 407C, 408C, 408D
Habitat: Chprl, LCFrs / inland marine sands
Life Form: Perennial herb
Blooming: March-July
Notes: Seriously threatened by residential development and sand mining. See *Leaflets of Western Botany* 2(5):73 (1938) for original description.
Status Report: 1986

Erythronium citrinum

See *Erythronium citrinum* var. *citrinum*

Erythronium citrinum Wats. var. *citrinum*

"lemon-colored fawn lily" Liliaceae
CNPS List: 4 **R-E-D Code:** 1-1-1 **State/Fed. Status:** CEQA?
Distribution: DNT, HUM, SIS, TRI, OR
Habitat: Chprl, LCFrs / usually serpentinite
Life Form: Perennial herb (bulbiferous)
Blooming: March-April
Notes: See *E. citrinum* in *The Jepson Manual.*

Erythronium citrinum Wats. var. *roderickii* Shevock & Allen

"Scott Mtns. fawn lily" Liliaceae
CNPS List: 1B **R-E-D Code:** 3-1-3 **State/Fed. Status:** CEQA
Distribution: TRI
Quads: 683B
Habitat: LCFrs (serpentinite)
Life Form: Perennial herb (bulbiferous)
Blooming: May
Notes: Endemic to the Scott Mtns. Not in *The Jepson Manual.* See *Phytologia* 71(2):101-103 (1991) for original description.

Erythronium cliftonii

Considered but rejected: Not published

Erythronium grandiflorum ssp. *pusaterii*

See *Erythronium pusaterii*

Erythronium helenae Appleg.

"St. Helena fawn lily" Liliaceae
CNPS List: 4 **R-E-D Code:** 1-2-3 **State/Fed. Status:** CEQA?
Distribution: LAK, NAP, SON
Habitat: Chprl, CmWld, LCFrs, VFGrs / volcanic or serpentinite
Life Form: Perennial herb (bulbiferous)
Blooming: March-May
Notes: Approximately fifteen NAP Co. occurrences are all near Mt. St. Helena; rare in SON Co. Threatened by horticultural collecting, road construction, and geothermal development. See *Contributions from the Dudley Herbarium* 1:188 (1933) for original description.

Erythronium hendersonii Wats.

"Henderson's fawn lily" Liliaceae
CNPS List: 2 **R-E-D Code:** 3-1-1 **State/Fed. Status:** CEQA
Distribution: DNT, SIS, OR
Quads: 718C, 734A, 736C, 736D, 740D
Habitat: LCFrs
Life Form: Perennial herb (bulbiferous)
Blooming: April-July
Notes: Forms hybrid swarms with *E. citrinum* in OR.

Erythronium howellii Wats.

"Howell's fawn lily" Liliaceae
CNPS List: 4 **R-E-D Code:** 1-1-2 **State/Fed. Status:** /C3c
Distribution: DNT, OR
Habitat: LCFrs
Life Form: Perennial herb (bulbiferous)
Blooming: April-May
Notes: Endangered in OR. See *E. citrinum* in *The Jepson Manual.*

Erythronium klamathense Appleg.

"Klamath fawn lily" Liliaceae
CNPS List: 4 **R-E-D Code:** 1-1-1 **State/Fed. Status:** CEQA?
Distribution: SHA, SIS, OR
Habitat: Medws, UCFrs
Life Form: Perennial herb (bulbiferous)
Blooming: April-July

Erythronium pluriflorum Shevock, Bartel & G. Allen

"Shuteye Peak fawn lily" Liliaceae
CNPS List: 1B **R-E-D Code:** 2-1-3 **State/Fed. Status:** CEQA
Distribution: MAD
Quads: 417B, 417C, 417D
Habitat: Medws, SCFrs, UCFrs / granitic
Life Form: Perennial herb (bulbiferous)
Blooming: May-July
Notes: Occurrences highly localized; endemic to Chiquito Ridge in the San Joaquin River watershed. See *Madroño* 37(4):261-273 (1990) for original description.

Erythronium pusaterii (Munz & J.T. Howell) Shevock, Bartel & G. Allen

"Hocket Lakes fawn lily" Liliaceae
CNPS List: 1B **R-E-D Code:** 3-1-3 **State/Fed. Status:** /C3c
Distribution: TUL
Quads: 308A, 308D, 331C
Habitat: SCFrs (granitic or metamorphic)
Life Form: Perennial herb (bulbiferous)
Blooming: May-July
Notes: Known from fewer than five occurrences. Most occurrences are relatively inaccessible. Protected in part at Slate Mtn. BA (USFS). See *Leaflets of Western Botany* 10(7):104-105 (1964) for original description, and *Madroño* 37(4):261-273 (1990) for revised nomenclature.
Status Report: 1977

Erythronium tuolumnense Appleg.

"Tuolumne fawn lily" Liliaceae
CNPS List: 1B **R-E-D Code:** 2-2-3 **State/Fed. Status:** /C2
Distribution: TUO
Quads: 457B, 458A, 474C, 475C, 475D
Habitat: BUFrs, Chprl, LCFrs
Life Form: Perennial herb (bulbiferous)
Blooming: March-May
Notes: Threatened by logging, vehicles, horticultural collecting, and reforestation with herbicides.
Status Report: 1980

Eschscholzia covillei

Considered but rejected: Too common; a synonym of *E. minutiflora* ssp. *covillei*

Eschscholzia hypecoides Benth.

"San Benito poppy" Papaveraceae
CNPS List: 4 **R-E-D Code:** 1-1-3 **State/Fed. Status:** CEQA?
Distribution: FRE, IMP, MEN, MNT, SBT, SLO
Habitat: Chprl, CmWld, VFGrs / serpentinite clay
Life Form: Annual herb
Blooming: March-June

Eschscholzia lemmonii Greene ssp. *kernensis* (Munz) C. Clark

"Tejon poppy" Papaveraceae
CNPS List: 1B **R-E-D Code:** 3-3-3 **State/Fed. Status:** CEQA
Distribution: KRN
Quads: 189B, 214D, 216C, 238C, 242D
Habitat: VFGrs
Life Form: Annual herb
Blooming: March-April
Notes: Probably threatened by grazing and non-native plants. See *Aliso* 4:90 (1958) for original description, and *Madroño* 33(3):224 (1986) for revised nomenclature.

Eschscholzia lemmonii ssp. *lemmonii*

Considered but rejected: Too common

Eschscholzia lobbii

Considered but rejected: Too common

Eschscholzia minutiflora ssp. *minutiflora*

Considered but rejected: Too common

Eschscholzia minutiflora Wats. ssp. *twisselmannii* C. Clark & Faull

"Red Rock poppy" Papaveraceae
CNPS List: 1B **R-E-D Code:** 3-2-3 **State/Fed. Status:** /C2
Distribution: KRN
Quads: 234B, 235A, 235C, 236D
Habitat: MDScr (volcanic tuff)
Life Form: Annual herb
Blooming: March-May
Notes: Known only from the Rand and El Paso Mtns. of the western Mojave Desert. Threatened by vehicles. See *E. minutiflora* in *The Jepson Manual.* See *Madroño* 38(2):73-79 (1991) for original description.

Eschscholzia procera Greene

"Kernville poppy" Papaveraceae
CNPS List: 3 **R-E-D Code:** ?-?-3 **State/Fed. Status:** /C2
Distribution: KRN
Quads: 212B, 260B
Habitat: CmWld (sandy floodplain)
Life Form: Perennial herb
Blooming: June-August
Notes: Move to List 1B? Taxonomic problem. Threatened by urbanization. See *E. californica* in *The Jepson Manual.*

Eschscholzia ramosa Greene

"island poppy" Papaveraceae
CNPS List: 4 **R-E-D Code:** 1-1-2 **State/Fed. Status:** /C3c
Distribution: SBR, SCM, SCT, SCZ, SMI, SNI, SRO, GU
Habitat: CBScr, Chprl
Life Form: Annual herb
Blooming: March-April
Notes: See *Bulletin of the California Academy of Sciences* 1:182 (1885) for original description.

Eschscholzia rhombipetala Greene

"diamond-petaled California poppy" Papaveraceae
CNPS List: 1A **Last Seen:** 1950 **State/Fed. Status:** /C2
Distribution: ALA*, CCA*, COL*, SLO*, STA*
Quads: 243B*, 244C?*, 424B*, 443C*, 445A*, 463C*, 464A*, 481D*, 563D*
Habitat: VFGrs (clay)
Life Form: Annual herb
Blooming: March-April
Notes: Historical occurrence for SJQ Co. from literature lacks documentation; occurrence from La Panza, SLO Co. (244C) probably misidentified *E. lemmonii* ssp. *lemmonii*. Field surveys needed. Threatened by agriculture and grazing.

Escobaria vivipara (Nutt.) F. Buxb. var. *alversonii* (Coult.) D. Hunt

"foxtail cactus" Cactaceae
CNPS List: 1B **R-E-D Code:** 2-2-2 **State/Fed. Status:** /C2
Distribution: IMP, RIV, SBD, AZ
Quads: 12D, 58A, 61A, 61B, 61D, 62A, 63A, 75A, 75B, 75C, 75D, 76B, 77A, 77B, 78B, 78C, 79D, 95C, 96D, 101C, 102D, 127B, 153B, 178B
Habitat: MDScr, SDScr
Life Form: Shrub (stem succulent)
Blooming: May-June
Notes: Threatened by horticultural collecting. USFWS uses the name *Coryphantha vivipara* var. *alversonii.*

Escobaria vivipara (Nutt.) F. Buxb. var. *rosea* (Clokey) D. Hunt

"viviparous foxtail cactus" Cactaceae
CNPS List: 1B **R-E-D Code:** 3-2-2 **State/Fed. Status:** /C3c
Distribution: SBD, AZ, NV
Quads: 200A, 200B, 225D, 226A, 249D
Habitat: MDScr, PJWld / carbonate
Life Form: Shrub (stem succulent)
Blooming: May-June
Notes: Known in CA from fewer than twenty occurrences. Threatened by horticultural collecting. See *Madroño* 7:75 (1943) for original description.

Eucnide rupestris (Baill.) Thomps. & Ernst

"rock nettle" Loasaceae
CNPS List: 2 **R-E-D Code:** 3-2-1 **State/Fed. Status:** CEQA
Distribution: IMP, SDG, AZ, BA
Quads: 17C, 18C
Habitat: SDScr
Life Form: Annual herb
Blooming: December-April

Eupatorium shastense

See *Ageratina shastensis*

Euphorbia arizonica

See *Chamaesyce arizonica*

Euphorbia exstipulata Engelm. var. *exstipulata*

"Clark Mtn. spurge" Euphorbiaceae
CNPS List: 2 **R-E-D Code:** 3-3-1 **State/Fed. Status:** CEQA
Distribution: SBD, AZ, ++
Quads: 249D
Habitat: MDScr (rocky)
Life Form: Annual herb
Blooming: September
Notes: Known in CA only from Clark Mtn. Threatened by mining.

Euphorbia hooveri

See *Chamaesyce hooveri*

Euphorbia misera Benth.

"cliff spurge" Euphorbiaceae
CNPS List: 2 **R-E-D Code:** 2-2-1 **State/Fed. Status:** CEQA
Distribution: ORA, RIV, SCM, SCT, SDG, BA, GU
Quads: 11B, 11D, 22B, 52B, 71D, 83B, SCMC, SCMS, SCTN
Habitat: CBScr, CoScr / rocky
Life Form: Shrub
Blooming: January-August
Notes: Threatened by development. Only RIV Co. population reduced to a few plants by habitat disturbance and frost damage.

Euphorbia ocellata var. *rattanii*

See *Chamaesyce ocellata* ssp. *rattanii*

Euphorbia parishii

Considered but rejected: A synonym of *Chamaesyce parishii*; a common taxon

Euphorbia pediculifera

Considered but rejected: A synonym of *Chamaesyce pediculifera*; a common taxon

Euphorbia platyspermum

See *Chamaesyce platysperma*

Fendlerella utahensis (Wats.) Heller

"yerba desierto" Hydrangeaceae
CNPS List: 4 **R-E-D Code:** 1-1-1 **State/Fed. Status:** CEQA?
Distribution: INY, SBD, AZ, NV+
Habitat: MDScr, PJWld / carbonate
Life Form: Shrub (deciduous)
Blooming: June-August

Ferocactus acanthodes var. *acanthodes*

Considered but rejected: Too common; a synonym of *F. cylindraceus* var. *cylindraceus*

Ferocactus viridescens (T. & G.) Britt. & Rose

"San Diego barrel cactus" Cactaceae
CNPS List: 2 **R-E-D Code:** 1-3-1 **State/Fed. Status:** /C2
Distribution: SDG, BA
Quads: 10B, 10C, 11A, 11B, 11D, 21B, 21C, 22A, 22B, 22C, 22D, 35C, 35D
Habitat: Chprl, CoScr, VFGrs, VnPls
Life Form: Shrub (stem succulent)
Blooming: May-June
Notes: Seriously threatened by urbanization, vehicles, and horticultural collecting.
Status Report: 1977

Festuca arizonica

Considered but rejected: Not documented in CA

Fimbristylis spadicea

See *Fimbristylis thermalis*

Fimbristylis thermalis Wats.

"hot-springs fimbristylis" Cyperaceae
CNPS List: 2 **R-E-D Code:** 2-2-1 **State/Fed. Status:** /C3b
Distribution: INY, KRN*, MNO, SBD, NV, AZ
Quads: 107A, 324A, 346B, 413B, 413D, 432C
Habitat: Medws (alkaline, near hot springs)
Life Form: Perennial herb (rhizomatous)
Blooming: July-September
Notes: Need historical quads for KRN Co. See *Intermountain Flora* 6:88 (1977) for revised nomenclature.

Forsellesia pungens var. *glabra*

See *Glossopetalon pungens*

Forsellesia stipulifera

Considered but rejected: A synonym of *Glossopetalon spinescens*; a common taxon

Frankenia palmeri Wats.

"Palmer's frankenia" Frankeniaceae
CNPS List: 2 **R-E-D Code:** 3-3-1 **State/Fed. Status:** CEQA
Distribution: SDG, BA, SO
Quads: 11A, 11B, 11D, 22C
Habitat: CoDns, MshSw (coastal salt), Plyas
Life Form: Perennial herb
Blooming: May-July
Notes: Seriously threatened by development.

Frasera neglecta

See *Swertia neglecta*

Frasera puberulenta

Considered but rejected: Too common; a synonym of *Swertia puberulenta*

Frasera tubulosa

Considered but rejected: Too common; a synonym of *Swertia tubulosa*

Frasera umpquaensis

See *Swertia fastigiata*

Fraxinus trifoliata

Considered but rejected: Taxonomic problem

Fremontodendron californicum ssp. *napensis*

Considered but rejected: A synonym of *F. californicum*; a common taxon

Fremontodendron californicum ssp. *obispoense*

Considered but rejected: A synonym of *F. californicum*; a common taxon

Fremontodendron decumbens R. Lloyd

"Pine Hill flannelbush" Sterculiaceae
CNPS List: 1B **R-E-D Code:** 3-2-3 **State/Fed. Status:** CR/C1
Distribution: ELD, NEV
Quads: 510B, 511A, 542A
Habitat: Chprl, CmWld / gabbroic or serpentinite
Life Form: Shrub (evergreen)
Blooming: April-June
Notes: Known from fewer than ten occurrences in the Pine Hill area (ELD Co.), and one near Grass Valley (NEV Co.) where plant occurs on serpentinite. See *F. californicum* ssp. *decumbens* in *The Jepson Manual.* See *Brittonia* 17:382 (1965) for original description, *Fremontia* 13(1):3-6 (1985) for species account, and *Systematic Botany* 16(1):3-20 (1991) for revised nomenclature and taxonomic treatment.
Status Report: 1993

Fremontodendron mexicanum A. Davids.

"Mexican flannelbush" Sterculiaceae
CNPS List: 1B **R-E-D Code:** 3-2-2 **State/Fed. Status:** CR/C2
Distribution: IMP, ORA, SDG, BA
Quads: 7A, 10A, 10B*, 10C, 10D, 11B*, 19B
Habitat: CCFrs, Chprl, CmWld / gabbroic or serpentinite
Life Form: Shrub (evergreen)
Blooming: March-June
Notes: Known from fewer than twenty occurrences. Need quads for ORA Co. See *Bulletin of the Southern California Academy of Sciences* 16:50 (1917) for original description, and *Systematic Botany* 16(1):3-20 (1991) for taxonomic treatment.
Status Report: 1987

Fritillaria affinis var. *tristulis*

Not yet published; see *F. lanceolata* var. *tristulis*

Fritillaria agrestis Greene

"stinkbells" Liliaceae
CNPS List: 4 **R-E-D Code:** 1-2-3 **State/Fed. Status:** /C3c
Distribution: ALA, CCA, FRE, KRN, MEN, MNT, MPA, PLA, SAC, SBA, SBT, SLO, SMT, STA, TUO
Habitat: Chprl, CmWld, VFGrs / clay, sometimes serpentinite
Life Form: Perennial herb (bulbiferous)
Blooming: March-April
Notes: Threatened by grazing and development.

Fritillaria biflora var. *biflora*

Considered but rejected: Too common

Fritillaria biflora Lindl. var. *ineziana* Jeps.

"Hillsborough chocolate lily" Liliaceae
CNPS List: 1B **R-E-D Code:** 3-3-3 **State/Fed. Status:** CEQA
Distribution: SMT
Quads: 448C
Habitat: CmWld, VFGrs / serpentinite
Life Form: Perennial herb (bulbiferous)
Blooming: March-April
Notes: Endemic to the Hillsborough area. See *Flora of California* 1(6):306-307 (1922) by W.L. Jepson for original description.

Fritillaria brandegei Eastw.

"Greenhorn fritillary" Liliaceae
CNPS List: 1B **R-E-D Code:** 1-2-3 **State/Fed. Status:** /C3c
Distribution: KRN, TUL
Quads: 212C, 261A, 261B, 261C, 261D, 285A, 285B, 285C, 285D, 307B, 308A, 308B, 308C, 308D
Habitat: LCFrs (granitic)
Life Form: Perennial herb (bulbiferous)
Blooming: April-June
Notes: Threatened by logging and grazing by cattle and deer. See *Bulletin of the Torrey Botanical Club* 30:484 (1903) for original description.

Fritillaria eastwoodiae MacFarlane

"Butte County fritillary" Liliaceae
CNPS List: 1B **R-E-D Code:** 2-2-3 **State/Fed. Status:** /C2
Distribution: BUT, SHA, TEH, YUB
Quads: 558D, 559A, 574C, 575A, 575B, 575D, 576A, 576B, 577A, 591B, 591C, 591D, 592A, 592B, 592C, 592D, 626B, 627A, 627B, 645A, 645B, 645C, 645D
Habitat: Chprl, CmWld, LCFrs (openings) / sometimes serpentinite
Life Form: Perennial herb (bulbiferous)
Blooming: March-May
Notes: Taxonomic questions resolved; not a hybrid between *F. micrantha* and *F. recurva*. Threatened on private lands by logging and development. See *Leaflets of Western Botany* 1:55 (1933) for original description as *F. phaeanthera*, and *Madroño* 25(2):93-100 (1978) for revised nomenclature.
Status Report: 1977

Fritillaria falcata (Jeps.) D. E. Beetle

"talus fritillary" Liliaceae
CNPS List: 1B **R-E-D Code:** 3-3-3 **State/Fed. Status:** /C2
Distribution: ALA, MNT, SBT, SCL, STA
Quads: 339B, 339C, 343C, 425B, 426A, 445D
Habitat: Chprl, CmWld, LCFrs / often on talus, serpentinite
Life Form: Perennial herb (bulbiferous)
Blooming: March-May
Notes: Threatened by vehicles. See *Flora of California* 1(6):309 (1922) by W.L. Jepson for original description, and *Madroño* 7:133-159 (1944) for revised nomenclature.
Status Report: 1979

Fritillaria glauca

Considered but rejected: Too common

Fritillaria grayana

See *Fritillaria roderickii*

Fritillaria lanceolata Pursh. var. *tristulis* Grant

"Marin checker lily" Liliaceae
CNPS List: 1B **R-E-D Code:** 3-3-3 **State/Fed. Status:** CEQA
Distribution: MRN
Quads: 467A, 467B, 484C, 485C, 485D
Habitat: CBScr, CoPrr, CoScr
Life Form: Perennial herb (bulbiferous)
Blooming: February-April
Notes: Known from fewer than ten extant occurrences. Some occurrences threatened by grazing, and all by small size. Plants seem not to set seed, but to reproduce by offsets. See *F. affinis* var. *tristulis* in *The Jepson Manual*. See *Flora of California* 1:(6)308 (1922) by W.L. Jepson for original description.

Fritillaria liliacea Lindl.

"fragrant fritillary" Liliaceae
CNPS List: 1B **R-E-D Code:** 1-2-3 **State/Fed. Status:** /C2
Distribution: ALA, CCA, MNT, MRN, SBT, SCL, SFO, SMT, SOL, SON
Quads: 366C*, 386B, 406B, 407B, 427D, 429A, 429C, 447A, 448C, 448D, 464B, 465C, 466A*, 484A, 484C, 484D, 485A, 485B, 485C, 485D, 498D, 501A, 501B, 501C*, 502B
Habitat: CoPrr, CoScr, VFGrs / often serpentinite
Life Form: Perennial herb (bulbiferous)
Blooming: February-April
Notes: Threatened by grazing and loss of habitat to agriculture and urban development. Quite variable.
Status Report: 1979

Fritillaria ojaiensis A. Davids.

"Ojai fritillary" Liliaceae
CNPS List: 1B **R-E-D Code:** 3-2-3 **State/Fed. Status:** /C2
Distribution: SBA, SLO?, VEN
Quads: 140A, 140B, 141A, 166D, 167C, 168C, 168D
Habitat: BUFrs (mesic), Chprl, LCFrs / rocky
Life Form: Perennial herb (bulbiferous)
Blooming: March-May
Notes: Known from approximately five occurrences. Does plant occur in SLO Co.? Closely related to *F. affinis*.

Fritillaria phaeanthera

See *Fritillaria eastwoodiae*

Fritillaria pluriflora Benth.

"adobe-lily" Liliaceae
CNPS List: 1B **R-E-D Code:** 1-2-3 **State/Fed. Status:** /C2
Distribution: BUT, COL, GLE, LAK, NAP, PLU, SOL, TEH, YOL
Quads: 497C, 498A*, 498C*, 515D, 531B, 532A, 532B, 532C, 532D, 547B, 547C, 547D, 563B, 563C, 563D, 564D, 576A, 576C*, 577A*, 579A, 580A, 580D*, 591A, 593B, 593C*, 593D, 594A, 595D, 596C, 596D, 610B, 612D
Habitat: Chprl, CmWld, VFGrs / often adobe
Life Form: Perennial herb (bulbiferous)
Blooming: February-April
Notes: Threatened by grazing, vehicles, and horticultural collecting.
Status Report: 1990

Fritillaria purdyi Eastw.

"Purdy's fritillary" Liliaceae
CNPS List: 4 **R-E-D Code:** 1-1-3 **State/Fed. Status:** CEQA?
Distribution: COL, GLE, HUM, LAK, MEN, NAP, TEH, TRI, YOL
Habitat: Chprl, VFGrs / serpentinite
Life Form: Perennial herb (bulbiferous)
Blooming: March-June

Fritillaria roderickii Knight

"Roderick's fritillary" Liliaceae
CNPS List: 1B **R-E-D Code:** 3-2-3 **State/Fed. Status:** CE/C3b
Distribution: MEN
Quads: 537B*, 537C, 537D, 551C, 569A
Habitat: CBScr, CoPrr, VFGrs
Life Form: Perennial herb (bulbiferous)
Blooming: March-May
Notes: Known from fewer than ten occurrences. SON Co. plants are introduced. Threatened by road maintenance, residential development, and erosion. Taxonomic validity has been questioned; further study needed. A synonym of *F. biflora* var. *biflora* in *The Jepson Manual.* USFWS uses the name *F. grayana.* See *Four Seasons* 2(2):14-16 (1967) for original description.
Status Report: 1988

Fritillaria striata Eastw.

"striped adobe-lily" Liliaceae
CNPS List: 1B **R-E-D Code:** 3-3-3 **State/Fed. Status:** CT/C1
Distribution: KRN, TUL
Quads: 213C, 239A, 261C, 262B, 262D, 309B*, 309C, 310A*, 310D
Habitat: CmWld, VFGrs / adobe
Life Form: Perennial herb (bulbiferous)
Blooming: February-April
Notes: Known from fewer than twenty occurrences. Threatened by citriculture, urbanization, and grazing. See *Proceedings of the California Academy of Sciences* IV 20:136 (1931) for original description.
Status Report: 1985

Fritillaria viridea Kell.

"San Benito fritillary" Liliaceae
CNPS List: 4 **R-E-D Code:** 1-2-3 **State/Fed. Status:** /C2
Distribution: MNT, SBT, SLO
Habitat: Chprl (serpentinite)
Life Form: Perennial herb (bulbiferous)
Blooming: March-May
Notes: Much more common than previously thought in SBT Co.; plants from MNT Co. may be another taxon. Threatened by vehicles and expansion of mining.

Fritillaria striata

Galium andrewsii Gray ssp. *gatense* (Dempster) Dempster & Steb.

"serpentine bedstraw" Rubiaceae
CNPS List: 4 **R-E-D Code:** 1-2-3 **State/Fed. Status:** CEQA?
Distribution: ALA?, CCA, FRE, MNT, SBT, SCL, SLO
Habitat: Chprl, CmWld, LCFrs / serpentinite
Life Form: Perennial herb
Blooming: April-July
Notes: Does plant occur in ALA Co? Threatened by vehicles in the Clear Creek area near San Benito Mtn. See *Brittonia* 10:186 (1958) for original description, and *Flora of California* 4(2):35-36 (1979) by W.L. Jepson for taxonomic treatment.

Galium angustifolium Nutt. ssp. *borregoense* Dempster

"Borrego bedstraw" Rubiaceae
CNPS List: 1B **R-E-D Code:** 3-2-3 **State/Fed. Status:** CR/C2
Distribution: SDG
Quads: 32B, 32D, 47C
Habitat: SDScr (rocky)
Life Form: Perennial herb
Blooming: March
Notes: Known from fewer than ten occurrences. See *Madroño* 21(2):88 (1971) for original description, and *Flora of California* 4(2):23-24 (1979) by W.L. Jepson for taxonomic treatment.
Status Report: 1986

Galium angustifolium Nutt. ssp. *gabrielense* (Munz & Jtn.) Dempster & Steb.

"San Antonio Canyon bedstraw" Rubiaceae
CNPS List: 4 **R-E-D Code:** 1-1-3 **State/Fed. Status:** CEQA?
Distribution: LAX, SBD
Habitat: Chprl, LCFrs / granitic, sandy or rocky
Life Form: Perennial herb
Blooming: April-August
Notes: See *Bulletin of the Torrey Botanical Club* 51:299 (1924) for original description, *Madroño* 21(2):90 (1971) for revised nomenclature, and *Flora of California* 4(2):23-25 (1979) by W.L. Jepson for taxonomic treatment.

Galium angustifolium Nutt. ssp. *gracillimum* Dempster & Steb.

"slender bedstraw" Rubiaceae
CNPS List: 4 **R-E-D Code:** 1-2-3 **State/Fed. Status:** CEQA?
Distribution: RIV, SBD
Habitat: JTWld, SDScr / rocky
Life Form: Perennial herb
Blooming: April-June
Notes: See *Madroño* 21(2):90 (1971) for original description, and *Flora of California* 4(2):23-26 (1979) by W.L. Jepson for taxonomic treatment.

Galium angustifolium Nutt. ssp. *jacinticum* Dempster & Steb.

"San Jacinto Mtns. bedstraw" Rubiaceae
CNPS List: 1B **R-E-D Code:** 3-1-3 **State/Fed. Status:** CEQA
Distribution: RIV
Quads: 84D
Habitat: LCFrs
Life Form: Perennial herb
Blooming: June-August
Notes: Known only from the Lake Fulmor and Black Mtn. area of the San Jacinto Mtns. See *Madroño* 21(2):92 (1971) for original description, and *Flora of California* 4(2):23-26 (1979) by W.L. Jepson for taxonomic treatment.

Galium angustifolium Nutt. ssp. *onycense* (Dempster) Dempster & Steb.

"Onyx Peak bedstraw" Rubiaceae
CNPS List: 4 **R-E-D Code:** 1-1-3 **State/Fed. Status:** CEQA?
Distribution: KRN
Habitat: CmWld (granitic, rocky)
Life Form: Perennial herb
Blooming: April-July
Notes: Known only from the Onyx Pk. area. See *Madroño* 21(2):94 (1971) for original description, and *Flora of California* 4(2):26 (1979) by W.L. Jepson for taxonomic treatment.

Galium buxifolium Greene

"box bedstraw" Rubiaceae
CNPS List: 1B **R-E-D Code:** 3-2-3 **State/Fed. Status:** CR/C1
Distribution: SCZ, SMI, SRO
Quads: SCZB, SCZC, SMIE, SROE
Habitat: CBScr, CCFrs, CoScr / rocky
Life Form: Shrub (deciduous)
Blooming: March-July
Notes: Known from approximately twenty occurrences. Threatened by feral herbivores. See *Bulletin of the California Academy of Sciences* 2:150 (1886) for original description, and *Flora of California* 4(2):20-21 (1979) by W.L. Jepson for taxonomic treatment.
Status Report: 1979

Galium californicum H. & A. ssp. *luciense* Dempster & Steb.

"Cone Peak bedstraw" Rubiaceae
CNPS List: 1B **R-E-D Code:** 3-1-3 **State/Fed. Status:** /C2
Distribution: MNT
Quads: 296B, 296D, 319C, 320B, 320D, 343C
Habitat: BUFrs, CmWld, LCFrs
Life Form: Perennial herb
Blooming: March-July
Notes: Known from fewer than ten occurrences. See *Madroño* 18(4):107 (1965) for original description, and *Flora of California* 4(2):39-40 (1979) by W.L. Jepson for taxonomic treatment.
Status Report: 1977

Galium californicum H. & A. ssp. *miguelense* (Greene) Dempster & Steb.

"San Miguel Island bedstraw" Rubiaceae
CNPS List: 4 **R-E-D Code:** 1-2-3 **State/Fed. Status:** /C3c
Distribution: SMI, SRO
Habitat: CmWld
Life Form: Perennial herb
Blooming: March-July
Notes: Threatened by feral herbivores and cattle grazing on SRO Isl. See *Pittonia* 1:34 (1887) for original description, and *University of California Publications in Botany* 46:32 (1968) and *Flora of California* 4(2):39-42 (1979) by W.L. Jepson for taxonomic treatments.

Galium californicum H. & A. ssp. *primum* Dempster & Steb.

"California bedstraw" Rubiaceae
CNPS List: 1B **R-E-D Code:** 3-2-3 **State/Fed. Status:** /C2
Distribution: RIV, SBD
Quads: 67A, 107D
Habitat: Chprl, LCFrs / granitic, sandy
Life Form: Perennial herb
Blooming: May-July
Notes: Known from only three occurrences. See *University of California Publications in Botany* 46:30 (1968) for original description, and *Flora of California* 4(2):39-41 (1979) by W.L. Jepson for taxonomic treatment.
Status Report: 1977

Galium californicum H. & A. ssp. *sierrae* Dempster & Steb.

"El Dorado bedstraw" Rubiaceae
CNPS List: 1B **R-E-D Code:** 3-2-3 **State/Fed. Status:** CR/C2
Distribution: ELD
Quads: 510B, 511A, 527D
Habitat: Chprl, CmWld, LCFrs / gabbroic
Life Form: Perennial herb
Blooming: May-June
Notes: Known from fewer than ten occurrences. Threatened by development, vehicles, and recreational activities. See *University of California Publications in Botany* 46:30 (1968) for original description, and *Flora of California* 4(2):39-40 (1979) by W.L. Jepson for taxonomic treatment.
Status Report: 1990

Galium catalinense Gray ssp. *acrispum* Dempster & Steb.

"San Clemente Island bedstraw" Rubiaceae
CNPS List: 1B **R-E-D Code:** 3-2-3 **State/Fed. Status:** CE/C2
Distribution: SCM
Quads: SCMC, SCMN, SCMS
Habitat: VFGrs
Life Form: Shrub (deciduous)
Blooming: March-May
Notes: Possibly threatened by Navy activities. Feral herbivores removed from SCM Isl., and vegetation recovering. See *University of California Publications in Botany* 64:11 (1973) for original description, and *Flora of California* 4(2):20 (1979) by W.L. Jepson for taxonomic treatment.
Status Report: 1978

Galium catalinense Gray ssp. *catalinense*

"Santa Catalina Island bedstraw" Rubiaceae
CNPS List: 4 **R-E-D Code:** 1-2-3 **State/Fed. Status:** CEQA?
Distribution: SCT
Habitat: Chprl, CoScr
Life Form: Shrub (deciduous)
Blooming: February-July
Notes: Threatened by feral herbivores. See *University of California Publications in Botany* 64:11 (1973) and *Flora of California* 4(2):20 (1979) by W.L. Jepson for taxonomic treatments.

Galium clementis Eastw.

"Santa Lucia bedstraw" Rubiaceae
CNPS List: 4 **R-E-D Code:** 1-1-3 **State/Fed. Status:** /C3c
Distribution: MNT
Habitat: LCFrs, UCFrs / granitic or serpentinite, sandy
Life Form: Perennial herb
Blooming: May-July
Notes: See *Leaflets of Western Botany* 1:56 (1933) for original description, and *Flora of California* 4(2):38 (1979) by W.L. Jepson for taxonomic treatment.

Galium cliftonsmithii (Dempster) Dempster & Steb.

"Santa Barbara bedstraw" Rubiaceae
CNPS List: 4 **R-E-D Code:** 1-1-3 **State/Fed. Status:** CEQA?
Distribution: LAX, MNT, SBA, SLO, VEN
Habitat: CmWld
Life Form: Perennial herb
Blooming: May-July
Notes: See *Brittonia* 10:183 (1958) for original description, and *Flora of California* 4(2):44 (1979) by W.L. Jepson for taxonomic treatment.

Galium glabrescens (Ehrend.) Dempster & Ehrend. ssp. *modocense* Dempster & Ehrend.

"Modoc bedstraw" Rubiaceae
CNPS List: 1B **R-E-D Code:** 2-2-3 **State/Fed. Status:** /C3c
Distribution: MOD
Quads: 673B, 690B, 690C, 707C, 724A
Habitat: GBScr (gravelly or rocky)
Life Form: Perennial herb
Blooming: July
Notes: See *Brittonia* 17:333 (1965) for original description, and *Flora of California* 4(2):34-35 (1979) by W.L. Jepson for taxonomic treatment.
Status Report: 1977

Galium grande McClat.

"San Gabriel bedstraw" Rubiaceae
CNPS List: 1B **R-E-D Code:** 3-1-3 **State/Fed. Status:** /C2
Distribution: LAX
Quads: 109B, 110A, 163D
Habitat: BUFrs, Chprl, CmWld, LCFrs
Life Form: Shrub (deciduous)
Blooming: January-July
Notes: Known from fewer than ten occurrences. See *Erythea* 2:124 (1894) for original description, and *Flora of California* 4(2):47 (1979) by W.L. Jepson for taxonomic treatment.
Status Report: 1977

Galium hardhamiae Dempster

"Hardham's bedstraw" Rubiaceae
CNPS List: 1B **R-E-D Code:** 2-1-3 **State/Fed. Status:** /C3c
Distribution: MNT, SLO
Quads: 270C, 271A, 271B, 295C, 296B, 296D
Habitat: CCFrs (serpentinite)
Life Form: Perennial herb
Blooming: April-October
Notes: Known from fewer than twenty occurrences. See *Madroño* 16(5):166 (1962) for original description, and *Flora of California* 4(2):35 (1979) by W.L. Jepson for taxonomic treatment.
Status Report: 1977

Galium hilendiae Dempster & Ehrend. ssp. *carneum* (Hilend. & Howell) Dempster & Ehrend.

"Panamint Mtns. bedstraw" Rubiaceae
CNPS List: 4 **R-E-D Code:** 1-1-3 **State/Fed. Status:** CEQA?
Distribution: INY
Habitat: MDScr, PJWld / gravelly or rocky
Life Form: Perennial herb
Blooming: May-August
Notes: See *Leaflets of Western Botany* 1:136 (1934) for original description, *Brittonia* 17:311 (1963) for revised nomenclature, and *Flora of California* 4(2):29 (1979) by W.L. Jepson for taxonomic treatment.

Galium hilendiae Dempster & Ehrend. ssp. *kingstonense* (Dempster) Dempster & Ehrend.

"Kingston Mtns. bedstraw" Rubiaceae
CNPS List: 1B **R-E-D Code:** 3-1-2 **State/Fed. Status:** /C2
Distribution: SBD, NV
Quads: 250B, 274C
Habitat: PJWld (rocky)
Life Form: Perennial herb
Blooming: June
Notes: Known in CA from fewer than five occurrences in the Kingston Mtns. Threatened in NV. See *Brittonia* 10:190 (1958) for original description, and *Flora of California* 4(2):29 (1979) by W.L. Jepson for taxonomic treatment.
Status Report: 1977

Galium hypotrichium Gray ssp. *tomentellum* Ehrend.

"Telescope Peak bedstraw" Rubiaceae
CNPS List: 1B **R-E-D Code:** 3-1-3 **State/Fed. Status:** /C3c
Distribution: INY
Quads: 302A
Habitat: SCFrs (talus)
Life Form: Perennial herb
Blooming: June-August
Notes: Known from only one extant occurrence on Telescope Pk. in the Panamint Mtns. See *Contributions from the Dudley Herbarium* 5:12 (1956) for original description, and *Flora of California* 4(2):31-32 (1979) by W.L. Jepson for taxonomic treatment.

Galium jepsonii Hilend. & Howell

"Jepson's bedstraw" Rubiaceae
CNPS List: 4 **R-E-D Code:** 1-1-3 **State/Fed. Status:** CEQA?
Distribution: LAX, SBD
Habitat: UCFrs (granitic, sandy or gravelly)
Life Form: Perennial herb (rhizomatous)
Blooming: July-August
Notes: See *Leaflets of Western Botany* 1:135 (1934) for original description, and *Flora of California* 4(2):22 (1979) by W.L. Jepson for taxonomic treatment.

Galium johnstonii Dempster & Steb.

"Johnston's bedstraw" Rubiaceae
CNPS List: 4 **R-E-D Code:** 1-1-3 **State/Fed. Status:** CEQA?
Distribution: LAX, SBD, SDG
Habitat: LCFrs
Life Form: Perennial herb
Blooming: June-July
Notes: See *Madroño* 21(2):84 (1971) for original description, and *Flora of California* 4(2):22-23 (1979) by W.L. Jepson for taxonomic treatments.

Galium munzii Hilend. & Howell

"Munz's bedstraw" Rubiaceae
CNPS List: 4 **R-E-D Code:** 1-1-1 **State/Fed. Status:** CEQA?
Distribution: SBD, AZ, NV+
Habitat: GBScr, PJWld, UCFrs
Life Form: Perennial herb
Blooming: May-July
Notes: See *Leaflets of Western Botany* 1:135 (1934) for original description, and *Flora of California* 4(2):30-31 (1979) by W.L. Jepson for taxonomic treatment.

Galium munzii ssp. *munzii*

See *Galium munzii*

Galium nuttallii Gray ssp. *insulare* Ferris

"Nuttall's island bedstraw" Rubiaceae
CNPS List: 4 **R-E-D Code:** 1-1-3 **State/Fed. Status:** CEQA?
Distribution: SCT, SCZ, SRO
Habitat: Chprl, CmWld, LCFrs
Life Form: Perennial herb
Blooming: March-June
Notes: See *Contributions from the Dudley Herbarium* 4:338 (1955) for original description, and *Flora of California* 4(2):43-44 (1979) by W.L. Jepson for taxonomic treatment.

Galium oreganum Britt.

"Oregon bedstraw" Rubiaceae
CNPS List: 3 **R-E-D Code:** 3-?-1 **State/Fed. Status:** CEQA?
Distribution: DNT?, SIS?, OR, WA+
Quads: 738A?
Habitat: LCFrs, NCFrs
Life Form: Perennial herb (rhizomatous)
Blooming: May-September
Notes: Move to List 2? Location, rarity, and endangerment information needed. Apparently known in CA from only a single 1951 collection from SIS Co., near the DNT Co. line. Common in OR. See *Bulletin of the Torrey Botanical Club* 21:31 (1894) for original description, and *Flora of California* 4(2):16-17 (1979) by W.L. Jepson for taxonomic treatment.

Galium serpenticum Dempster ssp. *scotticum* Dempster & Ehrend.

"Scott Mtn. bedstraw" Rubiaceae
CNPS List: 1B **R-E-D Code:** 2-2-3 **State/Fed. Status:** /C3c
Distribution: SIS, TRI
Quads: 683B, 684A, 684B, 700A, 700B, 700C, 701B, 701D, 717B
Habitat: LCFrs (serpentinite)
Life Form: Perennial herb
Blooming: June-July
Notes: Threatened by logging. See *Brittonia* 17:326 (1965) for original description, and *Flora of California* 4(2):33-34 (1979) by W.L. Jepson for taxonomic treatment.
Status Report: 1977

Galium serpenticum Dempster ssp. *warnerense* Dempster & Ehrend.

"Warner Mtns. bedstraw" Rubiaceae
CNPS List: 1B **R-E-D Code:** 3-2-2 **State/Fed. Status:** /C3c
Distribution: MOD, OR
Quads: 724B, 725A
Habitat: Medws, SCFrs
Life Form: Perennial herb (rhizomatous)
Blooming: June
Notes: Known in CA from only three occurrences in the vicinity of Mt. Bidwell in the Warner Mtns. On review list in OR. See *Brittonia* 17:325 (1965) for original description, and *Flora of California* 4(2):33-34 (1979) by W.L. Jepson for taxonomic treatment.

Galium wrightii Gray

"Wright's bedstraw" Rubiaceae
CNPS List: 2 **R-E-D Code:** 3-1-1 **State/Fed. Status:** CEQA
Distribution: SBD, AZ, BA, NM, SO, TX+
Quads: 249D
Habitat: PJWld (carbonate)
Life Form: Perennial herb
Blooming: June-October
Notes: Known in CA from only three occurrences near Clark Mtn. See *Flora of California* 4(2):21 (1979) by W.L. Jepson for taxonomic treatment.

Galvezia speciosa (Nutt.) Gray

"island snapdragon" Scrophulariaceae
CNPS List: 1B **R-E-D Code:** 2-2-2 **State/Fed. Status:** /C2
Distribution: SBR*, SCM, SCT, GU
Quads: SBRA*, SCMC, SCMS, SCTE
Habitat: CoScr
Life Form: Shrub
Blooming: February-May
Notes: Feral herbivores removed from SCM Isl., and vegetation recovering. See *Madroño* 21(5):380 (1972) for distributional information.

Gaura coccinea

Considered but rejected: Too common

Gentiana affinis var. *affinis*

Considered but rejected: Not in CA; name apparently misapplied to *G. affinis* var. *ovata*; a common taxon

Gentiana affinis Griseb. var. *parvidentata* Kusnez.

"small-toothed prairie gentian" Gentianaceae
CNPS List: 3 **R-E-D Code:** ?-?-1 **State/Fed. Status:** CEQA?
Distribution: MOD, ID, OR+
Quads: unknown
Habitat: GBScr, LCFrs
Life Form: Perennial herb
Blooming: July-September
Notes: Move to List 2? Location, rarity, and endangerment information needed. Searches of major herbaria in 1992 failed to identify confirmed CA localities. See *G. affinis* var. *ovata* in *The Jepson Manual.*

Gentiana aquatica

See *Gentiana fremontii*

Gentiana bisetaea

See *Gentiana setigera*

Gentiana fremontii Torr.

"moss gentian" Gentianaceae
CNPS List: 2 **R-E-D Code:** 3-1-1 **State/Fed. Status:** CEQA
Distribution: SBD, OR, ++
Quads: 105A
Habitat: Medws (mesic), UCFrs
Life Form: Annual herb
Blooming: June-August
Notes: Not *G. aquatica*, previously misapplied to *G. fremontii.*
Status Report: 1980

Gentiana newberryi

Considered but rejected: Too common

Gentiana plurisetosa C. Mason

"Klamath gentian" Gentianaceae
CNPS List: 4 **R-E-D Code:** 1-1-2 **State/Fed. Status:** /C2
Distribution: DNT, HUM, SIS, TRI, OR
Habitat: LCFrs, Medws, UCFrs / mesic
Life Form: Perennial herb
Blooming: July-September
Notes: Equals *G. setigera* of the *Inventory* fourth edition (except the Red Mtn. occurrence in MEN Co., which is now *G. setigera* — formerly called *G. bisetaea*!). Candidate for state listing in OR. See *Madroño* 37(4):289-292 (1990) for original description, and 36(1):49-50 (1989) for revised nomenclature of *G. setigera.*

Gentiana prostrata Haenke

"pygmy gentian" Gentianaceae
CNPS List: 4 **R-E-D Code:** 1-1-1 **State/Fed. Status:** CEQA?
Distribution: INY, MNO, NV, OR, ++
Habitat: AlpBR, Medws (mesic)
Life Form: Perennial herb
Blooming: July-August
Notes: Endangered in OR.

Gentiana setigera Gray

"Mendocino gentian" Gentianaceae
CNPS List: 1B **R-E-D Code:** 3-3-2 **State/Fed. Status:** /C2
Distribution: DNT, MEN, OR
Quads: 600B, 739B, 739C
Habitat: LCFrs, Medws / mesic
Life Form: Perennial herb
Blooming: August-September
Notes: Threatened by mining activities and wetland alteration. Called *G. bisetaea* in the *Inventory* fourth edition, but this name is a later synonym of *G. setigera*; see *G. plurisetosa* for plant called *G. setigera* in the fourth edition. Candidate for state listing in OR. See *Madroño* 36(1):49-50 (1989) for revised nomenclature.

Gentiana tenella

Considered but rejected: A synonym of *Gentianella tenella* ssp. *tenella*; a common taxon

Geraea viscida (Gray) Blake

"sticky geraea" Asteraceae
CNPS List: 2 **R-E-D Code:** 2-1-1 **State/Fed. Status:** CEQA
Distribution: SDG, BA
Quads: 7B, 8A, 8B, 8C, 19D
Habitat: Chprl (often in disturbed areas)
Life Form: Perennial herb
Blooming: May-June

Geum aleppicum Jacq.

"Aleppo avens" Rosaceae
CNPS List: 2 **R-E-D Code:** 3-2-1 **State/Fed. Status:** CEQA
Distribution: LAS, MOD, SIS, OR, ++
Quads: 622B, 639B, 640A, 677B, 698B, 698C, 699D
Habitat: GBScr, LCFrs, Medws
Life Form: Perennial herb
Blooming: June-August
Notes: Circumboreal distribution; plants in CA may be Pleistocene relicts.

Gilia aliquanta ssp. *aliquanta*

Considered but rejected: Too common

Gilia aliquanta ssp. *breviloba*

Considered but rejected: Too common

Gilia australis

Considered but rejected: Too common

Gilia campanulata

Considered but rejected: Too common

Gilia caruifolia Abrams

"caraway-leaved gilia" Polemoniaceae
CNPS List: 4 **R-E-D Code:** 1-1-1 **State/Fed. Status:** CEQA?
Distribution: RIV, SDG, BA
Habitat: Chprl, LCFrs / sandy
Life Form: Annual herb
Blooming: May-August

Gilia inyoensis

Considered but rejected: Too common

Gilia latiflora (Gray) Gray ssp. *cuyamensis* A. & V. Grant

"Cuyama gilia" Polemoniaceae
CNPS List: 4 **R-E-D Code:** 1-1-3 **State/Fed. Status:** CEQA?
Distribution: KRN, LAX, SBA, VEN
Habitat: PJWld
Life Form: Annual herb
Blooming: April-May

Gilia maculata Parish

"Little San Bernardino Mtns. gilia" Polemoniaceae
CNPS List: 1B **R-E-D Code:** 3-2-3 **State/Fed. Status:** /C1
Distribution: RIV, SBD
Quads: 65C, 82B, 83A, 83B, 102B, 103A, 103D
Habitat: DeDns, JTWld, MDScr, SDScr / sandy
Life Form: Annual herb
Blooming: April-May
Notes: Known from fewer than ten occurrences near Joshua Tree NM. Threatened by development, vehicles, and dumping.

Gilia nevinii Gray

"Nevin's gilia" Polemoniaceae
CNPS List: 4 **R-E-D Code:** 1-1-2 **State/Fed. Status:** CEQA?
Distribution: ANA, SBR, SCM, SCT, SCZ, SNI*, SRO, GU
Habitat: CBScr, VFGrs
Life Form: Annual herb
Blooming: March-May
Notes: Last seen on SNI Isl. in 1901.

Gilia ripleyi Barneby

"Ripley's gilia" Polemoniaceae
CNPS List: 2 **R-E-D Code:** 3-1-1 **State/Fed. Status:** /C3c
Distribution: INY, NV
Quads: 302D, 325B, 349C, 368D, 372C, 390B, 411D
Habitat: MDScr (carbonate)
Life Form: Perennial herb
Blooming: May-June
Notes: Easily confused with *G. latifolia.* See *Leaflets of Western Botany* 3:129 (1942) for original description.

Gilia tenuiflora Benth. ssp. *arenaria* (Benth.) A. & V. Grant

"sand gilia" Polemoniaceae
CNPS List: 1B **R-E-D Code:** 3-2-3 **State/Fed. Status:** CT/FE
Distribution: MNT
Quads: 366A, 366C, 366D, 387D
Habitat: CoDns, CoScr
Life Form: Annual herb
Blooming: April-May
Notes: Seriously threatened by coastal development, foot traffic, and non-native plants. Intergrades with ssp. *tenuiflora* near the Salinas River mouth.
Status Report: 1985

Gilia tenuiflora Benth. ssp. *hoffmannii* (Eastw.) A. & V. Grant

"Hoffmann's slender-flowered gilia" Polemoniaceae
CNPS List: 1B **R-E-D Code:** 3-1-3 **State/Fed. Status:** /C1
Distribution: SRO
Quads: SROE, SRON
Habitat: CoDns, CoScr
Life Form: Annual herb
Blooming: April
Notes: Known from only two occurrences at East Pt. and near Carrington Pt.

Gilmania luteola (Cov.) Cov.

"golden carpet" Polygonaceae
CNPS List: 1B **R-E-D Code:** 3-1-3 **State/Fed. Status:** /C3c
Distribution: INY
Quads: 324A, 324D, 346C
Habitat: ChScr (alkaline barrens)
Life Form: Annual herb
Blooming: March-April
Notes: Known from approximately five occurrences in Death Valley NM. See *Contributions from the U.S. National Herbarium* 4:190 (1893) for original description, *Fremontia* 11(4):25-26 (1984) for species account, and *Phytologia* 66(3):243-244 (1989) for taxonomic treatment.

Githopsis diffusa Gray ssp. *filicaulis* (Ewan) Morin

"Mission Canyon bluecup" Campanulaceae
CNPS List: 1B **R-E-D Code:** 3-3-2 **State/Fed. Status:** /C2
Distribution: RIV, SDG, BA
Quads: 20B*, 21D*, 21A, 22D, 67C
Habitat: Chprl (mesic, disturbed areas)
Life Form: Annual herb
Blooming: May
Notes: Known in CA from fewer than five occurrences. Reduced by development, which remains a threat. See *Rhodora* 41:312 (1939) for original description, and *Systematic Botany* 8(4):436-468 (1983) for taxonomic treatment.
Status Report: 1977

Githopsis filicaulis

See *Githopsis diffusa* ssp. *filicaulis*

Githopsis latifolia

Considered but rejected: Not native and not naturalized in CA; a synonym of *Legousia speculum-veneris*

Githopsis pulchella

Considered but rejected: Too common

Glossopetalon pungens Bdg.

"pungent glossopetalon" Crossosomataceae
CNPS List: 1B **R-E-D Code:** 3-2-2 **State/Fed. Status:** /C2
Distribution: SBD, NV
Quads: 249D
Habitat: Chprl, PJWld / carbonate
Life Form: Shrub (deciduous)
Blooming: May-June
Notes: Known in CA from only one occurrence in Forsellesia Cyn. in the Clark Mtns. On watch list in NV. See *Aliso* 9:171-178 (1978) for discussion of transfer to Crossosomataceae. USFWS uses the name *G. pungens* var. *glabra.* See *American Midland Naturalist* 27(2):503-504 (1942) for original description.
Status Report: 1977

Glyceria grandis Wats.

"American manna grass" Poaceae
CNPS List: 2 **R-E-D Code:** 3-1-1 **State/Fed. Status:** CEQA
Distribution: HUM, MEN, MNO, PLA, ++
Quads: 470B, 487C, 505D, 537B, 538B
Habitat: BgFns, Medws, MshSw (streambanks and lake margins)
Life Form: Perennial herb (rhizomatous)
Blooming: June-August

Glyceria grandis var. *grandis*

See *Glyceria grandis*

Goodmania luteola (Parry) Reveal & Ertter

"golden goodmania" Polygonaceae
CNPS List: 4 **R-E-D Code:** 1-2-2 **State/Fed. Status:** CEQA?
Distribution: FRE, INY, KRN, LAX, MAD, MNO, TUL, NV
Habitat: MDScr, Medws, Plyas, VFGrs / alkaline or clay
Life Form: Annual herb
Blooming: April-August
Notes: May be threatened by groundwater lowering and trampling by cattle. See *Bulletin of the Torrey Botanical Club* 10:23 (1883) for original description, *Brittonia* 28:427-429 (1976) for revised nomenclature, and *Phytologia* 66(4):389 (1989) for taxonomic treatment.

Gratiola heterosepala Mason & Bacig.

"Boggs Lake hedge-hyssop" Scrophulariaceae
CNPS List: 1B **R-E-D Code:** 1-2-2 **State/Fed. Status:** CE/C3c
Distribution: FRE, LAK, LAS, MAD, MOD, PLA, SAC, SHA, SJQ, SOL, TEH, OR
Quads: 398A, 398D, 495B, 495D, 496A, 498D, 511C, 512B*, 527C, 528A, 528D, 533D, 534A, 594B, 628A, 628B, 628D, 643B, 661A, 661C, 678B, 678D, 690C
Habitat: MshSw (lake margins), VnPls
Life Form: Annual herb
Blooming: April-June
Notes: Threatened by agriculture, development, grazing, and vehicles. Candidate for state listing in OR. Lassen NF has adopted species management guidelines. See *Madroño* 12:150-152 (1954) for original description.
Status Report: 1987

Grindelia camporum var. *parviflora*

Considered but rejected: A synonym of *G. camporum* var. *camporum*; a common taxon

Grindelia fraxino-pratensis Reveal & Beatley

"Ash Meadows gumplant" Asteraceae
CNPS List: 1B **R-E-D Code:** 3-2-2 **State/Fed. Status:** /FT
Distribution: INY, NV
Quads: 322C, 322D
Habitat: Medws (mesic clay)
Life Form: Perennial herb
Blooming: June-October
Notes: Known in CA from only two extant occurrences in Carson Slough in the Amargosa Desert. Threatened by water diversion, habitat alteration, and non-native plants. State-listed as Critically Endangered in NV. See *Bulletin of the Torrey Botanical Club* 98:332 (1971) for original description.

Grindelia hallii

See *Grindelia hirsutula* var. *hallii*

Grindelia hirsutula H. & A. var. *hallii* (Steyerm.) M.A. Lane

"San Diego gumplant" Asteraceae
CNPS List: 1B **R-E-D Code:** 2-2-3 **State/Fed. Status:** /C3c
Distribution: SDG
Quads: 9B?, 10D?, 19B, 19C, 20A, 22B, 33B, 33C, 33D
Habitat: Chprl, LCFrs, Medws, VFGrs
Life Form: Perennial herb
Blooming: July-October
Notes: Threatened by grazing and road maintenance. See *Annals of the Missouri Botanical Garden* 21:229 (1934) for original description, and *Novon* 2(3):215-217 (1992) for revised nomenclature.

Grindelia hirsutula H. & A. var. *maritima* (Greene) M.A. Lane

"San Francisco gumplant" Asteraceae
CNPS List: 1B **R-E-D Code:** 2-2-3 **State/Fed. Status:** /C2
Distribution: MNT, MRN, SCR, SFO, SLO, SMT
Quads: 271D, 272A, 344B, 366C, 366D, 387B, 387D, 387E, 408C, 409B, 409D, 429C, 448B, 448C, 466C, 467A, 467B, 467D, 467E, 485B, 485C, 485D
Habitat: CBScr, CoScr, VFGrs / sandy, serpentinite
Life Form: Perennial herb
Blooming: August-September
Notes: Most collections are old; need current information on distribution and rarity. Threatened by coastal development and non-native plants. See *Pittonia* 2:289 (1892) for original description, and *Novon* 2(3):215-217 (1992) for revised nomenclature.

Grindelia humilis

See *Grindelia stricta* var. *angustifolia*

Grindelia latifolia ssp. *latifolia*

Considered but rejected: A hybrid and synonym of *G. stricta* var. *platyphylla* and *G. camporum* var. *camporum*; common taxa

Grindelia maritima

See *Grindelia hirsutula* var. *maritima*

Grindelia paludosa

Considered but rejected: A hybrid

Grindelia stricta DC. var. *angustifolia* (Gray) M.A. Lane

"marsh gumplant" Asteraceae
CNPS List: 4 **R-E-D Code:** 1-1-3 **State/Fed. Status:** CEQA?
Distribution: ALA, CCA, MNT, MRN, NAP, SCL, SFO, SMT, SOL, SON
Habitat: MshSw (coastal salt)
Life Form: Perennial herb
Blooming: August-October
Notes: Rare in MNT Co. Hybridizes with *G. camporum* var. *camporum*. See *Novon* 2(3):215-217 (1992) for revised nomenclature.

Grindelia stricta ssp. *blakei*

Considered but rejected: A synonym of *G. stricta* var. *stricta*; a common taxon

Gutierrezia californica

Considered but rejected: Too common

Habenaria elegans* var. *maritima
Considered but rejected: A synonym of *Piperia elegans*; a common taxon

***Hackelia amethystina* J.T. Howell**
"amethyst stickseed" Boraginaceae
CNPS List: 4 **R-E-D Code:** 1-1-3 **State/Fed. Status:** CEQA?
Distribution: GLE, LAK, MEN, PLU, TEH, TRI
Habitat: LCFrs (openings, disturbed areas), Medws
Life Form: Perennial herb
Blooming: June-July
Notes: See *Leaflets of Western Botany* 3:125 (1942) for original description, and *Memoirs of the New York Botanical Garden* 26(1):213-214 (1976) for taxonomic treatment.

Hackelia bella
Considered but rejected: Too common

***Hackelia brevicula* (Jeps.) Gentry**
"Poison Canyon stickseed" Boraginaceae
CNPS List: 1B **R-E-D Code:** 3-1-3 **State/Fed. Status:** /C2
Distribution: INY, MNO
Quads: 412A, 412B, 431C, 431D
Habitat: SCFrs
Life Form: Perennial herb
Blooming: July
Notes: Known from six occurrences in the White Mtns. See *Southwestern Naturalist* 19(2):139-146 (1974) for revised nomenclature, and *Memoirs of the New York Botanical Garden* 26(1):205-206 (1976) for taxonomic treatment.

***Hackelia cusickii* (Piper) Brand**
"Cusick's stickseed" Boraginaceae
CNPS List: 4 **R-E-D Code:** 1-1-1 **State/Fed. Status:** CEQA?
Distribution: LAS, MOD, SIS, NV, OR
Habitat: AlpBR, PJWld (rocky loam), SCFrs
Life Form: Perennial herb
Blooming: April-July
Notes: See *Memoirs of the New York Botanical Garden* 26(1):203-204 (1976) for taxonomic treatment.

Hackelia longituba
Considered but rejected: A synonym of *H. velutina*; a common taxon

Hackelia mundula
Considered but rejected: Too common

***Hackelia sharsmithii* Jtn.**
"Sharsmith's stickseed" Boraginaceae
CNPS List: 2 **R-E-D Code:** 2-1-1 **State/Fed. Status:** /C3c
Distribution: FRE, INY, TUL, NV
Quads: 329A, 329B, 329D, 330A, 352A, 352D, 373B, 373C, 373D
Habitat: AlpBR, SCFrs / granitic
Life Form: Perennial herb
Blooming: July-September
Notes: See *Journal of the Arnold Arboretum* 20:398 (1939) for original description, and *Memoirs of the New York Botanical Garden* 26(1):149-150 (1976) for taxonomic treatment.

Hackelia velutina
Considered but rejected: Too common

***Halimolobos virgata* (Nutt.) Schulz**
"virgate halimolobos" Brassicaceae
CNPS List: 2 **R-E-D Code:** 3-1-1 **State/Fed. Status:** /C3c
Distribution: INY, MNO, NV, ++
Quads: 412A, 412B, 412C
Habitat: Medws, PJWld
Life Form: Perennial herb
Blooming: June-July
Notes: See *Contributions from the Gray Herbarium* 3(8):241-288 (1943) for taxonomic treatment and discussion of the genus.

Halodule wrightii
Considered but rejected: Not native

Haplopappus brickellioides
Considered but rejected: A synonym of *Hazardia brickellioides*; a common taxon

Haplopappus canus
See *Hazardia cana*

Haplopappus detonsus
See *Hazardia detonsa*

Haplopappus eastwoodiae
See *Ericameria fasciculata*

Haplopappus eximius
See *Tonestus eximius*

Haplopappus gilmanii
See *Ericameria gilmanii*

Haplopappus junceus
See *Machaeranthera juncea*

Haplopappus lucidus
See *Pyrrocoma lucida*

Haplopappus lyallii
See *Tonestus lyallii*

Haplopappus ophitidis
See *Ericameria ophitidis*

Haplopappus parishii
Considered but rejected: A synonym of *Ericameria parishii* var. *parishii*; a common taxon

Haplopappus peirsonii
Considered but rejected: Too common; a synonym of *Tonestus peirsonii*

Haplopappus propinquus
Considered but rejected: A synonym of *Ericameria brachylepis*; a common taxon

Haplopappus racemosus* ssp. *congestus
See *Pyrrocoma racemosa* var. *congesta*

Haplopappus uniflorus ssp. *gossypinus*

See *Pyrrocoma uniflora* var. *gossypina*

Harpagonella palmeri Gray

"Palmer's grapplinghook" Boraginaceae
CNPS List: 2 **R-E-D Code:** 1-2-1 **State/Fed. Status:** /C2
Distribution: LAX*, ORA, RIV, SCT, SDG, AZ, BA, SO
Quads: 9B, 10C, 20C, 22B, 22C, 22D, 35B, 35C, 36A, 36D, 50A, 51B, 51C, 52A, 52B, 67C, 68B, 68C, 68D, 69A, 69B, 70D, 86C, 86D, 110B*, 138A*
Habitat: Chprl, CoScr, VFGrs / clay
Life Form: Annual herb
Blooming: March-April
Notes: Need quads for SCT Isl. Very inconspicuous and easily overlooked. Threatened by development.

Harpagonella palmeri var. *palmeri*

See *Harpagonella palmeri*

Hastingsia bracteosa

Considered but rejected: Not in CA

Hazardia brickellioides

Considered but rejected: Too common

Hazardia cana (Gray) Greene

"San Clemente Island hazardia" Asteraceae
CNPS List: 1B **R-E-D Code:** 3-2-2 **State/Fed. Status:** /C2
Distribution: SCM, GU
Quads: SCMC, SCMS
Habitat: CBScr, CoScr, RpFrs
Life Form: Shrub (evergreen)
Blooming: June-September
Notes: Known in CA from fewer than twenty occurrences. Feral herbivores removed from SCM Isl., and vegetation recovering. See *Proceedings of the American Academy of Arts and Sciences* 11:75 (1876) for original description, *Aliso* 5(3):343 (1963) for species account, and *Madroño* 26(3):105-127 (1979) for taxonomic treatment.
Status Report: 1979

Hazardia detonsa (Greene) Greene

"island hazardia" Asteraceae
CNPS List: 4 **R-E-D Code:** 1-1-3 **State/Fed. Status:** CEQA?
Distribution: ANA, SCZ, SRO
Habitat: CoScr (rocky)
Life Form: Shrub (evergreen)
Blooming: July-November
Notes: See *Madroño* 26(3):105-127 (1979) for revised nomenclature.

Hazardia orcuttii (Gray) Greene

"Orcutt's hazardia" Asteraceae
CNPS List: 1B **R-E-D Code:** 3-3-2 **State/Fed. Status:** /C2
Distribution: SDG, BA
Quads: 35C, 36D
Habitat: Chprl, CoScr
Life Form: Shrub (evergreen)
Blooming: August-October
Notes: Known in CA from only one occurrence in Lux Cyn. Threatened by urbanization. See *Madroño* 26(3):105-127 (1979) for revised nomenclature and 28(1):38 (1981) for first CA record.

Hecastocleis shockleyi

Considered but rejected: Too common

Heleocharis parvula

See *Eleocharis parvula*

Helianthella castanea Greene

"Diablo helianthella" Asteraceae
CNPS List: 1B **R-E-D Code:** 3-2-3 **State/Fed. Status:** /C2
Distribution: ALA, CCA, MRN*, SFO*, SMT
Quads: 446B, 447A, 448B, 463C, 464A, 464B, 464C, 464D, 465A, 465B, 465C*, 465D, 467A*, 482C
Habitat: BUFrs, Chprl, CmWld, CoScr, RpWld, VFGrs
Life Form: Perennial herb
Blooming: April-June
Notes: Threatened by urbanization, grazing, and fire suppression.

Helianthemum greenei Rob.

"island rush-rose" Cistaceae
CNPS List: 1B **R-E-D Code:** 3-2-3 **State/Fed. Status:** /C2
Distribution: SCT, SCZ, SMI*, SRO
Quads: SCTN, SCZB, SCZC, SMIW*
Habitat: Chprl (rocky)
Life Form: Shrub (evergreen)
Blooming: March-May
Notes: Known from fewer than twenty occurrences. Need quads for SRO Isl. Last collected on SMI Isl. in 1939. Threatened by feral herbivores on SCZ and SRO islands; recovering on SCT Isl.

Helianthemum suffrutescens Schreib.

"Bisbee Peak rush-rose" Cistaceae
CNPS List: 3 **R-E-D Code:** 2-2-3 **State/Fed. Status:** /C3c
Distribution: AMA, CAL, ELD, SAC, TUO
Quads: 458B, 476B, 493C, 494B, 494C, 495A, 510B, 527D
Habitat: Chprl (often serpentinite, gabbroic, or Ione soil)
Life Form: Shrub (evergreen)
Blooming: April-May
Notes: Move to List 1B? Threatened by mining in the Ione area, development, and vehicles. Herbarium material indicates taxon is conspecific with *H. scoparium*, but field studies reveal differences; definitive taxonomic study needed. A synonym of *H. scoparium* in *The Jepson Manual*.

Helianthus exilis Gray

"serpentine sunflower" Asteraceae
CNPS List: 4 **R-E-D Code:** 1-2-3 **State/Fed. Status:** /C3c
Distribution: COL, GLE, LAK, NAP, SHA, SIS, SON, TEH, TRI
Habitat: Chprl, CmWld / serpentinite seeps
Life Form: Annual herb
Blooming: July-November
Notes: Difficult to distinguish from *H. bolanderi*; see that plant in *The Jepson Manual*.
Status Report: 1977

Helianthus niveus (Benth.) Bdg. ssp. *tephrodes* (Gray) Heiser

"Algodones Dunes sunflower" Asteraceae
CNPS List: 1B **R-E-D Code:** 3-2-2 **State/Fed. Status:** CE/C2
Distribution: IMP, AZ, SO
Quads: 2A, 2B, 13B, 13C, 14A, 14B, 14D, 27C, 28D
Habitat: DeDns
Life Form: Perennial herb
Blooming: September-May
Notes: Threatened by vehicles. See *Report on the U.S. and Mexican Boundary Survey*, p. 90 (1859) by W. Emory for original description, and *Memoirs of the Torrey Botanical Club* 22(3):39-43 (1969) for revised nomenclature.
Status Report: 1987

Helianthus nuttallii T. & G. ssp. *parishii* (Gray) Heiser

"Los Angeles sunflower" Asteraceae
CNPS List: 1A **Last Seen:** 1937 **State/Fed. Status:** /C2*
Distribution: LAX*, ORA*, SBD*
Quads: 71B*, 105B*, 105C*, 107D*, 110B*, 110C*, 111D*, 163C*
Habitat: MshSw (coastal salt and freshwater)
Life Form: Perennial herb (rhizomatous)
Blooming: August-October
Notes: See *Proceedings of the American Academy of Arts and Sciences* 14:7 (1883) for original description, and *Memoirs of the Torrey Botanical Club* 22(3):147-152 (1969) for taxonomic treatment.
Status Report: 1977

Hemitomes congestum

Considered but rejected: Too common

Hemizonia arida Keck

"Red Rock tarplant" Asteraceae
CNPS List: 1B **R-E-D Code:** 3-2-3 **State/Fed. Status:** CR/C1
Distribution: KRN
Quads: 235B, 235C
Habitat: MDScr (clay)
Life Form: Annual herb
Blooming: April-November
Notes: Known from fewer than ten occurrences near Red Rock Cyn. in the Mojave Desert. Threatened by vehicles. See *Aliso* 4(1):101-104 (1958) for original description.
Status Report: 1986

Hemizonia australis

See *Hemizonia parryi* ssp. *australis*

Hemizonia calyculata

See *Hemizonia congesta* ssp. *calyculata*

Hemizonia clementina Bdg.

"island tarplant" Asteraceae
CNPS List: 4 **R-E-D Code:** 1-1-3 **State/Fed. Status:** CEQA?
Distribution: ANA, SBR, SCM, SCT, SNI
Habitat: CBScr, VFGrs
Life Form: Shrub (deciduous)
Blooming: April-July

Hemizonia congesta

See *Hemizonia congesta* ssp. *leucocephala*

Hemizonia congesta DC. ssp. *calyculata* Babc. & Hall

"Mendocino tarplant" Asteraceae
CNPS List: 4 **R-E-D Code:** 1-1-3 **State/Fed. Status:** CEQA?
Distribution: LAK, MEN
Habitat: CmWld, VFGrs
Life Form: Annual herb
Blooming: July-October

Hemizonia congesta DC. ssp. *leucocephala* (Tanowitz) Keil

"Hayfield tarplant" Asteraceae
CNPS List: 3 **R-E-D Code:** ?-?-3 **State/Fed. Status:** CEQA?
Distribution: MEN, MRN, SON
Quads: 484D, 485B, 502A, 502B, 502C, 502D, 518D, 519A, 568B
Habitat: CoScr, VFGrs
Life Form: Annual herb
Blooming: April-October
Notes: Move to List 1B? Precise location, rarity, and endangerment information needed. Intergrades with ssp. *congesta*; dried plants may be indistinguishable. Many herbarium collections are misidentified. Threatened by agriculture and urban development. See *Bulletin of the Torrey Botanical Club* 110:15 (1983) for original description, and *Phytologia* 73(3):259-260 (1992) for revised nomenclature.

Hemizonia congesta DC. ssp. *tracyi* Babc. & Hall

"Tracy's tarplant" Asteraceae
CNPS List: 4 **R-E-D Code:** 1-1-3 **State/Fed. Status:** CEQA?
Distribution: HUM, MEN, TRI
Habitat: CoPrr, NCFrs
Life Form: Annual herb
Blooming: May-October

Hemizonia conjugens Keck

"Otay tarplant" Asteraceae
CNPS List: 1B **R-E-D Code:** 3-3-2 **State/Fed. Status:** CE/C2
Distribution: SDG, BA
Quads: 10B, 10C, 11A, 11B, 11D
Habitat: CoScr, VFGrs / clay
Life Form: Annual herb
Blooming: May-June
Notes: Threatened by residential development and illegal dumping. See *Aliso* 4(1):109 (1958) for original description, and *Madroño* 25:159 (1978) for information on distribution and taxonomy.
Status Report: 1987

Hemizonia floribunda Gray

"Tecate tarplant" Asteraceae
CNPS List: 1B **R-E-D Code:** 2-2-2 **State/Fed. Status:** /C2
Distribution: SDG, BA
Quads: 7B, 7C, 8A, 8D, 9B*, 9C, 9D, 19C, 19D
Habitat: Chprl, CoScr
Life Form: Annual herb
Blooming: August-October
Notes: Threatened by development. See *Proceedings of the American Academy of Arts and Sciences* 11:79 (1876) for original description.
Status Report: 1977

Hemizonia halliana Keck

"Hall's tarplant" Asteraceae
CNPS List: 1B **R-E-D Code:** 3-3-3 **State/Fed. Status:** CEQA
Distribution: MNT, SBT, SLO
Quads: 244A, 268A, 268B, 292D, 315C, 316D, 339B, 340C, 362D
Habitat: ChScr, CmWld, VFGrs / clay
Life Form: Annual herb
Blooming: April-May
Notes: Field searches in 1990 and 1992 failed to locate the plant in SBT Co., and status in MNT Co. is uncertain. Severely threatened by non-native plants in SLO Co. Appears only in unusually wet years.

Hemizonia increscens (Keck) Tanowitz ssp. *villosa* Tanowitz

"Gaviota tarplant" Asteraceae
CNPS List: 1B **R-E-D Code:** 3-3-3 **State/Fed. Status:** CE/C1
Distribution: SBA
Quads: 144B
Habitat: CBScr, CoScr, VFGrs
Life Form: Annual herb
Blooming: June-September
Notes: Known from only four extant occurrences at Gaviota. Seriously threatened by development. See *Systematic Botany* 7(3):314-339 (1982) for original description.
Status Report: 1988

Hemizonia laevis

See *Hemizonia pungens* ssp. *laevis*

Hemizonia minthornii Jeps.

"Santa Susana tarplant" Asteraceae
CNPS List: 1B **R-E-D Code:** 2-2-3 **State/Fed. Status:** CR/C2
Distribution: LAX, VEN
Quads: 112B, 112C, 113A, 113C, 113D, 138C, 138D
Habitat: Chprl, CoScr / rocky
Life Form: Shrub (deciduous)
Blooming: July-November
Notes: Threatened by development. See *Manual of the Flowering Plants of California*, p. 1092 (1925) by W.L. Jepson for original description.
Status Report: 1987

Hemizonia mohavensis Keck

"Mojave tarplant" Asteraceae
CNPS List: 1A **Last Seen:** 1933 **State/Fed. Status:** CE/C2*
Distribution: RIV*, SBD*
Quads: 84A*, 132C*
Habitat: RpScr
Life Form: Annual herb
Blooming: July-September
Notes: Recent attempts to rediscover plant at the type locality in SBD Co. have been unsuccessful. RIV Co. record may be erroneous. See *Madroño* 3:9-10 (1935) for original description.
Status Report: 1988

Hemizonia multicaulis ssp. *multicaulis*

Considered but rejected: A synonym of *H. congesta* ssp. *congesta*; a common taxon

Hemizonia multicaulis ssp. *vernalis*

Considered but rejected: A synonym of *H. congesta* ssp. *congesta*; a common taxon

Hemizonia pallida

Considered but rejected: Too common

Hemizonia parryi Greene ssp. *australis* Keck

"southern tarplant" Asteraceae
CNPS List: 1B **R-E-D Code:** 3-3-2 **State/Fed. Status:** /C2
Distribution: LAX, ORA, SBA, SCT?, SDG, BA
Quads: 34C, 34D, 35D, 70D, 71A, 71B, 72A, 88C*, 89C*, 89D, 90A*, 90B*, 90D, 111B*, 111C*, 111D*, 143A, 143B, SCTE?
Habitat: MshSw (estuary margins), VFGrs (vernally mesic), VnPls
Life Form: Annual herb
Blooming: June-November
Notes: Does plant occur on SCT Isl.? Many ORA Co. occurrences recently extirpated. Many historical occurrences also extirpated; need information. Population fragmentation a serious problem, and plant continues to be threatened by urbanization, vehicles, and foot traffic.

Hemizonia parryi Greene ssp. *congdonii* (Rob. & Greenm.) Keck

"Congdon's tarplant" Asteraceae
CNPS List: 1B **R-E-D Code:** 3-3-3 **State/Fed. Status:** /C1
Distribution: ALA*, CCA*, MNT, SCL(*?), SCR*, SLO, SOL*
Quads: 246C, 342A?, 365A, 365B, 386B, 387A*, 427B*, 427D(*?), 446A*, 446B*, 446C*, 447A*, 447B*, 465A*, 482C*
Habitat: VFGrs (alkaline)
Life Form: Annual herb
Blooming: June-November
Notes: Nearly extirpated from S.F. Bay Area; need information on historical and present distribution. Severely threatened by development. See *Botanical Gazette* 22:169 (1896) for original description, and *Madroño* 3(1):15 (1935) for revised nomenclature.

Hemizonia pentactis

Considered but rejected: Too common

Hemizonia pungens (H. & A.) T. & G. ssp. *laevis* Keck

"smooth tarplant" Asteraceae
CNPS List: 1B **R-E-D Code:** 2-3-3 **State/Fed. Status:** /C2
Distribution: RIV, SBD, SDG*
Quads: 21C*, 32A*, 32B*, 36A*, 36B*, 47D*, 50B*, 68A, 68C, 68D, 69A, 69D, 84C, 85A, 85C, 85D, 86A, 86B, 86D, 87A, 87B, 87C, 87D, 106B*, 106C, 107A*, 107D*, 108C*
Habitat: ChScr, Medws, Plyas, RpWld, VFGrs / alkaline
Life Form: Annual herb
Blooming: April-September
Notes: Many historical occurrences may be extirpated; need information. Frequently confused with other *Hemizonia* species such as *H. parryi* ssp. *australis* in ORA and LAX counties, and *H. pungens* ssp. *pungens* in KRN Co. Threatened by agriculture, urbanization, and flood control projects.

Hemizonia tracyi

See *Hemizonia congesta* ssp. *tracyi*

Herissantia crispa (L.) Brizicky

"curly herissantia" Malvaceae
CNPS List: 2 **R-E-D Code:** 3-1-1 **State/Fed. Status:** CEQA
Distribution: SDG, AZ, BA, TX, SO, ++
Quads: 7A, 32C
Habitat: SDScr
Life Form: Annual/Perennial herb
Blooming: August-September
Notes: Known in CA from only two occurrences at Mountain Springs Grade and Vallecito Wash.

Hesperevax acaulis var. *acaulis*

Considered but rejected: Too common

Hesperevax acaulis var. *ambusticola*

Considered but rejected: Too common

Hesperevax sparsiflora (Gray) Greene var. *brevifolia* (Gray) Morefield

"short-leaved evax" Asteraceae
CNPS List: 4 **R-E-D Code:** 1-2-1 **State/Fed. Status:** CEQA?
Distribution: HUM, MEN, MRN, SCR, SFO, SON, OR
Habitat: CBScr, CoDns
Life Form: Annual herb
Blooming: April-June
Notes: On review list in OR. See *Synoptical Flora of North America* 1(2):229 (1884) for original description, and *Systematic Botany* 17:293-310 (1992) for revised nomenclature.

Hesperolinon adenophyllum (Gray) Small

"glandular western flax" Linaceae
CNPS List: 1B **R-E-D Code:** 2-2-3 **State/Fed. Status:** /C2
Distribution: HUM*, LAK, MEN
Quads: 533B, 533C, 534A, 534B, 534D, 548B?, 549A, 549B, 549C, 549D, 550A, 565A, 565B, 565C, 565D, 566A, 566C, 566D, 567A, 567B, 582D, 583C, 654C*
Habitat: Chprl, VFGrs / serpentinite
Life Form: Annual herb
Blooming: May-August
Notes: Threatened in LAK Co. by geothermal development. See *Proceedings of the American Academy of Arts and Sciences* 8:624-625 (1873) for original description, and *University of California Publications in Botany* 32:235-314 (1961) for taxonomic treatment.
Status Report: 1978

Hesperolinon bicarpellatum (H.K. Sharsm.) H.K. Sharsm.

"two-carpellate western flax" Linaceae
CNPS List: 1B **R-E-D Code:** 2-2-3 **State/Fed. Status:** /C2
Distribution: LAK, NAP, SON
Quads: 499D, 500A, 516A, 516B, 516C, 516D, 517A, 517B, 533B, 533C, 533D, 549A
Habitat: Chprl (serpentinite)
Life Form: Annual herb
Blooming: May-July
Notes: Previously confused with *H. serpentinum*, which also occurs on serpentinite soils in the same general area. Threatened by development and grazing. See *University of California Publications in Botany* 32:235-314 (1961) for taxonomic treatment.

Hesperolinon breweri (Gray) Small

"Brewer's western flax" Linaceae
CNPS List: 1B **R-E-D Code:** 2-2-3 **State/Fed. Status:** /C2
Distribution: CCA, NAP, SOL
Quads: 464A, 464B, 464C, 464D, 499A, 499B, 499C, 515D
Habitat: Chprl, CmWld, VFGrs / mostly serpentinite
Life Form: Annual herb
Blooming: May-July
Notes: Need information on status of occurrences. Threatened by development, and several occurrences threatened by construction of Los Vaqueros Reservoir. See *University of California Publications in Botany* 32:235-314 (1961) for taxonomic treatment.

Hesperolinon congestum (Gray) Small

"Marin western flax" Linaceae
CNPS List: 1B **R-E-D Code:** 3-3-3 **State/Fed. Status:** CT/PT
Distribution: MRN, SFO, SMT
Quads: 429A, 448D, 466B, 466C, 467B, 484C*, 484D
Habitat: Chprl, VFGrs / serpentinite
Life Form: Annual herb
Blooming: May-July
Notes: Known from fewer than twenty occurrences. Protected in part at Ring Mtn. Preserve (TNC), MRN Co. Threatened by development and foot traffic. See *Proceedings of the American Academy of Arts and Sciences* 6:521 (1865) for original description, and *University of California Publications in Botany* 32:235-314 (1961) for taxonomic treatment.
Status Report: 1977

Hesperolinon didymocarpum H.K. Sharsm.

"Lake County western flax" Linaceae
CNPS List: 1B **R-E-D Code:** 3-2-3 **State/Fed. Status:** CE/C1
Distribution: LAK
Quads: 533B, 533C, 533D
Habitat: Chprl, CmWld, VFGrs / serpentinite
Life Form: Annual herb
Blooming: May-July
Notes: Known from fewer than ten occurrences in the Big Cyn. area near Middletown. Threatened by grazing. See *University of California Publications in Botany* 32:235-314 (1961) for original description.
Status Report: 1987

Hesperolinon drymarioides (Curran) Small

"drymaria-like western flax" Linaceae
CNPS List: 1B **R-E-D Code:** 3-2-3 **State/Fed. Status:** /C2
Distribution: COL, GLE, LAK, NAP, YOL
Quads: 532C, 532D, 548A, 564A, 564B, 564D, 565B, 565C, 565D
Habitat: CCFrs, Chprl, CmWld, VFGrs / serpentinite
Life Form: Annual herb
Blooming: May-August
Notes: Known from approximately twenty occurrences. Threatened by mining and vehicles. See *Bulletin of the California Academy of Sciences* 1:152 (1885) for original description, and *University of California Publications in Botany* 32:235-314 (1961) for taxonomic treatment.
Status Report: 1979

Hesperolinon serpentinum N. McCarten

"Napa western flax" Linaceae
CNPS List: 1B **R-E-D Code:** 3-2-3 **State/Fed. Status:** CEQA
Distribution: LAK, NAP
Quads: 516B
Habitat: Chprl (serpentinite)
Life Form: Annual herb
Blooming: May-July
Notes: Known from approximately six occurrences; one protected at ACEC (BLM), rest on private land and unprotected. Threatened by brush clearing and grading. See *Madroño* (1994, in review) for original description.

Hesperolinon tehamense H.K. Sharsm.

"Tehama County western flax" Linaceae
CNPS List: 1B **R-E-D Code:** 3-1-3 **State/Fed. Status:** /C2
Distribution: GLE, TEH
Quads: 580C, 580D, 596B, 596C
Habitat: Chprl, CmWld / serpentinite
Life Form: Annual herb
Blooming: May-July
Notes: Known from fewer than ten occurrences, mostly on Mendocino NF and BLM lands where locally abundant. See *University of California Publications in Botany* 32:298 (1961) for original description.

Hesperomecon linearis

Considered but rejected: Too common; a synonym of *Meconella linearis*

Heterotheca villosa (Pursh.) Shinn. var. *shevockii* Semple

"Shevock's hairy golden-aster" Asteraceae
CNPS List: 1B **R-E-D Code:** 2-1-3 **State/Fed. Status:** CEQA
Distribution: KRN
Quads: 261C, 261D
Habitat: Chprl, CmWld / sandy
Life Form: Perennial herb
Blooming: August-November
Notes: Endemic to lower Kern River Cyn. in the Greenhorn Mtns. See *Phytologia* 73(6):453 (1992) for original description.

Heuchera abramsii Rydb.

"Abrams's alumroot" Saxifragaceae
CNPS List: 4 **R-E-D Code:** 1-1-3 **State/Fed. Status:** CEQA?
Distribution: LAX, SBD
Habitat: UCFrs (rocky)
Life Form: Perennial herb (rhizomatous)
Blooming: July-August

Heuchera alpestris

See *Heuchera parishii*

Heuchera brevistaminea Wiggins

"Mt. Laguna alumroot" Saxifragaceae
CNPS List: 1B **R-E-D Code:** 3-1-3 **State/Fed. Status:** /C3c
Distribution: SDG
Quads: 19B, 19C, 20A
Habitat: BUFrs, Chprl, CmWld, RpFrs / rocky
Life Form: Perennial herb (rhizomatous)
Blooming: May-July
Notes: Known from fewer than ten occurrences. Relatively unthreatened since it occurs on steep, rocky sites. See *Contributions from the Dudley Herbarium* 1:100 (1929) for original description.
Status Report: 1979

Heuchera chlorantha Piper

"green-flowered alumroot" Saxifragaceae
CNPS List: 2 **R-E-D Code:** 3-1-1 **State/Fed. Status:** CEQA
Distribution: DNT, OR, ++
Quads: 740C
Habitat: Medws
Life Form: Perennial herb (rhizomatous)
Blooming: June-July
Notes: Known in CA only from Crescent City area. Not in *The Jepson Manual.* See *Contributions from the U.S. National Herbarium* 16:205-206 (1913) for original description.

Heuchera duranii Bacig.

"Duran's alumroot" Saxifragaceae
CNPS List: 1B **R-E-D Code:** 2-1-2 **State/Fed. Status:** /C3c
Distribution: INY, MNO, NV
Quads: 412A, 412B, 431C, 450D, 470A, 487C, 488A
Habitat: AlpBR, SCFrs (rocky) / carbonate
Life Form: Perennial herb (rhizomatous)
Blooming: July-August
Notes: May be inseparable from *H. parvifolia,* common in NV. Needs study.

Heuchera elegans Abrams

"urn-flowered alumroot" Saxifragaceae
CNPS List: 4 **R-E-D Code:** 1-1-3 **State/Fed. Status:** CEQA?
Distribution: LAX, SBD
Habitat: LCFrs, UCFrs / rocky
Life Form: Perennial herb (rhizomatous)
Blooming: May-June

Heuchera hirsutissima Rosend., Butt. & Lak.

"shaggy-haired alumroot" Saxifragaceae
CNPS List: 1B **R-E-D Code:** 3-1-3 **State/Fed. Status:** CEQA
Distribution: RIV
Quads: 65C, 65D, 83C, 83D
Habitat: SCFrs, UCFrs / rocky
Life Form: Perennial herb (rhizomatous)
Blooming: May-July
Notes: Very restricted, but habitat relatively unthreatened.

Heuchera leptomeria var. *peninsularis*

See *Heuchera rubescens* var. *versicolor*

Heuchera maxima Greene

"island alumroot" Saxifragaceae
CNPS List: 1B **R-E-D Code:** 2-2-3 **State/Fed. Status:** /C2
Distribution: ANA, SCZ, SRO
Quads: ANAC, SCZA, SCZB, SCZC, SRON
Habitat: CBScr
Life Form: Perennial herb (rhizomatous)
Blooming: February-April
Notes: Known from fewer than twenty occurrences. Potentially threatened by seabird rookery on ANA Isl., and feral herbivores on SCZ and SRO islands.

Heuchera merriamii

Considered but rejected: Too common

Heuchera parishii Rydb.

"Parish's alumroot" Saxifragaceae
CNPS List: 1B **R-E-D Code:** 2-1-3 **State/Fed. Status:** CEQA
Distribution: RIV, SBD
Quads: 83C, 105A, 105B, 105D, 106A
Habitat: AlpBR, LCFrs, SCFrs, UCFrs / rocky
Life Form: Perennial herb (rhizomatous)
Blooming: June-August
Notes: Includes *H. alpestris.*

Heuchera rubescens Torr. var. *versicolor* (Greene) M.G. Stewart

"San Diego County alumroot" Saxifragaceae
CNPS List: 2 **R-E-D Code:** 3-1-1 **State/Fed. Status:** CEQA
Distribution: SDG, BA, TX+
Quads: 20A, 48D
Habitat: Chprl, LCFrs / rocky
Life Form: Perennial herb (rhizomatous)
Blooming: May-June

Hibiscus californicus

See *Hibiscus lasiocarpus*

Hibiscus lasiocarpus Cav.

"rose-mallow" Malvaceae
CNPS List: 2 **R-E-D Code:** 2-2-1 **State/Fed. Status:** CEQA
Distribution: BUT, CCA, COL, GLE, SAC, SJQ, SOL, SUT, YOL, ++
Quads: 462A, 462B, 463A, 463D, 479B, 479C, 480A, 480B, 480C, 480D, 496C, 497D, 513D, 529B, 529C, 544C, 545A, 545B, 545D, 561A, 561B, 561C, 561D, 576B, 576C, 577B, 577C, 577D
Habitat: MshSw (freshwater)
Life Form: Perennial herb (rhizomatous, emergent)
Blooming: August-September
Notes: Most occurrences are very small. Seriously threatened by development, agriculture, recreation, and channelization of the Sacramento River and its tributaries. Widespread but also threatened in eastern NA. CA plants apparently unique in rhizomatous-stoloniferous propagation.
Status Report: 1977

Hierochloe odorata (L.) Wahl.

"vanilla-grass" Poaceae
CNPS List: 2 **R-E-D Code:** 3-1-1 **State/Fed. Status:** CEQA
Distribution: SIS, OR, WA, ++
Quads: 732C
Habitat: Medws (mesic)
Life Form: Perennial herb (rhizomatous)
Blooming: April-July
Notes: Known in CA from only a single confirmed occurrence on Ball Mtn. Unconfirmed specimens at the California Academy of Sciences have been recorded from Yosemite NP, San Pedro Valley (SMT Co.), and Fern Cyn. (MEN Co.); need verification.

Hollisteria lanata

Considered but rejected: Too common

Holocarpha macradenia (DC.) Greene

"Santa Cruz tarplant" Asteraceae
CNPS List: 1B **R-E-D Code:** 2-3-3 **State/Fed. Status:** CE/C1
Distribution: ALA*, CCA, MNT, MRN*, SCR
Quads: 386B, 386C, 387A, 387B, 407C, 408D, 447A*, 465B, 466A, 466D*, 467A*
Habitat: CoPrr, VFGrs / often clay
Life Form: Annual herb
Blooming: June-October
Notes: All CCA Co. occurrences are introduced; last remaining natural population in the S.F. Bay Area extirpated in 1993. Seriously threatened by urbanization, agriculture, non-native plants, and lack of appropriate ecological disturbance. See *Fremontia* 5(4):15-16 (1978) for species account.
Status Report: 1987

Holocarpha virgata (Gray) Keck ssp. *elongata* Keck

"graceful tarplant" Asteraceae
CNPS List: 4 **R-E-D Code:** 1-2-3 **State/Fed. Status:** /C2
Distribution: ORA, RIV, SDG
Habitat: Chprl?, CmWld, CoScr, VFGrs
Life Form: Annual herb
Blooming: August-November
Notes: Known in RIV Co. only from Santa Rosa Plateau. See *Aliso* 4:111 (1958) for original description.

Hordeum intercedens Nevski

"vernal barley" Poaceae
CNPS List: 3 **R-E-D Code:** ?-2-2 **State/Fed. Status:** CEQA?
Distribution: ANA, FRE?, KNG, LAX, MNO, RIV, SBA, SBR, SBT, SCM, SCT, SCZ, SDG, SMT, SNI, SRO, VEN, BA
Quads: 9D, 85C, 113B, 142B, 336C, 385D, ANAC, SBRA, SCZA, SNIC
Habitat: VFGrs (saline flats and depressions), VnPls
Life Form: Annual herb
Blooming: March-June
Notes: Move to List 1B? Location and rarity information needed, especially quads for LAX, MNO, and SMT counties, and SCM, SCT, and SRO islands. Does plant occur in FRE Co.? Most mainland occurrences have been extirpated by development; others are threatened. Previously confused with *H. pusillum*. See *Nordic Journal of Botany* 2:307-321 (1982) for taxonomic treatment.

Horkelia bolanderi Gray

"Bolander's horkelia" Rosaceae
CNPS List: 1B **R-E-D Code:** 3-2-3 **State/Fed. Status:** /C2
Distribution: COL*, LAK, MEN?
Quads: 533B, 533C, 534A, 534B
Habitat: LCFrs, Medws (edges, vernally mesic), VFGrs (edge habitats)
Life Form: Perennial herb
Blooming: June-August
Notes: Known from only three extant occurrences. Does plant occur in MEN Co.? Need historical quads for COL Co. Threatened by vehicles, and possibly by development. See *Systematic Botany* 18(1):137-144 (1993) for taxonomic treatment.

Horkelia bolanderi var. *bolanderi*

See *Horkelia bolanderi*

Horkelia congesta Hook. ssp. *nemorosa* Keck

"Josephine horkelia" Rosaceae
CNPS List: 2 **R-E-D Code:** 3-3-1 **State/Fed. Status:** CEQA
Distribution: DNT, OR
Quads: 739A
Habitat: NCFrs (clay, serpentinite seeps)
Life Form: Perennial herb
Blooming: May-July
Notes: Known in CA from only one occurrence at Shelly Creek.

Horkelia cuneata Lindl. ssp. *sericea* (Gray) Keck

"Kellogg's horkelia" Rosaceae
CNPS List: 1B **R-E-D Code:** 3-3-3 **State/Fed. Status:** /C2
Distribution: ALA*, MRN*, MNT, SBA, SCR, SFO*, SLO, SMT
Quads: 170B*, 171A, 196A*, 196B*, 196D*, 220C, 221D, 366A, 366C*, 366D, 386C*, 387B*, 408C, 408D, 409D, 448B, 465C*, 466C*, 466D*, 467D*, 485C*
Habitat: CCFrs, CoScr
Life Form: Perennial herb
Blooming: April-September
Notes: Historical occurrences need field surveys. Threatened by coastal development. Occurrence on Mt. San Bruno probably last remaining location in S.F. Bay. Remaining plants less distinct from ssp. *cuneata* than those formerly occurring near San Francisco.

Horkelia hendersonii Howell

"Henderson's horkelia" Rosaceae
CNPS List: 1B **R-E-D Code:** 3-3-2 **State/Fed. Status:** /C2
Distribution: SIS, OR
Quads: 735B
Habitat: UCFrs (granitic)
Life Form: Perennial herb
Blooming: June-August
Notes: Known in CA from only one occurrence. Candidate for state listing in OR.

Horkelia hispidula Rydb.

"White Mtns. horkelia" Rosaceae
CNPS List: 1B **R-E-D Code:** 3-1-3 **State/Fed. Status:** CEQA
Distribution: INY, MNO
Quads: 412B, 431B, 431C, 431D
Habitat: AlpDS, GBScr, SCFrs
Life Form: Perennial herb
Blooming: June-August
Notes: See *North American Flora* 22:278 (1908) for original description.

Horkelia marinensis (Elmer) Crum

"Point Reyes horkelia" Rosaceae
CNPS List: 1B **R-E-D Code:** 3-1-3 **State/Fed. Status:** /C2
Distribution: MEN, MRN, SCR, SMT
Quads: 387E, 408C?, 408D, 448C, 485C, 568B, 569A, 585A, 585D
Habitat: CoDns, CoPrr, CoScr
Life Form: Perennial herb
Blooming: May-September
Notes: Known from fewer than twenty occurrences. Historical occurrences need field surveys. See *Systematic Botany* 18(1):137-144 (1993) for distributional information.

Horkelia parryi Greene

"Parry's horkelia" Rosaceae
CNPS List: 1B **R-E-D Code:** 2-2-3 **State/Fed. Status:** /C2
Distribution: AMA, CAL, ELD, MPA
Quads: 439B, 476A, 494C, 495A, 509B, 509D, 510A
Habitat: Chprl, CmWld / especially Ione formation
Life Form: Perennial herb
Blooming: April-June
Notes: Potentially threatened by clay mining.

Horkelia sericata Wats.

"Howell's horkelia" Rosaceae
CNPS List: 4 **R-E-D Code:** 1-1-1 **State/Fed. Status:** CEQA?
Distribution: DNT, HUM?, OR
Habitat: Chprl, LCFrs / serpentinite or clay
Life Form: Perennial herb
Blooming: May-July
Notes: Does plant occur in HUM Co.?

Horkelia tenuiloba (Torr.) Gray

"thin-lobed horkelia" Rosaceae
CNPS List: 1B **R-E-D Code:** 2-2-3 **State/Fed. Status:** CEQA
Distribution: MEN, MRN, SON
Quads: 467A, 467B, 502A, 502B, 518B, 518C, 520D, 536C, 537B, 537C, 537D
Habitat: Chprl (mesic openings)
Life Form: Perennial herb
Blooming: May-July
Notes: Historical occurrences need field surveys. See *Report of the Pacific Railroad Expedition* 4(5):84 (1857) for original description, and *Systematic Botany* 18(1):137-144 (1993) for taxonomic treatment.

Horkelia truncata Rydb.

"Ramona horkelia" Rosaceae
CNPS List: 1B **R-E-D Code:** 3-1-2 **State/Fed. Status:** /C3c
Distribution: SDG, BA
Quads: 20B, 20C, 21A, 34A, 34B, 34D, 35B, 49D, 51A, 51B
Habitat: Chprl, CmWld / clay
Life Form: Perennial herb
Blooming: May-June
Notes: Known in CA from fewer than twenty occurrences. Possibly threatened by chaparral management.

Horkelia tularensis (J.T. Howell) Munz

"Kern Plateau horkelia" Rosaceae
CNPS List: 1B **R-E-D Code:** 3-1-3 **State/Fed. Status:** /C2
Distribution: TUL
Quads: 306B, 306C, 307D
Habitat: UCFrs (rocky)
Life Form: Perennial herb
Blooming: July-August
Notes: Known from approximately ten occurrences. Potentially threatened by mining and recreation. Protected in part in a BA (USFS) which includes the type locality. See *Leaflets of Western Botany* 10(13):254-255 (1966) for original description.
Status Report: 1979

Horkelia wilderae Parish

"Barton Flats horkelia" Rosaceae
CNPS List: 1B **R-E-D Code:** 3-3-3 **State/Fed. Status:** /C2
Distribution: SBD
Quads: 105A, 105B
Habitat: LCFrs, UCFrs
Life Form: Perennial herb
Blooming: May-September
Notes: Known from fewer than ten occurrences in the Barton Flats area. Threatened by logging and recreational activities. See *Botanical Gazette* 38:460 (1904) for original description.
Status Report: 1979

Horkelia yadonii Ertter

"Santa Lucia horkelia" Rosaceae
CNPS List: 4 **R-E-D Code:** 1-2-3 **State/Fed. Status:** CEQA?
Distribution: MNT, SBA, SLO
Habitat: Chprl, CmWld, Medws / granitic, sandy
Life Form: Perennial herb (rhizomatous)
Blooming: June-July
Notes: Possibly threatened by vehicles and recreational activities. Confused with *H. cuneata* ssp. *sericea*, *H. rydbergii*, and *H. tenuiloba*. See *Systematic Botany* 18(1):139 (1993) for original description.

Howellia aquatilis Gray

"water howellia" Campanulaceae
CNPS List: 1A **Last Seen:** 1928 **State/Fed. Status:** /PT
Distribution: MEN*, ID, OR*, WA+
Quads: 597B*
Habitat: MshSw (freshwater)
Life Form: Annual herb (aquatic)
Blooming: May-August
Notes: Known in CA from only a single collection near Howard Lake in 1927; rediscovery attempt in 1979 was unsuccessful. Possibly extirpated by cattle grazing and trampling. Endangered in ID, and state-listed as Endangered in WA. Inadvertently omitted from *The Jepson Manual*. See *Proceedings of the American Academy of Arts and Sciences* 15:43 (1879) for original description, *Wasmann Journal of Biology* 33:90-91 (1975) for the CA record, and *Conservation Biology* 2(3):275-282 (1988) for ecological information.
Status Report: 1980

Hulsea brevifolia Gray

"short-leaved hulsea" Asteraceae
CNPS List: 4 **R-E-D Code:** 1-1-3 **State/Fed. Status:** CEQA?
Distribution: FRE, MAD, MPA, TUL, TUO
Habitat: UCFrs (granitic or volcanic)
Life Form: Perennial herb
Blooming: June-August

Hulsea californica T. & G.

"San Diego sunflower" Asteraceae
CNPS List: 1B **R-E-D Code:** 2-1-3 **State/Fed. Status:** /C3c
Distribution: SDG
Quads: 8C, 19C, 19B, 20A, 20D, 33D, 48D, 49D
Habitat: LCFrs, UCFrs / openings
Life Form: Perennial herb
Blooming: April-June

Hulsea inyoensis

See *Hulsea vestita* ssp. *inyoensis*

Hulsea mexicana Rydb.

"Mexican hulsea" Asteraceae
CNPS List: 2 **R-E-D Code:** 3-1-1 **State/Fed. Status:** CEQA
Distribution: SDG, BA
Quads: 7C
Habitat: Chprl (often on burned or disturbed areas)
Life Form: Annual/Perennial herb
Blooming: April-June
Notes: Known in U.S. from only a single collection near Jacumba on Table Mtn.; is it more widespread? Common in BA. Easily confused with *H. californica.* See *North American Flora* 34:41 (1914) for original description.

Hulsea vestita Gray ssp. *callicarpha* (Hall) Wilken

"beautiful hulsea" Asteraceae
CNPS List: 4 **R-E-D Code:** 1-2-3 **State/Fed. Status:** CEQA?
Distribution: RIV, SDG
Habitat: Chprl, LCFrs / rocky or gravelly
Life Form: Perennial herb
Blooming: May-October
Notes: Intergrades with *H. californica.* See *University of California Publications in Botany* 1:129 (1902) for original description.

Hulsea vestita Gray ssp. *gabrielensis* Wilken

"San Gabriel Mtns. sunflower" Asteraceae
CNPS List: 4 **R-E-D Code:** 1-1-3 **State/Fed. Status:** CEQA?
Distribution: LAX, SBD, VEN
Habitat: LCFrs, UCFrs / rocky
Life Form: Perennial herb
Blooming: May-July
Notes: See *Madroño* 24(1):48-55 (1977) for original description.

Hulsea vestita Gray ssp. *inyoensis* (Keck) Wilken

"Inyo hulsea" Asteraceae
CNPS List: 2 **R-E-D Code:** 2-2-1 **State/Fed. Status:** /C3c
Distribution: INY, MNO, NV
Quads: 348C, 369C, 372A, 433C
Habitat: GBScr, PJWld / rocky
Life Form: Perennial herb
Blooming: April-June
Notes: Type locality degraded by grading and talus removal in 1989. On watch list in NV.

Hulsea vestita Gray ssp. *parryi* (Gray) Wilken

"Parry's sunflower" Asteraceae
CNPS List: 4 **R-E-D Code:** 1-1-3 **State/Fed. Status:** CEQA?
Distribution: KRN, RIV?, SBD, VEN
Habitat: LCFrs, UCFrs / rocky
Life Form: Perennial herb
Blooming: April-August
Notes: Does plant occur in RIV Co.? See *Proceedings of the American Academy of Arts and Sciences* 12:59 (1876) for original description.

Hystrix californica

See *Elymus californicus*

Iliamna bakeri (Jeps.) Wiggins

"Baker's globe mallow" Malvaceae
CNPS List: 4 **R-E-D Code:** 1-1-1 **State/Fed. Status:** CEQA?
Distribution: LAS, MOD, SHA, SIS?, OR
Habitat: Chprl, PJWld / volcanic
Life Form: Perennial herb
Blooming: July-August
Notes: Does plant occur in SIS Co.? See *Manual of the Flowering Plants of California*, p. 635 (1925) by W.L. Jepson for original description.

Iliamna latibracteata Wiggins

"California globe mallow" Malvaceae
CNPS List: 4 **R-E-D Code:** 1-1-2 **State/Fed. Status:** CEQA?
Distribution: DNT, HUM, OR
Habitat: NCFrs (mesic)
Life Form: Perennial herb
Blooming: June-July
Notes: Endangered in OR.

Ipomopsis effusa (Gray) Moran

"Baja California ipomopsis" Polemoniaceae
CNPS List: 2 **R-E-D Code:** 3-3-1 **State/Fed. Status:** CEQA
Distribution: IMP, BA
Quads: 6A
Habitat: Chprl, SDScr (alluvial fan)
Life Form: Annual herb
Blooming: April-June
Notes: Known in CA only from Pinto Wash, along the Mexican border. See *Madroño* 24(3):141-159 (1977) for taxonomic treatment, and *Aliso* 11(4):589-598 (1987) for first CA record.

Ipomopsis polycladon

Considered but rejected: Too common

Ipomopsis tenuifolia (Gray) V. Grant

"slender-leaved ipomopsis" Polemoniaceae
CNPS List: 2 **R-E-D Code:** 2-1-1 **State/Fed. Status:** CEQA
Distribution: IMP, SDG, BA
Quads: 7A, 7B
Habitat: Chprl, PJWld, SDScr / gravelly or rocky
Life Form: Perennial herb
Blooming: March-May

Iris bracteata Wats.

"Siskiyou iris" Iridaceae
CNPS List: 4 **R-E-D Code:** 1-1-1 **State/Fed. Status:** CEQA?
Distribution: DNT, OR
Habitat: LCFrs (serpentinite)
Life Form: Perennial herb (rhizomatous)
Blooming: May-June

Iris chrysophylla

Considered but rejected: Too common

Iris hartwegii Baker ssp. *columbiana* Lenz

"Tuolumne iris" Iridaceae
CNPS List: 4 **R-E-D Code:** 1-1-3 **State/Fed. Status:** CEQA?
Distribution: TUO
Habitat: CmWld, LCFrs
Life Form: Perennial herb (rhizomatous)
Blooming: May-June
Notes: Possibly threatened by logging.

Iris innominata Henders.

"Del Norte County iris" Iridaceae
CNPS List: 4 **R-E-D Code:** 1-1-1 **State/Fed. Status:** CEQA?
Distribution: DNT, OR
Habitat: LCFrs (serpentinite)
Life Form: Perennial herb (rhizomatous)
Blooming: May-June

Iris munzii Foster

"Munz's iris" Iridaceae
CNPS List: 4 **R-E-D Code:** 1-1-3 **State/Fed. Status:** CEQA?
Distribution: TUL
Habitat: CmWld
Life Form: Perennial herb (rhizomatous)
Blooming: March-April

Iris tenax Dougl. ssp. *klamathensis* Lenz

"Orleans iris" Iridaceae
CNPS List: 4 **R-E-D Code:** 1-1-3 **State/Fed. Status:** /C3c
Distribution: HUM, SIS
Habitat: LCFrs (often in disturbed areas)
Life Form: Perennial herb (rhizomatous)
Blooming: April-May
Notes: See *Aliso* 4(1):64-66 (1958) for original description.
Status Report: 1977

Isocoma arguta Greene

"Carquinez goldenbush" Asteraceae
CNPS List: 1B **R-E-D Code:** 3-3-3 **State/Fed. Status:** /C2
Distribution: CCA, SOL
Quads: 481A, 481B, 498D
Habitat: VFGrs (alkaline)
Life Form: Shrub
Blooming: August-December
Notes: Undocumented in CCA Co.; need quads. Probably threatened by development and agriculture. See *Manual of the Botany of the Region of San Francisco Bay*, p. 175 (1894) by E. Greene for original description, and *Phytologia* 70(2):69-114 (1991) for taxonomic treatment.

Isocoma menziesii (H. & A.) Nesom var. *decumbens* (Greene) Nesom

"decumbent goldenbush" Asteraceae
CNPS List: 1B **R-E-D Code:** 2-2-2 **State/Fed. Status:** CEQA
Distribution: ORA, SCM, SCT, SDG, BA
Quads: 10A, 10B, 10D, 11A, 11D, 36D, 71D
Habitat: CoScr (sandy, often in disturbed areas)
Life Form: Shrub
Blooming: April-November
Notes: Need quads for SCM and SCT islands. Equals *Haplopappus venetus* ssp. *furfuraceus*, a synonym of *I. menziesii* var. *menziesii* in *The Jepson Manual.* See *Leaflets of Botanical Observation and Criticism* 1:172 (1906) for original description, and *Phytologia* 70(2):69-114 (1991) for taxonomic treatment.

Isocoma menziesii (H. & A.) Nesom var. *diabolica* Nesom

"Satan's goldenbush" Asteraceae
CNPS List: 4 **R-E-D Code:** 1-2-3 **State/Fed. Status:** CEQA?
Distribution: SBT, SCL
Habitat: CmWld
Life Form: Shrub
Blooming: August-October
Notes: Not in *The Jepson Manual.* See *Phytologia* 70(2):69-114 (1991) for original description.

Isoetes bolanderi var. *pygmaea*

Considered but rejected: Too common

Isoetes howellii

Considered but rejected: Too common

Isoetes nuttallii

Considered but rejected: Too common

Isoetes orcuttii

Considered but rejected: Too common

Iva hayesiana Gray

"San Diego marsh-elder" Asteraceae
CNPS List: 2 **R-E-D Code:** 2-2-1 **State/Fed. Status:** /C2
Distribution: SDG, BA
Quads: 10A, 10B, 10C, 11D, 22B, 35C
Habitat: Plyas
Life Form: Perennial herb
Blooming: April-September
Notes: Threatened by waterway channelization and coastal development.

Iva nevadensis

Considered but rejected: Too common

Ivesia aperta

See *Ivesia aperta* var. *aperta*

Ivesia aperta (J.T. Howell) Munz var. *aperta*

"Sierra Valley ivesia" Rosaceae
CNPS List: 1B **R-E-D Code:** 2-2-2 **State/Fed. Status:** /C2
Distribution: LAS, PLU, SIE, NV
Quads: 570B, 570D, 571A, 571B, 571D, 586B, 587C, 587D, 602C, 603D
Habitat: GBScr, LCFrs, Medws (xeric), PJWld / usually volcanic
Life Form: Perennial herb
Blooming: June-August
Notes: Threatened by residential development, agriculture, grazing, and vehicles. Interim management guide prepared by USFS in 1992. See *Leaflets of Western Botany* 9:233-242 (1962) for original description, and *Brittonia* 40(4):398-399 (1988) for revised nomenclature.
Status Report: 1990

Ivesia aperta (J.T. Howell) Munz var. *canina* Ertter

"Dog Valley ivesia" Rosaceae
CNPS List: 1B **R-E-D Code:** 3-3-3 **State/Fed. Status:** /C1
Distribution: SIE
Quads: 570D
Habitat: LCFrs (openings), Medws (xeric) / volcanic
Life Form: Perennial herb
Blooming: June-August
Notes: Endemic to Dog Valley. Threatened by reservoir and recreation development, and possibly by vehicle use and grazing. Interim management guide prepared by USFS in 1992. See *Brittonia* 40(4):398-399 (1988) for original description.

Ivesia argyrocoma (Rydb.) Rydb.

"silver-haired ivesia" Rosaceae
CNPS List: 1B **R-E-D Code:** 2-2-2 **State/Fed. Status:** /C2
Distribution: SBD, BA
Quads: 104B, 105A, 105B, 126D, 131C, 131D, 132C, 132D
Habitat: Medws (alkaline), PbPln, UCFrs
Life Form: Perennial herb
Blooming: June-August
Notes: Threatened by development, grazing, and vehicles. Populations in BA possibly taxonomically distinct.

Ivesia arizonica (J.T. Howell) Ertter var. *arizonica*

"yellow ivesia" Rosaceae
CNPS List: 3 **R-E-D Code:** ?-?-1 **State/Fed. Status:** CEQA?
Distribution: INY, AZ, NV, UT
Quads: 368D
Habitat: PJWld, SCFrs, UCFrs / carbonate
Life Form: Perennial herb
Blooming: May-August
Notes: Move to List 2? Precise location, rarity, and endangerment information needed. Known in CA only from the Inyo, Grapevine, and Funeral Mtns.

Ivesia baileyi

See *Ivesia baileyi* var. *baileyi*

Ivesia baileyi Wats. var. *baileyi*

"Bailey's ivesia" Rosaceae
CNPS List: 2 **R-E-D Code:** 3-2-1 **State/Fed. Status:** CEQA
Distribution: LAS, PLU, ID, NV, OR
Quads: 586B, 586C, 587A, 620B, 620C
Habitat: LCFrs (volcanic)
Life Form: Perennial herb
Blooming: June-August
Notes: Known in CA from fewer than ten occurrences. See *Madroño* 32(2):123 (1985) for discussion of CA occurrences.

Ivesia baileyi Wats. var. *beneolens* (Nelson & J.F. Macbr.) Ertter

"Owyhee ivesia" Rosaceae
CNPS List: 2 **R-E-D Code:** 3-1-1 **State/Fed. Status:** CEQA
Distribution: MOD, ID, NV, OR
Quads: 708D
Habitat: PJWld, UCFrs / volcanic
Life Form: Perennial herb
Blooming: June
Notes: Known in CA from only three occurrences in the Warner Mtns. See *Botanical Gazette* 55:374 (1913) for original description, and *Systematic Botany* 14:231-244 (1989) for revised nomenclature.

Ivesia callida (Hall) Rydb.

"Tahquitz ivesia" Rosaceae
CNPS List: 1B **R-E-D Code:** 3-1-3 **State/Fed. Status:** CR/C2
Distribution: RIV
Quads: 83C
Habitat: UCFrs (granitic)
Life Form: Perennial herb
Blooming: July-September
Notes: Rediscovered in 1980 by K. Berg, M. Hamilton, and T. Krantz; known from only two occurrences in the San Jacinto Wilderness Area. See *University of California Publications in Botany* 1:86 (1902) for original description, and *Fremontia* 11(1):13-15 (1983) for discussion of rediscovery.
Status Report: 1987

Ivesia campestris (M.E. Jones) Rydb.

"field ivesia" Rosaceae
CNPS List: 4 **R-E-D Code:** 1-1-3 **State/Fed. Status:** CEQA?
Distribution: FRE, INY, TUL
Habitat: Medws (edges), SCFrs, UCFrs
Life Form: Perennial herb
Blooming: June-August
Notes: Possibly threatened by grazing. See *Proceedings of the California Academy of Sciences* II 5:679 (1895) for original description, and *North American Flora* 22:285 (1908) for revised nomenclature.

Ivesia corymbosa

See *Ivesia paniculata*

Ivesia jaegeri Munz & Jtn.

"Jaeger's ivesia" Rosaceae
CNPS List: 1B **R-E-D Code:** 3-1-2 **State/Fed. Status:** /C2
Distribution: SBD, NV
Quads: 249D
Habitat: PJWld, UCFrs / carbonate
Life Form: Perennial herb
Blooming: June-July
Notes: Known in CA from only two occurrences near Clark Mtn. On watch list in NV.

Ivesia kingii

See *Ivesia kingii* var. *kingii*

Ivesia kingii Wats. var. *kingii*

"alkali ivesia" Rosaceae
CNPS List: 1B **R-E-D Code:** 3-1-2 **State/Fed. Status:** CEQA
Distribution: INY, MNO, NV, UT
Quads: 413B, 432C, 434A, 451A, 451B, 451D
Habitat: GBScr, Medws, Plyas / mesic alkaline clay
Life Form: Perennial herb
Blooming: June-July
Notes: Known in CA from fewer than ten occurrences. Possibly threatened by grazing.

Ivesia longibracteata Ertter

"Castle Crags ivesia" Rosaceae
CNPS List: 1B **R-E-D Code:** 3-1-3 **State/Fed. Status:** /C2
Distribution: SHA
Quads: 682A
Habitat: LCFrs (granitic crevices)
Life Form: Perennial herb
Blooming: June
Notes: Known only from Castle Crags. See *Systematic Botany* 14:233 (1989) for original description.

Ivesia muirii

Considered but rejected: Too common

Ivesia paniculata T.W. Nelson & J.P. Nelson

"Ash Creek ivesia" Rosaceae
CNPS List: 1B **R-E-D Code:** 2-1-3 **State/Fed. Status:** /C2
Distribution: LAS
Quads: 659A, 675A, 675C, 676D
Habitat: GBScr, PJWld, UCFrs / volcanic
Life Form: Perennial herb
Blooming: June-July
Notes: Known from fewer than twenty occurrences in the Ash Valley region. Protected at Ash Valley RNA (BLM). See *Brittonia* 33(2):165-167 (1981) for original description, and *Fremontia* 16(1):15-17 (1988) for brief species account and BLM management plans.

Ivesia patellifera (J.T. Howell) Ertter

"Kingston Mtns. ivesia" Rosaceae
CNPS List: 1B **R-E-D Code:** 3-1-3 **State/Fed. Status:** /C2
Distribution: SBD
Quads: 250B, 274C, 274D
Habitat: PJWld (granitic)
Life Form: Perennial herb
Blooming: June-October
Notes: See *Leaflets of Western Botany* 4:173 (1945) for original description, and *Systematic Botany* 14(2):231-244 (1989) for revised nomenclature.

Ivesia pickeringii Gray

"Pickering's ivesia" Rosaceae
CNPS List: 1B **R-E-D Code:** 3-2-3 **State/Fed. Status:** /C2
Distribution: SIS, TRI
Quads: 683B, 683D, 699B, 700C, 701D
Habitat: LCFrs, Medws / mesic, clay, generally serpentinite seeps
Life Form: Perennial herb
Blooming: July-August
Notes: Known from fewer than twenty occurrences. Threatened by grazing, logging, mining, and road maintenance. Klamath NF has adopted species management guidelines. See *Proceedings of the American Academy of Arts and Sciences* 6:531 (1865) for original description.
Status Report: 1979

Ivesia purpurascens ssp. *purpurascens*

Considered but rejected: Too common; a synonym of *Horkeliella purpurascens*

Ivesia pygmaea

Considered but rejected: Too common

Ivesia sericoleuca (Rydb.) Rydb.

"Plumas ivesia" Rosaceae
CNPS List: 1B **R-E-D Code:** 1-2-3 **State/Fed. Status:** /C2
Distribution: LAS, NEV, PLA, PLU, SIE
Quads: 554A, 554B, 554C, 554D, 555A, 570C, 570D, 571B, 571C, 571D, 572D, 586B, 587B, 587C, 587D, 602C, 603C, 603D, 622D
Habitat: GBScr, LCFrs, Medws, VnPls / vernally mesic, usually volcanic
Life Form: Perennial herb
Blooming: May-September
Notes: Threatened by development, grazing, and vehicles. Interim management guide prepared by the USFS in 1992.

Ivesia unguiculata Gray

"Yosemite ivesia" Rosaceae
CNPS List: 1B **R-E-D Code:** 3-2-3 **State/Fed. Status:** CEQA
Distribution: FRE, MAD, MPA
Quads: 376A, 395B, 395C, 395D, 396B, 396C, 396D, 397D, 415B*, 416C, 416D, 417A, 418D, 436D, 437A, 437B
Habitat: Medws, SCFrs
Life Form: Perennial herb
Blooming: June-August
Notes: Historical occurrences need field surveys. Threatened by grazing.

Ivesia webberi Gray

"Webber's ivesia" Rosaceae
CNPS List: 1B **R-E-D Code:** 3-3-2 **State/Fed. Status:** /C2
Distribution: PLU, SIE, NV
Quads: 570A, 570B, 570D, 586C, 589B, 605C
Habitat: GBScr (volcanic ash), LCFrs
Life Form: Perennial herb
Blooming: May-June
Notes: Known from approximately ten occurrences over its range; in CA from only Sierra Valley and Dog Valley. Threatened by vehicles, and potentially by development and grazing. On watch list in NV. See *Proceedings of the American Academy of Arts and Sciences* 10:71 (1874) for original description.
Status Report: 1990

Jamesia americana T. & G. var. *rosea* C. Schneider

"rosy-petalled cliffbush" Philadelphaceae
CNPS List: 4 **R-E-D Code:** 1-1-2 **State/Fed. Status:** CEQA?
Distribution: FRE, INY, MNO, TUL, NV
Habitat: AlpBR, PJWld, SCFrs / granitic or carbonate
Life Form: Shrub (deciduous)
Blooming: May-September
Notes: Scattered but never abundant. See *Brittonia* 41(4):335-350 (1989) for taxonomic treatment.

Jepsonia heterandra Eastw.

"foothill jepsonia" Saxifragaceae
CNPS List: 4 **R-E-D Code:** 1-1-3 **State/Fed. Status:** CEQA?
Distribution: AMA, CAL, ELD, MPA, STA, TUO
Habitat: CmWld, LCFrs / rocky, metamorphic
Life Form: Perennial herb
Blooming: August-December
Notes: See *Bulletin of the Torrey Botanical Club* 32:201 (1905) for original description, and *Brittonia* 21:286-298 (1969) for taxonomic treatment.

Jepsonia malvaefolia

See *Jepsonia malvifolia*

Jepsonia malvifolia (Greene) Small

"island jepsonia" Saxifragaceae
CNPS List: 4 **R-E-D Code:** 1-1-2 **State/Fed. Status:** /C2
Distribution: SCM, SCT, SCZ, SNI, SRO, GU
Habitat: CoScr
Life Form: Perennial herb
Blooming: November-December

Jepsonia parryi

Considered but rejected: Too common

Juglans californica Wats. var. *californica*

"Southern California black walnut" Juglandaceae
CNPS List: 4 **R-E-D Code:** 1-2-3 **State/Fed. Status:** CEQA?
Distribution: LAX, ORA, RIV, SBA, SBD, SDG, VEN
Habitat: Chprl, CmWld, CoScr / alluvial
Life Form: Tree (deciduous)
Blooming: March-May
Notes: Walnut forest is a much fragmented, declining natural community, rare in ORA, RIV, SBD, and SDG counties. Threatened by urbanization, grazing, and possibly by lack of natural reproduction. See *Proceedings of the American Academy of Arts and Sciences* 10:349 (1875) for original description, and *Southern California Botanists Special Publication No. 3*, pp. 42-54 (1990) for additional information.

Juglans californica Wats. var. *hindsii* Jeps.

"Northern California black walnut" Juglandaceae
CNPS List: 1B **R-E-D Code:** 3-3-3 **State/Fed. Status:** /C2
Distribution: CCA, NAP, SAC*, SOL*, YOL*
Quads: 465D, 480A*, 480B*, 499B
Habitat: RpFrs, RpWld
Life Form: Tree (deciduous)
Blooming: April-May
Notes: Only two of three native stands are still extant. Widely naturalized in cismontane CA. Threatened by urbanization and conversion to agriculture. Formerly cultivated as rootstock for *J. regia*, with which it hybridizes readily. See *Flora of California* 1(2):365 (1909) by W.L. Jepson for original description, and *Madroño* 17(1):1-32 (1963) for discussion of origin.
Status Report: 1978

Juglans hindsii

See *Juglans californica* var. *hindsii*

Juncus abjectus

See *Juncus hemiendytus* var. *abjectus*

Juncus acutus L. ssp. *leopoldii* (Parl.) Snog.

"southwestern spiny rush" Juncaceae
CNPS List: 4 **R-E-D Code:** 1-2-1 **State/Fed. Status:** CEQA?
Distribution: IMP?, LAX, ORA, SBA, SDG, SLO, VEN, AZ?, BA, ++
Habitat: CoDns (mesic), Medws (alkaline seeps), MshSw (coastal salt)
Life Form: Perennial herb (rhizomatous)
Blooming: May-June
Notes: Threatened by urbanization and flood control.

Juncus acutus var. *sphaerocarpus*

See *Juncus acutus* ssp. *leopoldii*

Juncus bufonius var. *congdonii*

Considered but rejected: A synonym of *J. bufonius* var. *congestus*; a common taxon

Juncus chlorocephalus

Considered but rejected: Too common

Juncus cooperi

Considered but rejected: Too common

Juncus cyperoides
Considered but rejected: Not native

Juncus diffusissimus
Considered but rejected: Not native

***Juncus dudleyi* Wieg.**
"Dudley's rush" Juncaceae
CNPS List: 2 **R-E-D Code:** 3-1-1 **State/Fed. Status:** CEQA
Distribution: HUM, SIS, TRI, AZ, OR, ++
Quads: 649B, 703B, 704D, 736C
Habitat: LCFrs (mesic)
Life Form: Perennial herb
Blooming: July-August

***Juncus duranii* Ewan**
"Duran's rush" Juncaceae
CNPS List: 4 **R-E-D Code:** 1-1-3 **State/Fed. Status:** CEQA?
Distribution: LAX, RIV, SBD
Habitat: LCFrs, Medws, UCFrs / mesic
Life Form: Perennial herb (rhizomatous)
Blooming: July-August

Juncus falcatus* var. *falcatus
Considered but rejected: Too common

***Juncus hemiendytus* F.J. Herm. var. *abjectus* (F.J. Herm.) Ertter**
"Center Basin rush" Juncaceae
CNPS List: 4 **R-E-D Code:** 1-1-1 **State/Fed. Status:** CEQA?
Distribution: ALP, LAS, MNO, NEV, PLU, SIE, TUL, TUO, ID, NV, OR
Habitat: SCFrs
Life Form: Annual herb
Blooming: June-July
Notes: On review list in OR. See *Memoirs of the New York Botanical Garden* 39:72-78 (1986) for revised treatment.

Juncus leiospermus
See *Juncus leiospermus* var. *leiospermus*

***Juncus leiospermus* F.J. Herm. var. *ahartii* Ertter**
"Ahart's dwarf rush" Juncaceae
CNPS List: 1B **R-E-D Code:** 3-1-3 **State/Fed. Status:** /C1
Distribution: BUT, CAL, PLA, SAC
Quads: 477C, 511C, 528A, 560A, 560D
Habitat: VnPls
Life Form: Annual herb
Blooming: March-May
Notes: Known from only five occurrences. See *Memoirs of the New York Botanical Garden* 39:46-51 (1986) for original description.

Juncus leiospermus* F.J. Herm. var. *leiospermus
"Red Bluff dwarf rush" Juncaceae
CNPS List: 1B **R-E-D Code:** 3-2-3 **State/Fed. Status:** /C3c
Distribution: BUT, SHA, TEH
Quads: 560D, 576A, 576D, 592B, 593A, 593D, 595A, 610B*, 611B, 611D, 612A, 628B, 628C, 629B, 629C, 646C, 647D, 662B
Habitat: Chprl, CmWld, VFGrs, VnPls / vernally mesic
Life Form: Annual herb
Blooming: March-May
Notes: See *Leaflets of Western Botany* 5:113 (1948) for original description, and *Memoirs of the New York Botanical Garden* 39:46-51 (1986) for revised treatment.

Juncus marginatus* Rostkov var. *marginatus
"red-anthered rush" Juncaceae
CNPS List: 2 **R-E-D Code:** 3-2-1 **State/Fed. Status:** CEQA
Distribution: NEV, AZ, ++
Quads: 541B
Habitat: MshSw
Life Form: Perennial herb (rhizomatous)
Blooming: July
Notes: Distribution poorly documented; to be expected elsewhere in CA.

Juncus mertensianus* var. *duranii
See *Juncus duranii*

***Juncus nodosus* L.**
"knotted rush" Juncaceae
CNPS List: 4 **R-E-D Code:** 1-1-1 **State/Fed. Status:** CEQA?
Distribution: INY, TUL?, ++
Habitat: Medws (mesic), MshSw (lake margins)
Life Form: Perennial herb (rhizomatous)
Blooming: July-September
Notes: Does plant occur in TUL Co.?

***Juncus regelii* Buch.**
"Regel's rush" Juncaceae
CNPS List: 2 **R-E-D Code:** 3-1-1 **State/Fed. Status:** CEQA
Distribution: SIS, TRI, OR, WA+
Quads: 667C, 738C
Habitat: Medws, UCFrs / mesic
Life Form: Perennial herb (rhizomatous)
Blooming: August

***Juncus supiniformis* Engelm.**
"hair-leaved rush" Juncaceae
CNPS List: 4 **R-E-D Code:** 1-1-2 **State/Fed. Status:** CEQA?
Distribution: DNT, HUM, MEN, OR+
Habitat: MshSw (freshwater, near coast)
Life Form: Perennial herb (rhizomatous)
Blooming: April-June

Juncus tenuis* var. *dudleyi
See *Juncus dudleyi*

Juniperus communis* var. *jackii
Considered but rejected: A synonym of *J. communis*; a common taxon

Juniperus communis var. *montana*

Considered but rejected: A synonym of *J. communis*; a common taxon

Kallstroemia californica

Considered but rejected: Too common

Kobresia bellardii (All.) Degl.

"seep kobresia" Cyperaceae
CNPS List: 2 **R-E-D Code:** 3-1-1 **State/Fed. Status:** CEQA
Distribution: MNO, ID, OR, ++
Quads: 434C, 434D, 471D
Habitat: AlpBR (mesic), Medws (carbonate), SCFrs
Life Form: Perennial herb (rhizomatous)
Blooming: August
Notes: Known in CA only from Convict Basin. On review list in ID, and endangered in OR. See *Madroño* 17(4):93-109 (1964) and 40(1):66-67 (1993) for first and second CA reports respectively.

Kobresia myosuroides

See *Kobresia bellardii*

Koeberlinia spinosa

See *Koeberlinia spinosa* ssp. *tenuispina*

Koeberlinia spinosa Zucc. ssp. *tenuispina* (Kearn. & Peebles) E. Murray

"crown-of-thorns" Koeberliniaceae
CNPS List: 2 **R-E-D Code:** 3-2-1 **State/Fed. Status:** CEQA
Distribution: IMP, AZ, SO+
Quads: 12B, 27A, 27B, 42C, 43D
Habitat: RpWld, SDScr
Life Form: Shrub (deciduous)
Blooming: May-July
Notes: Known in CA from fewer than ten occurrences. Threatened by mining.

Lagophylla minor

Considered but rejected: Too common

Larrea tridentata var. *arenaria*

Considered but rejected: A synonym of *L. tridentata*; a common taxon

Lasthenia burkei (Greene) Greene

"Burke's goldfields" Asteraceae
CNPS List: 1B **R-E-D Code:** 3-3-3 **State/Fed. Status:** CE/FE
Distribution: LAK, MEN, SON
Quads: 502A, 518A, 518D, 533A, 533B, 550B
Habitat: Medws (mesic), VnPls
Life Form: Annual herb
Blooming: April-June
Notes: Threatened by agriculture, urbanization, and grazing. See *Bulletin of the California Academy of Sciences* 2(6):151 (1887) for original description, and *American Journal of Botany* 56(9):1042-1047 (1969) for information on origin and relationships.
Status Report: 1988

Lasthenia conjugens Greene

"Contra Costa goldfields" Asteraceae
CNPS List: 1B **R-E-D Code:** 3-3-3 **State/Fed. Status:** /C1
Distribution: ALA*, CCA*, MEN*, NAP, SBA*, SCL*, SOL
Quads: 142A*, 143A*, 427D*, 447D*, 463C*, 465A*, 481B, 481D*, 482A, 483A, 498C, 499B*, 499C*, 500D*, 517D*, 537B*
Habitat: VFGrs (mesic), VnPls
Life Form: Annual herb
Blooming: March-June
Notes: Known from only four occurrences after comprehensive 1993 surveys. Many historical occurrences extirpated by development; also threatened by overgrazing.
Status Report: 1979

Lasthenia coronaria

Considered but rejected: Too common

Lasthenia glabrata Lindl. ssp. *coulteri* (Gray) Ornduff

"Coulter's goldfields" Asteraceae
CNPS List: 1B **R-E-D Code:** 2-3-2 **State/Fed. Status:** /C2
Distribution: KRN*, LAX*, ORA*, RIV, SBA, SBD*, SDG, SLO, SRO, TUL?, VEN, BA
Quads: 11A, 11D, 22B, 22C, 36B, 36D, 50B, 68C, 68D, 69A, 71B*, 71D*, 72A*, 84C?, 85A, 85C, 85D, 86B, 88C*, 89A*, 89D*, 90A*, 90B*, 90D*, 102A*, 110A*, 110B*, 114B, 114D, 141D, 142A, 142B, 143A, 171A, 212A*, 212B*, 212C*, 217D, 218A, 247D, SROE, SRON
Habitat: MshSw (coastal salt), Plyas, VnPls
Life Form: Annual herb
Blooming: February-June
Notes: Known to have declined significantly by 1966, and now seriously threatened by urbanization and agricultural development. Does plant occur in TUL Co.? See *Synoptical Flora of North America* 1(2):324 (1884) for original description, and *University of California Publications in Botany* 40:1-92 (1966) for taxonomic treatment.

Lasthenia leptalea (Gray) Ornduff

"Salinas Valley goldfields" Asteraceae
CNPS List: 4 **R-E-D Code:** 1-1-3 **State/Fed. Status:** /C3c
Distribution: INY, KRN, MNT, SLO
Habitat: CmWld, VFGrs
Life Form: Annual herb
Blooming: April
Notes: See *Proceedings of the American Academy of Arts and Sciences* 6:546 (1865) for original description, and *University of California Publications in Botany* 40:63-66 (1969) for revised nomenclature.

Lasthenia maritima

Considered but rejected: Too common

Lasthenia minor ssp. *maritima*

Considered but rejected: A synonym of *L. maritima*; a common taxon

Lasthenia glabrata ssp. *coulteri*

Lathyrus biflorus T.W. Nelson & J.P. Nelson

"two-flowered pea" Fabaceae
CNPS List: 1B **R-E-D Code:** 3-1-3 **State/Fed. Status:** /C1
Distribution: HUM
Quads: 634A
Habitat: LCFrs (serpentinite)
Life Form: Perennial herb
Blooming: June-August
Notes: Known from only one occurrence in The Lassics area. See *Brittonia* 35(2):180-183 (1983) for original description.

Lathyrus delnorticus C.L. Hitchc.

"Del Norte pea" Fabaceae
CNPS List: 4 **R-E-D Code:** 1-1-2 **State/Fed. Status:** CEQA?
Distribution: DNT, SIS, OR
Habitat: LCFrs, NCFrs / often serpentinite
Life Form: Perennial herb
Blooming: June-July
Notes: On watch list in OR.

Lathyrus glandulosus Broich

"sticky pea" Fabaceae
CNPS List: 4 **R-E-D Code:** 1-1-3 **State/Fed. Status:** CEQA?
Distribution: HUM, MEN
Habitat: CmWld
Life Form: Perennial herb (rhizomatous)
Blooming: April-June
Notes: See *Madroño* 33(2):136-143 (1986) for original description.

Lathyrus hitchcockianus

Considered but rejected: Not in CA

Lathyrus jepsonii ssp. *jepsonii*

See *Lathyrus jepsonii* var. *jepsonii*

Lathyrus jepsonii Greene var. *jepsonii*

"Delta tule pea" Fabaceae
CNPS List: 1B **R-E-D Code:** 2-2-3 **State/Fed. Status:** /C2
Distribution: ALA, CCA, FRE, MRN, NAP, SAC, SBT, SCL, SJQ, SOL
Quads: 341A, 397A, 427A, 428A, 428B, 446C, 462A, 462B, 463A, 465A*, 465B*, 479B, 479C, 480B, 480D, 481B, 481C, 481D, 482A, 482C, 482D, 483A, 496C, 497C, 497D, 498C, 498D
Habitat: MshSw (freshwater and brackish)
Life Form: Perennial herb
Blooming: May-June
Notes: Threatened by agriculture and water diversions. See *Pittonia* 2:158 (1890) for original description.
Status Report: 1977

Lathyrus palustris L.

"marsh pea" Fabaceae
CNPS List: 2 **R-E-D Code:** 2-2-1 **State/Fed. Status:** CEQA
Distribution: DNT, HUM, OR, WA, ++
Quads: 618D, 672C, 689A, 689C, 723B, 740B, 740C
Habitat: BgFns, LCFrs, MshSw, NCFrs / mesic
Life Form: Perennial herb
Blooming: May-August
Notes: See *University of Washington Publications in Botany* 15:13 (1952) for taxonomic treatment.

Lathyrus splendens Kell.

"pride-of-California" Fabaceae
CNPS List: 4 **R-E-D Code:** 1-1-2 **State/Fed. Status:** CEQA?
Distribution: SDG, BA
Habitat: Chprl
Life Form: Perennial herb
Blooming: April-June

Lathyrus sulphureus Gray var. *argillaceus* Jeps.

"dubious pea" Fabaceae
CNPS List: 3 **R-E-D Code:** 3-?-3 **State/Fed. Status:** CEQA?
Distribution: NEV?, PLA, SHA, TEH
Quads: 527A, 630D, 647C
Habitat: CmWld, LCFrs, UCFrs
Life Form: Perennial herb
Blooming: April
Notes: Move to List 1B? Location, rarity, and endangerment information needed; does plant occur in NEV Co.? Fewer than ten specimens exist in CA herbaria. Taxonomy poorly understood; see *L. sulphureus* in *The Jepson Manual*. See *Flora of California* 2(1):393 (1936) by W.L. Jepson for original description.

Lathyrus vestitus var. *alefeldii*

Considered but rejected: Too common

Lathyrus vestitus* var. *ochropetalus

Considered but rejected: Too common

Lavatera assurgentiflora* Kell. ssp. *assurgentiflora

"island mallow" Malvaceae
CNPS List: 1B **R-E-D Code:** 3-3-3 **State/Fed. Status:** /C1
Distribution: ANA, SMI, SNI*, SRO
Quads: ANAC, SMIE, SMIW, SNIC*, SRON
Habitat: CBScr
Life Form: Shrub (evergreen)
Blooming: May-September
Notes: Known from fewer than ten native occurrences. Rare at all occurrences, and seriously threatened by grazing; reduced to one plant on ANA Isl. Reintroduced into native habitat on SMI Isl. (SMIE); cultivated plants grow on SNI Isl. but native occurrence extirpated. May not be native to SRO Isl.; plants on mainland and Todos Santos Isl. (BA) most likely planted. See *L. assurgentiflora* in *The Jepson Manual.* See *Proceedings of a Multidisciplinary Symposium: The California Islands*, pp. 157-158 (1980) for species account.
Status Report: 1979

***Lavatera assurgentiflora* Kell. ssp. *glabra* Philbrick**

"southern island mallow" Malvaceae
CNPS List: 1B **R-E-D Code:** 3-3-3 **State/Fed. Status:** /C1
Distribution: SCM, SCT
Quads: SCMC, SCMN, SCMS, SCTN, SCTW
Habitat: CBScr
Life Form: Shrub (evergreen)
Blooming: May-September
Notes: Known from approximately ten native occurrences. Feral herbivores removed from SCM Isl.; possibly still a threat on SCT Isl. See *L. assurgentiflora* in *The Jepson Manual.* See *Proceedings of a Multidisciplinary Symposium: The California Islands*, pp. 157-158 (1980) for original description.

***Layia carnosa* (Nutt.) T. & G.**

"beach layia" Asteraceae
CNPS List: 1B **R-E-D Code:** 3-3-3 **State/Fed. Status:** CE/FE
Distribution: HUM, MNT, MRN, SBA*, SFO*
Quads: 171A*, 171B*, 171C*, 366C, 466C*, 485B, 485C, 637D, 654B, 655A, 672A*, 672B, 672C, 689D*
Habitat: CoDns
Life Form: Annual herb
Blooming: May-July
Notes: Threatened by coastal development, vehicles, and non-native plants. Protected in part at Manila Dunes ACEC and Mattole Beach ACEC (both BLM), HUM Co.
Status Report: 1990

Layia chrysanthemoides* ssp. *maritima

Considered but rejected: A synonym of *L. chrysanthemoides*; a common taxon

***Layia discoidea* (Keck) Keck**

"rayless layia" Asteraceae
CNPS List: 1B **R-E-D Code:** 2-3-3 **State/Fed. Status:** /C2
Distribution: FRE, SBT
Quads: 339B, 339C, 339D, 340D
Habitat: Chprl, CmWld, LCFrs / serpentinite, talus and alluvial terraces
Life Form: Annual herb
Blooming: May
Notes: Threatened by vehicles in the New Idria area. Similar to *L. glandulosa.* See *Aliso* 4:101-104 (1958) for original description.
Status Report: 1977

***Layia heterotricha* (DC.) H. & A.**

"pale-yellow layia" Asteraceae
CNPS List: 1B **R-E-D Code:** 3-3-3 **State/Fed. Status:** /C2
Distribution: FRE*, KNG*, KRN*, MNT*, SBA, SBT(*?), SLO*, VEN(*?)
Quads: 165A(*?), 165B(*?), 166A(*?), 166B*, 166D*, 190C*, 190D(*?), 191C, 192C, 211C*, 212A*, 212B*, 212D*, 217D*, 243A*, 243B*, 244C*, 244D*, 267B*, 291B*, 294B*, 294C*, 315A*, 315C*, 318A(*?), 318B*, 340C*, 362D(*?)
Habitat: CmWld, PJWld, VFGrs / alkaline or clay
Life Form: Annual herb
Blooming: March-June
Notes: Recent searches of historical occurrences were largely unsuccessful. Threatened by agricultural conversion and previous construction of San Antonio Reservoir, and possibly by overgrazing.

***Layia jonesii* Gray**

"Jones's layia" Asteraceae
CNPS List: 1B **R-E-D Code:** 3-2-3 **State/Fed. Status:** /C2
Distribution: MNT, SLO
Quads: 221B, 246C, 246D, 247A, 247B, 247D, 294B, 366C
Habitat: Chprl, VFGrs / clay or serpentinite
Life Form: Annual herb
Blooming: March-May

***Layia leucopappa* Keck**

"Comanche Point layia" Asteraceae
CNPS List: 1B **R-E-D Code:** 3-3-3 **State/Fed. Status:** /C1
Distribution: KRN
Quads: 213C, 214A, 214D, 215D, 239D
Habitat: ChScr, VFGrs
Life Form: Annual herb
Blooming: May-April
Notes: Reduced by agriculture; also threatened by development and grazing.

***Layia munzii* Keck**

"Munz's tidy-tips" Asteraceae
CNPS List: 1B **R-E-D Code:** 2-2-3 **State/Fed. Status:** CEQA
Distribution: FRE, KRN, SLO
Quads: 192A, 218A, 243C, 244A, 265C, 265D, 268A, 337A, 359C, 360B, 360C, 381C, 383D
Habitat: ChScr, VFGrs (alkaline clay)
Life Form: Annual herb
Blooming: March-April
Notes: Historical occurrences need field surveys. Similar to *L. jonesii* and *L. leucopappa.*

Layia septentrionalis Keck

"Colusa layia" Asteraceae
CNPS List: 1B **R-E-D Code:** 2-2-3 **State/Fed. Status:** CEQA
Distribution: COL, LAK, MEN, NAP, SON, SUT, TEH, YOL
Quads: 501A, 516A, 516B, 516C, 516D, 517A, 531B, 532A, 532D, 533A, 533B, 534A, 534B, 534D, 535A, 535D, 545A, 546D, 547C, 548A, 548C, 548D, 549C, 549D, 550D, 596B
Habitat: Chprl, CmWld, VFGrs / sandy, serpentinite
Life Form: Annual herb
Blooming: April-May
Notes: Threatened by development. See *Aliso* 4(1):106 (1958) for original description.

Layia ziegleri

Considered but rejected: A synonym of *L. platyglossa*; a common taxon

Legenere limosa (Greene) McVaugh

"legenere" Campanulaceae
CNPS List: 1B **R-E-D Code:** 2-3-3 **State/Fed. Status:** /C2
Distribution: LAK, NAP, PLA, SAC, SMT, SOL, SON*, STA*, TEH
Quads: 428C, 460C*, 481B, 483A, 495A, 496A, 496B, 496D, 498C*, 498D, 499D*, 501D*, 511C, 512B, 528D, 533C, 534A, 628D
Habitat: VnPls
Life Form: Annual herb
Blooming: May-June
Notes: Many historical occurrences extirpated. Threatened by grazing and development. See *Pittonia* 2:81 (1890) for original description, *North American Flora* 32(1):13-14 (1943) for revised nomenclature, and *Wasmann Journal of Biology* 33(1-2):91 (1975) for distributional information.
Status Report: 1977

Lembertia congdonii (Gray) Greene

"San Joaquin woollythreads" Asteraceae
CNPS List: 1B **R-E-D Code:** 3-2-3 **State/Fed. Status:** /FE
Distribution: FRE, KNG*, KRN, SBA, SBT, SLO, TUL*
Quads: 192A*, 192B*, 192D, 217B, 217C, 217D, 218A, 218C, 218D, 239A*, 239B*, 239D, 240A*, 240B*, 240C, 241B, 242B*, 243A, 265B*, 265C, 265D*, 266D*, 287B*, 290B, 290C, 291A*, 291B*, 313C*, 314A*, 314B*, 314C, 314D, 315A*, 315C, 315D, 338A*, 338B*, 338D*, 360B*, 360C*, 361B, 361C, 361D
Habitat: ChScr, VFGrs
Life Form: Annual herb
Blooming: March-May
Notes: About half of historical occurrences extirpated; now known from approximately 30 occurrences. Seriously threatened by agricultural conversion, energy development, and urbanization. See *Proceedings of the American Academy of Arts and Sciences* 19:20 (1883) for original description.

Lepechinia cardiophylla Epl.

"heart-leaved pitcher sage" Lamiaceae
CNPS List: 1B **R-E-D Code:** 3-2-2 **State/Fed. Status:** /C2
Distribution: ORA, RIV, SDG, BA
Quads: 21B, 69B, 70A, 87C, 87D
Habitat: CCFrs, Chprl, CmWld
Life Form: Shrub
Blooming: April-July
Notes: Known in CA from fewer than ten occurrences. Threatened by development.

Lepechinia fragrans (Greene) Epl.

"fragrant pitcher sage" Lamiaceae
CNPS List: 4 **R-E-D Code:** 1-2-3 **State/Fed. Status:** CEQA?
Distribution: LAX, SCT, SCZ, SRO, VEN?
Habitat: Chprl
Life Form: Shrub
Blooming: March-May
Notes: Known in the Santa Monica Mtns. from near Triunfo Pass, and may cross into VEN Co. Threatened in the Santa Monica Mtns. by urbanization; threatened in the San Gabriel Mtns. by fire management and habitat alteration.

Lepechinia ganderi Epl.

"Gander's pitcher sage" Lamiaceae
CNPS List: 1B **R-E-D Code:** 3-1-2 **State/Fed. Status:** /C2
Distribution: SDG, BA
Quads: 10A, 10B, 10C, 10D
Habitat: CCFrs, Chprl, CoScr, VFGrs
Life Form: Shrub
Blooming: June-July
Notes: Known in CA from fewer than ten occurrences.

Lepidium flavum Torr. var. *felipense* C.L. Hitchc.

"Borrego Valley pepper-grass" Brassicaceae
CNPS List: 1B **R-E-D Code:** 3-2-3 **State/Fed. Status:** /C2
Distribution: SDG
Quads: 20A?, 32B, 32C, 33A, 47C
Habitat: PJWld, SDScr / sandy
Life Form: Annual herb
Blooming: March-May
Notes: Known from fewer than ten occurrences. Threatened by vehicles. See *Madroño* 3:299 (1936) for original description.

Lepidium jaredii

See *Lepidium jaredii* ssp. *jaredii*

Lepidium jaredii Bdg. ssp. *album* Hoov.

"Panoche pepper-grass" Brassicaceae
CNPS List: 1B **R-E-D Code:** 3-2-3 **State/Fed. Status:** /C2
Distribution: FRE, SBT, SLO
Quads: 267B, 336A*, 338B*, 339A, 339B, 340A, 362A*, 362D*
Habitat: VFGrs (alluvial fans, washes)
Life Form: Annual herb
Blooming: February-June
Notes: Recent surveys (1991) discovered a number of extant occurrences. Threatened by gravel mining and cattle grazing. Not in *The Jepson Manual.* See *Leaflets of Western Botany* 10:345 (1966) for original description.

Lepidium jaredii Bdg. ssp. *jaredii*

"Jared's pepper-grass" Brassicaceae
CNPS List: 1B **R-E-D Code:** 3-2-3 **State/Fed. Status:** /C2
Distribution: KRN, SLO
Quads: 217C, 218A, 267A, 269A*
Habitat: VFGrs (alkaline, adobe)
Life Form: Annual herb
Blooming: March-May
Notes: Known only from Soda Lake on the Carrizo Plain (SLO Co.) and Devil's Den (KRN Co.).

Lepidium latipes Hook. var. *heckardii* Roll.

"Heckard's pepper-grass" Brassicaceae
CNPS List: 1B **R-E-D Code:** 3-2-3 **State/Fed. Status:** CEQA
Distribution: YOL
Quads: 513B, 513C
Habitat: VFGrs (alkaline flats)
Life Form: Annual herb
Blooming: April-May

Lepidium latipes var. *latipes*

Considered but rejected: Too common

Lepidium virginicum L. var. *robinsonii* (Thell.) Hitchc.

"Robinson's pepper-grass" Brassicaceae
CNPS List: 1B **R-E-D Code:** 3-2-2 **State/Fed. Status:** CEQA
Distribution: LAX, ORA, RIV, SBA*, SBD, SCZ, SDG, BA
Quads: 8C, 11B, 34D, 71B, 87D, 88C, 108C, 109A, 110A, 160D, SCZB, SCZC
Habitat: Chprl, CoScr
Life Form: Annual herb
Blooming: January-July
Notes: Need historical quads for SBA Co. Threatened by erosion and feral herbivores on SCZ Isl. See *Madroño* 3(7):265-320 (1936) for taxonomic treatment.

Lepidospartum squamatum var. *palmeri*

Considered but rejected: A synonym of *L. squamatum*; a common taxon

Leptodactylon californicum H. & A. ssp. *tomentosum* Gordon

"fuzzy prickly phlox" Polemoniaceae
CNPS List: 4 **R-E-D Code:** 1-2-3 **State/Fed. Status:** CEQA?
Distribution: SBA, SLO
Habitat: CoDns
Life Form: Shrub (deciduous)
Blooming: March-August
Notes: See *L. californicum* in *The Jepson Manual*. See *Madroño* 37(1):28-42 (1990) for original description and discussion.

Leptodactylon jaegeri (Munz) Wherry

"San Jacinto prickly phlox" Polemoniaceae
CNPS List: 1B **R-E-D Code:** 2-2-3 **State/Fed. Status:** /C3c
Distribution: RIV
Quads: 83C
Habitat: SCFrs, UCFrs / granitic
Life Form: Perennial herb
Blooming: July-August
Notes: Known from fewer than twenty occurrences in the San Jacinto Mtns.

Lesquerella bernardina

See *Lesquerella kingii* ssp. *bernardina*

Lesquerella kingii (Wats.) Wats. ssp. *bernardina* (Munz) Munz

"San Bernardino Mtns. bladderpod" Brassicaceae
CNPS List: 1B **R-E-D Code:** 3-3-3 **State/Fed. Status:** /PE
Distribution: SBD
Quads: 105A, 131C, 131D
Habitat: LCFrs, PJWld / often carbonate
Life Form: Perennial herb
Blooming: May-June
Notes: Known from only five occurrences in the Big Bear Valley area. Threatened by development and carbonate mining. See *Fremontia* 16(1):20-21 (1988) for discussion of mining threats.

Lessingia arachnoidea Greene

"Crystal Springs lessingia" Asteraceae
CNPS List: 1B **R-E-D Code:** 3-2-3 **State/Fed. Status:** /C2
Distribution: SMT, SON?
Quads: 429A, 448C, 448D, 502B?, 518C?
Habitat: CmWld, CoScr, VFGrs / serpentinite, often roadsides
Life Form: Annual herb
Blooming: July-October
Notes: Known only from Crystal Springs Reservoir (SMT Co.); occurrences from SON Co. need taxonomic verification. See *Leaflets of Botanical Observation and Criticism* 2:29 (1910) for original description.

Lessingia germanorum Cham.

"San Francisco lessingia" Asteraceae
CNPS List: 1B **R-E-D Code:** 3-3-3 **State/Fed. Status:** CE/C1
Distribution: SFO, SMT
Quads: 448B, 466C
Habitat: CoScr (remnant dunes)
Life Form: Annual herb
Blooming: August-November
Notes: Known from only four occurrences at the Presidio of San Francisco, and one on San Bruno Mtn. (SMT Co.). Threatened by urbanization, base-closure activities, trampling, and non-native plants.
Status Report: 1990

Lessingia germanorum var. *germanorum*

See *Lessingia germanorum*

Lessingia germanorum var. *tenuis*

See *Lessingia tenuis*

Lessingia glandulifera Gray var. *tomentosa* (Greene) Ferris

"Warner Springs lessingia" Asteraceae
CNPS List: 2 **R-E-D Code:** 3-1-1 **State/Fed. Status:** /C2
Distribution: SDG, BA
Quads: 33A, 33B, 48C
Habitat: Chprl (sandy)
Life Form: Annual herb
Blooming: October
Notes: Known in CA from fewer than five occurrences.

Lessingia hololeuca Greene

"woolly-headed lessingia" Asteraceae
CNPS List: 3 **R-E-D Code:** ?-?-3 **State/Fed. Status:** CEQA?
Distribution: ALA, MNT, MRN, NAP, SCL, SMT, SOL, SON, YOL
Quads: 317D, 364B, 365A, 406D, 407B, 408B?, 427A, 428B, 428C?, 448C, 466B?, 467A, 481B, 483A, 484A, 484C, 484D, 485A, 501A, 502B, 503A, 503B, 514A, 517A, 517D
Habitat: CoScr, LCFrs, VFGrs / clay, serpentinite
Life Form: Annual herb
Blooming: June-October
Notes: Move to List 4? Need location, rarity, and endangerment information. Probably more widespread in the southern Sacramento Valley, southern North Coast Ranges, and northern S.F. Bay. Possibly threatened by grazing. See *Flora Franciscana*, p. 377 (1897) by E. Greene for original description, and *University of California Publications in Botany* 16:40 (1929) for taxonomic treatment.

Lessingia micradenia var. *arachnoidea*

See *Lessingia arachnoidea*

Lessingia micradenia Greene var. *glabrata* (Keck) Ferris

"smooth lessingia" Asteraceae
CNPS List: 1B **R-E-D Code:** 3-2-3 **State/Fed. Status:** /C2
Distribution: SCL
Quads: 406B, 406C, 406D, 407A?, 407B, 407D
Habitat: Chprl (serpentinite, often roadsides)
Life Form: Annual herb
Blooming: August-November
Notes: See *Aliso* 4:105 (1958) for original description, and *Contributions from the Dudley Herbarium* 5:101 (1958) for revised nomenclature.

Lessingia micradenia Greene var. *micradenia*

"Tamalpais lessingia" Asteraceae
CNPS List: 1B **R-E-D Code:** 3-2-3 **State/Fed. Status:** /C2
Distribution: MRN
Quads: 467A, 467B
Habitat: Chprl, VFGrs / usually serpentinite, often roadsides
Life Form: Annual herb
Blooming: June-September
Notes: Known only from Mt. Tamalpais. See *Leaflets of Botanical Observation and Criticism* 2:28 (1910) for original description, and *University of California Publications in Botany* 16:39-40 (1929) for taxonomic treatment.

Lessingia occidentalis (Hall) M.A. Lane

"western lessingia" Asteraceae
CNPS List: 4 **R-E-D Code:** 1-1-3 **State/Fed. Status:** /C3c
Distribution: FRE, MNT, SBT
Habitat: Chprl, CmWld, CoScr, VFGrs / serpentinite
Life Form: Annual herb
Blooming: May-November
Notes: See *Novon* 2(3):213-214 (1992) for revised nomenclature.

Lessingia tenuis (Gray) Cov.

"spring lessingia" Asteraceae
CNPS List: 4 **R-E-D Code:** 1-1-3 **State/Fed. Status:** CEQA?
Distribution: SBA, SLO, VEN
Habitat: LCFrs
Life Form: Annual herb
Blooming: May-June

Lewisia brachycalyx Gray

"short-sepaled lewisia" Portulacaceae
CNPS List: 4 **R-E-D Code:** 1-1-1 **State/Fed. Status:** CEQA?
Distribution: SBD, SDG, BA, ++
Habitat: LCFrs, Medws / mesic
Life Form: Perennial herb
Blooming: May-June

Lewisia cantelovii J.T. Howell

"Cantelow's lewisia" Portulacaceae
CNPS List: 1B **R-E-D Code:** 2-2-3 **State/Fed. Status:** /C3c
Distribution: BUT, NEV, PLU, SHA, SIE
Quads: 557A, 557B, 557C, 558C, 558D, 573D, 574D, 575A, 589C, 590D, 591A, 591C, 591D, 606C, 607D, 665B
Habitat: BUFrs, Chprl, CmWld, LCFrs / mesic, granitic, sometimes serpentinite seeps
Life Form: Perennial herb
Blooming: May-June
Notes: Threatened by horticultural collecting and road maintenance. Does not include *L. serrata*. See *Leaflets of Western Botany* 3(6):139 (1942) for original description.

Lewisia cantelowii

See *Lewisia cantelovii*

Lewisia congdonii (Rydb.) J.T. Howell

"Congdon's lewisia" Portulacaceae
CNPS List: 1B **R-E-D Code:** 3-1-3 **State/Fed. Status:** CR/C3c
Distribution: FRE, MPA
Quads: 375C, 437C, 438A
Habitat: Chprl, CmWld, LCFrs, UCFrs / granitic, mesic
Life Form: Perennial herb
Blooming: April-June
Notes: Known from fewer than ten occurrences. Most occurrences are relatively inaccessible. See *North American Flora* 21:328 (1932) for original description.
Status Report: 1985

Lewisia cotyledon var. *cotyledon*

Considered but rejected: Too common

Lewisia cotyledon var. *fimbriata*

Considered but rejected: Not published and taxonomic problem

Lewisia cotyledon (Wats.) Rob. var. *heckneri* (Mort.) Munz

"Heckner's lewisia" Portulacaceae
CNPS List: 1B **R-E-D Code:** 2-2-3 **State/Fed. Status:** /C2
Distribution: HUM, SIS, TRI
Quads: 650A, 650B, 659B, 667C, 668A, 668B, 668D, 669C, 683C, 684C, 685D, 703C, 737B
Habitat: LCFrs, NCFrs / rocky
Life Form: Perennial herb
Blooming: May-July
Notes: Range overlaps with other varieties; identification difficult.

Lewisia cotyledon (Wats.) Rob. var. *howellii* (Wats.) Jeps.

"Howell's lewisia" Portulacaceae
CNPS List: 3 **R-E-D Code:** 2-2-2 **State/Fed. Status:** /C2
Distribution: DNT, HUM, SHA, SIS, TRI, OR
Quads: 666B, 669B, 669C, 681C, 681D, 683D, 702C, 703B, 703C, 717B, 719A, 719B, 720A, 720B, 720C, 721A, 721C, 721D, 734C, 735A, 736A, 736C, 737A, 736D, 737C, 737D, 738D
Habitat: Chprl, CmWld / rocky
Life Form: Perennial herb
Blooming: April-June
Notes: Move to List 4? Taxonomic problems. Identification difficult. On watch list in OR. See *Proceedings of the American Academy of Arts and Sciences* 23:262 (1888) for original description.

Lewisia disepala Rydb.

"Yosemite lewisia" Portulacaceae
CNPS List: 1B **R-E-D Code:** 2-1-3 **State/Fed. Status:** /C3c
Distribution: KRN, MAD, MPA, TUL
Quads: 259A, 283C, 284D, 307B, 308D, 416B, 417D, 437A, 437B, 454C, 455D
Habitat: LCFrs, PJWld, UCFrs / granitic sand
Life Form: Perennial herb
Blooming: April-June
Notes: See *North American Flora* 21:328 (1932) for original description, and *Madroño* 37:63 (1990) for first KRN Co. record.

Lewisia leana

Considered but rejected: Too common

Lewisia longipetala (Piper) Clay

"long-petaled lewisia" Portulacaceae
CNPS List: 1B **R-E-D Code:** 3-1-3 **State/Fed. Status:** /C2
Distribution: ELD, FRE, NEV, PLA
Quads: 394B, 523B, 523C, 523D, 539A, 555A, 555D
Habitat: AlpBR, SCFrs (mesic, rocky)
Life Form: Perennial herb
Blooming: July-August
Notes: Known from fewer than twenty occurrences. Possibly threatened by horticultural collecting. Interim management guidelines prepared by the USFS in 1992. See *Contributions from the U.S. National Herbarium* 16:207 (1918) for original description.
Status Report: 1979

Lewisia oppositifolia (Wats.) Rob.

"opposite-leaved lewisia" Portulacaceae
CNPS List: 1B **R-E-D Code:** 2-2-2 **State/Fed. Status:** /C3c
Distribution: DNT, OR
Quads: 739A, 739B, 739C, 739D, 740A, 740D
Habitat: LCFrs (mesic)
Life Form: Perennial herb
Blooming: April-May
Notes: Threatened by logging and mining. CA plants possibly hybrids with *C. nevadensis*. On watch list in OR. See *Proceedings of the American Academy of Arts and Sciences* 20:355 (1885) for original description.

Lewisia pygmaea ssp. *longipetala*

See *Lewisia longipetala*

Lewisia serrata Heckard & Steb.

"saw-toothed lewisia" Portulacaceae
CNPS List: 1B **R-E-D Code:** 3-3-3 **State/Fed. Status:** /C2
Distribution: ELD, PLA
Quads: 524C, 525A, 525D, 540C, 540D
Habitat: BUFrs, LCFrs, RpFrs
Life Form: Perennial herb
Blooming: May-June
Notes: Known from approximately ten occurrences. Threatened by horticultural collecting and small hydroelectric power projects. See *L. cantelovii* in *The Jepson Manual*. See *Brittonia* 26:305 (1974) for original description.
Status Report: 1979

Lewisia sierrae

Considered but rejected: Too common; a synonym of *L. pygmaea*

Lewisia stebbinsii Gankin & Hildreth

"Stebbins's lewisia" Portulacaceae
CNPS List: 1B **R-E-D Code:** 3-2-3 **State/Fed. Status:** /C2
Distribution: MEN, TRI
Quads: 581B, 581C, 613C
Habitat: LCFrs, UCFrs / gravelly, sometimes serpentinite
Life Form: Perennial herb
Blooming: May-June
Notes: Known from approximately ten occurrences. Threatened by timber and range management activities, road maintenance, and vehicles. See *Four Seasons* 2(4):12-14 (1968) for original description.
Status Report: 1979

Lilaeopsis masonii Math. & Const.

"Mason's lilaeopsis" Apiaceae
CNPS List: 1B **R-E-D Code:** 2-2-3 **State/Fed. Status:** CR/C2
Distribution: ALA, CCA, MRN, NAP, SAC, SJQ, SOL
Quads: 462B, 463A, 463D, 479B, 480A, 480B, 480C, 480D, 481B, 481C, 481D, 482A, 482D, 483A, 483D, 496C, 497C, 498D, 500D
Habitat: MshSw (brackish or freshwater), RpScr
Life Form: Perennial herb
Blooming: April-October
Notes: Threatened by development, flood control projects, recreation, erosion, levee maintenance, and agriculture. See *Madroño* 24:81 (1977) for original description.
Status Report: 1988

Lilium bolanderi Wats.

"Bolander's lily" Liliaceae
CNPS List: 4 **R-E-D Code:** 1-2-1 **State/Fed. Status:** /C3c
Distribution: DNT, HUM, SIS, OR
Habitat: Chprl, LCFrs / serpentinite
Life Form: Perennial herb (bulbiferous)
Blooming: June-July
Notes: Hybridizes with *L. pardalinum* sspp., *L. rubescens*, and *L. washingtonianum* ssp. *purpurascens*. See *Proceedings of the American Academy of Arts and Sciences* 20:377 (1885) for original description.

Lilium fairchildii

See *Lilium humboldtii* ssp. *ocellatum*

Lilium humboldtii Roezl & Leichtl. ssp. *humboldtii*

"Humboldt lily" Liliaceae
CNPS List: 4 **R-E-D Code:** 1-2-3 **State/Fed. Status:** CEQA?
Distribution: AMA, BUT, CAL, ELD, FRE, MAD, MPA, NEV, PLA, TUO, YUB
Habitat: Chprl, LCFrs / openings
Life Form: Perennial herb (bulbiferous)
Blooming: June-July
Notes: Threatened by urbanization.

Lilium humboldtii Roezl & Leichtl. ssp. *ocellatum* (Kell.) Thorne

"ocellated Humboldt lily" Liliaceae
CNPS List: 4 **R-E-D Code:** 1-2-3 **State/Fed. Status:** /C2
Distribution: LAX, ORA, RIV, SBA, SBD, SCZ, SDG, SLO, SRO, VEN
Habitat: Chprl, CmWld, LCFrs / openings
Life Form: Perennial herb (bulbiferous)
Blooming: April-July
Notes: Localized below 5500 feet elevation. Threatened by development and horticultural collecting on the mainland, and by feral herbivores on SCZ and SRO islands. Includes *L. humboldtii* var. *bloomerianum* and *L. fairchildii*. See *Proceedings of the California Academy of Sciences* I 5:88 (1873) for original description.

Lilium kelloggii Purdy

"Kellogg's lily" Liliaceae
CNPS List: 4 **R-E-D Code:** 1-1-2 **State/Fed. Status:** CEQA?
Distribution: DNT, HUM, OR
Habitat: LCFrs, NCFrs / openings, roadsides
Life Form: Perennial herb (bulbiferous)
Blooming: June-July
Notes: Endangered in OR. See *Garden* 59:330 (1901) for original description.

Lilium maritimum Kell.

"coast lily" Liliaceae
CNPS List: 1B **R-E-D Code:** 2-3-3 **State/Fed. Status:** /C1
Distribution: MEN, MRN*, SFO?*, SMT*, SON
Quads: 448D*, 485C*, 520B, 520D, 537A, 537B, 537C, 537D, 552B, 553A, 568C, 568D, 569A, 569D, 585A, 585D
Habitat: BUFrs, CCFrs, CoPrr, CoScr, NCFrs
Life Form: Perennial herb (bulbiferous)
Blooming: May-July
Notes: Did this plant occur in SFO Co.? Populations along Highway 1 are routinely disturbed by road maintenance; also threatened by urbanization and horticultural collecting. Hybridizes with *L. pardalinum* ssp. *pardalinum*. See *Proceedings of the American Academy of Arts and Sciences* 6:140 (1875) for original description.

Lilium occidentale Purdy

"western lily" Liliaceae
CNPS List: 1B **R-E-D Code:** 3-3-2 **State/Fed. Status:** CE/PE
Distribution: DNT, HUM, OR
Quads: 654B, 655A, 672C, 672D, 723B, 740C
Habitat: BgFns, CBScr, CoPrr, CoScr, MshSw (freshwater), NCFrs (openings)
Life Form: Perennial herb (bulbiferous)
Blooming: June-July
Notes: Most CA occurrences under DFG management or voluntarily protected by landowners. Threatened by development, grazing, and horticultural collecting. State-listed as Endangered in OR. See *Erythea* 5:103-105 (1897) for original description.
Status Report: 1987

Lilium pardalinum Kell. ssp. *pitkinense* (Beane & Vollmer) M. Skinner

"Pitkin Marsh lily" Liliaceae
CNPS List: 1B **R-E-D Code:** 3-3-3 **State/Fed. Status:** CE/C1
Distribution: SON
Quads: 502A, 502D
Habitat: CmWld (mesic), MshSw (freshwater)
Life Form: Perennial herb (bulbiferous)
Blooming: June-July
Notes: Known from only two occurrences near Sebastopol. Most of marsh habitat has been destroyed; also threatened by horticultural collecting, grazing, and competition from other plants. State-listed as *L. pitkinense*.
Status Report: 1993

Lilium pardalinum ssp. *vollmeri*

Lilium pardalinum Kell. ssp. *vollmeri* (Eastw.) M. Skinner

"Vollmer's lily" Liliaceae
CNPS List: 4 **R-E-D Code:** 1-1-1 **State/Fed. Status:** /C3c
Distribution: DNT, HUM, SIS, OR
Habitat: BgFns, Medws (mesic)
Life Form: Perennial herb (bulbiferous)
Blooming: July-August
Notes: Forms hybrid swarms with ssp. *wigginsii* in western SIS Co. See *Leaflets of Western Botany* 5:120-122 (1948) for original description.
Status Report: 1980

Lilium pardalinum Kell. ssp. *wigginsii* (Beane & Vollmer) M. Skinner

"Wiggins's lily" Liliaceae
CNPS List: 4 **R-E-D Code:** 1-1-2 **State/Fed. Status:** /C3c
Distribution: DNT, SIS, OR
Habitat: BgFns, LCFrs, Medws / mesic
Life Form: Perennial herb (bulbiferous)
Blooming: June-August
Notes: Forms hybrid swarms with ssp. *vollmeri* in western SIS Co. On watch list in OR. See *Contributions from the Dudley Herbarium* 4(8):355 (1955) for original description.
Status Report: 1979

Lilium parryi Wats.

"lemon lily" Liliaceae
CNPS List: 1B **R-E-D Code:** 2-2-2 **State/Fed. Status:** /C2
Distribution: LAX, RIV, SBD, SDG, AZ
Quads: 20A*, 49C*, 49D, 65C, 66B, 83C, 84B, 84D, 105A, 105B, 105D, 106A, 107B, 108B, 109A, 131C, 131D, 134C, 134D, 135C, 135D
Habitat: LCFrs, Medws, RpFrs, UCFrs / mesic
Life Form: Perennial herb (bulbiferous)
Blooming: July-August
Notes: Most occurrences in LAX Co. are very small. Nearly extirpated from SDG Co., where known from only a few plants at Palomar Mtn. SP. Threatened by horticultural collecting, water diversion, and grazing. Includes *L. parryi* var. *kessleri*. Angeles NF has adopted species management guidelines.

Lilium parryi var. *kessleri*

See *Lilium parryi*

Lilium parryi var. *parryi*

See *Lilium parryi*

Lilium pitkinense

See *Lilium pardalinum* ssp. *pitkinense*

Lilium rubescens Wats.

"redwood lily" Liliaceae
CNPS List: 4 **R-E-D Code:** 1-2-3 **State/Fed. Status:** CEQA?
Distribution: DNT, HUM, LAK, MEN, NAP, SCR*, SHA, SIS, SON, TRI
Habitat: Chprl, LCFrs, UCFrs / sometimes serpentinite
Life Form: Perennial herb (bulbiferous)
Blooming: June-July
Notes: Increasingly rare in southern portion of range. Threatened by urbanization, horticultural collection, and grazing. See *Proceedings of the American Academy of Arts and Sciences* 14:256 (1879) for original description.

Lilium vollmeri

See *Lilium pardalinum* ssp. *vollmeri*

Lilium washingtonianum var. *minus*

Considered but rejected: A synonym of *L. washingtonianum* ssp. *washingtonianum*; a common taxon

Lilium washingtonianum Kell. ssp. *purpurascens* (Stearn) M. Skinner

"purple-flowered Washington lily" Liliaceae
CNPS List: 4 **R-E-D Code:** 1-1-1 **State/Fed. Status:** CEQA?
Distribution: DNT, HUM, SIS, TRI, OR
Habitat: Chprl, LCFrs, UCFrs / serpentinite
Life Form: Perennial herb (bulbiferous)
Blooming: June-August
Notes: Possibly threatened by logging and related activities, but often abundant in clear cuts.

Lilium washingtonianum var. *purpurascens*

See *Lilium washingtonianum* ssp. *purpurascens*

Lilium wigginsii

See *Lilium pardalinum* ssp. *wigginsii*

Limnanthes bakeri Howell

"Baker's meadowfoam" Limnanthaceae
CNPS List: 1B **R-E-D Code:** 3-3-3 **State/Fed. Status:** CR/C2
Distribution: MEN
Quads: 550B, 567A, 583B, 598C, 599A
Habitat: Medws, MshSw (freshwater), VFGrs (vernally mesic), VnPls
Life Form: Annual herb
Blooming: April-May
Notes: Known from fewer than twenty occurrences. Threatened by development, grazing, and road construction. See *Leaflets of Western Botany* 3(9):206 (1943) for original description.
Status Report: 1988

Limnanthes douglasii R. Br. ssp. *sulphurea* (C.T. Mason) C.T. Mason

"Point Reyes meadowfoam" Limnanthaceae
CNPS List: 1B **R-E-D Code:** 3-2-3 **State/Fed. Status:** CE/C2
Distribution: MRN, SMT
Quads: 409A, 485C
Habitat: CoPrr, Medws (mesic), MshSw (freshwater), VnPls
Life Form: Annual herb
Blooming: March-May
Notes: Known from approximately ten occurrences. Threatened by grazing, trampling, and non-native plants. See *University of California Publications in Botany* 25:477 (1952) for original description.
Status Report: 1988

Limnanthes douglasii var. *sulphurea*

See *Limnanthes douglasii* ssp. *sulphurea*

Limnanthes floccosa Howell ssp. *bellingeriana* (Peck) Arroyo

"Bellinger's meadowfoam" Limnanthaceae
CNPS List: 1B **R-E-D Code:** 3-2-2 **State/Fed. Status:** /C2
Distribution: SHA, OR
Quads: 646A, 679C, 679D
Habitat: CmWld, Medws / mesic
Life Form: Annual herb
Blooming: April-June
Notes: Known in CA from fewer than five occurrences near Ingot and Canyon Creek. Candidate for state listing in OR. See *L. floccosa* ssp. *floccosa* in *The Jepson Manual*. See *Proceedings of the Biological Society of Washington* 50:93-94 (1937) for original description.
Status Report: 1979

Limnanthes floccosa Howell ssp. *californica* Arroyo

"Butte County meadowfoam" Limnanthaceae
CNPS List: 1B **R-E-D Code:** 3-3-3 **State/Fed. Status:** CE/FE
Distribution: BUT
Quads: 576C, 577A, 593C, 593D
Habitat: VFGrs (mesic), VnPls
Life Form: Annual herb
Blooming: March-May
Notes: Known from fewer than twenty occurrences. Threatened by urbanization, road construction, grazing, non-native plants, and agriculture. See *Brittonia* 25:187 (1973) for original description.
Status Report: 1987

Limnanthes floccosa Howell ssp. *floccosa*

"woolly meadowfoam" Limnanthaceae
CNPS List: 2 **R-E-D Code:** 2-2-1 **State/Fed. Status:** CEQA
Distribution: BUT, LAK, SHA, SIS, TEH, TRI, OR
Quads: 534A, 593C, 593D, 595D, 610A, 628C, 628D, 629B, 631C, 646B, 646C, 650C, 679C, 679D, 680C
Habitat: CmWld, VFGrs / vernally mesic
Life Form: Annual herb
Blooming: March-June
Notes: Threatened by grazing. See *Brittonia* 25:177-193 (1973) for revised nomenclature.

Limnanthes floccosa var. *floccosa*

See *Limnanthes floccosa* ssp. *floccosa*

Limnanthes gracilis Howell ssp. *parishii* (Jeps.) Beauch.

"Parish's meadowfoam" Limnanthaceae
CNPS List: 1B **R-E-D Code:** 2-2-3 **State/Fed. Status:** CE/C2
Distribution: RIV, SDG
Quads: 19B, 19C, 20A, 33C, 33D, 49D, 69D
Habitat: Medws (vernally mesic), VnPls
Life Form: Annual herb
Blooming: April-June
Notes: Threatened by grazing and recreational development. See *Flora of California* 2(1):411 (1936) by W.L. Jepson for original description, and *University of California Publications in Botany* 25:490 (1952) for revised nomenclature.
Status Report: 1987

Limnanthes gracilis var. *parishii*

See *Limnanthes gracilis* ssp. *parishii*

Limnanthes vinculans Ornduff

"Sebastopol meadowfoam" Limnanthaceae
CNPS List: 1B **R-E-D Code:** 2-3-3 **State/Fed. Status:** CE/FE
Distribution: NAP, SON
Quads: 500A, 501B, 502A, 502B*, 502D
Habitat: Medws (mesic), VnPls
Life Form: Annual herb
Blooming: April-May
Notes: Only NAP Co. occurrence (500A) may be introduced; protected in part at Napa River ER (DFG). Threatened by urbanization, agriculture, and grazing. See *Brittonia* 21:11-14 (1969) for original description.
Status Report: 1987

Limosella subulata Ives

"delta mudwort" Scrophulariaceae
CNPS List: 2 **R-E-D Code:** 2-3-1 **State/Fed. Status:** CEQA
Distribution: CCA, MRN?, SAC, SJQ, SOL, OR, ++
Quads: 463A, 480B, 480C, 480D, 481C, 481D, 485C?, 498D
Habitat: MshSw
Life Form: Perennial herb (stoloniferous)
Blooming: May-August
Notes: Known in CA from several occurrences in the Delta; occurrence at Point Reyes NS (485C) needs verification. Also found on the Atlantic Coast, where threatened by habitat destruction. Native to CA, although *The Jepson Manual* states otherwise.

Linanthus acicularis Greene

"bristly linanthus" Polemoniaceae
CNPS List: 4 **R-E-D Code:** 1-2-3 **State/Fed. Status:** CEQA?
Distribution: ALA, CCA?, FRE, HUM, LAK, MEN, MRN, NAP, SMT, SON
Habitat: Chprl, CmWld, CoPrr
Life Form: Annual herb
Blooming: April-July
Notes: Historical occurrences need verification. Does plant occur in CCA Co.? See *Pittonia* 2:259 (1892) for original description.

Linanthus ambiguus (Rattan) Greene

"serpentine linanthus" Polemoniaceae
CNPS List: 4 **R-E-D Code:** 1-2-3 **State/Fed. Status:** CEQA?
Distribution: ALA, CCA, MER, SBT, SCL, SCR, SJQ, SMT, STA
Habitat: CmWld, CoScr, VFGrs / usually serpentinite
Life Form: Annual herb
Blooming: March-June
Notes: To be expected in other adjacent counties. See *Botanical Gazette* 11:339 (1886) for original description.

Linanthus arenicola (M.E. Jones) Jeps. & Bailey

"sand linanthus" Polemoniaceae
CNPS List: 2 **R-E-D Code:** 1-2-1 **State/Fed. Status:** /C3c
Distribution: INY, SBD, NV
Quads: 129C, 172D, 177D, 181C, 181D, 182A, 182D, 201C, 203B, 204D, 256B, 325B, 347C, 389C
Habitat: DeDns, JTWld, MDScr / sandy
Life Form: Annual herb
Blooming: March-April
Notes: Threatened by mining.

Linanthus bellus (Gray) Greene

"desert beauty" Polemoniaceae
CNPS List: 2 **R-E-D Code:** 2-1-1 **State/Fed. Status:** /C3c
Distribution: SDG, BA
Quads: 7B, 8A, 8D
Habitat: Chprl (sandy)
Life Form: Annual herb
Blooming: April-May

Linanthus concinnus Mlkn.

"San Gabriel linanthus" Polemoniaceae
CNPS List: 1B **R-E-D Code:** 3-2-3 **State/Fed. Status:** /C2
Distribution: LAX, SBD
Quads: 108B, 134B, 134C, 135A, 136D
Habitat: LCFrs, UCFrs / rocky
Life Form: Annual herb
Blooming: May-July
Notes: Infrequently collected in recent years; looked for but not seen in 1991. Field work needed. Threatened by recreational activities. See *University of California Publications in Botany* 2:53 (1904) for original description.

Linanthus floribundus (Gray) Mlkn. ssp. *hallii* (Jeps.) Mason

"Santa Rosa Mtns. linanthus" Polemoniaceae
CNPS List: 1B **R-E-D Code:** 3-1-3 **State/Fed. Status:** CEQA
Distribution: RIV, SDG
Quads: 46C, 47A, 47B, 48A, 65C
Habitat: SDScr
Life Form: Perennial herb
Blooming: May-July
Notes: Known only from the Santa Rosa Mtns.

Linanthus grandiflorus (Benth.) Greene

"large-flower linanthus" Polemoniaceae
CNPS List: 4 **R-E-D Code:** 1-2-3 **State/Fed. Status:** CEQA?
Distribution: ALA, KRN, MAD, MER, MNT, MRN, SBA*, SCL, SCR, SFO, SLO, SMT, SON
Habitat: CBScr, CCFrs, CmWld, CoDns, CoPrr, CoScr, VFGrs
Life Form: Annual herb
Blooming: April-July
Notes: Many historical occurrences extirpated by development; need information. Other taxa often misidentified as *L. grandiflorus.* See *Pittonia* 2:260 (1892) for original description.

Linanthus harknessii ssp. *condensatus*

Considered but rejected: A synonym of *L. harknessii*; a common taxon

Linanthus killipii Mason

"Baldwin Lake linanthus" Polemoniaceae
CNPS List: 1B **R-E-D Code:** 2-2-3 **State/Fed. Status:** /C2
Distribution: SBD
Quads: 104B, 105A, 126D, 130C, 131D
Habitat: Medws (alkaline), PbPln, PJWld, UCFrs
Life Form: Annual herb
Blooming: May-July
Notes: Threatened by urbanization and vehicles. See *Madroño* 9:251-252 (1948) for original description.
Status Report: 1979

Linanthus maculatus

See *Gilia maculata*

Linanthus nudatus

Considered but rejected: Too common

Linanthus nuttallii (Gray) Mlkn. ssp. *howellii* Nels. & Patterson

"Mt. Tedoc linanthus" Polemoniaceae
CNPS List: 1B **R-E-D Code:** 3-1-3 **State/Fed. Status:** /C2
Distribution: TEH
Quads: 613B, 631C
Habitat: LCFrs (serpentinite)
Life Form: Perennial herb
Blooming: May-August
Notes: Known from only four occurrences in the Mt. Tedoc region of the Klamath Mtns. See *Madroño* 32(2):102-105 (1985) for original description.

Linanthus oblanceolatus (Brand) Jeps.

"Sierra Nevada linanthus" Polemoniaceae
CNPS List: 4 **R-E-D Code:** 1-1-3 **State/Fed. Status:** CEQA?
Distribution: FRE, INY, TUL
Habitat: SCFrs
Life Form: Annual herb
Blooming: July-August

Linanthus orcuttii (Parry & Gray) Jeps.

"Orcutt's linanthus" Polemoniaceae
CNPS List: 1B **R-E-D Code:** 3-1-2 **State/Fed. Status:** /C2
Distribution: LAX*, RIV, SDG, BA
Quads: 19B, 20A, 20D, 48D, 49B, 49C, 49D, 110A*
Habitat: Chprl, LCFrs / openings
Life Form: Annual herb
Blooming: May-June
Notes: See *Madroño* 24(3):150-151 (1977) for taxonomic treatment.

Linanthus orcuttii ssp. *pacificus*

See *Linanthus orcuttii*

Linanthus pygmaeus (Brand) J.T. Howell ssp. *pygmaeus*

"pygmy linanthus" Polemoniaceae
CNPS List: 1B **R-E-D Code:** 3-2-2 **State/Fed. Status:** CEQA
Distribution: SCM, GU
Quads: SCMC, SCMN
Habitat: CoScr, VFGrs
Life Form: Annual herb
Blooming: April
Notes: Feral herbivores removed from SCM Isl., and vegetation recovering. See *Pflanzenreich* 4(250):134 (1907) for original description.

Linanthus rattanii (Gray) Greene

"Rattan's linanthus" Polemoniaceae
CNPS List: 4 **R-E-D Code:** 1-1-3 **State/Fed. Status:** CEQA?
Distribution: COL, GLE, LAK, MEN, TEH
Habitat: CmWld, LCFrs
Life Form: Annual herb
Blooming: May-July

Linanthus serrulatus Greene

"Madera linanthus" Polemoniaceae
CNPS List: 1B **R-E-D Code:** 2-2-3 **State/Fed. Status:** CEQA
Distribution: FRE, KRN, MAD, MPA, TUL
Quads: 261A, 309C, 332B, 354C, 376C, 380A, 396B, 398C, 399A, 419B, 420A
Habitat: CmWld, LCFrs
Life Form: Annual herb
Blooming: April-May

Listera caurina

Considered but rejected: Too common

Listera cordata (L.) R. Br.

"heart-leaved twayblade" Orchidaceae
CNPS List: 4 **R-E-D Code:** 1-2-1 **State/Fed. Status:** CEQA?
Distribution: DNT, HUM, SIS, NV, OR, WA, ++
Habitat: BgFns, LCFrs, NCFrs
Life Form: Perennial herb
Blooming: March-July
Notes: Easily overlooked. Threatened by grazing and logging. Includes *L. cordata* var. *nephrophylla.* See *Fremontia* 17(3):26-27 (1989) for species account.

Listera cordata var. *nephrophylla*

See *Listera cordata*

Lithophragma maximum Bacig.

"San Clemente Island woodland star" Saxifragaceae
CNPS List: 1B **R-E-D Code:** 3-3-3 **State/Fed. Status:** CE/C1
Distribution: SCM
Quads: SCMC, SCMS
Habitat: CBScr, CoScr
Life Form: Perennial herb
Blooming: April
Notes: Known from approximately five occurrences. Threatened by erosion. Feral herbivores removed from SCM Isl., and vegetation recovering. See *Aliso* 5:349-350 (1963) for original description.
Status Report: 1979

Lithophragma rupicola

Considered but rejected: A synonym of *L. tenellum*; a common taxon

Lithophragma tenellum

Considered but rejected: Too common

Lobelia cardinalis ssp. *cardinalis*

Considered but rejected: Not in CA; name misapplied to *L. cardinalis* var. *pseudosplendens*; a common taxon

Loeflingia pusilla

Considered but rejected: A synonym of *L. squarrosa* var. *squarrosa*; a common taxon

Loeflingia squarrosa ssp. *artemisiarum*

See *Loeflingia squarrosa* var. *artemisiarum*

Loeflingia squarrosa Nutt. var. *artemisiarum* (Barneby & Twisselm.) R. Dorn

"sagebrush loeflingia" Caryophyllaceae
CNPS List: 1B **R-E-D Code:** 2-2-2 **State/Fed. Status:** /C3c
Distribution: INY, KRN, LAX, RIV, OR, WY+
Quads: 185C, 185D, 186C, 392C, 393A
Habitat: DeDns, GBScr, SDScr / sandy
Life Form: Annual herb
Blooming: April-May
Notes: Need quads for RIV Co. See *Madroño* 20(8):406 (1970) for original description.

Lomatium ciliolatum

See *Lomatium ciliolatum* var. *hooveri*

Lomatium ciliolatum Jeps. var. *hooveri* Math. & Const.

"Hoover's lomatium" Apiaceae
CNPS List: 4 **R-E-D Code:** 1-1-3 **State/Fed. Status:** CEQA?
Distribution: COL, LAK, NAP, YOL
Habitat: Chprl, CmWld / serpentinite
Life Form: Perennial herb
Blooming: May-July

Lomatium congdonii Coult. & Rose

"Congdon's lomatium" Apiaceae
CNPS List: 1B **R-E-D Code:** 2-2-3 **State/Fed. Status:** /C2
Distribution: MPA*, TUO
Quads: 419B*, 458B, 458C, 459A, 459D
Habitat: Chprl, CmWld / serpentinite
Life Form: Perennial herb
Blooming: April-June
Notes: Known from fewer than twenty occurrences. Threatened by vehicles and mining. Protected at Red Hills ACEC (BLM), TUO Co. See *Contributions from the U.S. National Herbarium* 7:232 (1900) for original description.
Status Report: 1978

Lomatium engelmannii Math.

"Engelmann's lomatium" Apiaceae
CNPS List: 4 **R-E-D Code:** 1-1-2 **State/Fed. Status:** CEQA?
Distribution: MEN, SIS, TRI, OR
Habitat: LCFrs, UCFrs / serpentinite
Life Form: Perennial herb
Blooming: June-August
Notes: Endangered in OR.

Lomatium foeniculaceum (Nutt.) Coult. & Rose ssp. *inyoense* (Math. & Const.) Theobald

"Inyo lomatium" Apiaceae
CNPS List: 4 **R-E-D Code:** 1-1-1 **State/Fed. Status:** /C3c
Distribution: INY, ID, NV
Habitat: SCFrs (carbonate)
Life Form: Perennial herb
Blooming: June-July
Notes: Common along crest of the Inyo Mtns. May be a form induced by high-altitude conditions.

Lomatium hendersonii (Coult. & Rose) Coult. & Rose

"Henderson's lomatium" Apiaceae
CNPS List: 2 **R-E-D Code:** 3-1-1 **State/Fed. Status:** CEQA
Distribution: LAS, MOD, ID, NV, OR
Quads: 622C, 707C
Habitat: GBScr, PJWld / rocky, clay
Life Form: Perennial herb
Blooming: March-June
Notes: Easily confused with *L. canbyi*. Inadvertently omitted from *The Jepson Manual.* See *Botanical Gazette* 13:210 (1888) for original description, and *Systematic Botany Monographs* 4:1-55 (1984) for taxonomic treatment.

Lomatium howellii (Wats.) Jeps.

"Howell's lomatium" Apiaceae
CNPS List: 4 **R-E-D Code:** 1-1-1 **State/Fed. Status:** /C3c
Distribution: DNT, SIS, OR
Habitat: Chprl, LCFrs / serpentinite
Life Form: Perennial herb
Blooming: April-June
Notes: See *Proceedings of the American Academy of Arts and Sciences* 20:369 (1885) for original description.
Status Report: 1978

Lomatium insulare (Eastw.) Munz

"San Nicolas Island lomatium" Apiaceae
CNPS List: 1B **R-E-D Code:** 2-2-2 **State/Fed. Status:** /C2
Distribution: SCM*, SNI, GU
Quads: SNIC
Habitat: CBScr (sandy)
Life Form: Perennial herb
Blooming: February-April
Notes: Presumed extinct on SCM Isl.; last seen there in 1918 (need quad). Threatened by military development and road construction, and possibly by non-native plants.

Lomatium martindalei (Coult. & Rose) Coult. & Rose

"Coast Range lomatium" Apiaceae
CNPS List: 2 **R-E-D Code:** 2-1-1 **State/Fed. Status:** CEQA
Distribution: DNT, SIS, OR, WA+
Quads: 704A, 722D, 739A, 739C, 739D
Habitat: CBScr, LCFrs, Medws
Life Form: Perennial herb
Blooming: May-June

Lomatium parvifolium (H. & A.) Jeps.

"small-leaved lomatium" Apiaceae
CNPS List: 4 **R-E-D Code:** 1-2-3 **State/Fed. Status:** CEQA?
Distribution: MNT, SCR, SLO
Habitat: CCFrs, Chprl / serpentinite
Life Form: Perennial herb
Blooming: February-June
Notes: Rare in SCR Co.

Lomatium parvifolium var. *parvifolium*

See *Lomatium parvifolium*

Lomatium peckianum Math. & Const.

"Peck's lomatium" Apiaceae
CNPS List: 2 **R-E-D Code:** 2-2-1 **State/Fed. Status:** /C3c
Distribution: SIS, OR
Quads: 699A, 699B, 717A, 717B, 733B, 733D, 734A, 734B, 734C, 734D, 735A, 735D
Habitat: CmWld, LCFrs, PJWld / volcanic
Life Form: Perennial herb
Blooming: April-June
Notes: Threatened by development. See *Bulletin of the Torrey Botanical Club* 69:155 (1942) for original description.
Status Report: 1978

Lomatium ravenii Math. & Const.

"Raven's lomatium" Apiaceae
CNPS List: 4 **R-E-D Code:** 1-2-1 **State/Fed. Status:** /C3c
Distribution: LAS, ID, NV, OR, UT
Habitat: GBScr (adobe, alkaline)
Life Form: Perennial herb
Blooming: April-June
Notes: Occasionally threatened by grazing. Endangered in OR. See *Bulletin of the Torrey Botanical Club* 86(6):379 (1959) for original description.
Status Report: 1977

Lomatium repostum (Jeps.) Math.

"Napa lomatium" Apiaceae
CNPS List: 4 **R-E-D Code:** 1-1-3 **State/Fed. Status:** CEQA?
Distribution: LAK, NAP, SOL, SON
Habitat: Chprl, CmWld / serpentinite
Life Form: Perennial herb
Blooming: April-May

Lomatium rigidum (M.E. Jones) Jeps.

"stiff lomatium" Apiaceae
CNPS List: 4 **R-E-D Code:** 1-1-3 **State/Fed. Status:** /C3c
Distribution: INY
Habitat: GBScr, PJWld / rocky, near streams
Life Form: Perennial herb
Blooming: April-May
Notes: Common and widespread along east side of the Sierra in INY Co.

Lomatium shevockii Hartman & Const.

"Owens Peak lomatium" Apiaceae
CNPS List: 1B **R-E-D Code:** 3-1-3 **State/Fed. Status:** /C1
Distribution: KRN
Quads: 258B
Habitat: LCFrs, UCFrs / rocky
Life Form: Perennial herb
Blooming: April-May
Notes: Known from only two occurrences in the Owens Pk. and Mt. Jenkins area. See *Madroño* 35(2):121-125 (1988) for original description.

Lomatium stebbinsii Schlessman & Const.

"Stebbins's lomatium" Apiaceae
CNPS List: 1B **R-E-D Code:** 3-2-3 **State/Fed. Status:** /C2
Distribution: AMA, CAL, TUO
Quads: 474A, 474C, 492A, 492B
Habitat: Chprl, LCFrs / gravelly, volcanic clay
Life Form: Perennial herb
Blooming: March-April
Notes: Known from fewer than twenty occurrences. Threatened by vehicles. See *Madroño* 26(1):41 (1979) for original description.

Lomatium torreyi

Considered but rejected: Too common

Lomatium tracyi Math. & Const.

"Tracy's lomatium" Apiaceae
CNPS List: 4 **R-E-D Code:** 1-1-2 **State/Fed. Status:** CEQA?
Distribution: HUM, SHA, SIS, TEH, TRI, OR
Habitat: LCFrs, UCFrs / serpentinite
Life Form: Perennial herb
Blooming: May-June
Notes: Endangered in OR.

Lonicera cauriana

Considered but rejected: Too common

Lophochlaena californica var. *davyi*

See *Pleuropogon davyi*

Lophochlaena refracta var. *hooverianus*

See *Pleuropogon hooverianus*

Lophochlaena refracta var. *refracta*

See *Pleuropogon refractus*

Lotus argophyllus ssp. *adsurgens*

See *Lotus argophyllus* var. *adsurgens*

Lotus argophyllus (Gray) Greene var. *adsurgens* (Dunkle) Isley

"San Clemente Island bird's-foot trefoil" Fabaceae
CNPS List: 1B **R-E-D Code:** 3-3-3 **State/Fed. Status:** CE/C1
Distribution: SCM
Quads: SCMC, SCMS
Habitat: CBScr, CoScr / rocky
Life Form: Perennial herb
Blooming: April-June
Notes: Known from fewer than ten occurrences. Threatened by Navy activities. Feral herbivores removed from SCM Isl., and vegetation recovering. State-listed as *L. argophyllus* ssp. *adsurgens.* See *Bulletin of the Southern California Academy of Sciences* 39:175 (1940) for original description.
Status Report: 1987

Lotus argophyllus var. *argenteus*

Considered but rejected: Too common

Lotus argophyllus ssp. *niveus*

See *Lotus argophyllus* var. *niveus*

Lotus argophyllus (Gray) Greene var. *niveus* (Greene) Ottley

"Santa Cruz Island bird's-foot trefoil" Fabaceae
CNPS List: 1B **R-E-D Code:** 2-2-3 **State/Fed. Status:** CE/C2
Distribution: SCZ
Quads: SCZA, SCZB, SCZC, SCZD
Habitat: Chprl, CoScr / rocky
Life Form: Perennial herb
Blooming: April-July
Notes: Threatened by feral herbivores. State-listed as *L. argophyllus* ssp. *niveus.*
Status Report: 1979

Lotus argyraeus (Greene) Greene var. *multicaulis* (Ottley) Isely

"scrub lotus" Fabaceae
CNPS List: 1B **R-E-D Code:** 3-1-3 **State/Fed. Status:** CEQA
Distribution: SBD
Quads: 200A, 224C, 225D
Habitat: PJWld (granitic)
Life Form: Perennial herb
Blooming: April-June
Notes: Known only from the New York Mtns. See *University of California Publications in Botany* 10:211 (1923) for original description, and *Memoirs of the New York Botanical Garden* 25:128-206 (1981) for taxonomic treatment.

Lotus argyraeus (Greene) Greene var. *notitius* Isely

"Providence Mtns. lotus" Fabaceae
CNPS List: 1B **R-E-D Code:** 2-1-3 **State/Fed. Status:** CEQA
Distribution: SBD
Quads: 176A
Habitat: PJWld
Life Form: Perennial herb
Blooming: May-August
Notes: Known only from the Providence Mtns. See *Brittonia* 30(4):466-472 (1978) for original description, and *Memoirs of the New York Botanical Garden* 25:128-206 (1981) for taxonomic treatment.

Lotus crassifolius (Benth.) Greene var. *otayensis* Moran

"Otay Mtn. lotus" Fabaceae
CNPS List: 1B **R-E-D Code:** 3-3-3 **State/Fed. Status:** /C2
Distribution: SDG
Quads: 10D
Habitat: Chprl (often in disturbed areas)
Life Form: Perennial herb
Blooming: May-August
Notes: Known only from Otay Mtn. Threatened by road maintenance. See *Brittonia* 30(4):466-467 (1978) for original description.

Lotus cupreus

See *Lotus oblongifolius* var. *cupreus*

Lotus dendroideus (Greene) Greene var. *dendroideus*

"island broom" Fabaceae
CNPS List: 4 **R-E-D Code:** 1-2-3 **State/Fed. Status:** CEQA?
Distribution: ANA, SCT, SCZ, SRO
Habitat: CBScr, CCFrs, CoScr, CmWld
Life Form: Shrub
Blooming: February-August
Notes: Threatened by feral herbivores. See *Brittonia* 30:467 (1978) for revised nomenclature.

Lotus dendroideus (Greene) Greene var. *traskiae* (Noddin) Isely

"San Clemente Island lotus" Fabaceae
CNPS List: 1B **R-E-D Code:** 3-3-3 **State/Fed. Status:** CE/FE
Distribution: SCM
Quads: SCMC, SCMN, SCMS
Habitat: CBScr, CoScr, VFGrs
Life Form: Shrub
Blooming: February-August
Notes: Known from fewer than twenty occurrences. Threatened by Navy activities. Feral herbivores removed from SCM Isl., and vegetation recovering. Occasionally hybridizes with *L. argophyllus* var. *argenteus.* USFWS uses the name *L. dendroideus* ssp. *traskiae.* See *Brittonia* 30:466-472 (1978) for revised nomenclature.
Status Report: 1979

Lotus dendroideus (Greene) Greene var. *veatchii* (Greene) Isely

"San Miguel Island deerweed" Fabaceae
CNPS List: 4 **R-E-D Code:** 1-1-2 **State/Fed. Status:** CEQA?
Distribution: SMI, BA
Habitat: CBScr, CoScr
Life Form: Shrub
Blooming: March-June
Notes: See *Bulletin of the California Academy of Sciences* 1:83 (1886) for original description, and *Brittonia* 30:467 (1978) for revised nomenclature.

Lotus haydonii (Orcutt) Greene

"pygmy lotus" Fabaceae
CNPS List: 4 **R-E-D Code:** 1-1-2 **State/Fed. Status:** CEQA?
Distribution: IMP, SDG, BA
Habitat: PJWld, SDScr / rocky
Life Form: Perennial herb
Blooming: January-June
Notes: See *West American Scientist* 6:63 (1889) for original description, and *Memoirs of the New York Botanical Garden* 25:128-206 (1981) for taxonomic treatment.

Lotus neo-incanus

Considered but rejected: Too common; a synonym of *L. incanus*

Lotus nuttallianus Greene

"Nuttall's lotus" Fabaceae
CNPS List: 1B **R-E-D Code:** 3-3-2 **State/Fed. Status:** /C2
Distribution: SDG, BA
Quads: 11B, 11D, 22B, 22C, 36B, 36D
Habitat: CoDns, CoScr
Life Form: Annual herb
Blooming: March-June
Notes: Declining precipitously; now known in CA from fewer than ten occurrences. Threatened by development, non-native plants, and land management activities, particularly by U.S. Navy at Silver Strand and Imperial Beach.

Lotus oblongifolius (Benth.) Greene var. *cupreus* (Greene) Ottley

"copper-flowered bird's-foot trefoil" Fabaceae
CNPS List: 4 **R-E-D Code:** 1-1-3 **State/Fed. Status:** CEQA?
Distribution: TUL
Habitat: Medws (edges), UCFrs / mesic
Life Form: Perennial herb
Blooming: July-August

Lotus rubriflorus H.K. Sharsm.

"red-flowered lotus" Fabaceae
CNPS List: 1B **R-E-D Code:** 3-3-3 **State/Fed. Status:** /C2
Distribution: COL, STA, TEH
Quads: 425A, 425B, 564D, 628D
Habitat: CmWld, VFGrs
Life Form: Annual herb
Blooming: April-June
Notes: Known from only four disjunct occurrences. Threatened by development, and probably by grazing and non-native plants. Field work needed. See *Madroño* 2:56 (1941) for original description.
Status Report: 1988

Lotus scoparius var. *dendroideus*

See *Lotus dendroideus* var. *dendroideus*

Lotus scoparius ssp. *traskiae*

See *Lotus dendroideus* var. *traskiae*

Lotus yollabolliensis Munz

"Yolla Bolly Mtns. bird's-foot trefoil" Fabaceae
CNPS List: 4 **R-E-D Code:** 1-1-3 **State/Fed. Status:** CEQA?
Distribution: HUM, TRI
Habitat: UCFrs
Life Form: Perennial herb
Blooming: June-August

Ludwigia repens var. *stipitata*

Considered but rejected: Too common; a synonym of *L. repens*

Lupinus abramsii

See *Lupinus albifrons* var. *abramsii*

Lupinus adsurgens var. *undulatus*

Considered but rejected: Too common

Lupinus albifrons Benth. var. *abramsii* (C.P. Smith) Hoov.

"Abrams's lupine" Fabaceae
CNPS List: 3 **R-E-D Code:** 3-2-3 **State/Fed. Status:** CEQA?
Distribution: MNT
Quads: 320B
Habitat: LCFrs
Life Form: Perennial herb
Blooming: April-June
Notes: Move to List 1B? Possibly more widespread, but only specimens from 320B match the type; plants from SLO Co. are probably var. *albifrons.*

Lupinus antoninus Eastw.

"Anthony Peak lupine" Fabaceae
CNPS List: 1B **R-E-D Code:** 3-1-3 **State/Fed. Status:** /C2
Distribution: LAK, MEN, TEH, TRI
Quads: 549A, 581C, 597C, 613C
Habitat: LCFrs, UCFrs
Life Form: Perennial herb
Blooming: May-July
Notes: Known from only four occurrences. See *Leaflets of Western Botany* 3(9):202 (1943) for original description.

Lupinus cervinus Kell.

"Santa Lucia lupine" Fabaceae
CNPS List: 4 **R-E-D Code:** 1-1-3 **State/Fed. Status:** /C3c
Distribution: MNT, SLO
Habitat: LCFrs
Life Form: Perennial herb
Blooming: May-June

Lupinus citrinus

See *Lupinus citrinus* var. *citrinus*

Lupinus citrinus Kell. var. *citrinus*

"orange lupine" Fabaceae
CNPS List: 1B **R-E-D Code:** 1-2-3 **State/Fed. Status:** /C2
Distribution: FRE, MAD
Quads: 377A, 397A, 397B, 397C, 397D, 398B, 418D
Habitat: Chprl, CmWld, LCFrs / granitic
Life Form: Annual herb
Blooming: April-July
Notes: Threatened by development, road widening, vehicles, grazing, and logging. See *Proceedings of the California Academy of Sciences* I 7:93 (1877) for original description, and *Madroño* 28(3):184 for first MAD Co. record.
Status Report: 1979

Lupinus citrinus Kell. var. *deflexus* (Congd.) Jeps.

"Mariposa lupine" Fabaceae
CNPS List: 1B **R-E-D Code:** 3-2-3 **State/Fed. Status:** CT/C1
Distribution: MPA
Quads: 419B
Habitat: Chprl, CmWld / granitic
Life Form: Annual herb
Blooming: April-May
Notes: Known from only four occurrences near Mariposa Creek. Threatened by development and grazing. State-listed as *L. deflexus*. See *Muhlenbergia* 1:38 (1904) for original description.
Status Report: 1990

Lupinus constancei T.W. Nelson & J.P. Nelson

"The Lassics lupine" Fabaceae
CNPS List: 1B **R-E-D Code:** 3-2-3 **State/Fed. Status:** /C2
Distribution: HUM, TRI
Quads: 634D
Habitat: LCFrs (serpentinite)
Life Form: Perennial herb
Blooming: July
Notes: Known from approximately five occurrences. See *Brittonia* 35(2):180-183 (1983) for original description.

Lupinus covillei

Considered but rejected: Too common

Lupinus croceus Eastw. var. *pilosellus* (Eastw.) Munz

"saffron-flowered lupine" Fabaceae
CNPS List: 4 **R-E-D Code:** 1-1-3 **State/Fed. Status:** CEQA?
Distribution: SHA, SIS, TRI
Habitat: LCFrs
Life Form: Perennial herb
Blooming: May-August
Notes: See *L. croceus* in *The Jepson Manual.*

Lupinus culbertsonii ssp. *culbertsonii*

See *Lupinus lepidus* var. *culbertsonii*

Lupinus culbertsonii ssp. *hypolasius*

Considered but rejected: A synonym of *L. lepidus* var. *ramosus*; a common taxon

Lupinus dalesiae Eastw.

"Quincy lupine" Fabaceae
CNPS List: 1B **R-E-D Code:** 2-2-3 **State/Fed. Status:** /C3c
Distribution: BUT, PLU, SIE, YUB
Quads: 559A, 573A, 573B, 574A, 574B, 588B, 588C, 589A, 589B, 589C, 589D, 590A, 590B, 590D, 605C, 606A, 606B, 606C, 606D
Habitat: LCFrs, UCFrs / often in disturbed areas
Life Form: Perennial herb
Blooming: May-July
Notes: Possibly threatened by logging and roadside maintenance. See *Leaflets of Western Botany* 2:266 (1940) for original description.

Lupinus dedeckerae

See *Lupinus padre-crowleyi*

Lupinus deflexus

See *Lupinus citrinus* var. *deflexus*

Lupinus densiflorus var. *glareosus*

Considered but rejected: A synonym of *L. microcarpus* var. *horizontalis*; a common taxon

Lupinus duranii Eastw.

"Mono Lake lupine" Fabaceae
CNPS List: 1B **R-E-D Code:** 2-2-3 **State/Fed. Status:** /C2
Distribution: MNO
Quads: 434B, 435A, 452A, 452B, 452C, 453A, 453D, 470C
Habitat: GBScr, SCFrs, UCFrs / pumice flats
Life Form: Perennial herb
Blooming: May-July
Notes: Threatened by vehicles.
Status Report: 1980

Lupinus elatus Jtn.

"silky lupine" Fabaceae
CNPS List: 4 **R-E-D Code:** 1-1-3 **State/Fed. Status:** CEQA?
Distribution: LAX, VEN
Habitat: LCFrs, UCFrs
Life Form: Perennial herb
Blooming: June-August
Notes: Similar to *L. adsurgens* and *L. andersonii*; needs taxonomic study.

Lupinus excubitus M.E. Jones var. *johnstonii* C.P. Smith

"interior bush lupine" Fabaceae
CNPS List: 4 **R-E-D Code:** 1-1-3 **State/Fed. Status:** CEQA?
Distribution: KRN, LAX
Habitat: LCFrs (decomposed granitic)
Life Form: Shrub
Blooming: May-July

Lupinus excubitus M.E. Jones var. *medius* (Jeps.) Munz

"Mountain Springs bush lupine" Fabaceae
CNPS List: 1B **R-E-D Code:** 2-1-3 **State/Fed. Status:** /C2
Distribution: IMP, SDG
Quads: 7A, 7B, 7C, 18C, 19A, 19B, 19D
Habitat: PJWld, SDScr
Life Form: Shrub
Blooming: March-April
Notes: Possibly threatened by vehicles.

Lupinus eximius Davy

"San Mateo tree lupine" Fabaceae
CNPS List: 3 **R-E-D Code:** 2-2-3 **State/Fed. Status:** /C2
Distribution: SMT, SON?
Quads: 448C, 448D, 503D?
Habitat: Chprl, CoScr
Life Form: Shrub (evergreen)
Blooming: April-July
Notes: Move to List 1B? SON Co. plants need taxonomic confirmation. Identification is very difficult; study needed. See *L. arboreus* in *The Jepson Manual*. USFWS uses the name *L. arboreus* var. *eximius*. See *Erythea* 3:116 (1895) for original description.

Lupinus gracilentus Greene

"slender lupine" Fabaceae
CNPS List: 4 **R-E-D Code:** 1-1-3 **State/Fed. Status:** CEQA?
Distribution: INY, MPA
Habitat: SCFrs
Life Form: Perennial herb
Blooming: July-August

Lupinus guadalupensis Greene

"Guadalupe Island lupine" Fabaceae
CNPS List: 1B **R-E-D Code:** 3-2-2 **State/Fed. Status:** /C2
Distribution: SCM, GU
Quads: SCMC, SCMN, SCMS
Habitat: CoScr (sandy or gravelly)
Life Form: Annual herb
Blooming: February-April
Notes: Threatened by Navy activities. Feral herbivores removed from SCM Isl., and vegetation recovering. See *Bulletin of the California Academy of Sciences* 1:184 (1885) for original description, and *Aliso* 5:327 (1963) for distributional information.
Status Report: 1979

Lupinus holmgrenanus C.P. Smith

"Holmgren's lupine" Fabaceae
CNPS List: 2 **R-E-D Code:** 2-1-1 **State/Fed. Status:** /C3b
Distribution: INY, NV
Quads: 410C
Habitat: PJWld (volcanic)
Life Form: Perennial herb
Blooming: May-June

Lupinus horizontalis

Considered but rejected: Too common; a synonym of *L. microcarpus* var. *horizontalis*

Lupinus humboldtii

See *Lupinus constancei*

Lupinus hypolasius

Considered but rejected: A synonym of *L. lepidus* var. *ramosus*; a common taxon

Lupinus inyoensis

Considered but rejected: A synonym of *L. argenteus* var. *heteranthus*; a common taxon

Lupinus lapidicola Heller

"Mt. Eddy lupine" Fabaceae
CNPS List: 4 **R-E-D Code:** 1-1-3 **State/Fed. Status:** CEQA?
Distribution: DNT, HUM, SIS, TRI
Habitat: LCFrs, SCFrs, UCFrs / granitic or serpentinite
Life Form: Perennial herb
Blooming: July
Notes: See *Bulletin of the Torrey Botanical Club* 51:306 (1924) for original description.

Lupinus latifolius var. *dudleyi*

Considered but rejected: Taxonomic problem, but needs reevaluation

Lupinus lepidus Douglas var. *culbertsonii* (Greene) C.P. Smith

"Hockett Meadows lupine" Fabaceae
CNPS List: 1B **R-E-D Code:** 3-1-3 **State/Fed. Status:** /C2
Distribution: FRE, MNO, TUL
Quads: 331B, 331C, 331D, 352C, 434D
Habitat: Medws, UCFrs (mesic)
Life Form: Perennial herb
Blooming: July-August
Notes: Known from fewer than ten occurrences. Need quads for FRE Co. See *Madroño* 22(4):169-177 (1973) for taxonomic treatment.

Lupinus lobbii

Considered but rejected: Too common; a synonym of *L. lepidus* var. *lobbii*

Lupinus ludovicianus Greene

"San Luis Obispo County lupine" Fabaceae
CNPS List: 1B **R-E-D Code:** 3-2-3 **State/Fed. Status:** /C2
Distribution: SLO
Quads: 195B, 220A, 220B, 221A*, 221B, 245C, 245D, 246D, 294C
Habitat: Chprl, CmWld / carbonate
Life Form: Perennial herb
Blooming: April-July
Notes: Known from fewer than twenty occurrences. Threatened by grazing and trampling. Official flower of SLO Co. See *Bulletin of the California Academy of Sciences* 1:184 (1885) for original description.
Status Report: 1977

Lupinus magnificus M.E. Jones var. *glarecola* M.E. Jones

"Coso Mtns. lupine" Fabaceae
CNPS List: 4 **R-E-D Code:** 1-1-3 **State/Fed. Status:** CEQA?
Distribution: INY
Habitat: GBScr (granitic)
Life Form: Perennial herb
Blooming: April-June
Notes: Appears after fires. See *L. magnificus* in *The Jepson Manual*.

Lupinus magnificus M.E. Jones var. *hesperius* (Heller) C.P. Smith

"McGee Meadows lupine" Fabaceae
CNPS List: 4 **R-E-D Code:** 1-1-3 **State/Fed. Status:** CEQA?
Distribution: INY
Habitat: GBScr, UCFrs / sandy
Life Form: Perennial herb
Blooming: April-June
Notes: See *L. magnificus* in *The Jepson Manual.* See *Muhlenbergia* 2:212 (1906) for original description.

Lupinus magnificus M.E. Jones var. *magnificus*

"Panamint Mtns. lupine" Fabaceae
CNPS List: 1B **R-E-D Code:** 3-1-3 **State/Fed. Status:** /C2
Distribution: INY
Quads: 302A, 302C, 302D, 348C
Habitat: GBScr, MDScr, UCFrs
Life Form: Perennial herb
Blooming: April-June
Notes: Known from approximately ten, mostly historical, occurrences. Field surveys needed. See *Contributions to Western Botany* 8:26 (1898) for original description.

Lupinus milo-bakeri C.P. Smith

"Milo Baker's lupine" Fabaceae
CNPS List: 1B **R-E-D Code:** 2-3-3 **State/Fed. Status:** CT/C2
Distribution: COL, MEN
Quads: 547B, 547C, 583C, 598C
Habitat: CmWld (often along roadsides), VFGrs
Life Form: Annual herb
Blooming: June-September
Notes: Known from fewer than twenty occurrences. Threatened by urbanization, road widening, and herbicide application; many populations seriously affected by CalTrans spraying in 1984. Needs taxonomic study; see *L. luteolus* in *The Jepson Manual.* See *Four Seasons* 1(3):8-9 (1965) for differentiation from *L. luteolus.*
Status Report: 1985

Lupinus montigenus

Considered but rejected: A synonym of *L. argenteus* var. *montigenus*; a common taxon

Lupinus nipomensis Eastw.

"Nipomo Mesa lupine" Fabaceae
CNPS List: 1B **R-E-D Code:** 3-3-3 **State/Fed. Status:** CE/C1
Distribution: SLO
Quads: 221D
Habitat: CoDns
Life Form: Annual herb
Blooming: March-May
Notes: Known from only five occurrences on Nipomo Mesa, each with a small number of plants. Threatened by coastal development. See *Leaflets of Western Botany* 2(10):186-188 (1939) for original description.
Status Report: 1988

Lupinus padre-crowleyi C.P. Smith

"Father Crowley's lupine" Fabaceae
CNPS List: 1B **R-E-D Code:** 3-2-3 **State/Fed. Status:** CR/C2
Distribution: INY, MNO, TUL
Quads: 306B, 329A, 352A, 393B, 393D, 414B
Habitat: GBScr, RpFrs, RpScr, UCFrs / decomposed granitic
Life Form: Perennial herb
Blooming: July-August
Notes: Known from fewer than twenty occurrences. This name predates *L. dedeckerae* (both described independently from the same type locality).
Status Report: 1987

Lupinus peirsonii Mason

"Peirson's lupine" Fabaceae
CNPS List: 4 **R-E-D Code:** 1-1-3 **State/Fed. Status:** /C3c
Distribution: LAX
Habitat: JTWld, PJWld, UCFrs / gravelly
Life Form: Perennial herb
Blooming: April-May

Lupinus pratensis var. *eriostachyus*

Considered but rejected: Too common

Lupinus sericatus Kell.

"Cobb Mtn. lupine" Fabaceae
CNPS List: 1B **R-E-D Code:** 2-2-3 **State/Fed. Status:** /C3c
Distribution: COL, LAK, NAP, SON
Quads: 500B, 516B, 516C, 517A, 517B, 517D, 533A, 533C, 534A, 534D, 535C, 547C
Habitat: Chprl, CmWld, LCFrs
Life Form: Perennial herb
Blooming: March-June
Notes: Threatened by geothermal development, logging, and road widening; will colonize disturbed sites. See *Fremontia* 13(3):21-22 (1985) for account of reestablishment project in The Geysers geothermal area.

Lupinus sericeus

Considered but rejected: Taxonomic problem

Lupinus spectabilis Hoov.

"shaggyhair lupine" Fabaceae
CNPS List: 1B **R-E-D Code:** 2-2-3 **State/Fed. Status:** /C2
Distribution: MPA, TUO
Quads: 439B, 439C, 439D, 457C, 458B, 458D
Habitat: Chprl, CmWld / serpentinite
Life Form: Annual herb
Blooming: April-May
Notes: Threatened by mining, grazing, and road construction. See *Leaflets of Western Botany* 2(8):131 (1938) for original description.
Status Report: 1980

Lupinus sublanatus

Considered but rejected: A synonym of *L. argenteus* var. *argenteus*; a common taxon

Lupinus tidestromii Greene

"Tidestrom's lupine" Fabaceae
CNPS List: 1B **R-E-D Code:** 3-3-3 **State/Fed. Status:** CE/FE
Distribution: MNT, MRN, SON
Quads: 366A, 366C, 485C, 503B, 503D
Habitat: CoDns
Life Form: Perennial herb (rhizomatous)
Blooming: May-June
Notes: Seriously threatened by coastal development, trampling, and non-native plants. Includes *L. tidestromii* var. *layneae.* Only MNT Co. plants are state-listed Endangered as var. *tidestromii.* See *Erythea* 3:17 (1895) for original description.
Status Report: 1985

Lupinus tidestromii var. *layneae*

See *Lupinus tidestromii*

Lupinus tidestromii var. *tidestromii*

See *Lupinus tidestromii*

Lupinus tracyi Eastw.

"Tracy's lupine" Fabaceae
CNPS List: 4 **R-E-D Code:** 1-1-2 **State/Fed. Status:** /C3c
Distribution: DNT, HUM, SIS, TRI, OR
Habitat: UCFrs
Life Form: Perennial herb
Blooming: July
Notes: Endangered in OR. See *Leaflets of Western Botany* 2(15):268 (1940) for original description.
Status Report: 1977

Lycium brevipes Benth var. *hassei* (Greene) Hitchc.

"Santa Catalina Island desert-thorn" Solanaceae
CNPS List: 1B **R-E-D Code:** 3-3-3 **State/Fed. Status:** /C3a
Distribution: LAX, SCM*, SCT*
Quads: 73A, SCMN*, SCTE*
Habitat: CBScr, CoScr
Life Form: Shrub (deciduous)
Blooming: June
Notes: Rediscovered on the Palos Verdes Peninsula (LAX Co.) by A. Sanders in 1976. Only a few plants are known; the occurrence was diminished by grading in the 1930's. Only two wild specimens were known from the Channel Islands; one died in 1908, the other in 1936. See *Pittonia* 1:222 (1888) for original description, and *Annals of the Missouri Botanical Garden* 19:256-257 (1932) for taxonomic treatment.
Status Report: 1979

Lycium hassei

See *Lycium brevipes* var. *hassei*

Lycium parishii Gray

"Parish's desert-thorn" Solanaceae
CNPS List: 2 **R-E-D Code:** 2-1-1 **State/Fed. Status:** CEQA
Distribution: IMP, RIV, SBD*, SDG, AZ, SO
Quads: 18C, 80D, 107A*
Habitat: CoScr, SDScr
Life Form: Shrub
Blooming: March-April

Lycium verrucosum Eastw.

"San Nicolas Island desert-thorn" Solanaceae
CNPS List: 1A **Last Seen:** 1901 **State/Fed. Status:** /C3a
Distribution: SNI*
Quads: SNIC*
Habitat: CoScr
Life Form: Shrub
Blooming: April
Notes: Known only from the type collection. Perhaps a form of *L. brevipes.* See *Proceedings of the California Academy of Sciences* III 1:111 (1898) for original description, and *Annals of the Missouri Botanical Garden* 19:257-258 (1932) for taxonomic treatment.
Status Report: 1979

Lycopodiella inundata (L.) Holub

"bog club-moss" Lycopodiaceae
CNPS List: 2 **R-E-D Code:** 3-2-1 **State/Fed. Status:** CEQA
Distribution: HUM, NEV, ID, OR, WA, ++
Quads: 557C, 689A, 689C, 689D
Habitat: BgFns (coastal), MshSw (lake margins), LCFrs (mesic)
Life Form: Perennial herb (rhizomatous)
Fertile: September
Notes: Endangered in ID and OR, and state-listed as Sensitive in WA. See *Fremontia* 12(2):11-14 (1984) for information on NEV Co. occurrence.

Lycopodium clavatum L.

"running-pine" Lycopodiaceae
CNPS List: 2 **R-E-D Code:** 2-1-1 **State/Fed. Status:** CEQA
Distribution: HUM, ++
Quads: 652A, 672A, 672D, 689A, 689C, 689D
Habitat: MshSw, NCFrs (mesic)
Life Form: Perennial herb (rhizomatous)
Fertile: July-August

Lycopodium inundatum

See *Lycopodiella inundata*

Lycopus uniflorus Michx.

"northern bugleweed" Lamiaceae
CNPS List: 4 **R-E-D Code:** 1-1-1 **State/Fed. Status:** CEQA?
Distribution: DNT?, HUM, NEV, PLU, TUO, ++
Habitat: BgFns
Life Form: Perennial herb
Blooming: July-September
Notes: Does plant occur in DNT Co.? See *Wasmann Journal of Biology* 33(1-2):93 (1975) for distributional information.

Lycurus phleoides

See *Lycurus phleoides* var. *phleoides*

Lycurus phleoides HBK. var. *phleoides*

"wolftail" Poaceae
CNPS List: 2 **R-E-D Code:** 3-2-1 **State/Fed. Status:** CEQA
Distribution: SBD, AZ, ++
Quads: 200A
Habitat: JTWld, PJWld
Life Form: Perennial herb
Blooming: August-September
Notes: Threatened by grazing.

Lyonothamnus floribundus Gray ssp. *asplenifolius* (Greene) Raven

"Santa Cruz Island ironwood" Rosaceae
CNPS List: 1B **R-E-D Code:** 2-2-3 **State/Fed. Status:** /C2
Distribution: SCM, SCZ, SRO
Quads: SCMS, SCZA, SCZB, SCZC, SCZD, SRON
Habitat: BUFrs, Chprl, CmWld
Life Form: Tree (evergreen)
Blooming: June-July
Notes: Threatened by grazing on SCZ and SRO islands. Feral herbivores removed from SCM Isl., and vegetation recovering.

Lyonothamnus floribundus Gray ssp. *floribundus*

"Santa Catalina Island ironwood" Rosaceae
CNPS List: 1B **R-E-D Code:** 3-2-3 **State/Fed. Status:** /C2
Distribution: SCT
Quads: SCTN, SCTW
Habitat: BUFrs, Chprl, CmWld
Life Form: Tree (evergreen)
Blooming: May-June
Notes: Feral animal populations increasingly under control; stump-sprouting and seedlings recently noted.

Lyrocarpa coulteri Hook. & Harv. var. *palmeri* (Wats.) Roll.

"Coulter's lyrepod" Brassicaceae
CNPS List: 4 **R-E-D Code:** 1-1-1 **State/Fed. Status:** CEQA?
Distribution: IMP, SDG, BA
Habitat: SDScr
Life Form: Perennial herb
Blooming: December-April

Lythrum californicum

Considered but rejected: Too common

Machaeranthera ammophila

Considered but rejected: A synonym of *M. arida*; a common taxon

Machaeranthera asteroides (Torr.) Greene var. *lagunensis* (Keck) Turner

"Laguna Mtns. aster" Asteraceae
CNPS List: 2 **R-E-D Code:** 3-3-1 **State/Fed. Status:** CR/C2
Distribution: SDG, BA
Quads: 19C
Habitat: CmWld, LCFrs
Life Form: Perennial herb
Blooming: July-August
Notes: Known in CA from approximately five occurrences in the Wooded Hill area of Mt. Laguna. Threatened by grazing and recreational activities. State-listed as *M. lagunensis.* See *Brittonia* 9:238 (1957) for original description, and *Phytologia* 62(3):231-233 (1987) for revised nomenclature.
Status Report: 1987

Machaeranthera canescens ssp. *ziegleri*

See *Machaeranthera canescens* var. *ziegleri*

Machaeranthera canescens (Pursh) Gray var. *ziegleri* (Munz) Turner

"Ziegler's aster" Asteraceae
CNPS List: 1B **R-E-D Code:** 3-2-3 **State/Fed. Status:** /C3c
Distribution: RIV
Quads: 65C
Habitat: LCFrs, UCFrs
Life Form: Perennial herb
Blooming: July-October
Notes: Known only from the Santa Rosa Mtns. Threatened by grazing. See *Aliso* 7:65-66 (1969) for original description, and *Phytologia* 62(3):257 (1987) for revised nomenclature.

Machaeranthera cognata

See *Xylorhiza cognata*

Machaeranthera juncea (Greene) Shinn.

"rush-like bristleweed" Asteraceae
CNPS List: 4 **R-E-D Code:** 1-1-1 **State/Fed. Status:** CEQA?
Distribution: SDG, AZ, BA, SO
Habitat: Chprl, CoScr
Life Form: Perennial herb
Blooming: June-October
Notes: See *Bulletin of the California Academy of Sciences* 1:190 (1885) for original description, and *Phytologia* 68(6):439-465 (1990) for revised nomenclature.

Machaeranthera lagunensis

See *Machaeranthera asteroides* var. *lagunensis*

Machaeranthera leucanthemifolia

Considered but rejected: Too common; a synonym of *M. canescens* var. *leucanthemifolia*

Madia anomala

Considered but rejected: Too common

Madia doris-nilesiae T.W. Nelson & J.P. Nelson

"Niles's madia" Asteraceae
CNPS List: 1B **R-E-D Code:** 3-3-3 **State/Fed. Status:** CEQA
Distribution: TRI
Quads: 633A, 651B, 651C, 651D
Habitat: LCFrs (serpentinite)
Life Form: Annual herb
Blooming: June-July
Notes: Threatened by logging, recreation, and road maintenance. See *Brittonia* 37(4):394-396 (1985) for original description.

Madia hallii Keck

"Hall's madia" Asteraceae
CNPS List: 1B **R-E-D Code:** 2-2-3 **State/Fed. Status:** /C2
Distribution: COL, LAK, NAP, YOL
Quads: 516B, 517A, 532C, 532D, 533A, 533D, 547B, 547C
Habitat: Chprl (serpentinite)
Life Form: Annual herb
Blooming: April-May
Notes: Threatened by mining in NAP Co.

Madia nutans (Greene) Keck

"nodding madia" Asteraceae
CNPS List: 4 **R-E-D Code:** 1-1-3 **State/Fed. Status:** CEQA?
Distribution: NAP, SON
Habitat: Chprl, CmWld
Life Form: Annual herb
Blooming: April-May

Madia radiata Kell.

"showy madia" Asteraceae
CNPS List: 1B **R-E-D Code:** 2-3-3 **State/Fed. Status:** CEQA
Distribution: CCA*, FRE, KNG, KRN, MNT, SBT, SJQ, SLO
Quads: 267B, 268A, 291B, 291C, 315D, 316D, 339A, 339B, 340A, 361C, 362D, 444D, 464A*, 481D*
Habitat: CmWld, VFGrs
Life Form: Annual herb
Blooming: March-May
Notes: Apparently occurs as very scattered populations at only a few locations on private land. Threatened by grazing and non-native plants.

Madia stebbinsii T.W. Nelson & J.P. Nelson

"Stebbins's madia" Asteraceae
CNPS List: 1B **R-E-D Code:** 3-2-3 **State/Fed. Status:** /C3c
Distribution: SHA, TEH, TRI
Quads: 596A, 596B, 632C, 632D
Habitat: Chprl, LCFrs / serpentinite
Life Form: Annual herb
Blooming: May-June
Notes: Possibly threatened by road maintenance. See *Brittonia* 32(3):323-325 (1980) for original description.

Madia subspicata

Considered but rejected: Too common

Madia yosemitana Gray

"Yosemite madia" Asteraceae
CNPS List: 3 **R-E-D Code:** ?-2-3 **State/Fed. Status:** CEQA?
Distribution: AMA, FRE, MAD?, MPA, TUL, TUO
Quads: 353C, 397C, 397D, 419B, 436B, 437A, 455A, 455B, 455D, 456A, 490D, 492A
Habitat: LCFrs, Medws
Life Form: Annual herb
Blooming: April-July
Notes: Move to List 4? Easily overlooked; location and rarity information needed. Does plant occur in MAD Co.? See *Proceedings of the American Academy of Arts and Sciences* 17:219 (1881-2) for original description.

Mahonia higginsiae

See *Berberis fremontii*

Mahonia nervosa var. *mendocinensis*

Considered but rejected: A synonym of *Berberis nervosa*; a common taxon

Mahonia nevinii

See *Berberis nevinii*

Mahonia pinnata ssp. *insularis*

See *Berberis pinnata* ssp. *insularis*

Mahonia sonnei

Considered but rejected: A synonym of *Berberis aquifolium* var. *repens*; a common taxon

Malacothamnus abbottii (Eastw.) Kearn.

"Abbott's bush mallow" Malvaceae
CNPS List: 1B **R-E-D Code:** 3-3-3 **State/Fed. Status:** /C1
Distribution: MNT
Quads: 294A, 295D
Habitat: RpScr
Life Form: Shrub (deciduous)
Blooming: June-October
Notes: Rediscovered in 1990 by D. Mitchell near Sargent Creek; now known from about five extended populations. Threatened by housing development, grazing, energy development, and road construction. See *Leaflets of Western Botany* 1:213-222 (1936) for original description.
Status Report: 1977

Malacothamnus aboriginum (Rob.) Greene

"Indian Valley bush mallow" Malvaceae
CNPS List: 1B **R-E-D Code:** 2-2-3 **State/Fed. Status:** CEQA
Distribution: FRE, MNT, SBT
Quads: 293A, 294A, 295D, 315B, 315C, 316B, 339C, 340B, 340D, 341A, 341B, 341D, 362C, 363A, 363B, 363C, 364A?, 364C, 385C, 385D
Habitat: Chprl, CmWld / rocky
Life Form: Shrub (deciduous)
Blooming: April-October
Notes: Appears in abundance after fires. See *Synoptical Flora of North America* 1(1):311 (1897) for original description.

Malacothamnus arcuatus (Greene) Greene

"arcuate bush mallow" Malvaceae
CNPS List: 4 **R-E-D Code:** 1-1-3 **State/Fed. Status:** CEQA?
Distribution: SCL, SCR, SMT
Habitat: Chprl
Life Form: Shrub (evergreen)
Blooming: April-July
Notes: Rare in SCR Co. A synonym of *M. fasciculatus* in *The Jepson Manual.*

Malacothamnus clementinus (Munz & Jtn.) Kearn.

"San Clemente Island bush mallow" Malvaceae
CNPS List: 1B **R-E-D Code:** 3-3-3 **State/Fed. Status:** CE/FE
Distribution: SCM
Quads: SCMC, SCMS
Habitat: VFGrs
Life Form: Shrub (deciduous)
Blooming: March-August
Notes: Known from six occurrences. Threatened by Navy activities. Feral herbivores removed from SCM Isl., and vegetation recovering. See *Bulletin of the Torrey Botanical Club* 51:296 (1924) for original description, and *Leaflets of Western Botany* 6(6):127-128 (1951) for revised nomenclature.
Status Report: 1987

Malacothamnus davidsonii (Rob.) Greene

"Davidson's bush mallow" Malvaceae
CNPS List: 1B **R-E-D Code:** 2-2-3 **State/Fed. Status:** /C2
Distribution: LAX, MNT, SLO
Quads: 111A, 111B, 137C, 137D, 294C
Habitat: Chprl, CoScr, RpWld
Life Form: Shrub (deciduous)
Blooming: June-September
Notes: Need quads for MNT Co. Threatened by urbanization in LAX Co. Intergrades with *M. fasciculatus.*

Malacothamnus fasciculatus (T. & G.) Greene var. *nesioticus* (Rob.) Kearn.

"Santa Cruz Island bush mallow" Malvaceae
CNPS List: 1B **R-E-D Code:** 3-3-3 **State/Fed. Status:** CE/C1
Distribution: SCZ
Quads: SCZA, SCZB
Habitat: Chprl
Life Form: Shrub (deciduous)
Blooming: June-July
Notes: Known from only two occurrences near Christi Ranch and in Central Valley. Threatened by feral pigs, erosion, and non-native plants. A synonym of *M. fasciculatus* in *The Jepson Manual.*
Status Report: 1988

Malacothamnus gracilis (Eastw.) Kearn.

"slender bush mallow" Malvaceae
CNPS List: 4 **R-E-D Code:** 1-1-3 **State/Fed. Status:** CEQA?
Distribution: SLO
Habitat: Chprl
Life Form: Shrub (deciduous)
Blooming: June-October
Notes: A synonym of *M. jonesii* in *The Jepson Manual.*

Malacothamnus hallii (Eastw.) Kearn.

"Hall's bush mallow" Malvaceae
CNPS List: 1B **R-E-D Code:** 3-2-3 **State/Fed. Status:** CEQA
Distribution: ALA?, CCA, MER, SCL
Quads: 384A, 404C, 405D, 406B, 406C, 407A, 407B, 427D, 464A, 464B, 464C
Habitat: Chprl
Life Form: Shrub (evergreen)
Blooming: May-September
Notes: Does plant occur in ALA Co.? A synonym of *M. fasciculatus* in *The Jepson Manual.*

Malacothamnus helleri (Eastw.) Kearn.

"Heller's bush mallow" Malvaceae
CNPS List: 4 **R-E-D Code:** 1-1-3 **State/Fed. Status:** CEQA?
Distribution: COL, GLE, LAK, NAP, TEH, YOL
Habitat: Chprl (sandstone)
Life Form: Shrub (deciduous)
Blooming: June-August
Notes: A synonym of *M. fremontii* in *The Jepson Manual.*

Malacothamnus jonesii (Munz) Kearn.

"Jones's bush mallow" Malvaceae
CNPS List: 4 **R-E-D Code:** 1-1-3 **State/Fed. Status:** CEQA?
Distribution: MNT, SLO
Habitat: Chprl, CmWld
Life Form: Shrub (deciduous)
Blooming: May-July

Malacothamnus mendocinensis (Eastw.) Kearn.

"Mendocino bush mallow" Malvaceae
CNPS List: 1A **Last Seen:** 1938 **State/Fed. Status:** /C2*
Distribution: MEN*
Quads: 550C*, 551D*
Habitat: CmWld
Life Form: Shrub (deciduous)
Blooming: May-June
Notes: Known from only two historical collections. A synonym of *M. fasciculatus* in *The Jepson Manual.* See *Leaflets of Western Botany* 2:188 (1939) for original description and 6(6):133-134 (1951) for revised nomenclature.
Status Report: 1977

Malacothamnus niveus (Eastw.) Kearn.

"San Luis Obispo County bush mallow" Malvaceae
CNPS List: 4 **R-E-D Code:** 1-1-3 **State/Fed. Status:** CEQA?
Distribution: MNT, SBA, SLO
Habitat: Chprl
Life Form: Shrub (deciduous)
Blooming: May-July
Notes: A synonym of *M. jonesii* in *The Jepson Manual.*

Malacothamnus palmeri (Wats.) Greene var. *involucratus* (Rob.) Kearn.

"Carmel Valley bush mallow" Malvaceae
CNPS List: 1B **R-E-D Code:** 1-2-3 **State/Fed. Status:** /C2
Distribution: MNT, SLO
Quads: 270C, 295A, 295B, 295D, 318C, 318D, 343B, 365C, 365D, 366D
Habitat: Chprl, CmWld
Life Form: Shrub (deciduous)
Blooming: May-August
Notes: Threatened by development in MNT Co. A synonym of *M. palmeri* in *The Jepson Manual.* See *Synoptical Flora of North America* 1(1):310 (1897) for original description, and *Leaflets of Western Botany* 6(6):121 (1951) for revised nomenclature.
Status Report: 1977

Malacothamnus palmeri (Wats.) Greene var. *lucianus* Kearn.

"Arroyo Seco bush mallow" Malvaceae
CNPS List: 1B **R-E-D Code:** 3-2-3 **State/Fed. Status:** /C2
Distribution: MNT
Quads: 319B, 344D
Habitat: Chprl, Medws
Life Form: Shrub (deciduous)
Blooming: May-August
Notes: Known from only three occurrences: two near Big Sur, and one in Arroyo Seco in the Santa Lucia Mtns. A synonym of *M. palmeri* in *The Jepson Manual.* See *Leaflets of Western Botany* 7(12):289-290 (1955) for original description.
Status Report: 1979

Malacothamnus palmeri (Wats.) Greene var. *palmeri*

"Santa Lucia bush mallow" Malvaceae
CNPS List: 4 **R-E-D Code:** 1-1-3 **State/Fed. Status:** /C3c
Distribution: MNT?, SLO
Habitat: Chprl
Life Form: Shrub (deciduous)
Blooming: May-July
Notes: MNT Co. plants need confirmation. A synonym of *M. palmeri* in *The Jepson Manual.*

Malacothamnus parishii (Eastw.) Kearn.

"Parish's bush mallow" Malvaceae
CNPS List: 1A **Last Seen:** 1895 **State/Fed. Status:** /C2
Distribution: SBD*
Quads: 106C*
Habitat: Chprl, CoScr
Life Form: Shrub (deciduous)
Blooming: June-July
Notes: Historical occurrence in San Bernardino Valley was extirpated by urbanization. RIV Co. record from the Santa Rosa Mtns. probably erroneous, based on a misidentification of *Sphaeralcea.* A synonym of *M. fasciculatus* in *The Jepson Manual.*

Malacothrix foliosa Gray

"leafy malacothrix" Asteraceae
CNPS List: 4 **R-E-D Code:** 1-2-2 **State/Fed. Status:** CEQA?
Distribution: ANA, SBR, SCM, SNI, BA
Habitat: CoScr
Life Form: Annual herb
Blooming: March-July
Notes: See *Madroño* 21:386-388 (1992) for SBR Isl. distribution.

Malacothrix incana (Nutt.) T. & G.

"dunedelion" Asteraceae
CNPS List: 4 **R-E-D Code:** 1-1-3 **State/Fed. Status:** CEQA?
Distribution: SBA, SCZ*, SLO, SMI, SNI, SRO, VEN
Habitat: CoDns
Life Form: Perennial herb
Blooming: April-August
Notes: Last collected on SCZ Isl in the 1880's.

Malacothrix indecora Greene

"Santa Cruz Island cliff-aster" Asteraceae
CNPS List: 1B **R-E-D Code:** 3-3-3 **State/Fed. Status:** /C1
Distribution: SCZ, SMI*
Quads: SCZA, SCZB*, SMIE*
Habitat: CBScr, Chprl, CoDns
Life Form: Annual herb
Blooming: April-September
Notes: Known from only one occurrence with less than 50 individuals at Black Point. Threatened by feral herbivores and possibly by erosion. See *Bulletin of the California Academy of Sciences* 2:152 (1886) for original description.

Malacothrix saxatilis (Nutt.) T. & G. var. *arachnoidea* (McGregor) E. Williams

"Carmel Valley cliff-aster" Asteraceae
CNPS List: 1B **R-E-D Code:** 3-2-3 **State/Fed. Status:** /C2
Distribution: MNT, SBA
Quads: 167C, 342C, 343A, 343B, 366D
Habitat: Chprl
Life Form: Perennial herb (rhizomatous)
Blooming: June-December
Notes: Known from approximately ten occurrences.

Malacothrix squalida Greene

"island malacothrix" Asteraceae
CNPS List: 1B **R-E-D Code:** 3-3-3 **State/Fed. Status:** /C2
Distribution: ANA, SCZ
Quads: ANAC, SCZC
Habitat: Chprl, CmWld
Life Form: Annual herb
Blooming: April-July
Notes: Last seen on SCZ Isl. in 1968 and on ANA Isl. in 1986. Threatened by feral herbivores on SCZ Isl., and by seabird-caused erosion on ANA Isl. Field surveys needed. See *Bulletin of the California Academy of Sciences* 2:152 (1886) for original description.

Malaxis brachypoda

See *Malaxis monophyllos* ssp. *brachypoda*

Malaxis monophyllos (L.) Swartz ssp. *brachypoda* (Gray) A. Löve & D. Löve

"adder's-mouth" Orchidaceae
CNPS List: 2 **R-E-D Code:** 3-3-1 **State/Fed. Status:** CEQA
Distribution: RIV*, SBD, ++
Quads: 83C*, 105A
Habitat: BgFns, Medws, UCFrs / mesic
Life Form: Perennial herb (bulbiferous)
Blooming: June-August
Notes: Rediscovered in 1989 by R. Coleman at South Fork Meadows in the San Bernardino Mtns. Historically known from Tahquitz Valley in the San Jacinto Mtns. Threatened by grazing and trampling. USFWS uses the name *M. brachypoda.* See *Native Orchids of the United States and Canada* (1975) by Luer for nomenclature, and *Fremontia* 18(1):19-21 (1990) for account of rediscovery.

Malperia tenuis Wats.

"brown turbans" Asteraceae
CNPS List: 2 **R-E-D Code:** 3-1-1 **State/Fed. Status:** CEQA
Distribution: IMP, SDG, BA
Quads: 5B, 18C, 31D
Habitat: SDScr (sandy)
Life Form: Annual herb
Blooming: March-April

Malvastrum kernense

See *Eremalche kernensis*

Marina orcuttii (Wats.) Barneby var. *orcuttii*

"California marina" Fabaceae
CNPS List: 1B **R-E-D Code:** 3-1-2 **State/Fed. Status:** /C2
Distribution: RIV, BA
Quads: 65C
Habitat: Chprl, PJWld, SDScr
Life Form: Perennial herb
Blooming: May-October
Notes: Known in CA from only two occurrences at Deep Cyn. in the Santa Rosa Mtns. See *Memoirs of the New York Botanical Garden* 27:85 (1977) for revised nomenclature.

Marsilea oligospora Goodd.

"Nelson's pepperwort" Marsileaceae
CNPS List: 3 **R-E-D Code:** ?-?-1 **State/Fed. Status:** CEQA?
Distribution: BUT, ELD, LAS, MOD, NEV, PLU, SHA, SIE, TRI, TUL, NV, OR, ++
Quads: 523A, 554B, 587B, 587C, 607C, 644D, 726A, 727B
Habitat: MshSw, VnPls / muddy
Life Form: Annual/Perennial herb (rhizomatous)
Fertile: July-August
Notes: Move to List 4? Location, rarity, and endangerment information needed, especially quads for LAS, SIE, TRI, and TUL counties. See *Botanical Gazette* 33:66 (1902) for original description, and *Systematic Botany Monographs* 11:1-87 (1986) for taxonomic treatment.

Matelea parvifolia (Torr.) Woodson

"spearleaf" Asclepiadaceae
CNPS List: 2 **R-E-D Code:** 3-1-1 **State/Fed. Status:** CEQA
Distribution: RIV, SBD, SDG, AZ, BA, NV, TX+
Quads: 32B, 61A, 61C, 63A, 65B, 65C, 102C, 201C
Habitat: MDScr, SDScr / rocky
Life Form: Perennial herb
Blooming: March-May
Status Report: 1979

Maurandya antirrhiniflora Willd. ssp. *antirrhiniflora*

"violet twining snapdragon" Scrophulariaceae
CNPS List: 2 **R-E-D Code:** 3-1-1 **State/Fed. Status:** CEQA
Distribution: SBD, AZ, SO, TX, ++
Quads: 175B, 176A, 201C
Habitat: JTWld, MDScr / carbonate
Life Form: Perennial herb
Blooming: April-May
Notes: Known in CA from fewer than ten occurrences in the Providence Mtns. See *Systematic Botany Monographs* 5:48-49 (1985) for original description.

Maurandya petrophila Cov. & Mort.

"rock lady" Scrophulariaceae
CNPS List: 1B **R-E-D Code:** 3-2-3 **State/Fed. Status:** CR/C2
Distribution: INY
Quads: 368A, 368C, 368D
Habitat: MDScr (carbonate)
Life Form: Perennial herb
Blooming: April-June
Notes: USFWS uses the name *Holmgrenanthe petrophila.* See *Journal of the Washington Academy of Sciences* 25:291 (1935) for original description, and *Systematic Botany Monographs* 5:54-56 (1985) for taxonomic treatment.
Status Report: 1987

Meconella californica

Considered but rejected: Too common

Meconella denticulata

Considered but rejected: Too common

Meconella oregana

Considered but rejected: Thought not to be in CA, but needs reconsideration; now apparently known from SCL Co.

Melica inflata

Considered but rejected: Too common; a synonym of *M. bulbosa* var. *inflata*

Melica spectabilis Scribn.

"purple onion grass" Poaceae
CNPS List: 4 **R-E-D Code:** 1-1-1 **State/Fed. Status:** CEQA?
Distribution: DNT, HUM, MEN, OR, ++
Habitat: LCFrs, Medws, UCFrs / mesic
Life Form: Perennial herb (rhizomatous)
Blooming: May-July

Mentzelia hirsutissima Wats.

"hairy stickleaf" Loasaceae
CNPS List: 2 **R-E-D Code:** 2-1-1 **State/Fed. Status:** /C3c
Distribution: IMP, SDG, BA
Quads: 5B, 7A, 18C, 19B
Habitat: SDScr
Life Form: Annual herb
Blooming: April-May

Mentzelia hirsutissima var. *stenophylla*

See *Mentzelia hirsutissima*

Mentzelia leucophylla

Considered but rejected: Not in CA; misidentification of *M. oreophila*

Mentzelia reflexa

Considered but rejected: Too common

Menyanthes trifoliata

Considered but rejected: Too common

Mertensia bella Piper

"Oregon lungwort" Boraginaceae
CNPS List: 2 **R-E-D Code:** 3-2-1 **State/Fed. Status:** CEQA
Distribution: SIS, ID, OR+
Quads: 681A, 736B
Habitat: Medws, UCFrs / mesic
Life Form: Perennial herb
Blooming: May-July
Notes: Sensitive in ID. See *Proceedings of the Biological Society of Washington* 31:76 (1918) for original description.

Micropus amphibolus Gray

"Mt. Diablo cottonweed" Asteraceae
CNPS List: 4 **R-E-D Code:** 1-1-3 **State/Fed. Status:** CEQA?
Distribution: ALA, CCA, LAK, MNT, MRN, NAP, SCR, SON
Habitat: BUFrs, CmWld, VFGrs
Life Form: Annual herb
Blooming: April-May

Microseris borealis (Bong.) Sch.-Bip.

"northern microseris" Asteraceae
CNPS List: 2 **R-E-D Code:** 3-3-1 **State/Fed. Status:** CEQA
Distribution: HUM, MEN*, OR, WA, ++
Quads: 569D*, 671A
Habitat: BgFns, LCFrs, Medws / mesic
Life Form: Perennial herb
Blooming: June-September
Notes: Known in CA from only two occurrences. State-listed as Sensitive in WA.

Microseris decipiens

See *Stebbinsoseris decipiens*

Microseris douglasii (DC.) Sch.-Bip. var. *platycarpha* (Gray) Chambers

"small-flowered microseris" Asteraceae
CNPS List: 4 **R-E-D Code:** 1-2-2 **State/Fed. Status:** CEQA?
Distribution: LAX, ORA, RIV, SCM, SCT, SDG, BA
Habitat: CmWld, CoScr, VFGrs / clay
Life Form: Annual herb
Blooming: March-May
Notes: Common in coastal grassy areas in SDG Co., but this habitat is declining. See *Contributions from the Dudley Herbarium* 4(7):296-298 (1955) for taxonomic treatment.

Microseris laciniata ssp. *siskiyouensis*

Considered but rejected: Too common

Mimulus acutidens Greene

"Kings River monkeyflower" Scrophulariaceae
CNPS List: 3 **R-E-D Code:** ?-?-3 **State/Fed. Status:** CEQA?
Distribution: FRE, MAD, TUL
Quads: 418A
Habitat: CmWld, LCFrs
Life Form: Annual herb
Blooming: April-July
Notes: Move to List 4? Location, rarity, and endangerment information needed, especially quads for FRE and TUL counties. Many misidentifications of Sequoia-Kings Canyon NP plants. A synonym of *M. inconspicuus* in *The Jepson Manual.*

Mimulus arenarius

Considered but rejected: Too common; perhaps a synonym of *M. floribundus*

Mimulus aridus (Abrams) Grant

"low bush monkeyflower" Scrophulariaceae
CNPS List: 4 **R-E-D Code:** 1-1-2 **State/Fed. Status:** /C3c
Distribution: IMP, SDG, BA
Habitat: Chprl
Life Form: Shrub (evergreen)
Blooming: April-July
Notes: See *M. aurantiacus* in *The Jepson Manual.*

Mimulus barbatus

Considered but rejected: Too common; perhaps a synonym of *M. montioides*

Mimulus bifidus ssp. *bifidus*

Considered but rejected: Too common

Mimulus biolettii

See *Mimulus filicaulis*

Mimulus brachiatus Penn.

"serpentine monkeyflower" Scrophulariaceae
CNPS List: 3 **R-E-D Code:** ?-?-3 **State/Fed. Status:** CEQA?
Distribution: COL, LAK
Quads: 564D
Habitat: Chprl (serpentinite)
Life Form: Annual herb
Blooming: May-July
Notes: Move to List 4? Location, rarity, and endangerment information needed. A synonym of *M. layneae* in *The Jepson Manual.*

Mimulus brandegei Penn.

"Santa Cruz Island monkeyflower" Scrophulariaceae
CNPS List: 1A **Last Seen:** 1932 **State/Fed. Status:** /C2*
Distribution: SCZ*
Quads: SCZA*
Habitat: VFGrs (rocky)
Life Form: Annual herb
Blooming: May
Notes: Recent searches have not rediscovered only known historical occurrence, which may have been extirpated by grazing. May be conspecific with *M. latifolius*, which is rare on arroyo beds and cliffs on GU Isl. A synonym of *M. latifolius* in *The Jepson Manual.*
Status Report: 1979

Mimulus clevelandii Bdg.

"Cleveland's bush monkeyflower" Scrophulariaceae
CNPS List: 4 **R-E-D Code:** 1-2-2 **State/Fed. Status:** CEQA?
Distribution: ORA, RIV, SDG, BA
Habitat: Chprl, LCFrs / often in disturbed areas
Life Form: Perennial herb
Blooming: May-July
Notes: Threatened by recreational activities.

Mimulus diffusus Grant

"Palomar monkeyflower" Scrophulariaceae
CNPS List: 4 **R-E-D Code:** 1-1-1 **State/Fed. Status:** CEQA?
Distribution: ORA, RIV, SDG, BA
Habitat: Chprl, LCFrs
Life Form: Annual herb
Blooming: April-June
Notes: A synonym of *M. palmeri* in *The Jepson Manual*; probably indistinct from it. See *Annals of the Missouri Botanical Garden* 11:254 (1925) for original description.

Mimulus dudleyi

Considered but rejected: Too common; perhaps a synonym of *M. floribundus*

Mimulus exiguus Gray

"San Bernardino Mtns. monkeyflower" Scrophulariaceae
CNPS List: 1B **R-E-D Code:** 2-2-2 **State/Fed. Status:** /C2
Distribution: SBD, BA
Quads: 105A, 105B, 131C, 131D
Habitat: Medws, PbPln, UCFrs / mesic
Life Form: Annual herb
Blooming: June-July
Notes: Threatened by development and vehicles.

Mimulus filicaulis Wats.

"slender-stemmed monkeyflower" Scrophulariaceae
CNPS List: 1B **R-E-D Code:** 2-2-3 **State/Fed. Status:** /C2
Distribution: MPA, TUO
Quads: 419B, 438A, 438B, 438C, 438D, 439A, 456A, 456C, 456D, 457C, 457D
Habitat: CmWld, LCFrs, Medws, UCFrs / vernally mesic
Life Form: Annual herb
Blooming: April-August
Notes: Threatened by logging and reforestation with herbicides, and possibly by grazing. Includes *M. biolettii*. See *Proceedings of the American Academy of Arts and Sciences* 26:125 (1891) for original description, and *Changing Seasons* 1(3):3-5 (1981) for taxonomic discussion.

Mimulus flemingii Munz

"island bush monkeyflower" Scrophulariaceae
CNPS List: 4 **R-E-D Code:** 1-1-3 **State/Fed. Status:** CEQA?
Distribution: ANA, SCM, SCZ, SRO
Habitat: CBScr
Life Form: Shrub (evergreen)
Blooming: March-July
Notes: See *M. aurantiacus* in *The Jepson Manual*.

Mimulus glabratus Kunth. ssp. *utahensis* Penn.

"Utah monkeyflower" Scrophulariaceae
CNPS List: 2 **R-E-D Code:** 3-2-1 **State/Fed. Status:** CEQA
Distribution: MNO, INY, NV+
Quads: 326A, 453A, 453B
Habitat: Medws, PJWld
Life Form: Perennial herb (rhizomatous)
Blooming: April
Notes: Known in CA from fewer than ten occurrences. Threatened by the dewatering of Mono Lake. See *M. guttatus* in *The Jepson Manual*.

Mimulus glaucescens Greene

"shield-bracted monkeyflower" Scrophulariaceae
CNPS List: 4 **R-E-D Code:** 1-1-3 **State/Fed. Status:** CEQA?
Distribution: BUT, COL, LAK, TEH
Habitat: CmWld, VFGrs / serpentinite seeps
Life Form: Annual herb
Blooming: March-May

Mimulus gracilipes Rob.

"slender-stalked monkeyflower" Scrophulariaceae
CNPS List: 4 **R-E-D Code:** 1-1-3 **State/Fed. Status:** CEQA?
Distribution: FRE, MPA
Habitat: Chprl (often in burns and disturbed areas)
Life Form: Annual herb
Blooming: April-June
Notes: See *Madroño* 28(1):41 (1981) for range extension information.

Mimulus grayi Grant

"Gray's monkeyflower" Scrophulariaceae
CNPS List: 4 **R-E-D Code:** 1-1-3 **State/Fed. Status:** CEQA?
Distribution: FRE, MAD, MPA, TUL
Habitat: LCFrs, UCFrs / mesic
Life Form: Annual herb
Blooming: May-July
Notes: A synonym of *M. inconspicuus* in *The Jepson Manual*.

Mimulus guttatus ssp. *arenicola*

Considered but rejected: Too common; perhaps a synonym of *M. guttatus*

Mimulus inconspicuus Gray

"small-flowered monkeyflower" Scrophulariaceae
CNPS List: 4 **R-E-D Code:** 1-1-3 **State/Fed. Status:** CEQA?
Distribution: AMA, BUT, CAL, MPA, TUO
Habitat: Chprl, CmWld, LCFrs / mesic
Life Form: Annual herb
Blooming: May-June
Notes: Does not include *M. acutidens* or *M. grayi*.

Mimulus laciniatus Gray

"cut-leaved monkeyflower" Scrophulariaceae
CNPS List: 4 **R-E-D Code:** 1-1-3 **State/Fed. Status:** CEQA?
Distribution: AMA, BUT, FRE, MAD, MPA, PLU, TUL, TUO
Habitat: LCFrs, UCFrs / mesic, granitic
Life Form: Annual herb
Blooming: May-July
Notes: See *Proceedings of the American Academy of Arts and Sciences* 11:98 (1876) for original description.

Mimulus microphyllus Benth.

"small-leaved monkeyflower" Scrophulariaceae
CNPS List: 4 **R-E-D Code:** 1-1-3 **State/Fed. Status:** CEQA?
Distribution: KRN
Habitat: Medws (mesic)
Life Form: Annual herb
Blooming: May-August
Notes: See *M. guttatus* in *The Jepson Manual*.

Mimulus mohavensis Lemmon

"Mojave monkeyflower" Scrophulariaceae
CNPS List: 1B **R-E-D Code:** 2-2-3 **State/Fed. Status:** /C1
Distribution: SBD
Quads: 130B, 132A, 155B, 158A, 181A, 181C, 181D, 182A*, 182D
Habitat: JTWld, MDScr / gravelly
Life Form: Annual herb
Blooming: April-June
Notes: Most historical occurrences in the Barstow area have been extirpated or impacted. Threatened by development, mining, nonnative plants, and vehicles. See *Botanical Gazette* 9:142 (1884) for original description.
Status Report: 1988

Mimulus norrisii Heckard & Shevock

"Kaweah monkeyflower" Scrophulariaceae
CNPS List: 1B **R-E-D Code:** 3-1-3 **State/Fed. Status:** /C3c
Distribution: TUL
Quads: 332A, 332B, 332D, 353C, 354D
Habitat: Chprl, CmWld / carbonate
Life Form: Annual herb
Blooming: March-May
Notes: Known from fewer than ten occurrences in the Kaweah River drainage. See *Madroño* 32(3):179-185 (1985) for original description.

Mimulus nudatus Greene

"bare monkeyflower" Scrophulariaceae
CNPS List: 4 **R-E-D Code:** 1-1-3 **State/Fed. Status:** CEQA?
Distribution: LAK, MEN, NAP
Habitat: Chprl (serpentinite seeps)
Life Form: Annual herb
Blooming: May-June

Mimulus parviflorus

See *Mimulus flemingii*

Mimulus pictus (Curran) Gray

"calico monkeyflower" Scrophulariaceae
CNPS List: 1B **R-E-D Code:** 2-2-3 **State/Fed. Status:** /C3c
Distribution: KRN, TUL
Quads: 212B, 213A, 213C, 238B, 238C, 238D, 239B, 259B, 260A, 261C, 261D, 262A, 262D, 286C, 286D, 309C, 310A
Habitat: BUFrs, CmWld / granitic
Life Form: Annual herb
Blooming: April-May
Notes: Threatened by grazing. See *Bulletin of the California Academy of Sciences* 1:106 (1885) for original description.

Mimulus pulchellus (Greene) Grant

"pansy monkeyflower" Scrophulariaceae
CNPS List: 4 **R-E-D Code:** 1-1-3 **State/Fed. Status:** CEQA?
Distribution: CAL, MPA, TUO
Habitat: LCFrs, Medws / vernally mesic
Life Form: Annual herb
Blooming: May-July

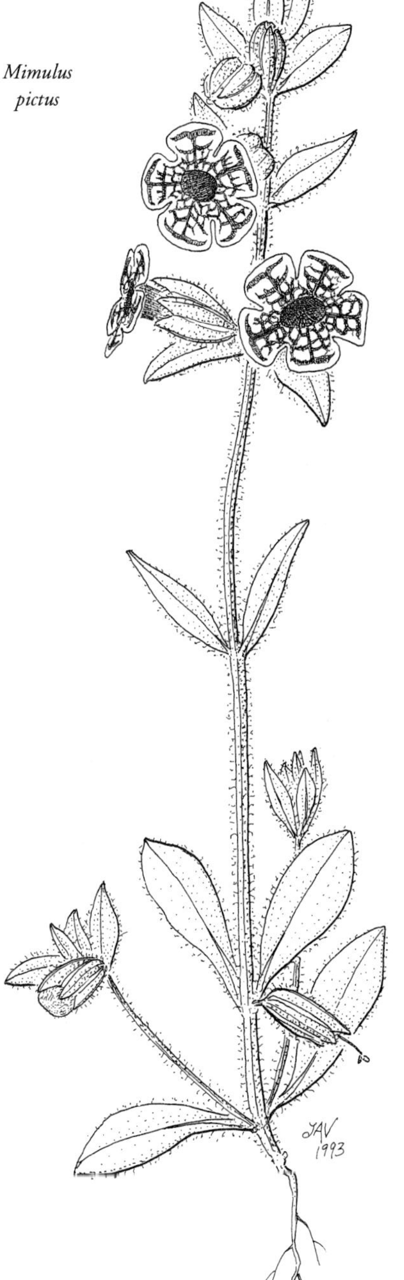

Mimulus purpureus Grant

"purple monkeyflower" Scrophulariaceae
CNPS List: 2 **R-E-D Code:** 2-3-1 **State/Fed. Status:** /C2
Distribution: SBD, BA
Quads: 105A, 105B, 131C, 131D
Habitat: Medws, PbPln, UCFrs
Life Form: Annual herb
Blooming: May-June
Notes: Known in CA from fewer than twenty occurrences. Threatened by development and vehicles. See *Annals of the Missouri Botanical Garden* 11:255 (1925) for original description.
Status Report: 1980

Mimulus purpureus var. *purpureus*

See *Mimulus purpureus*

Mimulus pygmaeus Grant

"Egg Lake monkeyflower" Scrophulariaceae
CNPS List: 1B **R-E-D Code:** 2-1-2 **State/Fed. Status:** /C2
Distribution: LAS, MOD, PLU, OR
Quads: 606B, 641A, 641B, 641C, 642A, 642B, 643A, 661D, 678B, 679C, 695C, 695D, 708D, 711D
Habitat: GBScr (clay), LCFrs, Medws / vernally mesic
Life Form: Annual herb
Blooming: May-June
Notes: Rediscovered in 1980. Candidate for state listing in OR. Lassen NF has adopted species management guidelines. See *Annals of the Missouri Botanical Garden* 11:312 (1925) for original description.
Status Report: 1977

Mimulus rattanii Gray ssp. *decurtatus* (Grant) Penn.

"Santa Cruz County monkeyflower" Scrophulariaceae
CNPS List: 4 **R-E-D Code:** 1-2-3 **State/Fed. Status:** CEQA?
Distribution: SCR
Habitat: Chprl, LCFrs / margins
Life Form: Annual herb
Blooming: May-July
Notes: Field work needed. Threatened by sand mining. A synonym of *M. rattanii* in *The Jepson Manual.*

Mimulus rupicola Cov. & Grant

"rock midget" Scrophulariaceae
CNPS List: 4 **R-E-D Code:** 1-1-3 **State/Fed. Status:** /C3c
Distribution: INY
Habitat: MDScr (carbonate)
Life Form: Annual herb
Blooming: March-May

Mimulus shevockii Heckard & Bacig.

"Kelso Creek monkeyflower" Scrophulariaceae
CNPS List: 1B **R-E-D Code:** 3-2-3 **State/Fed. Status:** /C1
Distribution: KRN
Quads: 259C, 260B, 260D
Habitat: JTWld, PJWld / sandy, granitic
Life Form: Annual herb
Blooming: April-May
Notes: Known from seven occurrences; some are affected by residential development. See *Madroño* 33(4):271-277 (1986) for original description.

Mimulus subsecundus Gray

"one-sided monkeyflower" Scrophulariaceae
CNPS List: 4 **R-E-D Code:** 1-1-3 **State/Fed. Status:** CEQA?
Distribution: FRE, MNT, SBT, SLO
Habitat: LCFrs
Life Form: Annual herb
Blooming: May-July
Notes: A synonym of *M. fremontii* in *The Jepson Manual.*

Mimulus traskiae Grant

"Santa Catalina Island monkeyflower" Scrophulariaceae
CNPS List: 1A **Last Seen:** 1904 **State/Fed. Status:** /C2*
Distribution: SCT*
Quads: SCTE*
Habitat: CoScr?
Life Form: Annual herb
Blooming: March-May
Notes: Known only from the type collection near Avalon. Possibly extirpated by grazing. See *Field Museum of Natural History* 5:226-227 (1923) for original description.
Status Report: 1979

Mimulus whipplei Grant

"Whipple's monkeyflower" Scrophulariaceae
CNPS List: 1A **Last Seen:** 1854 **State/Fed. Status:** /C2*
Distribution: CAL*
Quads: 475B*
Habitat: LCFrs
Life Form: Annual herb
Blooming: May
Notes: Known only from the type collection. Many recent searches have not rediscovered this plant. See *M. guttatus* in *The Jepson Manual.* See *Annals of the Missouri Botanical Garden* 11:184-185 (1924) for original description.
Status Report: 1977

Mimulus whitneyi

Considered but rejected: Too common

Minuartia decumbens T.W. Nelson & J.P. Nelson

"The Lassics sandwort" Caryophyllaceae
CNPS List: 1B **R-E-D Code:** 3-2-3 **State/Fed. Status:** /C2
Distribution: HUM, SHA, TEH, TRI
Quads: 632D, 634D
Habitat: LCFrs, UCFrs / serpentinite
Life Form: Perennial herb
Blooming: July
Notes: Known from only two occurrences. See *Brittonia* 33(2):162-164 (1981) for original description.

Minuartia howellii (Wats.) Mattf.

"Howell's sandwort" Caryophyllaceae
CNPS List: 4 **R-E-D Code:** 1-1-2 **State/Fed. Status:** /C3c
Distribution: DNT, OR
Habitat: Chprl, LCFrs / serpentinite
Life Form: Perennial herb
Blooming: May-July
Status Report: 1979

Minuartia obtusiloba (Rydb.) House

"alpine sandwort" Caryophyllaceae
CNPS List: 4 **R-E-D Code:** 1-1-1 **State/Fed. Status:** CEQA?
Distribution: FRE, INY, MNO, TUO, OR, ++
Habitat: AlpBR, AlpDS, SCFrs / granitic, metamorphic
Life Form: Perennial herb
Blooming: July-August

Minuartia rosei (Maguire & Barneby) McNeill

"Peanut sandwort" Caryophyllaceae
CNPS List: 4 **R-E-D Code:** 1-2-3 **State/Fed. Status:** /C2
Distribution: SHA, TEH, TRI
Habitat: LCFrs (serpentinite)
Life Form: Perennial herb
Blooming: May-July
Notes: Possibly threatened by logging. See *Rhodora* 82:499 (1980) for taxonomic treatment.
Status Report: 1979

Minuartia stolonifera T.W. Nelson & J.P. Nelson

"Scott Mtn. sandwort" Caryophyllaceae
CNPS List: 1B **R-E-D Code:** 3-1-3 **State/Fed. Status:** CEQA
Distribution: SIS
Quads: 700C
Habitat: LCFrs (serpentinite)
Life Form: Perennial herb
Blooming: May-July
Notes: Known only from the Scott Mtn. Pass area. See *Brittonia* 43(1):17-19 (1991) for original description.

Mirabilis tenuiloba Wats.

"slender-lobed four o'clock" Nyctaginaceae
CNPS List: 4 **R-E-D Code:** 1-1-1 **State/Fed. Status:** CEQA?
Distribution: IMP, RIV, SDG, BA
Habitat: SDScr
Life Form: Perennial herb
Blooming: March-May

Monardella antonina

See *Monardella antonina* ssp. *antonina*

Monardella antonina Hardham ssp. *antonina*

"San Antonio Hills monardella" Lamiaceae
CNPS List: 3 **R-E-D Code:** ?-?-3 **State/Fed. Status:** /C3c
Distribution: ALA?, CCA?, MNT, SBT?, SCL?
Quads: 295A, 319A, 344B, 362D?, 385C?, 464D?
Habitat: Chprl, CmWld
Life Form: Perennial herb (rhizomatous)
Blooming: June-August
Notes: Move to List 4? Easily confused with *M. villosa* ssp. *villosa*, which may be the taxon occurring in ALA, CCA, SBT, and SCL counties; needs clarification.

Monardella antonina Hardham ssp. *benitensis* (Hardham) Jokerst

"San Benito monardella" Lamiaceae
CNPS List: 4 **R-E-D Code:** 1-1-3 **State/Fed. Status:** /C3c
Distribution: MNT, SBT, SLO
Habitat: Chprl (serpentinite)
Life Form: Perennial herb (rhizomatous)
Blooming: June
Notes: See *Leaflets of Western Botany* 8(3):55 (1956) for original description, and *Phytologia* 72(1):9-16 (1992) for revised nomenclature.

Monardella beneolens Shevock, Ertter & Jokerst

"sweet-smelling monardella" Lamiaceae
CNPS List: 1B **R-E-D Code:** 3-1-3 **State/Fed. Status:** CEQA
Distribution: INY, KRN, TUL
Quads: 258B, 329A, 329B, 329D
Habitat: AlpBR, SCFrs, UCFrs / granitic
Life Form: Perennial herb (rhizomatous)
Blooming: July-September
Notes: Known from only three occurrences on public lands on the eastern Sierran crest. Highly restricted and endemic, but remoteness of occurrences limits disturbance. Hybridizes with *M. linoides* ssp. *linoides* and *M. odoratissima* ssp. *pallida*. See *Madroño* 36(4):271-279 (1989) for original description.

Monardella benitensis

See *Monardella antonina* ssp. *benitensis*

Monardella candicans Benth.

"Sierra monardella" Lamiaceae
CNPS List: 4 **R-E-D Code:** 1-1-3 **State/Fed. Status:** CEQA?
Distribution: AMA, CAL, ELD, FRE, KRN, MAD, MPA, NEV, PLA, SJQ, STA, TUL, TUO?
Habitat: Chprl, CmWld, LCFrs / sandy or gravelly
Life Form: Annual herb
Blooming: April-July
Notes: Does plant occur in TUO Co.? See *Annals of the Missouri Botanical Garden* 12:1-106 (1925) for taxonomic treatment.

Monardella cinerea Abrams

"gray monardella" Lamiaceae
CNPS List: 4 **R-E-D Code:** 1-1-3 **State/Fed. Status:** CEQA?
Distribution: LAX, SBD
Habitat: LCFrs, UCFrs
Life Form: Perennial herb (rhizomatous)
Blooming: July-August

Monardella crispa Elmer

"crisp monardella" Lamiaceae
CNPS List: 1B **R-E-D Code:** 2-2-3 **State/Fed. Status:** /C2
Distribution: SBA, SLO
Quads: 171A, 171C, 196A, 196B, 196D, 221D
Habitat: CoDns, CoScr
Life Form: Perennial herb (rhizomatous)
Blooming: April-August
Notes: Threatened by vehicles. Hybridizes with *M. frutescens*.

Monardella diaboli

Considered but rejected: Not published

Monardella douglasii Benth. ssp. *venosa* (Torr.) Jokerst

"veiny monardella" Lamiaceae
CNPS List: 1B **R-E-D Code:** 3-3-3 **State/Fed. Status:** /C2
Distribution: BUT, SUT*, TUO*
Quads: 459A*, 544D*, 576A*, 576B, 577A*, 593D*
Habitat: VFGrs (heavy clay)
Life Form: Annual herb
Blooming: May
Notes: Rediscovered in 1992 by B. Castro. Threatened by development of wastewater treatment plant.

Monardella douglasii var. *venosa*

See *Monardella douglasii* ssp. *venosa*

Monardella follettii (Jeps.) Jokerst

"Follett's monardella" Lamiaceae
CNPS List: 1B **R-E-D Code:** 3-1-3 **State/Fed. Status:** CEQA
Distribution: NEV, PLU
Quads: 542A, 590A, 606C, 606D
Habitat: LCFrs (rocky, serpentinite)
Life Form: Shrub
Blooming: June
Notes: Known from approximately three occurrences. See *Flora of California* 3(2):437 (1943) by W.L. Jepson for original description, and *Phytologia* 72(1):9-16 (1992) for revised nomenclature.

Monardella frutescens (Hoov.) Jokerst

"San Luis Obispo monardella" Lamiaceae
CNPS List: 1B **R-E-D Code:** 2-2-3 **State/Fed. Status:** /C2
Distribution: SBA, SLO
Quads: 171A, 171C, 171D, 196A, 196B, 196D, 221D, 247D
Habitat: CoDns
Life Form: Perennial herb (rhizomatous)
Blooming: May-July
Notes: Threatened by coastal development and vehicles. Hybridizes with *M. crispa*. See *Leaflets of Western Botany* 5:179-182 (1949) for original description, and *Phytologia* 72(1):9-16 (1992) for revised nomenclature.
Status Report: 1977

Monardella hypoleuca Gray ssp. *lanata* (Abrams) Munz

"felt-leaved monardella" Lamiaceae
CNPS List: 1B **R-E-D Code:** 2-2-2 **State/Fed. Status:** /C3c
Distribution: ORA, SDG, BA
Quads: 9B, 9C, 10A, 10D, 20A, 20B, 20C, 21A, 21D, 34A, 34C, 35C, 49D, 51A, 69B, 69C, 70A
Habitat: Chprl, CmWld
Life Form: Perennial herb (rhizomatous)
Blooming: June-July
Notes: Historical occurrences need field surveys.
Status Report: 1979

Monardella leucocephala Gray

"Merced monardella" Lamiaceae
CNPS List: 1A **Last Seen:** 1941 **State/Fed. Status:** /C2*
Distribution: MER*, STA*
Quads: 422D*, 423A*, 440B*, 441A*
Habitat: VFGrs (sandy)
Life Form: Annual herb
Blooming: July-August
Notes: May have been extirpated by agriculture. High potential for rediscovery; field work needed. See *Proceedings of the American Academy of Arts and Sciences* 7:385 (1867) for original description, and *Annals of the Missouri Botanical Garden* 12:90-91 (1925) for taxonomic treatment.
Status Report: 1988

Monardella linoides Gray ssp. *oblonga* (Greene) Abrams

"flax-like monardella" Lamiaceae
CNPS List: 1B **R-E-D Code:** 3-1-3 **State/Fed. Status:** /C2
Distribution: KRN, TUL, VEN
Quads: 189C, 190C, 190D, 239B, 262D, 285A
Habitat: LCFrs, UCFrs
Life Form: Perennial herb (rhizomatous)
Blooming: June-August
Notes: Known from fewer than twenty occurrences. Possibly indistinct from ssp. *linoides*.

Monardella linoides Gray ssp. *viminea* (Greene) Abrams

"willowy monardella" Lamiaceae
CNPS List: 1B **R-E-D Code:** 2-3-2 **State/Fed. Status:** CE/C2
Distribution: SDG, BA
Quads: 10B, 10C, 21B, 22A, 22B, 22C, 22D
Habitat: CCFrs, Chprl, RpFrs, RpScr, RpWld
Life Form: Perennial herb
Blooming: June-August
Notes: Threatened by road improvements, vehicles, non-native plants, and urbanization. See *Pittonia* 5:85 (1902) for original description.
Status Report: 1987

Monardella macrantha Gray ssp. *hallii* Abrams

"Hall's monardella" Lamiaceae
CNPS List: 1B **R-E-D Code:** 2-1-3 **State/Fed. Status:** /C3c
Distribution: ORA, RIV, SBD, SDG
Quads: 20A, 20D, 33D, 48C, 49A, 49B, 49C, 49D, 70A, 83C, 86A, 105C, 106B, 106D
Habitat: BUFrs, Chprl, CmWld, LCFrs, VFGrs
Life Form: Perennial herb (rhizomatous)
Blooming: June-August
Notes: Intermediates to ssp. *macrantha* are common. See *Muhlenbergia* 8:26-44 (1912) for original description.
Status Report: 1979

Monardella nana Gray ssp. *leptosiphon* (Torr.) Abrams

"San Felipe monardella" Lamiaceae
CNPS List: 2 **R-E-D Code:** 3-2-1 **State/Fed. Status:** /C2
Distribution: SDG, BA
Quads: 48D, 49A, 49D
Habitat: Chprl, LCFrs
Life Form: Perennial herb (rhizomatous)
Blooming: June-July
Notes: Known in CA from fewer than twenty occurrences. Threatened by woodcutting, habitat conversion, and grazing. See *Report of the U.S. and Mexican Boundary Survey*, p. 129 (1859) by W. Emory for original description, and *Muhlenbergia* 8:31 (1912) for revised nomenclature.

Monardella palmeri Gray

"Palmer's monardella" Lamiaceae
CNPS List: 4 **R-E-D Code:** 1-1-3 **State/Fed. Status:** /C3c
Distribution: MNT, SLO
Habitat: Chprl, CmWld / serpentinite
Life Form: Perennial herb (rhizomatous)
Blooming: June-August

Monardella pringlei Gray

"Pringle's monardella" Lamiaceae
CNPS List: 1A **Last Seen:** 1921 **State/Fed. Status:** /C2*
Distribution: RIV*, SBD*
Quads: 107C*, 107D*
Habitat: CoScr
Life Form: Annual herb
Blooming: May-June
Notes: Historically known only from the vicinity of Colton. Habitat lost to urbanization. See *Proceedings of the American Academy of Arts and Sciences* 19:96 (1883) for original description, and *Annals of the Missouri Botanical Garden* 12:86-87 (1925) for taxonomic treatment.
Status Report: 1977

Monardella purpurea

Considered but rejected: Too common

Monardella robisonii Epl.

"Robison's monardella" Lamiaceae
CNPS List: 1B **R-E-D Code:** 3-1-3 **State/Fed. Status:** /C2
Distribution: RIV, SBD
Quads: 81A, 100B, 102C, 103B, 103D, 104D
Habitat: PJWld
Life Form: Perennial herb (rhizomatous)
Blooming: April-July
Notes: Known from fewer than twenty occurrences. May occur in BA; need confirmation. Closely related to and possibly a variety of *M. linoides.*
Status Report: 1979

Monardella scelerata

Considered but rejected: Not published

Monardella stebbinsii Hardham & Bartel

"Stebbins's monardella" Lamiaceae
CNPS List: 1B **R-E-D Code:** 3-1-3 **State/Fed. Status:** /C3c
Distribution: PLU
Quads: 606C
Habitat: BUFrs, Chprl, LCFrs / serpentinite
Life Form: Perennial herb (rhizomatous)
Blooming: July-September
Notes: Known from only four occurrences along the North Fork of the Feather River. Possibly threatened by mining. See *Aliso* 12(4):693-699 (1990) for original description.
Status Report: 1979

Monardella undulata Benth.

"curly-leaved monardella" Lamiaceae
CNPS List: 4 **R-E-D Code:** 1-2-3 **State/Fed. Status:** CEQA?
Distribution: MNT, MRN, SBA, SCR, SFO, SLO, SMT, SON
Habitat: Chprl, CoDns, CoScr, LCFrs (ponderosa pine sandhills)
Life Form: Annual herb
Blooming: May-July
Notes: Threatened by coastal development, sand mining, and non-native plants.

Monardella undulata var. *frutescens*

See *Monardella frutescens*

Monardella undulata var. *undulata*

See *Monardella undulata*

Monardella villosa Benth. ssp. *globosa* (Greene) Jokerst

"robust monardella" Lamiaceae
CNPS List: 1B **R-E-D Code:** 3-2-3 **State/Fed. Status:** CEQA
Distribution: ALA, CCA, HUM, LAK, MRN, NAP, SMT, SON
Quads: 428C, 465B, 465C, 482C, 482D, 502C, 518D, 635C
Habitat: Chprl (openings), CmWld
Life Form: Perennial herb (rhizomatous)
Blooming: June-July
Notes: Need quads for LAK, MRN, and NAP counties. See *Pittonia* 5:82 (1902) for original description, and *Phytologia* 72(1):9-16 (1992) for revised nomenclature.

Monardella viridis Jeps. ssp. *saxicola* (Jtn.) Ewan

"rock monardella" Lamiaceae
CNPS List: 4 **R-E-D Code:** 1-2-3 **State/Fed. Status:** /C3c
Distribution: LAX, SBD
Habitat: Chprl, LCFrs
Life Form: Perennial herb (rhizomatous)
Blooming: June-September
Notes: Threatened by development.

Monardella viridis Jeps. ssp. *viridis*

"green monardella" Lamiaceae
CNPS List: 4 **R-E-D Code:** 1-1-3 **State/Fed. Status:** CEQA?
Distribution: LAK, NAP, SOL, SON
Habitat: BUFrs, Chprl, CmWld
Life Form: Perennial herb (rhizomatous)
Blooming: July-September
Notes: Hybridizes with *M. villosa* ssp. *villosa.*

Monotropa hypopithys

Considered but rejected: Too common

Monotropa uniflora L.

"Indian-pipe" Ericaceae
CNPS List: 2 **R-E-D Code:** 2-2-1 **State/Fed. Status:** CEQA
Distribution: DNT, HUM, OR, WA, ++
Quads: 672C, 740C, 740D
Habitat: BUFrs, NCFrs
Life Form: Perennial herb (saprophytic)
Blooming: June-July

Montia funstonii

Considered but rejected: A synonym of *M. fontana*; a common taxon

Montia howellii Wats.

"Howell's montia" Portulacaceae
CNPS List: 1A **Last Seen:** 1933 **State/Fed. Status:** /C2
Distribution: DNT?*, HUM*, TRI*, OR, WA+
Quads: 617A*, 617D*, 635A*, 670A*, 672C*
Habitat: Medws, NCFrs, VnPls / vernally mesic
Life Form: Annual herb
Blooming: March-May
Notes: Known in CA from seven collections. Did plant occur in DNT Co.? To be looked for in wet, disturbed sites. Known from about 30 sites in OR, WA, and British Columbia. Candidate for state listing in OR. Sometimes mistaken for *M. fontana* or *M. dichotoma.* See *Proceedings of the American Academy of Arts and Sciences* 18:191 (1883) for original description.

Montia saxosa

Considered but rejected: A synonym of *Claytonia saxosa*; a common taxon

Mucronea californica Benth.

"California spineflower" Polygonaceae
CNPS List: 4 **R-E-D Code:** 1-2-3 **State/Fed. Status:** CEQA?
Distribution: KRN, LAX, MNT, ORA, RIV, SBA, SBD, SDG, SLO, VEN
Habitat: Chprl, CmWld, CoDns, CoScr, VFGrs / sandy
Life Form: Annual herb
Blooming: March-August
Notes: Rare in southern California. Many herbarium records old. Threatened by aggregate mining, vehicles, flood control modification, and water percolation projects. Includes *Chorizanthe californica* var. *suskdorfii.* See *Phytologia* 66(3):203-205 (1989) for revised nomenclature.

Muhlenbergia appressa C. Goodd.

"appressed muhly" Poaceae
CNPS List: 2 **R-E-D Code:** 2-2-1 **State/Fed. Status:** CEQA
Distribution: SBD, SCM, AZ, BA
Quads: 176A, SCMC, SCMS
Habitat: CoScr, MDScr, VFGrs / rocky
Life Form: Annual herb
Blooming: April-May
Notes: Has this taxon been poorly collected in CA? See *Journal of the Washington Academy of Sciences* 31:504 (1914) for original description, and *Madroño* 35(4):353 (1988) for discussion of SCM Isl. records.

Muhlenbergia arsenei Hitchc.

"tough muhly" Poaceae
CNPS List: 2 **R-E-D Code:** 2-1-1 **State/Fed. Status:** CEQA
Distribution: SBD, AZ, BA, NM, NV, UT
Quads: 200B, 225D, 249D
Habitat: PJWld (rocky, carbonate)
Life Form: Perennial herb (rhizomatous)
Blooming: August-October
Notes: Known in CA only from the Clark and New York Mtns.

Muhlenbergia californica Vasey

"California muhly" Poaceae
CNPS List: 1B **R-E-D Code:** 2-2-3 **State/Fed. Status:** CEQA
Distribution: LAX, RIV, SBD
Quads: 83B, 83C, 108D, 110A, 135A, 135D, 136D
Habitat: Chprl, CoScr, LCFrs, Medws / mesic, seeps and streambanks
Life Form: Perennial herb (rhizomatous)
Blooming: July-September
Notes: See *Botanical Gazette* 7:92 (1882) for original description.

Muhlenbergia fragilis Swall.

"delicate muhly" Poaceae
CNPS List: 2 **R-E-D Code:** 3-1-1 **State/Fed. Status:** CEQA
Distribution: SBD, AZ, BA, NM, SO, TX+
Quads: 225D, 249D
Habitat: PJWld (carbonate, gravelly)
Life Form: Annual herb
Blooming: October
Notes: Known in CA only from the Clark and New York Mtns. See *Contributions from the U.S. National Herbarium* 29:206 (1947) for original description, and *Madroño* 35(4):353 (1988) for first CA record.

Muhlenbergia pauciflora Buckl.

"few-flowered muhly" Poaceae
CNPS List: 2 **R-E-D Code:** 3-1-1 **State/Fed. Status:** CEQA
Distribution: SBD, AZ, ++
Quads: 200A, 225D
Habitat: PJWld (rocky)
Life Form: Perennial herb (rhizomatous)
Blooming: September
Notes: Known in CA only from the New York Mtns. See *Madroño* 35(4):353-359 (1988) for first CA records.

Muilla clevelandii (Wats.) Hoov.

"San Diego goldenstar" Liliaceae
CNPS List: 1B **R-E-D Code:** 2-2-2 **State/Fed. Status:** /C2
Distribution: SDG, BA
Quads: 10A, 10B, 10C, 10D, 21B, 21C, 21D, 22A, 22B, 22C, 22D, 33D, 35C
Habitat: Chprl, CoScr, VFGrs, VnPls
Life Form: Perennial herb (bulbiferous)
Blooming: May
Notes: Threatened by urbanization, road construction, vehicles, and illegal dumping.

Muilla coronata Greene

"crowned muilla" Liliaceae
CNPS List: 4 **R-E-D Code:** 1-2-2 **State/Fed. Status:** /C3c
Distribution: INY, KRN, LAX, SBD, TUL, NV
Habitat: JTWld, MDScr, PJWld
Life Form: Perennial herb (bulbiferous)
Blooming: March-April
Notes: See *Pittonia* 1:165 (1888) for original description, and *Aliso* 10(4):621-627 (1984) for taxonomic treatment.
Status Report: 1978

Muilla transmontana

Considered but rejected: Too common

Munroa squarrosa (Nutt.) Torr.

"false buffalo-grass" Poaceae
CNPS List: 2 **R-E-D Code:** 3-2-1 **State/Fed. Status:** CEQA
Distribution: SBD, AZ, NV, ++
Quads: 200A, 249D
Habitat: PJWld (gravelly or rocky)
Life Form: Annual herb
Blooming: October
Notes: Known in CA only from the Clark and New York Mtns. May appear only after heavy summer rains.

Munzothamnus blairii

See *Stephanomeria blairii*

Myosotis laxa

Considered but rejected: Too common

Myosurus minimus L. ssp. *apus* (Greene) G.R. Campbell

"little mousetail" Ranunculaceae
CNPS List: 3 **R-E-D Code:** 2-3-2 **State/Fed. Status:** /C2
Distribution: ALA, BUT, CCA, COL, KRN, RIV, SBD, SDG, SOL, STA, BA, OR
Quads: 10A, 10B, 10C, 11D, 22B, 22D, 34C, 68A, 68B, 68C, 69A, 69D, 86A, 86D, 131B, 423B, 424A, 445B, 463D?, 498D, 561D, 562C
Habitat: VnPls (alkaline)
Life Form: Annual herb
Blooming: March-June
Notes: Move to List 1B? Need quads for KRN Co. Reduced by vernal pool habitat loss; threatened by vehicles, grazing, and agriculture. Endangered in OR. Taxonomic problems; distinguishing between this taxon and *M. sessilis* (= *M. minimus* ssp. *apus* var. *sessiliflorus* in *A California Flora* (1959) by P. Munz) is difficult; are both rare? May be a stabilized hybrid between *M. minimus* and *M. sessilis*, at least in the Central Valley; see *Evolution* 13:151-174 (1959) for details. See *M. minimus* in *The Jepson Manual.* See *Bulletin of the California Academy of Sciences* 1:277 (1885) for original description.

Myrica hartwegii

Considered but rejected: Too common

Myriophyllum quitense

Considered but rejected: Not native

Nama dichotomum (Ruiz, Lopez & Pav.) Choisy var. *dichotomum*

"forked purple mat" Hydrophyllaceae
CNPS List: 2 **R-E-D Code:** 3-1-1 **State/Fed. Status:** CEQA
Distribution: SBD, AZ, NM, TX+
Quads: 225D
Habitat: PJWld (granitic or carbonate)
Life Form: Annual herb
Blooming: September-October
Notes: Known in CA only from the New York Mtns.

Nama stenocarpum Gray

"mud nama" Hydrophyllaceae
CNPS List: 2 **R-E-D Code:** 3-2-1 **State/Fed. Status:** CEQA
Distribution: IMP, LAX*, ORA, SCM, SDG, AZ, BA+
Quads: 1A, 1B, 10B, 11A, 36A, 71D, 72A, 111C*, SCMN
Habitat: MshSw (lake margins, riverbanks)
Life Form: Annual/Perennial herb
Blooming: January-July
Notes: See *Proceedings of the American Academy of Arts and Sciences* 10:331 (1875) for original description, and *American Journal of Botany* 20:415-430, 518-534 (1933) for taxonomic treatment.

Nasturtium gambellii

See *Rorippa gambellii*

Navarretia eriocephala Mason

"hoary navarretia" Polemoniaceae
CNPS List: 4 **R-E-D Code:** 1-1-3 **State/Fed. Status:** CEQA?
Distribution: AMA, CAL, ELD, PLA, SAC
Habitat: CmWld, VFGrs
Life Form: Annual herb
Blooming: May-June
Notes: Intergrades somewhat with *N. heterandra.* See *Madroño* 8:196-197 (1946) for original description.

Navarretia fossalis Moran

"spreading navarretia" Polemoniaceae
CNPS List: 1B **R-E-D Code:** 2-3-2 **State/Fed. Status:** /C1
Distribution: RIV, SDG, BA
Quads: 10B, 10C, 11A, 11D, 22B, 22C, 34C, 35B, 35C, 36D, 51C, 68A, 85C, 85D
Habitat: ChScr, MshSw (assorted shallow freshwater), VnPls
Life Form: Annual herb
Blooming: April-June
Notes: Threatened by agriculture, road construction, grazing, and urbanization. See *Madroño* 24(3):155-159 (1977) for original description.

Navarretia heterandra Mason

"Tehama navarretia" Polemoniaceae
CNPS List: 4 **R-E-D Code:** 1-1-2 **State/Fed. Status:** CEQA?
Distribution: BUT, COL, LAK, SHA, TEH, TRI, YUB, OR*
Habitat: VFGrs (mesic), VnPls
Life Form: Annual herb
Blooming: May-June
Notes: To be expected elsewhere; need information. See *Madroño* 8:197 (1946) for original description.

Navarretia heterodoxa ssp. *rosulata*

See *Navarretia rosulata*

Navarretia jaredii Eastw.

"Paso Robles navarretia" Polemoniaceae
CNPS List: 4 **R-E-D Code:** 1-1-3 **State/Fed. Status:** CEQA?
Distribution: MNT, SLO
Habitat: Chprl?, CmWld, VFGrs / clay, serpentinite
Life Form: Annual herb
Blooming: April-June

Navarretia jepsonii Jeps.

"Jepson's navarretia" Polemoniaceae
CNPS List: 4 **R-E-D Code:** 1-1-3 **State/Fed. Status:** CEQA?
Distribution: COL, GLE, LAK, NAP, TEH, YOL
Habitat: Chprl, CmWld, VFGrs / serpentinite
Life Form: Annual herb
Blooming: May-June

Navarretia leucocephala Benth. ssp. *bakeri* (Mason) Day

"Baker's navarretia" Polemoniaceae
CNPS List: 1B **R-E-D Code:** 2-2-3 **State/Fed. Status:** CEQA
Distribution: COL, LAK, MEN, MRN, NAP, SOL, SON, TEH
Quads: 484A, 498D, 501A, 501B, 502A, 517C, 517D, 518D, 533A, 533B, 533D, 566C, 567A, 583C, 594C
Habitat: CmWld, LCFrs, Medws (mesic), VFGrs, VnPls
Life Form: Annual herb
Blooming: May-July
Notes: May be more widespread; need information. Need quads for COL Co. *N. leucocephala* on review list in OR. See *Madroño* 8(6):198 (1946) for original description, and *Novon* 3(4):331-340 (1993) for revised nomenclature.

Navarretia leucocephala Benth. ssp. *pauciflora* (Mason) Day

"few-flowered navarretia" Polemoniaceae
CNPS List: 1B **R-E-D Code:** 3-3-3 **State/Fed. Status:** CT/C1
Distribution: LAK, NAP
Quads: 499B, 500A, 533A, 533B, 533C, 534A
Habitat: VnPls (volcanic ash flow)
Life Form: Annual herb
Blooming: June
Notes: Threatened by erosion, grazing, vehicles, and recreation. Intergrades rarely with ssp. *plieantha.* State-listed as *N. pauciflora.* See *Madroño* 8:200 (1946) for original description, and *Novon* 3(4):331-340 (1993) for revised nomenclature.
Status Report: 1990

Navarretia leucocephala Benth. ssp. *plieantha* (Mason) Day

"many-flowered navarretia" Polemoniaceae
CNPS List: 1B **R-E-D Code:** 3-2-3 **State/Fed. Status:** CE/C1
Distribution: LAK, SON
Quads: 501A, 517C, 518D, 533B, 533C, 533D, 534A
Habitat: VnPls (volcanic ash flow)
Life Form: Annual herb
Blooming: May-June
Notes: Known from fewer than ten occurrences. Threatened by grazing, development, and vehicles. Protected in part at Loch Lomond ER (DFG). State-listed as *N. plieantha*; USFWS also uses this name. See *Madroño* 8:199 (1946) for original description, and *Novon* 3(4):331-340 (1993) for revised nomenclature.
Status Report: 1987

Navarretia mitracarpa ssp. *jaredii*

See *Navarretia jaredii*

Navarretia myersii P.S. Allen & Day

"pincushion navarretia" Polemoniaceae
CNPS List: 1B **R-E-D Code:** 3-3-3 **State/Fed. Status:** CEQA
Distribution: AMA, LAK, MER, SAC
Quads: 421A, 494B, 511B, 533D
Habitat: VnPls
Life Form: Annual herb
Blooming: May
Notes: Known from six occurrences. Threatened by development. See *Novon* 3(4):337-340 (1993) for original description.

Navarretia nigelliformis Greene ssp. *radians* (J.T. Howell) Day

"shining navarretia" Polemoniaceae
CNPS List: 1B **R-E-D Code:** 2-2-3 **State/Fed. Status:** CEQA
Distribution: FRE, MER, MNT, SBT, SLO
Quads: 268A, 269B, 269C, 269D, 292B, 292D, 315C, 364D, 384A, 384D
Habitat: CmWld, VFGrs, VnPls
Life Form: Annual herb
Blooming: May-June
Notes: Similar to *N. heterandra.* See *Leaflets of Western Botany* 2(8):136 (1938) for original description, and *Novon* 3(4):339 (1993) for revised nomenclature.

Navarretia pauciflora

See *Navarretia leucocephala* ssp. *pauciflora*

Navarretia peninsularis Greene

"Baja navarretia" Polemoniaceae
CNPS List: 1B **R-E-D Code:** 2-2-2 **State/Fed. Status:** CEQA
Distribution: KRN, SBA, SBD, SDG, BA
Quads: 20A, 105A, 131C, 131D, 167B, 238A
Habitat: LCFrs (mesic)
Life Form: Annual herb
Blooming: June-August
Notes: To be looked for in RIV Co. How common is plant in BA? Threatened in SBD Co. by gold-panning and vehicles. Similar to *N. breweri.* See *Pittonia* 1:136 (1887) for original description.

Navarretia plieantha

See *Navarretia leucocephala* ssp. *plieantha*

Navarretia prolifera Greene ssp. *lutea* (Brand) Mason

"yellow bur navarretia" Polemoniaceae
CNPS List: 4 **R-E-D Code:** 1-1-3 **State/Fed. Status:** /C3c
Distribution: ELD, PLA
Habitat: Chprl, CmWld
Life Form: Annual herb
Blooming: May-July
Notes: Eldorado NF has adopted species management guidelines.
Status Report: 1979

Navarretia prolifera ssp. *prolifera*

Considered but rejected: Too common

Navarretia rosulata Brand

"Marin County navarretia" Polemoniaceae
CNPS List: 1B **R-E-D Code:** 2-2-3 **State/Fed. Status:** CEQA
Distribution: MRN, NAP
Quads: 467A, 467B, 516D
Habitat: CCFrs, Chprl / serpentinite
Life Form: Annual herb
Blooming: June-July

Navarretia setiloba Cov.

"Piute Mtns. navarretia" Polemoniaceae
CNPS List: 1B **R-E-D Code:** 3-3-3 **State/Fed. Status:** /C1
Distribution: KRN, TUL
Quads: 189A*, 238D, 239D*, 260C, 261B, 261D, 262D, 285C
Habitat: CmWld, PJWld, VFGrs / clay or gravelly loam
Life Form: Annual herb
Blooming: April-June
Notes: Known from fewer than twenty occurrences. Many historical occurrences have been searched without success. Threatened by residential development at Bodfish, KRN Co. See *Contributions from the U.S. National Herbarium* 4:153 (1893) for original description.
Status Report: 1977

Navarretia subuligera Greene

"awl-leaved navarretia" Polemoniaceae
CNPS List: 4 **R-E-D Code:** 1-1-2 **State/Fed. Status:** CEQA?
Distribution: AMA, BUT, DNT, LAK, MEN, MOD, NAP?, SHA, TEH, OR
Habitat: CmWld, LCFrs / rocky, mesic
Life Form: Annual herb
Blooming: May-August
Notes: Does plant occur in NAP Co.?

Nemacaulis denudata Nutt. var. *denudata*

"coast woolly-heads" Polygonaceae
CNPS List: 2 **R-E-D Code:** 2-2-1 **State/Fed. Status:** CEQA
Distribution: LAX, ORA, SCT, SDG, BA
Quads: 11A, 11B, 11D, 22B, 22C, 36A, 36B, 36D, 71B, 72A, 73A, 89C*, 90D*
Habitat: CoDns
Life Form: Annual herb
Blooming: April-September
Notes: Need quads for SCT Isl. Much reduced by development in coastal dunes. Intergrades with var. *gracilis* at some localities. See *Madroño* 27(2):101-109 (1980) and *Phytologia* 66(4):390-91 (1989) for taxonomic treatments.

Nemacaulis denudata Nutt. var. *gracilis* Goodm. & Benson

"slender woolly-heads" Polygonaceae
CNPS List: 2 **R-E-D Code:** 2-2-1 **State/Fed. Status:** CEQA
Distribution: RIV, SDG, AZ, BA, SO
Quads: 11B, 11D, 65A, 83D
Habitat: CoDns, DeDns, SDScr
Life Form: Annual herb
Blooming: March-May
Notes: Threatened by urbanization near Palm Springs (RIV Co.) and along coast. Intergrades with var. *denudata* at some coastal localities. See *Aliso* 4:89 (1958) for original description, and *Madroño* 27(2):101-109 (1980) and *Phytologia* 66(4):390-91 (1989) for taxonomic treatments.

Nemacladus gracilis Eastw.

"slender nemacladus" Campanulaceae
CNPS List: 4 **R-E-D Code:** 1-1-3 **State/Fed. Status:** CEQA?
Distribution: FRE, KNG, KRN, LAX, MER
Habitat: CmWld, VFGrs
Life Form: Annual herb
Blooming: March-May

Nemacladus montanus

Considered but rejected: Too common

Nemacladus twisselmannii J.T. Howell

"Twisselmann's nemacladus" Campanulaceae
CNPS List: 1B **R-E-D Code:** 3-2-3 **State/Fed. Status:** CR/C2
Distribution: KRN, TUL
Quads: 283B, 284C, 284D
Habitat: UCFrs (sandy, granitic)
Life Form: Annual herb
Blooming: July
Notes: Known from only two occurrences. See *Leaflets of Western Botany* 10(3-4):45-46 (1963) for original description.
Status Report: 1979

Nemophila parviflora Benth. var. *quercifolia* (Eastw.) Chandl.

"oak-leaved nemophila" Hydrophyllaceae
CNPS List: 4 **R-E-D Code:** 1-1-2 **State/Fed. Status:** CEQA?
Distribution: FRE, KRN, MAD, TUL, OR
Habitat: CmWld, LCFrs
Life Form: Annual herb
Blooming: May-June

Neostapfia colusana (Davy) Davy

"Colusa grass" Poaceae
CNPS List: 1B **R-E-D Code:** 1-3-3 **State/Fed. Status:** CE/PT
Distribution: COL*, MER, SOL, STA, YOL
Quads: 401B*, 402A, 402B, 420C, 421A, 421C, 421D, 422C, 422D, 441A, 441B, 441C, 441D, 442A*, 459C, 460A, 481D, 497B, 498D, 562A*
Habitat: VnPls
Life Form: Annual herb
Blooming: May-July
Notes: Threatened by agriculture, overgrazing, flood control, and non-native plants. See *Erythea* 6:110-113 (1898) for original description, and *Fremontia* 4(3):22-23 (1976) for species account and habitat information.
Status Report: 1986

Neviusia cliftonii Shevock, Ertter & D.W. Taylor

"Shasta snow-wreath" Rosaceae
CNPS List: 1B **R-E-D Code:** 3-2-3 **State/Fed. Status:** CEQA
Distribution: SHA
Quads: 664A, 664B, 664D
Habitat: LCFrs (carbonate)
Life Form: Shrub (deciduous)
Blooming: May
Notes: Known from fewer than ten occurrences near Lake Shasta. Potentially threatened by mining. See *Novon* 2(4):285-289 (1992) for original description, and *Fremontia* 22(3):3-13 (1993) for species account and information about discovery.

Nicolletia occidentalis

Considered but rejected: Too common

Nitrophila mohavensis Munz & Roos

"Amargosa nitrophila" Chenopodiaceae
CNPS List: 1B **R-E-D Code:** 3-3-2 **State/Fed. Status:** CE/FE
Distribution: INY, NV
Quads: 275C, 322C, 322D
Habitat: Plyas (mesic, clay)
Life Form: Perennial herb
Blooming: May-October
Notes: Known in CA from fewer than five occurrences near Carson Slough in the Amargosa Desert. Severely threatened by water diversion and habitat alteration. State-listed as Critically Endangered in NV. See *Aliso* 3:112-114 (1955) for original description.
Status Report: 1987

Nolina interrata Gentry

"Dehesa nolina" Liliaceae
CNPS List: 1B **R-E-D Code:** 3-3-2 **State/Fed. Status:** CE/C1
Distribution: SDG, BA
Quads: 10A, 21D
Habitat: Chprl (gabbroic or serpentinite)
Life Form: Perennial herb
Blooming: June-July
Notes: Known in CA from approximately ten occurrences. Threatened by residential development and horticultural collecting. See *Madroño* 8:179-184 (1946) for original description and 22:214 (1973) for second CA location.
Status Report: 1987

Notholaena californica

Considered but rejected: Too common

Notholaena cochisensis

See *Astrolepis cochisensis*

Notholaena limitanea var. *limitanea*

See *Argyrochosma limitanea* var. *limitanea*

Oenothera avita ssp. *eurekensis*

See *Oenothera californica* ssp. *eurekensis*

Oenothera caespitosa Nutt. ssp. *crinita* (Rydb.) Munz

"caespitose evening-primrose" Onagraceae
CNPS List: 4 **R-E-D Code:** 1-2-1 **State/Fed. Status:** CEQA?
Distribution: INY, SBD, NV+
Habitat: PJWld, SCFrs, SDScr
Life Form: Perennial herb (rhizomatous)
Blooming: June-September
Notes: Threatened by cattle grazing.

Oenothera californica (Wats.) Wats. ssp. *eurekensis* (Munz & Roos) W. Klein

"Eureka Dunes evening-primrose" Onagraceae
CNPS List: 1B **R-E-D Code:** 3-2-3 **State/Fed. Status:** CR/FE
Distribution: INY
Quads: 390C, 391A, 391D
Habitat: DeDns
Life Form: Perennial herb (rhizomatous)
Blooming: April-July
Notes: Known from fewer than five occurrences in the Eureka Dunes. Recovery plan has been completed by BLM; still threatened by vehicle trespass and non-native plants. State-listed as *O. avita* ssp. *eurekensis*. See *Aliso* 3(2):118 (1955) for original description, and *Biological Conservation* 46:217-242 (1988) for population biology.
Status Report: 1977

Oenothera deltoides Torr. & Frem. ssp. *howellii* (Munz) W. Klein

"Antioch Dunes evening-primrose" Onagraceae
CNPS List: 1B **R-E-D Code:** 3-3-3 **State/Fed. Status:** CE/FE
Distribution: CCA, SAC
Quads: 480C, 481C, 481D
Habitat: InDns
Life Form: Perennial herb
Blooming: March-September
Notes: Known from seven occurrences. Seriously threatened by mining, agriculture, and industrial development. Recovery work in progress. See *Aliso* 2:81 (1949) for original description, *Four Seasons* 3(1):2-4 (1969) for threat information, and *Biological Conservation* 65:257-278 (1993) for population biology.
Status Report: 1988

Oenothera heterochroma

Considered but rejected: A synonym of *Camissonia heterochroma*; a common taxon

Oenothera hookeri ssp. *wolfii*

See *Oenothera wolfii*

Oenothera wolfii (Munz) Raven, Dietrich & Stubbe

"Wolf's evening-primrose" Onagraceae
CNPS List: 1B **R-E-D Code:** 3-3-2 **State/Fed. Status:** /C1
Distribution: DNT, HUM, TRI, OR
Quads: 637B, 683C, 689D, 704D, 706D, 723B, 723D, 740B, 740C
Habitat: CBScr, CoDns, CoPrr, LCFrs / sandy, usually mesic
Life Form: Perennial herb
Blooming: June-October
Notes: Known from approximately twenty occurrences. Threatened by road maintenance, foot traffic, and hybridization with non-native *Oenothera* spp. Candidate for state listing in OR. See *Aliso* 2:16 (1949) for original description, and *Systematic Botany* 4:242-252 (1979) for revised nomenclature.

Oenothera xylocarpa

Considered but rejected: Too common

Ophioglossum californicum Prantl

"California adder's-tongue" Ophioglossaceae
CNPS List: 4 **R-E-D Code:** 1-2-2 **State/Fed. Status:** /C3c
Distribution: AMA, BUT, MER, MNT*, MPA, ORA, SBD*, SDG, STA, TUO, BA
Habitat: Chprl, VFGrs, VnPls (margins)
Life Form: Perennial herb (rhizomatous)
Fertile: December-May
Status Report: 1979

Ophioglossum lusitanicum ssp. *californicum*

See *Ophioglossum californicum*

Ophioglossum pusillum Raf.

"northern adder's-tongue" Ophioglossaceae
CNPS List: 1A **Last Seen:** 1894 **State/Fed. Status:** CEQA
Distribution: ELD*, SIS*, AZ, OR, WA, ++
Quads: 699D*
Habitat: MshSw (margins), VFGrs (mesic)
Life Form: Perennial herb (rhizomatous)
Fertile: July
Notes: Need historical quads from ELD Co. Endangered in OR, and state-listed as Threatened in WA.
Status Report: 1980

Ophioglossum vulgatum

See *Ophioglossum pusillum*

Opuntia basilaris Engelm. & Bigel. var. *brachyclada* (Griffiths) Munz

"short-joint beavertail" Cactaceae
CNPS List: 1B **R-E-D Code:** 3-2-3 **State/Fed. Status:** /C2
Distribution: LAX, SBD
Quads: 133B, 133C, 133D, 134A, 134B, 134D, 135A, 135B, 136A, 138A, 161C, 161D, 175B, 176A
Habitat: Chprl, JTWld, MDScr, PJWld
Life Form: Shrub (stem succulent)
Blooming: April-June
Notes: Threatened by urbanization, mining, horticultural collecting, grazing, and vehicles. Angeles NF has adopted species management guidelines.

Opuntia basilaris Engelm. & Bigel. var. *treleasei* (Coult.) Toumey

"Bakersfield cactus" Cactaceae
CNPS List: 1B **R-E-D Code:** 3-3-3 **State/Fed. Status:** CE/FE
Distribution: KRN
Quads: 214A, 214C, 214D, 215D, 238C, 238D*, 239A, 239B, 239C*, 239D, 240A*, 263D
Habitat: ChScr, VFGrs (sandy)
Life Form: Shrub (stem succulent)
Blooming: May
Notes: Threatened by energy development, agricultural conversion, grazing, vehicles, and especially urbanization in the Bakersfield area. Known by 1991 to be extirpated at eleven of thirty-five occurrences. USFWS uses the name *O. treleasei*. See *Contributions from the U.S. National Herbarium* 3:434-435 (1896) for original description, *Wasmann Journal of Biology* 25:289-290 (1967) for species account, and *Madroño* 39(1):79 (1992) for information on a new population.
Status Report: 1990

Opuntia bigelovii var. *hoffmannii*

Considered but rejected: A hybrid (*O.* x*fosbergii*)

Opuntia curvospina Griffiths

"curved-spine beavertail" Cactaceae
CNPS List: 2 **R-E-D Code:** 2-2-1 **State/Fed. Status:** CEQA
Distribution: SBD, AZ, NV
Quads: 201A?, 224C, 225A, 225D
Habitat: Chprl, MDScr, PJWld
Life Form: Shrub (stem succulent)
Blooming: April-June
Notes: Stabilized hybrid between *O. phaeacantha* and *O. chlorotica*; see the latter in *The Jepson Manual*. See *Bulletin of the Torrey Botanical Club* 43:88-89 (1916) for original description, and *Systematic Botany* 5(4):408-418 (1980) for discussion of hybrid origin.

Opuntia fragilis (Nutt.) Haw.

"brittle prickly-pear" Cactaceae
CNPS List: 2 **R-E-D Code:** 3-3-1 **State/Fed. Status:** CEQA
Distribution: SIS(*?), AZ, OR, NV, ++
Quads: 716B(*?)
Habitat: PJWld (volcanic)
Life Form: Shrub (stem succulent)
Blooming: April-July
Notes: Known in CA only from Shasta Valley, where not collected since 1941; still extant? Field surveys needed. Probably reduced by grazing.

Opuntia munzii C.B. Wolf

"Munz's cholla" Cactaceae
CNPS List: 1B **R-E-D Code:** 3-1-3 **State/Fed. Status:** /C2
Distribution: IMP, RIV
Quads: 27D, 42B, 42C, 43A, 43D
Habitat: SDScr (sandy or gravelly)
Life Form: Shrub (stem succulent)
Blooming: May
Notes: Known from fewer than ten occurrences in the Chocolate Mtns. Of hybrid origin, but stabilized; perhaps only reproducing vegetatively. Relatively unthreatened as many occurrences are inaccessible; some populations threatened by military activities.

Opuntia parryi Engelm. var. *serpentina* (Engelm.) L. Benson

"snake cholla" Cactaceae
CNPS List: 1B **R-E-D Code:** 3-3-2 **State/Fed. Status:** /C2
Distribution: SDG, BA
Quads: 10C, 11A, 11B, 11D
Habitat: Chprl, CoScr
Life Form: Shrub (stem succulent)
Blooming: April-May
Notes: Threatened by development. A synonym of *O. parryi* in *The Jepson Manual*.

Opuntia phaeacantha var. *major*

Considered but rejected: Too common and taxonomic problem

Opuntia phaeacantha var. *mojavensis*

Considered but rejected: A synonym of *O. phaeacantha*; a common taxon

Opuntia pulchella Engelm.

"beautiful cholla" Cactaceae
CNPS List: 2 **R-E-D Code:** 2-2-1 **State/Fed. Status:** CEQA
Distribution: INY, MNO, AZ?, NV, UT+
Quads: 411B, 411C, 412D, 431A
Habitat: DeDns, GBScr?, MDScr
Life Form: Shrub (stem succulent)
Blooming: May-June
Notes: Threatened by grazing impacts. See *Madroño* 32(2):123 (1985) for first CA records.

Opuntia treleasei
See *Opuntia basilaris* var. *treleasei*

Opuntia wigginsii L. Benson

"Wiggins's cholla" Cactaceae
CNPS List: 3 **R-E-D Code:** 3-1-2 **State/Fed. Status:** /C3b
Distribution: IMP, RIV, SDG, AZ
Quads: 18A, 18B, 30C, 32C, 40B
Habitat: SDScr
Life Form: Shrub (stem succulent)
Blooming: March
Notes: Apparently a sporadic hybrid between *O. ramosissima* and *O. echinocarpa*; needs further study. See *O. ramosissima* in *The Jepson Manual*.

Opuntia wolfii (Benson) Baker

"Wolf's cholla" Cactaceae
CNPS List: 4 **R-E-D Code:** 1-1-3 **State/Fed. Status:** CEQA?
Distribution: IMP, SDG, BA?
Habitat: SDScr
Life Form: Shrub (stem succulent)
Blooming: April-May
Notes: Plant is locally common within range. May occur in BA; need confirmation. See *Cactus and Succulent Journal* 41:33 (1969) for original description, and *Madroño* 39(2):98-113 (1992) for revised nomenclature.

Orcuttia californica Vasey

"California Orcutt grass" Poaceae
CNPS List: 1B **R-E-D Code:** 3-3-2 **State/Fed. Status:** CE/FE
Distribution: LAX*, RIV, SDG, VEN, BA
Quads: 9B, 10C, 11D, 22C, 22D, 68A, 68B, 68C*, 68D, 69D, 89A*, 89B*, 89D*, 90A*, 139D
Habitat: VnPls
Life Form: Annual herb
Blooming: April-June
Notes: Known from fewer than twenty occurrences. Seriously threatened by agriculture, development, non-native plants, grazing, and vehicles. See *Bulletin of the Torrey Botanical Club* 13:219 (1886) for original description, and *American Journal of Botany* 69:1082-1095 (1982) for taxonomic treatment.
Status Report: 1987

Orcuttia californica var. *californica*
See *Orcuttia californica*

Orcuttia californica var. *inaequalis*
See *Orcuttia inaequalis*

Orcuttia californica var. *viscida*
See *Orcuttia viscida*

Orcuttia greenei
See *Tuctoria greenei*

Orcuttia inaequalis Hoov.

"San Joaquin Valley Orcutt grass" Poaceae
CNPS List: 1B **R-E-D Code:** 2-3-3 **State/Fed. Status:** CE/PE
Distribution: FRE, MAD, MER, STA*, TUL*
Quads: 333B*, 356B*, 378B*, 379A, 379D*, 398D, 400B, 400D*, 420C, 421A, 421B, 421D, 422B*, 441B*, 441C*, 442A*, 442D*
Habitat: VnPls
Life Form: Annual herb
Blooming: May-September
Notes: Seriously threatened by agriculture, development, overgrazing, and non-native plants. See *Madroño* 3:227-230 (1936) for original description, and *American Journal of Botany* 69:1082-1095 (1982) for taxonomic treatment.
Status Report: 1986

Orcuttia mucronata
See *Tuctoria mucronata*

Orcuttia pilosa Hoov.

"hairy Orcutt grass" Poaceae
CNPS List: 1B **R-E-D Code:** 2-3-3 **State/Fed. Status:** CE/PE
Distribution: BUT, GLE, MAD, MER, STA, TEH
Quads: 379A, 379B, 379C, 380A, 399C, 400D*, 421B*, 421C, 441A*, 441B*, 441C, 441D, 562B, 576B, 593B, 594A
Habitat: VnPls
Life Form: Annual herb
Blooming: May-August
Notes: Seriously threatened by agriculture, urbanization, overgrazing, non-native plants, and trampling. See *Bulletin of the Torrey Botanical Club* 68:149-156 (1941) for original description, and *American Journal of Botany* 69:1082-1095 (1982) for taxonomic treatment.
Status Report: 1987

Orcuttia tenuis Hitchc.

"slender Orcutt grass" Poaceae
CNPS List: 1B **R-E-D Code:** 2-3-3 **State/Fed. Status:** CE/PT
Distribution: LAK, PLU, SAC, SHA, SIS, TEH
Quads: 496A, 533D, 534A, 593B, 594A, 606B, 628A, 628B, 628C, 628D, 629A, 646C, 647D, 662B, 678A, 678B, 679D
Habitat: VnPls
Life Form: Annual herb
Blooming: May-July
Notes: Seriously threatened by agriculture, overgrazing, non-native plants, and residential development. Species management guidelines adopted by Lassen NF (USFS) and BLM. See *American Journal of Botany* 21:131 (1934) for original description and 69:1082-1095 (1982) for taxonomic treatment.
Status Report: 1987

Orcuttia viscida (Hoov.) J. Reeder

"Sacramento Orcutt grass" Poaceae
CNPS List: 1B **R-E-D Code:** 3-3-3 **State/Fed. Status:** CE/PE
Distribution: SAC
Quads: 495D, 511B, 511C
Habitat: VnPls
Life Form: Annual herb
Blooming: May-June
Notes: Known from seven occurrences. Seriously threatened by agriculture, urbanization, grazing, vehicles, and non-native plants. See *Bulletin of the Torrey Botanical Club* 68(3):155 (1941) for original description, and *American Journal of Botany* 69:1082-1095 (1982) for taxonomic treatment.
Status Report: 1977

Oreonana clementis

Considered but rejected: Too common

Oreonana purpurascens Shevock & Const.

"purple mountain-parsley" Apiaceae
CNPS List: 1B **R-E-D Code:** 2-2-3 **State/Fed. Status:** /C3c
Distribution: TUL
Quads: 308A, 308D, 331B, 331C, 331D, 354A
Habitat: BUFrs, SCFrs, UCFrs
Life Form: Perennial herb
Blooming: May-June
Notes: Threatened by road construction, logging, and trampling. See *Madroño* 26(3):128-134 (1979) for original description, and *Fremontia* 9(3):22-25 (1981) for species account.

Oreonana vestita (Wats.) Jeps.

"woolly mountain-parsley" Apiaceae
CNPS List: 4 **R-E-D Code:** 1-1-3 **State/Fed. Status:** CEQA?
Distribution: LAX, SBD
Habitat: SCFrs, UCFrs / scree
Life Form: Perennial herb
Blooming: June-July
Notes: See *Fremontia* 9(3):22-25 (1981) for species account.

Ornithostaphylos oppositifolia (Parry) Small

"Baja California birdbrush" Ericaceae
CNPS List: 2 **R-E-D Code:** 3-3-1 **State/Fed. Status:** CEQA
Distribution: SDG, BA
Quads: 11D
Habitat: Chprl
Life Form: Shrub (evergreen)
Blooming: January-April
Notes: Known in CA from only one occurrence west of San Ysidro near the Mexican border. Seriously threatened by aggregate mining operation.

Orobanche parishii (Jeps.) Heckard ssp. *brachyloba* Heckard

"short-lobed broom-rape" Orobanchaceae
CNPS List: 1B **R-E-D Code:** 2-2-2 **State/Fed. Status:** /C1
Distribution: SCT, SCZ, SDG, SLO, SMI, SNI, SRO, BA
Quads: 11B*, 36D, 221D, SCTS, SCZA, SMIE, SMIW, SNIC, SROW
Habitat: CBScr, CoDns, CoScr / sandy
Life Form: Perennial herb (parasitic)
Blooming: May-August
Notes: To be expected elsewhere; need information. Parasitic on shrubs such as *Isocoma menziesii.* See *Madroño* 22:68 (1973) for original description.

Orobanche valida

See *Orobanche valida* ssp. *valida*

Orobanche valida Jeps. ssp. *howellii* Heckard & Collins

"Howell's broomrape" Orobanchaceae
CNPS List: 4 **R-E-D Code:** 1-1-3 **State/Fed. Status:** CEQA?
Distribution: GLE, LAK, MEN, NAP, SON, TEH
Habitat: Chprl (serpentinite or volcanic)
Life Form: Perennial herb (parasitic)
Blooming: June-September
Notes: Generally parasitic on *Garrya* spp. See *Madroño* 29(2):95-100 (1982) for original description.

Orobanche valida Jeps. ssp. *valida*

"Rock Creek broomrape" Orobanchaceae
CNPS List: 1B **R-E-D Code:** 3-2-3 **State/Fed. Status:** /C2
Distribution: LAX
Quads: 108B, 135A
Habitat: Chprl, PJWld / granitic
Life Form: Perennial herb (parasitic)
Blooming: May-July
Notes: Known from only two occurrences in the San Gabriel Mtns. Parasitic on various chaparral shrubs. See *Madroño* 1:255-256 (1929) for original description, *Fremontia* 11(1):16-18 (1983) for discussion of rediscovery (1979), and *Madroño* 29(2):95-100 (1982) for taxonomic treatment.
Status Report: 1977

Orochaenactis thysanocarpha

Considered but rejected: Too common

Orthocarpus campestris var. *succulentus*

See *Castilleja campestris* ssp. *succulenta*

Orthocarpus castillejoides var. *humboldtiensis*

See *Castilleja ambigua* ssp. *humboldtiensis*

Orthocarpus cuspidatus
See *Orthocarpus cuspidatus* ssp. *cuspidatus*

Orthocarpus cuspidatus Greene ssp. *cuspidatus*
"Siskiyou Mtns. orthocarpus" Scrophulariaceae
CNPS List: 4 **R-E-D Code:** 1-1-2 **State/Fed. Status:** CEQA?
Distribution: HUM, SIS, OR
Habitat: LCFrs, UCFrs
Life Form: Annual herb
Blooming: June-August
Notes: On watch list in OR. Intergrades with ssp. *copelandii.* See *Systematic Botany* 17(4):560-582 (1992) for taxonomic treatment.

Orthocarpus floribundus
See *Triphysaria floribunda*

Orthocarpus lasiorhynchus
See *Castilleja lasiorhyncha*

Orthocarpus pachystachyus Gray
"Shasta orthocarpus" Scrophulariaceae
CNPS List: 1A **Last Seen:** 1913 **State/Fed. Status:** /C3a
Distribution: SIS*
Quads: 698B*, 717B*, 717D*
Habitat: GBScr, VFGrs
Life Form: Annual herb
Blooming: May
Notes: Location information vague, but area of historical distribution is farmed and grazed.
Status Report: 1978

Orthocarpus purpurascens var. *pallidus*
Considered but rejected: Too common; a synonym of *Castilleja exserta* ssp. *exserta*

Orthocarpus succulentus
See *Castilleja campestris* ssp. *succulenta*

Oryctes nevadensis Wats.
"Nevada oryctes" Solanaceae
CNPS List: 1B **R-E-D Code:** 3-3-2 **State/Fed. Status:** /C2
Distribution: INY, NV
Quads: 351A, 351D, 372B, 372D, 393A, 413A, 413B, 413C, 413D
Habitat: ChScr, MDScr / sandy
Life Form: Annual herb
Blooming: April-June
Notes: Known in CA from fewer than ten occurrences in Owens Valley. Seriously threatened by grazing, trampling, and vehicles. On watch list in NV. See *Botany of the King Expedition,* p. 274 (1871) for original description.

Oryzopsis kingii
Considered but rejected: Too common; a synonym of *Ptilagrostis kingii*

Oryzopsis micrantha
See *Piptatherum micranthum*

Oxystylis lutea
Considered but rejected: Too common

Oxytheca caryophylloides Parry
"chickweed oxytheca" Polygonaceae
CNPS List: 4 **R-E-D Code:** 1-1-3 **State/Fed. Status:** CEQA?
Distribution: LAX, RIV, SBD, TUL, VEN
Habitat: LCFrs (sandy)
Life Form: Annual herb
Blooming: July-September
Notes: See *Phytologia* 66(4):387-388 (1989) for taxonomic treatment.

Oxytheca emarginata Hall
"white-margined oxytheca" Polygonaceae
CNPS List: 4 **R-E-D Code:** 1-1-3 **State/Fed. Status:** CEQA?
Distribution: RIV
Habitat: LCFrs, PJWld
Life Form: Annual herb
Blooming: April-August
Notes: See *Phytologia* 66(4):388 (1989) for taxonomic treatment.

Oxytheca parishii Parry var. *abramsii* (McGregor) Munz
"Abrams's oxytheca" Polygonaceae
CNPS List: 1B **R-E-D Code:** 2-2-3 **State/Fed. Status:** CEQA
Distribution: SBA, VEN
Quads: 164C, 165D, 166A, 167B, 168A, 168B, 190C
Habitat: Chprl (sandy or shale)
Life Form: Annual herb
Blooming: June-August
Notes: See *Bulletin of the Torrey Botanical Club* 36:605 (1909) for original description, and *Brittonia* 32(1):70-102 (1980) and *Phytologia* 66(4):386-87 (1989) for taxonomic treatments.

Oxytheca parishii Parry var. *cienegensis* Ertter
"Cienega Seca oxytheca" Polygonaceae
CNPS List: 1B **R-E-D Code:** 3-1-3 **State/Fed. Status:** /C2
Distribution: SBD
Quads: 105A
Habitat: UCFrs (sandy, granitic)
Life Form: Annual herb
Blooming: June-September
Notes: Field work needed. See *Brittonia* 32:70-102 (1980) for original description, and *Phytologia* 66(4):386-87 (1989) for taxonomic treatment.

Oxytheca parishii Parry var. *goodmaniana* Ertter
"Cushenbury oxytheca" Polygonaceae
CNPS List: 1B **R-E-D Code:** 3-3-3 **State/Fed. Status:** /PE
Distribution: SBD
Quads: 131C, 131D
Habitat: PJWld (carbonate)
Life Form: Annual herb
Blooming: May-September
Notes: Threatened by carbonate mining. See *Brittonia* 32:70-102 (1980) for original description, and *Phytologia* 66(4):386-87 (1989) for taxonomic treatment.

Oxytheca watsonii T. & G.

"Watson's oxytheca" Polygonaceae
CNPS List: 2 **R-E-D Code:** 3-2-1 **State/Fed. Status:** CEQA
Distribution: INY, NV
Quads: 327A, 327B, 327C, 348C
Habitat: JTWld, MDScr / sandy
Life Form: Annual herb
Blooming: May-July
Notes: Known in CA from fewer than five occurrences. See *Madroño* 29(4):273-274 (1982) for first CA occurrences, and *Phytologia* 66(4):385-86 (1989) for taxonomic treatment.

Oxytropis deflexa (Pallas) DC. var. *sericea* T. & G.

"blue pendent-pod oxytrope" Fabaceae
CNPS List: 2 **R-E-D Code:** 3-3-1 **State/Fed. Status:** CEQA
Distribution: MNO, NV, OR, WA, ++
Quads: 431C
Habitat: Medws, UCFrs
Life Form: Perennial herb
Blooming: June-August
Notes: Known in CA from only five extant occurrences in the White Mtns. Threatened by cattle grazing.

Palafoxia arida Turner & Morris var. *gigantea* (M.E. Jones) Turner & Morris

"giant Spanish-needle" Asteraceae
CNPS List: 1B **R-E-D Code:** 2-1-2 **State/Fed. Status:** /C2
Distribution: IMP, AZ
Quads: 1B, 2A, 2B, 3A, 13B, 13C, 13D, 14A, 14B, 14D, 15A, 27C, 27D, 28D
Habitat: DeDns
Life Form: Annual/Perennial herb
Blooming: February-May
Notes: See *Contributions to Western Botany* 18:79 (1933) for original description, and *Rhodora* 78:604-605 (1976) for taxonomic treatment.
Status Report: 1977

Palafoxia linearis var. *gigantea*

See *Palafoxia arida* var. *gigantea*

Panicum shastense

Considered but rejected: A hybrid

Panicum thermale

See *Dichanthelium lanuginosum* var. *thermale*

Papaver californicum

Considered but rejected: Too common

Paronychia ahartii Ertter

"Ahart's paronychia" Caryophyllaceae
CNPS List: 1B **R-E-D Code:** 3-2-3 **State/Fed. Status:** /C2
Distribution: BUT, SHA, TEH
Quads: 560D, 593B, 593D, 594C, 610A, 611D, 627A, 628B, 628C, 628D, 646C, 646D
Habitat: CmWld, VFGrs, VnPls
Life Form: Annual herb
Blooming: April-June
Notes: Known only from the northern Sacramento Valley. Threatened by habitat loss, and possibly by grazing and trampling. See *Madroño* 32(2):87-90 (1985) for original description.

Paronychia franciscana

Considered but rejected: Not native

Parvisedum leiocarpum (H.K. Sharsm.) Clausen

"Lake County stonecrop" Crassulaceae
CNPS List: 1B **R-E-D Code:** 3-3-3 **State/Fed. Status:** CE/C1
Distribution: LAK
Quads: 533B, 533C, 533D
Habitat: CmWld, VFGrs, VnPls / vernally mesic depressions in rock outcrops
Life Form: Annual herb
Blooming: April-May
Notes: Known from approximately five occurrences. Extremely vulnerable to trampling; also threatened by grazing and development. See *Madroño* 5:192-194 (1940) for original description, and *Cactus and Succulent Journal* 18:58 (1946) for revised nomenclature.
Status Report: 1990

Parvisedum pentandrum

Considered but rejected: Too common

Pectocarya palmeri

See *Harpagonella palmeri*

Pedicularis bracteosa Benth. var. *flavida* (Pennell) Cronq.

"yellowish lousewort" Scrophulariaceae
CNPS List: 4 **R-E-D Code:** 1-1-1 **State/Fed. Status:** CEQA?
Distribution: SIS, TRI, OR
Habitat: UCFrs (mesic)
Life Form: Perennial herb
Blooming: July-August
Notes: See *Bulletin of the Torrey Botanical Club* 61:445 (1934) for original description.

Pedicularis centranthera Gray

"dwarf lousewort" Scrophulariaceae
CNPS List: 2 **R-E-D Code:** 3-1-1 **State/Fed. Status:** CEQA
Distribution: LAS, AZ, NV, OR, ++
Quads: 639D
Habitat: GBScr (alluvial)
Life Form: Perennial herb
Blooming: May-June
Notes: Known in CA from only two occurrences. Possibly threatened by agricultural development. On review list in OR. See *Report on the U.S. and Mexican Boundary Survey*, p. 120 (1859) by W. Emory for original description.

Pedicularis contorta Benth.

"curved-beak lousewort" Scrophulariaceae
CNPS List: 4 **R-E-D Code:** 1-1-1 **State/Fed. Status:** CEQA?
Distribution: SIS, TRI, OR, WA
Habitat: BgFns, Medws, LCFrs, UCFrs / mesic
Life Form: Perennial herb
Blooming: July-August

Pedicularis crenulata Benth.

"scalloped-leaved lousewort" Scrophulariaceae
CNPS List: 2 **R-E-D Code:** 3-1-1 **State/Fed. Status:** CEQA
Distribution: MNO, NV, WY+
Quads: 434D
Habitat: Medws (mesic)
Life Form: Perennial herb
Blooming: June-July
Notes: Known in CA from only one occurrence near Convict Creek. See *Madroño* 28(3):86 (1981) for this record.

Pedicularis dudleyi Elmer

"Dudley's lousewort" Scrophulariaceae
CNPS List: 1B **R-E-D Code:** 3-2-3 **State/Fed. Status:** CR/C2
Distribution: MNT, SCR*, SLO, SMT
Quads: 271B, 272A, 320B, 344D, 387B*, 428C, 429A
Habitat: Chprl (maritime), NCFrs, VFGrs
Life Form: Perennial herb
Blooming: April-June
Notes: Known from fewer than fifteen occurrences. Threatened by trampling, and potentially by development. Plants from Arroyo de la Cruz (SLO Co.) are different and warrant further study. See *Botanical Gazette* 41:316-317 (1906) for original description.
Status Report: 1988

Pedicularis flavida

See *Pedicularis bracteosa* var. *flavida*

Pedicularis howellii Gray

"Howell's lousewort" Scrophulariaceae
CNPS List: 4 **R-E-D Code:** 1-1-2 **State/Fed. Status:** /C3c
Distribution: SIS, OR
Habitat: UCFrs (often serpentinite)
Life Form: Perennial herb
Blooming: June-August
Notes: Endangered in OR. See *Proceedings of the American Academy of Arts and Sciences* 20:307 (1885) for original description.
Status Report: 1979

Pellaea truncata Goodd.

"cliff brake" Pteridaceae
CNPS List: 2 **R-E-D Code:** 2-1-1 **State/Fed. Status:** CEQA
Distribution: SBD, AZ, BA, NV, SO, ++
Quads: 225D
Habitat: PJWld (volcanic or granitic)
Life Form: Perennial herb (rhizomatous)
Fertile: April-June
Notes: Need quad for occurrence in the Providence Mtns. See *Madroño* 25(1):56 (1978) for species account.

Penstemon albomarginatus M.E. Jones

"white-margined beardtongue" Scrophulariaceae
CNPS List: 1B **R-E-D Code:** 3-2-2 **State/Fed. Status:** /C2
Distribution: SBD, AZ, NV
Quads: 150C, 153B, 154A, 179D
Habitat: DeDns (stabilized), MDScr
Life Form: Perennial herb
Blooming: March-May
Notes: Known in CA from approximately four occurrences; two have not been seen in many years. One confirmed extant occurrence was damaged by military activities in 1989. Threatened in NV.

Penstemon barnebyi N. Holmgren

"Barneby's beardtongue" Scrophulariaceae
CNPS List: 2 **R-E-D Code:** 3-3-1 **State/Fed. Status:** CEQA
Distribution: MNO, NV
Quads: 431A, 431B
Habitat: GBScr, PJWld / carbonate, gravelly
Life Form: Perennial herb
Blooming: May
Notes: Known in CA from only one occurrence along Busher Creek in the White Mtns. Possibly threatened by grazing. See *Brittonia* 31:226 (1979) for original description, and *Madroño* 35(2):164-165 (1988) for the CA record.

Penstemon calcareus Bdg.

"limestone beardtongue" Scrophulariaceae
CNPS List: 2 **R-E-D Code:** 2-1-1 **State/Fed. Status:** /C3c
Distribution: INY, SBD, NV
Quads: 175B, 176A, 201C, 348A, 368A, 368D, 369B, 390C, 390D
Habitat: JTWld, MDScr, PJWld / carbonate
Life Form: Perennial herb
Blooming: April-May

Penstemon californicus (Munz & Jtn.) Keck

"California beardtongue" Scrophulariaceae
CNPS List: 1B **R-E-D Code:** 3-2-2 **State/Fed. Status:** /C3c
Distribution: RIV, BA
Quads: 49A, 65C, 66A, 66B, 66C, 66D, 67A
Habitat: Chprl, LCFrs, PJWld / sandy
Life Form: Perennial herb
Blooming: May-June
Notes: Known in CA from fewer than twenty occurrences. Threatened by grazing. See *Bulletin of the Southern California Academy of Sciences* 23:31 (1924) for original description.
Status Report: 1979

Penstemon cinereus Piper

"gray beardtongue" Scrophulariaceae
CNPS List: 4 **R-E-D Code:** 1-1-1 **State/Fed. Status:** CEQA?
Distribution: MOD, SIS, NV, OR
Habitat: GBScr, PJWld / volcanic, gravelly
Life Form: Perennial herb
Blooming: May-August
Notes: Intergrades with *P. humilis* var. *humilis*; see that taxon in *The Jepson Manual*.

Penstemon cinicola Keck

"ash beardtongue" Scrophulariaceae
CNPS List: 4 **R-E-D Code:** 1-2-1 **State/Fed. Status:** /C3c
Distribution: LAS, MOD, SIS, OR
Habitat: LCFrs, Medws / volcanic, sandy
Life Form: Perennial herb
Blooming: June-August
Notes: See *Carnegie Institute of Washington Publication* 520:294 (1940) for original description.

Penstemon clevelandii ssp. *connatus*

See *Penstemon clevelandii* var. *connatus*

Penstemon clevelandii Gray var. *connatus* (Munz & Jtn.) N. Holmgren

"San Jacinto beardtongue" Scrophulariaceae
CNPS List: 4 **R-E-D Code:** 1-1-1 **State/Fed. Status:** CEQA?
Distribution: IMP, RIV, SDG, BA
Habitat: Chprl
Life Form: Perennial herb
Blooming: March-May

Penstemon confusus ssp. *confusus*

Considered but rejected: A synonym of *P. confusus*; not in CA

Penstemon confusus ssp. *patens*

Considered but rejected: A synonym of *P. patens*; a common taxon

Penstemon filiformis (Keck) Keck

"thread-leaved beardtongue" Scrophulariaceae
CNPS List: 1B **R-E-D Code:** 2-1-3 **State/Fed. Status:** /C2
Distribution: SHA, TRI
Quads: 665B, 666A, 666B, 667A, 667B, 667C, 667D, 682B, 682C, 682D, 683B, 683C, 683D
Habitat: CmWld, LCFrs / rocky
Life Form: Perennial herb
Blooming: June-July
Notes: Possibly threatened by logging and recreation. Confused with *P. laetus* var. *sagittatus*. See *University of California Publications in Botany* 16:394 (1932) for original description.

Penstemon floridus ssp. *floridus*

Considered but rejected: Too common

Penstemon fruticiformis ssp. *amargosae*

See *Penstemon fruticiformis* var. *amargosae*

Penstemon fruticiformis Cov. var. *amargosae* (Keck) N. Holmgren

"Death Valley beardtongue" Scrophulariaceae
CNPS List: 4 **R-E-D Code:** 1-1-2 **State/Fed. Status:** /C2
Distribution: INY, SBD, NV
Habitat: MDScr
Life Form: Perennial herb
Blooming: April-June
Notes: Threatened in NV. See *American Midland Naturalist* 18:801 (1937) for original description.

Penstemon heterodoxus Gray var. *shastensis* (Keck) N. Holmgren

"Shasta beardtongue" Scrophulariaceae
CNPS List: 4 **R-E-D Code:** 1-1-3 **State/Fed. Status:** CEQA?
Distribution: MOD, SHA, SIS
Habitat: LCFrs, Medws, UCFrs
Life Form: Perennial herb
Blooming: June-August
Notes: See *American Midland Naturalist* 33:165 (1945) for original description.

Penstemon monoensis

Considered but rejected: Too common

Penstemon neotericus

Considered but rejected: Too common

Penstemon newberryi ssp. *sonomensis*

See *Penstemon newberryi* var. *sonomensis*

Penstemon newberryi Gray var. *sonomensis* (Greene) Jeps.

"Sonoma beardtongue" Scrophulariaceae
CNPS List: 1B **R-E-D Code:** 3-1-3 **State/Fed. Status:** CEQA
Distribution: LAK, NAP, SON
Quads: 500A, 500B, 501A, 516B, 517A, 517B, 533C
Habitat: Chprl (rocky)
Life Form: Perennial herb
Blooming: May-July

Penstemon papillatus J.T. Howell

"Inyo beardtongue" Scrophulariaceae
CNPS List: 4 **R-E-D Code:** 1-1-3 **State/Fed. Status:** /C3c
Distribution: INY, KRN, MNO
Habitat: PJWld, SCFrs / rocky, granitic
Life Form: Perennial herb
Blooming: June-July

Penstemon personatus Keck

"closed-throated beardtongue" Scrophulariaceae
CNPS List: 1B **R-E-D Code:** 2-2-3 **State/Fed. Status:** /C2
Distribution: BUT, PLU
Quads: 589A, 590A, 590B, 590C, 590D, 591B, 591C, 591D, 592A, 605C, 606D
Habitat: LCFrs, UCFrs / metavolcanic
Life Form: Perennial herb
Blooming: July-September
Notes: Threatened by logging activities. Plumas NF has adopted species management guidelines. See *Madroño* 3:248-250 (1935) for original description.
Status Report: 1977

Penstemon purpusii Bdg.

"Snow Mtn. beardtongue" Scrophulariaceae
CNPS List: 4 **R-E-D Code:** 1-1-3 **State/Fed. Status:** CEQA?
Distribution: COL, GLE, HUM, LAK, MEN, SHA, TEH, TRI
Habitat: LCFrs, UCFrs / rocky, often serpentinite
Life Form: Perennial herb
Blooming: June-August

Penstemon rattanii ssp. *kleei*

See *Penstemon rattanii* var. *kleei*

Penstemon rattanii Gray var. *kleei* (Greene) Gray

"Santa Cruz Mtns. beardtongue" Scrophulariaceae
CNPS List: 1B **R-E-D Code:** 2-2-3 **State/Fed. Status:** CEQA
Distribution: SCL, SCR
Quads: 387B, 406C, 407D, 408B, 408C, 408D
Habitat: Chprl, LCFrs, NCFrs
Life Form: Perennial herb
Blooming: May-June

Penstemon scapoides

Considered but rejected: Too common

Penstemon shastensis

See *Penstemon heterodoxus* var. *shastensis*

Penstemon stephensii Bdg.

"Stephens's beardtongue" Scrophulariaceae
CNPS List: 1B **R-E-D Code:** 2-2-3 **State/Fed. Status:** /C2
Distribution: INY, SBD
Quads: 176A, 176C, 200C, 250A, 250B, 274C, 274D, 275A, 298D
Habitat: MDScr, PJWld / carbonate
Life Form: Perennial herb
Blooming: April-June
Notes: Threatened by limestone mining.

Penstemon thurberi Torr.

"Thurber's beardtongue" Scrophulariaceae
CNPS List: 4 **R-E-D Code:** 1-2-1 **State/Fed. Status:** CEQA?
Distribution: IMP, RIV, SBD, SDG, AZ, BA, NV
Habitat: Chprl, JTWld, PJWld, SDScr / sandy or gravelly
Life Form: Perennial herb
Blooming: May-July

Penstemon tracyi Keck

"Tracy's beardtongue" Scrophulariaceae
CNPS List: 1B **R-E-D Code:** 3-1-3 **State/Fed. Status:** /C3c
Distribution: TRI
Quads: 667B, 667C, 668A, 668B, 684C, 685D
Habitat: UCFrs (rocky)
Life Form: Perennial herb
Blooming: June-August
Notes: Known from fewer than ten occurrences. See *American Midland Naturalist* 23:603-605 (1940) for original description.
Status Report: 1979

Penstemon venustus

Considered but rejected: Not native

Pentachaeta bellidiflora Greene

"white-rayed pentachaeta" Asteraceae
CNPS List: 1B **R-E-D Code:** 3-3-3 **State/Fed. Status:** CE/PE
Distribution: MRN*, SCR*, SMT
Quads: 387B*, 408A*, 408B*, 408C*, 408D*, 429A, 448B*, 448C*, 448D*, 466B*, 467A*, 467D*
Habitat: VFGrs (often serpentinite)
Life Form: Annual herb
Blooming: March-May
Notes: Known from only one extended occurrence bisected by Highway 280; historical occurrences lost to development. See *Bulletin of the California Academy of Sciences* 1:86 (1885) for original description, and *University of California Publications in Botany* 65:1-41 (1973) for taxonomic treatment.

Pentachaeta exilis (Gray) Gray ssp. *aeolica* Van Horn & Ornduff

"slender pentachaeta" Asteraceae
CNPS List: 1B **R-E-D Code:** 3-2-3 **State/Fed. Status:** /C1
Distribution: MNT, SBT
Quads: 319C, 339C, 340D
Habitat: CmWld, VFGrs
Life Form: Annual herb
Blooming: April-May
Notes: Known from approximately five occurrences near The Indians (MNT Co.) and Hernandez (SBT Co.). See *University of California Publications in Botany* 65:1-41 (1973) for taxonomic treatment.

Pentachaeta lyonii Gray

"Lyon's pentachaeta" Asteraceae
CNPS List: 1B **R-E-D Code:** 3-3-3 **State/Fed. Status:** CE/PE
Distribution: LAX, SCT*, VEN
Quads: 73A*, 90D*, 112C, 113A, 113B, 113D, SCTN*
Habitat: Chprl (openings), VFGrs
Life Form: Annual herb
Blooming: March-August
Notes: Known from fewer than twenty extant occurrences. Threatened by development and recreational activities. See *Synoptical Flora of North America* 1(2):446 (1884) for original description, and *University of California Publications in Botany* 65:1-41 (1973) for taxonomic treatment.
Status Report: 1989

Perideridia bacigalupii Chuang & Const.

"Bacigalupi's yampah" Apiaceae
CNPS List: 4 **R-E-D Code:** 1-2-3 **State/Fed. Status:** /C3c
Distribution: AMA, BUT, CAL, MAD*, MPA, NEV, TUO
Habitat: Chprl, LCFrs / serpentinite
Life Form: Perennial herb
Blooming: June-August
Notes: See *University of California Publications in Botany* 55:36-38 (1969) for original description.
Status Report: 1978

Perideridia gairdneri (H. & A.) Math. ssp. *gairdneri*

"Gairdner's yampah" Apiaceae
CNPS List: 4 **R-E-D Code:** 1-2-3 **State/Fed. Status:** /C2
Distribution: DNT, HUM, KRN, LAS, LAX*, MEN, MNT, MOD, MRN, NAP, ORA*, SBT, SCL, SCR, SDG*, SIS, SLO, SMT(*?), SOL, SON, TRI
Habitat: BUFrs, Chprl, VFGrs, VnPls / mesic
Life Form: Perennial herb
Blooming: June-October
Notes: Endangered in the southern portion of its range; status of occurrences uncertain. Can be relatively common locally, especially in northern counties. Is plant extant in SMT Co.? Threatened by agriculture and urban development. See *University of California Publications in Botany* 55:1-74 (1969) for taxonomic treatment.

Perideridia leptocarpa Chuang & Const.

"narrow-seeded yampah" Apiaceae
CNPS List: 4 **R-E-D Code:** 1-1-2 **State/Fed. Status:** /C3c
Distribution: SIS, OR
Habitat: LCFrs (serpentinite)
Life Form: Perennial herb
Blooming: June-August
Notes: Taxonomic questions; possibly belongs in *P. oregana*. See *University of California Publications in Botany* 55:51-54 (1969) for original description.
Status Report: 1978

Perideridia parishii (Coult. & Rose) Nels. & Macbr. ssp. *parishii*

"Parish's yampah" Apiaceae
CNPS List: 2 **R-E-D Code:** 2-2-1 **State/Fed. Status:** CEQA
Distribution: SBD, AZ, NM, NV
Quads: 105A, 105B, 106A, 106B, 131C, 132C
Habitat: LCFrs, Medws, UCFrs
Life Form: Perennial herb
Blooming: June-July
Notes: See *Botanical Gazette* 12:157 (1887) for original description, and *University of California Publications in Botany* 55:1-74 (1969) for taxonomic treatment.
Status Report: 1980

Perideridia pringlei (Coult. & Rose) Nels. & Macbr.

"adobe yampah" Apiaceae
CNPS List: 4 **R-E-D Code:** 1-1-3 **State/Fed. Status:** /C3c
Distribution: KRN, LAX, MNT, NEV, SBA, SLO, TUL, VEN
Habitat: Chprl, CmWld, CoScr / serpentinite
Life Form: Perennial herb
Blooming: April-July

Perityle inyoensis (Ferris) Powell

"Inyo rock daisy" Asteraceae
CNPS List: 1B **R-E-D Code:** 3-2-3 **State/Fed. Status:** /C2
Distribution: INY
Quads: 327C, 350D
Habitat: PJWld (rocky)
Life Form: Perennial herb
Blooming: July-August
Notes: Known from fewer than ten occurrences. Threatened by proposed mining at Cerro Gordo Mine.

Perityle megalocephala var. *intricata*

Considered but rejected: A synonym of *P. megalocephala* var. *oligophylla*; a common taxon

Perityle megalocephala var. *oligophylla*

Considered but rejected: Too common

Perityle villosa (Blake) Shinners

"Hanaupah rock daisy" Asteraceae
CNPS List: 1B **R-E-D Code:** 3-1-3 **State/Fed. Status:** /C2
Distribution: INY
Quads: 302A(*?), 368B, 369B
Habitat: PJWld (rocky)
Life Form: Perennial herb
Blooming: June
Notes: Known from fewer than five extant occurrences. Has been searched for but not rediscovered in Hanaupah Cyn. Collected in 1980 on Mt. Palmer in the Grapevine Mtns.

Petalonyx gilmanii

See *Petalonyx thurberi* ssp. *gilmanii*

Petalonyx thurberi Gray ssp. *gilmanii* (Munz) Davis & Thompson

"Death Valley sandpaper-plant" Loasaceae
CNPS List: 1B **R-E-D Code:** 3-2-3 **State/Fed. Status:** /C2
Distribution: INY
Quads: 303A, 323C, 326B, 326C, 327D, 348A, 348D, 389C
Habitat: DeDns, MDScr
Life Form: Shrub (evergreen)
Blooming: May-September
Notes: Known from fewer than twenty occurrences. Distinctiveness from ssp. *thurberi* needs study.

Peteria thompsoniae Wats.

"spine-noded milk vetch" Fabaceae
CNPS List: 2 **R-E-D Code:** 3-1-1 **State/Fed. Status:** /C3c
Distribution: INY, AZ, ID, NV, UT+
Quads: 274C
Habitat: MDScr (bajadas)
Life Form: Perennial herb
Blooming: May-June
Notes: Known in CA from only one occurrence in California Valley. Endangered in ID. See *Madroño* 34(4):381 (1987) for the CA record.

Petradoria discoidea

See *Chrysothamnus gramineus*

Petunia parviflora

Considered but rejected: Too common

Phacelia amabilis Const.

"Saline Valley phacelia" Hydrophyllaceae
CNPS List: 3 **R-E-D Code:** 3-1-3 **State/Fed. Status:** /C2*
Distribution: INY
Quads: 347A, 350A*
Habitat: RpScr, SCFrs
Life Form: Annual herb
Blooming: April-May
Notes: Move to List 1B? Rediscovered in the mid-1980's by M. DeDecker above Mud Cyn., growing within a population of *P. crenulata*; probably a variant of the latter. Repeated searches of the type locality have been unsuccessful.

Phacelia anelsonii Macbr.

"Aven Nelson's phacelia" Hydrophyllaceae
CNPS List: 2 **R-E-D Code:** 2-1-1 **State/Fed. Status:** /C3c
Distribution: INY, SBD, NV+
Quads: 224C, 226A, 327A
Habitat: JTWld, PJWld / carbonate
Life Form: Annual herb
Blooming: April-May
Notes: See *Contributions from the Gray Herbarium* 49:26 (1917) for original description.

Phacelia argentea Nels. & Macbr.

"sand dune phacelia" Hydrophyllaceae
CNPS List: 1B **R-E-D Code:** 3-3-2 **State/Fed. Status:** /C2
Distribution: DNT, OR
Quads: 740B, 740C
Habitat: CoDns
Life Form: Perennial herb
Blooming: June-August
Notes: Known in CA from only four occurrences near Lake Earl and the Smith River Dunes. Threatened by coastal development, vehicles, and non-native plants. Candidate for state listing in OR. See *Botanical Gazette* 61:34 (1916) for original description.

Phacelia calthifolia

Considered but rejected: Too common

Phacelia ciliata Benth. var. *opaca* J.T. Howell

"Merced phacelia" Hydrophyllaceae
CNPS List: 1B **R-E-D Code:** 3-1-3 **State/Fed. Status:** /C2
Distribution: MER
Quads: 400B, 401A, 420C, 421C, 421D
Habitat: VFGrs (clay)
Life Form: Annual herb
Blooming: February-May
Notes: Known from fewer than ten extant occurrences. See *P. ciliata* in *The Jepson Manual*. See *Leaflets of Western Botany* 1:221 (1936) for original description.
Status Report: 1979

Phacelia cinerea Eastw.

"ashy phacelia" Hydrophyllaceae
CNPS List: 1A **Last Seen:** 1901 **State/Fed. Status:** /C2*
Distribution: SNI*
Quads: SNIC*
Habitat: Medws (mesic)
Life Form: Perennial herb
Blooming: March-April
Notes: Known only from the type collection. Rediscovery attempt in 1977 and others since 1983 have been unsuccessful. Not in *The Jepson Manual*. See *Contributions from the Gray Herbarium* 49:26 (1917) for original description.
Status Report: 1979

Phacelia cookei Const. & Heckard

"Cooke's phacelia" Hydrophyllaceae
CNPS List: 1B **R-E-D Code:** 3-3-3 **State/Fed. Status:** /C2
Distribution: SIS
Quads: 698B, 715C
Habitat: GBScr, LCFrs / sandy
Life Form: Annual herb
Blooming: June-July
Notes: Known from approximately five occurrences. Threatened by non-native plants, and possibly by fire suppression. See *Brittonia* 22(1):25-30 (1970) for original description.
Status Report: 1977

Phacelia dalesiana J.T. Howell

"Scott Mtn. phacelia" Hydrophyllaceae
CNPS List: 1B **R-E-D Code:** 1-2-3 **State/Fed. Status:** /C2
Distribution: SHA, SIS, TRI
Quads: 681A, 682B, 682C, 683A, 683B, 683D, 684A, 684D, 699C, 700C, 700D
Habitat: LCFrs, Medws, SCFrs, UCFrs / serpentinite
Life Form: Perennial herb
Blooming: May-June
Notes: Threatened by logging and grazing. See *Leaflets of Western Botany* 2:51 (1937) for original description.

Phacelia divaricata var. *insularis*

See *Phacelia insularis* var. *insularis*

Phacelia exilis (Gray) G.J. Lee

"Transverse Range phacelia" Hydrophyllaceae
CNPS List: 4 **R-E-D Code:** 1-1-3 **State/Fed. Status:** CEQA?
Distribution: KRN, LAX, SBD, TUL, VEN
Habitat: LCFrs, Medws, UCFrs / sandy or gravelly
Life Form: Annual herb
Blooming: May-August
Notes: Known from fewer than ten occurrences, but can be locally common. Difficult to separate from *P. mohavensis*. See *Synoptical Flora of North America* 2(1):165 (1878) for original description, and *Systematic Botany* 13(1):16-20 (1988) for revised nomenclature.

Phacelia floribunda Greene

"many-flowered phacelia" Hydrophyllaceae
CNPS List: 1B **R-E-D Code:** 3-2-2 **State/Fed. Status:** /C2
Distribution: SBR, SCM, GU
Quads: SBRA, SCMC, SCMN, SCMS
Habitat: CoScr
Life Form: Annual herb
Blooming: March-May
Notes: Known from fewer than twenty occurrences over its range. Feral herbivores removed from SBR and SCM islands, and vegetation recovering.

Phacelia greenei J.T. Howell

"Scott Valley phacelia" Hydrophyllaceae
CNPS List: 1B **R-E-D Code:** 2-2-3 **State/Fed. Status:** /C2
Distribution: SIS
Quads: 700C, 700D, 701A, 701C, 701D, 717B, 717C, 718C, 718D
Habitat: CCFrs, LCFrs, SCFrs, UCFrs / serpentinite
Life Form: Annual herb
Blooming: May-June
Notes: Possibly threatened by mining.
Status Report: 1989

Phacelia grisea

Considered but rejected: Too common

Phacelia insularis Munz var. *continentis* J.T. Howell

"North Coast phacelia" Hydrophyllaceae
CNPS List: 1B **R-E-D Code:** 3-2-3 **State/Fed. Status:** /C2
Distribution: MEN, MRN
Quads: 485C, 485D, 569A, 585D
Habitat: CBScr, CoDns
Life Form: Annual herb
Blooming: March-May
Notes: Known from approximately seven occurrences. Threatened by foot traffic, non-native plants, and grazing. See *American Midland Naturalist* 33:474 (1945) for original description.

Phacelia insularis Munz var. *insularis*

"northern Channel Islands phacelia" Hydrophyllaceae
CNPS List: 1B **R-E-D Code:** 3-2-3 **State/Fed. Status:** /C1
Distribution: SMI, SRO
Quads: SMIW, SROE, SRON
Habitat: CoDns, VFGrs
Life Form: Annual herb
Blooming: March-April
Notes: Known from fewer than five occurrences. Threatened by feral herbivores on SRO Isl. See *Bulletin of the Southern California Academy of Sciences* 31:113 (1932) for original description, and *American Midland Naturalist* 33:472-474 (1945) for taxonomic treatment.
Status Report: 1979

Phacelia inundata J.T. Howell

"playa phacelia" Hydrophyllaceae
CNPS List: 2 **R-E-D Code:** 2-1-1 **State/Fed. Status:** CEQA
Distribution: LAS, MOD, NV, OR
Quads: 640B, 640C, 712A, 728C
Habitat: GBScr, LCFrs, Plyas / alkaline
Life Form: Annual herb
Blooming: May-July
Notes: On review list in OR. See *Leaflets of Western Botany* 4:15 (1944) for original description, and *Madroño* 28(3):121-132 (1981) for taxonomic treatment.

Phacelia inyoensis (Macbr.) J.T. Howell

"Inyo phacelia" Hydrophyllaceae
CNPS List: 4 **R-E-D Code:** 1-1-3 **State/Fed. Status:** CEQA?
Distribution: INY, MNO
Habitat: Medws (alkaline)
Life Form: Annual herb
Blooming: May-August
Notes: See *Leaflets of Western Botany* 4:16 (1944) for original description, and *Madroño* 28(3):121-132 (1981) for taxonomic treatment.

Phacelia ixodes

Considered but rejected: Not in CA

Phacelia leonis J.T. Howell

"Siskiyou phacelia" Hydrophyllaceae
CNPS List: 1B **R-E-D Code:** 2-1-2 **State/Fed. Status:** CEQA
Distribution: SIS, TRI, OR*
Quads: 667A, 683B, 684C, 684D, 686D, 738A, 738D
Habitat: UCFrs (sometimes serpentinite)
Life Form: Annual herb
Blooming: June-July
Notes: On review list in OR.

Phacelia longipes

Considered but rejected: Too common

Phacelia lyonii

Considered but rejected: Too common

Phacelia marcescens

Considered but rejected: Too common

Phacelia mohavensis Gray

"Mojave phacelia" Hydrophyllaceae
CNPS List: 4 **R-E-D Code:** 1-1-3 **State/Fed. Status:** CEQA?
Distribution: LAX, SBD
Habitat: CmWld, LCFrs, Medws (xeric), PJWld / sandy or gravelly, often dry streambeds
Life Form: Annual herb
Blooming: April-August
Notes: Difficult to separate from *P. exilis*. See *Synoptical Flora of North America* 2(1):164 (1878) for original description, and *Systematic Botany* 13(1):16-20 (1988) for additional information.

Phacelia monoensis Halse

"Mono County phacelia" Hydrophyllaceae
CNPS List: 1B **R-E-D Code:** 3-3-2 **State/Fed. Status:** /C2
Distribution: MNO, NV
Quads: 469B, 470B, 486C, 487C, 487D, 488A
Habitat: GBScr, PJWld / clay, often roadsides
Life Form: Annual herb
Blooming: June-July
Notes: Known in CA from fewer than twenty occurrences. Threatened by vehicles and grazing. Threatened in NV. See *Madroño* 28(3):124-125 (1981) for original description, and *Intermountain Flora* 4:177 (1984) for alternate taxonomic treatment.

Phacelia mustelina Cov.

"Death Valley round-leaved phacelia" Hydrophyllaceae
CNPS List: 1B **R-E-D Code:** 2-1-2 **State/Fed. Status:** /C3c
Distribution: INY, SBD, NV
Quads: 232B, 302A, 325B, 325D, 368A, 368B, 368D, 369B, 389D
Habitat: MDScr, PJWld / carbonate or volcanic
Life Form: Annual herb
Blooming: May-July
Notes: On watch list in NV. See *Journal of the Washington Academy of Sciences* 27(5):196 (1937) for original description.

Phacelia nashiana Jeps.

"Charlotte's phacelia" Hydrophyllaceae
CNPS List: 1B **R-E-D Code:** 1-2-3 **State/Fed. Status:** /C2
Distribution: INY, KRN, TUL
Quads: 211A, 235A, 235B, 235C, 236D, 258B, 258C, 259A, 259C, 259D, 282A, 282B, 282C, 283D, 306A, 306D
Habitat: JTWld, MDScr, PJWld / granitic
Life Form: Annual herb
Blooming: March-June
Notes: Threatened by vehicles and grazing. See *Flora of California* 3(2):276-277 (1943) by W.L. Jepson for original description.
Status Report: 1989

Phacelia novenmillensis Munz

"Nine Mile Canyon phacelia" Hydrophyllaceae
CNPS List: 1B **R-E-D Code:** 3-2-3 **State/Fed. Status:** /C2
Distribution: INY, KRN, TUL
Quads: 258B, 282C, 283D, 306C, 307D
Habitat: BUFrs, CmWld, PJWld, UCFrs
Life Form: Annual herb
Blooming: May
Notes: Threatened by grazing and recreation. See *Aliso* 3(2):122-124 (1955) for original description.

Phacelia orogenes Brand

"mountain phacelia" Hydrophyllaceae
CNPS List: 4 **R-E-D Code:** 1-1-3 **State/Fed. Status:** /C3c
Distribution: FRE, TUL
Habitat: Medws, PJWld, SCFrs / mesic
Life Form: Annual herb
Blooming: July-August

Phacelia parishii Gray

"Parish's phacelia" Hydrophyllaceae
CNPS List: 2 **R-E-D Code:** 3-3-1 **State/Fed. Status:** /C2
Distribution: SBD, NV
Quads: 131B*, 181A*, 205C
Habitat: MDScr, Plyas
Life Form: Annual herb
Blooming: April-July
Notes: Rediscovered in CA in 1989 by M. Bagley. Threatened by military activities, and potentially by expansion of Ft. Irwin. See *Proceedings of the American Academy of Arts and Sciences* 19:88 (1883) for original description, and *American Midland Naturalist* 29:16-17 (1943) for taxonomic treatment.
Status Report: 1979

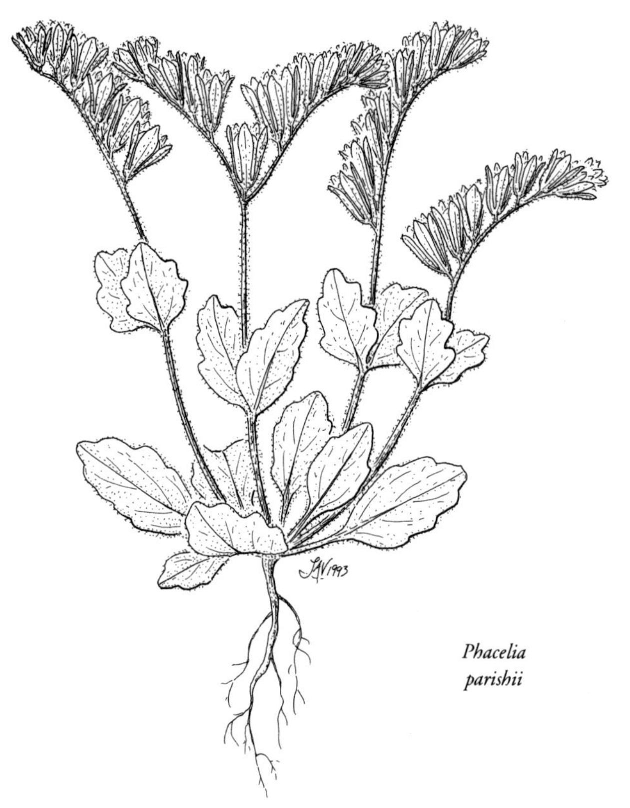

Phacelia parishii

Phacelia peirsoniana

Considered but rejected: Too common

Phacelia phacelioides (Benth.) Brand

"Mt. Diablo phacelia" Hydrophyllaceae
CNPS List: 1B **R-E-D Code:** 2-2-3 **State/Fed. Status:** /C2
Distribution: ALA?, CCA, SBT, SCL, STA
Quads: 339B, 425B, 425C, 426C, 426D, 444D, 464B, 464C
Habitat: Chprl, CmWld / rocky
Life Form: Annual herb
Blooming: April-May
Notes: Does plant occur in ALA Co.? Possibly threatened by foot traffic and trail construction.

Phacelia platyloba

Considered but rejected: Too common

Phacelia pringlei
Considered but rejected: Too common

Phacelia pulchella Gray var. *gooddingii* (Brand) J.T. Howell
"Goodding's phacelia" Hydrophyllaceae
CNPS List: 2 **R-E-D Code:** 3-1-1 **State/Fed. Status:** CEQA
Distribution: INY, AZ, NV, UT
Quads: 273C
Habitat: MDScr (clay, often alkaline)
Life Form: Annual herb
Blooming: April-June
Notes: See *Pflanzenreich* 4(251):120 (1913) for original description.

Phacelia sericea
See *Phacelia sericea* var. *ciliosa*

Phacelia sericea (Graham) Gray var. *ciliosa* Rydb.
"blue alpine phacelia" Hydrophyllaceae
CNPS List: 2 **R-E-D Code:** 2-1-1 **State/Fed. Status:** CEQA
Distribution: MOD, SIS, AZ, NV, OR+
Quads: 690B, 690C, 700A
Habitat: UCFrs (rocky)
Life Form: Perennial herb
Blooming: June-August
Notes: Known in CA only from the Warner Mtns. (MOD Co.) and China Mtn. (SIS Co.).

Phacelia stebbinsii Const. & Heckard
"Stebbins's phacelia" Hydrophyllaceae
CNPS List: 1B **R-E-D Code:** 2-2-3 **State/Fed. Status:** /C2
Distribution: ELD, PLA
Quads: 523C, 524A, 524B, 524C, 525A, 525B, 525D, 539B, 539C, 539D, 540C, 540D
Habitat: CmWld, LCFrs, Medws
Life Form: Annual herb
Blooming: June
Notes: See *Brittonia* 22(1):25-30 (1970) for original description.
Status Report: 1979

Phacelia stellaris Brand
"Brand's phacelia" Hydrophyllaceae
CNPS List: 1B **R-E-D Code:** 3-3-2 **State/Fed. Status:** CEQA
Distribution: LAX(*?), SDG, BA
Quads: 11B*, 11D, 22C*, 89B*, 90B*, 90C*, 110D(*?)
Habitat: CoDns, CoScr
Life Form: Annual herb
Blooming: March-June
Notes: Historical occurrences extirpated by development. Some plants intermediate to *P. douglasii* var. *douglasii* occur in SE Western Transverse Range foothills. See *Pflanzenreich* 4(251):123 (1913) for original description.

Phacelia suaveolens Greene ssp. *keckii* (Munz & Jtn.) Thorne
"Santiago Peak phacelia" Hydrophyllaceae
CNPS List: 1B **R-E-D Code:** 3-1-3 **State/Fed. Status:** /C2
Distribution: ORA, RIV
Quads: 70A, 87D
Habitat: CCFrs, Chprl
Life Form: Annual herb
Blooming: May-June
Notes: Known from only two occurrences near Santiago Pk. and Pleasants Pk. in the Santa Ana Mtns. See *Bulletin of the Torrey Botanical Club* 51:298 (1924) for original description, and *Aliso* 9(2):189-196 (1978) for revised nomenclature.

Phacelia vallicola
Considered but rejected: Too common

Phalacroseris bolanderi
Considered but rejected: Too common

Phaseolus filiformis Benth.
"slender-stem bean" Fabaceae
CNPS List: 2 **R-E-D Code:** 3-1-1 **State/Fed. Status:** CEQA
Distribution: RIV, AZ, BA+
Quads: 64C
Habitat: SDScr
Life Form: Annual herb
Blooming: April
Notes: Known in CA from only one occurrence in the Coachella Valley.

Phaseolus wrightii
See *Phaseolus filiformis*

Phlox adsurgens
Considered but rejected: Too common

Phlox azurea
Considered but rejected: A synonym of *P. diffusa*; a common taxon

Phlox bryoides
See *Phlox muscoides*

Phlox dispersa C.W. Sharsm.
"High Sierra phlox" Polemoniaceae
CNPS List: 4 **R-E-D Code:** 1-1-3 **State/Fed. Status:** CEQA?
Distribution: INY, TUL
Habitat: AlpBR (granitic)
Life Form: Perennial herb (stoloniferous)
Blooming: July-August
Notes: See *Aliso* 4:128 (1958) for original description.

Phlox dolichantha Gray

"Big Bear Valley phlox" Polemoniaceae
CNPS List: 1B **R-E-D Code:** 2-2-3 **State/Fed. Status:** /C2
Distribution: SBD
Quads: 104B, 105A, 105B, 131C, 131D
Habitat: PbPln, UCFrs (openings)
Life Form: Perennial herb
Blooming: May-July
Notes: Threatened by urbanization and recreation. See *Proceedings of the American Academy of Arts and Sciences* 22:310 (1887) for original description.

Phlox hirsuta E. Nels.

"Yreka phlox" Polemoniaceae
CNPS List: 1B **R-E-D Code:** 3-2-3 **State/Fed. Status:** CE/C1
Distribution: SIS
Quads: 717A, 717B
Habitat: LCFrs, UCFrs / serpentinite talus
Life Form: Perennial herb
Blooming: April-June
Notes: Known from only two occurrences on Soap Creek Ridge and on Juniper Terrace near Yreka. Threatened on private land by development and vehicles.
Status Report: 1987

Phlox muscoides Nutt.

"moss phlox" Polemoniaceae
CNPS List: 2 **R-E-D Code:** 2-1-1 **State/Fed. Status:** CEQA
Distribution: SHA, SIS, NV, OR, ++
Quads: 626A, 730C, 731A, 731C, 731D
Habitat: AlpBR
Life Form: Perennial herb
Blooming: July-August

Pholisma arenarium

Considered but rejected: Too common

Pholisma sonorae (Gray) Yatskievych

"sand food" Lennoaceae
CNPS List: 1B **R-E-D Code:** 2-2-2 **State/Fed. Status:** /C2
Distribution: IMP, AZ, BA, SO
Quads: 2A, 2B, 13A, 14B, 15C, 26C, 27D, 30D
Habitat: DeDns
Life Form: Perennial herb (parasitic)
Blooming: April-May
Notes: Threatened by vehicles and military activities. Parasitic on *Eriogonum, Tiquilia, Ambrosia,* and *Pluchea* spp. See *Memoirs of the American Academy of Arts and Sciences* 5:327 (1854), *Desert Plants* 2(3):188-196 (1980) for species account, and *Systematic Botany* 11(4):531-548 (1986) for revised treatment.
Status Report: 1977

Pholistoma auritum (Lindl.) Lilja var. *arizonicum* (M.E. Jones) Const.

"Arizona pholistoma" Hydrophyllaceae
CNPS List: 2 **R-E-D Code:** 3-1-1 **State/Fed. Status:** CEQA
Distribution: SBD, AZ
Quads: 121D
Habitat: MDScr
Life Form: Annual herb
Blooming: March
Notes: Known in CA from only one occurrence in the Whipple Mtns.

Pholistoma racemosum

Considered but rejected: Too common

Physalis crassifolia

Considered but rejected: Too common

Physalis greenei

Considered but rejected: A synonym of *P. crassifolia*; a common taxon

Physalis lobata Torr.

"lobed ground-cherry" Solanaceae
CNPS List: 2 **R-E-D Code:** 3-1-1 **State/Fed. Status:** CEQA
Distribution: SBD, AZ, NV, ++
Quads: 100A, 100D, 149A, 199B
Habitat: MDScr (decomposed granitic), Plyas
Life Form: Perennial herb
Blooming: September-January
Notes: See *Madroño* 26(2):101 (1979) for first CA records.

Picea breweriana

Considered but rejected: Too common

Picea engelmannii Engelm.

"Engelmann spruce" Pinaceae
CNPS List: 2 **R-E-D Code:** 2-2-1 **State/Fed. Status:** CEQA
Distribution: SHA, SIS, TRI, OR, WA, ++
Quads: 667A, 679C, 680C, 701C, 702D
Habitat: UCFrs
Life Form: Tree (evergreen)
Blooming: N/A
Notes: Threatened by logging.

Pilostyles thurberi Gray

"Thurber's pilostyles" Rafflesiaceae
CNPS List: 4 **R-E-D Code:** 1-1-1 **State/Fed. Status:** /C3c
Distribution: IMP, RIV, SDG, AZ, BA, NV+
Habitat: SDScr
Life Form: Perennial herb (parasitic)
Blooming: January
Notes: Grows inside the stems of *Psorothamnus*, especially *P. emoryi*; flowers on the stems of its host. See *Fremontia* 5(3):20-22 (1977) for species account, and *Madroño* 26(4):189 (1979) for distributional information.

Pilularia americana

Considered but rejected: Too common

Pinguicula macroceras
See *Pinguicula vulgaris* ssp. *macroceras*

Pinguicula vulgaris L. ssp. *macroceras* (Link) Calder & R. Taylor
"horned butterwort" Lentibulariaceae
CNPS List: 2 **R-E-D Code:** 1-2-1 **State/Fed. Status:** CEQA
Distribution: DNT, SIS, OR, WA, ++
Quads: 738A, 738D, 739B, 739C, 740C, 740D
Habitat: BgFns (serpentinite)
Life Form: Perennial herb (carnivorous)
Blooming: April-June
Notes: Threatened by horticultural collecting.

Pinus aristata
See *Pinus longaeva*

Pinus balfouriana
Considered but rejected: Too common

Pinus contorta Loud. ssp. *bolanderi* (Parl.) Vasey
"Bolander's beach pine" Pinaceae
CNPS List: 1B **R-E-D Code:** 1-2-3 **State/Fed. Status:** /C2
Distribution: MEN
Quads: 552B, 568C, 569A, 569D
Habitat: CCFrs (podzol-like soil)
Life Form: Tree (evergreen)
Blooming: N/A
Notes: Known only from the white sand pine barrens along the Mendocino coast. Threatened by development and vehicles.

Pinus edulis Engelm.
"two-needle pinyon pine" Pinaceae
CNPS List: 3 **R-E-D Code:** 3-1-1 **State/Fed. Status:** CEQA?
Distribution: SBD, AZ+
Quads: 200A, 200B, 225C
Habitat: PJWld
Life Form: Tree (evergreen)
Blooming: N/A
Notes: CA plants may be a form of *P. monophylla*, or hybrids of that plant; needs study.

Pinus lambertiana
Considered but rejected: Too common

Pinus longaeva D.K. Bailey
"bristlecone pine" Pinaceae
CNPS List: 4 **R-E-D Code:** 1-1-1 **State/Fed. Status:** CEQA?
Distribution: INY, MNO, NV+
Habitat: SCFrs (carbonate)
Life Form: Tree (evergreen)
Blooming: N/A
Notes: The name *P. aristata* has been misapplied to CA material. See *Annals of the Missouri Botanical Garden* 57:210-249 (1970) for revised treatment.

Pinus quadrifolia
Considered but rejected: Too common

Pinus radiata D. Don
"Monterey pine" Pinaceae
CNPS List: 1B **R-E-D Code:** 3-2-2 **State/Fed. Status:** /C2
Distribution: MNT, SCR, SLO, SMT, BA, GU
Quads: 271D, 344B, 366C, 386C, 387A, 387B, 387D, 408C, 409A, 409D
Habitat: CCFrs
Life Form: Tree (evergreen)
Blooming: N/A
Notes: Only three native stands in CA; introduced in many areas. Threatened by genetic contamination, development, and fragmentation, especially at Del Monte Forest (MNT Co.) and in SLO Co.; threatened by feral goats on GU Isl. Plants from BA (Cedros Isl.) and GU Isl. are genetically distinct. See *Fremontia* 18(2):15-21 (1990) for discussion of genetic conservation work.

Pinus remorata
Considered but rejected: A synonym of *P. muricata*; a common taxon

Pinus torreyana
See *P. torreyana* sspp. *insularis* and *torreyana*

Pinus torreyana Carr. ssp. *insularis* Haller
"Santa Rosa Isl. Torrey pine" Pinaceae
CNPS List: 1B **R-E-D Code:** 3-2-3 **State/Fed. Status:** /C2
Distribution: SRO
Quads: SROE, SRON
Habitat: CCFrs
Life Form: Tree (evergreen)
Blooming: N/A
Notes: Endemic to SRO Isl. See *Systematic Botany* 11(1):39-50 (1986) original description.

Pinus torreyana Carr. ssp. *torreyana*
"Torrey pine" Pinaceae
CNPS List: 1B **R-E-D Code:** 3-2-3 **State/Fed. Status:** /C2
Distribution: SDG
Quads: 22B, 22C
Habitat: CCFrs, Chprl / sandstone
Life Form: Tree (evergreen)
Blooming: N/A
Notes: Threatened by development. Seriously attacked by the five-spined ips bark beetle at Torrey Pines SR, but biological control apparently has contained this infestation. See *American Forests* 91(10):26-29 (1985) for species account, and *Systematic Botany* 11(1):39-50 (1986) for taxonomic treatment.

Pinus washoensis
Considered but rejected: Too common

Piperia candida Morgan & Ackerman
"white-flowered rein orchid" Orchidaceae
CNPS List: 4 **R-E-D Code:** 1-1-1 **State/Fed. Status:** CEQA?
Distribution: DNT, HUM, MEN, SCR, SIS, SMT, SON, TRI, OR, WA+
Habitat: LCFrs, NCFrs / sometimes serpentinite
Life Form: Perennial herb
Blooming: May-August
Notes: Difficult to identify from herbarium material. See *Lindleyana* 5(4):205-211 (1990) for original description.

Piperia elegans

Considered but rejected: Too common

Piperia elongata ssp. *michaelii*

See *Piperia michaelii*

Piperia michaelii (Greene) Rydb.

"Michael's rein orchid" Orchidaceae
CNPS List: 4 **R-E-D Code:** 1-2-3 **State/Fed. Status:** CEQA?
Distribution: ALA, CCA, HUM, MNT, MRN, SBT, SCR, SCZ, SFO, SLO, SMT
Habitat: CBScr, CCFrs, CmWld, LCFrs
Life Form: Perennial herb
Blooming: May-August
Notes: To be expected in the Sierra Nevada foothills, Western Transverse Ranges, and northern South Coast; need information. See *Bulletin of the California Academy of Sciences* 1:282 (1885) for original description, and *Botany Journal of the Linnean Society* 75:245-270 (1977) for revised nomenclature.

Piperia yadonii Morgan & Ackerman

"Yadon's rein orchid" Orchidaceae
CNPS List: 1B **R-E-D Code:** 3-3-3 **State/Fed. Status:** /C1
Distribution: MNT
Quads: 344B, 366A, 366C, 366D, 386B?, 386C
Habitat: CBScr, CCFrs, Chprl (maritime) / sandy
Life Form: Perennial herb
Blooming: May-August
Notes: Threatened by urbanization, recreational development, and non-native plants; only about 1000 plants remain. See *Lindleyana* 5(4):205-211 (1990) for original description.

Piptatherum micranthum (Trin. & Rupr.) Barkworth

"small-flowered rice grass" Poaceae
CNPS List: 2 **R-E-D Code:** 2-1-1 **State/Fed. Status:** CEQA
Distribution: INY, MNO, SBD, ++
Quads: 249D, 250B, 412A, 412B, 431D
Habitat: PJWld (gravelly, carbonate)
Life Form: Perennial herb
Blooming: June-September
Notes: Threatened in ID. See *Phytologia* 74(1):1-25 (1993) for revised nomenclature.

Pityopus californicus (Eastw.) Copel. f.

"California pinefoot" Ericaceae
CNPS List: 4 **R-E-D Code:** 1-2-1 **State/Fed. Status:** /C3c
Distribution: DNT, FRE, HUM, LAK, MEN, MRN, NAP, SIS, SON, TUL, OR
Habitat: BUFrs, LCFrs, NCFrs, UCFrs
Life Form: Perennial herb (saprophytic)
Blooming: May-July
Notes: Threatened by logging.

Plagiobothrys chorisianus (Chamb.) Jtn. var. *chorisianus*

"Choris's popcorn-flower" Boraginaceae
CNPS List: 3 **R-E-D Code:** 2-2-3 **State/Fed. Status:** CEQA?
Distribution: SCR, SFO, SMT
Quads: 409A, 409D, 446C, 448B, 466C
Habitat: Chprl, CoPrr, CoScr / mesic
Life Form: Annual herb
Blooming: April-June
Notes: Move to List 1B? Taxonomic work needed; differences from var. *hickmanii* may be environmentally induced.

Plagiobothrys diffusus (Greene) Jtn.

"San Francisco popcorn-flower" Boraginaceae
CNPS List: 1B **R-E-D Code:** 3-3-3 **State/Fed. Status:** CE/C2*
Distribution: SCR, SFO*
Quads: 387B, 407C, 408D, 466C*
Habitat: CoPrr, VFGrs?
Life Form: Annual herb
Blooming: April-June
Notes: Known from six occurrences. Threatened by development and non-native plants. Identification difficult; taxonomic work needed. See *P. reticulatus* var. *rossianorum* in *The Jepson Manual*. See *Pittonia* 1:14 (1887) for original description, and *Contributions from the Arnold Arboretum* 3:77 (1932) for revised nomenclature.
Status Report: 1988

Plagiobothrys distantiflorus

Considered but rejected: Too common

Plagiobothrys glaber (Gray) Jtn.

"hairless popcorn-flower" Boraginaceae
CNPS List: 1A **Last Seen:** 1954 **State/Fed. Status:** /C3a
Distribution: ALA*, MER*, MRN*, SBT*, SCL*
Quads: 384A*, 385C*, 407B*, 427C*, 427D*, 445B*, 447A*, 467A*
Habitat: Medws (alkaline), MshSw (coastal salt)
Life Form: Annual herb
Blooming: April-May
Notes: All collections since 1930's located in the Hollister area; plant should be looked for there. Is it a variety or ecotype of *P. stipitatus*? See *Proceedings of the American Academy of Arts and Sciences* 17:227 (1882) for original description.

Plagiobothrys glomeratus Gray

"Mammoth popcorn-flower" Boraginaceae
CNPS List: 2 **R-E-D Code:** 2-2-1 **State/Fed. Status:** CEQA
Distribution: MNO, NV
Quads: 434B?, 487B, 488A, 488C, 488D?
Habitat: GBScr, PJWld
Life Form: Annual herb
Blooming: July
Notes: Known in CA only from the Mammoth and Sweetwater Mtns. Threatened by mining. Not in *The Jepson Manual*. See *Proceedings of the American Academy of Arts and Sciences* 20:286 (1885) for original description.

Plagiobothrys glyptocarpus (Piper) Jtn. var. *modestus* Jtn.

"Cedar Crest popcorn-flower" Boraginaceae
CNPS List: 3 **R-E-D Code:** 3-?-3 **State/Fed. Status:** /C2
Distribution: NEV
Quads: 542A
Habitat: CmWld
Life Form: Annual herb
Blooming: April-May
Notes: Move to List 1A? Apparently not collected since 1937; field surveys needed. May be a minor variant or hybrid. Position in the *P. glyptocarpus*/*P. distantiflorus* complex needs study.

Plagiobothrys hystriculus (Piper) Jtn.

"bearded popcorn-flower" Boraginaceae
CNPS List: 1A **Last Seen:** 1892 **State/Fed. Status:** /C3a
Distribution: SOL*
Quads: 481D*
Habitat: VFGrs (mesic), VnPls
Life Form: Annual herb
Blooming: April-May
Notes: Known only from the type collection in the Montezuma Hills; all other reports have been misidentifications. Easily confused with *P. acanthocarpus*, *P. trachycarpus*, and others.

Plagiobothrys lithocaryus (Greene) Jtn.

"Mayacamas popcorn-flower" Boraginaceae
CNPS List: 1A **Last Seen:** 1899 **State/Fed. Status:** CEQA
Distribution: LAK*, MEN?*
Quads: 549C*, 566D?*
Habitat: Chprl?, CmWld, VFGrs / mesic
Life Form: Annual herb
Blooming: April-May
Notes: Known only from the type collection by Curran (Lakeport, LAK Co.) in 1884 and an uncertain collection by Purpus (Potter Valley, MEN Co.) in 1899; should be looked for in these areas.

Plagiobothrys mollis (Gray) Jtn. var. *vestitus* (Greene) Jtn.

"Petaluma popcorn-flower" Boraginaceae
CNPS List: 1A **Last Seen:** 1888 **State/Fed. Status:** /C2*
Distribution: SON*
Quads: 484*
Habitat: MshSw? (coastal salt), VFGrs (mesic)
Life Form: Perennial herb
Blooming: June-July
Notes: Known only from the type collection near Petaluma. Field work needed.

Plagiobothrys myosotoides (Lehm.) Brand

"forget-me-not popcorn-flower" Boraginaceae
CNPS List: 4 **R-E-D Code:** 1-1-1 **State/Fed. Status:** CEQA?
Distribution: FRE, SCL, TUL, SA
Habitat: Chprl
Life Form: Annual herb
Blooming: April-May
Notes: Identification uncertain. Relationship to *P. torreyi* complex needs study. More SA specimens needed for comparison with CA material.

Plagiobothrys salsus (Bdg.) Jtn.

"desert popcorn-flower" Boraginaceae
CNPS List: 2 **R-E-D Code:** 3-2-1 **State/Fed. Status:** CEQA
Distribution: INY, MOD, NV, OR
Quads: 299B, 322C, 322D, 707B, 707C
Habitat: Plyas (alkaline)
Life Form: Annual herb
Blooming: May-August
Notes: Endangered in OR, but more widespread in NV. See *Botanical Gazette* 27:452 (1899) for original description.

Plagiobothrys scriptus

Considered but rejected: Too common

Plagiobothrys strictus (Greene) Jtn.

"Calistoga popcorn-flower" Boraginaceae
CNPS List: 1B **R-E-D Code:** 3-3-3 **State/Fed. Status:** CT/C1
Distribution: NAP
Quads: 517D
Habitat: BUFrs, Medws, VFGrs / alkaline areas near thermal springs
Life Form: Annual herb
Blooming: March-June
Notes: Known from only three extant occurrences near Calistoga. Threatened by urbanization and viticulture.
Status Report: 1990

Plagiobothrys uncinatus J.T. Howell

"hooked popcorn-flower" Boraginaceae
CNPS List: 1B **R-E-D Code:** 2-2-3 **State/Fed. Status:** /C2
Distribution: MNT, SBT, SCL, SLO
Quads: 270B, 295B?, 295C, 296A?, 318B, 319C, 341B, 341C, 343A, 343D, 363C, 425B
Habitat: Chprl (sandy), CmWld, VFGrs
Life Form: Annual herb
Blooming: May
Notes: Field surveys needed in Gabilan and Santa Lucia ranges to determine status.

Platystemon californicus Benth. var. *ciliatus* Dunkle

"Santa Barbara Island cream cups" Papaveraceae
CNPS List: 1B **R-E-D Code:** 3-1-3 **State/Fed. Status:** /C2
Distribution: SBR
Quads: SBRA
Habitat: CBScr
Life Form: Annual herb
Blooming: March-May
Notes: Known from only one extended occurrence. See *P. californicus* in *The Jepson Manual*. See *Madroño* 21(5):366-367 (1972) for species account.

Plectritis eichleriana

Considered but rejected: A synonym of *P. macrocera*; a common taxon

Plectritis jepsonii

Considered but rejected: A synonym of *P. macrocera*; a common taxon

Plectritis macrocera ssp. *macrocera*

Considered but rejected: Too common; a synonym of *P. macrocera*

Pleuropogon californicus var. *davyi*

Not published; see *Pleuropogon davyi*

Pleuropogon davyi L. Benson

"Davy's semaphore grass" Poaceae
CNPS List: 4 **R-E-D Code:** 1-1-3 **State/Fed. Status:** CEQA?
Distribution: LAK, MEN
Habitat: CmWld, LCFrs, Medws
Life Form: Perennial herb (rhizomatous)
Blooming: March-May
Notes: See *P. californicus* in *The Jepson Manual.*

Pleuropogon hooverianus (L. Benson) J.T. Howell

"North Coast semaphore grass" Poaceae
CNPS List: 1B **R-E-D Code:** 3-2-3 **State/Fed. Status:** CR/C2
Distribution: MEN, MRN, SON
Quads: 467A, 484C, 501C, 502A, 502B, 567A, 568D, 584A
Habitat: BUFrs, Medws, NCFrs, VnPls / mesic
Life Form: Perennial herb (rhizomatous)
Blooming: May-August
Notes: Known from approximately twelve occurrences. See *American Journal of Botany* 28:360 (1941) for original description, *Leaflets of Western Botany* 4(10):247 (1946) for revised nomenclature, and *Taxon* 27(4):375 (1978) for alternate nomenclature.
Status Report: 1988

Pleuropogon refractus (Gray) Benth.

"nodding semaphore grass" Poaceae
CNPS List: 4 **R-E-D Code:** 1-2-1 **State/Fed. Status:** CEQA?
Distribution: DNT, HUM, MEN, MRN, OR, WA+
Habitat: LCFrs, Medws, NCFrs, RpFrs / mesic
Life Form: Perennial herb (rhizomatous)
Blooming: May-August

Poa abbreviata R. Br. ssp. *marshii* Soreng

"Marsh's blue grass" Poaceae
CNPS List: 2 **R-E-D Code:** 3-1-1 **State/Fed. Status:** CEQA
Distribution: MNO, ID, NV
Quads: 431B
Habitat: AlpBR
Life Form: Perennial herb
Blooming: June
Notes: Known in CA from only one occurrence along Perry Aiken Creek in the White Mtns. Not in *The Jepson Manual.* On review list in ID. See *Phytologia* 71(5):390-413 (1991) for original description.

Poa abbreviata R. Br. ssp. *pattersonii* (Vasey) A. Löve, D. Löve & Kapoor

"Patterson's blue grass" Poaceae
CNPS List: 2 **R-E-D Code:** 3-1-1 **State/Fed. Status:** CEQA
Distribution: MNO, ID, NV, ++
Quads: 431B, 431C
Habitat: AlpBR
Life Form: Perennial herb
Blooming: July
Notes: Known in CA from fewer than five occurrences in the White Mtns. See *P. pattersonii* in *The Jepson Manual.* See *Contributions from the U.S. National Herbarium* 1:275 (1893) for original description, *Madroño* 35(2):165 (1988) for first CA record, and *Phytologia* 71(5):390-413 (1991) for taxonomic treatment.

Poa atropurpurea Scribn.

"San Bernardino blue grass" Poaceae
CNPS List: 1B **R-E-D Code:** 2-2-3 **State/Fed. Status:** /C1
Distribution: SBD, SDG
Quads: 19B, 19C, 49D, 104B, 105A, 105B, 131C, 131D
Habitat: Medws (mesic)
Life Form: Perennial herb (rhizomatous)
Blooming: April-June
Notes: Threatened by development, grazing, and vehicles.
Status Report: 1979

Poa fibrata

Considered but rejected: A hybrid, sometimes stabilized, between *P. pratensis* and *P. secunda* ssp. *juncifolia*

Poa napensis Beetle

"Napa blue grass" Poaceae
CNPS List: 1B **R-E-D Code:** 3-3-3 **State/Fed. Status:** CE/C1
Distribution: NAP
Quads: 517D
Habitat: Medws (alkaline, near hot springs)
Life Form: Perennial herb
Blooming: May-August
Notes: Known from only two occurrences in the Calistoga area. Threatened by development. See *Leaflets of Western Botany* 4:289 (1946) for original description.
Status Report: 1987

Poa pattersonii

See *Poa abbreviata* ssp. *pattersonii*

Poa piperi Hitchc.

"Piper's blue grass" Poaceae
CNPS List: 4 **R-E-D Code:** 1-1-2 **State/Fed. Status:** /C3c
Distribution: DNT, SIS, OR
Habitat: Chprl, LCFrs / serpentinite
Life Form: Perennial herb (rhizomatous)
Blooming: April-May
Notes: Endangered in OR. See *Illustrated Flora of the Pacific States* 1:201 (1923) by L. Abrams for original description.

Poa rhizomata Hitchc.

"timber blue grass" Poaceae
CNPS List: 4 **R-E-D Code:** 1-1-1 **State/Fed. Status:** /C3c
Distribution: DNT, SIS, TRI, OR
Habitat: LCFrs (serpentinite)
Life Form: Perennial herb (rhizomatous)
Blooming: April-May
Notes: On review list in OR.

Poa sierrae

Considered but rejected: Too common

Poa tenerrima

Considered but rejected: Too common

Podistera nevadensis (Gray) Wats.

"Sierra podistera" Apiaceae
CNPS List: 4 **R-E-D Code:** 1-2-3 **State/Fed. Status:** CEQA?
Distribution: ALP, ELD, MNO, PLA, SBD*, TUO
Habitat: AlpBR
Life Form: Perennial herb
Blooming: July-September

Pogogyne abramsii J.T. Howell

"San Diego mesa mint" Lamiaceae
CNPS List: 1B **R-E-D Code:** 2-3-3 **State/Fed. Status:** CE/FE
Distribution: SDG
Quads: 22A, 22B, 22C, 22D
Habitat: VnPls
Life Form: Annual herb
Blooming: April-June
Notes: Seriously threatened by vehicles, dumping, road maintenance, and urbanization of San Diego mesas. See *Proceedings of the California Academy of Sciences* IV 20:119-120 (1931) for original description.
Status Report: 1988

Pogogyne clareana J.T. Howell

"Santa Lucia mint" Lamiaceae
CNPS List: 1B **R-E-D Code:** 3-2-3 **State/Fed. Status:** CE/C2
Distribution: MNT
Quads: 295B, 296A, 296D
Habitat: RpWld
Life Form: Annual herb
Blooming: May-June
Notes: Known from approximately five occurrences near Ft. Hunter Liggett. Possibly threatened by road maintenance and military activities. See *Four Seasons* 4(3):22 (1973) for original description.
Status Report: 1988

Pogogyne douglasii ssp. *minor*

Considered but rejected: A synonym of *P. douglasii* ssp. *douglasii*; a common taxon

Pogogyne douglasii Benth. ssp. *parviflora* (Benth.) J.T. Howell

"Douglas's pogogyne" Lamiaceae
CNPS List: 3 **R-E-D Code:** 1-2-3 **State/Fed. Status:** /C3c
Distribution: BUT?, LAK, MEN, NAP, SAC?, SON
Quads: 500B, 501A, 501B, 502A, 516B, 516C, 516D, 517A, 517B, 517C, 517D, 518D, 532D, 533B, 534A, 549B, 550B, 567A, 582B, 583C
Habitat: Chprl (serpentinite), MshSw (vernal freshwater), VFGrs, VnPls
Life Form: Annual herb
Blooming: May-June
Notes: Move to List 4? Does plant occur in BUT and SAC counties? Many new occurrences found recently, but threatened by urbanization and agriculture; some populations have been extirpated. Taxonomic questions; is ssp. distinct? See *P. douglasii* in *The Jepson Manual.* See *Proceedings of the California Academy of Sciences* IV 20:117 (1931) for revised nomenclature.

Pogogyne douglasii ssp. *ramosa*

Considered but rejected: A synonym of *P. douglasii* ssp. *douglasii*; a common taxon

Pogogyne floribunda Jokerst

"profuse-flowered pogogyne" Lamiaceae
CNPS List: 1B **R-E-D Code:** 2-2-3 **State/Fed. Status:** CEQA
Distribution: LAS, MOD, SHA
Quads: 660C, 662B, 709B, 710D, 711A, 711B, 711D, 726C, 726D, 728C, 728D
Habitat: VnPls
Life Form: Annual herb
Blooming: June-August
Notes: Similar to *P. zizyphoroides.* See *Aliso* 13(2):347-353 (1992) for original description.

Pogogyne nudiuscula Gray

"Otay Mesa mint" Lamiaceae
CNPS List: 1B **R-E-D Code:** 3-3-2 **State/Fed. Status:** CE/FE
Distribution: SDG, BA
Quads: 10C, 11B*, 11D, 22B*, 22C*
Habitat: VnPls
Life Form: Annual herb
Blooming: May-June
Notes: Known in CA from six occurrences on Otay Mesa. Seriously threatened by urbanization, grazing, vehicles, and trash dumping. See *Botany of California* 1:597 (1876) for original description.
Status Report: 1993

Pogogyne serpylloides ssp. *intermedia*

Considered but rejected: Too common; a synonym of *P. serpylloides*

Polemonium chartaceum Mason

"Mason's sky pilot" Polemoniaceae
CNPS List: 1B **R-E-D Code:** 3-1-2 **State/Fed. Status:** /C3c
Distribution: MNO, SIS, TRI, NV
Quads: 431B, 432A, 450D, 488A, 699C, 700C?
Habitat: AlpBR, SCFrs / serpentinite, granitic, or volcanic
Life Form: Perennial herb
Blooming: June-August
Notes: Disjunct occurrences in SIS and TRI counties may be taxonomically distinct from transmontane plants; currently under study. Rare in NV. Probably related to both *P. eximium* and *P. elegans.* See *Fremontia* 22(2):24-26 (1993) for information about ongoing biosystematic study.

Poliomintha incana (Torr.) Gray

"frosted mint" Lamiaceae
CNPS List: 1A **Last Seen:** 1938 **State/Fed. Status:** CEQA
Distribution: SBD*, AZ, NM, TX, UT+
Quads: 131D*
Habitat: LCFrs (mesic)
Life Form: Shrub
Blooming: June-July
Notes: Known in CA from only a single historical collection at Cushenbury Springs. See *Botany of the U.S. and Mexican Boundary Survey,* 2(1):130 (1858) by W. Emory for original description, and *Proceedings of the American Academy of Arts and Sciences* 7:247-296 (1870) for taxonomic treatment.

Polyctenium fremontii var. *fremontii*

Considered but rejected: Too common

Polygala cornuta ssp. *fishiae*

See *Polygala cornuta* var. *fishiae*

Polygala cornuta Kell. var. *fishiae* (Parry) Jeps.

"Fish's milkwort" Polygalaceae
CNPS List: 4 **R-E-D Code:** 1-1-2 **State/Fed. Status:** CEQA?
Distribution: LAX, ORA, RIV, SBA, SDG, VEN, BA
Habitat: Chprl, CmWld, RpWld
Life Form: Shrub (deciduous)
Blooming: May-August
Notes: Includes *P. cornuta* var. *pollardii.*

Polygala cornuta var. *pollardii*

See *Polygala cornuta* var. *fishiae*

Polygala heterorhyncha (Barneby) T. Wendt

"notch-beaked milkwort" Polygalaceae
CNPS List: 1B **R-E-D Code:** 2-1-2 **State/Fed. Status:** CEQA
Distribution: INY, NV
Quads: 346A, 346B, 367C
Habitat: MDScr (alkaline)
Life Form: Perennial herb
Blooming: April-May
Notes: Known in CA only from the Funeral Mtns. On watch list in NV. See *Leaflets of Western Botany* 3:194 (1943) for original description, and *Journal of the Arnold Arboretum* 60:504-514 (1979) for revised nomenclature.

Polygala subspinosa Wats.

"spiny milkwort" Polygalaceae
CNPS List: 2 **R-E-D Code:** 2-2-1 **State/Fed. Status:** CEQA
Distribution: LAS, AZ, NM, NV, UT+
Quads: 620A, 620B, 621A, 638C, 638D, 639A
Habitat: GBScr
Life Form: Perennial herb
Blooming: June-July
Notes: See *American Naturalist* 7:299 (1873) for original description.

Polygala subspinosa var. *subspinosa*

See *Polygala subspinosa*

Polygonum bidwelliae Wats.

"Bidwell's knotweed" Polygonaceae
CNPS List: 4 **R-E-D Code:** 1-1-3 **State/Fed. Status:** /C3c
Distribution: BUT, SHA, TEH
Habitat: Chprl, CmWld, VFGrs / volcanic
Life Form: Annual herb
Blooming: April-June

Polygonum esotericum

See *Polygonum polygaloides* ssp. *esotericum*

Polygonum fusiforme

Considered but rejected: A synonym of *P. persicaria*; a common, non-native taxon

Polygonum marinense Mertens & Raven

"Marin knotweed" Polygonaceae
CNPS List: 3 **R-E-D Code:** 3-3-3 **State/Fed. Status:** /C2
Distribution: MRN, NAP, SON
Quads: 467A, 483A, 484A, 485C*, 485D, 503D
Habitat: MshSw (coastal salt)
Life Form: Annual herb
Blooming: June-August
Notes: Move to List 1B? Known from fewer than ten occurrences. Taxonomic questions warrant immediate study. Threatened by coastal development.

Polygonum montereyense

Considered but rejected: A synonym of *P. arenastrum*; a common, non-native taxon

Polygonum patulum

Considered but rejected: Not native

Polygonum polygaloides Meisn. ssp. *esotericum* (Wheeler) Hickman

"Modoc County knotweed" Polygonaceae
CNPS List: 1B **R-E-D Code:** 3-3-3 **State/Fed. Status:** CEQA
Distribution: MOD, SIE
Quads: 571B, 708A, 726C
Habitat: GBScr (mesic), VnPls
Life Form: Annual herb
Blooming: July-August
Notes: Known from fewer than five occurrences. Possibly threatened by trampling. Intermediates to ssp. *confertiflorum* occur more broadly on Modoc Plateau, to OR. See *Madroño* 31(4):249-252 (1984) for revised nomenclature.

Polypodium hesperium

Considered but rejected: Too common

Polystichum dudleyi

Considered but rejected: Too common

Polystichum kruckebergii W.H. Wagner

"Kruckeberg's sword fern" Dryopteridaceae
CNPS List: 4 **R-E-D Code:** 1-1-1 **State/Fed. Status:** /C3c
Distribution: ALP, BUT, PLU, SBD, SHA, SIE, SIS, TUO, ID, OR, ++
Habitat: SCFrs, UCFrs / rocky
Life Form: Perennial herb (rhizomatous)
Fertile: June-August
Notes: Sensitive in ID, and on watch list in OR. See *American Fern Journal* 56:4 (1966) for original description, and *Pteridologia* 1:1-64 (1979) for taxonomic treatment.

Polystichum lonchitis (L.) Roth

"holly fern" Dryopteridaceae
CNPS List: 3 **R-E-D Code:** ?-?-1 **State/Fed. Status:** CEQA?
Distribution: ALP, ELD, PLU?, SIS, TRI?, AZ, ID, NV, OR, WA+
Quads: 506B, 523A, 735B, 737A
Habitat: SCFrs, UCFrs / granitic
Life Form: Perennial herb (rhizomatous)
Fertile: June-September
Notes: Move to List 2? Location, rarity, and endangerment information needed. Does plant occur in PLU and TRI counties?

Populus angustifolia James

"narrow-leaved cottonwood" Salicaceae
CNPS List: 2 **R-E-D Code:** 3-2-1 **State/Fed. Status:** CEQA
Distribution: INY, SBD, AZ, ID, NV, OR, ++
Quads: 105A, 106C, 106D, 346B, 351C, 373A, 412A
Habitat: RpFrs
Life Form: Tree (deciduous)
Blooming: March-April
Notes: On watch list in OR.

Portulaca halimoides L.

"desert portulaca" Portulacaceae
CNPS List: 4 **R-E-D Code:** 1-2-1 **State/Fed. Status:** CEQA?
Distribution: RIV, SBD, AZ, BA+
Habitat: JTWld (sandy)
Life Form: Annual herb
Blooming: September
Notes: Fairly common in the Granite, Providence, and New York Mtns. following unusually heavy summer rains in 1990, but otherwise known in CA from only three records: in Joshua Tree NM, Valley Wells, and the Little San Bernardino Mtns. See *Madroño* 36(4):281-282 (1989) for revised nomenclature.

Portulaca mundula

See *Portulaca halimoides*

Potamogeton epihydrus Raf. ssp. *nuttallii* (Cham. & Schldl.) Calder & R. Taylor

"Nuttall's pondweed" Potamogetonaceae
CNPS List: 2 **R-E-D Code:** 2-2-1 **State/Fed. Status:** CEQA
Distribution: ELD, MEN, MOD, MPA, PLU, OR, WA, ++
Quads: 437A, 437B, 523C, 567A, 598C, 599D, 603D, 692A
Habitat: MshSw (assorted shallow freshwater)
Life Form: Perennial herb (rhizomatous, aquatic)
Blooming: July-August
Notes: Hybridizes with *P. nodosus.*

Potamogeton filiformis Pers.

"slender-leaved pondweed" Potamogetonaceae
CNPS List: 2 **R-E-D Code:** 3-2-1 **State/Fed. Status:** CEQA
Distribution: LAS, MER, MNO, SCL*, AZ, NV, OR, ++
Quads: 403B, 428A*, 428B*, 453D
Habitat: MshSw (assorted shallow freshwater)
Life Form: Perennial herb (rhizomatous, aquatic)
Blooming: May-July
Notes: To be expected in the San Joaquin Valley, San Francisco Bay area, and the central high Sierra Nevada; need information. Need quads for LAS Co. On review list in OR.

Potamogeton foliosus Raf. var. *fibrillosus* (Fern.) R. Haynes & Reveal

"fibrous pondweed" Potamogetonaceae
CNPS List: 2 **R-E-D Code:** 3-1-1 **State/Fed. Status:** CEQA
Distribution: DNT, ID, OR*, WA+
Quads: 723B, 740C
Habitat: MshSw (assorted shallow freshwater)
Life Form: Perennial herb (rhizomatous, aquatic)
Blooming: unknown
Notes: To be expected in Great Basin areas of CA; need information. See *Rhodora* 75:76 (1973) for original description and 76:564-649 (1974) for taxonomic treatment.

Potamogeton praelongus Wulfen

"white-stemmed pondweed" Potamogetonaceae
CNPS List: 2 **R-E-D Code:** 3-1-1 **State/Fed. Status:** CEQA
Distribution: LAS, PLU, SHA, SIE, OR, WA, ++
Quads: 555B, 644D
Habitat: MshSw (deep water, lakes)
Life Form: Perennial herb (rhizomatous, aquatic)
Blooming: July-August
Notes: Need quads for LAS and PLU counties. Hybridizes with *P. richardsonii.*

Potamogeton robbinsii Oakes

"Robbins's pondweed" Potamogetonaceae
CNPS List: 2 **R-E-D Code:** 2-1-1 **State/Fed. Status:** CEQA
Distribution: ALP, INY, LAS, MAD, SIS, TUO, ++
Quads: 393A, 394A, 435A, 435D, 456A, 719C
Habitat: MshSw (deep water, lakes)
Life Form: Perennial herb (rhizomatous, aquatic)
Blooming: July-August
Notes: Need quads for LAS Co.

Potamogeton zosteriformis Fern.

"eel-grass pondweed" Potamogetonaceae
CNPS List: 2 **R-E-D Code:** 2-2-1 **State/Fed. Status:** CEQA
Distribution: CCA, LAK, LAS, MOD, SHA, OR, WA, ++
Quads: 480C, 480D, 533B, 620C, 662A, 674A, 692A, 693A, 693B
Habitat: MshSw (assorted freshwater)
Life Form: Annual herb (aquatic)
Blooming: June-July
Notes: To be expected in the Central Valley; need information.

Potentilla basaltica Tiehm & Ertter

"Black Rock potentilla" Rosaceae
CNPS List: 1B **R-E-D Code:** 3-1-2 **State/Fed. Status:** /C1
Distribution: LAS, NV
Quads: 675C
Habitat: Medws (alkaline, sandy, volcanic)
Life Form: Perennial herb
Blooming: June
Notes: Known in CA from only one occurrence in Ash Valley, and in NV from one occurrence. See *Brittonia* 36(3):228-231 (1984) for original description.

Potentilla concinna Richards.

"alpine cinquefoil" Rosaceae
CNPS List: 2 **R-E-D Code:** 3-1-1 **State/Fed. Status:** CEQA
Distribution: MNO, NV, ++
Quads: 431C
Habitat: AlpBR, Medws (rocky)
Life Form: Perennial herb
Blooming: June-July
Notes: Known in CA from only one occurrence near Tres Plumas Meadow in the White Mtns. See *Madroño* 35(2):165 (1988) for this record.

Potentilla concinna var. *divisa*

See *Potentilla concinna*

Potentilla cristae Ferlatte & Strother

"crested potentilla" Rosaceae
CNPS List: 1B **R-E-D Code:** 3-1-3 **State/Fed. Status:** CEQA
Distribution: SIS, TRI
Quads: 699C, 699D, 700D, 719D, 738D
Habitat: AlpBR, SCFrs / seasonally mesic, often serpentinite seeps, gravelly or rocky
Life Form: Perennial herb
Blooming: August-September
Notes: See *Madroño* 37(3):190-194 (1990) for original description.

Potentilla glandulosa Lindl. ssp. *ewanii* Keck

"Ewan's cinquefoil" Rosaceae
CNPS List: 1B **R-E-D Code:** 3-1-3 **State/Fed. Status:** CEQA
Distribution: LAX
Quads: 135D, 160B
Habitat: LCFrs (near seeps and springs)
Life Form: Perennial herb
Blooming: June-July
Notes: Endemic to the Dawson Saddle area of the San Gabriel Mtns. See *Carnegie Institution of Washington Publication* 520:47 (1940) for original description.

Potentilla hickmanii Eastw.

"Hickman's cinquefoil" Rosaceae
CNPS List: 1B **R-E-D Code:** 3-3-3 **State/Fed. Status:** CE/C1
Distribution: MNT, SMT*, SON*
Quads: 366C, 448C*, 502D*
Habitat: CBScr, CCFrs, Medws (vernally mesic), MshSw (freshwater)
Life Form: Perennial herb
Blooming: April-August
Notes: Known from only two extant occurrences on the Monterey Peninsula, where seriously threatened by urbanization and recreational activities. See *Bulletin of the Torrey Botanical Club* 29:77-78 (1902) for original description, and *Fremontia* 21(1):25-29 (1993) for species account.
Status Report: 1987

Potentilla hickmanii var. *hickmanii*

See *Potentilla hickmanii*

Potentilla hickmanii var. *uliginosa*

Considered but rejected: Not published

Potentilla morefieldii Ertter

"Morefield's cinquefoil" Rosaceae
CNPS List: 1B **R-E-D Code:** 2-1-3 **State/Fed. Status:** CEQA
Distribution: MNO
Quads: 431C, 432A, 450D
Habitat: AlpBR (carbonate)
Life Form: Perennial herb
Blooming: July-August
Notes: See *Brittonia* 44(4):429-435 (1992) for original description.

Potentilla multijuga Lehm.

"Ballona cinquefoil" Rosaceae
CNPS List: 1A **Last Seen:** 1890 **State/Fed. Status:** /C2*
Distribution: LAX*
Quads: 90B*
Habitat: Medws (brackish)
Life Form: Perennial herb
Blooming: June-August
Notes: Extirpated by development. Recent surveys have failed to relocate plant. See *Fremontia* 21(1):25-29 (1993) for species account.
Status Report: 1977

Potentilla newberryi Gray

"Newberry's cinquefoil" Rosaceae
CNPS List: 2 **R-E-D Code:** 2-1-1 **State/Fed. Status:** CEQA
Distribution: MNO?, MOD, SIS, NV, OR, WA
Quads: 712A, 715B, 728A, 728C
Habitat: MshSw (drying margins)
Life Form: Perennial herb
Blooming: May-August
Notes: Does plant occur in MNO Co.? Common near Plush, OR. See *Report of the Pacific Railroad Expedition* 6(3):72 (1857) for original description as *Ivesia gracilis*.

Potentilla patellifera

See *Ivesia patellifera*

Potentilla rimicola (Munz & Jtn.) Ertter

"cliff cinquefoil" Rosaceae
CNPS List: 1B **R-E-D Code:** 2-1-2 **State/Fed. Status:** CEQA
Distribution: RIV, BA
Quads: 83C
Habitat: SCFrs, UCFrs / granitic crevices
Life Form: Perennial herb
Blooming: July-September
Notes: Known in CA only from the San Jacinto Mtns., and in BA from the Sierra San Pedro Martir, where it may be rare. See *Bulletin of the Southern California Academy of Sciences* 24:18 (1925) for original description, and *Phytologia* 71(5):420-422 (1991) for revised nomenclature.

Potentilla tularensis

See *Horkelia tularensis*

Proboscidea althaeifolia (Benth.) Dcne.

"desert unicorn-plant" Martyniaceae
CNPS List: 4 **R-E-D Code:** 1-1-1 **State/Fed. Status:** CEQA?
Distribution: IMP, RIV, SDG, AZ, BA, SO
Habitat: SDScr
Life Form: Perennial herb
Blooming: May-August

Prunus fasciculata (Torr.) Gray var. *punctata* Jeps.

"sand almond" Rosaceae
CNPS List: 4 **R-E-D Code:** 1-1-3 **State/Fed. Status:** CEQA?
Distribution: SBA, SLO
Habitat: Chprl (maritime), CmWld, CoDns, CoScr / sandy
Life Form: Shrub (deciduous)
Blooming: March-April

Prunus lyonii

Considered but rejected: Too common

Pseudobahia bahiifolia (Benth.) Rydb.

"Hartweg's golden sunburst" Asteraceae
CNPS List: 1B **R-E-D Code:** 2-3-3 **State/Fed. Status:** CE/PE
Distribution: FRE, MAD, STA, SUT*, YUB*
Quads: 378B, 398C, 440B, 440C, 441A, 459C*, 544A*
Habitat: CmWld, VFGrs / clay
Life Form: Annual herb
Blooming: March-April
Notes: Known from fewer than twenty occurrences. Seriously threatened by development, agriculture, overgrazing, and trampling.
Status Report: 1986

Pseudobahia peirsonii Munz

"San Joaquin adobe sunburst" Asteraceae
CNPS List: 1B **R-E-D Code:** 2-3-3 **State/Fed. Status:** CE/PE
Distribution: FRE, KRN, TUL

Quads: 239A, 262A*, 262B, 262D, 286B, 287A*, 287D*, 309C, 310A*, 310D, 311A*, 333C*, 333D*, 356B, 356C*, 378D
Habitat: CmWld, VFGrs / adobe
Life Form: Annual herb
Blooming: March-April
Notes: Known from fewer than twenty occurrences. Seriously threatened by agriculture, grazing, development, road construction, and flood control activities. See *Aliso* 2:84 (1949) for original description.
Status Report: 1985

Psilocarphus brevissimus Nutt. var. *multiflorus* Cronq.

"delta woolly-marbles" Asteraceae
CNPS List: 4 **R-E-D Code:** 1-2-3 **State/Fed. Status:** CEQA?
Distribution: ALA, NAP, SCL, SJQ, SOL, STA, YOL
Habitat: VnPls
Life Form: Annual herb
Blooming: May-June
Notes: Does plant occur in CCA, SAC, or other counties? Similar to *P. elatior.* See *Research Studies of the State College of Washington* 18:80 (1950) for original description.

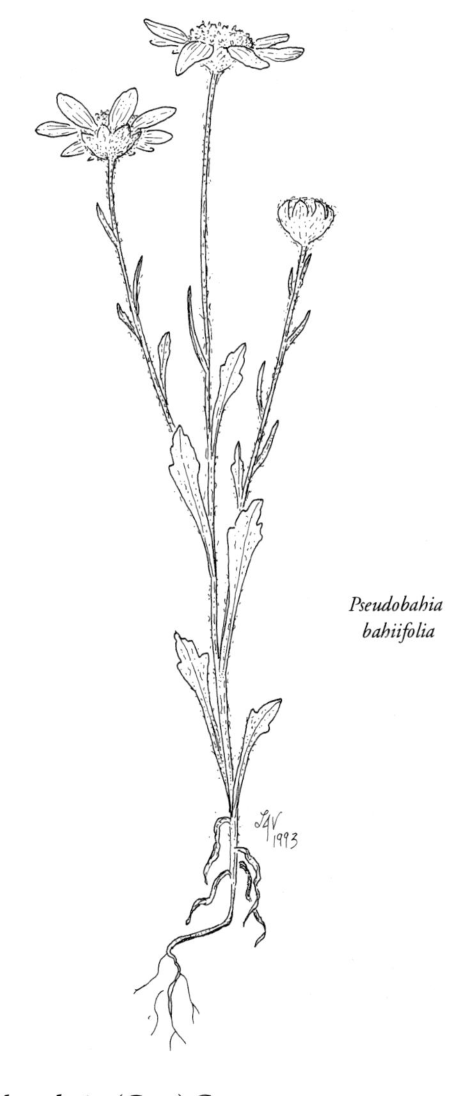

Pseudobahia bahiifolia

Psilocarphus elatior (Gray) Gray

"tall woolly-marbles" Asteraceae
CNPS List: 4 **R-E-D Code:** 1-1-1 **State/Fed. Status:** CEQA?
Distribution: LAS, MOD, OR, ++
Habitat: Medws, VFGrs / vernally mesic
Life Form: Annual herb
Blooming: May-August
Notes: Probably more widespread in northern CA; need information. Possibly a variant of *P. brevissimus* var. *brevissimus.* See *Proceedings of the American Academy of Arts and Sciences* 8:652 (1873) for original description.

Psilocarphus tenellus Nutt. var. *globiferus* (DC.) Morefield

"round woolly-marbles" Asteraceae
CNPS List: 4 **R-E-D Code:** 1-2-1 **State/Fed. Status:** CEQA?
Distribution: CAL, FRE, KRN, MER, MNT, MRN, SLO, STA, TUL, SA
Habitat: CoDns, VnPls
Life Form: Annual herb
Blooming: April-May
Notes: To be expected in other areas of the Sierra Nevada foothills, San Joaquin Valley, Central Coast, and S.F. Bay; need information. See *Madroño* 39(2):156 (1992) for revised nomenclature.

Psoralea rigida

See *Rupertia rigida*

Psorothamnus arborescens (Gray) Barneby var. *arborescens*

"Mojave indigobush" Fabaceae
CNPS List: 4 **R-E-D Code:** 1-1-1 **State/Fed. Status:** /C3c
Distribution: KRN, SBD, SO
Habitat: RpScr
Life Form: Shrub (deciduous)
Blooming: April-May
Notes: See *Memoirs of the New York Botanical Garden* 27:182 (1977) for taxonomic treatment.

Psorothamnus arborescens var. *simplicifolius*

Considered but rejected: Too common

Puccinellia californica (Beetle) Munz

"Sierra Nevada alkali-grass" Poaceae
CNPS List: 4 **R-E-D Code:** 1-1-3 **State/Fed. Status:** CEQA?
Distribution: FRE, TUL
Habitat: SCFrs (mesic)
Life Form: Perennial herb (rhizomatous)
Blooming: August-September
Notes: See *Torreyochloa pallida* var. *pauciflora* in *The Jepson Manual.*

Puccinellia howellii Davis

"Howell's alkali-grass" Poaceae
CNPS List: 1B **R-E-D Code:** 3-3-3 **State/Fed. Status:** /C1
Distribution: SHA
Quads: 648B
Habitat: Medws (mineralized)
Life Form: Perennial herb
Blooming: April-June
Notes: Known from only one occurrence. Threatened by highway runoff pollution and disturbance from the realignment of Highway 299. See *Madroño* 37(1):55-58 (1990) for original description.

Puccinellia lemmonii

Considered but rejected: Too common

Puccinellia parishii Hitchc.

"Parish's alkali-grass" Poaceae
CNPS List: 1B **R-E-D Code:** 3-3-2 **State/Fed. Status:** /C1
Distribution: KRN?, SBD, AZ, NM
Quads: 131B, 186D?
Habitat: Medws (alkaline)
Life Form: Annual herb
Blooming: April-May
Notes: Confirmed in 1992 at Rabbit Springs, SBD Co. (type locality, 131B). New occurrence also found in 1992 at Edwards AFB, KRN Co. (186D), but identity is questionable. Declining elsewhere; confirmed extant from only a few occurrences in AZ, and one in NM where state-listed as Endangered. Threatened by groundwater pumping, flood control, and grazing. See *Proceedings of the Biological Society of Washington* 41:157 (1928) for original description.
Status Report: 1979

Puccinellia pumila (Vasey) Hitchc.

"dwarf alkali-grass" Poaceae
CNPS List: 2 **R-E-D Code:** 3-2-1 **State/Fed. Status:** CEQA
Distribution: HUM, MEN, OR, WA, ++
Quads: 569A, 655A
Habitat: MshSw (coastal salt)
Life Form: Perennial herb
Blooming: July
Notes: Known in CA from only three occurrences. Is plant poorly known?

Purpusia saxosa

See *Ivesia arizonica* var. *arizonica*

Pyrrocoma lucida (Keck) Kartesz & Gandhi

"sticky pyrrocoma" Asteraceae
CNPS List: 1B **R-E-D Code:** 3-1-3 **State/Fed. Status:** CEQA
Distribution: PLU, SIE, YUB
Quads: 558A, 571D, 572A, 587C, 588A, 589B
Habitat: LCFrs (alkaline clay)
Life Form: Perennial herb
Blooming: July-September
Notes: See *Phytologia* 71(1):58-65 (1991) for revised nomenclature.

Pyrrocoma racemosa (Nutt.) T. & G. var. *congesta* (Greene) G. Brown & Keil

"Del Norte pyrrocoma" Asteraceae
CNPS List: 4 **R-E-D Code:** 1-1-1 **State/Fed. Status:** /C3c
Distribution: DNT, OR
Habitat: Chprl, LCFrs / serpentinite
Life Form: Perennial herb
Blooming: August-September
Notes: See *Pittonia* 3:23 (1898) for original description, and *Phytologia* 73(1):57-58 (1992) for revised nomenclature.

Pyrrocoma uniflora (Hook.) Greene var. *gossypina* (Greene) Kartesz & Gandhi

"Bear Valley pyrrocoma" Asteraceae
CNPS List: 1B **R-E-D Code:** 2-2-3 **State/Fed. Status:** /C2
Distribution: SBD
Quads: 105A, 105B, 131C, 131D
Habitat: Medws, PbPln
Life Form: Perennial herb
Blooming: July-September
Notes: Known from fewer than twenty occurrences. Threatened by grazing, development, and vehicles. See *Pittonia* 3:23 (1898) for original description, and *Phytologia* 71(1):58-65 (1991) for revised nomenclature.

Quercus dumosa Nutt.

"Nuttall's scrub oak" Fagaceae
CNPS List: 1B **R-E-D Code:** 2-3-2 **State/Fed. Status:** /C2
Distribution: ORA, SBA, SDG, BA
Quads: 10C, 11B, 22B, 35C, 36D, 52B, 70C, 70D?, 71D, 142A, 142B
Habitat: Chprl, CoScr / sandy, clay loam
Life Form: Shrub (evergreen)
Blooming: February-March
Notes: Threatened by development. Widespread scrub oak from much of cismontane CA, previously called *Q. dumosa*, is now *Q. berberidifolia.*

Quercus dumosa var. *kinselae*

Considered but rejected: Taxonomic problem

Quercus engelmannii Greene

"Engelmann oak" Fagaceae
CNPS List: 4 **R-E-D Code:** 1-2-2 **State/Fed. Status:** CEQA?
Distribution: LAX, ORA, RIV, SCT, SDG, BA
Habitat: Chprl, CmWld, RpWld, VFGrs
Life Form: Tree (deciduous)
Blooming: April-May
Notes: Only one tree remaining on SCT Isl. Threatened by development. Protected in part at the Santa Rosa Plateau Preserve (TNC), RIV Co. See *Fremontia* 18(3):26-35 (1990) for species account and ecological discussion.

Quercus lobata

Considered but rejected: Too common

Quercus macdonaldiana

Considered but rejected: A hybrid between *Q. berberidifolia* and *Q. lobata*

Quercus parvula Greene var. *parvula*

"Santa Cruz Island oak" Fagaceae
CNPS List: 4 **R-E-D Code:** 1-1-3 **State/Fed. Status:** /C3c
Distribution: SBA, SCZ
Habitat: CCFrs, Chprl, CmWld
Life Form: Tree (evergreen)
Blooming: March-June

Quercus tomentella Engelm.

"island oak" Fagaceae
CNPS List: 4 **R-E-D Code:** 1-2-2 **State/Fed. Status:** CEQA?
Distribution: ANA, SCM, SCT, SCZ, SRO, GU
Habitat: CCFrs, Chprl, CmWld
Life Form: Tree (evergreen)
Blooming: March-July
Notes: Threatened by trampling, erosion, grazing, feral herbivores, and introgression. Hybridizes with *Q. chrysolepis* except on SRO and GU islands. Populations on GU Isl. have declined dramatically in recent years. See *Transactions of the Academy of Science of St. Louis* 3:393 (1877) for original description.

Raillardella muirii

See *Raillardiopsis muirii*

Raillardella pringlei Greene

"showy raillardella" Asteraceae
CNPS List: 1B **R-E-D Code:** 2-2-3 **State/Fed. Status:** /C2
Distribution: SIS, TRI
Quads: 667A, 667B, 682B, 683A, 683B, 684A, 684C, 684D, 699C, 700A, 700D, 736A
Habitat: BgFns, Medws (mesic)
Life Form: Perennial herb (rhizomatous)
Blooming: July-September
Notes: Known from approximately twenty occurrences. Threatened by grazing. See *Bulletin of the Torrey Botanical Club* 9:17 (1882) for original description, and *Madroño* 25:138 (1978) for distributional information.
Status Report: 1979

Raillardella scabrida

See *Raillardiopsis scabrida*

Raillardiopsis muirii (Gray) Rydb.

"Muir's raillardella" Asteraceae
CNPS List: 1B **R-E-D Code:** 2-1-3 **State/Fed. Status:** /C3c
Distribution: FRE, KRN, MNT, TUL
Quads: 258B, 284D, 285D, 331C, 343C, 353C, 353D, 374B, 374C, 374D, 375A, 375B, 376A, 394C
Habitat: Chprl (montane), LCFrs, UCFrs
Life Form: Perennial herb (rhizomatous)
Blooming: July-August
Notes: Identity of Ventana Cones occurrence in the Santa Lucia Mtns. of MNT Co. has been confirmed. See *Botany of California* 1:618 (1876) for original description, *North American Flora* 34:320 (1927) for revised nomenclature, and *Madroño* 37(1):48 (1990) for revised description.
Status Report: 1979

Raillardiopsis scabrida (Eastw.) Rydb.

"scabrid raillardella" Asteraceae
CNPS List: 4 **R-E-D Code:** 1-1-3 **State/Fed. Status:** /C3c
Distribution: COL, LAK, MEN, SHA, TEH, TRI
Habitat: UCFrs (metamorphic)
Life Form: Perennial herb
Blooming: July-August
Notes: See *Bulletin of the Torrey Botanical Club* 32:216 (1905) for original description, and *North American Flora* 34:319-320 (1927) for revised nomenclature.
Status Report: 1979

Ranunculus alveolatus

Considered but rejected: A synonym of *R. bonariensis* var. *trisepalus*; a common taxon

Ranunculus bonariensis var. *trisepalus*

Considered but rejected: Too common

Ranunculus hydrocharoides Gray

"frog's-bit buttercup" Ranunculaceae
CNPS List: 1A **Last Seen:** 1874 **State/Fed. Status:** CEQA
Distribution: INY*, AZ, BA, NM+
Quads: 373D*, 394A*, 413C*
Habitat: MshSw (freshwater)
Life Form: Perennial herb (aquatic)
Blooming: June-July
Notes: Last seen in CA in 1874, but date uncertain since these Kellogg specimens may be mislabeled. See *Memoirs of the American Academy of Arts and Sciences* II 5:306 (1855) for original description.

Ranunculus lobbii Gray

"Lobb's aquatic buttercup" Ranunculaceae
CNPS List: 4 **R-E-D Code:** 1-2-3 **State/Fed. Status:** CEQA?
Distribution: ALA, CCA, MEN, MRN, NAP, SCL, SOL, SON
Habitat: CmWld, NCFrs, VFGrs, VnPls / mesic
Life Form: Annual herb (aquatic)
Blooming: March-May
Notes: Threatened by urbanization and agriculture.

Rhamnus pilosa

Considered but rejected: Too common

Rhus trilobata T. & G. var. *simplicifolia* (Greene) Barkley

"single-leaf skunkbrush" Anacardiaceae
CNPS List: 2 **R-E-D Code:** 3-1-1 **State/Fed. Status:** CEQA
Distribution: SDG, AZ, BA, ++
Quads: 31C, 32D
Habitat: Chprl, PJWld
Life Form: Shrub (deciduous)
Blooming: March-April
Notes: Not in *The Jepson Manual.* See *Annals of the Missouri Botanical Garden* 24:410-411 (1937) for revised nomenclature, and *Madroño* 34(2):171 (1987) for first CA records.

Rhynchospora alba (L.) Vahl

"white beaked-rush" Cyperaceae
CNPS List: 4 **R-E-D Code:** 1-1-1 **State/Fed. Status:** CEQA?
Distribution: DNT?, LAS, MEN, NEV?, PLU, SON, ID, ++
Habitat: BgFns, MshSw (freshwater)
Life Form: Perennial herb (rhizomatous)
Blooming: July-August
Notes: Does plant occur in DNT and NEV counties? Threatened in ID.

Rhynchospora californica Gale

"California beaked-rush" Cyperaceae
CNPS List: 1B **R-E-D Code:** 3-3-3 **State/Fed. Status:** /C2
Distribution: BUT, MPA, MRN, SON
Quads: 457D, 485C, 502A, 502D, 592C, 593D
Habitat: LCFrs, Medws (seeps), MshSw (freshwater)
Life Form: Perennial herb (rhizomatous)
Blooming: May-July
Notes: Known from fewer than ten occurrences. Threatened by marsh habitat loss. See *Rhodora* 46:272-273 (1944) for original description, and *Madroño* 33(2):150 (1986) for information on BUT Co. collection.
Status Report: 1977

Rhynchospora capitellata

Considered but rejected: Too common

Rhynchospora globularis

See *Rhynchospora globularis* var. *globularis*

Rhynchospora globularis (Chapm.) Small var. *globularis*

"round-headed beaked-rush" Cyperaceae
CNPS List: 2 **R-E-D Code:** 3-3-1 **State/Fed. Status:** CEQA
Distribution: SON, ++
Quads: 502A, 502D
Habitat: MshSw (freshwater)
Life Form: Perennial herb (rhizomatous)
Blooming: July-August
Notes: Seriously threatened by marsh habitat loss.

Rhynchospora glomerata var. *capitellata*

Considered but rejected: A synonym of *R. capitellata*; a common taxon

Rhynchospora glomerata var. *minor*

Considered but rejected: A synonym of *R. capitellata*; a common taxon

Ribes amarum McClat. var. *hoffmannii* Munz

"bitter gooseberry" Grossulariaceae
CNPS List: 3 **R-E-D Code:** ?-?-3 **State/Fed. Status:** CEQA?
Distribution: SBA
Quads: 142B, 169D
Habitat: Chprl
Life Form: Shrub (deciduous)
Blooming: March-April
Notes: Move to List 1B? Location, rarity, and endangerment information needed. See *R. amarum* in *The Jepson Manual.*

Ribes canthariforme Wiggins

"Moreno currant" Grossulariaceae
CNPS List: 1B **R-E-D Code:** 3-1-3 **State/Fed. Status:** /C2
Distribution: SDG
Quads: 9A, 9B, 10A, 20D, 21A, 21D
Habitat: Chprl
Life Form: Shrub (deciduous)
Blooming: February-April
Notes: Known from approximately ten occurrences. See *Contributions from the Dudley Herbarium* 1:101 (1929) for original description.

Ribes divaricatum Dougl. var. *parishii* (Heller) Jeps.

"Parish's gooseberry" Grossulariaceae
CNPS List: 1B **R-E-D Code:** 3-3-3 **State/Fed. Status:** /C2
Distribution: LAX, SBD*
Quads: 89A*, 107D*, 110D
Habitat: RpWld
Life Form: Shrub (deciduous)
Blooming: February-April
Notes: Possibly extinct; last documented in 1980 from two plants at Whittier Narrows Nature Center. Known from fewer than five historical occurrences; these extirpated by urbanization. Field surveys needed. See *Rhodora* 87:259-260 (1985) for taxonomic treatment.

Ribes divaricatum var. *pubiflorum*

Considered but rejected: Too common

Ribes hudsonianum Richards. var. *petiolare* (Dougl.) Jancz.

"western black currant" Grossulariaceae
CNPS List: 2 **R-E-D Code:** 3-1-1 **State/Fed. Status:** CEQA
Distribution: MOD, SIS, ID, OR, WA+
Quads: 691A, 732B
Habitat: RpScr
Life Form: Shrub (deciduous)
Blooming: May-July
Notes: Known in CA only from Shovel Creek (SIS Co.) and the Warner Mtns. (MOD Co.).

Ribes laxiflorum Pursh

"trailing black currant" Grossulariaceae
CNPS List: 4 **R-E-D Code:** 1-1-1 **State/Fed. Status:** CEQA?
Distribution: DNT?, HUM, OR, WA, ++
Habitat: NCFrs
Life Form: Shrub (deciduous)
Blooming: March-May
Notes: Does plant occur in DNT Co.?

Ribes marshallii Greene

"Marshall's gooseberry" Grossulariaceae
CNPS List: 4 **R-E-D Code:** 1-1-2 **State/Fed. Status:** CEQA?
Distribution: HUM, SIS, OR
Habitat: CCFrs
Life Form: Shrub (deciduous)
Blooming: June-July
Notes: Endangered in OR.

Ribes menziesii Pursh var. *ixoderme* Quick

"aromatic canyon gooseberry" Grossulariaceae
CNPS List: 4 **R-E-D Code:** 1-2-3 **State/Fed. Status:** CEQA?
Distribution: FRE, KRN, TUL
Habitat: Chprl
Life Form: Shrub (deciduous)
Blooming: April
Notes: Field work needed. See *R. menziesii* in *The Jepson Manual.* See *Madroño* 4:287 (1938) for original description.

Ribes sericeum Eastw.

"Santa Lucia gooseberry" Grossulariaceae
CNPS List: 4 **R-E-D Code:** 1-1-3 **State/Fed. Status:** CEQA?
Distribution: MNT, SLO
Habitat: CBScr, NCFrs
Life Form: Shrub (deciduous)
Blooming: February-April

Ribes thacherianum (Jeps.) Munz

"Santa Cruz Island gooseberry" Grossulariaceae
CNPS List: 1B **R-E-D Code:** 2-2-3 **State/Fed. Status:** /C2
Distribution: SCZ
Quads: SCZB, SCZC
Habitat: CCFrs, RpWld
Life Form: Shrub (deciduous)
Blooming: March-April
Notes: Possibly threatened by feral herbivores.

Ribes tularense (Cov.) Fedde

"Sequoia gooseberry" Grossulariaceae
CNPS List: 1B **R-E-D Code:** 3-1-3 **State/Fed. Status:** /C2
Distribution: TUL
Quads: 331B, 331D, 332A, 354D
Habitat: LCFrs, UCFrs
Life Form: Shrub (deciduous)
Blooming: May
Notes: Known from fewer than ten occurrences. Possibly threatened by logging and road improvements. See *North American Flora* 22:218 (1908) for original description.

Ribes viburnifolium Gray

"Santa Catalina Island currant" Grossulariaceae
CNPS List: 4 **R-E-D Code:** 1-1-2 **State/Fed. Status:** CEQA?
Distribution: SCT, SDG, BA
Habitat: Chprl
Life Form: Shrub (evergreen)
Blooming: February-April

Ribes victoris Greene

"Victor's gooseberry" Grossulariaceae
CNPS List: 4 **R-E-D Code:** 1-1-3 **State/Fed. Status:** CEQA?
Distribution: MEN, MRN, NAP, SOL, SON
Habitat: BUFrs, Chprl
Life Form: Shrub (deciduous)
Blooming: March-April

Robinia neomexicana

Considered but rejected: Not native and taxonomic problem, but needs reevaluation

Romneya coulteri Harv.

"Coulter's matilija poppy" Papaveraceae
CNPS List: 4 **R-E-D Code:** 1-2-3 **State/Fed. Status:** CEQA?
Distribution: LAX, ORA, RIV, SDG
Habitat: Chprl, CoScr / often in burned areas
Life Form: Perennial herb (rhizomatous)
Blooming: May-July
Notes: Threatened by urbanization, flood control, and road widening and maintenance.

Romneya trichocalyx

Considered but rejected: Too common

Rorippa columbiae (Robinson) Howell

"Columbia yellow cress" Brassicaceae
CNPS List: 1B **R-E-D Code:** 3-2-2 **State/Fed. Status:** /C2
Distribution: HUM, MOD, SIS, OR, WA+
Quads: 696C, 697A, 704D, 714B, 728A, 728D, 730D
Habitat: Medws, PJWld, Plyas
Life Form: Perennial herb (rhizomatous)
Blooming: May-September
Notes: Candidate for state listing in OR, and state-listed as Endangered in WA. See *Sida* 4:294-296 (1972) for taxonomic treatment.

Rorippa gambellii (Wats.) Roll. & Al-Shehbaz

"Gambel's water cress" Brassicaceae
CNPS List: 1B **R-E-D Code:** 3-3-2 **State/Fed. Status:** CT/FE
Distribution: LAX*, ORA*, SDG*, SLO, BA
Quads: 33D*, 107D*, 111D*, 221D
Habitat: MshSw (freshwater or brackish)
Life Form: Perennial herb (rhizomatous)
Blooming: April-June
Notes: Nearly extinct in U.S.; known in CA from only one or two extant occurrences. VEN Co. occurrence (141D) erroneous; probably misidentified *R. nasturtium-aquaticum.* Seriously threatened by habitat loss, and *Eucalyptus* may be altering hydrology at Black Lake Cyn. See *Journal of the Arnold Arboretum* 69:65-71 (1988) for revised nomenclature.
Status Report: 1991

Rorippa subumbellata Roll.

"Tahoe yellow cress" Brassicaceae
CNPS List: 1B **R-E-D Code:** 3-3-2 **State/Fed. Status:** CE/C1
Distribution: ELD, NEV*, PLA, NV
Quads: 522B, 523A, 538A*, 538B, 538C, 538D, 554C*
Habitat: LCFrs, Medws / decomposed granitic beaches
Life Form: Perennial herb (rhizomatous)
Blooming: June-September
Notes: Known in CA from fewer than ten extant occurrences around Lake Tahoe; over one-half of historical occurrences extirpated. Threatened by development, recreation, and trampling; recovery work underway. State-listed as Critically Endangered in NV. See *Contributions from the Dudley Herbarium* 3:177 (1941) for original description.
Status Report: 1991

Rosa minutifolia Engelm.

"small-leaved rose" Rosaceae
CNPS List: 2 **R-E-D Code:** 3-3-1 **State/Fed. Status:** CE/C2
Distribution: SDG, BA
Quads: 11D
Habitat: Chprl
Life Form: Shrub (deciduous)
Blooming: January-June
Notes: Known in CA from only one occurrence on Otay Mesa. Threatened by development and vehicles. See *Madroño* 33:150 (1986) for first CA record.
Status Report: 1989

Rubus glaucifolius Kell. var. *ganderi* (Bailey) Munz

"Cuyamaca raspberry" Rosaceae
CNPS List: 1B **R-E-D Code:** 3-1-3 **State/Fed. Status:** /C2
Distribution: SDG
Quads: 20A, 33D
Habitat: LCFrs
Life Form: Shrub (evergreen)
Blooming: June
Notes: Known from only two occurrences on Harrison Pk. and Middle Pk. in the Cuyamaca Mtns. Historical occurrences need field surveys. See *R. glaucifolius* in *The Jepson Manual.*

Rubus nivalis Dougl.

"snow dwarf bramble" Rosaceae
CNPS List: 2 **R-E-D Code:** 3-1-1 **State/Fed. Status:** CEQA
Distribution: DNT, ID, OR+
Quads: 738C
Habitat: NCFrs
Life Form: Vine (evergreen)
Blooming: June-August

Rubus pedatus

Considered but rejected: Not in CA; name misapplied to *R. lasiococcus*; a common taxon

Rupertia hallii (Rydb.) Grimes

"Hall's rupertia" Fabaceae
CNPS List: 1B **R-E-D Code:** 3-2-3 **State/Fed. Status:** CEQA
Distribution: BUT, TEH
Quads: 608C, 608D
Habitat: CmWld
Life Form: Perennial herb
Blooming: July-August
Notes: See *North American Flora* 24:11 (1919) for original description, and *Memoirs of the New York Botanical Garden* 61:1-114 (1990) for taxonomic treatment.

Rupertia rigida (Parish) Grimes

"Parish's rupertia" Fabaceae
CNPS List: 4 **R-E-D Code:** 1-1-2 **State/Fed. Status:** CEQA?
Distribution: RIV, SBD, SDG, BA
Habitat: Chprl, CmWld, LCFrs
Life Form: Perennial herb
Blooming: June-July
Notes: See *Systematic Botany* 14:233 (1989) for revised nomenclature.

Sagittaria rigida

Considered but rejected: Not native

Sagittaria sanfordii Greene

"Sanford's arrowhead" Alismataceae
CNPS List: 1B **R-E-D Code:** 2-2-3 **State/Fed. Status:** /C2
Distribution: BUT, DNT, FRE, KRN, MER, MRN, ORA*, SAC, SHA, SJQ, TEH, VEN*
Quads: 72A*, 141A*, 381D, 402C, 403D, 421C, 423D, 462D, 478B, 485C, 496A, 496B, 496D, 512C, 575B, 593B, 628A, 628D, 740C
Habitat: MshSw (assorted shallow freshwater)
Life Form: Perennial herb (rhizomatous, emergent)
Blooming: May-August
Notes: Mostly extirpated from the Central Valley. Need quads for FRE and KRN counties. Threatened by grazing, development, and channel alteration.

Salix bebbiana Sarg.

"gray willow" Salicaceae
CNPS List: 2 **R-E-D Code:** 2-1-1 **State/Fed. Status:** CEQA
Distribution: MOD, SIS, AZ, NV, OR, ++
Quads: 730B, 730C
Habitat: MshSw (streambanks and lake margins), RpScr
Life Form: Tree (deciduous)
Blooming: May
Notes: Need quads for MOD Co. See *Garden and Forest* 8:463 (1895) for original description.

Salix brachycarpa

See *Salix brachycarpa* ssp. *brachycarpa*

Salix brachycarpa Nutt. ssp. *brachycarpa*

"short-fruited willow" Salicaceae
CNPS List: 2 **R-E-D Code:** 3-1-1 **State/Fed. Status:** CEQA
Distribution: MNO, OR, ++
Quads: 434D, 453B, 454A
Habitat: AlpDS, Medws, SCFrs / carbonate
Life Form: Shrub (deciduous)
Blooming: June-July
Notes: On watch list in OR.

Salix delnortensis C. Schneider

"Del Norte willow" Salicaceae
CNPS List: 4 **R-E-D Code:** 1-1-2 **State/Fed. Status:** CEQA?
Distribution: DNT, SIS, OR
Habitat: RpFrs (serpentinite)
Life Form: Shrub (deciduous)
Blooming: April-May
Notes: Similar to *S. breweri*; may hybridize with *S. lasiolepis*. Endangered in OR.

Salix nivalis

See *Salix reticulata* ssp. *nivalis*

Salix parksiana

Considered but rejected: A synonym of *S. sessilifolia*; a common taxon

Salix reticulata L. ssp. *nivalis* (Hook.) A. Löve, D. Löve & Kapoor

"snow willow" Salicaceae
CNPS List: 2 **R-E-D Code:** 2-1-1 **State/Fed. Status:** CEQA
Distribution: MNO, ++
Quads: 453B, 453C, 454A
Habitat: AlpDS
Life Form: Shrub (deciduous)
Blooming: July-August

Salix tracyi

Considered but rejected: A synonym of *S. lasiolepis*; a common taxon

Salvia brandegei Munz

"Brandegee's sage" Lamiaceae
CNPS List: 1B **R-E-D Code:** 2-2-2 **State/Fed. Status:** /C3c
Distribution: SRO, BA
Quads: SROE, SRON
Habitat: CCFrs, Chprl, CoScr
Life Form: Shrub (evergreen)
Blooming: April-May
Notes: Fairly widespread on SRO Isl. Possibly threatened by feral herbivores.

Salvia clevelandii

Considered but rejected: Too common

Salvia columbariae var. *ziegleri*

Considered but rejected: A synonym of *S. columbariae*; a common taxon

Salvia dorrii var. *carnosa*

See *Salvia dorrii* var. *incana*

Salvia dorrii (Kellogg) Abrams var. *incana* (Benth.) Strachan

"fleshy sage" Lamiaceae
CNPS List: 3 **R-E-D Code:** ?-?-1 **State/Fed. Status:** CEQA?
Distribution: MOD, SIS, ID, OR, WA
Quads: 690B, 730D, 734A
Habitat: GBScr, PJWld
Life Form: Shrub (evergreen)
Blooming: May-July
Notes: Move to List 2? CA plants may be intermediate to var. *dorrii*. See *Brittonia* 34(2):151-169 (1982) for taxonomic treatment and 35(2):170 (1983) for revised nomenclature.

Salvia eremostachya Jeps.

"desert sage" Lamiaceae
CNPS List: 4 **R-E-D Code:** 1-1-1 **State/Fed. Status:** /C3c
Distribution: RIV, SDG, BA
Habitat: SDScr
Life Form: Shrub (evergreen)
Blooming: March-May

Salvia funerea M.E. Jones

"Death Valley sage" Lamiaceae
CNPS List: 4 **R-E-D Code:** 1-1-2 **State/Fed. Status:** /C3c
Distribution: INY, NV
Habitat: MDScr (carbonate)
Life Form: Shrub (evergreen)
Blooming: March-May
Notes: On watch list in NV.

Salvia greatae Bdg.

"Orocopia sage" Lamiaceae
CNPS List: 1B **R-E-D Code:** 2-1-3 **State/Fed. Status:** /C2
Distribution: IMP, RIV, SBD?
Quads: 44B, 45A, 61A, 61B, 62C, 63C, 63D, 151D?
Habitat: MDScr, SDScr
Life Form: Shrub (evergreen)
Blooming: March-April
Notes: SBD Co. report is questionable. See *Zoe* 5:229 (1906) for original description, and *Phytologia* 71(3):167-170 (1991) for nomenclatural correction.

Salvia munzii Epl.

"Munz's sage" Lamiaceae
CNPS List: 2 **R-E-D Code:** 2-2-1 **State/Fed. Status:** CEQA
Distribution: SDG, BA
Quads: 10B, 10C, 10D
Habitat: Chprl, CoScr
Life Form: Shrub (evergreen)
Blooming: February-April
Notes: Threatened by development.

Salvia vaseyi

Considered but rejected: Too common

Sanguisorba officinalis L.

"great burnet" Rosaceae
CNPS List: 2 **R-E-D Code:** 2-2-1 **State/Fed. Status:** CEQA
Distribution: DNT, HUM, MEN, OR, WA, ++
Quads: 553A, 653A, 671A?, 671D, 722A, 739A, 739B, 739C
Habitat: BgFns, BUFrs, Medws, MshSw, NCFrs, RpFrs / often serpentinite
Life Form: Perennial herb (rhizomatous)
Blooming: July-September
Notes: CA plants may be ssp. *microcephala.*

Sanicula hoffmannii (Munz) C.R. Bell

"Hoffmann's sanicle" Apiaceae
CNPS List: 4 **R-E-D Code:** 1-1-3 **State/Fed. Status:** /C3c
Distribution: MNT, SBA, SCR, SCZ, SLO, SMT, SRO
Habitat: BUFrs, Chprl, CoScr / often serpentinite or clay
Life Form: Perennial herb
Blooming: March-May

Sanicula maritima Wats.

"adobe sanicle" Apiaceae
CNPS List: 1B **R-E-D Code:** 3-3-3 **State/Fed. Status:** CR/C2
Distribution: ALA*, MNT, SFO*, SLO
Quads: 246C, 247D, 272A, 296B, 344D, 466C*, 466D*
Habitat: Chprl, CoPrr, Medws, VFGrs / clay, serpentinite
Life Form: Perennial herb
Blooming: April-May
Notes: Known from fewer than ten occurrences. Threatened by urbanization. See *Botany of California* 2:451 (1880) for original description, and *University of California Publications in Botany* 25:61-62 (1951) for taxonomic treatment.
Status Report: 1988

Sanicula peckiana Macbr.

"Peck's sanicle" Apiaceae
CNPS List: 4 **R-E-D Code:** 1-1-1 **State/Fed. Status:** /C3c
Distribution: DNT, OR
Habitat: Chprl, LCFrs / serpentinite
Life Form: Perennial herb
Blooming: March-June
Notes: On watch list in OR. See *Contributions from the Gray Herbarium* 59:28 (1919) for original description, and *University of California Publications in Botany* 25:64-65 (1951) for taxonomic treatment.
Status Report: 1979

Sanicula saxatilis Greene

"rock sanicle" Apiaceae
CNPS List: 1B **R-E-D Code:** 3-2-3 **State/Fed. Status:** CR/C2
Distribution: CCA, SCL
Quads: 406A, 426C, 426D, 464B, 464C
Habitat: BUFrs, Chprl, VFGrs / rocky
Life Form: Perennial herb
Blooming: April-May
Notes: Known from fewer than fifteen occurrences. Threatened by development. See *Erythea* 1:6 (1893) for original description, and *University of California Publications in Botany* 25:71-72 (1951) for taxonomic treatment.
Status Report: 1988

Sanicula tracyi Shan & Const.

"Tracy's sanicle" Apiaceae
CNPS List: 1B **R-E-D Code:** 1-2-3 **State/Fed. Status:** /C2
Distribution: BUT, DNT, HUM, TEH, TRI
Quads: 574C, 575B, 608A, 614B, 614C, 615A, 615B, 615C, 615D, 632C, 632D, 633B, 633C, 634A, 634B, 634C, 634D, 652B, 652C, 652D, 671A, 688A, 721C
Habitat: CmWld, LCFrs, UCFrs / openings
Life Form: Perennial herb
Blooming: April-July
Notes: Threatened by grazing, logging, and development. See *University of California Publications in Botany* 25:69 (1951) for original description.
Status Report: 1977

Sanvitalia abertii Gray

"Abert's sanvitalia" Asteraceae
CNPS List: 2 **R-E-D Code:** 3-2-1 **State/Fed. Status:** CEQA
Distribution: SBD, AZ, SO, TX
Quads: 224C, 249D
Habitat: PJWld (carbonate)
Life Form: Annual herb
Blooming: August-September
Notes: Known in CA only from the Clark and New York Mtns. Threatened by mining.

Satureja chandleri (Bdg.) Druce

"San Miguel savory" Lamiaceae
CNPS List: 4 **R-E-D Code:** 1-2-2 **State/Fed. Status:** /C3c
Distribution: ORA, RIV, SDG, BA
Habitat: Chprl, CmWld, CoScr, RpWld, VFGrs
Life Form: Perennial herb
Blooming: March-May
Notes: Threatened by residential development, agriculture, and recreational activities. See *Crossosoma* 12(1):12 (1986) for taxonomic treatment.

Saussurea americana D.C. Eat.

"American saw-wort" Asteraceae
CNPS List: 2 **R-E-D Code:** 3-2-1 **State/Fed. Status:** CEQA
Distribution: SIS, ID, OR, WA+
Quads: 735B, 736A
Habitat: LCFrs, Medws / mesic
Life Form: Perennial herb
Blooming: July-August
Notes: See *Botanical Gazette* 6:283 (1881) for original description.
Status Report: 1979

Saxifraga cespitosa L.

"tufted saxifrage" Saxifragaceae
CNPS List: 2 **R-E-D Code:** 3-1-1 **State/Fed. Status:** CEQA
Distribution: MOD, SIS, AZ, NV, OR, WA, ++
Quads: 690B, 719C
Habitat: Medws (mesic, rocky)
Life Form: Perennial herb
Blooming: June-September

Saxifraga fragarioides

Considered but rejected: Too common

Saxifraga howellii Greene

"Howell's saxifrage" Saxifragaceae
CNPS List: 4 **R-E-D Code:** 1-1-1 **State/Fed. Status:** CEQA?
Distribution: DNT, SIS, OR
Habitat: CmWld (sometimes serpentinite)
Life Form: Perennial herb (rhizomatous)
Blooming: March-May

Saxifraga integrifolia var. *integrifolia*

Considered but rejected: A synonym of *S. integrifolia*; a common taxon

Saxifraga nuttallii Small

"Nuttall's saxifrage" Saxifragaceae
CNPS List: 2 **R-E-D Code:** 3-3-1 **State/Fed. Status:** CEQA
Distribution: DNT, OR, WA
Quads: 740D
Habitat: NCFrs (mesic, rocky)
Life Form: Perennial herb (rhizomatous)
Blooming: May
Notes: Known in CA from only one occurrence along the Smith River. See *Bulletin of the Torrey Botanical Club* 23:368 (1896) for original description.

Saxifraga rufidula (Small) Macoun

"rusty saxifrage" Saxifragaceae
CNPS List: 2 **R-E-D Code:** 3-1-1 **State/Fed. Status:** CEQA
Distribution: SIS, OR, WA+
Quads: 720B
Habitat: UCFrs (mesic)
Life Form: Perennial herb (rhizomatous)
Blooming: March-July
Notes: Known in CA from only one occurrence in the Marble Mtns. See *North American Flora* 22:140 (1905) for original description.

Scheuchzeria palustris L. var. *americana* (Fern.) Hulten

"American scheuchzeria" Scheuchzeriaceae
CNPS List: 2 **R-E-D Code:** 3-3-1 **State/Fed. Status:** CEQA
Distribution: PLU, SIE*, ID, OR, WA, ++
Quads: 606B*, 625A, 625B
Habitat: BgFns, MshSw (lake margins)
Life Form: Perennial herb (rhizomatous, emergent)
Blooming: July
Notes: Rediscovered in 1988 by V. Oswald at Willow Lake. Need historical quads for SIE Co. Threatened by logging. Threatened in ID, and endangered in OR. See *Fremontia* 20(4):19-20 (1992) for information on rediscovery in CA.

Schoenolirion bracteosum

Considered but rejected: A synonym of *Hastingsia bracteosa*; not in CA

Scirpus clementis M.E. Jones

"Yosemite bulrush" Cyperaceae
CNPS List: 4 **R-E-D Code:** 1-1-3 **State/Fed. Status:** CEQA?
Distribution: FRE, INY, MAD, MNO, MPA, TUL, TUO
Habitat: AlpBR, Medws, SCFrs
Life Form: Perennial herb
Blooming: July-August

Scirpus heterochaetus Chase

"slender bulrush" Cyperaceae
CNPS List: 2 **R-E-D Code:** 3-1-1 **State/Fed. Status:** CEQA
Distribution: TEH, OR, WA, ++
Quads: 625C
Habitat: MshSw (lake margins), LCFrs
Life Form: Perennial herb (rhizomatous)
Blooming: August
Notes: Known in CA only from Wilson Lake. See *Rhodora* 6:70 (1904) for original description.

Scirpus pumilus Vahl.

"dwarf bulrush" Cyperaceae
CNPS List: 2 **R-E-D Code:** 3-2-1 **State/Fed. Status:** CEQA
Distribution: MNO, ++
Quads: 434D
Habitat: AlpDS? (mesic, carbonate)
Life Form: Perennial herb (rhizomatous)
Blooming: August
Notes: Known in CA only from Convict Creek Basin.

Scirpus rollandii

See *Scirpus pumilus*

Scirpus subterminalis Torr.

"water bulrush" Cyperaceae
CNPS List: 2 **R-E-D Code:** 2-1-1 **State/Fed. Status:** CEQA
Distribution: BUT, DNT, ELD, HUM, PLU, TEH, ID, OR, ++
Quads: 522C, 523D, 572B, 589C, 590D, 607C, 625A, 625B, 625C, 704C, 738B
Habitat: MshSw (montane lake margins)
Life Form: Perennial herb (rhizomatous)
Blooming: July-August
Notes: Sensitive in ID, and on review list in OR. See *Leaflets of Western Botany* 10:16 (1966) for first CA record.

Sclerocactus polyancistrus (Engelm. & Bigel.) Britt. & Rose

"Mojave fish-hook cactus" Cactaceae
CNPS List: 4 **R-E-D Code:** 1-2-2 **State/Fed. Status:** /C3c
Distribution: INY, KRN, SBD, NV
Habitat: JTWld, MDScr / carbonate
Life Form: Shrub (stem succulent)
Blooming: April-June
Notes: Threatened by horticultural collecting and herbivory by insects. Does not usually survive transplanting. See *Proceedings of the American Academy of Arts and Sciences* 3:272 (1856) for original description, *Cactus and Succulent Journal* 51:228-232 (1979) for ecological information, and *Fremontia* 10(4):23-24 (1983) for species account.

Scleropogon brevifolius Phil.

"burro grass" Poaceae
CNPS List: 2 **R-E-D Code:** 3-1-1 **State/Fed. Status:** CEQA
Distribution: SBD, AZ, NV, ++
Quads: 200A
Habitat: MDScr (decomposed granitic)
Life Form: Perennial herb (stoloniferous)
Blooming: October
Notes: Known in CA only from near Caruther's Cyn. in the New York Mtns.

Scopulophila rixfordii

Considered but rejected: Too common

Scrophularia atrata Penn.

"black-flowered figwort" Scrophulariaceae
CNPS List: 1B **R-E-D Code:** 2-2-3 **State/Fed. Status:** /C2
Distribution: SBA, SLO
Quads: 170B, 170C, 171A, 195C, 196A, 196B, 196D, 221A, 221B
Habitat: CCFrs, Chprl, CoDns, CoScr, RpScr
Life Form: Perennial herb
Blooming: April-June
Notes: Plants from south of Pt. Conception (143A, 143B, 144A) are probably hybrids with *S. californica* ssp. *floribunda*. Threatened by energy development and mining.

Scrophularia villosa Penn.

"Santa Catalina figwort" Scrophulariaceae
CNPS List: 1B **R-E-D Code:** 3-2-3 **State/Fed. Status:** /C2
Distribution: SCM, SCT
Quads: SCMC, SCMS, SCTE, SCTN, SCTW
Habitat: Chprl, CoScr
Life Form: Shrub
Blooming: April-August
Notes: Feral herbivores removed from SCM Isl. and vegetation recovering; still a threat on SCT Isl.

Scutellaria bolanderi Gray ssp. *austromontana* Epl.

"southern skullcap" Lamiaceae
CNPS List: 1B **R-E-D Code:** 2-2-3 **State/Fed. Status:** CEQA
Distribution: LAX(*?), RIV, SBD*, SDG
Quads: 20A, 20C, 20D, 33B, 49D, 66B, 69D, 110D(*?), 158D*
Habitat: Chprl, CmWld, LCFrs / mesic
Life Form: Perennial herb (rhizomatous)
Blooming: June-August
Notes: See *Madroño* 5(2):58 (1939) for original description.

Scutellaria galericulata L.

"marsh skullcap" Lamiaceae
CNPS List: 2 **R-E-D Code:** 2-2-1 **State/Fed. Status:** CEQA
Distribution: ELD, NEV, PLA, PLU, SHA, SIS?, OR, ++
Quads: 554C, 588D, 604A, 624C, 678C, 678D
Habitat: LCFrs, Medws (mesic), MshSw
Life Form: Perennial herb (rhizomatous)
Blooming: June-September
Notes: Need quads for ELD Co. Is SIS Co. occurrence a misidentification?

Scutellaria holmgreniorum Cronq.

"Holmgren's skullcap" Lamiaceae
CNPS List: 3 **R-E-D Code:** 3-1-2 **State/Fed. Status:** /C3c
Distribution: LAS, NV
Quads: 621B, 638A, 638B, 638C, 639A, 639D
Habitat: GBScr, PJWld / volcanic
Life Form: Perennial herb (rhizomatous)
Blooming: July
Notes: Move to List 1B? Taxonomic problems: a synonym of *S. nana* in *The Jepson Manual*, but quite likely distinct. On watch list in NV. See *Brittonia* 33(3):449-450 (1981) for original description.

Scutellaria lateriflora L.

"mad-dog skullcap" Lamiaceae
CNPS List: 2 **R-E-D Code:** 3-2-1 **State/Fed. Status:** CEQA
Distribution: INY, SJQ, NM, OR, ++
Quads: 480D
Habitat: Medws (mesic), MshSw
Life Form: Perennial herb (rhizomatous)
Blooming: July-September
Notes: Known in CA from only two occurrences. Need quad for occurrence in Saline Valley (INY Co.).

Sedum albomarginatum Clausen

"Feather River stonecrop" Crassulaceae
CNPS List: 1B **R-E-D Code:** 3-2-3 **State/Fed. Status:** /C3c
Distribution: BUT, PLU
Quads: 591C, 606B, 606C, 607C
Habitat: Chprl, LCFrs / serpentinite
Life Form: Perennial herb
Blooming: June
Notes: Known from fewer than twenty occurrences. Threatened by horticultural collecting, road construction, and new mining claims. See *Sedum of North America North of the Mexican Plateau,* pp. 424-433 (1975) by R. T. Clausen for original description.

Sedum divergens Wats.

"Cascade stonecrop" Crassulaceae
CNPS List: 2 **R-E-D Code:** 3-1-1 **State/Fed. Status:** CEQA
Distribution: SIS, OR, WA+
Quads: 738D
Habitat: AlpBR
Life Form: Perennial herb
Blooming: July-September
Notes: Known in CA from only one collection in the Siskiyou Wilderness; is it more common? See *Proceedings of the American Academy of Arts and Sciences* 17:372 (1882) for original description.

Sedum eastwoodiae (Britt.) Berger

"Red Mtn. stonecrop" Crassulaceae
CNPS List: 1B **R-E-D Code:** 3-2-3 **State/Fed. Status:** /C1
Distribution: MEN
Quads: 600B
Habitat: LCFrs (serpentinite)
Life Form: Perennial herb
Blooming: May-July
Notes: Known from only three occurrences on Red Mtn. Protected at Red Mtn. ACEC (BLM). See *Bulletin of the New York Botanical Garden* 3:31 (1903) for original description, and *Sedum of North America North of the Mexican Plateau,* pp. 398-403 (1975) by R. T. Clausen for taxonomic treatment.

Sedum laxum ssp. *eastwoodiae*

See *Sedum eastwoodiae*

Sedum laxum (Britt.) Berger ssp. *flavidum* Denton

"pale yellow stonecrop" Crassulaceae
CNPS List: 4 **R-E-D Code:** 1-1-3 **State/Fed. Status:** /C3c
Distribution: DNT, GLE, HUM, SHA, SIS, TEH, TRI
Habitat: BUFrs, Chprl, CmWld, LCFrs / serpentinite or volcanic
Life Form: Perennial herb
Blooming: May-July
Notes: Subspecies are somewhat indistinct. See *Brittonia* 30:233-238 (1978) for original description.

Sedum laxum (Britt.) Berger ssp. *heckneri* (Peck) Clausen

"Heckner's stonecrop" Crassulaceae
CNPS List: 4 **R-E-D Code:** 1-1-2 **State/Fed. Status:** /C3c
Distribution: DNT, HUM, SIS, TRI, OR
Habitat: LCFrs, UCFrs / serpentinite or gabbroic
Life Form: Perennial herb
Blooming: June-July
Notes: Subspecies are somewhat indistinct. Endangered in OR. See *Proceedings of the Biological Society of Washington* 50:121 (1937) for original description, and *Sedum of North America North of the Mexican Plateau,* pp. 391-393 (1975) by R. T. Clausen for taxonomic treatment.

Sedum niveum A. Davids.

"Davidson's stonecrop" Crassulaceae
CNPS List: 4 **R-E-D Code:** 1-2-2 **State/Fed. Status:** /C3c
Distribution: RIV, SBD, BA
Habitat: UCFrs (rocky)
Life Form: Perennial herb (rhizomatous)
Blooming: June-July
Notes: See *Bulletin of the Southern California Academy of Sciences* 20:53 (1921) for original description, and *Sedum of North America North of the Mexican Plateau,* pp. 178-179 (1975) by R. T. Clausen for taxonomic treatment.

Sedum oblanceolatum Clausen

"Applegate stonecrop" Crassulaceae
CNPS List: 1B **R-E-D Code:** 3-3-2 **State/Fed. Status:** /C2
Distribution: SIS, OR
Quads: 735B, 736A, 736B
Habitat: UCFrs (rocky)
Life Form: Perennial herb
Blooming: June-July
Notes: Known in CA from fewer than five occurrences. Threatened by logging and horticultural collecting. Protected at Botanical Special Interest Areas (USFS) on the Siskiyou Crest. Candidate for state listing in OR. See *Madroño* 39(4):310 (1992) for distributional information.
Status Report: 1989

Sedum obtusatum ssp. *paradisum*

See *Sedum paradisum*

Sedum paradisum (M. Denton) M. Denton

"Canyon Creek stonecrop" Crassulaceae
CNPS List: 1B **R-E-D Code:** 3-2-3 **State/Fed. Status:** /C2
Distribution: SHA, TRI
Quads: 648A, 667B, 668A
Habitat: BUFrs, Chprl, LCFrs, SCFrs / granitic
Life Form: Perennial herb
Blooming: May-June
Notes: Known from only five occurrences. See *Brittonia* 30:233-238 (1978) for original description.

Sedum pinetorum

Considered but rejected: Not in CA

Sedum purdyi

Considered but rejected: Too common; a synonym of *S. spathulifolium*

Sedum radiatum ssp. *depauperatum*

Considered but rejected: A synonym of *S. radiatum*; a common taxon

Sedum spathulifolium ssp. *purdyi*

Considered but rejected: Too common; a synonym of *S. spathulifolium*

Selaginella asprella Maxon

"bluish spike-moss" Selaginellaceae
CNPS List: 4 **R-E-D Code:** 1-1-2 **State/Fed. Status:** CEQA?
Distribution: KRN, LAX, ORA?, RIV, SBD, SDG, TUL, BA
Habitat: LCFrs, UCFrs, SCFrs / granitic
Life Form: Perennial herb (rhizomatous)
Fertile: July
Notes: Previously rejected as too common, but probably merely unthreatened. Does plant occur in ORA Co.? See *Smithsonian Miscellaneous Collections* 72(5):6-8 (1920) for original description.

Selaginella cinerascens A.A. Eat.

"ashy spike-moss" Selaginellaceae
CNPS List: 4 **R-E-D Code:** 1-2-1 **State/Fed. Status:** CEQA?
Distribution: ORA, SDG, BA
Habitat: Chprl, CoScr
Life Form: Perennial herb (rhizomatous)
Fertile: March
Notes: Still fairly widespread, but threatened by development.

Selaginella densa Rydb. var. *scopulorum* (Maxon) Tryon

"Rocky Mountain spike-moss" Selaginellaceae
CNPS List: 3 **R-E-D Code:** ?-?-1 **State/Fed. Status:** CEQA?
Distribution: DNT, SIS, AZ, OR, ++
Quads: unknown
Habitat: NCFrs, SCFrs, UCFrs / decomposed granitic
Life Form: Perennial herb (rhizomatous)
Fertile: August
Notes: Move to List 2? Location, rarity, and endangerment information needed. See *American Fern Journal* 11:36 (1921) for original description, and *Annals of the Missouri Botanical Garden* 42:67-69 (1955) for revised nomenclature.

Selaginella eremophila Maxon

"desert spike-moss" Selaginellaceae
CNPS List: 2 **R-E-D Code:** 2-2-1 **State/Fed. Status:** CEQA
Distribution: RIV, SDG, AZ, BA, NM, TX
Quads: 7A, 32C, 83C, 83D
Habitat: SDScr (gravelly or rocky)
Life Form: Perennial herb (rhizomatous)
Fertile: May-July
Notes: See *Smithsonian Miscellaneous Collections* 72(5):3 (1920) for original description.

Selaginella leucobryoides Maxon

"Mojave spike-moss" Selaginellaceae
CNPS List: 4 **R-E-D Code:** 1-1-3 **State/Fed. Status:** CEQA?
Distribution: INY, SBD
Habitat: MDScr, PJWld (rocky, usually carbonate)
Life Form: Perennial herb (rhizomatous)
Fertile: June
Notes: See *Smithsonian Miscellaneous Collections* 72(5):8 (1920) for original description.

Selinocarpus nevadensis (Standl.) Fowler & Turner

"desert wing-fruit" Nyctaginaceae
CNPS List: 2 **R-E-D Code:** 3-1-1 **State/Fed. Status:** CEQA
Distribution: INY, AZ, NV, UT
Quads: 274D
Habitat: JTWld, MDScr / shale
Life Form: Perennial herb
Blooming: June-September
Notes: Known in CA from only one occurrence in the Kingston Range. See *Madroño* 30(2):129 (1983) for this record.

Senecio aphanactis Greene

"rayless ragwort" Asteraceae
CNPS List: 2 **R-E-D Code:** 3-2-1 **State/Fed. Status:** CEQA
Distribution: CCA, FRE, LAX, MER, ORA, RIV, SBA, SCL, SCZ, SDG, SLO, SOL, SRO, VEN, BA
Quads: 10B, 86B?, 109D, 113B, 114A, 169D, 246C, 268C, 383A, 427D, 464B, 483D, SCZA, SRON
Habitat: CmWld, CoScr / alkaline
Life Form: Annual herb
Blooming: January-April
Notes: Rare in LAX, ORA, and RIV counties. Need quads for RIV Co. Not seen on SCZ Isl. between 1934 and 1991. See *Pittonia* 1:220 (1888) for original description, and *North American Flora* II 10:50-139 (1978) for taxonomic treatment.

Senecio bernardinus Greene

"San Bernardino ragwort" Asteraceae
CNPS List: 1B **R-E-D Code:** 2-2-3 **State/Fed. Status:** /C2
Distribution: SBD
Quads: 105A, 105B, 131C, 131D
Habitat: Medws (mesic, sometimes alkaline), PbPln, UCFrs
Life Form: Perennial herb
Blooming: May-July
Notes: Known from fewer than twenty occurrences. Threatened by development, grazing, and vehicles.
Status Report: 1980

Senecio blochmaniae

Considered but rejected: Too common

Senecio clarkianus

Considered but rejected: Too common

Senecio clevelandii

See *Senecio clevelandii* var. *clevelandii*

Senecio clevelandii Greene var. *clevelandii*

"Cleveland's ragwort" Asteraceae
CNPS List: 4 **R-E-D Code:** 1-1-3 **State/Fed. Status:** CEQA?
Distribution: COL, LAK, NAP, TRI, YOL
Habitat: Chprl (serpentinite seeps)
Life Form: Perennial herb
Blooming: June-July
Notes: See *Bulletin of the Torrey Botanical Club* 10:87 (1883) for original description.

Senecio clevelandii Greene var. *heterophyllus* Hoov.

"Red Hills ragwort" Asteraceae
CNPS List: 1B **R-E-D Code:** 3-3-3 **State/Fed. Status:** CEQA
Distribution: TUO
Quads: 458C
Habitat: CmWld (serpentinite seeps)
Life Form: Perennial herb
Blooming: June-July
Notes: Known only from the Red Hills. Possibly threatened by grazing. Not in *The Jepson Manual.* Protected at Red Hills ACEC (BLM). See *Leaflets of Western Botany* 2:132 (1938) for original description.

Senecio eurycephalus Gray var. *lewisrosei* (J.T. Howell) T.M. Barkley

"cut-leaved ragwort" Asteraceae
CNPS List: 1B **R-E-D Code:** 3-2-3 **State/Fed. Status:** CEQA
Distribution: BUT, LAS?, PLU
Quads: 575B, 591C, 592A, 592D, 606B, 606C, 607D, 623C?
Habitat: Chprl, CmWld, LCFrs / serpentinite
Life Form: Perennial herb
Blooming: April-July
Notes: Are plants from LAS Co. (623C) this taxon? Threatened by mining and road maintenance. See *Leaflets of Western Botany* 3(6):141-142 (1942) for original description.

Senecio foetidus

See *Senecio hydrophiloides*

Senecio ganderi Barkl. & Beauchamp

"Gander's ragwort" Asteraceae
CNPS List: 1B **R-E-D Code:** 3-2-3 **State/Fed. Status:** CR/C2
Distribution: SDG
Quads: 9B, 9C, 20A, 21A, 21D, 34A
Habitat: Chprl (burned areas, gabbroic outcrops)
Life Form: Perennial herb
Blooming: April-May
Notes: Known from fewer than fifteen occurrences. See *Brittonia* 26:106-108 (1974) for original description.
Status Report: 1987

Senecio hydrophiloides Rydb.

"sweet marsh ragwort" Asteraceae
CNPS List: 3 **R-E-D Code:** ?-?-1 **State/Fed. Status:** CEQA?
Distribution: BUT?, LAS, MOD, PLU, SIS, NV, OR, ++
Quads: 587B, 588B, 623C, 677A, 708A, 719C
Habitat: LCFrs, Medws / mesic
Life Form: Perennial herb
Blooming: May-July
Notes: Move to List 4? Location, rarity, and endangerment information needed. Does plant occur in BUT Co.? Includes *S. foetidus.*

Senecio ionophyllus Greene

"Tehachapi ragwort" Asteraceae
CNPS List: 4 **R-E-D Code:** 1-1-3 **State/Fed. Status:** CEQA?
Distribution: KRN, LAX, SBD
Habitat: LCFrs, UCFrs / granitic
Life Form: Perennial herb
Blooming: June-July

Senecio layneae Greene

"Layne's ragwort" Asteraceae
CNPS List: 1B **R-E-D Code:** 2-2-3 **State/Fed. Status:** CR/C2
Distribution: ELD, TUO
Quads: 458C, 510A, 510B, 511A, 526A, 527D
Habitat: Chprl, CmWld / serpentinite or gabbroic
Life Form: Perennial herb
Blooming: April-July
Notes: Threatened by urbanization, grazing, road construction, and vehicles. See *Bulletin of the Torrey Botanical Club* 10:87 (1883) for original description.
Status Report: 1990

Senecio ligulifolius

See *Senecio macounii*

Senecio lyonii

Considered but rejected: Too common

Senecio macounii Greene

"Siskiyou Mtns. ragwort" Asteraceae
CNPS List: 4 **R-E-D Code:** 1-1-1 **State/Fed. Status:** CEQA?
Distribution: DNT, SIS, OR, WA, +
Habitat: LCFrs (sometimes serpentinite, often in disturbed areas)
Life Form: Perennial herb
Blooming: June-July
Notes: See *Pittonia* 3:169 (1897) for original description, and *North American Flora* II 10:93-94 (1978) for taxonomic treatment.

Senecio pattersonensis Hoov.

"Mono ragwort" Asteraceae
CNPS List: 4 **R-E-D Code:** 1-1-3 **State/Fed. Status:** CEQA?
Distribution: MNO
Habitat: AlpBR
Life Form: Perennial herb (rhizomatous)
Blooming: July-August

Senecio streptanthifolius

Considered but rejected: Too common

Senna covesii (Gray) H. Irwin & Barneby

"Coves's cassia" Fabaceae
CNPS List: 2 **R-E-D Code:** 2-2-1 **State/Fed. Status:** CEQA
Distribution: IMP, RIV, SBD, SDG, AZ, BA
Quads: 19A, 32B, 32C, 47A, 61B, 64C, 121D
Habitat: SDScr (sandy)
Life Form: Perennial herb
Blooming: April-June
Notes: Need quads for IMP Co. and occurrence near "Turtle Mtns." (SBD Co.). Threatened by vehicles.

Sequoiadendron giganteum

Considered but rejected: Too common

Sibara deserti

Considered but rejected: Too common

Sibara filifolia (Greene) Greene

"Santa Cruz Island rock cress" Brassicaceae
CNPS List: 1B **R-E-D Code:** 3-3-3 **State/Fed. Status:** /C1
Distribution: SCM, SCT*, SCZ*
Quads: SCMS, SCTN*, SCZB*, SCZC*
Habitat: CoScr
Life Form: Annual herb
Blooming: April
Notes: Rediscovered in 1986 by M. Beauchamp and A. Kelly; known from only two extant occurrences. Feral herbivores removed from SCM Isl., and vegetation recovering.

Sibara rosulata

Considered but rejected: Too common

Sidalcea calycosa M.E. Jones ssp. *rhizomata* (Jeps.) Munz

"Point Reyes checkerbloom" Malvaceae
CNPS List: 1B **R-E-D Code:** 2-2-3 **State/Fed. Status:** CEQA
Distribution: MEN, MRN, SON
Quads: 467A, 484B, 485B, 485C, 485D, 502C, 503A, 520B, 537B, 552B, 553A
Habitat: MshSw (near coast)
Life Form: Perennial herb (rhizomatous)
Blooming: April-September
Notes: See *Manual of the Flowering Plants of California,* p. 629 (1925) by W.L. Jepson for original description.

Sidalcea covillei Greene

"Owens Valley checkerbloom" Malvaceae
CNPS List: 1B **R-E-D Code:** 2-3-3 **State/Fed. Status:** CE/C1
Distribution: INY
Quads: 305B*, 329D, 351A*, 351B, 351D, 372C, 393A, 393D, 413A, 413B, 413C, 413D, 414A
Habitat: GBScr, Medws (alkaline)
Life Form: Perennial herb
Blooming: April-June
Notes: Threatened by lowering of water table, non-native plants, and overgrazing. See *Cybele Columbiana* 1:35 (1914) for original description, and *Fremontia* 5(4):34-35 (1978), 6(3):26-27 (1978), and 8(4):16 (1981) for discussion of threats.
Status Report: 1987

Sidalcea hickmanii Greene ssp. *anomala* C.L. Hitchc.

"Cuesta Pass checkerbloom" Malvaceae
CNPS List: 1B **R-E-D Code:** 3-2-3 **State/Fed. Status:** CR/C2
Distribution: SLO
Quads: 246C
Habitat: CCFrs (serpentinite)
Life Form: Perennial herb
Blooming: May
Notes: Known from only three occurrences on Cuesta Ridge in the Los Padres NF.
Status Report: 1988

Sidalcea hickmanii Greene ssp. *hickmanii*

"Hickman's checkerbloom" Malvaceae
CNPS List: 1B **R-E-D Code:** 2-1-3 **State/Fed. Status:** /C3c
Distribution: MNT
Quads: 295B, 295C, 318C, 319A, 319B, 320A
Habitat: Chprl
Life Form: Perennial herb
Blooming: June-July
Notes: See *Fremontia* 6(2):8-14 (1978) for discussion of Marble-Cone fire and effects.

Sidalcea hickmanii Greene ssp. *parishii* (Rob.) C.L. Hitchc.

"Parish's checkerbloom" Malvaceae
CNPS List: 1B **R-E-D Code:** 3-2-3 **State/Fed. Status:** CR/C1
Distribution: SBA, SBD
Quads: 105B, 105C, 106D, 167B, 168A, 193A, 193B, 193D
Habitat: Chprl, LCFrs
Life Form: Perennial herb
Blooming: June-August
Notes: Threatened by urbanization, grazing, and road maintenance.
Status Report: 1987

Sidalcea hickmanii Greene ssp. *viridis* C.L. Hitchc.

"Marin checkerbloom" Malvaceae
CNPS List: 1B **R-E-D Code:** 3-1-3 **State/Fed. Status:** /C2
Distribution: MRN, NAP, SFO, SMT, SON
Quads: 409B, 466C, 467B, 499C, 503B
Habitat: Chprl (serpentinite)
Life Form: Perennial herb
Blooming: May-June
Notes: Possibly threatened by development.

Sidalcea keckii Wiggins

"Keck's checkerbloom" Malvaceae
CNPS List: 1B **R-E-D Code:** 3-3-3 **State/Fed. Status:** /C1
Distribution: TUL, FRE?*
Quads: 286D?*, 309C, 310D*, 377C*
Habitat: CmWld, VFGrs / serpentinite
Life Form: Annual herb
Blooming: April
Notes: Rediscovered in 1992 by J. Stebbins and K. Kirkpatrick near Mine Hill, TUL Co. Threatened by grazing and potential development. Recent work suggests FRE Co. record is probably erroneous. See *Contributions from the Dudley Herbarium* 3:55-56 (1940) for original description.

Sidalcea malachroides (H. & A.) Gray

"maple-leaved checkerbloom" Malvaceae
CNPS List: 1B **R-E-D Code:** 2-2-2 **State/Fed. Status:** CEQA
Distribution: HUM, MEN, MNT, SCL, SCR, OR
Quads: 344A, 344B, 344D, 366C, 387B, 427A, 537B, 537D, 552C, 585A, 585D, 601B, 618D, 635B, 635C, 636A, 637D, 654D, 671B, 671C, 672C, 672D
Habitat: BUFrs, CoPrr, NCFrs / often in disturbed areas
Life Form: Perennial herb
Blooming: May-August
Notes: How common is plant in HUM and MEN counties? Endangered in OR. See *Botany of Captain Beechey's Voyage Supplement,* p. 326 (1840) for original description, and *University of Washington Publications in Biology* 18:1-96 (1957) for taxonomic treatment.

Sidalcea malvaeflora ssp. *elegans*

Considered but rejected: A synonym of *S. malvaeflora* ssp. *asprella*; a common taxon

Sidalcea malvaeflora (DC.) Benth. ssp. *patula* C.L. Hitchc.

"Siskiyou checkerbloom" Malvaceae
CNPS List: 1B **R-E-D Code:** 3-2-2 **State/Fed. Status:** /C2
Distribution: DNT, HUM, OR
Quads: 654C, 654D, 655D, 672A, 672C(*?), 740B
Habitat: CBScr?, CoPrr, NCFrs
Life Form: Perennial herb (rhizomatous)
Blooming: May-June
Notes: Known in CA from fewer than ten occurrences, and three in OR, where plant is a candidate for state listing. See *University of Washington Publications in Biology* 18:1-96 (1957) for original description.

Sidalcea neomexicana Gray

"salt spring checkerbloom" Malvaceae
CNPS List: 2 **R-E-D Code:** 2-2-1 **State/Fed. Status:** CEQA
Distribution: LAX*, ORA, RIV, SBA, SBD, VEN, AZ, BA, NM, NV, SO, UT+
Quads: 85D, 89D*, 102A, 108C*, 131B, 141A, 165A, 166A, 190C
Habitat: Chprl, CoScr, LCFrs, MDScr, Plyas / alkaline, mesic
Life Form: Perennial herb
Blooming: March-June
Notes: Need quads for ORA and SBA counties. See *University of Washington Publications in Biology* 18:1-96 (1957) for taxonomic treatment.

Sidalcea oregana (T. & G.) Gray ssp. *eximia* (Greene) Hitchc.

"coast checkerbloom" Malvaceae
CNPS List: 1B **R-E-D Code:** 3-2-3 **State/Fed. Status:** CEQA
Distribution: HUM
Quads: 654B, 671A, 672A, 686C
Habitat: LCFrs, Medws, NCFrs
Life Form: Perennial herb
Blooming: June-August
Notes: Intergrades with sspp. *oregana* and *spicata*. See *University of Washington Publications in Biology* 18:1-96 (1957) for taxonomic treatment.

Sidalcea oregana (T. & G.) Gray ssp. *hydrophila* (Heller) Hitchc.

"marsh checkerbloom" Malvaceae
CNPS List: 1B **R-E-D Code:** 2-2-3 **State/Fed. Status:** CEQA
Distribution: GLE, LAK, MEN, NAP
Quads: 516C, 517B, 533C, 565A, 565B, 565C, 581A, 581C, 581D, 582B, 582D, 597C
Habitat: Medws, RpFrs / mesic
Life Form: Perennial herb
Blooming: July-August
Notes: Intergrades with ssp. *valida*. See *Muhlenbergia* 1:107 (1904) for original description, and *University of Washington Publications in Biology* 18:1-96 (1957) for taxonomic treatment.

Sidalcea oregana (Nutt.) Gray ssp. *valida* (Greene) C.L. Hitchc.

"Kenwood Marsh checkerbloom" Malvaceae
CNPS List: 1B **R-E-D Code:** 3-3-3 **State/Fed. Status:** CE/C1
Distribution: SON
Quads: 501A, 517B
Habitat: MshSw (freshwater)
Life Form: Perennial herb (rhizomatous)
Blooming: June-September
Notes: Known from only three occurrences: one in Knights Valley and two in Kenwood Marsh. Threatened by grazing and habitat alteration. See *Pittonia* 3:157-158 (1897) for original description, and *University of Washington Publications in Biology* 18:56-58 (1957) for revised nomenclature.
Status Report: 1988

Sidalcea pedata Gray

"bird-foot checkerbloom" Malvaceae
CNPS List: 1B **R-E-D Code:** 3-3-3 **State/Fed. Status:** CE/FE
Distribution: SBD
Quads: 105A*, 105B, 106A*, 131B, 131C, 131D
Habitat: Medws (mesic), PbPln
Life Form: Perennial herb
Blooming: May-August
Notes: Known from approximately ten occurrences. Seriously threatened by development, grazing, and vehicles. Protected in part at Baldwin Lake ER (DFG). See *Proceedings of the American Academy of Arts and Sciences* 22:288 (1887) for original description, and *Fremontia* 13(1):22-23 (1985) for species account.
Status Report: 1987

Sidalcea ranunculacea

Considered but rejected: Too common

Sidalcea robusta Roush

"Butte County checkerbloom" Malvaceae
CNPS List: 1B **R-E-D Code:** 2-2-3 **State/Fed. Status:** /C2
Distribution: BUT
Quads: 576A, 576B, 577A, 592B, 592C, 593D
Habitat: Chprl, CmWld
Life Form: Perennial herb (rhizomatous)
Blooming: April-June
Notes: Known from approximately twenty occurrences. Possibly threatened by residential development and fire suppression. See *Annals of the Missouri Botanical Garden* 18:205-207 (1931) for original description.
Status Report: 1979

Sidalcea setosa ssp. *setosa*

Considered but rejected: A synonym of *S. oregana* ssp. *spicata*; a common taxon

Sidalcea stipularis J.T. Howell & G. True

"Scadden Flat checkerbloom" Malvaceae
CNPS List: 1B **R-E-D Code:** 3-3-3 **State/Fed. Status:** CE/C1
Distribution: NEV
Quads: 541B, 542A
Habitat: MshSw (montane freshwater)
Life Form: Perennial herb (rhizomatous)
Blooming: July-August
Notes: Known from only two occurrences in Scadden Flat near Grass Valley. One occurrence voluntarily protected by landowner. Threatened by altered hydrology and non-native plants. See *Four Seasons* 4(4):20-22 (1974) for original description.

Silene aperta

Considered but rejected: Too common

Silene campanulata Wats. ssp. *campanulata*

"Red Mtn. catchfly" Caryophyllaceae
CNPS List: 1B **R-E-D Code:** 3-3-3 **State/Fed. Status:** CE/C1
Distribution: COL, MEN
Quads: 564D, 600B, 600C
Habitat: Chprl, LCFrs / serpentinite
Life Form: Perennial herb
Blooming: May-June
Notes: Known from fewer than ten occurrences near Cook Springs (COL Co.) and in the Red Mtn. area near Leggett (MEN Co.). Threatened by mining. Protected in part at Red Mtn. ACEC (BLM), MEN Co. See *Proceedings of the American Academy of Arts and Sciences* 10:341-342 (1875) for original description, and *University of Washington Publications in Biology* 13:22-23 (1947) for taxonomic treatment.
Status Report: 1988

Silene hookeri ssp. *pulverulenta*

Considered but rejected: A synonym of *S. hookeri*; a common taxon

Silene invisa C.L. Hitchc. & Maguire

"short-petaled campion" Caryophyllaceae
CNPS List: 4 **R-E-D Code:** 1-2-3 **State/Fed. Status:** /C3c
Distribution: ALP, AMA, ELD, LAS, NEV, PLA, PLU, SHA, SIE, TRI, TUO
Habitat: SCFrs, UCFrs / granitic
Life Form: Perennial herb
Blooming: July-August
Notes: Threatened by logging. See *University of Washington Publications in Biology* 13:31-32 (1947) for original description.
Status Report: 1988

Silene marmorensis Kruckeberg

"Marble Mtn. campion" Caryophyllaceae
CNPS List: 1B **R-E-D Code:** 2-2-3 **State/Fed. Status:** /C2
Distribution: HUM, SIS
Quads: 685B, 685D, 686A, 702C, 703A, 703B, 703C, 703D
Habitat: BUFrs, CmWld, LCFrs
Life Form: Perennial herb
Blooming: June
Notes: Need quads for HUM Co. Threatened by logging. Closely related to *S. bridgesii.* See *Madroño* 15(6):172-177 (1960) for original description.
Status Report: 1977

Silene occidentalis Wats. ssp. *longistipitata* Hitchc. & Maguire

"western campion" Caryophyllaceae
CNPS List: 3 **R-E-D Code:** ?-?-3 **State/Fed. Status:** /C2
Distribution: BUT, PLU, SHA, TEH
Quads: 607A, 607C, 608D, 662B
Habitat: Chprl, LCFrs
Life Form: Perennial herb
Blooming: July-August
Notes: Move to List 1B? Location, rarity, and endangerment information needed. Relationship to ssp. *occidentalis* needs study.

Silene suksdorfii Rob.

"Cascade alpine campion" Caryophyllaceae
CNPS List: 2 **R-E-D Code:** 2-1-1 **State/Fed. Status:** CEQA
Distribution: SHA, SIS, OR, WA
Quads: 625B, 626A, 643C, 644D, 698B, 698C
Habitat: AlpBR (volcanic)
Life Form: Perennial herb
Blooming: July-September
Notes: Known in CA only from Mt. Shasta and Mt. Lassen. On watch list in OR. See *Madroño* 34(1):29-40 (1987) for taxonomic discussion.

Silene verecunda Wats. ssp. *verecunda*

"San Francisco campion" Caryophyllaceae
CNPS List: 1B **R-E-D Code:** 3-2-3 **State/Fed. Status:** /C2
Distribution: SCR, SFO, SMT
Quads: 408B, 408C, 408D, 409D, 429A, 448B, 448C, 466C
Habitat: CBScr, Chprl, CoPrr, CoScr, VFGrs
Life Form: Perennial herb
Blooming: March-June
Notes: Known from fewer than twenty occurrences. Threatened by development. See *Proceedings of the American Academy of Arts and Sciences* 10:344 (1875) for original description, and *University of Washington Publications in Biology* 13:41-42 (1947) for taxonomic treatment.
Status Report: 1979

Sisyrinchium halophilum

Considered but rejected: Too common

Smelowskia ovalis M.E. Jones var. *congesta* Roll.

"Lassen Peak smelowskia" Brassicaceae
CNPS List: 1B **R-E-D Code:** 3-2-3 **State/Fed. Status:** /C2
Distribution: SHA
Quads: 625B, 626A, 643C
Habitat: AlpBR
Life Form: Perennial herb
Blooming: July-August
Notes: Known from six occurrences on Mt. Lassen. See *Rhodora* 40:302 (1938) for original description.
Status Report: 1977

Smilax jamesii Wallace

"English Peak greenbriar" Smilacaceae
CNPS List: 1B **R-E-D Code:** 2-1-3 **State/Fed. Status:** CEQA
Distribution: DNT, SHA, SIS, TRI
Quads: 662B, 667A, 667B, 668A, 680C, 702B, 702C, 702D, 718C, 719C, 721C
Habitat: BUFrs?, LCFrs, MshSw (streambanks and lake margins), NCFrs
Life Form: Perennial herb (rhizomatous)
Blooming: May-July
Notes: See *Brittonia* 31:416-421 (1979) for original description.

Solanum clokeyi Munz

"island nightshade" Solanaceae
CNPS List: 4 **R-E-D Code:** 1-2-3 **State/Fed. Status:** CEQA?
Distribution: SCZ, SRO
Habitat: CmWld
Life Form: Perennial herb
Blooming: March-July
Notes: Threatened by feral herbivores. A synonym of *S. wallacei* in *The Jepson Manual.*

Solanum tenuilobatum

Considered but rejected: A synonym of *S. xanti*; a common taxon

Solanum wallacei (Gray) Parish

"Wallace's nightshade" Solanaceae
CNPS List: 4 **R-E-D Code:** 1-2-2 **State/Fed. Status:** CEQA?
Distribution: SCT, GU
Habitat: Chprl, CmWld
Life Form: Perennial herb
Blooming: March-August
Notes: Threatened by feral herbivores. Occurrences reported from the mainland (SBA and SLO counties) need further study but are probably *S. xanti.* Plants from GU Isl. may be distinct, perhaps at the varietal level.

Solanum xanti var. *hoffmannii*

Considered but rejected: A synonym of *S. xanti*; a common taxon

Solanum xanti var. *obispoense*

Considered but rejected: A synonym of *S. xanti*; a common taxon

Solidago gigantea Ait.

"smooth goldenrod" Asteraceae
CNPS List: 2 **R-E-D Code:** 3-2-1 **State/Fed. Status:** CEQA
Distribution: MOD, PLU, AZ, NV, OR, ++
Quads: 605D, 724B, 724C
Habitat: Medws (mesic), MshSw (streambanks and lake margins)
Life Form: Perennial herb (rhizomatous)
Blooming: July-September
Notes: Known in CA from fewer than five occurrences; is it more common? Similar to *S. canadensis.*

Solidago guiradonis Gray

"Guirado's goldenrod" Asteraceae
CNPS List: 4 **R-E-D Code:** 1-2-3 **State/Fed. Status:** CEQA?
Distribution: FRE, SBT
Habitat: CmWld, VFGrs / serpentinite seeps
Life Form: Perennial herb (rhizomatous)
Blooming: September-October
Notes: Endemic to the San Benito Mtns. Threatened by vehicles. Frequently confused with *S. confinis.* See *Proceedings of the American Academy of Arts and Sciences* 6:543 (1865) for original description.

Solidago missouriensis

Considered but rejected: Not in CA; misidentification of *S. spectabilis*; a common taxon

Sparganium minimum

See *Sparganium natans*

Sparganium natans L.

"small bur-reed" Typhaceae
CNPS List: 4 **R-E-D Code:** 1-1-1 **State/Fed. Status:** CEQA?
Distribution: ELD, PLA, PLU, SHA, TUO, ID, OR, WA, ++
Habitat: MshSw (lake margins)
Life Form: Perennial herb (rhizomatous, emergent)
Blooming: August

Spartina gracilis Trin.

"alkali cord grass" Poaceae
CNPS List: 4 **R-E-D Code:** 1-2-1 **State/Fed. Status:** CEQA?
Distribution: INY, MNO, NV, OR, ++
Habitat: GBScr, Medws, MshSw / alkaline
Life Form: Perennial herb (rhizomatous)
Blooming: June-August
Notes: Threatened by grazing.

Sphaeralcea rusbyi ssp. *eremicola*

See *Sphaeralcea rusbyi* var. *eremicola*

Sphaeralcea rusbyi Gray var. *eremicola* (Jeps.) Kearn.

"Rusby's desert mallow" Malvaceae
CNPS List: 1B **R-E-D Code:** 3-2-3 **State/Fed. Status:** /C2
Distribution: INY, SBD
Quads: 248C, 249A, 249D, 325B, 348D
Habitat: JTWld, MDScr
Life Form: Perennial herb
Blooming: May-June
Notes: Known from ten occurrences in Death Valley NM (INY Co.) and near Clark Mtn. (SBD Co.). See *Manual of the Flowering Plants of California,* p. 635 (1925) by W.L. Jepson for original description.

Sphenopholis obtusata (Michx.) Scribn.

"prairie wedge grass" Poaceae
CNPS List: 2 **R-E-D Code:** 2-1-1 **State/Fed. Status:** CEQA
Distribution: AMA, FRE, INY, MNO, RIV, SBD, TUL, ++
Quads: 105A, 107D, 284B, 395B, 413A, 416D, 487C, 493B, 494D
Habitat: CmWld, Medws / mesic
Life Form: Perennial herb
Blooming: April-July
Notes: See *Rhodora* 8(92):137-146 (1906) for discussion of genus.
Status Report: 1980

Sphenopholis obtusata var. *obtusata*

See *Sphenopholis obtusata*

Stachys bergii

Considered but rejected: A synonym of *S. ajugoides* var. *rigida*; a common taxon

Stachys chamissonis var. *cooleyae*

Considered but rejected: A synonym of *S. chamissonis*; a common taxon

Stachys pycnantha

Considered but rejected: Too common

Stachys stebbinsii

Considered but rejected: A synonym of *S. rigida*; a common taxon

Stebbinsoseris decipiens (Chambers) Chambers

"Santa Cruz microseris" Asteraceae
CNPS List: 1B **R-E-D Code:** 2-2-3 **State/Fed. Status:** /C2
Distribution: MNT, MRN, SCR
Quads: 366D, 408B, 408C, 409D, 467A
Habitat: BUFrs, CCFrs, Chprl, CoPrr, CoScr / open areas, sometimes serpentinite
Life Form: Annual herb
Blooming: April-May
Notes: Known from fewer than twenty occurrences. Threatened by grazing. USFWS uses the name *Microseris decipiens*. See *Contributions from the Dudley Herbarium* 4:290-291 (1955) for original description, and *American Journal of Botany* 78(8):1015-1027 (1991) for revised nomenclature.
Status Report: 1977

Stellaria longifolia Willd.

"long-leaved starwort" Caryophyllaceae
CNPS List: 2 **R-E-D Code:** 3-2-1 **State/Fed. Status:** CEQA
Distribution: BUT, SHA, AZ, NM, OR, WA, ++
Quads: 607C, 662B
Habitat: Medws (mesic)
Life Form: Perennial herb (rhizomatous)
Blooming: May-July
Notes: Known in CA only from Goose Valley (SHA Co.) and near Jonesville (BUT Co.).

Stellaria obtusa Engelm.

"obtuse starwort" Caryophyllaceae
CNPS List: 2 **R-E-D Code:** 3-1-1 **State/Fed. Status:** CEQA
Distribution: BUT, GLE, HUM, TUO, ID, OR, WA, ++
Quads: 456A, 581A, 607C, 686C
Habitat: UCFrs (mesic)
Life Form: Perennial herb (rhizomatous)
Blooming: July
Notes: Occurrences are widely scattered.

Stenotus lanuginosus (Gray) Greene

"woolly stenotus" Asteraceae
CNPS List: 2 **R-E-D Code:** 3-2-1 **State/Fed. Status:** CEQA
Distribution: LAS, MOD, ID, OR, WA+
Quads: 659A, 675A, 676D, 727A
Habitat: GBScr, PJWld / gravelly loam
Life Form: Perennial herb
Blooming: May-July
Notes: Known in CA from fewer than five occurrences. See *Botany of the Wilkes Exploratory Expedition* 17:347 (1874) for original description, and *Carnegie Institution of Washington Publication* 389:171 (1928) for additional information.

Stephanomeria blairii Munz & Jtn.

"Blair's stephanomeria" Asteraceae
CNPS List: 1B **R-E-D Code:** 3-2-3 **State/Fed. Status:** /C2
Distribution: SCM
Quads: SCMC, SCMN, SCMS
Habitat: CBScr, CoScr / rocky
Life Form: Shrub
Blooming: July-September
Notes: Feral herbivores removed from SCM Isl., and vegetation recovering. See *Bulletin of the Torrey Botanical Club* 51:301 (1924) for original description.

Stipa arida

See *Achnatherum aridum*

Stipa diegoensis

See *Achnatherum diegoense*

Stipa latiglumis

Considered but rejected: Too common; a synonym of *Achnatherum latiglumis*

Stipa lemmonii var. *pubescens*

See *Achnatherum lemmonii* var. *pubescens*

Stipa lettermanii

Considered but rejected: Too common; a synonym of *Achnatherum lettermanii*

Stipa scribneri

Considered but rejected: Not in CA; material misidentified

Stipa stillmanii

Considered but rejected: Too common; a synonym of *Achnatherum stillmanii*

Streptanthus albidus Greene ssp. *albidus*

"Metcalf Canyon jewel-flower" Brassicaceae
CNPS List: 1B **R-E-D Code:** 3-3-3 **State/Fed. Status:** /PE
Distribution: SCL
Quads: 406A*, 406B, 406D, 407A*, 407B, 426C, 427D
Habitat: VFGrs (serpentinite)
Life Form: Annual herb
Blooming: April-July
Notes: Known from fewer than ten extant occurrences. Threatened by residential development and vehicles. See *Pittonia* 1:62 (1887) for original description, and *Madroño* 14(7):217-227 (1958) for taxonomic treatment.
Status Report: 1977

Streptanthus albidus Greene ssp. *peramoenus* (Greene) Kruckeberg

"most beautiful jewel-flower" Brassicaceae
CNPS List: 1B **R-E-D Code:** 2-2-3 **State/Fed. Status:** /C1
Distribution: ALA, CCA, SCL
Quads: 386A, 406A, 406B, 406D, 407A, 407B, 446C, 464B, 464C?, 465B, 465C, 465D, 466A
Habitat: Chprl, VFGrs / serpentinite
Life Form: Annual herb
Blooming: April-June
Notes: Historical occurrences need field surveys. Threatened by development and grazing. Similar plants from SLO Co. are likely *S. glandulosus* ssp. *glandulosus*, but further study needed.

Streptanthus barbatus

Considered but rejected: Too common

Streptanthus batrachopus Morrison

"Tamalpais jewel-flower" Brassicaceae
CNPS List: 1B **R-E-D Code:** 3-1-3 **State/Fed. Status:** /C2
Distribution: MRN
Quads: 467A, 467B
Habitat: CCFrs, Chprl / serpentinite
Life Form: Annual herb
Blooming: May-June
Notes: Known from fewer than ten occurrences in the Mt. Tamalpais area. Similar plants from the southern North Coast Ranges may be an undescribed new taxon. Intergrades with *S. barbiger.*

Streptanthus bernardinus (Greene) Parish

"Laguna Mtns. jewel-flower" Brassicaceae
CNPS List: 1B **R-E-D Code:** 2-1-2 **State/Fed. Status:** /C3c
Distribution: RIV, SBD, SDG, BA
Quads: 10B, 20A, 33D, 65C, 66D, 83C, 84D, 102A, 106A, 106C, 108A
Habitat: Chprl, LCFrs
Life Form: Perennial herb
Blooming: June-July
Notes: How common is this plant in BA?

Streptanthus brachiatus

See *Streptanthus brachiatus* ssp. *brachiatus*

Streptanthus brachiatus F.W. Hoffm. ssp. *brachiatus*

"Socrates Mine jewel-flower" Brassicaceae
CNPS List: 1B **R-E-D Code:** 3-2-3 **State/Fed. Status:** /C1
Distribution: LAK, NAP, SON
Quads: 517A, 517B, 533C, 534D
Habitat: CCFrs, Chprl / serpentinite
Life Form: Perennial herb
Blooming: June
Notes: See *Madroño* 36(1):36 (1989) for revised nomenclature.

Streptanthus brachiatus F.W. Hoffm. ssp. *hoffmanii* Dolan & LaPre

"Freed's jewel-flower" Brassicaceae
CNPS List: 1B **R-E-D Code:** 3-2-3 **State/Fed. Status:** /C1
Distribution: LAK, SON
Quads: 517B, 533C
Habitat: Chprl, CmWld / serpentinite
Life Form: Perennial herb
Blooming: May-July
Notes: See *Madroño* 36(1):36 (1989) for original description.

Streptanthus callistus Morrison

"Mt. Hamilton jewel-flower" Brassicaceae
CNPS List: 1B **R-E-D Code:** 3-1-3 **State/Fed. Status:** /C2
Distribution: SCL
Quads: 406A, 425C, 426D
Habitat: Chprl, CmWld
Life Form: Annual herb
Blooming: April-May
Notes: Known from approximately five occurrences in the Mt. Hamilton Range. See *Madroño* 4:205 (1938) for original description.
Status Report: 1977

Streptanthus campestris Wats.

"southern jewel-flower" Brassicaceae
CNPS List: 1B **R-E-D Code:** 2-1-2 **State/Fed. Status:** CEQA
Distribution: RIV, SBD, SDG, BA
Quads: 8A, 8B, 8C, 20D, 32C, 32D, 33D, 48A?, 65C, 83C, 105B
Habitat: Chprl, LCFrs, PJWld / rocky
Life Form: Perennial herb
Blooming: May-July
Notes: See *Proceedings of the American Academy of Arts and Sciences* 25:125 (1890) for original description.

Streptanthus cordatus Nutt. var. *piutensis* J.T. Howell

"Piute Mtns. jewel-flower" Brassicaceae
CNPS List: 1B **R-E-D Code:** 3-2-3 **State/Fed. Status:** /C2
Distribution: KRN
Quads: 211B, 212A, 260C, 261D
Habitat: BUFrs, CCFrs, PJWld / clay
Life Form: Perennial herb
Blooming: May-July
Notes: Known from approximately ten occurrences. See *Leaflets of Western Botany* 10:31 (1963) for original description.

Streptanthus drepanoides Kruckeberg & Morrison

"sickle-fruit jewel-flower" Brassicaceae
CNPS List: 4 **R-E-D Code:** 1-1-3 **State/Fed. Status:** CEQA?
Distribution: BUT, MEN, SHA, TEH, TRI
Habitat: Chprl, LCFrs / serpentinite
Life Form: Annual herb
Blooming: April-June
Notes: See *Madroño* 30(4):230-244 (1983) for original description, and *Fremontia* 14(2):19 (1986) for species account.

Streptanthus farnsworthianus J.T. Howell

"Farnsworth's jewel-flower" Brassicaceae
CNPS List: 4 **R-E-D Code:** 1-1-3 **State/Fed. Status:** /C3c
Distribution: FRE, KRN, MAD, TUL
Habitat: CmWld
Life Form: Annual herb
Blooming: May-June
Notes: See *Leaflets of Western Botany* 10(11):182-183 (1965) for original description.
Status Report: 1977

Streptanthus fenestratus (Greene) J.T. Howell

"Tehipite Valley jewel-flower" Brassicaceae
CNPS List: 1B **R-E-D Code:** 2-1-3 **State/Fed. Status:** /C3c
Distribution: FRE
Quads: 374B, 374C, 374D, 375A, 375C, 375D
Habitat: LCFrs, UCFrs
Life Form: Annual herb
Blooming: June-July
Notes: See *Leaflets of Botanical Observation and Criticism* 1:86 (1904) for original description, and *Leaflets of Western Botany* 9(12):184-185 (1961) for taxonomic treatment.
Status Report: 1979

Streptanthus glandulosus Hook. var. *hoffmanii* Kruckeberg

"secund jewel-flower" Brassicaceae
CNPS List: 1B **R-E-D Code:** 3-1-3 **State/Fed. Status:** /C2
Distribution: SON
Quads: 503B, 519A, 519B, 519C, 519D
Habitat: Chprl, CmWld, VFGrs (often serpentinite)
Life Form: Annual herb
Blooming: April-July
Notes: Historical occurrences need field surveys. See *S. glandulosus* ssp. *secundus* in *The Jepson Manual.* See *Madroño* 14(7):223 (1958) for original description.

Streptanthus glandulosus Hook. ssp. *pulchellus* (Greene) Kruckeberg

"Mt. Tamalpais jewel-flower" Brassicaceae
CNPS List: 1B **R-E-D Code:** 3-1-3 **State/Fed. Status:** /C3c
Distribution: MRN
Quads: 467A, 467B, 484C, 484D
Habitat: Chprl, VFGrs / serpentinite
Life Form: Annual herb
Blooming: May-July
Notes: Endemic to the Mt. Tamalpais area.

Streptanthus gracilis Eastw.

"alpine jewel-flower" Brassicaceae
CNPS List: 4 **R-E-D Code:** 1-1-3 **State/Fed. Status:** /C3c
Distribution: FRE, INY, TUL
Habitat: SCFrs, UCFrs / granitic
Life Form: Annual herb
Blooming: July-August

Streptanthus hispidus Gray

"Mt. Diablo jewel-flower" Brassicaceae
CNPS List: 1B **R-E-D Code:** 3-1-3 **State/Fed. Status:** /C2
Distribution: CCA
Quads: 464B, 464C
Habitat: Chprl, VFGrs / rocky
Life Form: Annual herb
Blooming: March-June
Notes: Known from fewer than fifteen occurrences in the Mt. Diablo area.

Streptanthus howellii Wats.

"Howell's jewel-flower" Brassicaceae
CNPS List: 1B **R-E-D Code:** 3-2-2 **State/Fed. Status:** /C3c
Distribution: DNT, OR
Quads: 739B, 739C, 740A, 740B, 740C
Habitat: LCFrs (serpentinite)
Life Form: Perennial herb
Blooming: July-August
Notes: Threatened by mining. On watch list in OR. See *Proceedings of the American Academy of Arts and Sciences* 20:353 (1885) for original description.

Streptanthus insignis Jeps. ssp. *lyonii* Kruckeberg & Morrison

"Arburua Ranch jewel-flower" Brassicaceae
CNPS List: 1B **R-E-D Code:** 3-2-3 **State/Fed. Status:** /C2
Distribution: MER
Quads: 383C, 384A, 384D
Habitat: CoScr (sometimes serpentinite)
Life Form: Annual herb
Blooming: April-May
Notes: Known from fewer than ten occurrences, where not recently reported. Threatened by grazing, and potentially by reservoir construction. See *Madroño* 30(4):230-244 (1985) for original description.

Streptanthus morrisonii

See *Streptanthus morrisonii* sspp. *elatus, hirtiflorus, kruckebergii,* and *morrisonii*

Streptanthus morrisonii F.W. Hoffm. ssp. *elatus* F.W. Hoffm.

"Three Peaks jewel-flower" Brassicaceae
CNPS List: 1B **R-E-D Code:** 3-2-3 **State/Fed. Status:** /C1
Distribution: LAK, NAP, SON
Quads: 516B, 517A, 517B, 519C, 519D, 532C, 532D, 533D
Habitat: Chprl (serpentinite)
Life Form: Perennial herb
Blooming: June-September
Notes: See *Streptanthus morrisonii* in *The Jepson Manual.* See *Madroño* 11(6):228 (1952) for original description and 36(1):33-40 (1989) for additional information.

Streptanthus morrisonii F.W. Hoffm. ssp. *hirtiflorus* F.W. Hoffm.

"Dorr's Cabin jewel-flower" Brassicaceae
CNPS List: 1B **R-E-D Code:** 3-2-3 **State/Fed. Status:** /C1
Distribution: SON
Quads: 519D
Habitat: Chprl, CCFrs / serpentinite
Life Form: Perennial herb
Blooming: June
Notes: Known from only one small occurrence in The Cedars. See *Streptanthus morrisonii* in *The Jepson Manual.* See *Madroño* 11(6):228 (1952) for original description and 36(1):33-40 (1989) for additional information.
Status Report: 1977

Streptanthus morrisonii F.W. Hoffm. ssp. *kruckebergii* Dolan & LaPre

"Kruckeberg's jewel-flower" Brassicaceae
CNPS List: 1B **R-E-D Code:** 3-2-3 **State/Fed. Status:** /C2
Distribution: LAK, NAP, SON
Quads: 517B, 532C, 532D, 533C
Habitat: CmWld (serpentinite)
Life Form: Perennial herb
Blooming: April-July
Notes: Possibly threatened by gold mining activities. See *Streptanthus morrisonii* in *The Jepson Manual.* See *Madroño* 36(1):38 (1989) for original description.

Streptanthus morrisonii F.W. Hoffm. ssp. *morrisonii*

"Morrison's jewel-flower" Brassicaceae
CNPS List: 1B **R-E-D Code:** 3-2-3 **State/Fed. Status:** /C2
Distribution: SON
Quads: 519C, 519D
Habitat: Chprl (serpentinite)
Life Form: Perennial herb
Blooming: May-September
Notes: See *Streptanthus morrisonii* in *The Jepson Manual.* See *Madroño* 11(6):225 (1952) for original description and 36(1):33-40 (1989) for additional information.

Streptanthus niger Greene

"Tiburon jewel-flower" Brassicaceae
CNPS List: 1B **R-E-D Code:** 3-3-3 **State/Fed. Status:** CE/PE
Distribution: MRN
Quads: 466B
Habitat: VFGrs (serpentinite)
Life Form: Annual herb
Blooming: May-June
Notes: Known from only three occurrences. Threatened by road construction, foot traffic, and development on the Tiburon Peninsula. See *Bulletin of the Torrey Botanical Club* 13:141 (1886) for original description, and *Madroño* 14(7):217-227 (1958) for taxonomic treatment.
Status Report: 1988

Streptanthus oliganthus Roll.

"Masonic Mtn. jewel-flower" Brassicaceae
CNPS List: 1B **R-E-D Code:** 2-2-2 **State/Fed. Status:** /C2
Distribution: INY, MNO, NV
Quads: 412C, 469B, 470C, 487A, 487B, 487C, 487D, 488A, 488B
Habitat: PJWld (volcanic or granitic)
Life Form: Perennial herb
Blooming: June-July
Notes: Known in CA from fewer than twenty occurrences. Threatened by mining, grazing, and vehicles. On watch list in NV. See *Contributions from the Dudley Herbarium* 3:372 (1946) for original description.

Streptanthus tortuosus var. *suffrutescens*

Considered but rejected: Too common

Stylocline amphibola

See *Micropus amphibolus*

Stylocline citroleum Morefield

"oil neststraw" Asteraceae
CNPS List: 1B **R-E-D Code:** 3-3-3 **State/Fed. Status:** /C2
Distribution: KRN, SDG*
Quads: 216B*, 239A*, 239B*, 241C, 242C*, 242D*
Habitat: ChScr, CoScr? / clay
Life Form: Annual herb
Blooming: April
Notes: Collected only once (1988) since 1935; plant is poorly known. Now appears to be restricted to oil-producing areas in the southern San Joaquin Valley. Need historical quads for SDG Co. Threatened by energy development and urbanization. See *Madroño* 39(2):123 (1992) for original description.

Stylocline masonii Morefield

"Mason neststraw" Asteraceae
CNPS List: 1B **R-E-D Code:** 3-3-3 **State/Fed. Status:** /C2
Distribution: KRN, LAX, MNT, SLO
Quads: 136B, 240B, 245D, 260A, 260B, 292D, 294C, 295D
Habitat: ChScr, PJWld / sandy
Life Form: Annual herb
Blooming: March-April
Notes: Collected only once (1990) since 1971. Threatened by development and habitat disturbance. See *Madroño* 39(2):117 (1992) for original description.

Stylocline micropoides

Considered but rejected: Too common

Stylocline sonorensis Wiggins

"mesquite neststraw" Asteraceae
CNPS List: 2 **R-E-D Code:** 3-3-1 **State/Fed. Status:** CEQA
Distribution: RIV(*?), AZ, SO
Quads: 62B(*?)
Habitat: SDScr (sandy)
Life Form: Annual herb
Blooming: April
Notes: Known in CA from only a single collection (1930) at Hayfields Dry Lake. Possibly extirpated after 1930 by development. See *Contributions from the Dudley Herbarium* 4:26 (1950) for original description, and *Madroño* 35(3):279 (1988) for the CA record.

Stylomecon heterophylla

Considered but rejected: Too common

Suaeda californica Wats.

"California seablite" Chenopodiaceae
CNPS List: 1B **R-E-D Code:** 3-3-3 **State/Fed. Status:** /PE
Distribution: ALA*, SCL*, SLO
Quads: 247B*, 247D, 428A*, 447B*
Habitat: MshSw (coastal salt)
Life Form: Shrub (evergreen)
Blooming: July-October
Notes: Formerly known from San Francisco Bay area, where extirpated by development; now extant only in Morro Bay. Remains from adobe bricks indicate plant may once have occurred along the Petaluma River, SON Co. (484A). Threatened by recreation, erosion, and alteration of marsh habitat. Often confused with *S. esteroa* and *S. taxifolia* in southern California, but does not occur there.

Suaeda esteroa Ferren & Whitmore

"estuary seablite" Chenopodiaceae
CNPS List: 4 **R-E-D Code:** 1-2-1 **State/Fed. Status:** CEQA?
Distribution: LAX, ORA, SBA, SDG, VEN, BA
Habitat: MshSw (coastal salt)
Life Form: Perennial herb
Blooming: July-October
Notes: See *Madroño* 30(3):181-190 (1983) for original description.

Suaeda taxifolia (Standl.) Standl.

"woolly seablite" Chenopodiaceae
CNPS List: 4 **R-E-D Code:** 1-2-1 **State/Fed. Status:** CEQA?
Distribution: ANA, LAX, ORA, SBA, SBR, SCM, SCT, SCZ, SDG, SLO, SNI, SRO, VEN, BA
Habitat: CBScr, MshSw (margins of coastal salt)
Life Form: Shrub (evergreen)
Blooming: January-December
Notes: See *North American Flora* 21:91 (1916) for original description.

Swallenia alexandrae (Swall.) Sderstrom & Decker

"Eureka Valley dune grass" Poaceae
CNPS List: 1B **R-E-D Code:** 3-2-3 **State/Fed. Status:** CR/FE
Distribution: INY
Quads: 390C, 391A, 391D
Habitat: DeDns
Life Form: Perennial herb (rhizomatous)
Blooming: April-June
Notes: Known from only four occurrences in the Eureka Valley. Recovery plan has been completed by BLM for the Eureka Dunes; still threatened by vehicle trespass. See *Journal of the Washington Academy of Sciences* 40(1):19-21 (1950) for original description, *Madroño* 17:88 (1963) for revised nomenclature, and *Biological Conservation* 46:217-242 (1988) for population biology.
Status Report: 1987

Swertia fastigiata Pursh

"clustered green-gentian" Gentianaceae
CNPS List: 2 **R-E-D Code:** 3-2-1 **State/Fed. Status:** CEQA
Distribution: TRI, ID, OR, WA
Quads: 632B, 632C, 633D
Habitat: Chprl, LCFrs, Medws, NCFrs
Life Form: Perennial herb
Blooming: June-July
Notes: Known in CA from fewer than ten occurrences in the Pickett Pk. area of South Fork Mtn. Candidate for state listing in OR. See *Madroño* 6:12 (1941) for original description, and *American Midland Naturalist* 26:14 (1941) for revised nomenclature.
Status Report: 1979

Swertia neglecta Hall (Jeps.)

"pine green-gentian" Gentianaceae
CNPS List: 4 **R-E-D Code:** 1-1-3 **State/Fed. Status:** CEQA?
Distribution: KRN, LAX, SBD, VEN
Habitat: LCFrs, PJWld
Life Form: Perennial herb
Blooming: May-July

Syntrichopappus lemmonii (Gray) Gray

"Lemmon's syntrichopappus" Asteraceae
CNPS List: 4 **R-E-D Code:** 1-1-3 **State/Fed. Status:** CEQA?
Distribution: KRN, LAX, MNT, RIV, SBD
Habitat: Chprl, JTWld / sandy or gravelly
Life Form: Annual herb
Blooming: April-May

Systenotheca vortriedei (Bdg.) Reveal & Hardham

"Vortriede's spineflower" Polygonaceae
CNPS List: 4 **R-E-D Code:** 1-1-3 **State/Fed. Status:** CEQA?
Distribution: MNT, SLO
Habitat: CmWld (sandy or serpentinite)
Life Form: Annual herb
Blooming: June-September
Notes: See *Great Basin Naturalist Memoirs* 2:169-190 (1978) for taxonomic revision, and *Phytologia* 66(2):83-88 (1989) for revised nomenclature.

Tanacetum camphoratum

Considered but rejected: Too common

Tanacetum douglasii
Considered but rejected: A synonym of *T. camphoratum*; a common taxon

Taraxacum californicum Munz & Jtn.
"California dandelion" Asteraceae
CNPS List: 1B **R-E-D Code:** 3-2-3 **State/Fed. Status:** /C1
Distribution: SBD
Quads: 104B, 105A, 105B, 105D, 131C, 131D
Habitat: Medws (mesic)
Life Form: Perennial herb
Blooming: May-August
Notes: Seriously threatened by development, grazing, vehicles, and hybridization with non-native *T. officinale*. See *Bulletin of the Torrey Botanical Club* 52:227-228 (1925) for original description.
Status Report: 1980

Tauschia glauca (Coult. & Rose) Math. & Const.
"glaucous tauschia" Apiaceae
CNPS List: 4 **R-E-D Code:** 1-1-1 **State/Fed. Status:** /C3c
Distribution: DNT, HUM, TRI, OR
Habitat: LCFrs (gravelly, serpentinite)
Life Form: Perennial herb
Blooming: April-June
Notes: See *Contributions from the U.S. National Herbarium* 3:321 (1895) for original description.
Status Report: 1979

Tauschia howellii (Coult. & Rose) Macbr.
"Howell's tauschia" Apiaceae
CNPS List: 1B **R-E-D Code:** 3-1-2 **State/Fed. Status:** /C2
Distribution: SIS, OR
Quads: 702B, 703D
Habitat: SCFrs, UCFrs / granitic, gravelly
Life Form: Perennial herb
Blooming: June-August
Notes: Known in CA from only three occurrences in the Marble Mtns. Wilderness. Candidate for state listing in OR. See *Contributions from the Gray Herbarium* 56:32 (1918) for revised nomenclature.
Status Report: 1979

Taxus brevifolia
Considered but rejected: Too common

Tetracoccus dioicus Parry
"Parry's tetracoccus" Euphorbiaceae
CNPS List: 1B **R-E-D Code:** 3-2-2 **State/Fed. Status:** /C2
Distribution: ORA, RIV, SDG, BA
Quads: 7B, 9D, 10A, 10C, 21A, 21D, 34D, 35B, 50A, 50B, 50D, 51A, 69C, 70D
Habitat: Chprl, CoScr
Life Form: Shrub (deciduous)
Blooming: April-May
Notes: Threatened by agriculture and development.

Tetracoccus ilicifolius Cov. & Gilman
"holly-leaved tetracoccus" Euphorbiaceae
CNPS List: 1B **R-E-D Code:** 3-1-3 **State/Fed. Status:** /C3c
Distribution: INY
Quads: 325D, 368D, 369C
Habitat: MDScr (carbonate)
Life Form: Shrub (deciduous)
Blooming: May-June
Notes: Known from fewer than ten occurrences in Death Valley NM.

Tetradymia argyraea Munz & Roos
"striped horsebrush" Asteraceae
CNPS List: 4 **R-E-D Code:** 1-1-1 **State/Fed. Status:** CEQA?
Distribution: RIV, SBD, AZ
Habitat: PJWld (rocky)
Life Form: Shrub (deciduous)
Blooming: August-September

Tetradymia stenolepis
Considered but rejected: Too common

Thelypodium brachycarpum Torr.
"short-podded thelypodium" Brassicaceae
CNPS List: 4 **R-E-D Code:** 1-2-2 **State/Fed. Status:** /C3c
Distribution: COL, LAK, NAP, SHA, SIS, TRI, OR
Habitat: Chprl, Medws / serpentinite, adobe, alkaline
Life Form: Perennial herb
Blooming: June-August
Notes: Endangered in OR. See *Contributions from the Gray Herbarium* 204:91-96 (1973) for taxonomic treatment.

Thelypodium flexuosum
Considered but rejected: Too common

Thelypodium jaegeri
See *Caulostramina jaegeri*

Thelypodium stenopetalum Wats.
"slender-petaled thelypodium" Brassicaceae
CNPS List: 1B **R-E-D Code:** 3-3-3 **State/Fed. Status:** CE/FE
Distribution: SBD
Quads: 105A, 105B, 131C, 131D
Habitat: Medws (mesic, alkaline)
Life Form: Perennial herb
Blooming: May-August
Notes: Known from eight occurrences. Seriously threatened by grazing, vehicles, and urbanization. Protected in part at Baldwin Lake ER (DFG). See *Proceedings of the American Academy of Arts and Sciences* 22:468 (1887) for original description, *Contributions from the Gray Herbarium* 204:114-115 (1973) for taxonomic treatment, and *Fremontia* 13(1):22-23 (1985) for species account.
Status Report: 1989

Thelypteris puberula (Baker) C. Morton var. *sonorensis* A.R. Smith

"Sonoran maiden fern" Thelypteridaceae
CNPS List: 2 **R-E-D Code:** 2-2-1 **State/Fed. Status:** CEQA
Distribution: LAX, RIV, SBA, AZ, BA, SO
Quads: 48B, 49A, 65D, 83D, 109B, 110A, 113D, 142A, 142B, 143A, 144A, 144B, 194A
Habitat: Medws (seeps and streams)
Life Form: Perennial herb (rhizomatous)
Fertile: January-September
Notes: See *University of California Publications in Botany* 59:91 (1971) for original description.

Thermopsis argentata

See *Thermopsis californica* var. *argentata*

Thermopsis californica Wats. var. *argentata* (Greene) Chen & Turner

"silvery false lupine" Fabaceae
CNPS List: 4 **R-E-D Code:** 1-1-3 **State/Fed. Status:** CEQA?
Distribution: LAS, LAX, MOD, SBA, SHA, SIS, VEN
Habitat: LCFrs, PJWld
Life Form: Perennial herb (rhizomatous)
Blooming: April-July
Notes: See *T. macrophylla* var. *argentata* in *The Jepson Manual.* See *Annals of the Missouri Botanical Garden* 81(3):x-x (1993, in press) for taxonomic treatment and revised nomenclature.

Thermopsis californica Wats. var. *semota* (Jeps.) Chen & Turner

"velvety false lupine" Fabaceae
CNPS List: 1B **R-E-D Code:** 2-2-3 **State/Fed. Status:** /C2
Distribution: SDG
Quads: 19B, 19C, 20A, 33C, 33D, 34A
Habitat: CmWld, LCFrs, VFGrs
Life Form: Perennial herb (rhizomatous)
Blooming: March-June
Notes: Threatened by grazing, trampling, and recreation. See *T. macrophylla* var. *semota* in *The Jepson Manual*; USFWS uses this name. See *Annals of the Missouri Botanical Garden* 81(3):x-x (1993, in press) for taxonomic treatment and revised nomenclature.

Thermopsis gracilis Howell

"slender false lupine" Fabaceae
CNPS List: 4 **R-E-D Code:** 1-1-1 **State/Fed. Status:** CEQA?
Distribution: DNT, HUM, SHA, SIS, TRI, OR
Habitat: CmWld, LCFrs, Medws, NCFrs
Life Form: Perennial herb (rhizomatous)
Blooming: March-July
Notes: A synonym of *T. macrophylla* var. *venosa* in *The Jepson Manual.* See *Erythea* 1:109 (1893) for original description, and *Annals of the Missouri Botanical Garden* 81(3):x-x (1993, in press) for taxonomic treatment.

Thermopsis macrophylla H. & A.

"Santa Ynez false lupine" Fabaceae
CNPS List: 1B **R-E-D Code:** 3-1-3 **State/Fed. Status:** CR/C2
Distribution: SBA
Quads: 142B, 167B, 168C, 168D
Habitat: Chprl (sandy granitic disturbed areas)
Life Form: Perennial herb (rhizomatous)
Blooming: April-June
Notes: Known from fewer than fifteen occurrences. Seems to respond well to properly timed fire. Equals *T. macrophylla* var. *agnina*; USFWS uses this name. See *T. macrophylla* var. *macrophylla* in *The Jepson Manual.* See *Annals of the Missouri Botanical Garden* 81(3):x-x (1993, in press) for taxonomic treatment.
Status Report: 1986

Thermopsis macrophylla var. *agnina*

See *Thermopsis macrophylla*

Thermopsis macrophylla var. *argentata*

See *Thermopsis californica* var. *argentata*

Thermopsis macrophylla var. *semota*

See *Thermopsis californica* var. *semota*

Thermopsis robusta Howell

"robust false lupine" Fabaceae
CNPS List: 1B **R-E-D Code:** 2-2-3 **State/Fed. Status:** CEQA
Distribution: DNT?, HUM, SIS
Quads: 671A, 703B, 704A, 704C, 704D, 721C
Habitat: NCFrs
Life Form: Perennial herb (rhizomatous)
Blooming: May-July
Notes: To be expected in DNT Co. Most collections are 35-40 years old. Potentially threatened by logging. Not in *The Jepson Manual*; a synonym of *T. macrophylla* in *A California Flora* (1959) by P. Munz. See *Erythea* 1:109 (1893) for original description, and *Annals of the Missouri Botanical Garden* 81(3):x-x (1993, in press) for taxonomic treatment.

Thlaspi californicum Wats.

"Kneeland Prairie pennycress" Brassicaceae
CNPS List: 1B **R-E-D Code:** 3-3-3 **State/Fed. Status:** /C1
Distribution: HUM, MEN
Quads: 581B, 581C, 653B, 687C*
Habitat: BUFrs (one extirpated occurrence), CoPrr / serpentinite
Life Form: Perennial herb
Blooming: May-June
Notes: Known from fewer than five extant occurrences. HUM Co. occurrence potentially threatened by road maintenance; voluntarily protected by landowner. USFWS uses the name *T. montanum* var. *californicum.* See *Proceedings of the American Academy of Arts and Sciences* 17:365 (1882) for original description, and *Memoirs of the New York Botanical Garden* 21(2):80-81 (1971) for revised nomenclature.
Status Report: 1988

Thlaspi montanum var. *californicum*

See *Thlaspi californicum*

Thysanocarpus conchuliferus Greene

"Santa Cruz Island fringepod" Brassicaceae
CNPS List: 1B **R-E-D Code:** 3-2-3 **State/Fed. Status:** /C1
Distribution: SCZ
Quads: SCZA, SCZB, SCZC
Habitat: Chprl, CmWld
Life Form: Annual herb
Blooming: March-April
Notes: Known from fewer than twenty occurrences. Threatened by grazing.

Tiarella trifoliata L. var. *trifoliata*

"trifoliate laceflower" Saxifragaceae
CNPS List: 3 **R-E-D Code:** ?-?-1 **State/Fed. Status:** CEQA?
Distribution: HUM, TRI, OR, ++
Quads: 671A
Habitat: LCFrs?, NCFrs
Life Form: Perennial herb (rhizomatous)
Blooming: June
Notes: Move to List 2? Location, rarity, and endangerment information needed, especially quads for TRI Co. See *Madroño* 18(5):152-160 (1966) for additional information.

Tiarella trifoliata var. *unifoliata*

Considered but rejected: Too common

Tonestus eximius (Hall) A. Nels. & Macbr.

"Tahoe tonestus" Asteraceae
CNPS List: 4 **R-E-D Code:** 1-1-1 **State/Fed. Status:** /C3c
Distribution: ALP, ELD, INY, NV
Habitat: SCFrs (granitic)
Life Form: Perennial herb (rhizomatous)
Blooming: July-August
Notes: Occurs in relatively inaccessible areas. See *Phytologia* 68(2):144-155 (1990) for revised nomenclature.

Tonestus lyallii (Gray) Nesom

"Lyall's tonestus" Asteraceae
CNPS List: 2 **R-E-D Code:** 3-1-1 **State/Fed. Status:** CEQA
Distribution: SIS, TRI, OR, ++
Quads: 667B, 668A, 685D
Habitat: AlpBR
Life Form: Perennial herb (rhizomatous)
Blooming: July-August
Notes: See *Madroño* 21(8):535 (1972) for first CA occurrence, and *Phytologia* 68(2):144-155 (1990) for revised nomenclature.

Townsendia parryi D.C. Eat.

"Parry's townsendia" Asteraceae
CNPS List: 2 **R-E-D Code:** 3-1-1 **State/Fed. Status:** CEQA
Distribution: MNO, OR, ++
Quads: 488A
Habitat: AlpBR
Life Form: Perennial herb
Blooming: May-August
Notes: Known in CA only from the Sweetwater Mtns. Endangered in OR.

Tracyina rostrata Blake

"beaked tracyina" Asteraceae
CNPS List: 1B **R-E-D Code:** 3-1-3 **State/Fed. Status:** /C3c
Distribution: HUM, LAK, SON
Quads: 535C, 549C, 616A, 616B, 616D
Habitat: CmWld, VFGrs
Life Form: Annual herb
Blooming: May-June
Notes: Known from fewer than twenty occurrences.
Status Report: 1977

Trichocoronis wrightii

See *Trichocoronis wrightii* var. *wrightii*

Trichocoronis wrightii (T. & G.) Gray var. *wrightii*

"Wright's trichocoronis" Asteraceae
CNPS List: 2 **R-E-D Code:** 3-3-1 **State/Fed. Status:** CEQA
Distribution: COL*, MER*, RIV, SJQ*, SUT*, TX, ++
Quads: 84B, 85A, 85C, 85D, 403D*, 462D*, 544C*, 545B*, 545C*, 545D*, 546D*
Habitat: Medws, MshSw, RpFrs, VnPls / alkaline
Life Form: Annual herb
Blooming: May-September
Notes: Extirpated in the Central Valley. Habitat lost to agriculture and urbanization. Confusion persists, but plant is probably native to CA. Rare in TX; perhaps more common in northern Mexico. Perhaps best treated as a full species distinct from TX plants.

Trichostema austromontanum ssp. *austromontanum*

Considered but rejected: Too common

Trichostema austromontanum Lewis ssp. *compactum* Lewis

"Hidden Lake bluecurls" Lamiaceae
CNPS List: 1B **R-E-D Code:** 3-3-3 **State/Fed. Status:** /C1
Distribution: RIV
Quads: 83C
Habitat: CCFrs, UCFrs / seasonally submerged lake margins
Life Form: Annual herb
Blooming: July-September
Notes: Known only from Hidden Lake in the San Jacinto Mtns. Threatened by recreational activities. See *Brittonia* 5:285 (1945) for original description.
Status Report: 1977

Trichostema lanatum

Considered but rejected: Too common

Trichostema micranthum Gray

"small-flowered bluecurls" Lamiaceae
CNPS List: 4 **R-E-D Code:** 1-1-1 **State/Fed. Status:** CEQA?
Distribution: RIV, SBD, BA
Habitat: LCFrs, Medws
Life Form: Annual herb
Blooming: July-September
Notes: See *Synoptical Flora of North America* 2(1):348 (1878) for original description.

Trichostema ovatum Curran

"San Joaquin bluecurls" Lamiaceae
CNPS List: 4 **R-E-D Code:** 1-2-3 **State/Fed. Status:** CEQA?
Distribution: FRE, KNG, KRN, TUL
Habitat: VFGrs
Life Form: Annual herb
Blooming: July-October

Trichostema rubisepalum Elmer

"Hernandez bluecurls" Lamiaceae
CNPS List: 4 **R-E-D Code:** 1-1-3 **State/Fed. Status:** CEQA?
Distribution: MPA, NAP, SBT, TUO
Habitat: BUFrs, Chprl, CmWld / volcanic or serpentinite
Life Form: Annual herb
Blooming: June-August

Tridens pilosus

See *Erioneuron pilosum*

Trientalis arctica Hook.

"arctic starflower" Primulaceae
CNPS List: 2 **R-E-D Code:** 3-2-1 **State/Fed. Status:** CEQA
Distribution: DNT, ID, OR, ++
Quads: 723B, 740C
Habitat: BgFns, Medws / coastal
Life Form: Perennial herb
Blooming: June-July
Notes: Known in CA from only three occurrences. Threatened by cattle grazing and trampling. Sensitive in ID. See *Leaflets of Western Botany* 10:333 (1966) for first CA records.

Trifolium amoenum Greene

"showy Indian clover" Fabaceae
CNPS List: 1B **R-E-D Code:** 3-3-3 **State/Fed. Status:** /C2*
Distribution: ALA*, MEN*, MRN*, NAP*, SCL*, SOL*, SON
Quads: 406D*, 445A*, 466B*, 467B*, 483A*, 484B*, 485C*, 485D*, 498C*, 500D*, 501A*, 501B*, 501C*, 502A*, 502B*, 502C*, 503A, 567B*, 567D*
Habitat: VFGrs (sometimes serpentinite)
Life Form: Annual herb
Blooming: April-June
Notes: Rediscovered in 1993 by P. Conners; only one plant found. Habitat lost to urbanization and agriculture. See *Flora Franciscana*, p. 27 (1891) by E. Greene for original description.
Status Report: 1977

Trifolium andersonii ssp. *beatleyae*

Considered but rejected: A synonym of *T. andersonii* var. *beatleyae*; a common taxon

Trifolium bolanderi Gray

"Bolander's clover" Fabaceae
CNPS List: 4 **R-E-D Code:** 1-1-3 **State/Fed. Status:** /C2
Distribution: FRE, MAD, MPA
Habitat: LCFrs, Medws, UCFrs / mesic
Life Form: Perennial herb
Blooming: June-August
Notes: See *Proceedings of the American Academy of Arts and Sciences* 7:335 (1868) for original description.
Status Report: 1979

Trifolium buckwestiorum Isely

"Santa Cruz clover" Fabaceae
CNPS List: 1B **R-E-D Code:** 3-3-3 **State/Fed. Status:** CEQA
Distribution: SCR
Quads: 407D, 408C, 409D
Habitat: BUFrs, CoPrr / margins
Life Form: Annual herb
Blooming: May,Oct
Notes: Known from about six very small occurrences; only one fully protected, others threatened by grazing, land clearing, and planting of non-native forage plants. See *Madroño* 39(2):90 (1992) for original description.

Trifolium dedeckerae

See *Trifolium macilentum* var. *dedeckerae*

Trifolium gracilentum T. & G. var. *palmeri* (Wats.) L.F. McDermott

"southern island clover" Fabaceae
CNPS List: 4 **R-E-D Code:** 1-2-2 **State/Fed. Status:** CEQA?
Distribution: SBR, SCM, SCT, SNI, GU
Habitat: CBScr, VFGrs
Life Form: Annual herb
Blooming: March-May
Notes: Rediscovered on SCT and SNI islands in 1978. Common on SNI Isl. in 1993.

Trifolium grayi

Considered but rejected: A synonym of *T. barbigerum* var. *andrewsii*; a common taxon

Trifolium howellii Wats.

"Howell's clover" Fabaceae
CNPS List: 4 **R-E-D Code:** 1-1-1 **State/Fed. Status:** CEQA?
Distribution: DNT, HUM, SIS, OR
Habitat: LCFrs, Medws, UCFrs / mesic
Life Form: Perennial herb
Blooming: July-August

Trifolium lemmonii Wats.

"Lemmon's clover" Fabaceae
CNPS List: 4 **R-E-D Code:** 1-1-2 **State/Fed. Status:** /C3c
Distribution: NEV, PLU, SIE, NV
Habitat: GBScr, LCFrs
Life Form: Perennial herb
Blooming: May-June
Notes: See *Proceedings of the American Academy of Arts and Sciences* 11:127 (1876) for original description, *Canadian Journal of Botany* 50:1975-2007 (1972) for taxonomic treatment, and *Four Seasons* 4:22-23 (1974) for discussion of rediscovery (1972).
Status Report: 1977

Trifolium macilentum Greene var. *dedeckerae* (Gillett) Barneby

"DeDecker's clover" Fabaceae
CNPS List: 1B **R-E-D Code:** 3-1-3 **State/Fed. Status:** /C3c
Distribution: INY, KRN, MNO, TUL
Quads: 283D, 306A, 307D, 329A, 329D, 352A, 373D, 412A, 431C, 431D, 488A, 488B
Habitat: LCFrs, PJWld, SCFrs, UCFrs / granitic crevices
Life Form: Perennial herb
Blooming: May-July
Notes: Known from fewer than twenty occurrences. Possibly threatened by mining and grazing. See *Madroño* 21(7):451-455 (1972) for original description.

Trifolium monoense

Considered but rejected: Too common; a synonym of *T. andersonii* var. *beatleyae*

Trifolium palmeri

See *Trifolium gracilentum* var. *palmeri*

Trifolium polyodon Greene

"Pacific Grove clover" Fabaceae
CNPS List: 1B **R-E-D Code:** 3-3-3 **State/Fed. Status:** CR/C1
Distribution: MNT
Quads: 366C
Habitat: CCFrs, CoPrr, Medws (mesic)
Life Form: Annual herb
Blooming: May-June
Notes: Known from only three occurrences on the Monterey Peninsula. Seriously threatened by urbanization and trampling. A synonym of *T. variegatum* (phase 4) in *The Jepson Manual.* See *Pittonia* 3:215 (1897) for original description.
Status Report: 1987

Trifolium trichocalyx Heller

"Monterey clover" Fabaceae
CNPS List: 1B **R-E-D Code:** 3-3-3 **State/Fed. Status:** CE/C1
Distribution: MNT
Quads: 366C
Habitat: CCFrs
Life Form: Annual herb
Blooming: April-June
Notes: Known from only two occurrences on the Monterey Peninsula. Seriously threatened by urbanization. Possible hybrid. See *Muhlenbergia* 1:55 (1904) for original description.
Status Report: 1987

Triglochin palustris

Considered but rejected: Too common

Triglochin striata

Considered but rejected: Too common

Trillium kurabayashii

Considered but rejected: A synonym of *T. angustipetalum*; a common taxon

Trillium ovatum Pursh ssp. *oettingeri* Munz & Thorne

"Salmon Mtns. wakerobin" Liliaceae
CNPS List: 4 **R-E-D Code:** 1-2-3 **State/Fed. Status:** /C3c
Distribution: SHA, SIS, TRI
Habitat: LCFrs (mesic)
Life Form: Perennial herb
Blooming: February-May
Notes: Klamath NF has adopted species management guidelines, but logging and grazing may threaten on private land in eastern part of range. See *Aliso* 8:15-17 (1973) for original description.

Trillium rivale

Considered but rejected: Too common

Trimorpha acris (L.) S.F. Gray var. *debilis* (Gray) Nesom

"northern daisy" Asteraceae
CNPS List: 2 **R-E-D Code:** 2-1-1 **State/Fed. Status:** CEQA
Distribution: PLU, SHA, SIS, OR, WA+
Quads: 625A, 626A, 698B, 713D
Habitat: AlpBR, Medws, SCFrs (volcanic)
Life Form: Perennial herb
Blooming: July-August
Notes: See *Synoptical Flora of North America* 1(2):220 (1884) for original description, and *Phytologia* 67(1):61-66 (1989) for taxonomic treatment.

Triphysaria floribunda (Benth.) Chuang & Heckard

"San Francisco owl's-clover" Scrophulariaceae
CNPS List: 1B **R-E-D Code:** 2-2-3 **State/Fed. Status:** /C2
Distribution: MRN, SFO, SMT
Quads: 448B, 448C, 466C, 485A, 485B, 485C, 485D
Habitat: CoPrr, VFGrs / serpentinite
Life Form: Annual herb
Blooming: April-May
Notes: Threatened by grazing and trampling. See *Systematic Botany* 16(4):644-666 (1991) for revised nomenclature.

Triteleia clementina Hoov.

"San Clemente Island triteleia" Liliaceae
CNPS List: 1B **R-E-D Code:** 3-2-3 **State/Fed. Status:** /C2
Distribution: SCM
Quads: SCMC, SCMS
Habitat: VFGrs (rocky)
Life Form: Perennial herb (bulbiferous)
Blooming: March-April
Notes: Known from fewer than twenty occurrences. Feral herbivores removed from SCM Isl., and vegetation recovering.

Triteleia crocea (Wood) Greene var. *crocea*

"yellow triteleia" Liliaceae
CNPS List: 4 **R-E-D Code:** 1-1-2 **State/Fed. Status:** CEQA?
Distribution: DNT, SHA, SIS, TRI, OR
Habitat: LCFrs (granitic or serpentinite)
Life Form: Perennial herb (bulbiferous)
Blooming: May-June
Notes: On watch list in OR.

Triteleia crocea (Wood) Greene var. *modesta* (Hall) Hoov.

"Trinity Mtns. triteleia" Liliaceae
CNPS List: 4 **R-E-D Code:** 1-1-3 **State/Fed. Status:** CEQA?
Distribution: SHA, SIS, TRI
Habitat: LCFrs (serpentinite)
Life Form: Perennial herb (bulbiferous)
Blooming: May-June

Triteleia dudleyi

Considered but rejected: Too common

Triteleia gracilis

Considered but rejected: Too common; a synonym of *T. montana*

Triteleia hendersonii Wats. var. *hendersonii*

"Henderson's triteleia" Liliaceae
CNPS List: 2 **R-E-D Code:** 3-2-1 **State/Fed. Status:** CEQA
Distribution: SIS, OR
Quads: 735B, 736A, 736C?
Habitat: CmWld
Life Form: Perennial herb (bulbiferous)
Blooming: May-July
Notes: Not in *The Jepson Manual*. See *Pittonia* 1:164 (1888) for original description.

Triteleia ixioides (Wats.) Greene ssp. *cookii* (Hoov.) Lenz

"Cook's triteleia" Liliaceae
CNPS List: 4 **R-E-D Code:** 1-1-3 **State/Fed. Status:** CEQA?
Distribution: SLO
Habitat: CCFrs, CmWld / serpentinite seeps
Life Form: Perennial herb (bulbiferous)
Blooming: May-June
Notes: See *Aliso* 8(8):273 (1975) for revised nomenclature.

Triteleia lugens

Considered but rejected: Too common

Tropidocarpum capparideum Greene

"caper-fruited tropidocarpum" Brassicaceae
CNPS List: 1A **Last Seen:** 1957 **State/Fed. Status:** /C2*
Distribution: ALA*, CCA*, GLE*, MNT*, SCL*, SJQ*
Quads: 318C*, 428B*, 428D*, 444B*, 445A*, 445B*, 463A*, 463C*, 463D*, 464B*, 578C*
Habitat: VFGrs (alkaline hills)
Life Form: Annual herb
Blooming: March-April
Notes: Recent attempts to rediscover this plant have been unsuccessful.
Status Report: 1988

Tuctoria greenei (Vasey) J. Reeder

"Greene's tuctoria" Poaceae
CNPS List: 1B **R-E-D Code:** 2-3-3 **State/Fed. Status:** CR/PE
Distribution: BUT, FRE*, MAD*, MER, SHA, SJQ*, STA*, TEH, TUL*
Quads: 333A*, 378C*, 378D*, 400B, 400D*, 420C, 421D, 441A*, 441B*, 441C*, 442A*, 460B*, 460C*, 560B, 576B, 593B, 594A, 661C
Habitat: VnPls
Life Form: Annual herb
Blooming: May-July
Notes: Threatened by agriculture, urbanization, and overgrazing. See *Botanical Gazette* 16:145-147 (1891) for original description, and *American Journal of Botany* 69:1082-1095 (1982) for taxonomic treatment.
Status Report: 1986

Tuctoria mucronata (Crampt.) J. Reeder

"Crampton's tuctoria" Poaceae
CNPS List: 1B **R-E-D Code:** 3-3-3 **State/Fed. Status:** CE/FE
Distribution: SOL, YOL
Quads: 497B, 498D
Habitat: VnPls
Life Form: Annual herb
Blooming: April-July
Notes: Known from only three occurrences: one at Olcott Lake at Jepson Prairie (TNC), one nearby on private land, and one south of Davis on DOD land. Possibly threatened by non-native plants. See *Madroño* 15:106-108 (1959) for original description, and *American Journal of Botany* 69:1082-1095 (1982) for revised taxonomy.
Status Report: 1991

Vaccinium coccineum Piper

"Siskiyou Mtns. huckleberry" Ericaceae
CNPS List: 3 **R-E-D Code:** 2-1-? **State/Fed. Status:** /C3b
Distribution: BUT, PLU, SIE, SIS, OR
Quads: 572C, 574A, 574B, 574C, 590A, 590C, 590D, 591D, 701C, 719C, 721B, 721C, 736C, 737A, 737C, 738A
Habitat: LCFrs, UCFrs / often serpentinite
Life Form: Shrub (deciduous)
Blooming: June-August
Notes: Move to List 4? Taxonomic work in progress. See *V. membranaceum* in *The Jepson Manual*. See *Proceedings of the Biological Society of Washington* 31:75-78 (1918) for original description.
Status Report: 1979

Vaccinium scoparium Leib.

"littleleaf huckleberry" Ericaceae
CNPS List: 2 **R-E-D Code:** 2-2-1 **State/Fed. Status:** CEQA
Distribution: SIS, OR, WA, ++
Quads: 682B, 684C, 701C, 702B, 719C
Habitat: SCFrs (rocky)
Life Form: Shrub (deciduous)
Blooming: June-August
Notes: See *Mazama* 1:196 (1897) for original description.

Vaccinium uliginosum ssp. *occidentale*

Considered but rejected: Too common

Vaccinium uliginosum* ssp. *uliginosum
Considered but rejected: Not confirmed in CA, but plants from Big Lagoon (HUM Co.) may be this taxon

Vahlodea atropurpurea
See *Deschampsia atropurpurea*

***Vancouveria chrysantha* Greene**
"Siskiyou inside-out-flower" Berberidaceae
CNPS List: 4 **R-E-D Code:** 2-1-2 **State/Fed. Status:** /C3c
Distribution: DNT, SIS, OR
Habitat: LCFrs (serpentinite)
Life Form: Perennial herb (rhizomatous)
Blooming: June
Notes: On watch list in OR.

***Veratrum fimbriatum* Gray**
"fringed false-hellebore" Liliaceae
CNPS List: 4 **R-E-D Code:** 1-1-3 **State/Fed. Status:** /C3c
Distribution: MEN, SON
Habitat: CoScr, Medws, NCFrs / mesic
Life Form: Perennial herb
Blooming: July-September

***Veratrum insolitum* Jeps.**
"Siskiyou false-hellebore" Liliaceae
CNPS List: 4 **R-E-D Code:** 1-1-1 **State/Fed. Status:** CEQA?
Distribution: DNT, HUM, SHA, SIS, TRI, OR, WA
Habitat: Chprl, LCFrs / clay
Life Form: Perennial herb
Blooming: June-August
Notes: On review list in OR, and state-listed as Sensitive in WA.

***Verbena californica* Mold.**
"California vervain" Verbenaceae
CNPS List: 1B **R-E-D Code:** 3-3-3 **State/Fed. Status:** CC/C1
Distribution: TUO
Quads: 458C
Habitat: CmWld, VFGrs / mesic, usually serpentinite seeps or creeks
Life Form: Perennial herb
Blooming: June-September
Notes: Known from nine occurrences in the Red Hills. Threatened by grazing, mining, development, recreation, and vehicles. Protected in part at Red Hills ACEC (BLM).

***Verbesina dissita* Gray**
"crownbeard" Asteraceae
CNPS List: 1B **R-E-D Code:** 3-3-2 **State/Fed. Status:** CT/PT
Distribution: ORA, BA
Quads: 70C, 71D
Habitat: Chprl (maritime), CoScr
Life Form: Perennial herb
Blooming: April-July
Notes: Known in CA from only two occurrences near southern Laguna Beach. Threatened by urbanization and fuelbreak clearing.
Status Report: 1990

***Veronica copelandii* Eastw.**
"Copeland's speedwell" Scrophulariaceae
CNPS List: 4 **R-E-D Code:** 1-1-3 **State/Fed. Status:** /C3c
Distribution: SIS, TRI
Habitat: Medws, SCFrs / serpentinite
Life Form: Perennial herb
Blooming: August
Notes: See *Botanical Gazette* 41:288 (1906) for original description.

***Veronica cusickii* Gray**
"Cusick's speedwell" Scrophulariaceae
CNPS List: 4 **R-E-D Code:** 1-1-1 **State/Fed. Status:** CEQA?
Distribution: ALP, AMA, MAD, MPA, PLA, SIE, TUO, OR, WA+
Habitat: AlpBR, Medws, SCFrs, UCFrs
Life Form: Perennial herb
Blooming: July-August

Viburnum ellipticum
Considered but rejected: Too common

***Viguiera laciniata* Gray**
"San Diego County viguiera" Asteraceae
CNPS List: 4 **R-E-D Code:** 1-2-1 **State/Fed. Status:** CEQA?
Distribution: SDG, BA, SO
Habitat: Chprl, CoScr
Life Form: Shrub
Blooming: February-June
Notes: Locally common in SDG Co. Threatened by development.

Viguiera reticulata
Considered but rejected: Too common

Viola adunca* var. *kirkii
Considered but rejected: A synonym of *V. adunca*; a common taxon

***Viola aurea* Kell.**
"golden violet" Violaceae
CNPS List: 2 **R-E-D Code:** 2-2-1 **State/Fed. Status:** CEQA
Distribution: KRN, MNO, SBD, SDG, NV
Quads: 49C, 211C, 243D, 487C
Habitat: GBScr, PJWld / sandy
Life Form: Perennial herb
Blooming: April-June
Notes: Rarely collected. Need quads for SBD Co. Probably more common in NV. Threatened by grazing. See *Proceedings of the California Academy of Natural Sciences* 2:185 (1862) for original description, and *Madroño* 12(1):8-18 (1953) for taxonomic treatment.

Viola cuneata
Considered but rejected: Too common

Viola hallii
Considered but rejected: Too common

Viola lanceolata* ssp. *occidentalis
See *Viola primulifolia* ssp. *occidentalis*

Viola langsdorfii (Regel) Fisch.

"Langsdorf's violet" Violaceae
CNPS List: 2 **R-E-D Code:** 3-3-1 **State/Fed. Status:** CEQA
Distribution: DNT, OR*, WA*, ++
Quads: 740C
Habitat: BgFns (coastal)
Life Form: Perennial herb
Blooming: June-July
Notes: Known in CA from only two occurrences at Lake Earl and Pt. St. George. Threatened by development. Not in *The Jepson Manual.* See *Madroño* 40(2):135 (1993) for first CA record.

Viola palustris L.

"marsh violet" Violaceae
CNPS List: 2 **R-E-D Code:** 3-2-1 **State/Fed. Status:** CEQA
Distribution: DNT, HUM, MEN, OR, WA, ++
Quads: 569A, 672C, 689A
Habitat: CoScr (mesic), BgFns (coastal)
Life Form: Perennial herb (rhizomatous)
Blooming: March-August
Notes: Known in CA from only a few occurrences; often overlooked and rarely collected. See *Madroño* 17(6):173-197 (1964) for taxonomic treatment.

Viola pinetorum Greene ssp. *grisea* (Jeps.) R.J. Little

"grey-leaved violet" Violaceae
CNPS List: 1B **R-E-D Code:** 3-1-3 **State/Fed. Status:** CEQA
Distribution: KRN, SBD, TUL
Quads: 105A, 131D, 329C, 355D
Habitat: SCFrs, UCFrs
Life Form: Perennial herb
Blooming: April
Notes: Currently known from only a few occurrences; rarely collected. See *Flora of California* 2(1):521 (1936) by W.L. Jepson for original description, and *Phytologia* 72(2):77-78 (1992) for revised nomenclature.

Viola primulifolia L. ssp. *occidentalis* (Gray) L.E. McKinney & R.J. Little

"western bog violet" Violaceae
CNPS List: 1B **R-E-D Code:** 2-2-2 **State/Fed. Status:** /C2
Distribution: DNT, OR
Quads: 722A, 739A, 739B, 739C
Habitat: BgFns (serpentinite), MshSw
Life Form: Perennial herb (rhizomatous)
Blooming: April-September
Notes: Threatened by mining, logging, road construction, and vehicles. Candidate for state listing in OR. See *Botanical Gazette* 11:255 (1886) for original description, and *Phytologia* 72(2):79 (1992) for revised nomenclature.
Status Report: 1980

Viola tomentosa Baker & Clausen

"woolly violet" Violaceae
CNPS List: 1B **R-E-D Code:** 2-2-3 **State/Fed. Status:** CEQA
Distribution: ELD, NEV, PLA, PLU, SIE
Quads: 508A, 508B, 523C, 524A, 524B, 524C, 524D, 539B, 539D, 540D, 555B, 556B, 556C, 557D, 573B, 573C, 574A, 590C, 616C
Habitat: LCFrs, SCFrs, UCFrs / gravelly
Life Form: Perennial herb
Blooming: May-August
Notes: Threatened by road building, vehicles, and proposed reservoir construction in national forests. See *Leaflets of Western Botany* 5:142 (1949) for original description, and *Madroño* 17(6):173-197 (1964) for taxonomic treatment.

Washingtonia filifera

Considered but rejected: Too common

Whitneya dealbata

Considered but rejected: Too common

Wislizenia refracta Engelm. ssp. *refracta*

"jackass-clover" Capparaceae
CNPS List: 2 **R-E-D Code:** 3-2-1 **State/Fed. Status:** CEQA
Distribution: RIV, SBD, AZ, NM, TX
Quads: 77C, 102A, 128C
Habitat: DeDns, MDScr, Plyas, SDScr
Life Form: Annual herb
Blooming: April-November

Woodsia plummerae Lemmon

"Plummer's woodsia" Dryopteridaceae
CNPS List: 2 **R-E-D Code:** 3-1-1 **State/Fed. Status:** CEQA
Distribution: SBD, ++
Quads: 225D
Habitat: PJWld (granitic)
Life Form: Perennial herb (rhizomatous)
Fertile: May-September
Notes: See *Madroño* 22(4):378 (1974) and 26(1):56 (1978) for distributional information.

Wyethia elata Hall

"Hall's wyethia" Asteraceae
CNPS List: 4 **R-E-D Code:** 1-1-3 **State/Fed. Status:** CEQA?
Distribution: FRE, MAD, MPA, TUL
Habitat: CmWld, LCFrs
Life Form: Perennial herb
Blooming: May-August

Wyethia invenusta
Considered but rejected: Too common

Wyethia longicaulis Gray
"Humboldt County wyethia" Asteraceae
CNPS List: 4 **R-E-D Code:** 1-1-3 **State/Fed. Status:** CEQA?
Distribution: HUM, MEN, TRI
Habitat: BUFrs, CoPrr, LCFrs
Life Form: Perennial herb
Blooming: May-July

Wyethia reticulata Greene
"El Dorado County mule ears" Asteraceae
CNPS List: 1B **R-E-D Code:** 2-2-3 **State/Fed. Status:** /C2
Distribution: ELD
Quads: 510B, 511A, 526C, 527D
Habitat: Chprl, CmWld, LCFrs / clay or gabbroic
Life Form: Perennial herb
Blooming: May-July
Notes: Threatened by development and vehicles. See *Bulletin of the California Academy of Sciences* 1:9 (1884) for original description.
Status Report: 1979

Xylorhiza cognata (H.M. Hall) T.J. Watson
"Mecca-aster" Asteraceae
CNPS List: 1B **R-E-D Code:** 2-2-3 **State/Fed. Status:** /C2
Distribution: RIV
Quads: 62C, 63A, 63B, 63C, 64A, 64B, 64D, 81C, 82C, 82D
Habitat: SDScr
Life Form: Perennial herb
Blooming: January-June
Notes: Known only from Indio Hills and Mecca Hills. Threatened by vehicles. See *Brittonia* 29:199-216 (1977) for revised nomenclature and description.

Xylorhiza orcuttii (Vasey & Rose) Greene
"Orcutt's woody-aster" Asteraceae
CNPS List: 1B **R-E-D Code:** 2-2-2 **State/Fed. Status:** /C2
Distribution: IMP, RIV, SDG, BA
Quads: 18A, 18B, 18C, 31A, 31B, 31C, 31D, 32A, 33D, 46C, 46D, 62B
Habitat: SDScr
Life Form: Perennial herb
Blooming: March-April
Notes: Threatened by vehicles.

Zauschneria septentrionalis
See *Epilobium septentrionale*

Zigadenus fremontii var. *inezianus*
Considered but rejected: Too common; a synonym of *Z. fremontii*

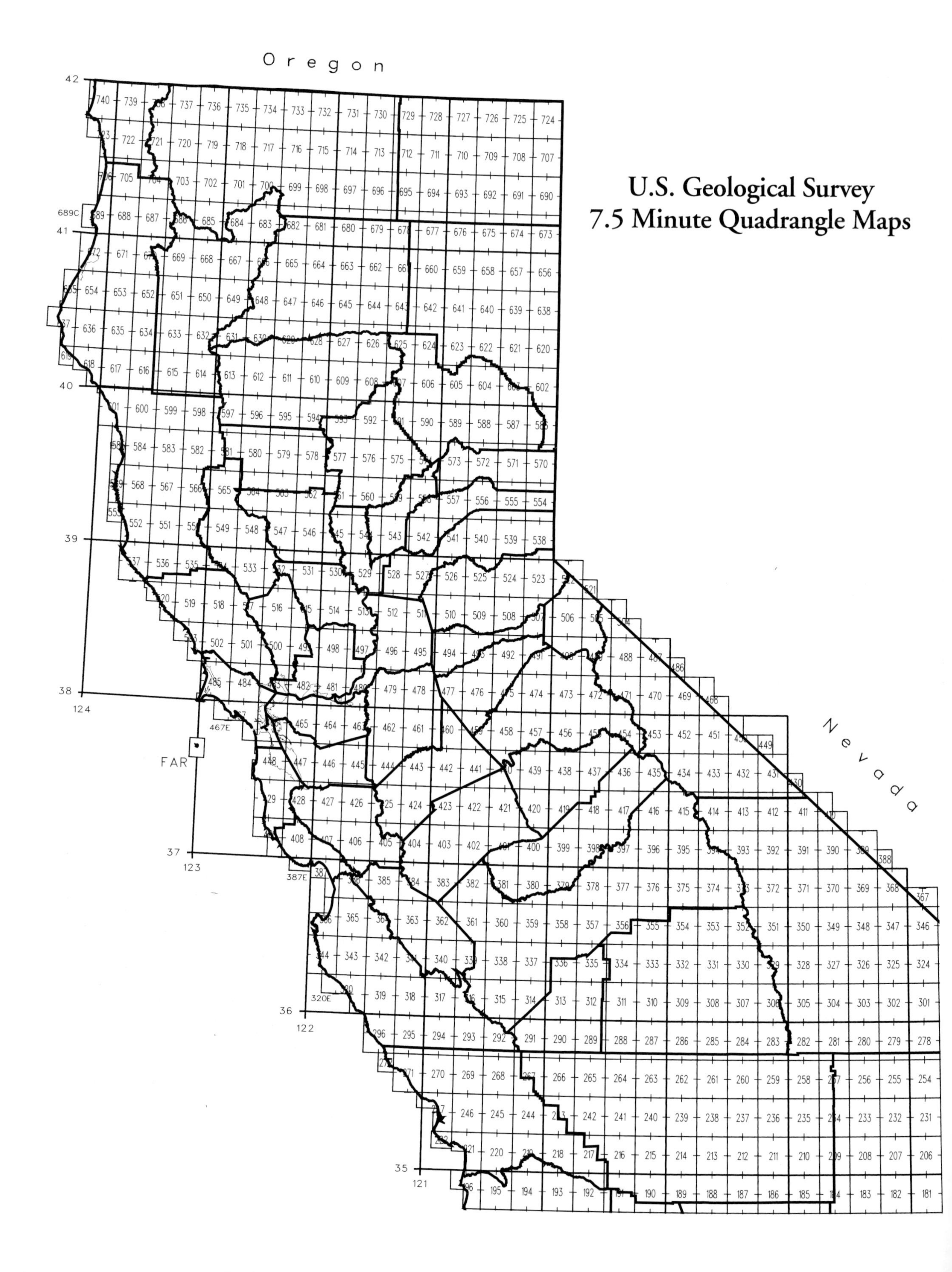
U.S. Geological Survey
7.5 Minute Quadrangle Maps
Oregon
Nevada
42
41
40
39
38
124
37
123
36
122
35
121
689C
467E
FAR
387E
320E

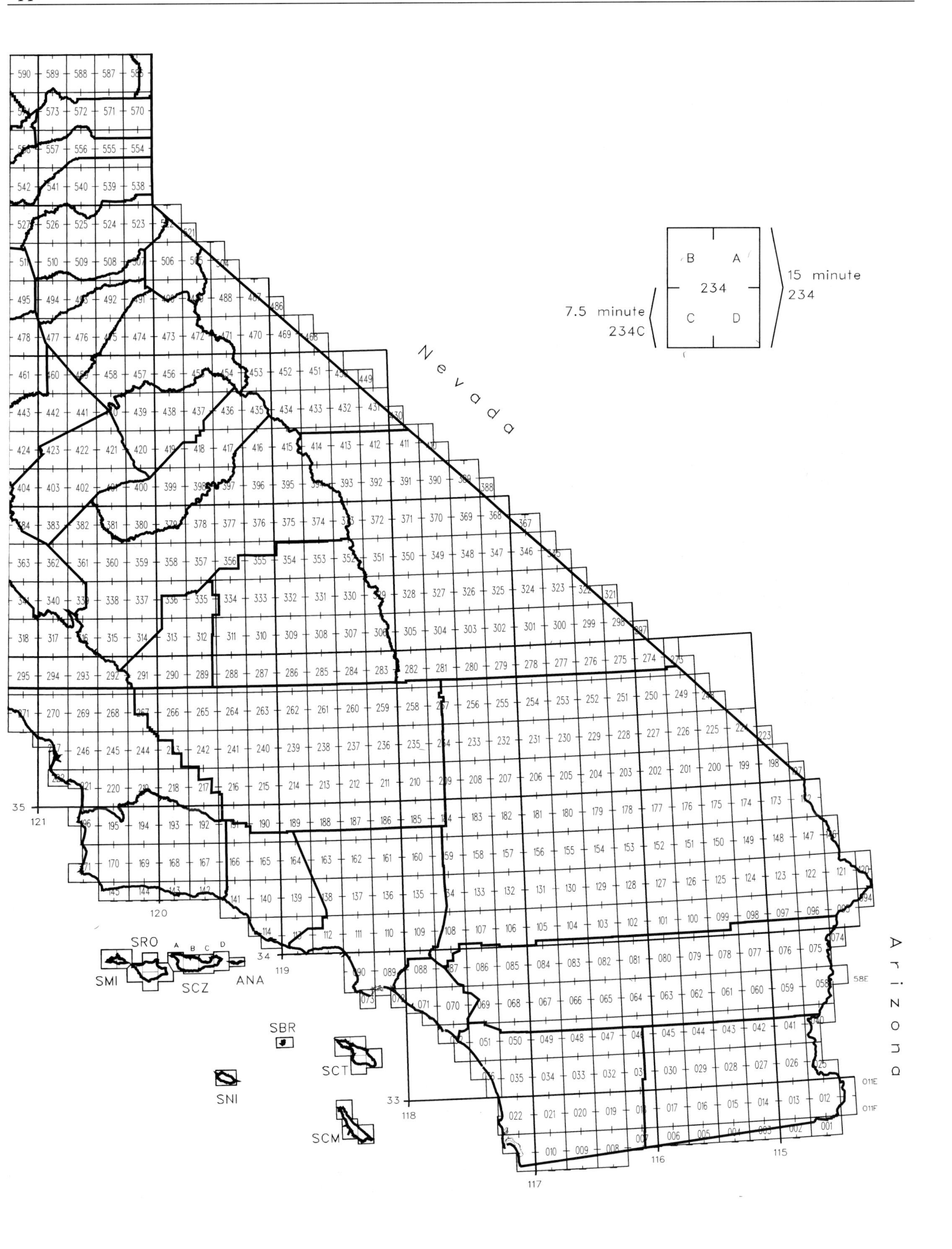

Nevada
Arizona
15 minute
234
7.5 minute
234C
B
A
C
D
SRO
SMI
SCZ
ANA
SBR
SCT
SNI
SCM
35
34
33
121
120
119
118
117
116
115

Appendix I: List of Quadrangle Maps

This index to 7.5 minute topographic quadrangle maps generally conforms to the quad code system used by the Department of Water Resources; the quad names are those used by the U.S. Geological Survey. Two lists are included below, the first sorted by quad code, the second by quad name.

There is some confusion regarding the names of certain 7.5 minute quads: the 1993 USGS *California Index to Topographic Map Coverage* indicates a preferred name for future map updates, but current actual maps indicate a different name. In these few cases we have listed the name on our current maps and indicated the alternative *Index* name in parentheses; it is probably best to order maps by the latter name. Also, there are apparently 20 quadrangles lacking maps at the 7.5 minute scale, and we understand that the USGS has no immediate plans to produce them. For these quads we have indicated the 15 minute map name and quadrant (NE, NW, SW, SE), along with the provisional name, now obsolete, previously used for that 7.5 minute quad.

We use a unique quad code system for the California islands, and frequent users will note that we have changed these quad codes from the fourth edition; now the first three digits of each match the island code given on the inside front cover.

Quadrangle Codes

Code	Quad Name
ANAC	Anacapa Island
FARA	Farallon Islands
SBRA	Santa Barbara Island
SCMC	San Clemente Island Central
SCMN	San Clemente Island North
SCMS	San Clemente Island South
SCTE	Santa Catalina East
SCTN	Santa Catalina North
SCTS	Santa Catalina South
SCTW	Santa Catalina West
SCZA	Santa Cruz Island A
SCZB	Santa Cruz Island B
SCZC	Santa Cruz Island C
SCZD	Santa Cruz Island D
SMIE	San Miguel Island East
SMIW	San Miguel Island West
SNIC	San Nicolas Island
SROE	Santa Rosa Island East
SRON	Santa Rosa Island North
SROS	Santa Rosa Island South
SROW	Santa Rosa Island West
001A	Yuma East
001B	Yuma West
002A	Grays Well NE
002B	Grays Well
003A	Midway Well
003B	Midway Well NW
004A	Bonds Corner
004B	Calexico
005A	Heber
005B	Mount Signal
006A	Yuha Basin
006B	Coyote Wells
007A	In-ko-pah Gorge
007B	Jacumba
008A	Live Oak Springs
008B	Cameron Corners
008C	Campo
008D	Tierra Del Sol
009A	Morena Reservoir
009B	Barrett Lake
009C	Tecate
009D	Potrero
010A	Dulzura
010B	Jamul Mountains
010C	Otay Mesa
010D	Otay Mountain
011A	National City
011B	Point Loma
011D	Imperial Beach
011E	Imperial Reservoir
011F	Laguna Dam
012A	Little Picacho Peak
012B	Picacho Peak
012C	Araz
012D	Bard
013A	Hedges
013B	Clyde
013C	Cactus
013D	Ogilby
014A	Glamis
014B	Glamis NW
014C	Glamis SW
014D	Glamis SE
015A	Holtville NE
015B	Alamorio
015C	Holtville West
015D	Holtville East
016A	Brawley
016B	Brawley NW
016C	Seeley
016D	El Centro
017A	Superstition Mtn.
017B	Plaster City NW
017C	Painted Gorge
017D	Plaster City
018A	Carrizo Mtn. NE
018B	Arroyo Tapiado
018C	Sweeney Pass
018D	Carrizo Mtn.
019A	Agua Caliente Springs
019B	Monument Peak
019C	Mount Laguna
019D	Sombrero Peak
020A	Cuyamaca Peak
020B	Tule Springs
020C	Viejas Mountain
020D	Descanso
021A	El Cajon Mtn.
021B	San Vicente Reservoir
021C	El Cajon
021D	Alpine
022A	Poway
022B	Del Mar
022C	La Jolla
022D	La Mesa
025B	Picacho NW
025C	Picacho SW
025D	Picacho
026A	Buzzards Peak
026B	Mt. Barrow
026C	Gables Wash
026D	Quartz Peak
027A	Blue Mountain
027B	Mammoth Wash
027C	Acolita
027D	East Of Acolita
028A	Tortuga
028B	Iris
028C	Wiest
028D	Amos
029A	Niland
029B	Obsidian Butte
029C	Calipatria SW (=Westmorland West)
029D	Westmorland (=Westmorland East)
030A	Kane Spring NE
030B	Kane Spring NW
030C	Harpers Well
030D	Kane Spring
031A	Shell Reef
031B	Borrego Mountain
031C	Harper Canyon
031D	Borrego Mountain SE
032A	Borrego Sink
032B	Tubb Canyon
032C	Earthquake Valley
032D	Whale Peak
033A	Ranchita
033B	Warners Ranch
033C	Santa Ysabel
033D	Julian
034A	Mesa Grande
034B	Rodriquez Mtn.
034C	San Pasqual
034D	Ramona
035A	Valley Center
035B	San Marcos
035C	Rancho Santa Fe
035D	Escondido
036A	San Luis Rey
036B	Oceanside
036D	Encinitas
040A	Mule Wash
040B	Palo Verde
040C	Cibola
041A	Thumb Peak
041B	Wileys Well
041C	West Of Palo Verde Peak
041D	Palo Verde Peak
042A	Little Chuckwalla Mts.
042B	Chuckwalla Spring
042C	Pegleg Well
042D	Little Mule Mts.
043A	Augustine Pass
043B	Iris Pass
043C	Iris Wash
043D	Lion Head Mtn.
044A	Frink NE
044B	Frink NW
044C	Frink
044D	Wister
045A	Durmid
045B	Salton
045C	Truckhaven
045D	Durmid SE
046A	Oasis
046B	Rabbit Peak
046C	Fonts Point
046D	Seventeen Palms
047A	Clark Lake NE
047B	Collins Valley
047C	Borrego Palm Canyon
047D	Clark Lake
048A	Bucksnort Mtn.

048B Beauty Mountain
048C Warner Springs
048D Hot Springs Mtn.
049A Aguanga
049B Vail Lake
049C Boucher Hill
049D Palomar Observatory
050A Pechanga
050B Temecula
050C Bonsall
050D Pala
051A Fallbrook
051B Margarita Peak
051C Las Pulgas Canyon
051D Morro Hill
052A San Clemente
052B Dana Point
052D San Onofre Bluff
058A Blythe NE
058B McCoy Wash
058C Ripley
058D Blythe
058E La Paz Mtn.
059A McCoy Peak
059B McCoy Spring
059C Hopkins Well
059D Roosevelt Mine
060A Ford Dry Lake
060B Sidewinder Well
060C Aztec Mines
060D East Of Aztec Mines
061A Corn Spring
061B Desert Center
061C Red Cloud Canyon
061D Pilot Mountain
062A Hayfield Spring
062B Hayfield
062C Red Canyon
062D East Of Red Canyon
063A Cottonwood Spring
063B Cottonwood Basin
063C Mortmar
063D Orocopia Canyon
064A Thermal Canyon
064B Indio
064C Valerie
064D Mecca
065A La Quinta
065B Rancho Mirage
065C Toro Peak
065D Martinez Mtn.
066A Palm View Peak
066B Idyllwild
066C Anza
066D Butterfly Peak
067A Blackburn Canyon
067B Hemet
067C Sage
067D Cahuilla Mtn.
068A Winchester
068B Romoland
068C Murrieta
068D Bachelor Mtn.
069A Lake Elsinore
069B Alberhill
069C Sitton Peak
069D Wildomar
070A Santiago Peak
070B El Toro
070C San Juan Capistrano
070D Canada Gobernadora
071A Tustin
071B Newport Beach
071D Laguna Beach
072A Seal Beach
073A San Pedro
074B Poston
075A Big Maria Mts. NE
075B Big Maria Mts. NW
075C Big Maria Mts. SW
075D Big Maria Mts. SE
076A Styx
076B Little Maria Mts.
076C Arlington Mine
076D Inca
077A Palen Pass
077B West Of Palen Pass
077C Palen Lake
077D Palen Mountains
078A Coxcomb Mts.
078B Pinto Wells
078C Victory Pass
078D East Of Victory Pass
079A Placer Canyon
079B San Bernardino Wash
079C Conejo Well
079D Buzzard Spring
080A Pinto Mountain
080B Fried Liver Wash
080C Washington Wash
080D Porcupine Wash
081A Malapai Hill
081B Keys View
081C West Berdoo Canyon
081D Rockhouse Canyon
082A East Deception Canyon
082B Seven Palms Valley
082C Cathedral City
082D Myoma
083A Desert Hot Springs
083B White Water
083C San Jacinto Peak
083D Palm Springs
084A Cabazon
084B Beaumont
084C San Jacinto
084D Lake Fulmor
085A El Casco
085B Sunnymead
085C Perris
085D Lakeview
086A Riverside East
086B Riverside West
086C Lake Mathews
086D Steele Peak
087A Corona North
087B Prado Dam
087C Black Star Canyon
087D Corona South
088A Yorba Linda
088B La Habra
088C Anaheim
088D Orange
089A Whittier
089B South Gate
089C Long Beach
089D Los Alamitos
090A Inglewood
090B Venice
090C Redondo Beach
090D Torrance
094B Cross Roads
095A Parker
095B Parker NW
095C Parker SW
095D Parker SE
096A Vidal Junction
096B Vidal NW
096C Grommet
096D Vidal
097A Horn Spring
097B Sablon
097C Arica Mountains
097D Rice
098A Danby Lake
098B Iron Mountains
098C Granite Pass
098D East Of Granite Pass
099A Cadiz Valley NE
099B Cadiz Valley NW
099C Cadiz Valley SW
099D Cadiz Valley SE
100A East Of Dale Lake
100B Dale Lake
100C New Dale
100D Clarks Pass
101A East Of Valley Mtn.
101B Valley Mtn.
101C Twentynine Palms Mtn.
101D Humbug Mountain
102A Twentynine Palms
102B Sunfair
102C Indian Cove
102D Queen Mtn.
103A Joshua Tree North
103B Yucca Valley North
103C Yucca Valley South
103D Joshua Tree South
104A Rimrock
104B Onyx Peak
104C Catclaw Flat
104D Morongo Valley
105A Moonridge
105B Big Bear Lake
105C Forest Falls
105D San Gorgonio Mtn.
106A Keller Peak
106B Harrison Mtn.
106C Redlands
106D Yucaipa
107A San Bernardino North
107B Devore
107C Fontana
107D San Bernardino South
108A Cucamonga Peak
108B Mt. Baldy
108C Ontario
108D Guasti
109A Glendora
109B Azusa
109C Baldwin Park
109D San Dimas
110A Mt. Wilson
110B Pasadena
110C Los Angeles
110D El Monte
111A Burbank
111B Van Nuys
111C Beverly Hills
111D Hollywood
112A Canoga Park
112B Calabasas
112C Malibu Beach
112D Topanga
113A Thousand Oaks
113B Newbury Park
113C Triunfo Pass
113D Point Dume
114A Camarillo
114B Oxnard
114D Point Mugu
120B Standard Wash
120C Gene Wash
121A Lake Havasu City South
121B Havasu Lake
121C Whipple Mts. SW
121D Whipple Wash
122A Savahia Peak NE
122B Savahia Peak NW
122C Savahia Peak SW
122D Savahia Peak
123A Mohawk Spring
123B West Of Mohawk Spring
123C Martins Well
123D Mopah Peaks
124A Wilhelm Spring
124B Sheep Camp Spring
124C Milligan
124D East Of Milligan
125A Cadiz Lake NE
125B Cadiz Lake NW
125C Cadiz Lake
125D Chubbuck
126A Calumet Mine
126B Bristol Lake NW
126C Bristol Lake SW
126D Calumet Mountains
127A Lead Mtn. NE
127B Lead Mountain
127C Lead Mountain SW
127D Cleghorn Lakes
128A Deadman Lake NE
128B Deadman Lake NW
128C Deadman Lake SW
128D Deadman Lake SE
129A Hidalgo Mtn.
129B Emerson Lake
129C Landers
129D Goat Mountain
130A Melville Lake
130B Old Woman Springs
130C Rattlesnake Canyon
130D Bighorn Canyon
131A Cougar Buttes

131B Lucerne Valley
131C Fawnskin
131D Big Bear City
132A Fifteenmile Valley
132B Apple Valley South
132C Lake Arrowhead
132D Butler Peak
133A Hesperia
133B Baldy Mesa
133C Cajon
133D Silverwood Lake
134A Phelan
134B Mescal Creek
134C Mount San Antonio
134D Telegraph Peak
135A Valyermo
135B Juniper Hills
135C Waterman Mtn.
135D Crystal Lake
136A Pacifico Mountain
136B Acton
136C Condor Peak
136D Chilao Flat
137A Agua Dulce
137B Mint Canyon
137C San Fernando
137D Sunland
138A Newhall
138B Val Verde
138C Santa Susana (= Simi Valley East)
138D Oat Mountain
139A Piru
139B Fillmore
139C Moorpark
139D Simi (= Simi Valley West)
140A Santa Paula Peak
140B Ojai
140C Saticoy
140D Santa Paula
141A Matilija
141B White Ledge Peak
141C Pitas Point
141D Ventura
142A Carpinteria
142B Santa Barbara
143A Goleta
143B Dos Pueblos Canyon
144A Tajiguas
144B Gaviota
145A Sacate
145B Point Conception
146B Topock
146C Castle Rock
147A Whale Mtn.
147B Monumental Pass
147C Snaggletooth
147D Chemehuevi Peak
148A Stepladder Mts. NE
148B Stepladder Mts. NW
148C Stepladder Mountains SW
148D Stepladder Mountains
149A Little Piute Mountains
149B Essex
149C Old Woman Statue
149D Painted Rock Wash
150A Danby
150B Castle Dome
150C Cadiz Summit
150D Skeleton Pass
151A Van Winkle Wash
151B Brown Buttes
151C Amboy
151D Cadiz
152A East Of Siberia
152B Siberia
152C Bagdad SW
152D Amboy Crater
153A Ash Hill
153B Ludlow
153C Morgans Well
153D Ludlow SE
154A Lavic Lake
154B Sunshine Peak
154C Galway Lake
154D Lavic SE
155A Silver Bell Mine
155B Camp Rock Mine
155C Fry Mountains
155D Iron Ridge
156A Ord Mountain
156B West Ord Mountain
156C White Horse Mtn.
156D Grand View Mine
157A Stoddard Well
157B Turtle Valley
157C Apple Valley North
157D Fairview Valley
158A Helendale
158B Victorville NW
158C Adelanto
158D Victorville
159A Shadow Mountains
159B Adobe Mountain
159C El Mirage
159D Shadow Mountains SE
160A Hi Vista
160B Alpine Butte
160C Littlerock
160D Lovejoy Buttes
161A Lancaster East
161B Lancaster West
161C Ritter Ridge
161D Palmdale
162A Del Sur
162B Lake Hughes
162C Green Valley
162D Sleepy Valley
163A Burnt Peak
163B Liebre Mtn.
163C Whitaker Peak
163D Warm Springs Mountain
164A Black Mtn.
164B Alamo Mountain
164C Devils Heart Peak
164D Cobblestone Mtn.
165A Lockwood Valley
165B San Guillermo
165C Lion Canyon
165D Topatopa Mountains
166A Reyes Peak
166B Rancho Nuevo Creek
166C Old Man Mountain
166D Wheeler Springs
167A Madulce Peak
167B Big Pine Mtn.
167C Little Pine Mtn.
167D Hildreth Peak
168A San Rafael Mtn.
168B Figueroa Mtn.
168C Lake Cachuma
168D San Marcos Pass
169A Los Olivos
169B Zaca Creek
169C Solvang
169D Santa Ynez
170A Los Alamos
170B Lompoc
170C Lompoc Hills
170D Santa Rosa Hills
171A Surf
171C Point Arguello
171D Tranquillon Mtn.
172B Needles NW
172C Needles SW
172D Needles
173A Bannock
173B Homer
173C West Of Flattop Mtn.
173D Flattop Mtn.
174A Goffs
174B Fenner Hills
174C Fenner
174D Fenner Spring
175A Desert Spring
175B Colton Well
175C West Of Blind Hills
175D Blind Hills
176A Fountain Peak
176B Kelso Dunes
176C Bighorn Basin
176D Van Winkle Spring
177A Glasgow
177B West Of Glasgow
177C West Of Budweiser Wash
177D Budweiser Wash
178A Broadwell Mesa
178B West Of Broadwell Mesa
178C Broadwell Lake
178D East Of Broadwell Lake
179A Hidden Valley East
179B Hidden Valley West
179C Hector
179D Sleeping Beauty
180A Manix
180B Harvard Hill
180C Newberry Springs
180D Troy Lake
181A Yermo
181B Nebo
181C Daggett
181D Minneola
182A Barstow
182B Hinkley
182C Hodge
182D Barstow SE
183A Twelve Gauge Lake
183B Kramer Hills
183C Astley Rancho
183D Wild Crossing
184A Kramer Junction
184B Leuhman Ridge
184C Jackrabbit Hill
184D Red Buttes
185A Rogers Lake North
185B Edwards
185C Redman
185D Rogers Lake South
186A Bissell
186B Soledad Mtn.
186C Rosamond
186D Rosamond Lake
187A Willow Springs
187B Tylerhorse Canyon
187C Fairmont Butte
187D Little Buttes
188A Liebre Twins
188B Winters Ridge
188C La Liebre Ranch
188D Neenach School
189A Pastoria Creek
189B Grapevine
189C Frazier Mtn.
189D Lebec
190A Pleito Hills
190B Eagle Rest Peak
190C Sawmill Mountain
190D Cuddy Valley
191A Santiago Creek
191B Ballinger Canyon
191C Cuyama Peak
191D Apache Canyon
192A Cuyama
192B New Cuyama
192C Salisbury Potrero
192D Fox Mountain
193A Peak Mountain
193B Bates Canyon
193C Bald Mountain
193D Hurricane Deck
194A Manzanita Mtn.
194B Tepusquet Canyon
194C Foxen Canyon
194D Zaca Lake
195A Twitchell Dam
195B Santa Maria
195C Orcutt
195D Sisquoc
196A Guadalupe
196B Point Sal
196D Casmalia
197C Mt. Manchester
198A Juniper Mine
198B West Of Juniper Mine
198C Homer Mtn.
198D East Of Homer Mtn.
199A East Of Grotto Hills
199B Grotto Hills
199C Hackberry Mountain
199D Signal Hill
200A Pinto Valley
200B Mid Hills
200C Columbia Mtn.
200D Woods Mountains

201A	Cima
201B	Marl Mountains
201C	Kelso
201D	Hayden
202A	Indian Spring
202B	Seventeenmile Point
202C	Cowhole Mountain
202D	Old Dad Mtn.
203A	Soda Lake North
203B	West Of Soda Lake
203C	Crucero Hill
203D	Soda Lake South
204A	Cronese Lakes
204B	Bitter Spring
204C	Dunn
204D	Cave Mountain
205A	East Of Langford Well
205B	Langford Well
205C	Alvord Mtn. West
205D	Alvord Mtn. East
206A	Paradise Range
206B	Williams Well
206C	Lane Mountain
206D	Coyote Lake
207A	Superior Lake
207B	Opal Mountain
207C	Water Valley
207D	Mud Hills
208A	Bird Spring
208B	Fremont Peak
208C	The Buttes
208D	Lockhart
209A	Boron NE
209B	Boron NW
209C	Boron
209D	Saddleback Mtn.
210A	Galileo Hill
210B	California City North
210C	California City South
210D	North Edwards
211A	Mojave NE
211B	Cache Peak
211C	Mojave
211D	Sanborn
212A	Tehachapi NE
212B	Tehachapi North
212C	Tehachapi South
212D	Monolith
213A	Keene
213B	Bear Mountain
213C	Tejon Ranch
213D	Cummings Mtn.
214A	Arvin
214B	Weed Patch
214C	Mettler
214D	Tejon Hills
215A	Conner
215B	Millux
215C	Conner SW
215D	Coal Oil Canyon
216A	Mouth Of Kern (=Buena Vista Lake Bed)
216B	Taft
216C	Maricopa
216D	Pentland
217A	Fellows
217B	Panorama Hills
217C	Wells Ranch
217D	Elkhorn Hills
218A	Painted Rock
218B	Chimineas Ranch
218C	Taylor Canyon
218D	Caliente Mtn.
219A	Branch Mtn.
219B	Los Machos Hills
219C	Chimney Canyon
219D	Miranda Pine Mtn.
220A	Caldwell Mesa
220B	Tar Spring Ridge
220C	Nipomo
220D	Huasna Peak
221A	Arroyo Grande NE
221B	Pismo Beach
221D	Oceano
222A	Port San Luis
223C	Tenmile Well
224A	Hopps Well
224B	Crescent Peak
224C	Castle Peaks
224D	Hart Peak
225A	Nipton
225B	Mineral Hill
225C	Joshua
225D	Ivanpah
226A	Mescal Range
226B	Valley Wells
226C	Cow Cove
226D	Cima Dome
227A	Solomons Knob
227B	Turquoise Mtn.
227C	Halloran Springs
227D	Granite Spring
228A	North Of Baker
228B	Silurian Valley
228C	West Of Baker
228D	Baker
229A	Red Pass Lake NE
229B	Red Pass Lake NW
229C	West Of Red Pass Lake
229D	Red Pass Lake
230A	Drinkwater Lake
230B	West Of Drinkwater Lake
230C	Fort Irwin
230D	Tiefort Mountains
231A	Nelson Lake
231B	West Of Nelson Lake
231C	Goldstone
231D	East Of Goldstone
232A	Eagle Crags
232B	Pilot Knob
232C	Slocum Mtn.
232D	Superior Valley
233A	Black Hills
233B	West Of Black Hills
233C	Cuddeback Lake
233D	Blackwater Well
234A	Klinker Mtn.
234B	El Paso Peaks
234C	Johannesburg
234D	Red Mountain
235A	Garlock
235B	Saltdale NW
235C	Cantil
235D	Saltdale SE
236A	Dove Spring
236B	Pinyon Mtn.
236C	Cross Mountain
236D	Cinco
237A	Claraville
237B	Piute Peak
237C	Loraine
237D	Emerald Mtn.
238A	Breckenridge Mtn.
238B	Mt. Adelaide
238C	Bena
238D	Oiler Peak
239A	Rio Bravo Ranch
239B	Oil Center
239C	Lamont
239D	Edison
240A	Oildale
240B	Rosedale
240C	Stevens
240D	Gosford
241A	Rio Bravo
241B	Buttonwillow
241C	East Elk Hills
241D	Tupman
242A	Lokern
242B	Belridge
242C	Reward
242D	West Elk Hills
243A	Carneros Rocks
243B	Las Yeguas Ranch
243C	Simmler
243D	McKittrick Summit
244A	La Panza NE
244B	La Panza Ranch
244C	La Panza
244D	California Valley
245A	Camatta Ranch
245B	Wilson Corner
245C	Santa Margarita Lake
245D	Pozo Summit
246A	Santa Margarita
246B	Atascadero
246C	San Luis Obispo
246D	Lopez Mtn.
247A	Morro Bay North
247B	Cayucos
247D	Morro Bay South
248B	State Line Pass
248C	Ivanpah Lake
248D	Desert
249A	Mesquite Lake
249B	Mesquite Mountains
249C	Pachalka Spring
249D	Clark Mtn.
250A	East Of Kingston Peak
250B	Kingston Peak
250C	Kingston Spring
250D	East Of Kingston Spring
251A	Valjean Hills
251B	Dumont Dunes
251C	Silurian Lake
251D	Silurian Hills
252A	Saddle Peak Hills
252B	Old Ibex Pass
252C	Avawatz Pass
252D	Sheep Creek Spring
253A	East Of Owl Lake
253B	Owl Lake
253C	Leach Lake
253D	East Of Leach Lake
254A	Quail Spring
254B	Hidden Spring
254C	West Of Leach Spring
254D	Leach Spring
255A	Wingate Pass
255B	Layton Spring
255C	Pilot Knob Valley West
255D	Pilot Knob Valley East
256A	Searles Lake
256B	Westend
256C	Spangler Hills East
256D	Christmas Canyon
257A	Lone Butte
257B	Ridgecrest North
257C	Ridgecrest South
257D	Spangler Hills West
258A	Inyokern
258B	Owens Peak
258C	Freeman Junction
258D	Inyokern SE
259A	Walker Pass
259B	Onyx
259C	Cane Canyon
259D	Horse Canyon
260A	Weldon
260B	Lake Isabella North
260C	Lake Isabella South
260D	Woolstalf Creek
261A	Alta Sierra
261B	Glennville
261C	Democrat Hot Springs
261D	Miracle Hot Springs
262A	Woody
262B	Sand Canyon
262C	Knob Hill
262D	Pine Mountain
263A	Deepwell Ranch
263B	McFarland
263C	Famoso
263D	North Of Oildale
264A	Pond
264B	Wasco NW
264C	Wasco SW
264D	Wasco
265A	Lost Hills NE
265B	Lost Hills NW
265C	Lost Hills
265D	Semitropic
266A	Antelope Plain
266B	Emigrant Hill
266C	Shale Point
266D	Blackwells Corner
267A	Sawtooth Ridge
267B	Orchard Peak
267C	Holland Canyon
267D	Packwood Creek
268A	Cholame
268B	Shandon
268C	Shedd Canyon
268D	Camatta Canyon

269A Estrella
269B Paso Robles
269C Templeton
269D Creston
270A Adelaida
270B Lime Mountain
270C Cypress Mountain
270D York Mountain
271A Pebblestone Shut-in
271B San Simeon
271C Pico Creek
271D Cambria
272A Piedras Blancas
273B Green Monster Mine
273C West Of Shenandoah Peak
273D Shenandoah Peak
274A Stump Spring, Nev.
274B Calvada Springs
274C Horse Thief Springs
274D Blackwater Mine
275A North Of Tecopa Pass
275B Resting Spring
275C Tecopa
275D Tecopa Pass
276A Shoshone
276B Salsberry Peak
276C Ibex Spring
276D Ibex Pass
277A Epaulet Peak
277B Shore Line Butte
277C Confidence Hills West
277D Confidence Hills East
278A Anvil Spring Canyon East
278B Anvil Spring Canyon West
278C Wingate Wash West
278D Wingate Wash East
279A Manly Peak
279B Manly Fall
279C Copper Queen Canyon
279D Sourdough Spring
280A Slate Range Crossing
280B Homewood Canyon
280C Trona West
280D Trona East
281A Mountain Springs Canyon
281B Airport Lake
281C White Hills
281D Burro Canyon
282A Volcano Peak
282B Little Lake
282C Ninemile Canyon
282D Pearsonville
283A Sacatar Canyon
283B Rockhouse Basin
283C White Dome
283D Lamont Peak
284A Sirretta Peak
284B Fairview
284C Kernville
284D Cannell Peak
285A Johnsondale
285B California Hot Springs
285C Posey
285D Tobias Peak
286A Gibbon Peak
286B Fountain Springs
286C Quincy School
286D White River
287A Ducor
287B Sausalito School
287C Delano East
287D Richgrove
288A Pixley
288B Alpaugh
288C Allensworth
288D Delano West
289A Hacienda Ranch NE
289B Hacienda Ranch NW
289C Lone Tree Well
289D Hacienda Ranch
290A Dudley Ridge
290B Los Viejos
290C Avenal Gap
290D West Camp
291A Kettleman Plain
291B Garza Peak
291C Tent Hills
291D Pyramid Hills
292A The Dark Hole
292B Parkfield
292C Cholame Hills
292D Cholame Valley
293A Stockdale Mountain
293B Valleton
293C San Miguel
293D Ranchito Canyon
294A Wunpost
294B Hames Valley
294C Tierra Redonda Mountain
294D Bradley
295A Williams Hill
295B Jolon
295C Burnett Peak
295D Bryson
296A Alder Peak
296B Cape San Martin
296C Villa Creek
296D Burro Mountain
297C Mound Spring
298A Sixmile Spring
298B Stewart Valley
298C Twelvemile Spring
298D Nopah Peak
299A Eagle Mtn.
299B West Of Eagle Mtn.
299C Deadman Pass
299D East Of Deadman Pass
300A Greenwater Canyon
300B Dantes View
300C Gold Valley
300D Funeral Peak
301A Badwater
301B Hanaupah Canyon
301C Galena Canyon
301D Mormon Point
302A Telescope Peak
302B Jail Canyon
302C Ballarat
302D Panamint
303A Maturango Peak NE
303B Revenue Canyon
303C Maturango Peak
303D Maturango Peak SE
304A China Gardens
304B Coso Peak
304C Petroglyph Canyon
304D Louisiana Butte
305A Upper Centennial Flat
305B Haiwee Reservoirs
305C Coso Junction
305D Cactus Peak
306A Haiwee Pass
306B Monache Mountain
306C Crag Peak
306D Long Canyon
307A Casa Vieja Meadows
307B Hockett Peak
307C Durrwood Creek
307D Bonita Meadows
308A Camp Nelson
308B Camp Wishon
308C Solo Peak
308D Sentinel Peak
309A Springville
309B Frazier Valley
309C Success Dam
309D Globe
310A Lindsay
310B Cairns Corner
310C Woodville
310D Porterville
311A Tulare
311B Paige
311C Taylor Weir
311D Tipton
312A Waukena
312B Guernsey
312C El Rico Ranch
312D Corcoran
313A Stratford
313B Westhaven
313C Kettleman City
313D Stratford SE
314A Huron
314B Guijarral Hills
314C Avenal
314D La Cima
315A Coalinga
315B Alcalde Hills
315C Curry Mountain
315D Kreyenhagen Hills
316A Sherman Peak
316B Priest Valley
316C Slack Canyon
316D Smith Mountain
317A Monarch Peak
317B Nattrass Valley
317C San Ardo
317D Pancho Rico Valley
318A San Lucas
318B Thompson Canyon
318C Cosio Knob
318D Espinosa Canyon
319A Reliz Canyon
319B Junipero Serra Peak
319C Cone Peak
319D Bear Canyon
320A Tassajara Hot Springs
320B Partington Ridge
320D Lopez Point
320E Pfeiffer Point
321C High Peak
322B Franklin Well
322C Death Valley Junction
322D Bole Spring
323A East Of Echo Canyon
323B Echo Canyon
323C Ryan
323D East Of Ryan
324A Furnace Creek
324B West Of Furnace Creek
324C Devils Speedway
324D Devils Golf Course
325A Tucki Wash
325B Emigrant Canyon
325C Emigrant Pass
325D Wildrose Peak
326A Panamint Butte
326B The Dunes
326C Panamint Springs
326D Nova Canyon
327A Lee Wash
327B Santa Rosa Flat
327C Talc City Hills
327D Darwin
328A Keeler
328B Owens Lake
328C Vermillion Canyon
328D Centennial Canyon
329A Bartlett
329B Cirque Peak
329C Templeton Mtn.
329D Olancha
330A Johnson Peak
330B Chagoopa Falls
330C Kern Lake
330D Kern Peak
331A Mineral King
331B Silver City
331C Moses Mtn.
331D Quinn Peak
332A Case Mountain
332B Kaweah
332C Chickencoop Canyon
332D Dennison Peak
333A Woodlake
333B Ivanhoe
333C Exeter
333D Rocky Hill
334A Monson
334B Traver
334C Goshen
334D Visalia
335A Burris Park
335B Laton
335C Hanford
335D Remnoy
336A Riverdale
336B Burrel
336C Vanguard
336D Lemoore
337A Five Points
337B Westside
337C Harris Ranch

337D Calflax
338A Tres Pecos Farms
338B Lillis Ranch
338C Joaquin Rocks
338D Domengine Ranch
339A Ciervo Mtn.
339B Idria
339C San Benito Mtn.
339D Santa Rita Peak
340A Hernandez Reservoir
340B Rock Spring Peak
340C Lonoak
340D Hepsedam Peak
341A Topo Valley
341B North Chalone Peak
341C Greenfield
341D Pinalito Canyon
342A Soledad
342B Palo Escrito Peak
342C Sycamore Flat
342D Paraiso Springs
343A Rana Creek
343B Carmel Valley
343C Ventana Cones
343D Chews Ridge
344A Mt. Carmel
344B Soberanes Point
344C Point Sur
344D Big Sur
345B Ashton
345C Lees Camp
345D Leeland
346A East Of Chloride City
346B Chloride City
346C Beatty Junction
346D Nevares Peak
347A Stovepipe Wells NE
347B Mesquite Flat
347C Stovepipe Wells
347D Grotto Canyon
348A East Of Sand Flat
348B Sand Flat
348C Harris Hill
348D Cottonwood Canyon
349A Ubehebe Peak
349B West Of Ubehebe Peak
349C Nelson Range
349D Jackass Canyon
350A Craig Canyon
350B New York Butte
350C Dolomite
350D Cerro Gordo Peak
351A Union Wash
351B Manzanar
351C Mt. Langley
351D Lone Pine
352A Mt. Williamson
352B Mt. Brewer
352C Mt. Kaweah
352D Mount Whitney
353A Sphinx Lakes
353B Mt. Silliman
353C Lodgepole
353D Triple Divide Peak
354A Muir Grove
354B General Grant Grove
354C Shadequarter Mtn.
354D Giant Forest
355A Miramonte
355B Tucker Mtn.
355C Stokes Mtn.
355D Auckland
356A Orange Cove North
356B Wahtoke
356C Reedley
356D Orange Cove South
357A Sanger
357B Malaga
357C Conejo
357D Selma
358A Fresno South
358B Kearney Park
358C Raisin
358D Caruthers
359A Kerman
359B Jamesan
359C San Joaquin
359D Helm
360A Tranquillity
360B Coit Ranch
360C Levis
360D Cantua Creek
361A Chaney Ranch
361B Chounet Ranch
361C Tumey Hills
361D Monocline Ridge
362A Mercy Hot Springs
362B Cerro Colorado
362C Llanada
362D Panoche
363A Panoche Pass
363B Cherry Peak
363C Bickmore Canyon
363D San Benito
364A Paicines
364B Mt. Harlan
364C Gonzales
364D Mount Johnson
365A Natividad
365B Salinas
365C Spreckels
365D Chualar
366A Marina
366C Monterey
366D Seaside
367C Daylight Pass
367D Gold Center
368A Wahguyhe Peak
368B Grapevine Peak
368C Fall Canyon
368D Thimble Peak
369A East Of Tin Mountain
369B Tin Mountain
369C White Top Mtn.
369D Dry Bone Canyon
370A Dry Mountain
370B Saline Peak
370C West Of Teakettle Junction
370D Teakettle Junction
371A East Of Waucoba Canyon
371B Waucoba Canyon
371C Pat Keyes Canyon
371D Lower Warm Springs
372A Mazourka Peak
372B Blackrock
372C Independence
372D Bee Springs Canyon
373A Aberdeen
373B Mt. Pinchot
373C Mt. Clarence King
373D Kearsarge Peak
374A Marion Peak
374B Slide Bluffs
374C Cedar Grove
374D The Sphinx
375A Tehipite Dome
375B Rough Spur
375C Hume
375D Wren Peak
376A Patterson Mtn.
376B Sacate Ridge
376C Luckett Mtn.
376D Verplank Ridge
377A Trimmer
377B Humphreys Station
377C Piedra
377D Pine Flat Dam
378A Academy
378B Friant
378C Clovis
378D Round Mountain
379A Lanes Bridge
379B Gregg
379C Herndon
379D Fresno North
380A Madera
380B Bonita Ranch
380C Gravelly Ford
380D Biola
381A Firebaugh NE
381B Poso Farm
381C Firebaugh
381D Mendota Dam
382A Oxalis
382B Dos Palos
382C Hammonds Ranch
382D Broadview Farms
383A Charleston School
383B Ortigalita Peak NW
383C Ortigalita Peak
383D Laguna Seca Ranch
384A Los Banos Valley
384B Mariposa Peak
384C Quien Sabe Valley
384D Ruby Canyon
385A Three Sisters
385B San Felipe
385C Hollister
385D Tres Pinos
386A Chittenden
386B Watsonville East
386C Prunedale
386D San Juan Bautista
387A Watsonville West
387B Soquel
387D Moss Landing
387E Santa Cruz
388C Bonnie Claire SW
389A Gold Mountain
389B West Of Gold Mtn.
389C Ubehebe Crater
389D Scottys Castle
390A Sand Spring
390B Hanging Rock Canyon
390C Last Chance Range SW
390D Last Chance Range SE
391A East Of Joshua Flats
391B Joshua Flats
391C Waucoba Spring
391D East Of Waucoba Spring
392A Cowhorn Valley
392B Uhlmeyer Spring
392C Tinemaha Reservoir
392D Waucoba Mtn.
393A Big Pine
393B Coyote Flat
393C Split Mtn.
393D Fish Springs
394A Mt. Thompson
394B Mt. Darwin
394C Mt. Goddard
394D North Palisade
395A Mt. Henry
395B Ward Mountain
395C Courtright Reservoir
395D Blackcap Mtn.
396A Dogtooth Peak
396B Huntington Lake
396C Dinkey Creek
396D Nelson Mtn.
397A Musick Mtn.
397B Cascadel Point
397C Auberry
397D Shaver Lake
398A North Fork
398B O'Neals
398C Millerton Lake West
398D Millerton Lake East
399A Knowles
399B Raymond
399C Daulton
399D Little Table Mtn.
400A Raynor Creek
400B Le Grand
400C Berenda
400D Kismet
401A Plainsburg
401B El Nido
401C Bliss Ranch
401D Chowchilla
402A Sandy Mush
402B Turner Ranch
402C Delta Ranch
402D Santa Rita Bridge
403A San Luis Ranch
403B Ingomar
403C Volta
403D Los Banos
404A Howard Ranch
404B Crevison Peak
404C Pacheco Pass
404D San Luis Dam
405A Mustang Peak
405B Mississippi Creek

405C Gilroy Hot Springs
405D Pacheco Peak
406A Mt. Sizer
406B Morgan Hill
406C Mt. Madonna
406D Gilroy
407A Santa Teresa Hills
407B Los Gatos
407C Laurel
407D Loma Prieta
408A Castle Rock Ridge
408B Big Basin
408C Davenport
408D Felton
409A Franklin Point
409B Pigeon Point
409D Ano Nuevo
410B Sylvania Mts.
410C Last Chance Mtn.
410D Tule Canyon
411A Sylvania Canyon
411B Chocolate Mtn.
411C Soldier Pass
411D Horse Thief Canyon
412A Crooked Creek
412B Blanco Mtn.
412C Westgard Pass
412D Deep Springs Lake
413A Laws
413B Fish Slough
413C Bishop
413D Poleta Canyon
414A Rovana
414B Mt. Morgan
414C Mount Tom
414D Tungsten Hills
415A Mt. Abbot
415B Graveyard Peak
415C Florence Lake
415D Mt. Hilgard
416A Sharktooth Peak
416B Balloon Dome
416C Kaiser Peak
416D Mt. Givens
417A Squaw Dome
417B Little Shuteye Peak
417C Shuteye Peak
417D Mammoth Pool Dam
418A White Chief Mtn.
418B Fish Camp
418C Bass Lake
418D Ahwahnee
419A Stumpfield Mtn.
419B Mariposa
419C Ben Hur
419D Horsecamp Mountain
420A Catheys Valley
420B Indian Gulch
420C Owens Reservoir
420D Illinois Hill
421A Haystack Mtn.
421B Yosemite Lake
421C Merced
421D Planada
422A Winton
422B Cressey
422C Arena
422D Atwater
423A Turlock
423B Hatch
423C Gustine
423D Stevinson
424A Crows Landing
424B Patterson
424C Orestimba Peak
424D Newman
425A Copper Mtn.
425B Mt. Boardman
425C Mt. Stakes
425D Wilcox Ridge
426A Eylar Mtn.
426B Mt. Day
426C Lick Observatory
426D Isabel Valley
427A Calaveras Reservoir
427B Milpitas
427C San Jose West
427D San Jose East
428A Mountain View
428B Palo Alto
428C Mindego Hill
428D Cupertino
429A Woodside
429B Half Moon Bay
429C San Gregorio
429D La Honda
430C Indian Garden Creek
431A Dyer
431B Juniper Mtn.
431C Mt. Barcroft
431D Station Peak
432A White Mtn. Peak
432B Hammil Valley
432C Chidago Canyon
432D Chalfant Valley
433A Banner Ridge
433B Watterson Canyon
433C Toms Place
433D Casa Diablo Mtn.
434A Whitmore Hot Springs
434B Old Mammoth
434C Bloody Mtn.
434D Convict Lake
435A Mammoth Mtn.
435B Mt. Ritter
435C Cattle Mtn.
435D Crystal Crag
436A Mt. Lyell
436B Merced Peak
436C Sing Peak
436D Timber Knob
437A Half Dome
437B El Capitan
437C Wawona
437D Mariposa Grove
438A El Portal
438B Kinsley
438C Feliciana Mtn.
438D Buckingham Mtn.
439A Buckhorn Peak
439B Coulterville
439C Hornitos
439D Bear Valley
440A Penon Blanco Peak
440B La Grange
440C Snelling
440D Merced Falls
441A Cooperstown
441B Paulsell
441C Montpelier
441D Turlock Lake
442A Waterford
442B Riverbank
442C Ceres
442D Denair
443A Salida
443B Ripon
443C Westley
443D Brush Lake
444A Vernalis
444B Tracy
444C Lone Tree Creek
444D Solyo
445A Midway
445B Altamont
445C Mendenhall Springs
445D Cedar Mtn.
446A Livermore
446B Dublin
446C Niles
446D La Costa Valley
447A Hayward
447B San Leandro
447C Redwood Point
447D Newark
448A Hunters Point
448B San Francisco South
448C Montara Mountain
448D San Mateo
449C Davis Mountain
450B Truman Meadows
450C Benton
450D Boundary Peak
451A River Spring
451B Indian Meadows
451C Glass Mountain
451D Benton Hot Springs
452A Cowtrack Mountain
452B Mono Mills
452C Crestview
452D Dexter Canyon
453A Lee Vining
453B Mount Dana
453C Koip Peak
453D June Lake
454A Tioga Pass
454B Falls Ridge
454C Tenaya Lake
454D Vogelsang Peak
455A Ten Lakes
455B Hetch Hetchy Reservoir
455C Tamarack Flat
455D Yosemite Falls
456A Lake Eleanor
456B Cherry Lake South
456C Ascension Mtn.
456D Ackerson Mtn.
457A Duckwall Mtn.
457B Tuolumne
457C Groveland
457D Jawbone Ridge
458A Standard
458B Sonora
458C Chinese Camp
458D Moccasin
459A New Melones Dam
459B Copperopolis
459C Knights Ferry
459D Keystone
460A Bachelor Valley
460B Farmington
460C Escalon
460D Oakdale
461A Peters
461B Stockton East
461C Manteca
461D Avena
462A Stockton West
462B Holt
462C Union Island
462D Lathrop
463A Woodward Island
463B Brentwood
463C Byron Hot Springs
463D Clifton Court Forebay
464A Antioch South
464B Clayton
464C Diablo
464D Tassajara
465A Walnut Creek
465B Briones Valley
465C Oakland East
465D Las Trampas Ridge
466A Richmond
466B San Quentin
466C San Francisco North
466D Oakland West
467A San Rafael
467B Bolinas
467D Point Bonita
467E Double Point
468B Anchorite Hills
468C West Of Huntoon Spring
468D Huntoon Spring
469A Cedar Hill
469B Kirkwood Spring
469C Sulphur Pond
469D Alameda Well
470A Bodie
470B Big Alkali
470C Lundy
470D Negit Island
471A Twin Lakes
471B Buckeye Ridge
471C Matterhorn Peak
471D Dunderberg Peak
472A Tower Peak
472B Emigrant Lake
472C Tiltill Mtn.
472D Piute Mtn.
473A Cooper Peak
473B Pinecrest
473C Cherry Lake North
473D Kibbie Lake

474A Strawberry
474B Crandall Peak
474C Twain Harte
474D Hull Creek
475A Stanislaus
475B Murphys
475C Columbia
475D Columbia SE
476A Calaveritas
476B San Andreas
476C Salt Spring Valley
476D Angels Camp
477A Valley Springs
477B Wallace
477C Valley Springs SW
477D Jenny Lind
478A Clements
478B Lockeford
478C Waterloo
478D Linden
479A Lodi North
479B Thornton
479C Terminous
479D Lodi South
480A Isleton
480B Rio Vista
480C Jersey Island
480D Bouldin Island
481A Birds Landing
481B Denverton
481C Honker Bay
481D Antioch North
482A Fairfield South
482B Cordelia
482C Benicia
482D Vine Hill
483A Cuttings Wharf
483B Sears Point
483C Petaluma Point
483D Mare Island
484A Petaluma River
484B Petaluma
484C San Geronimo
484D Novato
485A Point Reyes NE
485B Tomales
485C Drakes Bay
485D Inverness
486C Aurora
487A Conway Stage Station
487B Sweetwater Creek
487C Bridgeport
487D Dome Hill
488A Mt. Patterson
488B Chris Flat
488C Fales Hot Springs
488D Mt. Jackson
489A Lost Cannon Peak
489B Disaster Peak
489C Sonora Pass
489D Pickel Meadow
490A Dardanelles Cone
490B Spicer Meadows Res.
490C Donnell Lake
490D Dardanelle
491A Tamarack
491B Calaveras Dome
491C Boards Crossing
491D Liberty Hill
492A Garnet Hill
492B Devils Nose
492C Fort Mountain
492D Dorrington
493A West Point
493B Pine Grove
493C Mokelumne Hill
493D Rail Road Flat
494A Amador City
494B Irish Hill
494C Ione
494D Jackson
495A Carbondale
495B Sloughhouse
495C Clay
495D Goose Creek
496A Elk Grove
496B Florin
496C Bruceville
496D Galt
497A Clarksburg
497B Saxon
497C Liberty Island
497D Courtland
498A Dixon
498B Allendale
498C Elmira
498D Dozier
499A Mt. Vaca
499B Capell Valley
499C Mt. George
499D Fairfield North
500A Yountville
500B Rutherford
500C Sonoma
500D Napa
501A Kenwood
501B Santa Rosa
501C Cotati
501D Glen Ellen
502A Sebastopol
502B Camp Meeker
502C Valley Ford
502D Two Rock
503A Duncans Mills
503B Arched Rock
503D Bodega Head
504B Long Dry Canyon
504C Risue Canyon
504D Desert Creek Peak
505A Topaz Lake
505B Heenan Lake
505C Wolf Creek
505D Coleville
506A Markleeville
506B Carson Pass
506C Pacific Valley
506D Ebbetts Pass
507A Caples Lake
507B Tragedy Spring
507C Bear River Reservoir
507D Mokelumne Peak
508A Leek Spring Hill
508B Stump Spring, Calif. (= Old Iron Mountain)
508C Caldor
508D Peddler Hill
509A Sly Park
509B Camino
509C Aukum
509D Omo Ranch
510A Placerville
510B Shingle Springs
510C Latrobe
510D Fiddletown
511A Clarksville
511B Folsom
511C Buffalo Creek
511D Folsom SE
512A Citrus Heights
512B Rio Linda
512C Sacramento East
512D Carmichael
513A Taylor Monument
513B Grays Bend
513C Davis
513D Sacramento West
514A Woodland
514B Madison
514C Winters
514D Merritt
515A Esparto
515B Brooks
515C Lake Berryessa
515D Monticello Dam
516A Walter Springs
516B Aetna Springs
516C St. Helena
516D Chiles Valley
517A Detert Reservoir
517B Mount St. Helena
517C Mark West Springs
517D Calistoga
518A Jimtown
518B Geyserville
518C Guerneville
518D Healdsburg
519A Warm Springs Dam
519B Tombs Creek
519C Fort Ross
519D Cazadero
520A Annapolis
520B Stewarts Point
520D Plantation
521C Carters Station
522A Minden
522B South Lake Tahoe
522C Freel Peak
522D Woodfords
523A Emerald Bay
523B Rockbound Valley
523C Pyramid Peak
523D Echo Lake
524A Loon Lake
524B Robbs Peak
524C Riverton
524D Kyburz
525A Devil Peak
525B Tunnel Hill
525C Slate Mtn.
525D Pollack Pines
526A Georgetown
526B Greenwood
526C Coloma
526D Garden Valley
527A Auburn
527B Gold Hill
527C Rocklin
527D Pilot Hill
528A Lincoln
528B Sheridan
528C Pleasant Grove
528D Roseville
529A Nicolaus
529B Sutter Causeway
529C Knights Landing
529D Verona
530A Kirkville
530B Dunnigan
530C Zamora
530D Eldorado Bend
531A Wildwood School
531B Rumsey
531C Guinda
531D Bird Valley
532A Glascock Mtn.
532B Wilson Valley
532C Jericho Valley
532D Knoxville
533A Lower Lake
533B Clearlake Highlands
533C Whispering Pines
533D Middletown
534A Kelseyville
534B Highland Springs
534C Asti
534D The Geysers
535A Hopland
535B Yorkville
535C Big Foot Mtn.
535D Cloverdale
536A Ornbaun Valley
536B Zeni Ridge
536C McGuire Ridge
536D Gube Mountain
537A Eureka Hill
537B Point Arena
537C Saunders Reef
537D Gualala
538A Kings Beach
538B Tahoe City
538C Homewood
538D Meeks Bay
539A Granite Chief
539B Royal Gorge
539C Bunker Hill
539D Wentworth Springs
540A Duncan Peak
540B Westville
540C Michigan Bluff
540D Greek Store
541A Dutch Flat
541B Chicago Park
541C Colfax
541D Foresthill

Map	Name
542A	Grass Valley
542B	Rough And Ready
542C	Wolf
542D	Lake Combie
543A	Smartville
543B	Browns Valley
543C	Wheatland
543D	Camp Far West
544A	Yuba City
544B	Sutter
544C	Gilsizer Slough
544D	Olivehurst
545A	Sutter Buttes
545B	Meridian
545C	Grimes
545D	Tisdale Weir
546A	Colusa
546B	Williams
546C	Cortina Creek
546D	Arbuckle
547A	Manor Slough
547B	Leesville
547C	Wilbur Springs
547D	Salt Canyon
548A	Hough Springs
548B	Bartlett Springs
548C	Clearlake Oaks
548D	Benmore Canyon
549A	Bartlett Mtn.
549B	Upper Lake
549C	Lakeport
549D	Lucerne
550A	Cow Mountain
550B	Ukiah
550C	Elledge Peak
550D	Purdys Gardens
551A	Orrs Springs
551B	Bailey Ridge
551C	Philo
551D	Boonville
552A	Navarro
552B	Elk
552C	Mallo Pass Creek
552D	Cold Spring
553A	Albion
554A	Boca
554B	Hobart Mills
554C	Truckee
554D	Martis Peak
555A	Independence Lake
555B	Webber Peak
555C	Soda Springs
555D	Norden
556A	English Mtn.
556B	Graniteville
556C	Blue Canyon
556D	Cisco Grove
557A	Alleghany
557B	Pike
557C	North Bloomfield
557D	Washington
558A	Camptonville
558B	Challenge
558C	French Corral
558D	Nevada City
559A	Rackerby
559B	Bangor
559C	Loma Rica
559D	Oregon House
560A	Palermo
560B	Biggs
560C	Gridley
560D	Honcut
561A	West Of Biggs
561B	Butte City
561C	Sanborn Slough
561D	Pennington
562A	Princeton
562B	Logandale
562C	Maxwell
562D	Moulton Weir
563A	Logan Ridge
563B	Rail Canyon
563C	Lodoga
563D	Sites
564A	Stonyford
564B	St. John Mtn.
564C	Fouts Springs
564D	Gilmore Peak
565A	Crockett Peak
565B	Lake Pillsbury
565C	Elk Mountain
565D	Potato Hill
566A	Van Arsdale Reservoir
566B	Foster Mtn.
566C	Redwood Valley
566D	Potter Valley
567A	Willits
567B	Burbeck
567C	Greenough Ridge
567D	Laughlin Range
568A	Northspur
568B	Noyo Hill
568C	Mathison Peak
568D	Comptche
569A	Fort Bragg
569D	Mendocino
570A	Evans Canyon
570B	Loyalton
570C	Sardine Peak
570D	Dog Valley
571A	Antelope Valley
571B	Calpine
571C	Sattley
571D	Sierraville
572A	Clio
572B	Gold Lake
572C	Sierra City
572D	Haypress Valley
573A	Mt. Fillmore
573B	La Porte
573C	Goodyears Bar
573D	Downieville
574A	American House
574B	Cascade
574C	Clipper Mills
574D	Strawberry Valley
575A	Brush Creek
575B	Berry Creek
575C	Oroville Dam
575D	Forbestown
576A	Cherokee
576B	Hamlin Canyon
576C	Shippee
576D	Oroville
577A	Chico
577B	Ord Ferry
577C	Llano Seco
577D	Nelson
578A	Hamilton City
578B	Orland
578C	Willows
578D	Glenn
579A	Fruto NE
579B	Julian Rocks
579C	Fruto
579D	Stone Valley
580A	Chrome
580B	Alder Springs
580C	Felkner Hill
580D	Elk Creek
581A	Plaskett Meadows
581B	Plaskett Ridge
581C	Hull Mountain
581D	Kneecap Ridge
582A	Thatcher Ridge
582B	Jamison Ridge
582C	Brushy Mtn.
582D	Sanhedrin Mtn.
583A	Dos Rios
583B	Laytonville
583C	Longvale
583D	Willis Ridge
584A	Cahto Peak
584B	Lincoln Ridge
584C	Dutchmans Knoll
584D	Sherwood Peak
585A	Westport
585D	Inglenook
586A	Constantia
586B	Frenchman Lake
586C	Chilcoot
586D	Beckwourth Pass
587A	Dixie Mountain
587B	Crocker Mtn.
587C	Portola
587D	Reconnaissance Peak
588A	Grizzly Valley
588B	Mt. Ingalls
588C	Johnsville
588D	Blairsden
589A	Spring Garden
589B	Quincy
589C	Onion Valley
589D	Blue Nose Mtn.
590A	Meadow Valley
590B	Bucks Lake
590C	Haskins Valley
590D	Dogwood Peak
591A	Storrie
591B	Kimshew Point
591C	Pulga
591D	Soapstone Hill
592A	Stirling City
592B	Cohasset
592C	Paradise West
592D	Paradise East
593A	Campbell Mound
593B	Richardson Springs NW
593C	Nord
593D	Richardson Springs
594A	Vina
594B	Corning
594C	Kirkwood
594D	Foster Island
595A	Henleyville
595B	Flournoy
595C	Sehorn Creek
595D	Black Butte Dam
596A	Paskenta
596B	Riley Ridge
596C	Hall Ridge
596D	Newville
597A	Ball Mountain
597B	Buck Rock
597C	Mendocino Pass
597D	Log Spring
598A	Leech Lake Mtn.
598B	Bluenose Ridge
598C	Covelo East
598D	Newhouse Ridge
599A	Mina
599B	Updegraff Ridge
599C	Iron Peak
599D	Covelo West
600A	Bell Springs
600B	Noble Butte
600C	Leggett
600D	Tan Oak Park
601A	Piercy
601B	Bear Harbor
601C	Mistake Point
601D	Hales Grove
602A	Calneva Lake
602B	Herlong
602C	McKesick Peak
602D	Doyle
603A	Milford
603B	Stony Ridge
603C	Squaw Valley Peak
603D	Ferris Creek
604A	Antelope Lake
604B	Kettle Rock
604C	Genesee Valley
604D	Babcock Peak
605A	Moonlight Peak
605B	Greenville
605C	Crescent Mills
605D	Taylorsville
606A	Canyondam
606B	Almanor
606C	Caribou
606D	Twain
607A	Humbug Valley
607B	Humboldt Peak
607C	Jonesville
607D	Belden
608A	Onion Butte
608B	Barkley Mtn.
608C	Devils Parade Ground
608D	Butte Meadows
609A	Panther Spring
609B	Dewitt Peak
609C	Acorn Hollow

609D	Ishi Caves
610A	Tuscan Springs
610B	Red Bluff East
610C	Gerber
610D	Los Molinos
611A	Red Bluff West
611B	Blossom
611C	Red Bank
611D	West Of Gerber
612A	Oxbow Bridge
612B	Cold Fork
612C	Raglin Ridge
612D	Lowrey
613A	Yolla Bolly 15' NE (was Tomhead Mtn. 7.5')
613B	Yolla Bolly 15' NW (was North Yolla Bolly 7.5')
613C	Yolla Bolly 15' SW (was Solomon Peak 7.5')
613D	Yolla Bolly 15' SE (was South Yolla Bolly 7.5')
614A	Black Rock Mtn. 15' NE (was Black Rock Mtn. 7.5')
614B	Black Rock Mtn. 15' NW (was Swim Ridge 7.5')
614C	Black Rock Mtn. 15' SW (was Four Corners Rock 7.5')
614D	Black Rock Mtn. 15' SE (was Wrights Ridge 7.5')
615A	Shannon Butte
615B	Zenia
615C	Lake Mountain
615D	Long Ridge
616A	Alderpoint
616B	Fort Seward
616C	Harris
616D	Jewett Rock
617A	Miranda
617B	Ettersburg
617C	Briceland
617D	Garberville
618A	Honeydew
618B	Shubrick Peak
618D	Shelter Cove
619A	Cooskie Creek
620A	Bull Flat
620B	Little Mud Flat
620C	Wendel
620D	Spencer Creek
621A	Shaffer Mtn.
621B	Litchfield
621C	Standish
621D	Wendel Hot Springs
622A	Johnstonville
622B	Susanville
622C	Diamond Mtn.
622D	Janesville
623A	Roop Mountain
623B	Pegleg Mtn.
623C	Westwood East
623D	Fredonyer Pass
624A	Swain Mountain
624B	Red Cinder
624C	Chester
624D	Westwood West
625A	Mt. Harkness
625B	Reading Peak
625C	Childs Meadows
625D	Stover Mtn.
626A	Lassen Peak
626B	Grays Peak
626C	Lyonsville
626D	Mineral
627A	Manton
627B	Shingletown
627C	Inskip Hill
627D	Finley Butte
628A	Tuscan Buttes NE
628B	Balls Ferry
628C	Bend
628D	Dales
629A	Cottonwood
629B	Olinda
629C	Mitchell Gulch
629D	Hooker
630A	Ono 15' NE (was Ono 7.5')
630B	Ono 15' NW (was Tar Bully 7.5')
630C	Ono 15' SW (was Chickabally Mtn. 7.5')
630D	Ono 15' SE (was Rosewood 7.5')
631A	Chanchelulla Peak 15' NE (was Arbuckle Mtn. 7.5')
631B	Chanchelulla Peak 15' NW (was Chanchelulla Peak 7.5')
631C	Chanchelulla Peak 15' SW (was Platina 7.5')
631D	Chanchelulla Peak 15' SE (was Beegum 7.5')
632A	Dubakella Mtn. 15' NE (was Wildwood 7.5')
632B	Dubakella Mtn. 15' NW (was Dubakella Mtn. 7.5')
632C	Dubakella Mtn. 15' SW (was Smoky Creek 7.5')
632D	Dubakella Mtn. 15' SE (was Pony Buck Peak 7.5')
633A	Naufus Creek
633B	Sportshaven
633C	Ruth Reservoir
633D	Forest Glen
634A	Dinsmore
634B	Larabee Valley
634C	Blocksburg
634D	Black Lassic
635A	Bridgeville
635B	Redcrest
635C	Weott
635D	Myers Flat
636A	Scotia
636B	Taylor Peak
636C	Buckeye Mtn.
636D	Bull Creek
637A	Capetown
637B	Cape Mendocino
637D	Petrolia
638A	Al Shinn Canyon
638B	Shinn Mountain
638C	Five Springs
638D	Cherry Mtn.
639A	Snowstorm Mtn.
639B	West Of Snowstorm Mtn.
639C	Petes Valley
639D	Karlo
640A	Fredonyer Peak
640B	Troxel Point
640C	Gallatin Peak
640D	Tunnison Mtn.
641A	Spalding Tract
641B	Champs Flat
641C	Antelope Mtn.
641D	Pikes Point
642A	Harvey Mtn.
642B	Poison Lake
642C	Bogard Buttes
642D	Pine Creek Valley
643A	Swains Hole
643B	Old Station
643C	West Prospect Peak
643D	Prospect Peak
644A	Thousand Lakes Valley
644B	Jacks Backbone
644C	Viola
644D	Manzanita Lake
645A	Miller Mtn.
645B	Whitmore
645C	Inwood
645D	Hagaman Gulch
646A	Oak Run
646B	Bella Vista
646C	Palo Cedro
646D	Clough Gulch
647A	Project City
647B	Shasta Dam
647C	Redding
647D	Enterprise
648A	Whiskeytown
648B	French Gulch
648C	Shasta Bally
648D	Igo
649A	Lewiston
649B	Weaverville
649C	Hoosimbim Mtn.
649D	Bully Choop Mtn.
650A	Junction City
650B	Hayfork Bally
650C	Hayfork
650D	Hayfork Summit
651A	Big Bar
651B	Hyampom Mtn.
651C	Hyampom
651D	Halfway Ridge
652A	Sims Mountain
652B	Board Camp Mtn.
652C	Showers Mtn.
652D	Blake Mountain
653A	Mad River Buttes
653B	Iaqua Buttes
653C	Owl Creek
653D	Yager Junction
654A	McWhinney Creek
654B	Fields Landing
654C	Fortuna
654D	Hydesville
655A	Cannibal Island
655D	Ferndale
656A	Buckhorn Lake
656B	Dodge Reservoir
656C	Observation Peak
656D	Buckhorn Canyon
657A	Juniper Ridge
657B	McDonald Peak
657C	Termo
657D	Ravendale
658A	Anderson Mtn.
658B	Whitinger Mtn.
658C	Grasshopper Valley
658D	Cleghorn Flat
659A	Said Valley
659B	Silva Flat Reservoir
659C	Bullard Lake
659D	Sheepshead
660A	Dixie Peak
660B	Little Valley
660C	Corders Reservoir
660D	Straylor Lake
661A	Cable Mtn.
661B	Hogback Ridge
661C	Murken Bench
661D	Jellico
662A	Cassel
662B	Burney
662C	Burney Mtn. West
662D	Burney Mtn. East
663A	Chalk Mtn.
663B	Roaring Creek
663C	Montgomery Creek
663D	Hatchet Mtn. Pass
664A	Goose Gap
664B	Bollibokka Mtn.
664C	Minnesota Mtn.
664D	Devils Rock
665A	Hanland Peak
665B	Lamoine
665C	Bohemotash Mtn.
665D	O'Brien
666A	Damnation Peak
666B	Trinity Center
666C	Papoose Creek
666D	Schell Mtn.
667A	Covington Mill
667B	Siligo Peak
667C	Rush Creek Lakes
667D	Trinity Dam
668A	Mt. Hilton
668B	Thurston Peaks
668C	Helena
668D	Dedrick
669A	Jim Jam Ridge
669B	Denny
669C	Ironside Mtn.
669D	Del Loma
670A	Salyer
670B	Willow Creek
670C	Grouse Mtn.
670D	Hennessy Peak
671A	Lord-Ellis Summit
671B	Blue Lake
671C	Korbel
671D	Maple Creek

672A	Arcata North	688C	Panther Creek	705B	Ah Pah Ridge	722B	Cant Hook Mtn.
672B	Tyee City	688D	Hupa Mountain	705C	Holter Ridge	722C	Klamath Glen
672C	Eureka	689A	Rodgers Peak	705D	Johnsons	722D	Summit Valley
672D	Arcata South	689C	Trinidad	706A	Fern Canyon	723A	Childs Hill
673A	Snake Lake	689D	Crannell	706D	Orick	723B	Sister Rocks
673B	Emerson Peak	690A	Hansen Island	707A	Boyd Hot Spring	723D	Requa
673C	Boot Lake	690B	Warren Peak	707B	Lake City	724A	Lake Annie
673D	Little Hat Mtn.	690C	Eagle Peak	707C	Cedarville	724B	Mt. Bidwell
674A	Jess Valley	690D	Eagleville	707D	Leonards Hot Springs	724C	Fort Bidwell
674B	Tule Mountain	691A	Shields Creek	708A	Davis Creek	724D	Larkspur Hills
674C	Madeline	691B	Dorris Reservoir	708B	Lauer Reservoir	725A	Willow Ranch
674D	Cold Spring Mtn.	691C	Little Juniper Reservoir	708C	Surprise	725B	West Of Willow Ranch
675A	Likely	691D	Soup Creek	708D	Payne Peak	725C	McGinty Point
675B	Knox Mountain	692A	Alturas	709A	Dead Horse Reservoir	725D	Sugar Hill
675C	Ash Valley	692B	Rattlesnake Butte	709B	Whittemore Ridge	726A	Pease Flat
675D	Holbrook Canyon	692C	Graven Ridge	709C	Big Sage Reservoir	726B	Beaver Mtn.
676A	Ambrose Valley	692D	Infernal Caverns	709D	Mahogany Ridge	726C	South Mtn.
676B	Adin	693A	Canby	710A	Boles Meadows East	726D	McGinty Reservoir
676C	Letterbox Hill	693B	Washington Mtn.	710B	Boles Meadows West	727A	Weed Valley
676D	Lane Reservoir	693C	Adin Pass	710C	Ambrose	727B	Steele Swamp
677A	Big Swamp	693D	Hermit Butte	710D	Jacks Butte	727C	Pothole Valley
677B	Lookout	694A	Happy Camp Mtn.	711A	Lone Pine Butte	727D	Hager Basin
677C	Bieber	694B	Crank Mountain	711B	Rimrock Lake	728A	Sagebrush Butte
677D	Hog Valley	694C	Donica Mtn.	711C	Spaulding Butte	728B	Carr Butte
678A	Day	694D	Halls Canyon	711D	Knobcone Butte	728C	Double Head Mtn.
678B	Timbered Crater	695A	Hollenbeck	712A	Perez	728D	Pinnacle Lake
678C	Fall River Mills	695B	Border Mtn.	712B	Caldwell Butte	729A	Newell
678D	Pittville	695C	Whitehorse	712C	West Of Kephart	729B	Tulelake
679A	East Of Pondosa	695D	Egg Lake	712D	Kephart	729C	The Panhandle
679B	Pondosa	696A	Porcupine Butte	713A	Schonchin Butte	729D	Copic
679C	Burney Falls	696B	Snag Hill	713B	Bonita Butte	730A	Hatfield
679D	Dana	696C	Hambone	713C	Little Glass Mtn.	730B	Lower Klamath Lake
680A	Dead Horse Summit	696D	Indian Spring Mtn.	713D	Medicine Lake	730C	Mount Dome
680B	Grizzly Peak	697A	Horse Peak	714A	Sharp Mtn.	730D	Captain Jacks Stronghold
680C	Big Bend	697B	Rainbow Mtn.	714B	Bray	731A	Sheepy Lake
680D	Skunk Ridge	697C	Kinyon	714C	Tennant	731B	Dorris
681A	Lake McCloud	697D	Bartle	714D	Garner Mtn.	731C	Sheep Mountain
681B	Girard Ridge	698A	Ash Creek Butte	715A	Penoyar	731D	Red Rock Lakes
681C	Yellowjacket Mtn.	698B	Mt. Shasta	715B	Grass Lake	732A	Sams Neck
681D	Shoeinhorse Mtn.	698C	McCloud	715C	The Whaleback	732B	Secret Spring Mtn.
682A	Dunsmuir	698D	Elk Spring	715D	West Haight Mtn.	732C	Panther Rock
682B	Seven Lakes Basin	699A	Hotlum	716A	Solomons Temple	732D	Macdoel
682C	Chicken Hawk Hill	699B	Weed	716B	Little Shasta	733A	Copco
682D	Tombstone Mtn.	699C	Mount Eddy	716C	Lake Shastina	733B	Iron Gate Reservoir
683A	Mumbo Basin	699D	City Of Mount Shasta	716D	Juniper Flat	733C	Bogus Mountain
683B	Tangle Blue Lake	700A	China Mtn.	717A	Montague	733D	Dewey Gulch
683C	Carrville	700B	Gazelle Mtn.	717B	Yreka	734A	Hornbrook
683D	Whisky Bill Peak	700C	Scott Mountain	717C	Duzel Rock	734B	Cottonwood Peak
684A	Billys Peak	700D	South China Mtn.	717D	Gazelle	734C	Badger Mtn.
684B	Deadman Peak	701A	McConaughy Gulch	718A	Indian Creek Baldy	734D	Hawkinsville
684C	Caribou Lake	701B	Etna	718B	Russell Peak	735A	Buckhorn Bally
684D	Ycatapom Peak	701C	Eaton Peak	718C	Greenview	735B	Condrey Mtn.
685A	Grasshopper Ridge	701D	Callahan	718D	Fort Jones	735C	Horse Creek
685B	Cecilville	702A	Yellow Dog Peak	719A	Scott Bar	735D	McKinley Mtn.
685C	Cecil Lake	702B	English Peak	719B	Grider Valley	736A	Dutch Creek
685D	Thompson Peak	702C	Sawyers Bar	719C	Marble Mountain	736B	Kangaroo Mtn.
686A	Youngs Peak	702D	Tanners Peak	719D	Boulder Peak	736C	Seiad Valley
686B	Salmon Mtn.	703A	Medicine Mtn.	720A	Huckleberry Mtn.	736D	Hamburg
686C	Trinity Mtn.	703B	Somes Bar	720B	Clear Creek	737A	Figurehead Mtn.
686D	Dees Peak	703C	Orleans Mtn.	720C	Ukonom Mtn.	737B	Deadman Point
687A	Hopkins Butte	703D	Forks Of Salmon	720D	Ukonom Lake	737C	Happy Camp
687B	Weitchpec	704A	Bark Shanty Gulch	721A	Bear Peak	737D	Slater Butte
687C	Hoopa	704B	Lonesome Ridge	721B	Prescott Mtn.	738A	Polar Bear Mtn.
687D	Tish Tang Point	704C	Fish Lake	721C	Chimney Rock	738B	Broken Rib Mtn.
688A	French Camp Ridge	704D	Orleans	721D	Dillon Mtn.	738C	Devils Punchbowl
688B	Bald Hills	705A	Blue Creek Mtn.	722A	Ship Mountain	738D	Preston Peak

739A Shelly Creek Ridge
739B High Plateau Mtn.
739C Gasquet
739D Hurdygurdy Butte
740A High Divide
740B Smith River
740C Crescent City
740D Hiouchi

Quadrangle Names

373A Aberdeen
378A Academy
456D Ackerson Mtn.
027C Acolita
609C Acorn Hollow
136B Acton
270A Adelaida
158C Adelanto
676B Adin
693C Adin Pass
159B Adobe Mtn.
516B Aetna Springs
019A Agua Caliente Springs
137A Agua Dulce
049A Aguanga
705B Ah Pah Ridge
418D Ahwahnee
281B Airport Lake
638A Al Shinn Canyon
469D Alameda Well
164B Alamo Mountain
015B Alamorio
069B Alberhill
553A Albion
315B Alcalde Hills
296A Alder Peak
580B Alder Springs
616A Alderpoint
557A Alleghany
498B Allendale
288C Allensworth
606B Almanor
288B Alpaugh
021D Alpine
160B Alpine Butte
261A Alta Sierra
445B Altamont
692A Alturas
205D Alvord Mtn. East
205C Alvord Mtn. West
494A Amador City
151C Amboy
152D Amboy Crater
710C Ambrose
676A Ambrose Valley
574A American House
028D Amos
ANAC Anacapa Island
088C Anaheim
468B Anchorite Hills
658A Anderson Mtn.
476D Angels Camp
520A Annapolis
409D Ano Nuevo
604A Antelope Lake
641C Antelope Mtn.
266A Antelope Plain
571A Antelope Valley
481D Antioch North
464A Antioch South
278A Anvil Spring Canyon East
278B Anvil Spring Canyon West
066C Anza
191D Apache Canyon
157C Apple Valley North
132B Apple Valley South
012C Araz
546D Arbuckle
672A Arcata North
672D Arcata South
503B Arched Rock
422C Arena
097C Arica Mountains
076C Arlington Mine
221A Arroyo Grande NE
018B Arroyo Tapiado
214A Arvin
456C Ascension Mtn.
698A Ash Creek Butte
153A Ash Hill
675C Ash Valley
345B Ashton
534C Asti
183C Astley Rancho
246B Atascadero
422D Atwater
397C Auberry
527A Auburn
355D Auckland
043A Augustine Pass
509C Aukum
486C Aurora
252C Avawatz Pass
461D Avena
314C Avenal
290C Avenal Gap
060C Aztec Mines
109B Azusa
604D Babcock Peak
068D Bachelor Mtn.
460A Bachelor Valley
734C Badger Mtn.
301A Badwater
152C Bagdad SW
551B Bailey Ridge
228D Baker
688B Bald Hills
193C Bald Mountain
109C Baldwin Park
133B Baldy Mesa
597A Ball Mountain
302C Ballarat
191B Ballinger Canyon
416B Balloon Dome
628B Balls Ferry
559B Bangor
433A Banner Ridge
173A Bannock
012D Bard
704A Bark Shanty Gulch
608B Barkley Mtn.
009B Barrett Lake
182A Barstow
182D Barstow SE
697D Bartle
329A Bartlett
549A Bartlett Mtn.
548B Bartlett Springs
418C Bass Lake
193B Bates Canyon
319D Bear Canyon
601B Bear Harbor
213B Bear Mountain
721A Bear Peak
507C Bear River Reservoir
439D Bear Valley
346C Beatty Junction
084B Beaumont
048B Beauty Mountain
726B Beaver Mtn.
586D Beckwourth Pass
372D Bee Springs Canyon
607D Belden
600A Bell Springs
646B Bella Vista
242B Belridge
419C Ben Hur
238C Bena
628C Bend
482C Benicia
548D Benmore Canyon
450C Benton
451D Benton Hot Springs
400C Berenda
575B Berry Creek
111C Beverly Hills
363C Bickmore Canyon
677C Bieber
470B Big Alkali
651A Big Bar
408B Big Basin
131D Big Bear City
105B Big Bear Lake
680C Big Bend
535C Big Foot Mtn.
075A Big Maria Mts. NE
075B Big Maria Mts. NW
075D Big Maria Mts. SE
075C Big Maria Mts. SW
393A Big Pine
167B Big Pine Mtn.
709C Big Sage Reservoir
344D Big Sur
677A Big Swamp
560B Biggs
176C Bighorn Basin
130D Bighorn Canyon
684A Billys Peak
380D Biola
208A Bird Spring
531D Bird Valley
481A Birds Landing
413C Bishop
186A Bissell
204B Bitter Spring
595D Black Butte Dam
233A Black Hills
634D Black Lassic
164A Black Mtn.
614A Black Rock Mtn. 15' NE (was Black Rock Mtn. 7.5')
614B Black Rock Mtn. 15' NW (was Swim Ridge 7.5')
614D Black Rock Mtn. 15' SE (was Wrights Ridge 7.5')
614C Black Rock Mtn. 15' SW (was Four Corners Rock 7.5')
087C Black Star Canyon
067A Blackburn Canyon
395D Blackcap Mtn.
372B Blackrock
274D Blackwater Mine
233D Blackwater Well
266D Blackwells Corner
588D Blairsden
652D Blake Mountain
412B Blanco Mtn.
175D Blind Hills
401C Bliss Ranch
634C Blocksburg
434C Bloody Mtn.
611B Blossom
556C Blue Canyon
705A Blue Creek Mtn.
671B Blue Lake
027A Blue Mountain
589D Blue Nose Mtn.
598B Bluenose Ridge
058D Blythe
058A Blythe NE
652B Board Camp Mtn.
491C Boards Crossing
554A Boca
503D Bodega Head
470A Bodie
642C Bogard Buttes
733C Bogus Mountain
665C Bohemotash Mtn.
322D Bole Spring
710A Boles Meadows East
710B Boles Meadows West
467B Bolinas
664B Bollibokka Mtn.
004A Bonds Corner
713B Bonita Butte
307D Bonita Meadows
380B Bonita Ranch
388C Bonnie Claire SW
050C Bonsall
551D Boonville
673C Boot Lake
695B Border Mtn.
209C Boron
209A Boron NE
209B Boron NW
031B Borrego Mountain
031D Borrego Mountain SE
047C Borrego Palm Canyon
032A Borrego Sink
049C Boucher Hill
719D Boulder Peak
480D Bouldin Island
450D Boundary Peak

707A Boyd Hot Spring
294D Bradley
219A Branch Mtn.
016A Brawley
016B Brawley NW
714B Bray
238A Breckenridge Mtn.
463B Brentwood
617C Briceland
487C Bridgeport
635A Bridgeville
465B Briones Valley
126B Bristol Lake NW
126C Bristol Lake SW
382D Broadview Farms
178C Broadwell Lake
178A Broadwell Mesa
738B Broken Rib Mtn.
515B Brooks
151B Brown Buttes
543B Browns Valley
496C Bruceville
575A Brush Creek
443D Brush Lake
582C Brushy Mtn.
295D Bryson
597B Buck Rock
636C Buckeye Mtn.
471B Buckeye Ridge
735A Buckhorn Bally
656D Buckhorn Canyon
656A Buckhorn Lake
439A Buckhorn Peak
438D Buckingham Mtn.
590B Bucks Lake
048A Bucksnort Mtn.
177D Budweiser Wash
511C Buffalo Creek
636D Bull Creek
620A Bull Flat
659C Bullard Lake
649D Bully Choop Mtn.
539C Bunker Hill
111A Burbank
567B Burbeck
295C Burnett Peak
662B Burney
679C Burney Falls
662D Burney Mtn. East
662C Burney Mtn. West
163A Burnt Peak
336B Burrel
335A Burris Park
281D Burro Canyon
296D Burro Mountain
132D Butler Peak
561B Butte City
608D Butte Meadows
066D Butterfly Peak
241B Buttonwillow
079D Buzzard Spring
026A Buzzards Peak
463C Byron Hot Springs
084A Cabazon
661A Cable Mtn.
211B Cache Peak

013C Cactus
305D Cactus Peak
151D Cadiz
125C Cadiz Lake
125A Cadiz Lake NE
125B Cadiz Lake NW
150C Cadiz Summit
099A Cadiz Valley NE
099B Cadiz Valley NW
099D Cadiz Valley SE
099C Cadiz Valley SW
584A Cahto Peak
067D Cahuilla Mtn.
310B Cairns Corner
133C Cajon
112B Calabasas
491B Calaveras Dome
427A Calaveras Reservoir
476A Calaveritas
508C Caldor
712B Caldwell Butte
220A Caldwell Mesa
004B Calexico
337D Calflax
218D Caliente Mtn.
210B California City North
210C California City South
285B California Hot Springs
244D California Valley
029C Calipatria SW (=Westmorland West)
517D Calistoga
701D Callahan
602A Calneva Lake
571B Calpine
126A Calumet Mine
126D Calumet Mountains
274B Calvada Springs
114A Camarillo
268D Camatta Canyon
245A Camatta Ranch
271D Cambria
008B Cameron Corners
509B Camino
543D Camp Far West
502B Camp Meeker
308A Camp Nelson
155B Camp Rock Mine
308B Camp Wishon
593A Campbell Mound
008C Campo
558A Camptonville
070D Canada Gobernadora
693A Canby
259C Cane Canyon
284D Cannell Peak
655A Cannibal Island
112A Canoga Park
722B Cant Hook Mtn.
235C Cantil
360D Cantua Creek
606A Canyondam
637B Cape Mendocino
296B Cape San Martin
499B Capell Valley
637A Capetown

507A Caples Lake
730D Captain Jacks Stronghold
495A Carbondale
606C Caribou
684C Caribou Lake
343B Carmel Valley
512D Carmichael
243A Carneros Rocks
142A Carpinteria
728B Carr Butte
018D Carrizo Mtn.
018A Carrizo Mtn. NE
683C Carrville
506B Carson Pass
521C Carters Station
358D Caruthers
433D Casa Diablo Mtn.
307A Casa Vieja Meadows
574B Cascade
397B Cascadel Point
332A Case Mountain
196D Casmalia
662A Cassel
150B Castle Dome
224C Castle Peaks
146C Castle Rock
408A Castle Rock Ridge
104C Catclaw Flat
082C Cathedral City
420A Catheys Valley
435C Cattle Mtn.
204D Cave Mountain
247B Cayucos
519D Cazadero
685C Cecil Lake
685B Cecilville
374C Cedar Grove
469A Cedar Hill
445D Cedar Mtn.
707C Cedarville
328D Centennial Canyon
442C Ceres
362B Cerro Colorado
350D Cerro Gordo Peak
330B Chagoopa Falls
432D Chalfant Valley
663A Chalk Mtn.
558B Challenge
641B Champs Flat
631A Chanchelulla Peak 15' NE (was Arbuckle Mtn. 7.5')
631B Chanchelulla Peak 15' NW (was Chanchelulla Peak 7.5')
631D Chanchelulla Peak 15' SE (was Beegum 7.5')
631C Chanchelulla Peak 15'SW (was Platina 7.5')
361A Chaney Ranch
383A Charleston School
147D Chemehuevi Peak
576A Cherokee
473C Cherry Lake North
456B Cherry Lake South
638D Cherry Mtn.
363B Cherry Peak

624C Chester
343D Chews Ridge
541B Chicago Park
682C Chicken Hawk Hill
332C Chickencoop Canyon
577A Chico
432C Chidago Canyon
136D Chilao Flat
586C Chilcoot
723A Childs Hill
625C Childs Meadows
516D Chiles Valley
218B Chimineas Ranch
219C Chimney Canyon
721C Chimney Rock
304A China Gardens
700A China Mtn.
458C Chinese Camp
386A Chittenden
346B Chloride City
411B Chocolate Mtn.
268A Cholame
292C Cholame Hills
292D Cholame Valley
361B Chounet Ranch
401D Chowchilla
488B Chris Flat
256D Christmas Canyon
580A Chrome
365D Chualar
125D Chubbuck
042B Chuckwalla Spring
040C Cibola
339A Ciervo Mtn.
201A Cima
226D Cima Dome
236D Cinco
329B Cirque Peak
556D Cisco Grove
512A Citrus Heights
699D City Of Mount Shasta
237A Claraville
047D Clark Lake
047A Clark Lake NE
249D Clark Mtn.
100D Clarks Pass
497A Clarksburg
511A Clarksville
495C Clay
464B Clayton
720B Clear Creek
533B Clearlake Highlands
548C Clearlake Oaks
658D Cleghorn Flat
127D Cleghorn Lakes
478A Clements
463D Clifton Court Forebay
572A Clio
574C Clipper Mills
646D Clough Gulch
535D Cloverdale
378C Clovis
013B Clyde
215D Coal Oil Canyon
315A Coalinga
164D Cobblestone Mtn.

592B Cohasset
360B Coit Ranch
612B Cold Fork
552D Cold Spring
674D Cold Spring Mtn.
505D Coleville
541C Colfax
047B Collins Valley
526C Coloma
175B Colton Well
475C Columbia
200C Columbia Mtn.
475D Columbia SE
546A Colusa
568D Comptche
136C Condor Peak
735B Condrey Mtn.
319C Cone Peak
357C Conejo
079C Conejo Well
277D Confidence Hills East
277C Confidence Hills West
215A Conner
215C Conner SW
586A Constantia
434D Convict Lake
487A Conway Stage Station
473A Cooper Peak
441A Cooperstown
619A Cooskie Creek
733A Copco
729D Copic
425A Copper Mtn.
279C Copper Queen Canyon
459B Copperopolis
312D Corcoran
482B Cordelia
660C Corders Reservoir
061A Corn Spring
594B Corning
087A Corona North
087D Corona South
546C Cortina Creek
318C Cosio Knob
305C Coso Junction
304B Coso Peak
501C Cotati
629A Cottonwood
063B Cottonwood Basin
348D Cottonwood Canyon
734B Cottonwood Peak
063A Cottonwood Spring
131A Cougar Buttes
439B Coulterville
497D Courtland
395C Courtright Reservoir
598C Covelo East
599D Covelo West
667A Covington Mill
226C Cow Cove
550A Cow Mountain
202C Cowhole Mountain
392A Cowhorn Valley
452A Cowtrack Mountain
078A Coxcomb Mts.
393B Coyote Flat

206D Coyote Lake
006B Coyote Wells
306C Crag Peak
350A Craig Canyon
474B Crandall Peak
694B Crank Mountain
689D Crannell
740C Crescent City
605C Crescent Mills
224B Crescent Peak
422B Cressey
269D Creston
452C Crestview
404B Crevison Peak
587B Crocker Mtn.
565A Crockett Peak
204A Cronese Lakes
412A Crooked Creek
236C Cross Mountain
094B Cross Roads
424A Crows Landing
203C Crucero Hill
435D Crystal Crag
135D Crystal Lake
108A Cucamonga Peak
233C Cuddeback Lake
190D Cuddy Valley
213D Cummings Mtn.
428D Cupertino
315C Curry Mountain
483A Cuttings Wharf
192A Cuyama
191C Cuyama Peak
020A Cuyamaca Peak
270C Cypress Mountain
181C Daggett
100B Dale Lake
628D Dales
666A Damnation Peak
679D Dana
052B Dana Point
150A Danby
098A Danby Lake
300B Dantes View
490D Dardanelle
490A Dardanelles Cone
327D Darwin
399C Daulton
408C Davenport
513C Davis
708A Davis Creek
449C Davis Mountain
678A Day
367C Daylight Pass
709A Dead Horse Reservoir
680A Dead Horse Summit
128A Deadman Lake NE
128B Deadman Lake NW
128D Deadman Lake SE
128C Deadman Lake SW
299C Deadman Pass
684B Deadman Peak
737B Deadman Point
322C Death Valley Junction
668D Dedrick
412D Deep Springs Lake

263A Deepwell Ranch
686D Dees Peak
669D Del Loma
022B Del Mar
162A Del Sur
287C Delano East
288D Delano West
402C Delta Ranch
261C Democrat Hot Springs
442D Denair
332D Dennison Peak
669B Denny
481B Denverton
020D Descanso
248D Desert
061B Desert Center
504D Desert Creek Peak
083A Desert Hot Springs
175A Desert Spring
517A Detert Reservoir
525A Devil Peak
324D Devils Golf Course
164C Devils Heart Peak
492B Devils Nose
608C Devils Parade Ground
738C Devils Punchbowl
664D Devils Rock
324C Devils Speedway
107B Devore
733D Dewey Gulch
609B Dewitt Peak
452D Dexter Canyon
464C Diablo
622C Diamond Mtn.
721D Dillon Mtn.
396C Dinkey Creek
634A Dinsmore
489B Disaster Peak
587A Dixie Mountain
660A Dixie Peak
498A Dixon
656B Dodge Reservoir
570D Dog Valley
396A Dogtooth Peak
590D Dogwood Peak
350C Dolomite
487D Dome Hill
338D Domengine Ranch
694C Donica Mtn.
490C Donnell Lake
492D Dorrington
731B Dorris
691B Dorris Reservoir
382B Dos Palos
143B Dos Pueblos Canyon
583A Dos Rios
728C Double Head Mtn.
467E Double Point
236A Dove Spring
573D Downieville
602D Doyle
498D Dozier
485C Drakes Bay
230A Drinkwater Lake
369D Dry Bone Canyon
370A Dry Mountain

632A Dubakella Mtn. 15' NE (was Wildwood 7.5')
632B Dubakella Mtn. 15' NW (was Dubakella Mtn. 7.5')
632D Dubakella Mtn. 15' SE (was Pony Buck Peak 7.5')
632C Dubakella Mtn. 15' SW (was Smoky Creek 7.5')
446B Dublin
457A Duckwall Mtn.
287A Ducor
290A Dudley Ridge
010A Dulzura
251B Dumont Dunes
540A Duncan Peak
503A Duncans Mills
471D Dunderberg Peak
204C Dunn
530B Dunnigan
682A Dunsmuir
045A Durmid
045D Durmid SE
307C Durrwood Creek
736A Dutch Creek
541A Dutch Flat
584C Dutchmans Knoll
717C Duzel Rock
431A Dyer
232A Eagle Crags
299A Eagle Mtn.
690C Eagle Peak
190B Eagle Rest Peak
690D Eagleville
032C Earthquake Valley
082A East Deception Canyon
241C East Elk Hills
027D East Of Acolita
060D East Of Aztec Mines
178D East Of Broadwell Lake
346A East Of Chloride City
100A East Of Dale Lake
299D East Of Deadman Pass
323A East Of Echo Canyon
231D East Of Goldstone
098D East Of Granite Pass
199A East Of Grotto Hills
198D East Of Homer Mtn.
391A East Of Joshua Flats
250A East Of Kingston Peak
250D East Of Kingston Spring
205A East Of Langford Well
253D East Of Leach Lake
124D East Of Milligan
253A East Of Owl Lake
679A East Of Pondosa
062D East Of Red Canyon
323D East Of Ryan
348A East Of Sand Flat
152A East Of Siberia
369A East Of Tin Mountain
101A East Of Valley Mtn.
078D East Of Victory Pass
371A East Of Waucoba Canyon
391D East Of Waucoba Spring
701C Eaton Peak
506D Ebbetts Pass

323B	Echo Canyon
523D	Echo Lake
239D	Edison
185B	Edwards
695D	Egg Lake
021C	El Cajon
021A	El Cajon Mtn.
437B	El Capitan
085A	El Casco
016D	El Centro
159C	El Mirage
110D	El Monte
401B	El Nido
234B	El Paso Peaks
438A	El Portal
312C	El Rico Ranch
070B	El Toro
530D	Eldorado Bend
552B	Elk
580D	Elk Creek
496A	Elk Grove
565C	Elk Mountain
698D	Elk Spring
217D	Elkhorn Hills
550C	Elledge Peak
498C	Elmira
523A	Emerald Bay
237D	Emerald Mtn.
129B	Emerson Lake
673B	Emerson Peak
325B	Emigrant Canyon
266B	Emigrant Hill
472B	Emigrant Lake
325C	Emigrant Pass
036D	Encinitas
556A	English Mtn.
702B	English Peak
647D	Enterprise
277A	Epaulet Peak
460C	Escalon
035D	Escondido
515A	Esparto
318D	Espinosa Canyon
149B	Essex
269A	Estrella
701B	Etna
617B	Ettersburg
672C	Eureka
537A	Eureka Hill
570A	Evans Canyon
333C	Exeter
426A	Eylar Mtn.
499D	Fairfield North
482A	Fairfield South
187C	Fairmont Butte
284B	Fairview
157D	Fairview Valley
488C	Fales Hot Springs
368C	Fall Canyon
678C	Fall River Mills
051A	Fallbrook
454B	Falls Ridge
263C	Famoso
FARA	Farallon Islands
460B	Farmington
131C	Fawnskin
438C	Feliciana Mtn.
580C	Felkner Hill
217A	Fellows
408D	Felton
174C	Fenner
174B	Fenner Hills
174D	Fenner Spring
706A	Fern Canyon
655D	Ferndale
603D	Ferris Creek
510D	Fiddletown
654B	Fields Landing
132A	Fifteenmile Valley
168B	Figueroa Mtn.
737A	Figurehead Mtn.
139B	Fillmore
627D	Finley Butte
381C	Firebaugh
381A	Firebaugh NE
418B	Fish Camp
704C	Fish Lake
413B	Fish Slough
393D	Fish Springs
337A	Five Points
638C	Five Springs
173D	Flattop Mtn.
415C	Florence Lake
496B	Florin
595B	Flournoy
511B	Folsom
511D	Folsom SE
107C	Fontana
046C	Fonts Point
575D	Forbestown
060A	Ford Dry Lake
105C	Forest Falls
633D	Forest Glen
541D	Foresthill
703D	Forks Of Salmon
724C	Fort Bidwell
569A	Fort Bragg
230C	Fort Irwin
718D	Fort Jones
492C	Fort Mountain
519C	Fort Ross
616B	Fort Seward
654C	Fortuna
594D	Foster Island
566B	Foster Mtn.
176A	Fountain Peak
286B	Fountain Springs
564C	Fouts Springs
192D	Fox Mountain
194C	Foxen Canyon
409A	Franklin Point
322B	Franklin Well
189C	Frazier Mtn.
309B	Frazier Valley
623D	Fredonyer Pass
640A	Fredonyer Peak
522C	Freel Peak
258C	Freeman Junction
208B	Fremont Peak
688A	French Camp Ridge
558C	French Corral
648B	French Gulch
586B	Frenchman Lake
379D	Fresno North
358A	Fresno South
378B	Friant
080B	Fried Liver Wash
044C	Frink
044A	Frink NE
044B	Frink NW
579C	Fruto
579A	Fruto NE
155C	Fry Mountains
300D	Funeral Peak
324A	Furnace Creek
026C	Gables Wash
301C	Galena Canyon
210A	Galileo Hill
640C	Gallatin Peak
496D	Galt
154C	Galway Lake
617D	Garberville
526D	Garden Valley
235A	Garlock
714D	Garner Mtn.
492A	Garnet Hill
291B	Garza Peak
739C	Gasquet
144B	Gaviota
717D	Gazelle
700B	Gazelle Mtn.
120C	Gene Wash
354B	General Grant Grove
604C	Genesee Valley
526A	Georgetown
610C	Gerber
518B	Geyserville
354D	Giant Forest
286A	Gibbon Peak
564D	Gilmore Peak
406D	Gilroy
405C	Gilroy Hot Springs
544C	Gilsizer Slough
681B	Girard Ridge
014A	Glamis
014B	Glamis NW
014D	Glamis SE
014C	Glamis SW
532A	Glascock Mtn.
177A	Glasgow
451C	Glass Mountain
501D	Glen Ellen
109A	Glendora
578D	Glenn
261B	Glennville
309D	Globe
129D	Goat Mountain
174A	Goffs
367D	Gold Center
527B	Gold Hill
572B	Gold Lake
389A	Gold Mountain
300C	Gold Valley
231C	Goldstone
143A	Goleta
364C	Gonzales
573C	Goodyears Bar
495D	Goose Creek
664A	Goose Gap
240D	Gosford
334C	Goshen
156D	Grand View Mine
539A	Granite Chief
098C	Granite Pass
227D	Granite Spring
556B	Graniteville
189B	Grapevine
368B	Grapevine Peak
715B	Grass Lake
542A	Grass Valley
685A	Grasshopper Ridge
658C	Grasshopper Valley
380C	Gravelly Ford
692C	Graven Ridge
415B	Graveyard Peak
513B	Grays Bend
626B	Grays Peak
002B	Grays Well
002A	Grays Well NE
540D	Greek Store
273B	Green Monster Mine
162C	Green Valley
341C	Greenfield
567C	Greenough Ridge
718C	Greenview
605B	Greenville
300A	Greenwater Canyon
526B	Greenwood
379B	Gregg
719B	Grider Valley
560C	Gridley
545C	Grimes
680B	Grizzly Peak
588A	Grizzly Valley
096C	Grommet
347D	Grotto Canyon
199B	Grotto Hills
670C	Grouse Mtn.
457C	Groveland
196A	Guadalupe
537D	Gualala
108D	Guasti
536D	Gube Mountain
518C	Guerneville
312B	Guernsey
314B	Guijarral Hills
531C	Guinda
423C	Gustine
289D	Hacienda Ranch
289A	Hacienda Ranch NE
289B	Hacienda Ranch NW
199C	Hackberry Mountain
645D	Hagaman Gulch
727D	Hager Basin
306A	Haiwee Pass
305B	Haiwee Reservoirs
601D	Hales Grove
437A	Half Dome
429B	Half Moon Bay
651D	Halfway Ridge
596C	Hall Ridge
227C	Halloran Springs
694D	Halls Canyon
696C	Hambone

Code	Name
736D	Hamburg
294B	Hames Valley
578A	Hamilton City
576B	Hamlin Canyon
432B	Hammil Valley
382C	Hammonds Ranch
301B	Hanaupah Canyon
335C	Hanford
390B	Hanging Rock Canyon
665A	Hanland Peak
690A	Hansen Island
737C	Happy Camp
694A	Happy Camp Mtn.
031C	Harper Canyon
030C	Harpers Well
616C	Harris
348C	Harris Hill
337C	Harris Ranch
106B	Harrison Mtn.
224D	Hart Peak
180B	Harvard Hill
642A	Harvey Mtn.
590C	Haskins Valley
423B	Hatch
663D	Hatchet Mtn. Pass
730A	Hatfield
121B	Havasu Lake
734D	Hawkinsville
201D	Hayden
062B	Hayfield
062A	Hayfield Spring
650C	Hayfork
650B	Hayfork Bally
650D	Hayfork Summit
572D	Haypress Valley
421A	Haystack Mtn.
447A	Hayward
518D	Healdsburg
005A	Heber
179C	Hector
013A	Hedges
505B	Heenan Lake
668C	Helena
158A	Helendale
359D	Helm
067B	Hemet
595A	Henleyville
670D	Hennessy Peak
340D	Hepsedam Peak
602B	Herlong
693D	Hermit Butte
340A	Hernandez Reservoir
379C	Herndon
133A	Hesperia
455B	Hetch Hetchy Reservoir
160A	Hi Vista
129A	Hidalgo Mtn.
254B	Hidden Spring
179A	Hidden Valley East
179B	Hidden Valley West
740A	High Divide
321C	High Peak
739B	High Plateau Mtn.
534B	Highland Springs
167D	Hildreth Peak
182B	Hinkley
740D	Hiouchi
554B	Hobart Mills
307B	Hockett Peak
182C	Hodge
677D	Hog Valley
661B	Hogback Ridge
675D	Holbrook Canyon
267C	Holland Canyon
695A	Hollenbeck
385C	Hollister
111D	Hollywood
462B	Holt
705C	Holter Ridge
015D	Holtville East
015A	Holtville NE
015C	Holtville West
173B	Homer
198C	Homer Mtn.
538C	Homewood
280B	Homewood Canyon
560D	Honcut
618A	Honeydew
481C	Honker Bay
629D	Hooker
687C	Hoopa
649C	Hoosimbim Mtn.
687A	Hopkins Butte
059C	Hopkins Well
535A	Hopland
224A	Hopps Well
097A	Horn Spring
734A	Hornbrook
439C	Hornitos
259D	Horse Canyon
735C	Horse Creek
697A	Horse Peak
411D	Horse Thief Canyon
274C	Horse Thief Springs
419D	Horsecamp Mountain
048D	Hot Springs Mtn.
699A	Hotlum
548A	Hough Springs
404A	Howard Ranch
220D	Huasna Peak
720A	Huckleberry Mtn.
474D	Hull Creek
581C	Hull Mountain
607B	Humboldt Peak
101D	Humbug Mountain
607A	Humbug Valley
375C	Hume
377B	Humphreys Station
448A	Hunters Point
396B	Huntington Lake
468D	Huntoon Spring
688D	Hupa Mountain
739D	Hurdygurdy Butte
314A	Huron
193D	Hurricane Deck
651C	Hyampom
651B	Hyampom Mtn.
654D	Hydesville
653B	Iaqua Buttes
276D	Ibex Pass
276C	Ibex Spring
339B	Idria
066B	Idyllwild
648D	Igo
420D	Illinois Hill
011D	Imperial Beach
011E	Imperial Reservoir
007A	In-ko-pah Gorge
076D	Inca
372C	Independence
555A	Independence Lake
102C	Indian Cove
718A	Indian Creek Baldy
430C	Indian Garden Creek
420B	Indian Gulch
451B	Indian Meadows
202A	Indian Spring
696D	Indian Spring Mtn.
064B	Indio
692D	Infernal Caverns
585D	Inglenook
090A	Inglewood
403B	Ingomar
627C	Inskip Hill
485D	Inverness
645C	Inwood
258A	Inyokern
258D	Inyokern SE
494C	Ione
028B	Iris
043B	Iris Pass
043C	Iris Wash
494B	Irish Hill
733B	Iron Gate Reservoir
098B	Iron Mountains
599C	Iron Peak
155D	Iron Ridge
669C	Ironside Mtn.
426D	Isabel Valley
609D	Ishi Caves
480A	Isleton
333B	Ivanhoe
225D	Ivanpah
248C	Ivanpah Lake
349D	Jackass Canyon
184C	Jackrabbit Hill
644B	Jacks Backbone
710D	Jacks Butte
494D	Jackson
007B	Jacumba
302B	Jail Canyon
359B	Jamesan
582B	Jamison Ridge
010B	Jamul Mountains
622D	Janesville
457D	Jawbone Ridge
661D	Jellico
477D	Jenny Lind
532C	Jericho Valley
480C	Jersey Island
674A	Jess Valley
616D	Jewett Rock
669A	Jim Jam Ridge
518A	Jimtown
338C	Joaquin Rocks
234C	Johannesburg
330A	Johnson Peak
285A	Johnsondale
705D	Johnsons
622A	Johnstonville
588C	Johnsville
295B	Jolon
607C	Jonesville
225C	Joshua
391B	Joshua Flats
103A	Joshua Tree North
103D	Joshua Tree South
033D	Julian
579B	Julian Rocks
650A	Junction City
453D	June Lake
716D	Juniper Flat
135B	Juniper Hills
198A	Juniper Mine
431B	Juniper Mtn.
657A	Juniper Ridge
319B	Junipero Serra Peak
416C	Kaiser Peak
030D	Kane Spring
030A	Kane Spring NE
030B	Kane Spring NW
736B	Kangaroo Mtn.
639D	Karlo
332B	Kaweah
358B	Kearney Park
373D	Kearsarge Peak
328A	Keeler
213A	Keene
106A	Keller Peak
534A	Kelseyville
201C	Kelso
176B	Kelso Dunes
501A	Kenwood
712D	Kephart
359A	Kerman
330C	Kern Lake
330D	Kern Peak
284C	Kernville
604B	Kettle Rock
313C	Kettleman City
291A	Kettleman Plain
081B	Keys View
459D	Keystone
473D	Kibbie Lake
591B	Kimshew Point
538A	Kings Beach
250B	Kingston Peak
250C	Kingston Spring
438B	Kinsley
697C	Kinyon
530A	Kirkville
594C	Kirkwood
469B	Kirkwood Spring
400D	Kismet
722C	Klamath Glen
234A	Klinker Mtn.
581D	Kneecap Ridge
459C	Knights Ferry
529C	Knights Landing
262C	Knob Hill
711D	Knobcone Butte
399A	Knowles
675B	Knox Mountain
532D	Knoxville

453C Koip Peak
671C Korbel
183B Kramer Hills
184A Kramer Junction
315D Kreyenhagen Hills
524D Kyburz
314D La Cima
446D La Costa Valley
440B La Grange
088B La Habra
429D La Honda
022C La Jolla
188C La Liebre Ranch
022D La Mesa
244C La Panza
244A La Panza NE
244B La Panza Ranch
058E La Paz Mtn.
573B La Porte
065A La Quinta
071D Laguna Beach
011F Laguna Dam
383D Laguna Seca Ranch
724A Lake Annie
132C Lake Arrowhead
515C Lake Berryessa
168C Lake Cachuma
707B Lake City
542D Lake Combie
456A Lake Eleanor
069A Lake Elsinore
084D Lake Fulmor
121A Lake Havasu City South
162B Lake Hughes
260B Lake Isabella North
260C Lake Isabella South
086C Lake Mathews
681A Lake McCloud
615C Lake Mountain
565B Lake Pillsbury
716C Lake Shastina
549C Lakeport
085D Lakeview
665B Lamoine
239C Lamont
283D Lamont Peak
161A Lancaster East
161B Lancaster West
129C Landers
206C Lane Mountain
676D Lane Reservoir
379A Lanes Bridge
205B Langford Well
634B Larabee Valley
724D Larkspur Hills
051C Las Pulgas Canyon
465D Las Trampas Ridge
243B Las Yeguas Ranch
626A Lassen Peak
410C Last Chance Mtn.
390D Last Chance Range SE
390C Last Chance Range SW
462D Lathrop
335B Laton
510C Latrobe
708B Lauer Reservoir
567D Laughlin Range
407C Laurel
154A Lavic Lake
154D Lavic SE
413A Laws
255B Layton Spring
583B Laytonville
400B Le Grand
253C Leach Lake
254D Leach Spring
127B Lead Mountain
127C Lead Mountain SW
127A Lead Mtn. NE
189D Lebec
453A Lee Vining
327A Lee Wash
598A Leech Lake Mtn.
508A Leek Spring Hill
345D Leeland
345C Lees Camp
547B Leesville
600C Leggett
336D Lemoore
707D Leonards Hot Springs
676C Letterbox Hill
184B Leuhman Ridge
360C Levis
649A Lewiston
491D Liberty Hill
497C Liberty Island
426C Lick Observatory
163B Liebre Mtn.
188A Liebre Twins
675A Likely
338B Lillis Ranch
270B Lime Mountain
528A Lincoln
584B Lincoln Ridge
478D Linden
310A Lindsay
165C Lion Canyon
043D Lion Head Mtn.
621B Litchfield
187D Little Buttes
042A Little Chuckwalla Mts.
713C Little Glass Mtn.
673D Little Hat Mtn.
691C Little Juniper Reservoir
282B Little Lake
076B Little Maria Mts.
620B Little Mud Flat
042D Little Mule Mts.
012A Little Picacho Peak
167C Little Pine Mtn.
149A Little Piute Mountains
716B Little Shasta
417B Little Shuteye Peak
399D Little Table Mtn.
660B Little Valley
160C Littlerock
008A Live Oak Springs
446A Livermore
362C Llanada
577C Llano Seco
478B Lockeford
208D Lockhart
165A Lockwood Valley
353C Lodgepole
479A Lodi North
479D Lodi South
563C Lodoga
597D Log Spring
563A Logan Ridge
562B Logandale
242A Lokern
407D Loma Prieta
559C Loma Rica
170B Lompoc
170C Lompoc Hills
257A Lone Butte
351D Lone Pine
711A Lone Pine Butte
444C Lone Tree Creek
289C Lone Tree Well
704B Lonesome Ridge
089C Long Beach
306D Long Canyon
504B Long Dry Canyon
615D Long Ridge
583C Longvale
340C Lonoak
677B Lookout
524A Loon Lake
246D Lopez Mtn.
320D Lopez Point
237C Loraine
671A Lord-Ellis Summit
089D Los Alamitos
170A Los Alamos
110C Los Angeles
403D Los Banos
384A Los Banos Valley
407B Los Gatos
219B Los Machos Hills
610D Los Molinos
169A Los Olivos
290B Los Viejos
489A Lost Cannon Peak
265C Lost Hills
265A Lost Hills NE
265B Lost Hills NW
304D Louisiana Butte
160D Lovejoy Buttes
730B Lower Klamath Lake
533A Lower Lake
371D Lower Warm Springs
612D Lowrey
570B Loyalton
549D Lucerne
131B Lucerne Valley
376C Luckett Mtn.
153B Ludlow
153D Ludlow SE
470C Lundy
626C Lyonsville
732D Macdoel
653A Mad River Buttes
674C Madeline
380A Madera
514B Madison
167A Madulce Peak
709D Mahogany Ridge
357B Malaga
081A Malapai Hill
112C Malibu Beach
552C Mallo Pass Creek
435A Mammoth Mtn.
417D Mammoth Pool Dam
027B Mammoth Wash
180A Manix
279B Manly Fall
279A Manly Peak
547A Manor Slough
461C Manteca
627A Manton
351B Manzanar
644D Manzanita Lake
194A Manzanita Mtn.
671D Maple Creek
719C Marble Mountain
483D Mare Island
051B Margarita Peak
216C Maricopa
366A Marina
374A Marion Peak
419B Mariposa
437D Mariposa Grove
384B Mariposa Peak
517C Mark West Springs
506A Markleeville
201B Marl Mountains
065D Martinez Mtn.
123C Martins Well
554D Martis Peak
568C Mathison Peak
141A Matilija
471C Matterhorn Peak
303C Maturango Peak
303A Maturango Peak NE
303D Maturango Peak SE
562C Maxwell
372A Mazourka Peak
698C McCloud
701A McConaughy Gulch
059A McCoy Peak
059B McCoy Spring
058B McCoy Wash
657B McDonald Peak
263B McFarland
725C McGinty Point
726D McGinty Reservoir
536C McGuire Ridge
602C McKesick Peak
735D McKinley Mtn.
243D McKittrick Summit
654A McWhinney Creek
590A Meadow Valley
064D Mecca
713D Medicine Lake
703A Medicine Mtn.
538D Meeks Bay
130A Melville Lake
445C Mendenhall Springs
569D Mendocino
597C Mendocino Pass
381D Mendota Dam
421C Merced
440D Merced Falls

436B Merced Peak
362A Mercy Hot Springs
545B Meridian
514D Merritt
034A Mesa Grande
134B Mescal Creek
226A Mescal Range
347B Mesquite Flat
249A Mesquite Lake
249B Mesquite Mountains
214C Mettler
540C Michigan Bluff
200B Mid Hills
533D Middletown
445A Midway
003A Midway Well
003B Midway Well NW
603A Milford
645A Miller Mtn.
398D Millerton Lake East
398C Millerton Lake West
124C Milligan
215B Millux
427B Milpitas
599A Mina
428C Mindego Hill
522A Minden
626D Mineral
225B Mineral Hill
331A Mineral King
181D Minneola
664C Minnesota Mtn.
137B Mint Canyon
261D Miracle Hot Springs
355A Miramonte
617A Miranda
219D Miranda Pine Mtn.
405B Mississippi Creek
601C Mistake Point
629C Mitchell Gulch
458D Moccasin
123A Mohawk Spring
211C Mojave
211A Mojave NE
493C Mokelumne Hill
507D Mokelumne Peak
306B Monache Mountain
317A Monarch Peak
452B Mono Mills
361D Monocline Ridge
212D Monolith
334A Monson
717A Montague
448C Montara Mountain
366C Monterey
663C Montgomery Creek
515D Monticello Dam
441C Montpelier
019B Monument Peak
147B Monumental Pass
605A Moonlight Peak
105A Moonridge
139C Moorpark
123D Mopah Peaks
009A Morena Reservoir
406B Morgan Hill
153C Morgans Well
301D Mormon Point
104D Morongo Valley
247A Morro Bay North
247D Morro Bay South
051D Morro Hill
063C Mortmar
331C Moses Mtn.
387D Moss Landing
562D Moulton Weir
297C Mound Spring
453B Mount Dana
730C Mount Dome
699C Mount Eddy
364D Mount Johnson
019C Mount Laguna
134C Mount San Antonio
005B Mount Signal
517B Mount St. Helena
414C Mount Tom
352D Mount Whitney
281A Mountain Springs Canyon
428A Mountain View
216A Mouth Of Kern (=Buena Vista Lake Bed)
415A Mt. Abbot
238B Mt. Adelaide
108B Mt. Baldy
431C Mt. Barcroft
026B Mt. Barrow
724B Mt. Bidwell
425B Mt. Boardman
352B Mt. Brewer
344A Mt. Carmel
373C Mt. Clarence King
394B Mt. Darwin
426B Mt. Day
573A Mt. Fillmore
499C Mt. George
416D Mt. Givens
394C Mt. Goddard
625A Mt. Harkness
364B Mt. Harlan
395A Mt. Henry
415D Mt. Hilgard
668A Mt. Hilton
588B Mt. Ingalls
488D Mt. Jackson
352C Mt. Kaweah
351C Mt. Langley
436A Mt. Lyell
406C Mt. Madonna
197C Mt. Manchester
414B Mt. Morgan
488A Mt. Patterson
373B Mt. Pinchot
435B Mt. Ritter
698B Mt. Shasta
353B Mt. Silliman
406A Mt. Sizer
425C Mt. Stakes
394A Mt. Thompson
499A Mt. Vaca
352A Mt. Williamson
110A Mt. Wilson
207D Mud Hills
354A Muir Grove
040A Mule Wash
683A Mumbo Basin
661C Murken Bench
475B Murphys
068C Murrieta
397A Musick Mtn.
405A Mustang Peak
635D Myers Flat
082D Myoma
500D Napa
011A National City
365A Natividad
317B Nattrass Valley
633A Naufus Creek
552A Navarro
181B Nebo
172D Needles
172B Needles NW
172C Needles SW
188D Neenach School
470D Negit Island
577D Nelson
231A Nelson Lake
396D Nelson Mtn.
349C Nelson Range
558D Nevada City
346D Nevares Peak
192B New Cuyama
100C New Dale
459A New Melones Dam
350B New York Butte
447D Newark
180C Newberry Springs
113B Newbury Park
729A Newell
138A Newhall
598D Newhouse Ridge
424D Newman
071B Newport Beach
596D Newville
529A Nicolaus
029A Niland
446C Niles
282C Ninemile Canyon
220C Nipomo
225A Nipton
600B Noble Butte
298D Nopah Peak
593C Nord
555D Norden
557C North Bloomfield
341B North Chalone Peak
210D North Edwards
398A North Fork
228A North Of Baker
263D North Of Oildale
275A North Of Tecopa Pass
394D North Palisade
568A Northspur
326D Nova Canyon
484D Novato
568B Noyo Hill
665D O'Brien
398B O'Neals
646A Oak Run
460D Oakdale
465C Oakland East
466D Oakland West
046A Oasis
138D Oat Mountain
656C Observation Peak
029B Obsidian Butte
221D Oceano
036B Oceanside
013D Ogilby
239B Oil Center
240A Oildale
238D Oiler Peak
140B Ojai
329D Olancha
202D Old Dad Mtn.
252B Old Ibex Pass
434B Old Mammoth
166C Old Man Mountain
643B Old Station
130B Old Woman Springs
149C Old Woman Statue
629B Olinda
544D Olivehurst
509D Omo Ranch
608A Onion Butte
589C Onion Valley
630A Ono 15' NE (was Ono 7.5')
630B Ono 15' NW (was Tar Bully 7.5')
630D Ono 15' SE (was Rosewood 7.5')
630C Ono 15' SW (was Chickabally Mtn. 7.5')
108C Ontario
259B Onyx
104B Onyx Peak
207B Opal Mountain
088D Orange
356A Orange Cove North
356D Orange Cove South
267B Orchard Peak
195C Orcutt
577B Ord Ferry
156A Ord Mountain
559D Oregon House
424C Orestimba Peak
706D Orick
578B Orland
704D Orleans
703C Orleans Mtn.
536A Ornbaun Valley
063D Orocopia Canyon
576D Oroville
575C Oroville Dam
551A Orrs Springs
383C Ortigalita Peak
383B Ortigalita Peak NW
010C Otay Mesa
010D Otay Mountain
328B Owens Lake
258B Owens Peak
420C Owens Reservoir
653C Owl Creek
253B Owl Lake

382A Oxalis
612A Oxbow Bridge
114B Oxnard
249C Pachalka Spring
404C Pacheco Pass
405D Pacheco Peak
506C Pacific Valley
136A Pacifico Mountain
267D Packwood Creek
364A Paicines
311B Paige
017C Painted Gorge
218A Painted Rock
149D Painted Rock Wash
050D Pala
077C Palen Lake
077D Palen Mountains
077A Palen Pass
560A Palermo
083D Palm Springs
066A Palm View Peak
161D Palmdale
428B Palo Alto
646C Palo Cedro
342B Palo Escrito Peak
040B Palo Verde
041D Palo Verde Peak
049D Palomar Observatory
302D Panamint
326A Panamint Butte
326C Panamint Springs
317D Pancho Rico Valley
362D Panoche
363A Panoche Pass
217B Panorama Hills
688C Panther Creek
732C Panther Rock
609A Panther Spring
666C Papoose Creek
592D Paradise East
206A Paradise Range
592C Paradise West
342D Paraiso Springs
095A Parker
095B Parker NW
095D Parker SE
095C Parker SW
292B Parkfield
320B Partington Ridge
110B Pasadena
596A Paskenta
269B Paso Robles
189A Pastoria Creek
371C Pat Keyes Canyon
424B Patterson
376A Patterson Mtn.
441B Paulsell
708D Payne Peak
193A Peak Mountain
282D Pearsonville
726A Pease Flat
271A Pebblestone Shut-in
050A Pechanga
508D Peddler Hill
623B Pegleg Mtn.
042C Pegleg Well
561D Pennington
440A Penon Blanco Peak
715A Penoyar
216D Pentland
712A Perez
085C Perris
484B Petaluma
483C Petaluma Point
484A Petaluma River
461A Peters
639C Petes Valley
304C Petroglyph Canyon
637D Petrolia
320E Pfeiffer Point
134A Phelan
551C Philo
025D Picacho
025B Picacho NW
012B Picacho Peak
025C Picacho SW
489D Pickel Meadow
271C Pico Creek
377C Piedra
272A Piedras Blancas
601A Piercy
409B Pigeon Point
557B Pike
641D Pikes Point
527D Pilot Hill
232B Pilot Knob
255D Pilot Knob Valley East
255C Pilot Knob Valley West
061D Pilot Mountain
341D Pinalito Canyon
642D Pine Creek Valley
377D Pine Flat Dam
493B Pine Grove
262D Pine Mountain
473B Pinecrest
728D Pinnacle Lake
080A Pinto Mountain
200A Pinto Valley
078B Pinto Wells
236B Pinyon Mtn.
139A Piru
221B Pismo Beach
141C Pitas Point
678D Pittville
472D Piute Mtn.
237B Piute Peak
288A Pixley
079A Placer Canyon
510A Placerville
401A Plainsburg
421D Planada
520D Plantation
581A Plaskett Meadows
581B Plaskett Ridge
017D Plaster City
017B Plaster City NW
528C Pleasant Grove
190A Pleito Hills
537B Point Arena
171C Point Arguello
467D Point Bonita
145B Point Conception
113D Point Dume
011B Point Loma
114D Point Mugu
485A Point Reyes NE
196B Point Sal
344C Point Sur
642B Poison Lake
738A Polar Bear Mtn.
413D Poleta Canyon
525D Pollack Pines
264A Pond
679B Pondosa
696A Porcupine Butte
080D Porcupine Wash
222A Port San Luis
310D Porterville
587C Portola
285C Posey
381B Poso Farm
074B Poston
565D Potato Hill
727C Pothole Valley
009D Potrero
566D Potter Valley
022A Poway
245D Pozo Summit
087B Prado Dam
721B Prescott Mtn.
738D Preston Peak
316B Priest Valley
562A Princeton
647A Project City
643D Prospect Peak
386C Prunedale
591C Pulga
550D Purdys Gardens
291D Pyramid Hills
523C Pyramid Peak
254A Quail Spring
026D Quartz Peak
102D Queen Mtn.
384C Quien Sabe Valley
589B Quincy
286C Quincy School
331D Quinn Peak
046B Rabbit Peak
559A Rackerby
612C Raglin Ridge
563B Rail Canyon
493D Rail Road Flat
697B Rainbow Mtn.
358C Raisin
034D Ramona
343A Rana Creek
033A Ranchita
293D Ranchito Canyon
065B Rancho Mirage
166B Rancho Nuevo Creek
035C Rancho Santa Fe
692B Rattlesnake Butte
130C Rattlesnake Canyon
657D Ravendale
399B Raymond
400A Raynor Creek
625B Reading Peak
587D Reconnaissance Peak
611C Red Bank
610B Red Bluff East
611A Red Bluff West
184D Red Buttes
062C Red Canyon
624B Red Cinder
061C Red Cloud Canyon
234D Red Mountain
229D Red Pass Lake
229A Red Pass Lake NE
229B Red Pass Lake NW
731D Red Rock Lakes
635B Redcrest
647C Redding
106C Redlands
185C Redman
090C Redondo Beach
447C Redwood Point
566C Redwood Valley
356C Reedley
319A Reliz Canyon
335D Remnoy
723D Requa
275B Resting Spring
303B Revenue Canyon
242C Reward
166A Reyes Peak
097D Rice
593D Richardson Springs
593B Richardson Springs NW
287D Richgrove
466A Richmond
257B Ridgecrest North
257C Ridgecrest South
596B Riley Ridge
104A Rimrock
711B Rimrock Lake
241A Rio Bravo
239A Rio Bravo Ranch
512B Rio Linda
480B Rio Vista
058C Ripley
443B Ripon
504C Risue Canyon
161C Ritter Ridge
451A River Spring
442B Riverbank
336A Riverdale
086A Riverside East
086B Riverside West
524C Riverton
663B Roaring Creek
524B Robbs Peak
340B Rock Spring Peak
523B Rockbound Valley
283B Rockhouse Basin
081D Rockhouse Canyon
527C Rocklin
333D Rocky Hill
689A Rodgers Peak
034B Rodriquez Mtn.
185A Rogers Lake North
185D Rogers Lake South
068B Romoland
623A Roop Mountain
059D Roosevelt Mine

186C	Rosamond
186D	Rosamond Lake
240B	Rosedale
528D	Roseville
542B	Rough And Ready
375B	Rough Spur
378D	Round Mountain
414A	Rovana
539B	Royal Gorge
384D	Ruby Canyon
531B	Rumsey
667C	Rush Creek Lakes
718B	Russell Peak
633C	Ruth Reservoir
500B	Rutherford
323C	Ryan
097B	Sablon
283A	Sacatar Canyon
145A	Sacate
376B	Sacate Ridge
512C	Sacramento East
513D	Sacramento West
252A	Saddle Peak Hills
209D	Saddleback Mtn.
067C	Sage
728A	Sagebrush Butte
659A	Said Valley
443A	Salida
365B	Salinas
370B	Saline Peak
192C	Salisbury Potrero
686B	Salmon Mtn.
276B	Salsberry Peak
547D	Salt Canyon
476C	Salt Spring Valley
235B	Saltdale NW
235D	Saltdale SE
045B	Salton
670A	Salyer
732A	Sams Neck
476B	San Andreas
317C	San Ardo
363D	San Benito
339C	San Benito Mtn.
107A	San Bernardino North
107D	San Bernardino South
079B	San Bernardino Wash
052A	San Clemente
SCMC	San Clemente Island Central
SCMN	San Clemente Island North
SCMS	San Clemente Island South
109D	San Dimas
385B	San Felipe
137C	San Fernando
466C	San Francisco North
448B	San Francisco South
484C	San Geronimo
105D	San Gorgonio Mtn.
429C	San Gregorio
165B	San Guillermo
084C	San Jacinto
083C	San Jacinto Peak
359C	San Joaquin
427D	San Jose East
427C	San Jose West

386D	San Juan Bautista
070C	San Juan Capistrano
447B	San Leandro
318A	San Lucas
404D	San Luis Dam
246C	San Luis Obispo
403A	San Luis Ranch
036A	San Luis Rey
035B	San Marcos
168D	San Marcos Pass
448D	San Mateo
293C	San Miguel
SMIE	San Miguel Island East
SMIW	San Miguel Island West
SNIC	San Nicolas Island
052D	San Onofre Bluff
034C	San Pasqual
073A	San Pedro
466B	San Quentin
467A	San Rafael
168A	San Rafael Mtn.
271B	San Simeon
021B	San Vicente Reservoir
211D	Sanborn
561C	Sanborn Slough
262B	Sand Canyon
348B	Sand Flat
390A	Sand Spring
402A	Sandy Mush
357A	Sanger
582D	Sanhedrin Mtn.
142B	Santa Barbara
SBRA	Santa Barbara Island
SCTE	Santa Catalina East
SCTN	Santa Catalina North
SCTS	Santa Catalina South
SCTW	Santa Catalina West
387E	Santa Cruz
SCZA	Santa Cruz Island A
SCZB	Santa Cruz Island B
SCZC	Santa Cruz Island C
SCZD	Santa Cruz Island D
246A	Santa Margarita
245C	Santa Margarita Lake
195B	Santa Maria
140D	Santa Paula
140A	Santa Paula Peak
402D	Santa Rita Bridge
339D	Santa Rita Peak
501B	Santa Rosa
327B	Santa Rosa Flat
170D	Santa Rosa Hills
SROE	Santa Rosa Island East
SRON	Santa Rosa Island North
SROS	Santa Rosa Island South
SROW	Santa Rosa Island West
138C	Santa Susana (= Simi Valley East)
407A	Santa Teresa Hills
169D	Santa Ynez
033C	Santa Ysabel
191A	Santiago Creek
070A	Santiago Peak
570C	Sardine Peak
140C	Saticoy
571C	Sattley

537C	Saunders Reef
287B	Sausalito School
122D	Savahia Peak
122A	Savahia Peak NE
122B	Savahia Peak NW
122C	Savahia Peak SW
190C	Sawmill Mountain
267A	Sawtooth Ridge
702C	Sawyers Bar
497B	Saxon
666D	Schell Mtn.
713A	Schonchin Butte
636A	Scotia
719A	Scott Bar
700C	Scott Mountain
389D	Scottys Castle
072A	Seal Beach
256A	Searles Lake
483B	Sears Point
366D	Seaside
502A	Sebastopol
732B	Secret Spring Mtn.
016C	Seeley
595C	Sehorn Creek
736C	Seiad Valley
357D	Selma
265D	Semitropic
308D	Sentinel Peak
682B	Seven Lakes Basin
082B	Seven Palms Valley
046D	Seventeen Palms
202B	Seventeenmile Point
354C	Shadequarter Mtn.
159A	Shadow Mountains
159D	Shadow Mountains SE
621A	Shaffer Mtn.
266C	Shale Point
268B	Shandon
615A	Shannon Butte
416A	Sharktooth Peak
714A	Sharp Mtn.
648C	Shasta Bally
647B	Shasta Dam
397D	Shaver Lake
268C	Shedd Canyon
124B	Sheep Camp Spring
252D	Sheep Creek Spring
731C	Sheep Mountain
659D	Sheepshead
731A	Sheepy Lake
031A	Shell Reef
739A	Shelly Creek Ridge
618D	Shelter Cove
273D	Shenandoah Peak
528B	Sheridan
316A	Sherman Peak
584D	Sherwood Peak
691A	Shields Creek
510B	Shingle Springs
627B	Shingletown
638B	Shinn Mountain
722A	Ship Mountain
576C	Shippee
681D	Shoeinhorse Mtn.
277B	Shore Line Butte
276A	Shoshone

652C	Showers Mtn.
618B	Shubrick Peak
417C	Shuteye Peak
152B	Siberia
060B	Sidewinder Well
572C	Sierra City
571D	Sierraville
199D	Signal Hill
667B	Siligo Peak
251D	Silurian Hills
251C	Silurian Lake
228B	Silurian Valley
659B	Silva Flat Reservoir
155A	Silver Bell Mine
331B	Silver City
133D	Silverwood Lake
139D	Simi (= Simi Valley West)
243C	Simmler
652A	Sims Mountain
436C	Sing Peak
284A	Sirretta Peak
195D	Sisquoc
723B	Sister Rocks
563D	Sites
069C	Sitton Peak
298A	Sixmile Spring
150D	Skeleton Pass
680D	Skunk Ridge
316C	Slack Canyon
525C	Slate Mtn.
280A	Slate Range Crossing
737D	Slater Butte
179D	Sleeping Beauty
162D	Sleepy Valley
374B	Slide Bluffs
232C	Slocum Mtn.
495B	Sloughhouse
509A	Sly Park
543A	Smartville
316D	Smith Mountain
740B	Smith River
696B	Snag Hill
147C	Snaggletooth
673A	Snake Lake
440C	Snelling
639A	Snowstorm Mtn.
591D	Soapstone Hill
344B	Soberanes Point
203A	Soda Lake North
203D	Soda Lake South
555C	Soda Springs
411C	Soldier Pass
342A	Soledad
186B	Soledad Mtn.
308C	Solo Peak
227A	Solomons Knob
716A	Solomons Temple
169C	Solvang
444D	Solyo
019D	Sombrero Peak
703B	Somes Bar
500C	Sonoma
458B	Sonora
489C	Sonora Pass
387B	Soquel
691D	Soup Creek

279D Sourdough Spring
700D South China Mtn.
089B South Gate
522B South Lake Tahoe
726C South Mtn.
641A Spalding Tract
256C Spangler Hills East
257D Spangler Hills West
711C Spaulding Butte
620D Spencer Creek
353A Sphinx Lakes
490B Spicer Meadows Res.
393C Split Mtn.
633B Sportshaven
365C Spreckels
589A Spring Garden
309A Springville
417A Squaw Dome
603C Squaw Valley Peak
516C St. Helena
564B St. John Mtn.
458A Standard
120B Standard Wash
621C Standish
475A Stanislaus
248B State Line Pass
431D Station Peak
086D Steele Peak
727B Steele Swamp
148D Stepladder Mountains
148C Stepladder Mountains SW
148A Stepladder Mts. NE
148B Stepladder Mts. NW
240C Stevens
423D Stevinson
298B Stewart Valley
520B Stewarts Point
592A Stirling City
293A Stockdale Mountain
461B Stockton East
462A Stockton West
157A Stoddard Well
355C Stokes Mtn.
579D Stone Valley
603B Stony Ridge
564A Stonyford
591A Storrie
347C Stovepipe Wells
347A Stovepipe Wells NE
625D Stover Mtn.
313A Stratford
313D Stratford SE
474A Strawberry
574D Strawberry Valley
660D Straylor Lake
508B Stump Spring, Calif. (= Old Iron Mountain)
274A Stump Spring, Nev.
419A Stumpfield Mtn.
076A Styx
309C Success Dam
725D Sugar Hill
469C Sulphur Pond
722D Summit Valley
102B Sunfair
137D Sunland
085B Sunnymead
154B Sunshine Peak
207A Superior Lake
232D Superior Valley
017A Superstition Mtn.
171A Surf
708C Surprise
622B Susanville
544B Sutter
545A Sutter Buttes
529B Sutter Causeway
624A Swain Mountain
643A Swains Hole
018C Sweeney Pass
487B Sweetwater Creek
342C Sycamore Flat
411A Sylvania Canyon
410B Sylvania Mts.
216B Taft
538B Tahoe City
144A Tajiguas
327C Talc City Hills
491A Tamarack
455C Tamarack Flat
600D Tan Oak Park
683B Tangle Blue Lake
702D Tanners Peak
220B Tar Spring Ridge
464D Tassajara
320A Tassajara Hot Springs
218C Taylor Canyon
513A Taylor Monument
636B Taylor Peak
311C Taylor Weir
605D Taylorsville
370D Teakettle Junction
009C Tecate
275C Tecopa
275D Tecopa Pass
212A Tehachapi NE
212B Tehachapi North
212C Tehachapi South
375A Tehipite Dome
214D Tejon Hills
213C Tejon Ranch
134D Telegraph Peak
302A Telescope Peak
050B Temecula
269C Templeton
329C Templeton Mtn.
455A Ten Lakes
454C Tenaya Lake
223C Tenmile Well
714C Tennant
291C Tent Hills
194B Tepusquet Canyon
479C Terminous
657C Termo
582A Thatcher Ridge
208C The Buttes
292A The Dark Hole
326B The Dunes
534D The Geysers
729C The Panhandle
374D The Sphinx
715C The Whaleback
064A Thermal Canyon
368D Thimble Peak
318B Thompson Canyon
685D Thompson Peak
479B Thornton
644A Thousand Lakes Valley
113A Thousand Oaks
385A Three Sisters
041A Thumb Peak
668B Thurston Peaks
230D Tiefort Mountains
008D Tierra Del Sol
294C Tierra Redonda Mountain
472C Tiltill Mtn.
436D Timber Knob
678B Timbered Crater
369B Tin Mountain
392C Tinemaha Reservoir
454A Tioga Pass
311D Tipton
545D Tisdale Weir
687D Tish Tang Point
285D Tobias Peak
485B Tomales
519B Tombs Creek
682D Tombstone Mtn.
433C Toms Place
112D Topanga
165D Topatopa Mountains
505A Topaz Lake
341A Topo Valley
146B Topock
065C Toro Peak
090D Torrance
028A Tortuga
472A Tower Peak
444B Tracy
507B Tragedy Spring
360A Tranquillity
171D Tranquillon Mtn.
334B Traver
338A Tres Pecos Farms
385D Tres Pinos
377A Trimmer
689C Trinidad
666B Trinity Center
667D Trinity Dam
686C Trinity Mtn.
353D Triple Divide Peak
113C Triunfo Pass
280D Trona East
280C Trona West
640B Troxel Point
180D Troy Lake
554C Truckee
045C Truckhaven
450B Truman Meadows
032B Tubb Canyon
355B Tucker Mtn.
325A Tucki Wash
311A Tulare
410D Tule Canyon
674B Tule Mountain
020B Tule Springs
729B Tulelake
361C Tumey Hills
414D Tungsten Hills
525B Tunnel Hill
640D Tunnison Mtn.
457B Tuolumne
241D Tupman
423A Turlock
441D Turlock Lake
402B Turner Ranch
227B Turquoise Mtn.
157B Turtle Valley
628A Tuscan Buttes NE
610A Tuscan Springs
071A Tustin
606D Twain
474C Twain Harte
183A Twelve Gauge Lake
298C Twelvemile Spring
102A Twentynine Palms
101C Twentynine Palms Mtn.
471A Twin Lakes
195A Twitchell Dam
502D Two Rock
672B Tyee City
187B Tylerhorse Canyon
389C Ubehebe Crater
349A Ubehebe Peak
392B Uhlmeyer Spring
550B Ukiah
720D Ukonom Lake
720C Ukonom Mtn.
462C Union Island
351A Union Wash
599B Updegraff Ridge
305A Upper Centennial Flat
549B Upper Lake
049B Vail Lake
138B Val Verde
064C Valerie
251A Valjean Hills
293B Valleton
035A Valley Center
502C Valley Ford
101B Valley Mtn.
477A Valley Springs
477C Valley Springs SW
226B Valley Wells
135A Valyermo
566A Van Arsdale Reservoir
111B Van Nuys
176D Van Winkle Spring
151A Van Winkle Wash
336C Vanguard
090B Venice
343C Ventana Cones
141D Ventura
328C Vermillion Canyon
444A Vernalis
529D Verona
376D Verplank Ridge
158D Victorville
158B Victorville NW
078C Victory Pass
096D Vidal
096A Vidal Junction
096B Vidal NW
020C Viejas Mountain

296C Villa Creek
594A Vina
482D Vine Hill
644C Viola
334D Visalia
454D Vogelsang Peak
282A Volcano Peak
403C Volta
368A Wahguyhe Peak
356B Wahtoke
259A Walker Pass
477B Wallace
465A Walnut Creek
516A Walter Springs
395B Ward Mountain
519A Warm Springs Dam
163D Warm Springs Mountain
048C Warner Springs
033B Warners Ranch
690B Warren Peak
264D Wasco
264B Wasco NW
264C Wasco SW
557D Washington
693B Washington Mtn.
080C Washington Wash
207C Water Valley
442A Waterford
478C Waterloo
135C Waterman Mtn.
386B Watsonville East
387A Watsonville West
433B Watterson Canyon
371B Waucoba Canyon
392D Waucoba Mtn.
391C Waucoba Spring
312A Waukena
437C Wawona
649B Weaverville
555B Webber Peak
699B Weed
214B Weed Patch
727A Weed Valley
687B Weitchpec
260A Weldon
217C Wells Ranch
620C Wendel
621D Wendel Hot Springs
539D Wentworth Springs
635C Weott
081C West Berdoo Canyon
290D West Camp
242D West Elk Hills
715D West Haight Mtn.
228C West Of Baker
561A West Of Biggs
233B West Of Black Hills
175C West Of Blind Hills
178B West Of Broadwell Mesa
177C West Of Budweiser Wash
230B West Of Drinkwater Lake
299B West Of Eagle Mtn.
173C West Of Flattop Mtn.
324B West Of Furnace Creek
611D West Of Gerber
177B West Of Glasgow
389B West Of Gold Mtn.
468C West Of Huntoon Spring
198B West Of Juniper Mine
712C West Of Kephart
254C West Of Leach Spring
123B West Of Mohawk Spring
231B West Of Nelson Lake
077B West Of Palen Pass
041C West Of Palo Verde Peak
229C West Of Red Pass Lake
273C West Of Shenandoah Peak
639B West Of Snowstorm Mtn.
203B West Of Soda Lake
370C West Of Teakettle Junction
349B West Of Ubehebe Peak
725B West Of Willow Ranch
156B West Ord Mtn.
493A West Point
643C West Prospect Peak
256B Westend
412C Westgard Pass
313B Westhaven
443C Westley
029D Westmorland (=Westmorland East)
585A Westport
337B Westside
540B Westville
623C Westwood East
624D Westwood West
147A Whale Mtn.
032D Whale Peak
543C Wheatland
166D Wheeler Springs
121C Whipple Mts. SW
121D Whipple Wash
648A Whiskeytown
683D Whisky Bill Peak
533C Whispering Pines
163C Whitaker Peak
418A White Chief Mtn.
283C White Dome
281C White Hills
156C White Horse Mtn.
141B White Ledge Peak
432A White Mtn. Peak
286D White River
369C White Top Mtn.
083B White Water
695C Whitehorse
658B Whitinger Mtn.
645B Whitmore
434A Whitmore Hot Springs
709B Whittemore Ridge
089A Whittier
028C Wiest
547C Wilbur Springs
425D Wilcox Ridge
183D Wild Crossing
069D Wildomar
325D Wildrose Peak
531A Wildwood School
041B Wileys Well
124A Wilhelm Spring
546B Williams
295A Williams Hill
206B Williams Well
583D Willis Ridge
567A Willits
670B Willow Creek
725A Willow Ranch
187A Willow Springs
578C Willows
245B Wilson Corner
532B Wilson Valley
068A Winchester
255A Wingate Pass
278D Wingate Wash East
278C Wingate Wash West
514C Winters
188B Winters Ridge
422A Winton
044D Wister
542C Wolf
505C Wolf Creek
522D Woodfords
333A Woodlake
514A Woodland
200D Woods Mountains
429A Woodside
310C Woodville
463A Woodward Island
262A Woody
260D Woolstalf Creek
375D Wren Peak
294A Wunpost
653D Yager Junction
684D Ycatapom Peak
702A Yellow Dog Peak
681C Yellowjacket Mtn.
181A Yermo
613A Yolla Bolly 15' NE (was Tomhead Mtn. 7.5')
613B Yolla Bolly 15' NW (was North Yolla Bolly 7.5')
613D Yolla Bolly 15' SE (was South Yolla Bolly 7.5')
613C Yolla Bolly 15' SW (was Solomon Peak 7.5')
088A Yorba Linda
270D York Mountain
535B Yorkville
455D Yosemite Falls
421B Yosemite Lake
686A Youngs Peak
500A Yountville
717B Yreka
544A Yuba City
106D Yucaipa
103B Yucca Valley North
103C Yucca Valley South
006A Yuha Basin
001A Yuma East
001B Yuma West
169B Zaca Creek
194D Zaca Lake
530C Zamora
536B Zeni Ridge
615B Zenia

Appendix II: Plants of List 1A and List 3, The Lists of Uncertainty

The following two lists are presented to highlight our need to know more about the plants on List 1A (Plants Presumed Extinct in CA) and List 3 (Plants About Which We Need More Information). We hope our contributors will seek to rediscover the plants on List 1A, and consult the records for the information missing for each List 3 plant and try to provide it.

In previous editions we have included all the plants in the *Inventory* sorted by each of the five lists. The utility of this has now dwindled with the release of the CNPS *Electronic Inventory*, which allows preparation of customized lists such as the ones below. Users wishing to know the list placement for a plant are referred to the main body of the text; for printed summaries of the plants by list, please consult the *Electronic Inventory*.

List 1A

Total number of taxa: *34*

Arctostaphylos hookeri ssp. *franciscana*
Astragalus mojavensis var. *hemigyrus*
A. pycnostachyus var. *lanosissimus*
Calochortus monanthus
Carex livida
Castilleja uliginosa
Chorizanthe parryi var. *fernandina*
Dissanthelium californicum
Erigeron mariposanus
Eriogonum truncatum
Eschscholzia rhombipetala
Helianthus nuttallii ssp. *parishii*
Hemizonia mohavensis
Howellia aquatilis
Lycium verrucosum
Malacothamnus mendocinensis
M. parishii
Mimulus brandegei
M. traskiae
M. whipplei
Monardella leucocephala
M. pringlei
Montia howellii
Ophioglossum pusillum
Orthocarpus pachystachyus
Phacelia cinerea
Plagiobothrys glaber
P. hystriculus
P. lithocaryus
P. mollis var. *vestitus*
Poliomintha incana
Potentilla multijuga
Ranunculus hydrocharoides
Tropidocarpum capparideum

List 3

Total number of taxa: *47*

Achnatherum lemmonii var. *pubescens*
Agrostis hendersonii
Arenaria macradenia var. *kuschei*
Berberis fremontii
Calyptridium parryi var. *hesseae*
Calystegia atriplicifolia ssp. *buttensis*
Camissonia lewisii
Cardamine pachystigma var. *dissectifolia*
Chorizanthe parryi var. *parryi*
Dudleya alainae
Equisetum palustre
Erigeron biolettii
Eriogonum luteolum var. *caninum*
Eschscholzia procera
Galium oreganum
Gentiana affinis var. *parvidentata*
Helianthemum suffrutescens
Hemizonia congesta ssp. *leucocephala*
Hordeum intercedens
Ivesia arizonica var. *arizonica*
Lathyrus sulphureus var. *argillaceus*
Lessingia hololeuca
Lewisia cotyledon var. *howellii*
Lupinus albifrons var. *abramsii*
L. eximius
Madia yosemitana
Marsilea oligospora
Mimulus acutidens
M. brachiatus
Monardella antonina ssp. *antonina*
Myosurus minimus ssp. *apus*
Opuntia wigginsii
Phacelia amabilis
Pinus edulis
Plagiobothrys chorisianus var. *chorisianus*
P. glyptocarpus var. *modestus*
Pogogyne douglasii ssp. *parviflora*
Polygonum marinense
Polystichum lonchitis
Ribes amarum var. *hoffmannii*
Salvia dorrii var. *incana*
Scutellaria holmgreniorum
Selaginella densa var. *scopulorum*
Senecio hydrophiloides
Silene occidentalis ssp. *longistipitata*
Tiarella trifoliata var. *trifoliata*
Vaccinium coccineum

Appendix III: Plants by County and Island

The following lists include all the plants on CNPS Lists 1 through 4. Since we record occurrences in mainland counties and on islands separately, the county lists in Appendix III do not include island occurrences even if an island occurs in that county.

Counties

Alameda County

***Total number of taxa:* 64**

***List 1A:* 4; *List 1B:* 36; *List 2:* 0; *List 3:* 4; *List 4:* 20**

Acanthomintha lanceolata, Allium sharsmithae, Amsinckia grandiflora, A. lunaris, Androsace elongata ssp. *acuta, Arctostaphylos pallida, Aspidotis carlotta-halliae, Astragalus tener* var. *tener, Atriplex cordulata, A. coronata* var. *coronata, A. depressa, A. joaquiniana, Balsamorhiza macrolepis* var. *macrolepis, Blepharizonia plumosa* ssp. *plumosa, Calochortus umbellatus, Campanula exigua, Chorizanthe cuspidata* var. *cuspidata, C. robusta* var. *robusta, Cirsium fontinale* var. *campylon, Clarkia breweri, C. concinna* ssp. *automixa, C. franciscana, Cordylanthus maritimus* ssp. *palustris, C. mollis* ssp. *hispidus, C. palmatus, Cryptantha hooveri, Delphinium californicum* ssp. *interius, D. recurvatum, Dirca occidentalis, Eriogonum luteolum* var. *caninum, E. truncatum, Eriophyllum jepsonii, Eschscholzia rhombipetala, Fritillaria agrestis, F. falcata, F. liliacea, Galium andrewsii* ssp. *gatense, Grindelia stricta* var. *angustifolia, Helianthella castanea, Hemizonia parryi* ssp. *congdonii, Holocarpha macradenia, Horkelia cuneata* ssp. *sericea, Lasthenia conjugens, Lathyrus jepsonii* var. *jepsonii, Lessingia hololeuca, Lilaeopsis masonii, Linanthus acicularis, L. ambiguus, L. grandiflorus, Malacothamnus hallii, Micropus amphibolus, Monardella antonina* ssp. *antonina, M. villosa* ssp. *globosa, Myosurus minimus* ssp. *apus, Phacelia phacelioides, Piperia michaelii, Plagiobothrys glaber, Psilocarphus brevissimus* var. *multiflorus, Ranunculus lobbii, Sanicula maritima, Streptanthus albidus* ssp. *peramoenus, Suaeda californica, Trifolium amoenum,* and *Tropidocarpum capparideum.*

Alpine County

***Total number of taxa:* 19**

***List 1A:* 0; *List 1B:* 2; *List 2:* 5; *List 3:* 1; *List 4:* 11**

Agrostis humilis, Antennaria pulchella, Bolandra californica, Carex davyi, C. petasata, Chaenactis douglasii var. *alpina, Claytonia megarhiza, C. umbellata, Cryptantha crymophila, Draba asterophora* var. *asterophora, Eriogonum ovalifolium* var. *eximium, Juncus hemiendytus* var. *abjectus, Podistera nevadensis, Polystichum kruckebergii, P. lonchitis, Potamogeton robbinsii, Silene invisa, Tonestus eximius,* and *Veronica cusickii.*

Amador County

***Total number of taxa:* 26**

***List 1A:* 0; *List 1B:* 7; *List 2:* 1; *List 3:* 2; *List 4:* 16**

Arctostaphylos myrtifolia, Bolandra californica, Calochortus clavatus var. *avius, Carex davyi, Clarkia virgata, Eriogonum apricum* var. *apricum, E. apricum* var. *prostratum, E. tripodum, Eryngium pinnatisectum, Helianthemum suffrutescens, Horkelia parryi, Jepsonia heterandra, Lilium humboldtii* ssp. *humboldtii, Lomatium stebbinsii, Madia yosemitana, Mimulus inconspicuus, M. laciniatus, Monardella candicans, Navarretia eriocephala, N. myersii, N. subuligera, Ophioglossum californicum, Perideridia bacigalupii, Silene invisa, Sphenopholis obtusata,* and *Veronica cusickii.*

Butte County

***Total number of taxa:* 71**

***List 1A:* 0; *List 1B:* 28; *List 2:* 10; *List 3:* 9; *List 4:* 24**

Agrostis hendersonii, Allium jepsonii, A. sanbornii var. *sanbornii, Arctostaphylos mewukka* ssp. *truei, Astragalus pauperculus, A. tener* var. *ferrisiae, Atriplex cordulata, Azolla mexicana, Balsamorhiza macrolepis* var. *macrolepis, Botrychium ascendens, B. crenulatum, B. minganense, B. montanum, Calycadenia oppositifolia, Calystegia atriplicifolia* ssp. *buttensis, Cardamine pachystigma* var. *dissectifolia, Carex geyeri, C. vulpinoidea, Chamaesyce hooveri, Clarkia gracilis* ssp. *albicaulis, C. mildrediae, C. mosquinii* ssp. *mosquinii, C. mosquinii* ssp. *xerophila, Claytonia palustris, Corydalis caseana* ssp. *caseana, Cypripedium californicum, C. fasciculatum, Eleocharis quadrangulata, Erigeron inornatus* var. *calidipetris, E. petrophilus* var. *sierrensis, Fritillaria eastwoodiae, F. pluriflora, Hibiscus lasiocarpus, Juncus leiospermus* var. *ahartii, J. leiospermus* var. *leiospermus, Lewisia cantelovii, Lilium humboldtii* ssp. *humboldtii, Limnanthes floccosa* ssp. *californica, L. floccosa* ssp. *floccosa, Lupinus dalesiae, Marsilea oligospora, Mimulus glaucescens, M. inconspicuus, M. laciniatus, Monardella douglasii* ssp. *venosa, Myosurus minimus* ssp. *apus, Navarretia heterandra, N. subuligera, Ophioglossum californicum, Orcuttia pilosa, Paronychia ahartii, Penstemon personatus, Perideridia bacigalupii, Pogogyne douglasii* ssp. *parviflora, Polygonum bidwelliae, Polystichum kruckebergii, Rhynchospora californica, Rupertia hallii, Sagittaria sanfordii, Sanicula tracyi, Scirpus subterminalis, Sedum albomarginatum, Senecio eurycephalus* var. *lewisrosei, S. hydrophiloides, Sidalcea robusta, Silene occidentalis* ssp. *longistipitata, Stellaria longifolia, S. obtusa, Streptanthus drepanoides, Tuctoria greenei,* and *Vaccinium coccineum.*

Calaveras County

***Total number of taxa:* 26**

***List 1A:* 1; *List 1B:* 7; *List 2:* 0; *List 3:* 2; *List 4:* 16**

Agrostis hendersonii, Allium sanbornii var. *sanbornii, A. tribracteatum, Arctostaphylos myrtifolia, Bolandra californica, Calycadenia hooveri, Carex davyi, Ceanothus fresnensis, Clarkia virgata, Cryptantha mariposae, Delphinium hansenii* ssp. *ewanianum, Eryngium pinnatisectum, E. racemosum, Helianthemum suffrutescens, Horkelia parryi, Jepsonia heterandra, Juncus leiospermus* var. *ahartii, Lilium humboldtii* ssp. *humboldtii, Lomatium stebbinsii, Mimulus inconspicuus, M. pulchellus, M. whipplei, Monardella candicans, Navarretia eriocephala, Perideridia bacigalupii,* and *Psilocarphus tenellus* var. *globiferus.*

Colusa County

Total number of taxa: 53

List 1A: 1; List 1B: 23; List 2: 2; List 3: 3; List 4: 24

Allium fimbriatum var. *purdyi, Antirrhinum subcordatum, Arctostaphylos malloryi, Asclepias solanoana, Astragalus breweri, A. clevelandii, A. rattanii* var. *jepsonianus, A. rattanii* var. *rattanii, A. tener* var. *ferrisiae, Atriplex depressa, A. joaquiniana, Brodiaea coronaria* ssp. *rosea, Chamaesyce ocellata* ssp. *rattanii, Chlorogalum pomeridianum* var. *minus, Collomia diversifolia, Cordylanthus palmatus, Cryptantha excavata, Delphinium recurvatum, D. uliginosum, Epilobium nivium, Eriastrum brandegeae, Eriogonum luteolum* var. *caninum, E. nervulosum, E. strictum* var. *greenei, E. tripodum, Eschscholzia rhombipetala, Fritillaria pluriflora, F. purdyi, Helianthus exilis, Hesperolinon drymarioides, Hibiscus lasiocarpus, Horkelia bolanderi, Layia septentrionalis, Linanthus rattanii, Lomatium ciliolatum* var. *hooveri, Lotus rubriflorus, Lupinus milo-bakeri, L. sericatus, Madia hallii, Malacothamnus helleri, Mimulus brachiatus, M. glaucescens, Myosurus minimus* ssp. *apus, Navarretia heterandra, N. jepsonii, N. leucocephala* ssp. *bakeri, Neostapfia colusana, Penstemon purpusii, Raillardiopsis scabrida, Senecio clevelandii* var. *clevelandii, Silene campanulata* ssp. *campanulata, Thelypodium brachycarpum,* and *Trichocoronis wrightii* var. *wrightii.*

Contra Costa County

Total number of taxa: 66

List 1A: 3; List 1B: 36; List 2: 4; List 3: 3; List 4: 20

Amsinckia grandiflora, A. lunaris, Androsace elongata ssp. *acuta, Arabis blepharophylla, Arctostaphylos auriculata, A. manzanita* ssp. *laevigata, A. pallida, Aster lentus, Astragalus tener* var. *tener, Atriplex cordulata, A. coronata* var. *coronata, A. depressa, A. joaquiniana, Blepharizonia plumosa* ssp. *plumosa, Calandrinia breweri, Calochortus pulchellus, C. umbellatus, Campanula exigua, Collomia diversifolia, Convolvulus simulans, Cordylanthus mollis* ssp. *mollis, C. nidularius, Cryptantha hooveri, Delphinium californicum* ssp. *interius, D. recurvatum, Dirca occidentalis, Eleocharis parvula, Eriogonum luteolum* var. *caninum, E. nudum* var. *decurrens, E. truncatum, Eriophyllum jepsonii, Erysimum capitatum* ssp. *angustatum, Eschscholzia rhombipetala, Fritillaria agrestis, F. liliacea, Galium andrewsii* ssp. *gatense, Grindelia stricta* var. *angustifolia, Helianthella castanea, Hemizonia parryi* ssp. *congdonii, Hesperolinon breweri, Hibiscus lasiocarpus, Holocarpha macradenia, Isocoma arguta, Juglans californica* var. *hindsii, Lasthenia conjugens, Lathyrus jepsonii* var. *jepsonii, Lilaeopsis masonii, Limosella subulata, Linanthus acicularis, L. ambiguus, Madia radiata, Malacothamnus hallii, Micropus amphibolus, Monardella antonina* ssp. *antonina, M. villosa* ssp. *globosa, Myosurus minimus* ssp. *apus, Oenothera deltoides* ssp. *howellii, Phacelia phacelioides, Piperia michaelii, Potamogeton zosteriformis, Ranunculus lobbii, Sanicula saxatilis, Senecio aphanactis, Streptanthus albidus* ssp. *peramoenus, S. hispidus,* and *Tropidocarpum capparideum.*

Del Norte County

Total number of taxa: 115

List 1A: 1; List 1B: 20; List 2: 28; List 3: 3; List 4: 63

Abronia umbellata ssp. *breviflora, Allium siskiyouense, Antennaria suffrutescens, Arabis aculeolata, A. koehleri* var. *stipitata, A. mac-donaldiana, A. serpentinicola, Arctostaphylos hispidula, A. nortensis, Arnica cernua, A. spathulata, Asarum marmoratum, Asplenium trichomanes* ssp. *trichomanes, Boschniakia hookeri, Calamagrostis crassiglumis, C. foliosa, Cardamine nuttallii* var. *gemmata, Carex gigas, C. leptalea, C. praticola, Castilleja hispida* ssp. *brevilobata, C. miniata* ssp. *elata, Cochlearia officinalis* var. *arctica, Collomia tracyi, Cupressus nootkatensis, Cypripedium californicum, C. fasciculatum, C. montanum, Darlingtonia californica, Dicentra formosa* ssp. *oregana, Draba carnosula, Empetrum nigrum* ssp. *hermaphroditum, Epilobium oreganum, E. rigidum, Erigeron bloomeri* var. *nudatus, E. cervinus, Eriogonum hirtellum, E. nudum* var. *paralinum, E. pendulum, E. ternatum, Erythronium citrinum* var. *citrinum, E. hendersonii, E. howellii, Galium oreganum, Gentiana plurisetosa, G. setigera, Heuchera chlorantha, Horkelia congesta* ssp. *nemorosa, H. sericata, Iliamna latibracteata, Iris bracteata, I. innominata, Juncus supiniformis, Lathyrus delnorticus, L. palustris, Lewisia cotyledon* var. *howellii, L. oppositifolia, Lilium bolanderi, L. kelloggii, L. occidentale, L. pardalinum* ssp. *vollmeri, L. pardalinum* ssp. *wigginsii, L. rubescens, L. washingtonianum* ssp. *purpurascens, Listera cordata, Lomatium howellii, L. martindalei, Lupinus lapidicola, L. tracyi, Lycopus uniflorus, Melica spectabilis, Minuartia howellii, Monotropa uniflora, Montia howellii, Navarretia subuligera, Oenothera wolfii, Perideridia gairdneri* ssp. *gairdneri, Phacelia argentea, Pinguicula vulgaris* ssp. *macroceras, Piperia candida, Pityopus californicus, Pleuropogon refractus, Poa piperi, P. rhizomata, Potamogeton foliosus* var. *fibrillosus, Pyrrocoma racemosa* var. *congesta, Rhynchospora alba, Ribes laxiflorum, Rubus nivalis, Sagittaria sanfordii, Salix delnortensis, Sanguisorba officinalis, Sanicula peckiana, S. tracyi, Saxifraga howellii, S. nuttallii, Scirpus subterminalis, Sedum laxum* ssp. *flavidum, S. laxum* ssp. *heckneri, Selaginella densa* var. *scopulorum, Senecio macounii, Sidalcea malvaeflora* ssp. *patula, Smilax jamesii, Streptanthus howellii, Tauschia glauca, Thermopsis gracilis, T. robusta, Trientalis arctica, Trifolium howellii, Triteleia crocea* var. *crocea, Vancouveria chrysantha, Veratrum insolitum, Viola langsdorfii, V. palustris,* and *V. primulifolia* ssp. *occidentalis.*

El Dorado County

Total number of taxa: 46

List 1A: 1; List 1B: 18; List 2: 6; List 3: 3; List 4: 18

Allium sanbornii var. *congdonii, A. sanbornii* var. *sanbornii, Antennaria pulchella, Arctostaphylos nissenana, Bolandra californica, Botrychium ascendens, Calochortus clavatus* var. *avius, Calystegia stebbinsii, Carex davyi, C. limosa, Ceanothus roderickii, Chaenactis douglasii* var. *alpina, Chlorogalum grandiflorum, Clarkia virgata, Draba asterophora* var. *asterophora, D. asterophora* var. *macrocarpa, Epilobium oreganum, Erigeron petrophilus* var. *sierrensis, Eriogonum ovalifolium* var. *eximium, E. tripodum, Fremontodendron decumbens, Galium californicum* ssp. *sierrae, Helianthemum suffrutescens, Horkelia parryi, Jepsonia heterandra,*

Lewisia longipetala, L. serrata, Lilium humboldtii ssp. *humboldtii, Marsilea oligospora, Monardella candicans, Navarretia eriocephala, N. prolifera* ssp. *lutea, Ophioglossum pusillum, Phacelia stebbinsii, Podistera nevadensis, Polystichum lonchitis, Potamogeton epihydrus* ssp. *nuttallii, Rorippa subumbellata, Scirpus subterminalis, Scutellaria galericulata, Senecio layneae, Silene invisa, Sparganium natans, Tonestus eximius, Viola tomentosa,* and *Wyethia reticulata.*

Fresno County

Total number of taxa: *115*

List 1A: *0;* ***List 1B:*** *48;* ***List 2:*** *4;* ***List 3:*** *3;* ***List 4:*** *60*

Acanthomintha lanceolata, A. obovata ssp. *obovata, Amsinckia vernicosa* var. *furcata, Androsace elongata* ssp. *acuta, Angelica callii, Antennaria pulchella, Arabis bodiensis, Astragalus kentrophyta* var. *danaus, A. monoensis* var. *ravenii, Atriplex cordulata, A. coronata* var. *coronata, A. depressa, A. minuscula, A. vallicola, Botrychium minganense, Calyptridium pulchellum, Calystegia collina* ssp. *venusta, C. malacophylla* var. *berryi, Camissonia benitensis, C. sierrae* ssp. *alticola, Carex congdonii, C. tompkinsii, Carpenteria californica, Castilleja campestris* ssp. *succulenta, Caulanthus californicus, Ceanothus fresnensis, Chorizanthe biloba* var. *immemora, Clarkia breweri, Claytonia palustris, Cordylanthus palmatus, C. tenuis* ssp. *barbatus, Cryptantha rattanii, Delphinium gypsophilum* ssp. *gypsophilum, D. inopinum, D. recurvatum, Dicentra nevadensis, Draba sharsmithii, D. sierrae, Epilobium howellii, Eriastrum hooveri, Erigeron aequifolius, E. inornatus* var. *keilii, Eriogonum eastwoodianum, E. gossypinum, E. heermannii* var. *occidentale, E. nudum* var. *indictum, E. nudum* var. *murinum, E. nudum* var. *regivirum, E. polypodum, E. prattenianum* var. *avium, E. vestitum, Eriophyllum lanatum* var. *obovatum, Eryngium spinosepalum, Eschscholzia hypecoides, Fritillaria agrestis, Galium andrewsii* ssp. *gatense, Goodmania luteola, Gratiola heterosepala, Hackelia sharsmithii, Hordeum intercedens, Hulsea brevifolia, Ivesia campestris, I. unguiculata, Jamesia americana* var. *rosea, Lathyrus jepsonii* var. *jepsonii, Layia discoidea, L. heterotricha, L. munzii, Lembertia congdonii, Lepidium jaredii* ssp. *album, Lessingia occidentalis, Lewisia congdonii, L. longipetala, Lilium humboldtii* ssp. *humboldtii, Linanthus acicularis, L. oblanceolatus, L. serrulatus, Lupinus citrinus* var. *citrinus, L. lepidus* var. *culbertsonii, Madia radiata, M. yosemitana, Malacothamnus aboriginum, Mimulus acutidens, M. gracilipes, M. grayi, M. laciniatus, M. subsecundus, Minuartia obtusiloba, Monardella candicans, Navarretia nigelliformis* ssp. *radians, Nemacladus gracilis, Nemophila parviflora* var. *quercifolia, Orcuttia inaequalis, Phacelia orogenes, Pityopus californicus, Plagiobothrys myosotoides, Pseudobahia bahiifolia, P. peirsonii, Psilocarphus tenellus* var. *globiferus, Puccinellia californica, Raillardiopsis muirii, Ribes menziesii* var. *ixoderme, Sagittaria sanfordii, Scirpus clementis, Senecio aphanactis, Sidalcea keckii, Solidago guiradonis, Sphenopholis obtusata, Streptanthus farnsworthianus, S. fenestratus, S. gracilis, Trichostema ovatum, Trifolium bolanderi, Tuctoria greenei,* and *Wyethia elata.*

Glenn County

Total number of taxa: *32*

List 1A: *1;* ***List 1B:*** *15;* ***List 2:*** *2;* ***List 3:*** *0;* ***List 4:*** *14*

Antirrhinum subcordatum, Asclepias solanoana, Astragalus rattanii var. *jepsonianus, A. rattanii* var. *rattanii, Atriplex cordulata, A. depressa, A. joaquiniana, Brodiaea coronaria* ssp. *rosea, Calyptridium quadripetalum, Chamaesyce hooveri, C. ocellata* ssp. *rattanii, Collomia diversifolia, Epilobium nivium, Eriastrum brandegeae, Eriogonum nervulosum, Fritillaria pluriflora, F. purdyi, Hackelia amethystina, Helianthus exilis, Hesperolinon drymarioides, H. tehamense, Hibiscus lasiocarpus, Linanthus rattanii, Malacothamnus helleri, Navarretia jepsonii, Orcuttia pilosa, Orobanche valida* ssp. *howellii, Penstemon purpusii, Sedum laxum* ssp. *flavidum, Sidalcea oregana* ssp. *hydrophila, Stellaria obtusa,* and *Tropidocarpum capparideum.*

Humboldt County

Total number of taxa: *111*

List 1A: *1;* ***List 1B:*** *28;* ***List 2:*** *17;* ***List 3:*** *3;* ***List 4:*** *62*

Abronia umbellata ssp. *breviflora, Allium hoffmanii, A. siskiyouense, Antennaria suffrutescens, Arabis rigidissima* var. *rigidissima, Arctostaphylos canescens* ssp. *sonomensis, A. hispidula, Arnica cernua, A. spathulata, Astragalus agnicidus, A. rattanii* var. *rattanii, A. umbraticus, Bensoniella oregona, Boschniakia hookeri, Calamagrostis bolanderi, C. crassiglumis, C. foliosa, Carex geyeri, C. leptalea, C. praticola, Castilleja ambigua* ssp. *humboldtiensis, C. mendocinensis, Clarkia amoena* ssp. *whitneyi, Collinsia corymbosa, Collomia tracyi, Cordylanthus maritimus* ssp. *palustris, Cypripedium californicum, C. fasciculatum, C. montanum, Dicentra formosa* ssp. *oregana, Draba howellii, Eleocharis parvula, Empetrum nigrum* ssp. *hermaphroditum, Epilobium oreganum, E. septentrionale, Erigeron biolettii, E. decumbens* var. *robustior, E. petrophilus* var. *viscidulus, Erysimum menziesii* ssp. *eurekense, Erythronium citrinum* var. *citrinum, Fritillaria purdyi, Gentiana plurisetosa, Glyceria grandis, Hemizonia congesta* ssp. *tracyi, Hesperevax sparsiflora* var. *brevifolia, Hesperolinon adenophyllum, Horkelia sericata, Iliamna latibracteata, Iris tenax* ssp. *klamathensis, Juncus dudleyi, J. supiniformis, Lathyrus biflorus, L. glandulosus, L. palustris, Layia carnosa, Lewisia cotyledon* var. *heckneri, L. cotyledon* var. *howellii, Lilium bolanderi, L. kelloggii, L. occidentale, L. pardalinum* ssp. *vollmeri, L. rubescens, L. washingtonianum* ssp. *purpurascens, Linanthus acicularis, Listera cordata, Lomatium tracyi, Lotus yollabolliensis, Lupinus constancei, L. lapidicola, L. tracyi, Lycopodiella inundata, Lycopodium clavatum, Lycopus uniflorus, Melica spectabilis, Microseris borealis, Minuartia decumbens, Monardella villosa* ssp. *globosa, Monotropa uniflora, Montia howellii, Oenothera wolfii, Orthocarpus cuspidatus* ssp. *cuspidatus, Penstemon purpusii, Perideridia gairdneri* ssp. *gairdneri, Piperia candida, P. michaelii, Pityopus californicus, Pleuropogon refractus, Puccinellia pumila, Ribes laxiflorum, R. marshallii, Rorippa columbiae, Sanguisorba officinalis, Sanicula tracyi, Scirpus subterminalis, Sedum laxum* ssp. *flavidum, S. laxum* ssp. *heckneri, Sidalcea malachroides, S. malvaeflora* ssp. *patula, S. oregana* ssp. *eximia, Silene marmorensis, Stellaria obtusa, Tauschia glauca, Thermopsis gracilis, T. robusta, Thlaspi californicum, Tiarella trifoliata* var. *trifoliata, Tracyina rostrata, Trifolium howellii, Veratrum insolitum, Viola palustris,* and *Wyethia longicaulis.*

Imperial County

Total number of taxa: *52*

List 1A: *0;* ***List 1B:*** *13;* ***List 2:*** *17;* ***List 3:*** *1;* ***List 4:*** *21*

Allium parishii, Astragalus crotalariae, A. douglasii var. *perstrictus, A. insularis* var. *harwoodii, A. lentiginosus* var. *borreganus, A. magdalenae* var. *peirsonii, A. nutans, Bursera microphylla, Calliandra eriophylla, Carnegiea gigantea, Castela emoryi, Chaenactis carphoclinia* var. *peirsonii, Chamaesyce arizonica, C. platysperma, Colubrina californica, Condalia globosa* var. *pubescens, Croton wigginsii, Cryptantha costata, C. holoptera, Cynanchum utahense, Delphinium parishii* ssp. *subglobosum, Ditaxis clariana, Eschscholzia hypecoides, Escobaria vivipara* var. *alversonii, Eucnide rupestris, Fremontodendron mexicanum, Helianthus niveus* ssp. *tephrodes, Ipomopsis effusa, I. tenuifolia, Juncus acutus* ssp. *leopoldii, Koeberlinia spinosa* ssp. *tenuispina, Lotus haydonii, Lupinus excubitus* var. *medius, Lycium parishii, Lyrocarpa coulteri* var. *palmeri, Malperia tenuis, Mentzelia hirsutissima, Mimulus aridus, Mirabilis tenuiloba, Nama stenocarpum, Opuntia munzii, O. wigginsii, O. wolfii, Palafoxia arida* var. *gigantea, Penstemon clevelandii* var. *connatus, P. thurberi, Pholisma sonorae, Pilostyles thurberi, Proboscidea althaeifolia, Salvia greatae, Senna covesii,* and *Xylorhiza orcuttii.*

Inyo County

Total number of taxa: *160*

List 1A: *2;* ***List 1B:*** *50;* ***List 2:*** *46;* ***List 3:*** *2;* ***List 4:*** *60*

Abronia nana ssp. *covillei, Achnatherum aridum, Agave utahensis, Androstephium breviflorum, Antennaria pulchella, Arabis bodiensis, A. dispar, A. pulchra* var. *munciensis, A. pygmaea, A. shockleyi, Arctomecon merriamii, Astragalus argophyllus* var. *argophyllus, A. atratus* var. *mensanus, A. funereus, A. geyeri* var. *geyeri, A. gilmanii, A. kentrophyta* var. *danaus, A. kentrophyta* var. *elatus, A. lentiginosus* var. *kernensis, A. lentiginosus* var. *micans, A. lentiginosus* var. *piscinensis, A. lentiginosus* var. *sesquimetralis, A. mojavensis* var. *hemigyrus, A. monoensis* var. *ravenii, A. platytropis, A. preussii* var. *preussii, A. serenoi* var. *shockleyi, Blepharidachne kingii, Bouteloua trifida, Calochortus excavatus, C. panamintensis, Camissonia boothii* ssp. *boothii, Carex congdonii, C. eleocharis, C. incurviformis* var. *danaensis, Caulostramina jaegeri, Chaenactis douglasii* var. *alpina, Cheilanthes wootonii, Chenopodium simplex, Chrysothamnus gramineus, Cordylanthus eremicus* ssp. *eremicus, C. eremicus* ssp. *kernensis, C. tecopensis, Crepis runcinata* ssp. *hallii, Cryptantha costata, C. holoptera, C. roosiorum, C. scoparia, C. tumulosa, Cymopterus gilmanii, C. ripleyi, Dedeckera eurekensis, Delphinium inopinum, Draba californica, D. sharsmithii, D. sierrae, D. subumbellata, Dryopteris filix-mas, Dudleya calcicola, D. saxosa* ssp. *saxosa, Enceliopsis covillei, E. nudicaulis, Ericameria gilmanii, Erigeron calvus, E. uncialis* var. *uncialis, Eriogonum bifurcatum, E. contiguum, E. eremicola, E. gilmanii, E. hoffmannii* var. *hoffmannii, E. hoffmannii* var. *robustius, E. intrafractum, E. microthecum* var. *lapidicola, E. microthecum* var. *panamintense, E. puberulum, E. shockleyi* var. *shockleyi, Erioneuron pilosum, Fendlerella utahensis, Fimbristylis thermalis, Galium hilendiae* ssp. *carneum, G. hypotrichium* ssp. *tomentellum, Gentiana prostrata, Gilia ripleyi, Gilmania luteola, Goodmania luteola, Grindelia fraxino-pratensis, Hackelia brevicula, H. sharsmithii, Halimolobos virgata, Heuchera duranii, Horkelia hispidula, Hulsea vestita* ssp. *inyoensis, Ivesia arizonica* var. *arizonica, I. campestris, I. kingii* var. *kingii, Jamesia americana* var. *rosea, Juncus nodosus, Lasthenia leptalea, Linanthus arenicola, L. oblanceolatus, Loeflingia squarrosa* var. *artemisiarum, Lomatium foeniculaceum* ssp. *inyoense, L. rigidum, Lupinus gracilentus, L. holmgrenanus, L. magnificus* var. *glarecola, L. magnificus* var. *hesperius, L. magnificus* var. *magnificus, L. padre-crowleyi, Maurandya petrophila, Mimulus glabratus* ssp. *utahensis, M. rupicola, Minuartia obtusiloba, Monardella beneolens, Muilla coronata, Nitrophila mohavensis, Oenothera caespitosa* ssp. *crinita, O. californica* ssp. *eurekensis, Opuntia pulchella, Oryctes nevadensis, Oxytheca watsonii, Penstemon calcareus, P. fruticiformis* var. *amargosae, P. papillatus, P. stephensii, Perityle inyoensis, P. villosa, Petalonyx thurberi* ssp. *gilmanii, Peteria thompsoniae, Phacelia amabilis, P. anelsonii, P. inyoensis, P. mustelina, P. nashiana, P. novenmillensis, P. pulchella* var. *gooddingii, Phlox dispersa, Pinus longaeva, Piptatherum micranthum, Plagiobothrys salsus, Polygala heterorhyncha, Populus angustifolia, Potamogeton robbinsii, Ranunculus hydrocharoides, Salvia funerea, Scirpus clementis, Sclerocactus polyancistrus, Scutellaria lateriflora, Selaginella leucobryoides, Selinocarpus nevadensis, Sidalcea covillei, Spartina gracilis, Sphaeralcea rusbyi* var. *eremicola, Sphenopholis obtusata, Streptanthus gracilis, S. oliganthus, Swallenia alexandrae, Tetracoccus ilicifolius, Tonestus eximius,* and *Trifolium macilentum* var. *dedeckerae.*

Kern County

Total number of taxa: *122*

List 1A: *0;* ***List 1B:*** *63;* ***List 2:*** *2;* ***List 3:*** *2;* ***List 4:*** *55*

Allium shevockii, Amsinckia vernicosa var. *furcata, Androsace elongata* ssp. *acuta, Angelica callii, Antirrhinum ovatum, Astragalus ertterae, A. macrodon, A. subvestitus, Atriplex cordulata, A. coronata* var. *coronata, A. depressa, A. minuscula, A. tularensis, A. vallicola, Azolla mexicana, Calochortus palmeri* var. *palmeri, C. striatus, C. westonii, Calycadenia villosa, Camissonia integrifolia, C. kernensis* ssp. *kernensis, Canbya candida, Castilleja plagiotoma, Caulanthus californicus, Chorizanthe spinosa, Cirsium crassicaule, Clarkia exilis, C. tembloriensis* ssp. *calientensis, C. xantiana* ssp. *parviflora, Convolvulus simulans, Cordylanthus eremicus* ssp. *kernensis, C. mollis* ssp. *hispidus, Cupressus arizonica* ssp. *nevadensis, Cymopterus deserticola, Delphinium gypsophilum* ssp. *gypsophilum, D. hansenii* ssp. *ewanianum, D. inopinum, D. parryi* ssp. *purpureum, D. purpusii, D. recurvatum, Dudleya calcicola, Eremalche kernensis, Eriastrum hooveri, Ericameria gilmanii, Erigeron aequifolius, Eriogonum breedlovei* var. *breedlovei, E. breedlovei* var. *shevockii, E. gossypinum, E. kennedyi* var. *pinicola, E. nudum* var. *indictum, E. temblorense, Eriophyllum jepsonii, E. lanatum* var. *hallii, E. lanatum* var. *obovatum, Eschscholzia lemmonii* ssp. *kernensis, E. minutiflora* ssp. *twisselmannii, E. procera, Fimbristylis thermalis, Fritillaria agrestis, F. brandegei, F. striata, Galium angustifolium* ssp. *onycense, Gilia latiflora* ssp. *cuyamensis, Goodmania luteola, Hemizonia arida, Heterotheca villosa* var. *shevockii, Hulsea vestita* ssp. *parryi, Lasthenia glabrata* ssp. *coulteri, L. leptalea, Layia heterotricha, L. leucopappa, L. munzii, Lembertia congdonii, Lepidium jaredii* ssp. *jaredii, Lewisia disepala, Linanthus grandiflorus, L. serrulatus, Loeflingia squarrosa* var. *artemisiarum, Lomatium shevockii, Lupinus excubitus* var. *johnstonii, Madia radiata, Mimulus microphyllus, M. pictus, M. shevockii, Monardella beneolens, M. candicans, M. linoides* ssp. *oblonga, Mucronea californica, Muilla coronata, Myosurus minimus* ssp. *apus*, *N. peninsularis*, *N. setiloba*, *Nemacladus gracilis*, *N. twisselmannii*, *Nemophila parviflora* var. *quercifolia*, *Opuntia basilaris* var. *treleasei*, *Penstemon papillatus*, *Perideridia gairdneri* ssp. *gairdneri*, *P. pringlei*, *Phacelia exilis*, *P. nashiana*, *P. novenmillensis*, *Pseudobahia peirsonii*, *Psilocarphus tenellus* var. *globiferus*, *Psorothamnus arborescens* var. *arborescens*, *Puccinellia parishii*, *Raillardiopsis muirii*, *Ribes menziesii* var. *ixoderme*, *Sagittaria sanfordii*, *Sclerocactus polyancistrus*, *Selaginella asprella*, *Senecio ionophyllus*, *Streptanthus cordatus* var. *piutensis*, *S. farnsworthianus*, *Stylocline citroleum*, *S. masonii*, *Swertia neglecta*, *Syntrichopappus lemmonii*, *Trichostema ovatum*, *Trifolium macilentum* var. *dedeckerae*, *Viola aurea*, and *V. pinetorum* ssp. *grisea*.

Kings County

Total number of taxa: 16

List 1A: 0; *List 1B:* 8; *List 2:* 0; *List 3:* 1; *List 4:* 7

Amsinckia vernicosa var. *furcata*, *Atriplex cordulata*, *A. coronata* var. *coronata*, *A. vallicola*, *Caulanthus californicus*, *Cirsium crassicaule*, *Delphinium gypsophilum* ssp. *gypsophilum*, *D. recurvatum*, *Eriastrum hooveri*, *Eriogonum gossypinum*, *Hordeum intercedens*, *Layia heterotricha*, *Lembertia congdonii*, *Madia radiata*, *Nemacladus gracilis*, and *Trichostema ovatum*.

Lake County

Total number of taxa: 93

List 1A: 1; *List 1B:* 42; *List 2:* 3; *List 3:* 5; *List 4:* 42

Achnatherum lemmonii var. *pubescens*, *Allium fimbriatum* var. *purdyi*, *Amsinckia lunaris*, *Antirrhinum subcordatum*, *A. virga*, *Arabis oregana*, *Arctostaphylos canescens* ssp. *sonomensis*, *A. stanfordiana* ssp. *raichei*, *Asclepias solanoana*, *Astragalus breweri*, *A. clevelandii*, *A. rattanii* var. *jepsonianus*, *A. rattanii* var. *rattanii*, *Azolla mexicana*, *Brodiaea coronaria* ssp. *rosea*, *Calamagrostis ophitidis*, *Calyptridium quadripetalum*, *Calystegia collina* ssp. *oxyphylla*, *Carex comosa*, *C. confusus*, *C. divergens*, *Chlorogalum pomeridianum* var. *minus*, *Collomia diversifolia*, *Cordylanthus tenuis* ssp. *brunneus*, *Cryptantha clevelandii* var. *dissita*, *C. excavata*, *Delphinium uliginosum*, *Epilobium nivium*, *Equisetum palustre*, *Eriastrum brandegeae*, *Erigeron angustatus*, *Eriogonum luteolum* var. *caninum*, *E. nervulosum*, *E. tripodum*, *Eryngium constancei*, *Erythronium helenae*, *Fritillaria pluriflora*, *F. purdyi*, *Gratiola heterosepala*, *Hackelia amethystina*, *Helianthus exilis*, *Hemizonia congesta* ssp. *calyculata*, *Hesperolinon adenophyllum*, *H. bicarpellatum*, *H. didymocarpum*, *H. drymarioides*, *H. serpentinum*, *Horkelia bolanderi*, *Lasthenia burkei*, *Layia septentrionalis*, *Legenere limosa*, *Lilium rubescens*, *Limnanthes floccosa* ssp. *floccosa*, *Linanthus acicularis*, *L. rattanii*, *Lomatium ciliolatum* var. *hooveri*, *L. repostum*, *Lupinus antoninus*, *L. sericatus*, *Madia hallii*, *Malacothamnus helleri*, *Micropus amphibolus*, *Mimulus brachiatus*, *M. glaucescens*, *M. nudatus*, *Monardella villosa* ssp. *globosa*, *M. viridis* ssp. *viridis*, *Navarretia heterandra*, *N. jepsonii*, *N. leucocephala* ssp. *bakeri*, *N. leucocephala* ssp. *pauciflora*, *N. leucocephala* ssp. *plieantha*, *N. myersii*, *N. subuligera*, *Orcuttia tenuis*, *Orobanche valida* ssp. *howellii*, *Parvisedum leiocarpum*, *Penstemon newberryi* var. *sonomensis*, *P. purpusii*, *Pityopus californicus*, *Plagiobothrys lithocaryus*, *Pleuropogon davyi*, *Pogogyne douglasii* ssp. *parviflora*, *Potamogeton zosteriformis*, *Raillardiopsis scabrida*, *Senecio clevelandii* var. *clevelandii*, *Sidalcea oregana* ssp. *hydrophila*, *Streptanthus brachiatus* ssp. *brachiatus*, *S. brachiatus* ssp. *hoffmanii*, *S. morrisonii* ssp. *elatus*, *S. morrisonii* ssp. *kruckebergii*, *Thelypodium brachycarpum*, and *Tracyina rostrata*.

Lassen County

Total number of taxa: 56

List 1A: 0; *List 1B:* 13; *List 2:* 21; *List 3:* 3; *List 4:* 19

Allium atrorubens var. *atrorubens*, *Antennaria flagellaris*, *Arnica fulgens*, *A. sororia*, *Astragalus anxius*, *A. argophyllus* var. *argophyllus*, *A. inversus*, *A. pulsiferae* var. *pulsiferae*, *A. pulsiferae* var. *suksdorfii*, *Camissonia boothii* ssp. *alyssoides*, *C. minor*, *C. tanacetifolia* ssp. *quadriperforata*, *Carex lasiocarpa*, *C. petasata*, *Claytonia umbellata*, *Collomia tracyi*, *Cordylanthus capitatus*, *Corydalis caseana* ssp. *caseana*, *Dalea ornata*, *Dimeresia howellii*, *Drosera anglica*, *Erigeron elegantulus*, *Eriogonum nutans*, *E. prociduum*, *Geum aleppicum*, *Gratiola heterosepala*, *Hackelia cusickii*, *Iliamna bakeri*, *Ivesia aperta* var. *aperta*, *I. baileyi* var. *baileyi*, *I. paniculata*, *I. sericoleuca*, *Juncus hemiendytus* var. *abjectus*, *Lomatium hendersonii*, *L. ravenii*, *Marsilea oligospora*, *Mimulus pygmaeus*, *Pedicularis centranthera*, *Penstemon cinicola*, *Perideridia gairdneri* ssp. *gairdneri*, *Phacelia inundata*, *Pogogyne floribunda*, *Polygala subspinosa*, *Potamogeton filiformis*, *P. praelongus*, *P. robbinsii*, *P. zosteriformis*, *Potentilla basaltica*, *Psilocarphus elatior*, *Rhynchospora alba*, *Scutellaria holmgreniorum*, *Senecio eurycephalus* var. *lewisrosei*, *S. hydrophiloides*, *Silene invisa*, *Stenotus lanuginosus*, and *Thermopsis californica* var. *argentata*.

Los Angeles County

Total number of taxa: 133

List 1A: 4; *List 1B:* 59; *List 2:* 6; *List 3:* 3; *List 4:* 61

Abronia maritima, *Acanthomintha obovata* ssp. *cordata*, *Androsace elongata* ssp. *acuta*, *Aphanisma blitoides*, *Arctostaphylos gabrielensis*, *Arenaria paludicola*, *Aster greatae*, *Astragalus brauntonii*, *A. lentiginosus* var. *antonius*, *A. leucolobus*, *A. preussii* var. *laxiflorus*, *A. pycnostachyus* var. *lanosissimus*, *A. tener* var. *titi*, *Atriplex coulteri*, *A. pacifica*, *A. parishii*, *A. serenana* var. *davidsonii*, *Baccharis plummerae* ssp. *plummerae*, *Berberis nevinii*, *Botrychium crenulatum*, *Boykinia rotundifolia*, *Brodiaea filifolia*, *Calandrinia breweri*, *C. maritima*, *Calochortus catalinae*, *C. clavatus* var. *clavatus*, *C. clavatus* var. *gracilis*, *C. palmeri* var. *palmeri*, *C. plummerae*, *C. striatus*, *C. weedii* var. *intermedius*, *Calystegia peirsonii*, *C. sepium* ssp. *binghamiae*, *Camissonia lewisii*, *Canbya candida*, *Castilleja gleasonii*, *C. plagiotoma*, *Cercocarpus betuloides* var. *blancheae*, *Chamaebatia australis*, *Chorizanthe parryi* var. *fernandina*, *C. parryi* var. *parryi*, *C. procumbens*, *C. spinosa*, *Convolvulus simulans*, *Cordylanthus maritimus* ssp. *maritimus*, *Crossosoma californicum*, *Cymopterus deserticola*, *Dichondra occidentalis*, *Dithyrea maritima*, *Dodecahema leptoceras*, *Dudleya blochmaniae* ssp. *blochmaniae*, *D. cymosa* ssp. *crebrifolia*, *D. cymosa* ssp. *marcescens*, *D. cymosa* ssp. *ovatifolia*, *D. densiflora*, *D. multicaulis*, *D. virens*, *Erigeron breweri* var. *bisanctus*, *E. breweri* var. *jacinteus*, *Eriogonum kennedyi* var. *alpigenum*, *E. microthecum* var. *johnstonii*, *E. umbellatum* var. *minus*, *Erysimum insulare* ssp. *suffrutescens*, *Galium angustifolium* ssp. *gabrielense*, *G. cliftonsmithii*, *G. grande*, *G. jepsonii*, *G. johnstonii*, *Gilia latiflora* ssp. *cuyamensis*, *Goodmania luteola*, *Harpagonella palmeri*, *Helianthus nuttallii* ssp. *parishii*, *Hemizonia minthornii*, *H. parryi* ssp. *australis*, *Heuchera abramsii, H. elegans, Hordeum intercedens, Hulsea vestita* ssp. *gabrielensis, Juglans californica* var. *californica, Juncus acutus* ssp. *leopoldii, J. duranii, Lasthenia glabrata* ssp. *coulteri, Lepechinia fragrans, Lepidium virginicum* var. *robinsonii, Lilium humboldtii* ssp. *ocellatum, L. parryi, Linanthus concinnus, L. orcuttii, Loeflingia squarrosa* var. *artemisiarum, Lupinus elatus, L. excubitus* var. *johnstonii, L. peirsonii, Lycium brevipes* var. *hassei, Malacothamnus davidsonii, Microseris douglasii* var. *platycarpha, Monardella cinerea, M. viridis* ssp. *saxicola, Mucronea californica, Muhlenbergia californica, Muilla coronata, Nama stenocarpum, Nemacaulis denudata* var. *denudata, Nemacladus gracilis, Opuntia basilaris* var. *brachyclada, Orcuttia californica, Oreonana vestita, Orobanche valida* ssp. *valida, Oxytheca caryophylloides, Pentachaeta lyonii, Perideridia gairdneri* ssp. *gairdneri, P. pringlei, Phacelia exilis, P. mohavensis, P. stellaris, Polygala cornuta* var. *fishiae, Potentilla glandulosa* ssp. *ewanii, P. multijuga, Quercus engelmannii, Ribes divaricatum* var. *parishii, Romneya coulteri, Rorippa gambellii, Scutellaria bolanderi* ssp. *austromontana, Selaginella asprella, Senecio aphanactis, S. ionophyllus, Sidalcea neomexicana, Stylocline masonii, Suaeda esteroa, S. taxifolia, Swertia neglecta, Syntrichopappus lemmonii, Thelypteris puberula* var. *sonorensis,* and *Thermopsis californica* var. *argentata.*

Madera County

Total number of taxa: 46

List 1A: 0; List 1B: 21; List 2: 2; List 3: 2; List 4: 21

Atriplex cordulata, A. depressa, A. minuscula, Calycadenia hooveri, Calyptridium pulchellum, Carex congdonii, C. praticola, Castilleja campestris ssp. *succulenta, Ceanothus fresnensis, Clarkia australis, Collomia rawsoniana, Cordylanthus palmatus, Cryptantha hooveri, Cypripedium montanum, Delphinium gypsophilum* ssp. *gypsophilum, D. hansenii* ssp. *ewanianum, Eriogonum prattenianum* var. *avium, Eriophyllum nubigenum, Eryngium spinosepalum, Erythronium pluriflorum, Goodmania luteola, Gratiola heterosepala, Hulsea brevifolia, Ivesia unguiculata, Lewisia disepala, Lilium humboldtii* ssp. *humboldtii, Linanthus grandiflorus, L. serrulatus, Lupinus citrinus* var. *citrinus, Madia yosemitana, Mimulus acutidens, M. grayi, M. laciniatus, Monardella candicans, Nemophila parviflora* var. *quercifolia, Orcuttia inaequalis, O. pilosa, Perideridia bacigalupii, Potamogeton robbinsii, Pseudobahia bahiifolia, Scirpus clementis, Streptanthus farnsworthianus, Trifolium bolanderi, Tuctoria greenei, Veronica cusickii,* and *Wyethia elata.*

Marin County

Total number of taxa: 90

List 1A: 1; List 1B: 54; List 2: 4; List 3: 5; List 4: 26

Agrostis blasdalei, Alopecurus aequalis var. *sonomensis, Amsinckia lunaris, Arabis blepharophylla, Arctostaphylos hookeri* ssp. *montana, A. virgata, Aspidotis carlotta-halliae, Astragalus breweri, Blennosperma nanum* var. *robustum, Boschniakia hookeri, Calamagrostis crassiglumis, C. ophitidis, Calandrinia breweri, Calochortus tiburonensis, C. umbellatus, Campanula californica, Carex leptalea, Castilleja affinis* ssp. *neglecta, C. ambigua* ssp. *humboldtiensis, Ceanothus gloriosus* var. *gloriosus, C. gloriosus* var. *porrectus, C. masonii, Chorizanthe cuspidata* var. *cuspidata, C. cuspidata* var. *villosa, C. valida, Cirsium andrewsii, C. hydrophilum* var. *vaseyi, Clarkia concinna* ssp. *raichei, Collinsia corymbosa, Cordylanthus maritimus* ssp. *palustris, C. mollis* ssp. *mollis, Cypripedium californicum, Delphinium bakeri, Dichondra occidentalis, Dirca occidentalis, Elymus californicus, Erigeron biolettii, E. supplex, Eriogonum luteolum* var. *caninum, Erysimum franciscanum, Fritillaria lanceolata* var. *tristulis, F. liliacea, Grindelia hirsutula* var. *maritima, G. stricta* var. *angustifolia, Helianthella castanea, Hemizonia congesta* ssp. *leucocephala, Hesperevax sparsiflora* var. *brevifolia, Hesperolinon congestum, Holocarpha macradenia, Horkelia cuneata* ssp. *sericea, H. marinensis, H. tenuiloba, Lathyrus jepsonii* var. *jepsonii, Layia carnosa, Lessingia hololeuca, L. micradenia* var. *micradenia, Lilaeopsis masonii, Lilium maritimum, Limnanthes douglasii* ssp. *sulphurea, Limosella subulata, Linanthus acicularis, L. grandiflorus, Lupinus tidestromii, Micropus amphibolus, Monardella undulata, M. villosa* ssp. *globosa, Navarretia leucocephala* ssp. *bakeri, N. rosulata, Pentachaeta bellidiflora, Perideridia gairdneri* ssp. *gairdneri, Phacelia insularis* var. *continentis, Piperia michaelii, Pityopus californicus, Plagiobothrys glaber, Pleuropogon hooverianus, P. refractus, Polygonum marinense, Psilocarphus tenellus* var. *globiferus, Ranunculus lobbii, Rhynchospora californica, Ribes victoris, Sagittaria sanfordii, Sidalcea calycosa* ssp. *rhizomata, S. hickmanii* ssp. *viridis, Stebbinsoseris decipiens, Streptanthus batrachopus, S. glandulosus* ssp. *pulchellus, S. niger, Trifolium amoenum,* and *Triphysaria floribunda.*

Mariposa County

Total number of taxa: 58

List 1A: 1; List 1B: 25; List 2: 4; List 3: 1; List 4: 27

Agrostis humilis, Allium sanbornii var. *congdonii, A. yosemitense, Balsamorhiza macrolepis* var. *macrolepis, Bolandra californica, Calandrinia breweri, Calochortus clavatus* var. *avius, Calycadenia hooveri, Calyptridium pulchellum, Camissonia sierrae* ssp. *alticola, Carex congdonii, C. tompkinsii, Castilleja campestris* ssp. *succulenta, Ceanothus fresnensis, Clarkia australis, C. biloba* ssp. *australis, C. lingulata, C. rostrata, C. virgata, Claytonia megarhiza, Collomia rawsoniana, Cryptantha mariposae, Cypripedium montanum, Downingia pusilla, Erigeron mariposanus, Eriogonum tripodum, Eriophyllum congdonii, E. nubigenum, Fritillaria agrestis, Horkelia parryi, Hulsea brevifolia, Ivesia unguiculata, Jepsonia heterandra, Lewisia congdonii, L. disepala, Lilium humboldtii* ssp. *humboldtii, Linanthus serrulatus, Lomatium congdonii, Lupinus citrinus* var. *deflexus, L. gracilentus, L. spectabilis, Madia yosemitana, Mimulus filicaulis, M. gracilipes, M. grayi, M. inconspicuus, M. laciniatus, M. pulchellus, Monardella candicans, Ophioglossum californicum, Perideridia bacigalupii, Potamogeton epihydrus* ssp. *nuttallii, Rhynchospora californica, Scirpus clementis, Trichostema rubisepalum, Trifolium bolanderi, Veronica cusickii,* and *Wyethia elata.*

Mendocino County

Total number of taxa: 104

List 1A: 4; List 1B: 44; List 2: 9; List 3: 4; List 4: 43

Abronia umbellata ssp. *breviflora, Agrostis blasdalei, Antirrhinum virga, Arabis macdonaldiana, Arctostaphylos canescens* ssp. *sonomensis, A. mendocinoensis, A. stanfordiana* ssp. *raichei, Asclepias solanoana, Astragalus breweri, A. rattanii* var. *rattanii, Blennosperma nanum* var. *robustum, Boschniakia hookeri, Calamagrostis bolanderi, C. crassiglumis, C. foliosa, Calandrinia breweri, Campanula californica, Cardamine pachystigma* var. *dissectifolia*, *Carex californica*, *C. livida*, *Castilleja mendocinensis*, *Ceanothus confusus*, *C. foliosus* var. *vineatus*, *C. gloriosus* var. *gloriosus*, *Chorizanthe howellii*, *Clarkia amoena* ssp. *whitneyi*, *Collinsia corymbosa*, *Collomia diversifolia*, *Cupressus goveniana* ssp. *pigmaea*, *Cypripedium montanum*, *Epilobium nivium*, *E. oreganum*, *E. septentrionale*, *Erigeron biolettii*, *E. supplex*, *Eriogonum kelloggii*, *E. strictum* var. *greenei*, *Erysimum menziesii* ssp. *menziesii*, *Eschscholzia hypecoides*, *Fritillaria agrestis*, *F. purdyi*, *F. roderickii*, *Gentiana setigera*, *Glyceria grandis*, *Hackelia amethystina*, *Hemizonia congesta* ssp. *calyculata*, *H. congesta* ssp. *leucocephala*, *H. congesta* ssp. *tracyi*, *Hesperevax sparsiflora* var. *brevifolia*, *Hesperolinon adenophyllum*, *Horkelia bolanderi*, *H. marinensis*, *H. tenuiloba*, *Howellia aquatilis*, *Juncus supiniformis*, *Lasthenia burkei*, *L. conjugens*, *Lathyrus glandulosus*, *Layia septentrionalis*, *Lewisia stebbinsii*, *Lilium maritimum*, *L. rubescens*, *Limnanthes bakeri*, *Linanthus acicularis*, *L. rattanii*, *Lomatium engelmannii*, *Lupinus antoninus*, *L. milo-bakeri*, *Malacothamnus mendocinensis*, *Melica spectabilis*, *Microseris borealis*, *Mimulus nudatus*, *Navarretia leucocephala* ssp. *bakeri*, *N. subuligera*, *Orobanche valida* ssp. *howellii*, *Penstemon purpusii*, *Perideridia gairdneri* ssp. *gairdneri*, *Phacelia insularis* var. *continentis*, *Pinus contorta* ssp. *bolanderi*, *Piperia candida*, *Pityopus californicus*, *Plagiobothrys lithocaryus*, *Pleuropogon davyi*, *P. hooverianus*, *P. refractus*, *Pogogyne douglasii* ssp. *parviflora*, *Potamogeton epihydrus* ssp. *nuttallii*, *Puccinellia pumila*, *Raillardiopsis scabrida*, *Ranunculus lobbii*, *Rhynchospora alba*, *Ribes victoris*, *Sanguisorba officinalis*, *Sedum eastwoodiae*, *Sidalcea calycosa* ssp. *rhizomata*, *S. malachroides*, *S. oregana* ssp. *hydrophila*, *Silene campanulata* ssp. *campanulata*, *Streptanthus drepanoides*, *Thlaspi californicum*, *Trifolium amoenum*, *Veratrum fimbriatum*, *Viola palustris*, and *Wyethia longicaulis*.

Merced County

Total number of taxa: *43*

List 1A: *2;* ***List 1B:*** *22;* ***List 2:*** *5;* ***List 3:*** *1;* ***List 4:*** *13*

Acanthomintha lanceolata, *Agrostis hendersonii*, *Astragalus tener* var. *tener*, *Atriplex cordulata*, *A. coronata* var. *coronata*, *A. depressa*, *A. joaquiniana*, *A. minuscula*, *A. vallicola*, *Calycadenia hooveri*, *Castilleja campestris* ssp. *succulenta*, *Clarkia breweri*, *C. rostrata*, *Cordylanthus mollis* ssp. *hispidus*, *Cryptantha hooveri*, *C. rattanii*, *Delphinium gypsophilum* ssp. *gypsophilum*, *D. recurvatum*, *Downingia pusilla*, *Eleocharis quadrangulata*, *Eriogonum nudum* var. *indictum*, *E. vestitum*, *Eryngium racemosum*, *Linanthus ambiguus*, *L. grandiflorus*, *Malacothamnus hallii*, *Monardella leucocephala*, *Navarretia myersii*, *N. nigelliformis* ssp. *radians*, *Nemacladus gracilis*, *Neostapfia colusana*, *Ophioglossum californicum*, *Orcuttia inaequalis*, *O. pilosa*, *Phacelia ciliata* var. *opaca*, *Plagiobothrys glaber*, *Potamogeton filiformis*, *Psilocarphus tenellus* var. *globiferus*, *Sagittaria sanfordii*, *Senecio aphanactis*, *Streptanthus insignis* ssp. *lyonii*, *Trichocoronis wrightii* var. *wrightii*, and *Tuctoria greenei*.

Modoc County

Total number of taxa: *63*

List 1A: *0;* ***List 1B:*** *13;* ***List 2:*** *23;* ***List 3:*** *4;* ***List 4:*** *23*

Antennaria flagellaris, *Arabis cobrensis*, *A. microphylla* var. *microphylla*, *A. oregana*, *Arnica fulgens*, *A. sororia*, *Astragalus inversus*, *A. pulsiferae* var. *pulsiferae*, *Azolla mexicana*, *Botrychium crenulatum*, *B. lunaria*, *Calochortus greenei*, *C. longebarbatus* var. *longebarbatus*, *Camissonia minor*, *Campanula scabrella*, *Carex petasata*, *C. sheldonii*, *Chenopodium simplex*, *Claytonia megarhiza*, *C. umbellata*, *Cordylanthus capitatus*, *Cupressus bakeri*, *Cypripedium montanum*, *Delphinium stachydeum*, *Dimeresia howellii*, *Erigeron elegantulus*, *E. inornatus* var. *calidipetris*, *Eriogonum prociduum*, *E. umbellatum* var. *glaberrimum*, *Galium glabrescens* ssp. *modocense*, *G. serpenticum* ssp. *warnerense*, *Gentiana affinis* var. *parvidentata*, *Geum aleppicum*, *Gratiola heterosepala*, *Hackelia cusickii*, *Iliamna bakeri*, *Ivesia baileyi* var. *beneolens*, *Lomatium hendersonii*, *Marsilea oligospora*, *Mimulus pygmaeus*, *Navarretia subuligera*, *Penstemon cinereus*, *P. cinicola*, *P. heterodoxus* var. *shastensis*, *Perideridia gairdneri* ssp. *gairdneri*, *Phacelia inundata*, *P. sericea* var. *ciliosa*, *Plagiobothrys salsus*, *Pogogyne floribunda*, *Polygonum polygaloides* ssp. *esotericum*, *Potamogeton epihydrus* ssp. *nuttallii*, *P. zosteriformis*, *Potentilla newberryi*, *Psilocarphus elatior*, *Ribes hudsonianum* var. *petiolare*, *Rorippa columbiae*, *Salix bebbiana*, *Salvia dorrii* var. *incana*, *Saxifraga cespitosa*, *Senecio hydrophiloides*, *Solidago gigantea*, *Stenotus lanuginosus*, and *Thermopsis californica* var. *argentata*.

Mono County

Total number of taxa: *101*

List 1A: *0;* ***List 1B:*** *30;* ***List 2:*** *47;* ***List 3:*** *1;* ***List 4:*** *23*

Abronia nana ssp. *covillei*, *Achnatherum aridum*, *Allium atrorubens* var. *atrorubens*, *Antennaria pulchella*, *Arabis bodiensis*, *A. cobrensis*, *A. dispar*, *A. fernaldiana* var. *stylosa*, *A. microphylla* var. *microphylla*, *A. pinzlae*, *A. tiehmii*, *Arnica sororia*, *Astragalus argophyllus* var. *argophyllus*, *A. geyeri* var. *geyeri*, *A. johannis-howellii*, *A. kentrophyta* var. *danaus*, *A. lentiginosus* var. *piscinensis*, *A. monoensis* var. *monoensis*, *A. monoensis* var. *ravenii*, *A. oophorus* var. *lavinii*, *A. platytropis*, *A. pseudiodanthus*, *A. serenoi* var. *shockleyi*, *Blepharidachne kingii*, *Botrychium lunaria*, *Calochortus excavatus*, *Camissonia boothii* ssp. *boothii*, *Carex congdonii*, *C. eleocharis*, *C. incurviformis* var. *danaensis*, *C. norvegica*, *C. parryana* var. *hallii*, *C. petasata*, *C. praticola*, *C. tiogana*, *Claytonia megarhiza*, *C. umbellata*, *Crepis runcinata* ssp. *hallii*, *Cusickiella quadricostata*, *Dedeckera eurekensis*, *Draba asterophora* var. *asterophora*, *D. californica*, *D. cana*, *D. cruciata*, *D. incrassata*, *D. monoensis*, *D. subumbellata*, *Dryopteris filix-mas*, *Elymus scribneri*, *Epilobium howellii*, *Eriogonum beatleyae*, *E. nutans*, *E. ochrocephalum* var. *alexanderae*, *Fimbristylis thermalis*, *Gentiana prostrata*, *Glyceria grandis*, *Goodmania luteola*, *Hackelia brevicula*, *Halimolobos virgata*, *Heuchera duranii*, *Hordeum intercedens*, *H. hispidula*, *Hulsea vestita* ssp. *inyoensis*, *Ivesia kingii* var. *kingii*, *Jamesia americana* var. *rosea*, *Juncus hemiendytus* var. *abjectus*, *Kobresia bellardii*, *Lupinus duranii*, *L. lepidus* var. *culbertsonii*, *L. padre-crowleyi*, *Mimulus glabratus* ssp. *utahensis*, *Minuartia obtusiloba*, *Opuntia pulchella*, *Oxytropis deflexa* var. *sericea*, *Pedicularis crenulata*, *Penstemon barnebyi*, *P. papillatus*, *Phacelia inyoensis*, *P. monoensis*, *Pinus longaeva*, *Piptatherum micranthum*, *Plagiobothrys glomeratus*, *Poa abbreviata* ssp. *marshii*, *P. pattersonii*, *Podistera nevadensis*, *Polemonium chartaceum*, *Potamogeton filiformis*, *Potentilla concinna*, *P. morefieldii*, *P. newberryi*, *Salix brachycarpa* ssp. *brachycarpa*, *S. reticulata* ssp. *nivalis*, *Scirpus clementis*, *S. pumilus*, *Senecio pattersonensis*, *Spartina gracilis*, *Sphenopholis obtusata*, *Streptanthus oliganthus*, *Townsendia parryi*, *Trifolium macilentum* var. *dedeckerae*, and *Viola aurea*.

Monterey County

Total number of taxa: 137

List 1A: 1; List 1B: 67; List 2: 0; List 3: 4; List 4: 65

Abies bracteata, Acanthomintha lanceolata, A. obovata ssp. *obovata, Allium hickmanii, Antirrhinum ovatum, Arctostaphylos cruzensis, A. edmundsii, A. hookeri* ssp. *hookeri, A. hooveri, A. montereyensis, A. obispoensis, A. pajaroensis, A. pilosula, A. pumila, Aristocapsa insignis, Aspidotis carlotta-halliae, Astragalus macrodon, A. tener* var. *tener, A. tener* var. *titi, Atriplex coronata* var. *coronata, Bloomeria humilis, Calandrinia breweri, Calochortus weedii* var. *vestus, Calycadenia villosa, Calyptridium parryi* var. *hesseae, Calystegia collina* ssp. *venusta, Camissonia hardhamiae, Castilleja latifolia, Ceanothus cuneatus* var. *rigidus, Chlorogalum purpureum* var. *purpureum, Chorizanthe biloba* var. *immemora, C. douglasii, C. palmeri, C. pungens* var. *pungens, C. rectispina, C. robusta* var. *robusta, Cirsium occidentale* var. *compactum, Clarkia breweri, C. jolonensis, C. lewisii, Collinsia multicolor, Cordylanthus rigidus* ssp. *littoralis, Corethrogyne leucophylla, Cryptantha rattanii, Cupressus goveniana* ssp. *goveniana, C. macrocarpa, Delphinium gypsophilum* ssp. *gypsophilum, D. gypsophilum* ssp. *parviflorum, D. hutchinsoniae, D. umbraculorum, Elymus californicus, Eriastrum luteum, E. virgatum, Ericameria fasciculata, Eriogonum argillosum, E. butterworthianum, E. eastwoodianum, E. heermannii* var. *occidentale, E. nortonii, E. nudum* var. *indictum, E. temblorense, Erysimum ammophilum, E. menziesii* ssp. *menziesii, E. menziesii* ssp. *yadonii, Eschscholzia hypecoides, Fritillaria agrestis, F. falcata, F. liliacea, F. viridea, Galium andrewsii* ssp. *gatense, G. californicum* ssp. *luciense, G. clementis, G. cliftonsmithii, G. hardhamiae, Gilia tenuiflora* ssp. *arenaria, Grindelia hirsutula* var. *maritima, G. stricta* var. *angustifolia, Hemizonia halliana, H. parryi* ssp. *congdonii, Holocarpha macradenia, Horkelia cuneata* ssp. *sericea, H. yadonii, Lasthenia leptalea, Layia carnosa, L. heterotricha, L. jonesii, Lessingia hololeuca, L. occidentalis, Linanthus grandiflorus, Lomatium parvifolium, Lupinus albifrons* var. *abramsii, L. cervinus, L. tidestromii, Madia radiata, Malacothamnus abbottii, M. aboriginum, M. davidsonii, M. jonesii, M. niveus, M. palmeri* var. *involucratus, M. palmeri* var. *lucianus, M. palmeri* var. *palmeri, Malacothrix saxatilis* var. *arachnoidea, Micropus amphibolus, Mimulus subsecundus, Monardella antonina* ssp. *antonina, M. antonina* ssp. *benitensis, M. palmeri, M. undulata, Mucronea californica, Navarretia jaredii, N. nigelliformis* ssp. *radians, Ophioglossum californicum, Pedicularis dudleyi, Pentachaeta exilis* ssp. *aeolica, Perideridia gairdneri* ssp. *gairdneri, P. pringlei, Pinus radiata, Piperia michaelii, P. yadonii, Plagiobothrys uncinatus, Pogogyne clareana, Potentilla hickmanii, Psilocarphus tenellus* var. *globiferus, Raillardiopsis muirii, Ribes sericeum, Sanicula hoffmannii, S. maritima, Sidalcea hickmanii* ssp. *hickmanii, S. malachroides, Stebbinsoseris decipiens, Stylocline masonii, Syntrichopappus lemmonii, Systenotheca vortriedei, Trifolium polyodon, T. trichocalyx,* and *Tropidocarpum capparideum.*

Napa County

Total number of taxa: 87

List 1A: 0; List 1B: 41; List 2: 1; List 3: 5; List 4: 40

Antirrhinum virga, Arabis modesta, A. oregana, Asclepias solanoana, Aster lentus, Astragalus breweri, A. clarianus, A. clevelandii, A. rattanii var. *jepsonianus, A. tener* var. *tener, Atriplex joaquiniana, Balsamorhiza macrolepis* var. *macrolepis, Calamagrostis ophitidis, Calandrinia breweri, Calyptridium quadripetalum, Calystegia collina* ssp. *oxyphylla, Castilleja affinis* ssp. *neglecta, Ceanothus confusus, C. divergens, C. purpureus, C. sonomensis, Collomia diversifolia, Cordylanthus mollis* ssp. *mollis, C. tenuis* ssp. *brunneus, Cryptantha clevelandii* var. *dissita, Delphinium uliginosum, Downingia pusilla, Eleocharis parvula, Erigeron angustatus, E. biolettii, Eriogonum luteolum* var. *caninum, E. nervulosum, E. tripodum, Erythronium helenae, Fritillaria pluriflora, F. purdyi, Grindelia stricta* var. *angustifolia, Helianthus exilis, Hesperolinon bicarpellatum, H. breweri, H. drymarioides, H. serpentinum, Juglans californica* var. *hindsii, Lasthenia conjugens, Lathyrus jepsonii* var. *jepsonii, Layia septentrionalis, Legenere limosa, Lessingia hololeuca, Lilaeopsis masonii, Lilium rubescens, Limnanthes vinculans, Linanthus acicularis, Lomatium ciliolatum* var. *hooveri, L. repostum, Lupinus sericatus, Madia hallii, M. nutans, Malacothamnus helleri, Micropus amphibolus, Mimulus nudatus, Monardella villosa* ssp. *globosa, M. viridis* ssp. *viridis, Navarretia jepsonii, N. leucocephala* ssp. *bakeri, N. leucocephala* ssp. *pauciflora, N. rosulata, N. subuligera, Orobanche valida* ssp. *howellii, Penstemon newberryi* var. *sonomensis, Perideridia gairdneri* ssp. *gairdneri, Pityopus californicus, Plagiobothrys strictus, Poa napensis, Pogogyne douglasii* ssp. *parviflora, Polygonum marinense, Psilocarphus brevissimus* var. *multiflorus, Ranunculus lobbii, Ribes victoris, Senecio clevelandii* var. *clevelandii, Sidalcea hickmanii* ssp. *viridis, S. oregana* ssp. *hydrophila, Streptanthus brachiatus* ssp. *brachiatus, S. morrisonii* ssp. *elatus, S. morrisonii* ssp. *kruckebergii, Thelypodium brachycarpum, Trichostema rubisepalum,* and *Trifolium amoenum.*

Nevada County

Total number of taxa: 35

List 1A: 0; List 1B: 11; List 2: 5; List 3: 3; List 4: 16

Allium sanbornii var. *congdonii, A. sanbornii* var. *sanbornii, Azolla mexicana, Calystegia stebbinsii, Carex davyi, Claytonia megarhiza, Cypripedium fasciculatum, Darlingtonia californica, Drosera anglica, Erigeron miser, E. petrophilus* var. *sierrensis, Eriogonum umbellatum* var. *torreyanum, Fremontodendron decumbens, Ivesia sericoleuca, Juncus hemiendytus* var. *abjectus, J. marginatus* var. *marginatus, Lathyrus sulphureus* var. *argillaceus, Lewisia cantelovii, L. longipetala, Lilium humboldtii* ssp. *humboldtii, Lycopodiella inundata, Lycopus uniflorus, Marsilea oligospora, Monardella candicans, M. follettii, Perideridia bacigalupii, P. pringlei, Plagiobothrys glyptocarpus* var. *modestus, Rhynchospora alba, Rorippa subumbellata, Scutellaria galericulata, Sidalcea stipularis, Silene invisa, Trifolium lemmonii,* and *Viola tomentosa.*

Orange County

Total number of taxa: 69

List 1A: 3; List 1B: 33; List 2: 7; List 3: 1; List 4: 25

Abronia maritima, Aphanisma blitoides, Astragalus brauntonii, A. pycnostachyus var. *lanosissimus, Atriplex coulteri, A. pacifica, A. parishii, A. serenana* var. *davidsonii, Bergerocactus emoryi, Boykinia rotundifolia, Brodiaea filifolia, B. orcuttii, Calandrinia maritima, Calochortus catalinae, C. weedii* var. *intermedius, Calystegia sepium* ssp. *binghamiae, Camissonia lewisii, Chorizanthe parryi* var. *fernandina, C. procumbens,*

Comarostaphylis diversifolia ssp. *diversifolia, Cordylanthus maritimus* ssp. *maritimus, Cupressus forbesii, Dichondra occidentalis, Dudleya blochmaniae* ssp. *blochmaniae, D. cymosa* ssp. *ovatifolia, D. multicaulis, D. stolonifera, D. viscida, Eleocharis parvula, Eriastrum densifolium* ssp. *sanctorum, Euphorbia misera, Fremontodendron mexicanum, Harpagonella palmeri, Helianthus nuttallii* ssp. *parishii, Hemizonia parryi* ssp. *australis, Holocarpha virgata* ssp. *elongata, Isocoma menziesii* var. *decumbens, Juglans californica* var. *californica, Juncus acutus* ssp. *leopoldii, Lasthenia glabrata* ssp. *coulteri, Lepechinia cardiophylla, Lepidium virginicum* var. *robinsonii, Lilium humboldtii* ssp. *ocellatum, Microseris douglasii* var. *platycarpha, Mimulus clevelandii, M. diffusus, Monardella hypoleuca* ssp. *lanata, M. macrantha* ssp. *hallii, Mucronea californica, Nama stenocarpum, Nemacaulis denudata* var. *denudata, Ophioglossum californicum, Perideridia gairdneri* ssp. *gairdneri, Phacelia suaveolens* ssp. *keckii, Polygala cornuta* var. *fishiae, Quercus dumosa, Q. engelmannii, Romneya coulteri, Rorippa gambellii, Sagittaria sanfordii, Satureja chandleri, Selaginella asprella, S. cinerascens, Senecio aphanactis, Sidalcea neomexicana, Suaeda esteroa, S. taxifolia, Tetracoccus dioicus,* and *Verbesina dissita.*

Placer County

Total number of taxa: 36

List 1A: 0; *List 1B:* 15; *List 2:* 4; *List 3:* 2; *List 4:* 15

Allium sanbornii var. *congdonii, A. sanbornii* var. *sanbornii, Arabis rigidissima* var. *demota, Astragalus pauperculus, Balsamorhiza macrolepis* var. *macrolepis, Cardamine pachystigma* var. *dissectifolia, Carex davyi, C. sheldonii, Chlorogalum grandiflorum, Cordylanthus mollis* ssp. *hispidus, Corydalis caseana* ssp. *caseana, Downingia pusilla, Erigeron miser, Eriogonum tripodum, E. umbellatum* var. *torreyanum, Fritillaria agrestis, Glyceria grandis, Gratiola heterosepala, Ivesia sericoleuca, Juncus leiospermus* var. *ahartii, Lathyrus sulphureus* var. *argillaceus, Legenere limosa, Lewisia longipetala, L. serrata, Lilium humboldtii* ssp. *humboldtii, Monardella candicans, Navarretia eriocephala, N. prolifera* ssp. *lutea, Phacelia stebbinsii, Podistera nevadensis, Rorippa subumbellata, Scutellaria galericulata, Silene invisa, Sparganium natans, Veronica cusickii,* and *Viola tomentosa.*

Plumas County

Total number of taxa: 67

List 1A: 0; *List 1B:* 20; *List 2:* 16; *List 3:* 5; *List 4:* 26

Arabis constancei, A. microphylla var. *microphylla, Arctostaphylos mewukka* ssp. *truei, Arnica fulgens, Astragalus lentiformis, A. pulsiferae* var. *pulsiferae, A. pulsiferae* var. *suksdorfii, A. webberi, Azolla mexicana, Camissonia tanacetifolia* ssp. *quadriperforata, Carex geyeri, C. gigas, C. lasiocarpa, C. limosa, C. scoparia, C. sheldonii, Chenopodium simplex, Clarkia mildrediae, Claytonia palustris, Corallorhiza trifida, Corydalis caseana* ssp. *caseana, Cupressus bakeri, Cypripedium californicum, C. fasciculatum, C. montanum, Darlingtonia californica, Drosera anglica, Epilobium luteum, Erigeron inornatus* var. *calidipetris, E. petrophilus* var. *sierrensis, Fritillaria pluriflora, Hackelia amethystina, Ivesia aperta* var. *aperta, I. baileyi* var. *baileyi, I. sericoleuca, I. webberi, Juncus hemiendytus* var. *abjectus, Lewisia cantelovii, Lupinus dalesiae, Lycopus uniflorus, Marsilea oligospora, Mimulus laciniatus, M. pygmaeus, Monardella follettii, M. stebbinsii, Orcuttia tenuis, Penstemon personatus, Polystichum kruckebergii, P. lonchitis, Potamogeton epihydrus* ssp. *nuttallii, P. praelongus, Pyrrocoma lucida, Rhynchospora alba, Scheuchzeria palustris* var. *americana, Scirpus subterminalis, Scutellaria galericulata, Sedum albomarginatum, Senecio eurycephalus* var. *lewisrosei, S. hydrophiloides, Silene invisa, S. occidentalis* ssp. *longistipitata, Solidago gigantea, Sparganium natans, Trifolium lemmonii, Trimorpha acris* var. *debilis, Vaccinium coccineum,* and *Viola tomentosa.*

Riverside County

Total number of taxa: 154

List 1A: 2; *List 1B:* 72; *List 2:* 26; *List 3:* 4; *List 4:* 50

Acleisanthes longiflora, Allium munzii, A. parishii, Ambrosia pumila, Ammoselinum giganteum, Antirrhinum cyathiferum, Arabis johnstonii, Arctostaphylos peninsularis ssp. *peninsularis, A. rainbowensis, Astragalus bicristatus, A. crotalariae, A. insularis* var. *harwoodii, A. lentiginosus* var. *borreganus, A. lentiginosus* var. *coachellae, A. leucolobus, A. nutans, A. pachypus* var. *jaegeri, A. tricarinatus, Atriplex coronata* var. *notatior, A. coulteri, A. pacifica, A. parishii, A. serenana* var. *davidsonii, Ayenia compacta, Berberis nevinii, Boykinia rotundifolia, Brodiaea filifolia, B. orcuttii, Calochortus palmeri* var. *munzii, C. palmeri* var. *palmeri, C. plummerae, C. weedii* var. *intermedius, Carnegiea gigantea, Castela emoryi, Castilleja lasiorhyncha, Caulanthus simulans, Ceanothus cyaneus, C. ophiochilus, Chaenactis carphoclinia* var. *peirsonii, C. parishii, Chamaesyce arizonica, C. platysperma, Chorizanthe leptotheca, C. parryi* var. *parryi, C. polygonoides* var. *longispina, C. procumbens, C. xanti* var. *leucotheca, Colubrina californica, Comarostaphylis diversifolia* ssp. *diversifolia, Condalia globosa* var. *pubescens, Convolvulus simulans, Cryptantha costata, C. holoptera, Cynanchum utahense, Delphinium hesperium* ssp. *cuyamacae, Ditaxis californica, D. clariana, Dodecahema leptoceras, Dudleya multicaulis, D. saxosa* ssp. *saxosa, D. viscida, Erigeron breweri* var. *jacinteus, E. parishii, Eriogonum foliosum, Eryngium aristulatum* var. *parishii, Escobaria vivipara* var. *alversonii, Euphorbia misera, Galium angustifolium* ssp. *gracillimum, G. angustifolium* ssp. *jacinticum, G. californicum* ssp. *primum, Gilia caruifolia, G. maculata, Githopsis diffusa* ssp. *filicaulis, Harpagonella palmeri, Hemizonia mohavensis, H. pungens* ssp. *laevis, Heuchera hirsutissima, H. parishii, Holocarpha virgata* ssp. *elongata, Hordeum intercedens, Hulsea vestita* ssp. *callicarpha, H. vestita* ssp. *parryi, Ivesia callida, Juglans californica* var. *californica, Juncus duranii, Lasthenia glabrata* ssp. *coulteri, Lepechinia cardiophylla, Lepidium virginicum* var. *robinsonii, Leptodactylon jaegeri, Lilium humboldtii* ssp. *ocellatum, L. parryi, Limnanthes gracilis* ssp. *parishii, Linanthus floribundus* ssp. *hallii, L. orcuttii, Loeflingia squarrosa* var. *artemisiarum, Lycium parishii, Machaeranthera canescens* var. *ziegleri, Malaxis monophyllos* ssp. *brachypoda, Marina orcuttii* var. *orcuttii, Matelea parvifolia, Microseris douglasii* var. *platycarpha, Mimulus clevelandii, M. diffusus, Mirabilis tenuiloba, Monardella macrantha* ssp. *hallii, M. pringlei, M. robisonii, Mucronea californica, Muhlenbergia californica, Myosurus minimus* ssp. *apus, Navarretia fossalis, Nemacaulis denudata* var. *gracilis, Opuntia munzii, O. wigginsii, Orcuttia californica, Oxytheca caryophylloides, O. emarginata, Penstemon californicus, P. clevelandii* var. *connatus, P. thurberi, Phacelia suaveolens* ssp. *keckii, Phaseolus filiformis, Pilostyles thurberi, Polygala cornuta* var. *fishiae, Portulaca halimoides, Potentilla rimicola, Proboscidea althaeifolia, Quercus engelmannii, Romneya coulteri, Rupertia rigida, Salvia eremostachya, S. greatae, Satureja chandleri, Scutellaria bolanderi* ssp. *austromontana, Sedum niveum, Selaginella asprella, S. eremophila, Senecio aphanactis, Senna covesii, Sidalcea neomexicana, Sphenopholis obtusata, Streptanthus bernardinus, S. campestris, Stylocline sonorensis, Syntrichopappus lemmonii, Tetracoccus dioicus, Tetradymia argyraea, Thelypteris puberula* var. *sonorensis, Trichocoronis wrightii* var. *wrightii, Trichostema austromontanum* ssp. *compactum, T. micranthum, Wislizenia refracta* ssp. *refracta, Xylorhiza cognata,* and *X. orcuttii.*

Sacramento County

Total number of taxa: 21

List 1A: 0; List 1B: 13; List 2: 3; List 3: 2; List 4: 3

Aster lentus, Atriplex joaquiniana, Downingia pusilla, Eryngium pinnatisectum, Fritillaria agrestis, Gratiola heterosepala, Helianthemum suffrutescens, Hibiscus lasiocarpus, Juglans californica var. *hindsii, Juncus leiospermus* var. *ahartii, Lathyrus jepsonii* var. *jepsonii, Legenere limosa, Lilaeopsis masonii, Limosella subulata, Navarretia eriocephala, N. myersii, Oenothera deltoides* ssp. *howellii, Orcuttia tenuis, O. viscida, Pogogyne douglasii* ssp. *parviflora,* and *Sagittaria sanfordii.*

San Benito County

Total number of taxa: 61

List 1A: 1; List 1B: 19; List 2: 0; List 3: 3; List 4: 38

Acanthomintha lanceolata, A. obovata ssp. *obovata, Amsinckia vernicosa* var. *furcata, Antirrhinum ovatum, Aristocapsa insignis, Aspidotis carlotta-halliae, Astragalus clevelandii, A. macrodon, A. tener* var. *tener, Atriplex joaquiniana, Calochortus clavatus* var. *clavatus, Calyptridium parryi* var. *hesseae, Calystegia collina* ssp. *venusta, Camissonia benitensis, Campanula exigua, Chorizanthe biloba* var. *immemora, C. douglasii, C. palmeri, Clarkia breweri, C. lewisii, Convolvulus simulans, Cryptantha rattanii, Eriastrum hooveri, E. virgatum, Eriogonum argillosum, E. heermannii* var. *occidentale, E. nortonii, E. nudum* var. *indictum, E. vestitum, Eriophyllum jepsonii, Eryngium aristulatum* var. *hooveri, Eschscholzia hypecoides, Fritillaria agrestis, F. falcata, F. liliacea, F. viridea, Galium andrewsii* ssp. *gatense, Hemizonia halliana, Hordeum intercedens, Isocoma menziesii* var. *diabolica, Lathyrus jepsonii* var. *jepsonii, Layia discoidea, L. heterotricha, Lembertia congdonii, Lepidium jaredii* ssp. *album, Lessingia occidentalis, Linanthus ambiguus, Madia radiata, Malacothamnus aboriginum, Mimulus subsecundus, Monardella antonina* ssp. *antonina, M. antonina* ssp. *benitensis, Navarretia nigelliformis* ssp. *radians, Pentachaeta exilis* ssp. *aeolica, Perideridia gairdneri* ssp. *gairdneri, Phacelia phacelioides, Piperia michaelii, Plagiobothrys glaber, P. uncinatus, Solidago guiradonis,* and *Trichostema rubisepalum.*

San Bernardino County

Total number of taxa: 232

List 1A: 5; List 1B: 91; List 2: 61; List 3: 5; List 4: 70

Abronia nana ssp. *covillei, Achnatherum aridum, Agave utahensis, Allium nevadense, A. parishii, Androsace elongata* ssp. *acuta, Androstephium breviflorum, Antennaria marginata, Arabis breweri* var. *pecuniaria, A. dispar, A. parishii, A. pulchra* var. *munciensis, A. shockleyi, Arctomecon merriamii, Arenaria macradenia* var. *kuschei, A. paludicola, A. ursina, Argyrochosma limitanea* var. *limitanea, Astragalus albens, A. allochrous* var. *playanus, A. bicristatus, A. cimae* var. *cimae, A. jaegerianus, A. lentiginosus* var. *antonius, A. lentiginosus* var. *borreganus, A. lentiginosus* var. *sierrae, A. leucolobus, A. nutans, A. preussii* var. *preussii, A. tricarinatus, Astrolepis cochisensis, Atriplex coulteri, A. parishii, Ayenia compacta, Berberis fremontii, B. nevinii, Botrychium crenulatum, Bouteloua trifida, Boykinia rotundifolia, Brodiaea filifolia, Calandrinia breweri, Calochortus palmeri* var. *palmeri, C. plummerae, C. striatus, Camissonia boothii* ssp. *boothii, Canbya candida, Carex comosa, Carnegiea gigantea, Castela emoryi, Castilleja cinerea, C. lasiorhyncha, C. montigena, C. plagiotoma, Chamaesyce platysperma, Cheilanthes wootonii, Chorizanthe leptotheca, C. parryi* var. *parryi, C. procumbens, C. spinosa, C. xanti* var. *leucotheca, Chrysothamnus gramineus, Claytonia lanceolata* var. *peirsonii, Cordylanthus eremicus* ssp. *eremicus, C. parviflorus, C. tecopensis, Cryptantha clokeyi, C. costata, C. holoptera, C. tumulosa, Cymopterus deserticola, C. gilmanii, Cynanchum utahense, Ditaxis clariana, Dodecahema leptoceras, Dryopteris filix-mas, Dudleya abramsii* ssp. *affinis, D. multicaulis, D. saxosa* ssp. *saxosa, Echinocereus engelmannii* var. *howei, Enceliopsis nudicaulis, Enneapogon desvauxii, Eriastrum densifolium* ssp. *sanctorum, Erigeron breweri* var. *bisanctus, E. breweri* var. *jacinteus, E. parishii, E. uncialis* var. *uncialis, Eriodictyon angustifolium, Eriogonum bifurcatum, E. contiguum, E. ericifolium* var. *thornei, E. foliosum, E. heermannii* var. *floccosum, E. kennedyi* var. *alpigenum, E. kennedyi* var. *austromontanum, E. microthecum* var. *corymbosoides, E. microthecum* var. *johnstonii, E. ovalifolium* var. *vineum, E. umbellatum* var. *juniporinum, E. umbellatum* var. *minus, Erioneuron pilosum, Eriophyllum lanatum* var. *obovatum, E. mohavense, Escobaria vivipara* var. *alversonii, E. vivipara* var. *rosea, Euphorbia exstipulata* var. *exstipulata, Fendlerella utahensis, Fimbristylis thermalis, Galium angustifolium* ssp. *gabrielense, G. angustifolium* ssp. *gracillimum, G. californicum* ssp. *primum, G. hilendiae* ssp. *kingstonense, G. jepsonii, G. johnstonii, G. munzii, G. wrightii, Gentiana fremontii, Gilia maculata, Glossopetalon pungens, Helianthus nuttallii* ssp. *parishii, Hemizonia mohavensis, H. pungens* ssp. *laevis, Heuchera abramsii, H. elegans, H. parishii, Horkelia wilderae, Hulsea vestita* ssp. *gabrielensis, H. vestita* ssp. *parryi, Ivesia argyrocoma, I. jaegeri, I. patellifera, Juglans californica* var. *californica, Juncus duranii, Lasthenia glabrata* ssp. *coulteri, Lepidium virginicum* var. *robinsonii, Lesquerella kingii* ssp. *bernardina, Lewisia brachycalyx, Lilium humboldtii* ssp. *ocellatum, L. parryi, Linanthus arenicola, L. concinnus, L. killipii, Lotus argyraeus* var. *multicaulis, L. argyraeus* var. *notitius, Lycium parishii, Lycurus phleoides* var. *phleoides, Malacothamnus parishii, Malaxis monophyllos* ssp. *brachypoda, Matelea parvifolia, Maurandya antirrhiniflora* ssp. *antirrhiniflora, Mimulus exiguus, M. mohavensis, M. purpureus, Monardella cinerea, M. macrantha* ssp. *hallii, M. pringlei, M. robisonii, M. viridis* ssp. *saxicola, Mucronea californica, Muhlenbergia appressa, M. arsenei, M. californica, M. fragilis, M. pauciflora, Muilla coronata, Munroa squarrosa, Myosurus minimus* ssp. *apus, Nama dichotomum* var. *dichotomum, Navarretia peninsularis, Oenothera caespitosa* ssp. *crinita, Ophioglossum californicum, O. basilaris* var. *brachyclada, O. curvospina, Oreonana vestita, Oxytheca caryophylloides, O. parishii* var. *cienegensis, O. parishii* var. *goodmaniana, Pellaea truncata, Penstemon albomarginatus, P. calcareus, P. fruticiformis* var. *amargosae, P. stephensii,*

P. thurberi, Perideridia parishii ssp. *parishii, Phacelia anelsonii, P. exilis, P. mohavensis, P. mustelina, P. parishii, Phlox dolichantha, Pholistoma auritum* var. *arizonicum, Physalis lobata, Pinus edulis, Piptatherum micranthum, Poa atropurpurea, Podistera nevadensis, Poliomintha incana, Polystichum kruckebergii, Populus angustifolia, Portulaca halimoides, Psorothamnus arborescens* var. *arborescens, Puccinellia parishii, Pyrrocoma uniflora* var. *gossypina, Ribes divaricatum* var. *parishii, Rupertia rigida, Salvia greatae, Sanvitalia abertii, Sclerocactus polyancistrus, Scleropogon brevifolius, Scutellaria bolanderi* ssp. *austromontana, Sedum niveum, Selaginella asprella, S. leucobryoides, Senecio bernardinus, S. ionophyllus, Senna covesii, Sidalcea hickmanii* ssp. *parishii, S. neomexicana, S. pedata, Sphaeralcea rusbyi* var. *eremicola, Sphenopholis obtusata, Streptanthus bernardinus, S. campestris, Swertia neglecta, Syntrichopappus lemmonii, Taraxacum californicum, Tetradymia argyraea, Thelypodium stenopetalum, Trichostema micranthum, Viola aurea, V. pinetorum* ssp. *grisea, Wislizenia refracta* ssp. *refracta,* and *Woodsia plummerae.*

San Diego County

Total number of taxa: 210

List 1A: 1; *List 1B:* 97; *List 2:* 47; *List 3:* 6; *List 4:* 59

Abronia maritima, Acanthomintha ilicifolia, Achnatherum diegoense, Adolphia californica, Agave shawii, Ambrosia chenopodiifolia, A. pumila, Androsace elongata ssp. *acuta, Aphanisma blitoides, Arctostaphylos glandulosa* ssp. *crassifolia, A. otayensis, A. rainbowensis, Artemisia palmeri, Astragalus crotalariae, A. deanei, A. douglasii* var. *perstrictus, A. insularis* var. *harwoodii, A. lentiginosus* var. *borreganus, A. leucolobus, A. magdalenae* var. *peirsonii, A. oocarpus, A. tener* var. *titi, Atriplex coulteri, A. pacifica, A. parishii, A. serenana* var. *davidsonii, Ayenia compacta, Azolla mexicana, Baccharis vanessae, Berberis fremontii, B. nevinii, Bergerocactus emoryi, Boykinia rotundifolia, Brodiaea filifolia, B. orcuttii, Bursera microphylla, Calandrinia breweri, C. maritima, Calliandra eriophylla, Calochortus catalinae, C. dunnii, Camissonia lewisii, Carlowrightia arizonica, Castilleja lasiorhyncha, Caulanthus simulans, Ceanothus cyaneus, C. verrucosus, Chaenactis carphoclinia* var. *peirsonii, C. parishii, Chamaebatia australis, Chamaesyce arizonica, C. platysperma, Chorizanthe leptotheca, C. orcuttiana, C. parryi* var. *fernandina, C. polygonoides* var. *longispina, C. procumbens, Clarkia delicata, Colubrina californica, Comarostaphylis diversifolia* ssp. *diversifolia, Convolvulus simulans, Cordylanthus maritimus* ssp. *maritimus, C. orcuttianus, Coreopsis maritima, Corethrogyne filaginifolia* var. *incana, C. filaginifolia* var. *linifolia, Cryptantha costata, C. ganderi, C. holoptera, Cupressus forbesii, C. stephensonii, Cynanchum utahense, Delphinium hesperium* ssp. *cuyamacae, D. parishii* ssp. *subglobosum, Dichondra occidentalis, Ditaxis californica, Downingia concolor* var. *brevior, Dudleya alainae, D. attenuata* ssp. *orcuttii, D. blochmaniae* ssp. *blochmaniae, D. blochmaniae* ssp. *brevifolia, D. multicaulis, D. variegata, D. viscida, Ericameria cuneata* var. *macrocephala, E. palmeri* ssp. *palmeri, Eriogonum foliosum, Eryngium aristulatum* var. *parishii, Eucnide rupestris, Euphorbia misera, Ferocactus viridescens, Frankenia palmeri, Fremontodendron mexicanum, Galium angustifolium* ssp. *borregoense, G. johnstonii, Geraea viscida, Gilia caruifolia, Githopsis diffusa* ssp. *filicaulis, Grindelia hirsutula* var. *hallii, Harpagonella palmeri, Hazardia orcuttii, Hemizonia conjugens, H. floribunda, H. parryi* ssp. *australis, H. pungens* ssp. *laevis, Herissantia crispa, Heuchera brevistaminea, H. rubescens* var. *versicolor, Holocarpha virgata* ssp. *elongata, Hordeum intercedens, Horkelia truncata, Hulsea californica, H. mexicana, H. vestita* ssp. *callicarpha, Ipomopsis tenuifolia, Isocoma menziesii* var. *decumbens, Iva hayesiana, Juglans californica* var. *californica, Juncus acutus* ssp. *leopoldii, Lasthenia glabrata* ssp. *coulteri, Lathyrus splendens, Lepechinia cardiophylla, L. ganderi, Lepidium flavum* var. *felipense, L. virginicum* var. *robinsonii, Lessingia glandulifera* var. *tomentosa, Lewisia brachycalyx, Lilium humboldtii* ssp. *ocellatum, L. parryi, Limnanthes gracilis* ssp. *parishii, Linanthus bellus, L. floribundus* ssp. *hallii, L. orcuttii, Lotus crassifolius* var. *otayensis, L. haydonii, L. nuttallianus, Lupinus excubitus* var. *medius, Lycium parishii, Lyrocarpa coulteri* var. *palmeri, Machaeranthera asteroides* var. *lagunensis, M. juncea, Malperia tenuis, Matelea parvifolia, Mentzelia hirsutissima, Microseris douglasii* var. *platycarpha, Mimulus aridus, M. clevelandii, M. diffusus, Mirabilis tenuiloba, Monardella hypoleuca* ssp. *lanata, M. linoides* ssp. *viminea, M. macrantha* ssp. *hallii, M. nana* ssp. *leptosiphon, Mucronea californica, Muilla clevelandii, Myosurus minimus* ssp. *apus, Nama stenocarpum, Navarretia fossalis, N. peninsularis, Nemacaulis denudata* var. *denudata, N. denudata* var. *gracilis, Nolina interrata, Ophioglossum californicum, Opuntia parryi* var. *serpentina, O. wigginsii, O. wolfii, Orcuttia californica, Ornithostaphylos oppositifolia, Orobanche parishii* ssp. *brachyloba, Penstemon clevelandii* var. *connatus, P. thurberi, Perideridia gairdneri* ssp. *gairdneri, Phacelia stellaris, Pilostyles thurberi, Pinus torreyana* ssp. *torreyana, Poa atropurpurea, Pogogyne abramsii, P. nudiuscula, Polygala cornuta* var. *fishiae, Proboscidea althaeifolia, Quercus dumosa, Q. engelmannii, Rhus trilobata* var. *simplicifolia, Ribes canthariforme, R. viburnifolium, Romneya coulteri, Rorippa gambellii, Rosa minutifolia, Rubus glaucifolius* var. *ganderi, Rupertia rigida, Salvia eremostachya, S. munzii, Satureja chandleri, Scutellaria bolanderi* ssp. *austromontana, Selaginella asprella, S. cinerascens, S. eremophila, Senecio aphanactis, S. ganderi, Senna covesii, Streptanthus bernardinus, S. campestris, Stylocline citroleum, Suaeda esteroa, S. taxifolia, Tetracoccus dioicus, Thermopsis californica* var. *semota, Viguiera laciniata, Viola aurea,* and *Xylorhiza orcuttii.*

San Francisco County

Total number of taxa: 33

List 1A: 1; *List 1B:* 20; *List 2:* 1; *List 3:* 2; *List 4:* 9

Arabis blepharophylla, Arctostaphylos hookeri ssp. *franciscana, A. hookeri* ssp. *ravenii, Arenaria paludicola, Astragalus tener* var. *tener, Carex comosa, Chorizanthe cuspidata* var. *cuspidata, Cirsium andrewsii, C. occidentale* var. *compactum, Clarkia franciscana, Collinsia corymbosa, C. multicolor, Equisetum palustre, Erysimum franciscanum, Fritillaria liliacea, Grindelia hirsutula* var. *maritima, G. stricta* var. *angustifolia, Helianthella castanea, Hesperevax sparsiflora* var. *brevifolia, Hesperolinon congestum, Horkelia cuneata* ssp. *sericea, Layia carnosa, Lessingia germanorum, Lilium maritimum, Linanthus grandiflorus, Monardella undulata, Piperia michaelii, Plagiobothrys chorisianus* var. *chorisianus, P. diffusus, Sanicula maritima, Sidalcea hickmanii* ssp. *viridis, Silene verecunda* ssp. *verecunda,* and *Triphysaria floribunda.*

San Joaquin County

Total number of taxa: 30

List 1A: 1; List 1B: 16; List 2: 5; List 3: 0; List 4: 8

Amsinckia grandiflora, Androsace elongata ssp. *acuta, Aster lentus, Astragalus tener* var. *tener, Atriplex cordulata, A. coronata* var. *coronata, A. joaquiniana, Blepharizonia plumosa* ssp. *plumosa, Carex comosa, Cirsium crassicaule, Convolvulus simulans, Cordylanthus palmatus, Cryptantha hooveri, Delphinium californicum* ssp. *interius, D. gypsophilum* ssp. *gypsophilum, Eryngium racemosum, Gratiola heterosepala, Hibiscus lasiocarpus, Lathyrus jepsonii* var. *jepsonii, Lilaeopsis masonii, Limosella subulata, Linanthus ambiguus, Madia radiata, Monardella candicans, Psilocarphus brevissimus* var. *multiflorus, Sagittaria sanfordii, Scutellaria lateriflora, Trichocoronis wrightii* var. *wrightii, Tropidocarpum capparideum,* and *Tuctoria greenei.*

San Luis Obispo County

Total number of taxa: 151

List 1A: 1; List 1B: 75; List 2: 1; List 3: 0; List 4: 74

Abies bracteata, Abronia maritima, Acanthomintha obovata ssp. *cordata, A. obovata* ssp. *obovata, Agrostis hooveri, Allium hickmanii, Amsinckia vernicosa* var. *furcata, Androsace elongata* ssp. *acuta, Antirrhinum ovatum, Arctostaphylos cruzensis, A. hookeri* ssp. *hearstiorum, A. hooveri, A. luciana, A. morroensis, A. obispoensis, A. osoensis, A. pechoensis, A. pilosula, A. rudis, A. tomentosa* ssp. *daciticola, A. wellsii, Arenaria paludicola, Aristocapsa insignis, Aspidotis carlotta-halliae, Astragalus macrodon, Atriplex coronata* var. *coronata, A. vallicola, Baccharis plummerae* ssp. *glabrata, Bloomeria humilis, Calandrinia breweri, Calochortus catalinae, C. clavatus* ssp. *recurvifolius, C. clavatus* var. *clavatus, C. obispoensis, C. palmeri* var. *palmeri, C. simulans, C. weedii* var. *vestus, Calycadenia villosa, Calystegia subacaulis* ssp. *episcopalis, Camissonia hardhamiae, Carex obispoensis, Castilleja plagiotoma, Caulanthus californicus, Ceanothus cuneatus* var. *rigidus, C. hearstiorum, C. maritimus, Chlorogalum pomeridianum* var. *minus, C. purpureum* var. *reductum, Chorizanthe blakleyi, C. breweri, C. douglasii, C. palmeri, C. rectispina, Cirsium fontinale* var. *obispoense, C. loncholepis, C. occidentale* var. *compactum, C. rhothophilum, Clarkia speciosa* ssp. *immaculata, Convolvulus simulans, Cordylanthus maritimus* ssp. *maritimus, Corethrogyne leucophylla, Delphinium californicum* ssp. *interius, D. gypsophilum* ssp. *gypsophilum, D. gypsophilum* ssp. *parviflorum, D. parryi* ssp. *blochmaniae, D. recurvatum, D. umbraculorum, Dithyrea maritima, Dudleya abramsii* ssp. *bettinae, D. abramsii* ssp. *murina, D. blochmaniae* ssp. *blochmaniae, Eleocharis parvula, Eriastrum hooveri, E. luteum, Erigeron blochmaniae, E. sanctarum, Eriodictyon altissimum, Eriogonum gossypinum, E. nudum* var. *indictum, E. temblorense, Eryngium aristulatum* var. *hooveri, E. capitatum* ssp. *lompocense, E. insulare* ssp. *suffrutescens, Eschscholzia hypecoides, E. rhombipetala, Fritillaria agrestis, F. ojaiensis, F. viridea, Galium andrewsii* ssp. *gatense, G. cliftonsmithii, G. hardhamiae, Grindelia hirsutula* var. *maritima, Hemizonia halliana, H. parryi* ssp. *congdonii, H. cuneata* ssp. *sericea, H. yadonii, Juncus acutus* ssp. *leopoldii, Lasthenia glabrata* ssp. *coulteri, L. leptalea, Layia heterotricha, L. jonesii, L. munzii, Lembertia congdonii, Lepidium jaredii* ssp. *album, L. jaredii* ssp. *jaredii, Leptodactylon californicum* ssp. *tomentosum, Lessingia tenuis, Lilium humboldtii* ssp. *ocellatum, Linanthus grandiflorus, Lomatium parvifolium, Lupinus cervinus, L. ludovicianus, L. nipomensis, Madia radiata, Malacothamnus davidsonii, M. gracilis, M. jonesii, M. niveus, M. palmeri* var. *involucratus, M. palmeri* var. *palmeri, Malacothrix incana, Mimulus subsecundus, Monardella antonina* ssp. *benitensis, M. crispa, M. frutescens, M. palmeri, M. undulata, Mucronea californica, Navarretia jaredii, N. nigelliformis* ssp. *radians, Orobanche parishii* ssp. *brachyloba, Pedicularis dudleyi, Perideridia gairdneri* ssp. *gairdneri, P. pringlei, Pinus radiata, Piperia michaelii, Plagiobothrys uncinatus, Prunus fasciculata* var. *punctata, Psilocarphus tenellus* var. *globiferus, Ribes sericeum, Rorippa gambellii, Sanicula hoffmannii, S. maritima, Scrophularia atrata, Senecio aphanactis, Sidalcea hickmanii* ssp. *anomala, Stylocline masonii, Suaeda californica, S. taxifolia, Systenotheca vortriedei,* and *Triteleia ixioides* ssp. *cookii.*

San Mateo County

Total number of taxa: 59

List 1A: 0; List 1B: 32; List 2: 0; List 3: 6; List 4: 21

Acanthomintha duttonii, Arabis blepharophylla, Arctostaphylos andersonii, A. imbricata, A. montaraensis, A. regismontana, Calandrinia breweri, Calochortus umbellatus, Chorizanthe cuspidata var. *cuspidata, C. robusta* var. *robusta, Cirsium andrewsii, C. fontinale* var. *fontinale, Collinsia multicolor, Cordylanthus maritimus* ssp. *palustris, Cupressus abramsiana, Cypripedium fasciculatum, C. montanum, Dirca occidentalis, Elymus californicus, Equisetum palustre, Eriogonum luteolum* var. *caninum, Eriophyllum latilobum, Erysimum ammophilum, E. franciscanum, Fritillaria agrestis, F. biflora* var. *ineziana, F. liliacea, Grindelia hirsutula* var. *maritima, G. stricta* var. *angustifolia, Helianthella castanea, Hesperolinon congestum, Hordeum intercedens, Horkelia cuneata* ssp. *sericea, H. marinensis, Legenere limosa, Lessingia arachnoidea, L. germanorum, L. hololeuca, Lilium maritimum, Limnanthes douglasii* ssp. *sulphurea, Linanthus acicularis, L. ambiguus, L. grandiflorus, L. eximius, Malacothamnus arcuatus, Monardella undulata, M. villosa* ssp. *globosa, Pedicularis dudleyi, Pentachaeta bellidiflora, Perideridia gairdneri* ssp. *gairdneri, Pinus radiata, Piperia candida, P. michaelii, Plagiobothrys chorisianus* var. *chorisianus, Potentilla hickmanii, Sanicula hoffmannii, Sidalcea hickmanii* ssp. *viridis, Silene verecunda* ssp. *verecunda,* and *Triphysaria floribunda.*

Santa Barbara County

Total number of taxa: 87

List 1A: 0; List 1B: 41; List 2: 3; List 3: 2; List 4: 41

Abronia maritima, Acanthomintha obovata ssp. *cordata, Agrostis hooveri, Antirrhinum ovatum, Aphanisma blitoides, Arctostaphylos purissima, A. refugioensis, A. rudis, A. tomentosa* ssp. *eastwoodiana, Atriplex coulteri, A. serenana* var. *davidsonii, Baccharis plummerae* ssp. *plummerae, Boykinia rotundifolia, Calandrinia breweri, C. maritima, Calochortus catalinae, C. clavatus* var. *clavatus, C. weedii* var. *vestus, Calystegia collina* ssp. *venusta, C. sepium* ssp. *binghamiae, Caulanthus amplexicaulis* var. *barbarae, C. californicus, Chorizanthe blakleyi, C. palmeri, C. rectispina, Cirsium loncholepis, C. rhothophilum, Convolvulus simulans, Cordylanthus maritimus* ssp. *maritimus, C. rigidus* ssp. *littoralis,*

Delphinium parryi ssp. *blochmaniae, Dichondra occidentalis, Dithyrea maritima, Dudleya blochmaniae* ssp. *blochmaniae, Eriastrum hooveri, Erigeron blochmaniae, E. sanctarum, Eriodictyon capitatum, Eriophyllum lanatum* var. *hallii, Erysimum capitatum* ssp. *lompocense, E. insulare* ssp. *suffrutescens, Fritillaria agrestis, F. ojaiensis, Galium cliftonsmithii, Gilia latiflora* ssp. *cuyamensis, Hemizonia increscens* ssp. *villosa, H. parryi* ssp. *australis, Hordeum intercedens, Horkelia cuneata* ssp. *sericea, H. yadonii, Juglans californica* var. *californica, Juncus acutus* ssp. *leopoldii, Lasthenia conjugens, L. glabrata* ssp. *coulteri, Layia carnosa, L. heterotricha, Lembertia congdonii, Lepidium virginicum* var. *robinsonii, Leptodactylon californicum* ssp. *tomentosum, Lessingia tenuis, Lilium humboldtii* ssp. *ocellatum, Linanthus grandiflorus, Malacothamnus niveus, Malacothrix incana, M. saxatilis* var. *arachnoidea, Monardella crispa, M. frutescens, M. undulata, Mucronea californica, Navarretia peninsularis, Oxytheca parishii* var. *abramsii, Perideridia pringlei, Polygala cornuta* var. *fishiae, Prunus fasciculata* var. *punctata, Quercus dumosa, Q. parvula* var. *parvula, Ribes amarum* var. *hoffmannii, Sanicula hoffmannii, Scrophularia atrata, Senecio aphanactis, Sidalcea hickmanii* ssp. *parishii, S. neomexicana, Suaeda esteroa, S. taxifolia, Thelypteris puberula* var. *sonorensis, Thermopsis californica* var. *argentata*, and *T. macrophylla.*

Santa Clara County

Total number of taxa: 64

List 1A: 2; *List 1B:* 35; *List 2:* 2; *List 3:* 4; *List 4:* 21

Acanthomintha lanceolata, Allium sharsmithae, Arctostaphylos andersonii, Astragalus tener var. *tener, Atriplex joaquiniana, Azolla mexicana, Balsamorhiza macrolepis* var. *macrolepis, Calandrinia breweri, Calochortus umbellatus, Calyptridium parryi* var. *hesseae, Campanula exigua, C. sharsmithiae, Castilleja affinis* ssp. *neglecta, Ceanothus ferrisae, Chorizanthe cuspidata* var. *cuspidata, C. robusta* var. *robusta, Cirsium fontinale* var. *campylon, Clarkia breweri, C. concinna* ssp. *automixa, Cordylanthus maritimus* ssp. *palustris, Coreopsis hamiltonii, Cypripedium fasciculatum, Delphinium californicum* ssp. *interius, Dirca occidentalis, Dudleya setchellii, Eriastrum brandegeae, Eriogonum argillosum, E. luteolum* var. *caninum, Eriophyllum jepsonii, Eryngium aristulatum* var. *hooveri, Erysimum franciscanum, Fritillaria falcata, F. liliacea, Galium andrewsii* ssp. *gatense, Grindelia stricta* var. *angustifolia, Hemizonia parryi* ssp. *congdonii, Isocoma menziesii* var. *diabolica, Lasthenia conjugens, Lathyrus jepsonii* var. *jepsonii, Lessingia hololeuca, L. micradenia* var. *glabrata, Linanthus ambiguus, L. grandiflorus, Malacothamnus arcuatus, M. hallii, Monardella antonina* ssp. *antonina, Penstemon rattanii* var. *kleei, Perideridia gairdneri* ssp. *gairdneri, Phacelia phacelioides, Plagiobothrys glaber, P. myosotoides, P. uncinatus, Potamogeton filiformis, Psilocarphus brevissimus* var. *multiflorus, Ranunculus lobbii, Sanicula saxatilis, Senecio aphanactis, Sidalcea malachroides, Streptanthus albidus* ssp. *albidus, S. albidus* ssp. *peramoenus, S. callistus, Suaeda californica, Trifolium amoenum*, and *Tropidocarpum capparideum.*

Santa Cruz County

Total number of taxa: 58

List 1A: 0; *List 1B:* 30; *List 2:* 1; *List 3:* 2; *List 4:* 25

Agrostis blasdalei, Amsinckia lunaris, Arabis blepharophylla, Arctostaphylos andersonii, A. glutinosa, A. hookeri ssp. *hookeri, A. pajaroensis, A. regismontana, A. silvicola, Arenaria paludicola, Calandrinia breweri, Calochortus umbellatus, Calyptridium parryi* var. *hesseae, Campanula californica, Carex comosa, Castilleja latifolia, Ceanothus cuneatus* var. *rigidus, Chorizanthe pungens* var. *hartwegiana, C. pungens* var. *pungens, C. robusta* var. *hartwegii, C. robusta* var. *robusta, Collinsia multicolor, Corethrogyne leucophylla, Cupressus abramsiana, Cypripedium fasciculatum, Elymus californicus, Eriogonum nudum* var. *decurrens, Erysimum ammophilum, E. franciscanum, E. teretifolium, Grindelia hirsutula* var. *maritima, Hemizonia parryi* ssp. *congdonii, Hesperevax sparsiflora* var. *brevifolia, Holocarpha macradenia, Horkelia cuneata* ssp. *sericea, H. marinensis, Lilium rubescens, Linanthus ambiguus, L. grandiflorus, Lomatium parvifolium, Malacothamnus arcuatus, Micropus amphibolus, Mimulus rattanii* ssp. *decurtatus, Monardella undulata, Pedicularis dudleyi, Penstemon rattanii* var. *kleei, Pentachaeta bellidiflora, Perideridia gairdneri* ssp. *gairdneri, Pinus radiata, Piperia candida, P. michaelii, Plagiobothrys chorisianus* var. *chorisianus, P. diffusus, Sanicula hoffmannii, Sidalcea malachroides, Silene verecunda* ssp. *verecunda, Stebbinsoseris decipiens*, and *Trifolium buckwestiorum.*

Shasta County

Total number of taxa: 96

List 1A: 0; *List 1B:* 28; *List 2:* 14; *List 3:* 5; *List 4:* 49

Ageratina shastensis, Agrostis hendersonii, Allium hoffmanii, A. sanbornii var. *sanbornii, Amsinckia lunaris, Arctostaphylos klamathensis, A. malloryi, Arnica venosa, Asclepias solanoana, Asplenium septentrionale, Astragalus inversus, A. pauperculus, A. pulsiferae* var. *suksdorfii, Calochortus longebarbatus* var. *longebarbatus, Campanula scabrella, C. shetleri, Carex comosa, C. scoparia, C. vulpinoidea, Clarkia borealis* ssp. *arida, C. borealis* ssp. *borealis, Collomia diversifolia, C. larsenii, Corydalis caseana* ssp. *caseana, Cryptantha crinita, Cupressus bakeri, Cypripedium californicum, C. fasciculatum, Darlingtonia californica, Draba aureola, D. howellii, Epilobium oreganum, Ericameria ophitidis, Erigeron elegantulus, E. inornatus* var. *calidipetris, E. petrophilus* var. *viscidulus, Eriogonum congdonii, E. libertini, Erythronium klamathense, Fritillaria eastwoodiae, Gratiola heterosepala, Helianthus exilis, Iliamna bakeri, Ivesia longibracteata, Juncus leiospermus* var. *leiospermus, Lathyrus sulphureus* var. *argillaceus, Lewisia cantelovii, L. cotyledon* var. *howellii, Lilium rubescens, Limnanthes floccosa* ssp. *bellingeriana, L. floccosa* ssp. *floccosa, Lomatium tracyi, Lupinus croceus* var. *pilosellus, Madia stebbinsii, Marsilea oligospora, Minuartia decumbens, M. rosei, Navarretia heterandra, N. subuligera, Neviusia cliftonii, Orcuttia tenuis, Paronychia ahartii, Penstemon filiformis, P. heterodoxus* var. *shastensis, P. purpusii, Phacelia dalesiana, Phlox muscoides, Picea engelmannii, Pogogyne floribunda, Polygonum bidwelliae, Polystichum kruckebergii, Potamogeton praelongus, P. zosteriformis, Puccinellia howellii, Raillardiopsis scabrida, Sagittaria sanfordii, Scutellaria galericulata, Sedum laxum* ssp. *flavidum, S. paradisum, Silene invisa, S. occidentalis* ssp. *longistipitata, S. suksdorfii, Smelowskia ovalis* var. *congesta, Smilax jamesii, Sparganium natans, Stellaria longifolia, Streptanthus drepanoides, Thelypodium brachycarpum, Thermopsis californica* var. *argentata, T. gracilis, Trillium ovatum* ssp. *oettingeri, Trimorpha acris* var. *debilis, Triteleia crocea* var. *crocea, T. crocea* var. *modesta, Tuctoria greenei*, and *Veratrum insolitum.*

Sierra County

Total number of taxa: *32*

List 1A: *0;* ***List 1B:*** *14;* ***List 2:*** *4;* ***List 3:*** *2;* ***List 4:*** *12*

Arabis constancei, Asplenium trichomanes-ramosum, Astragalus pulsiferae var. *pulsiferae, A. webberi, Camissonia tanacetifolia* ssp. *quadriperforata, Carex geyeri, Corydalis caseana* ssp. *caseana, Cypripedium fasciculatum, C. montanum, Darlingtonia californica, Drosera anglica, Epilobium howellii, Erigeron petrophilus* var. *sierrensis, Eriogonum umbellatum* var. *torreyanum, Ivesia aperta* var. *aperta, I. aperta* var. *canina, I. sericoleuca, I. webberi, Juncus hemiendytus* var. *abjectus, Lewisia cantelovii, Lupinus dalesiae, Marsilea oligospora, Polygonum polygaloides* ssp. *esotericum, Polystichum kruckebergii, Potamogeton praelongus, Pyrrocoma lucida, Scheuchzeria palustris* var. *americana, Silene invisa, Trifolium lemmonii, Vaccinium coccineum, Veronica cusickii,* and *Viola tomentosa.*

Siskiyou County

Total number of taxa: *182*

List 1A: *3;* ***List 1B:*** *38;* ***List 2:*** *43;* ***List 3:*** *7;* ***List 4:*** *91*

Abies amabilis, A. lasiocarpa var. *lasiocarpa, Allium siskiyouense, Amsinckia lunaris, Androsace elongata* ssp. *acuta, A. filiformis, Arabis aculeolata, A. koehleri* var. *stipitata, A. modesta, A. oregana, A. rigidissima* var. *rigidissima, A. serpentinicola, Arctostaphylos klamathensis, Arnica cernua, A. spathulata, A. viscosa, Asarum marmoratum, Astragalus inversus, Balsamorhiza hookeri* var. *lanata, B. sericea, Botrychium pinnatum, Calochortus greenei, C. longebarbatus* var. *longebarbatus, C. monanthus, C. persistens, Campanula scabrella, C. shetleri, C. wilkinsiana, Carex geyeri, C. gigas, C. halliana, C. vulpinoidea, Castilleja hispida* ssp. *brevilobata, C. miniata* ssp. *elata, C. schizotricha, Chaenactis douglasii* var. *alpina, C. suffrutescens, Cirsium ciliolatum, Claytonia palustris, C. umbellata, Collomia larsenii, C. tracyi, Cordylanthus tenuis* ssp. *pallescens, Cupressus bakeri, C. nootkatensis, Cypripedium californicum, C. fasciculatum, C. montanum, Darlingtonia californica, Delphinium uliginosum, Deschampsia atropurpurea, Dicentra formosa* ssp. *oregana, Dimeresia howellii, Draba aureola, D. carnosula, D. howellii, D. pterosperma, Drosera anglica, Epilobium luteum, E. oreganum, E. rigidum, E. siskiyouense, Erigeron bloomeri* var. *nudatus, E. cervinus, E. elegantulus, E. inornatus* var. *calidipetris, E. petrophilus* var. *viscidulus, Eriogonum alpinum, E. congdonii, E. diclinum, E. hirtellum, E. siskiyouense, E. strictum* var. *greenei, E. ternatum, E. umbellatum* var. *humistratum, Erythronium citrinum* var. *citrinum, E. hendersonii, E. klamathense, Galium oreganum, G. serpenticum* ssp. *scotticum, Gentiana plurisetosa, Geum aleppicum, Hackelia cusickii, Helianthus exilis, Hierochloe odorata, Horkelia hendersonii, Iliamna bakeri, Iris tenax* ssp. *klamathensis, Ivesia pickeringii, Juncus dudleyi, J. regelii, Lathyrus delnorticus, Lewisia cotyledon* var. *heckneri, L. cotyledon* var. *howellii, Lilium bolanderi, L. pardalinum* ssp. *vollmeri, L. pardalinum* ssp. *wigginsii, L. rubescens, L. washingtonianum* ssp. *purpurascens, Limnanthes floccosa* ssp. *floccosa, Listera cordata, Lomatium engelmannii, L. howellii, L. martindalei, L. peckianum, L. tracyi, Lupinus croceus* var. *pilosellus, L. lapidicola, L. tracyi, Mertensia bella, Minuartia stolonifera, Ophioglossum pusillum, Opuntia fragilis, Orcuttia tenuis, Orthocarpus cuspidatus* ssp. *cuspidatus, O. pachystachyus, Pedicularis bracteosa* var. *flavida, P. contorta, P. howellii, Penstemon cinereus, P. cinicola, P. heterodoxus* var. *shastensis, Perideridia gairdneri* ssp. *gairdneri, P. leptocarpa, Phacelia cookei, P. dalesiana, P. greenei, P. leonis, P. sericea* var. *ciliosa, Phlox hirsuta, P. muscoides, Picea engelmannii, Pinguicula vulgaris* ssp. *macroceras, Piperia candida, Pityopus californicus, Poa piperi, P. rhizomata, Polemonium chartaceum, Polystichum kruckebergii, P. lonchitis, Potamogeton robbinsii, Potentilla cristae, P. newberryi, Raillardella pringlei, Ribes hudsonianum* var. *petiolare, R. marshallii, Rorippa columbiae, Salix bebbiana, S. delnortensis, Salvia dorrii* var. *incana, Saussurea americana, Saxifraga cespitosa, S. howellii, S. rufidula, Scutellaria galericulata, Sedum divergens, S. laxum* ssp. *flavidum, S. laxum* ssp. *heckneri, S. oblanceolatum, Selaginella densa* var. *scopulorum, Senecio hydrophiloides, S. macounii, Silene marmorensis, S. suksdorfii, Smilax jamesii, Tauschia howellii, Thelypodium brachycarpum, Thermopsis californica* var. *argentata, T. gracilis, T. robusta, Tonestus lyallii, Trifolium howellii, Trillium ovatum* ssp. *oettingeri, Trimorpha acris* var. *debilis, Triteleia crocea* var. *crocea, T. crocea* var. *modesta, T. hendersonii* var. *hendersonii, Vaccinium coccineum, V. scoparium, Vancouveria chrysantha, Veratrum insolitum,* and *Veronica copelandii.*

Solano County

Total number of taxa: *44*

List 1A: *2;* ***List 1B:*** *27;* ***List 2:*** *4;* ***List 3:*** *3;* ***List 4:*** *8*

Aster lentus, Astragalus tener var. *ferrisiae, A. tener* var. *tener, Atriplex cordulata, A. depressa, A. joaquiniana, Blepharizonia plumosa* ssp. *plumosa, Calochortus pulchellus, Ceanothus purpureus, Cirsium hydrophilum* var. *hydrophilum, Cordylanthus mollis* ssp. *hispidus, C. mollis* ssp. *mollis, Delphinium recurvatum, Downingia pusilla, Erigeron biolettii, Eriogonum truncatum, Fritillaria liliacea, F. pluriflora, Gratiola heterosepala, Grindelia stricta* var. *angustifolia, Hemizonia parryi* ssp. *congdonii, Hesperolinon breweri, Hibiscus lasiocarpus, Isocoma arguta, Juglans californica* var. *hindsii, Lasthenia conjugens, Lathyrus jepsonii* var. *jepsonii, Legenere limosa, Lessingia hololeuca, Lilaeopsis masonii, Limosella subulata, Lomatium repostum, Monardella viridis* ssp. *viridis, Myosurus minimus* ssp. *apus, Navarretia leucocephala* ssp. *bakeri, Neostapfia colusana, Perideridia gairdneri* ssp. *gairdneri, Plagiobothrys hystriculus, Psilocarphus brevissimus* var. *multiflorus, Ranunculus lobbii, Ribes victoris, Senecio aphanactis, Trifolium amoenum,* and *Tuctoria mucronata.*

Sonoma County

Total number of taxa: *122*

List 1A: *2;* ***List 1B:*** *67;* ***List 2:*** *5;* ***List 3:*** *8;* ***List 4:*** *40*

Abronia umbellata ssp. *breviflora, Agrostis blasdalei, Alopecurus aequalis* var. *sonomensis, Antirrhinum virga, A. blepharophylla, Arctostaphylos bakeri* ssp. *bakeri, A. bakeri* ssp. *sublaevis, A. canescens* ssp. *sonomensis, A. densiflora, A. hispidula, A. stanfordiana* ssp. *decumbens, Asclepias solanoana, Astragalus breweri, A. clarianus, A. rattanii* var. *rattanii, A. tener* var. *tener, Blennosperma bakeri, Calamagrostis bolanderi, C. crassiglumis, C. ophitidis, Calandrinia breweri, Calochortus raichei, Calyptridium quadripetalum, Calystegia collina* ssp. *oxyphylla, Campanula californica, Cardamine pachystigma* var. *dissectifolia, Carex albida, C. californica, C. comosa, Castilleja uliginosa, Ceanothus confusus, C. divergens, C. foliosus* var. *vineatus, C. gloriosus* var. *gloriosus, C.*

sonomensis, Chlorogalum pomeridianum var. *minus, Chorizanthe cuspidata* var. *cuspidata, C. cuspidata* var. *villosa, C. valida, Cirsium andrewsii, Clarkia imbricata, Collinsia corymbosa, Cordylanthus maritimus* ssp. *palustris, C. mollis* ssp. *mollis, C. tenuis* ssp. *brunneus, C. tenuis* ssp. *capillaris, Cupressus goveniana* ssp. *pigmaea, Cypripedium californicum, C. montanum, Delphinium bakeri, D. luteum, Dichanthelium lanuginosum* var. *thermale, Dirca occidentalis, Downingia pusilla, Eleocharis parvula, Elymus californicus, Erigeron angustatus, E. biolettii, E. serpentinus, E. supplex, Eriogonum luteolum* var. *caninum, E. nervulosum, E. ternatum, Erysimum franciscanum, Erythronium helenae, Fritillaria liliacea, Grindelia stricta* var. *angustifolia, Helianthus exilis, Hemizonia congesta* ssp. *leucocephala, Hesperevax sparsiflora* var. *brevifolia, Hesperolinon bicarpellatum, Horkelia tenuiloba, Lasthenia burkei, Layia septentrionalis, Legenere limosa, Lessingia arachnoidea, L. hololeuca, Lilium maritimum, L. pardalinum* ssp. *pitkinense, L. rubescens, Limnanthes vinculans, Linanthus acicularis, L. grandiflorus, Lomatium repostum, Lupinus eximius, L. sericatus, L. tidestromii, Madia nutans, Micropus amphibolus, Monardella undulata, M. villosa* ssp. *globosa, M. viridis* ssp. *viridis, Navarretia leucocephala* ssp. *bakeri, N. leucocephala* ssp. *plieantha, Orobanche valida* ssp. *howellii, Penstemon newberryi* var. *sonomensis, Perideridia gairdneri* ssp. *gairdneri, Piperia candida, Pityopus californicus, Plagiobothrys mollis* var. *vestitus, Pleuropogon hooverianus, Pogogyne douglasii* ssp. *parviflora, Polygonum marinense, Potentilla hickmanii, Ranunculus lobbii, Rhynchospora alba, R. californica, R. globularis* var. *globularis, Ribes victoris, Sidalcea calycosa* ssp. *rhizomata, S. hickmanii* ssp. *viridis, S. oregana* ssp. *valida, Streptanthus brachiatus* ssp. *brachiatus, S. brachiatus* ssp. *hoffmanii, S. glandulosus* var. *hoffmanii, S. morrisonii* ssp. *elatus, S. morrisonii* ssp. *hirtiflorus, S. morrisonii* ssp. *kruckebergii, S. morrisonii* ssp. *morrisonii, Tracyina rostrata, Trifolium amoenum,* and *Veratrum fimbriatum.*

Stanislaus County

Total number of taxa: *44*

List 1A: *2;* ***List 1B:*** *23;* ***List 2:*** *1;* ***List 3:*** *1;* ***List 4:*** *17*

Acanthomintha lanceolata, Allium sharsmithae, Astragalus tener var. *tener, Atriplex cordulata, A. coronata* var. *coronata, A. depressa, Blepharizonia plumosa* ssp. *plumosa, Bolandra californica, Calycadenia hooveri, Campanula exigua, C. sharsmithiae, Castilleja campestris* ssp. *succulenta, Chamaesyce hooveri, Cirsium fontinale* var. *campylon, Clarkia breweri, C. rostrata, Collomia diversifolia, Convolvulus simulans, Coreopsis hamiltonii, Cryptantha hooveri, Delphinium gypsophilum* ssp. *gypsophilum, Downingia pusilla, Eriophyllum jepsonii, Eryngium racemosum, E. spinosepalum, Eschscholzia rhombipetala, Fritillaria agrestis, F. falcata, Jepsonia heterandra, Legenere limosa, Linanthus ambiguus, Lotus rubriflorus, Monardella candicans, M. leucocephala, Myosurus minimus* ssp. *apus, Neostapfia colusana, Ophioglossum californicum, Orcuttia inaequalis, O. pilosa, Phacelia phacelioides, Pseudobahia bahiifolia, Psilocarphus brevissimus* var. *multiflorus, P. tenellus* var. *globiferus,* and *Tuctoria greenei.*

Sutter County

Total number of taxa: *6*

List 1A: *0;* ***List 1B:*** *4;* ***List 2:*** *2;* ***List 3:*** *0;* ***List 4:*** *0*

Astragalus tener var. *ferrisiae, Hibiscus lasiocarpus, Layia septentrionalis, Monardella douglasii* ssp. *venosa, Pseudobahia bahiifolia,* and *Trichocoronis wrightii* var. *wrightii.*

Tehama County

Total number of taxa: *79*

List 1A: *0;* ***List 1B:*** *32;* ***List 2:*** *9;* ***List 3:*** *3;* ***List 4:*** *35*

Achnatherum lemmonii var. *pubescens, Allium hoffmanii, A. sanbornii* var. *sanbornii, Androsace elongata* ssp. *acuta, Antirrhinum subcordatum, Arctostaphylos canescens* ssp. *sonomensis, Asclepias solanoana, Asplenium septentrionale, Astragalus pauperculus, A. rattanii* var. *jepsonianus, A. rattanii* var. *rattanii, Balsamorhiza macrolepis* var. *macrolepis, Botrychium ascendens, B. crenulatum, B. minganense, B. montanum, Brodiaea coronaria* ssp. *rosea, Calyptridium quadripetalum, Campanula wilkinsiana, Chamaesyce hooveri, C. ocellata* ssp. *rattanii, Chlorogalum pomeridianum* var. *minus, Claytonia palustris, Collomia tracyi, Corydalis caseana* ssp. *caseana, Cryptantha crinita, Cypripedium fasciculatum, C. montanum, Downingia pusilla, Eleocharis quadrangulata, Epilobium oreganum, Eriastrum brandegeae, Ericameria ophitidis, Eriogonum libertini, E. strictum* var. *greenei, E. ternatum, E. tripodum, Fritillaria eastwoodiae, F. pluriflora, F. purdyi, Gratiola heterosepala, Hackelia amethystina, Helianthus exilis, Hesperolinon tehamense, Juncus leiospermus* var. *leiospermus, Lathyrus sulphureus* var. *argillaceus, Layia septentrionalis, Legenere limosa, Limnanthes floccosa* ssp. *floccosa, Linanthus nuttallii* ssp. *howellii, L. rattanii, Lomatium tracyi, Lotus rubriflorus, Lupinus antoninus, Madia stebbinsii, Malacothamnus helleri, Mimulus glaucescens, Minuartia decumbens, M. rosei, Navarretia heterandra, N. jepsonii, N. leucocephala* ssp. *bakeri, N. subuligera, Orcuttia pilosa, O. tenuis, Orobanche valida* ssp. *howellii, Paronychia ahartii, Penstemon purpusii, Polygonum bidwelliae, Raillardiopsis scabrida, Rupertia hallii, Sagittaria sanfordii, Sanicula tracyi, Scirpus heterochaetus, S. subterminalis, Sedum laxum* ssp. *flavidum, Silene occidentalis* ssp. *longistipitata, Streptanthus drepanoides,* and *Tuctoria greenei.*

Trinity County

Total number of taxa: *117*

List 1A: *1;* ***List 1B:*** *32;* ***List 2:*** *9;* ***List 3:*** *4;* ***List 4:*** *71*

Allium hoffmanii, A. siskiyouense, Arabis macdonaldiana, A. modesta, A. oregana, A. rigidissima var. *rigidissima, Arctostaphylos klamathensis, A. malloryi, Arnica spathulata, A. venosa, A. viscosa, Asclepias solanoana, Astragalus rattanii* var. *rattanii, Balsamorhiza sericea, Calyptridium quadripetalum, Campanula scabrella, C. wilkinsiana, Carex gigas, C. hystricina, C. leptalea, C. vulpinoidea, Chaenactis suffrutescens, Clarkia borealis* ssp. *borealis, Collomia tracyi, Cypripedium californicum, C. fasciculatum, C. montanum, Darlingtonia californica, Deschampsia atropurpurea, Draba aureola, D. carnosula, D. howellii, Epilobium nivium, E. oreganum, E. septentrionale, E. siskiyouense, Eriastrum brandegeae, Ericameria ophitidis, Erigeron cervinus, E. decumbens* var.

robustior, E. petrophilus var. *viscidulus, Eriogonum alpinum, E. congdonii, E. diclinum, E. libertini, E. siskiyouense, E. strictum* var. *greenei, E. umbellatum* var. *humistratum, Erythronium citrinum* var. *citrinum, E. citrinum* var. *roderickii, Fritillaria purdyi, Galium serpenticum* ssp. *scotticum, Gentiana plurisetosa, Hackelia amethystina, Helianthus exilis, Hemizonia congesta* ssp. *tracyi, Ivesia pickeringii, Juncus dudleyi, J. regelii, Lewisia cotyledon* var. *heckneri, L. cotyledon* var. *howellii, L. stebbinsii, Lilium rubescens, L. washingtonianum* ssp. *purpurascens, Limnanthes floccosa* ssp. *floccosa, Lomatium engelmannii, L. tracyi, Lotus yollabolliensis, Lupinus antoninus, L. constancei, L. croceus* var. *pilosellus, L. lapidicola, L. tracyi, Madia doris-nilesiae, M. stebbinsii, Marsilea oligospora, Minuartia decumbens, M. rosei, Montia howellii, Navarretia heterandra, Oenothera wolfii, Pedicularis bracteosa* var. *flavida, P. contorta, Penstemon filiformis, P. purpusii, P. tracyi, Perideridia gairdneri* ssp. *gairdneri, Phacelia dalesiana, P. leonis, Picea engelmannii, Piperia candida, Poa rhizomata, Polemonium chartaceum, Polystichum lonchitis, Potentilla cristae, Raillardella pringlei, Raillardiopsis scabrida, Sanicula tracyi, Sedum laxum* ssp. *flavidum, S. laxum* ssp. *heckneri, S. paradisum, Senecio clevelandii* var. *clevelandii, Silene invisa, Smilax jamesii, Streptanthus drepanoides, Swertia fastigiata, Tauschia glauca, Thelypodium brachycarpum, Thermopsis gracilis, Tiarella trifoliata* var. *trifoliata, Tonestus lyallii, Trillium ovatum* ssp. *oettingeri, Triteleia crocea* var. *crocea, T. crocea* var. *modesta, Veratrum insolitum, Veronica copelandii,* and *Wyethia longicaulis.*

Tulare County

Total number of taxa: 111

List 1A: 0; *List 1B:* 52; *List 2:* 5; *List 3:* 3; *List 4:* 51

Abronia alpina, Angelica callii, Antennaria pulchella, Arabis bodiensis, A. dispar, A. pygmaea, Asplenium septentrionale, Astragalus lentiginosus var. *kernensis, A. shevockii, A. subvestitus, Atriplex cordulata, A. depressa, A. joaquiniana, A. minuscula, Azolla mexicana, Botrychium crenulatum, Brodiaea insignis, Calochortus westonii, Calystegia malacophylla* var. *berryi, Carex congdonii, C. incurviformis* var. *danaensis, Caulanthus californicus, Ceanothus fresnensis, Chamaesyce hooveri, Clarkia exilis, C. springvillensis, Claytonia palustris, Cordylanthus eremicus* ssp. *kernensis, Corydalis caseana* ssp. *caseana, Cupressus arizonica* ssp. *nevadensis, Delphinium inopinum, D. purpusii, D. recurvatum, Dicentra nevadensis, Draba cruciata, D. sharsmithii, Dudleya calcicola, D. cymosa* ssp. *costafolia, Eriastrum hooveri, Erigeron aequifolius, E. inornatus* var. *keilii, E. multiceps, Eriogonum breedlovei* var. *shevockii, E. nudum* var. *murinum, E. polypodum, E. twisselmannii, E. wrightii* var. *olanchense, Eriophyllum lanatum* var. *obovatum, Eryngium spinosepalum, Erythronium pusaterii, Fritillaria brandegei, F. striata, Goodmania luteola, Hackelia sharsmithii, Horkelia tularensis, Hulsea brevifolia, Iris munzii, Ivesia campestris, Jamesia americana* var. *rosea, Juncus hemiendytus* var. *abjectus, J. nodosus, Lasthenia glabrata* ssp. *coulteri, Lembertia congdonii, Lewisia disepala, Linanthus oblanceolatus, L. serrulatus, Lotus oblongifolius* var. *cupreus, Lupinus lepidus* var. *culbertsonii, L. padre-crowleyi, Madia yosemitana, Marsilea oligospora, Mimulus acutidens, M. grayi, M. laciniatus, M. norrisii, M. pictus, Monardella beneolens, M. candicans, M. linoides* ssp. *oblonga, Muilla coronata, Navarretia setiloba, Nemacladus twisselmannii, Nemophila parviflora* var. *quercifolia, Orcuttia inaequalis, Oreonana purpurascens, Oxytheca caryophylloides, Perideridia pringlei, Phacelia exilis, P. nashiana, P. novenmillensis, P. orogenes, Phlox dispersa, Pityopus californicus, Plagiobothrys myosotoides, Pseudobahia peirsonii, Psilocarphus tenellus* var. *globiferus, Puccinellia californica, Raillardiopsis muirii, Ribes menziesii* var. *ixoderme, R. tularense, Scirpus clementis, Selaginella asprella, Sidalcea keckii, Sphenopholis obtusata, Streptanthus farnsworthianus, S. gracilis, Trichostema ovatum, Trifolium macilentum* var. *dedeckerae, Tuctoria greenei, Viola pinetorum* ssp. *grisea,* and *Wyethia elata.*

Tuolumne County

Total number of taxa: 64

List 1A: 0; *List 1B:* 20; *List 2:* 7; *List 3:* 2; *List 4:* 35

Agrostis humilis, Allium jepsonii, A. sanbornii var. *congdonii, A. tribracteatum, A. tuolumnense, A. yosemitense, Antennaria pulchella, Arctostaphylos nissenana, Astragalus kentrophyta* var. *danaus, Bolandra californica, Botrychium lunaria, Brodiaea pallida, Carex congdonii, C. davyi, C. incurviformis* var. *danaensis, C. praticola, Ceanothus fresnensis, Chaenactis douglasii* var. *alpina, Chlorogalum grandiflorum, Clarkia australis, C. biloba* ssp. *australis, C. virgata, Claytonia megarhiza, Cryptantha crymophila, C. mariposae, Cypripedium montanum, Draba asterophora* var. *asterophora, Eriogonum tripodum, Eriophyllum nubigenum, Eryngium pinnatisectum, Erythronium tuolumnense, Fritillaria agrestis, Helianthemum suffrutescens, Hulsea brevifolia, Iris hartwegii* ssp. *columbiana, Jepsonia heterandra, Juncus hemiendytus* var. *abjectus, Lilium humboldtii* ssp. *humboldtii, Lomatium congdonii, L. stebbinsii, Lupinus spectabilis, Lycopus uniflorus, Madia yosemitana, Mimulus filicaulis, M. inconspicuus, M. laciniatus, M. pulchellus, Minuartia obtusiloba, Monardella candicans, M. douglasii* ssp. *venosa, Ophioglossum californicum, Perideridia bacigalupii, Podistera nevadensis, Polystichum kruckebergii, Potamogeton robbinsii, Scirpus clementis, Senecio clevelandii* var. *heterophyllus, S. layneae, Silene invisa, Sparganium natans, Stellaria obtusa, Trichostema rubisepalum, Verbena californica,* and *Veronica cusickii.*

Ventura County

Total number of taxa: 64

List 1A: 1; *List 1B:* 25; *List 2:* 2; *List 3:* 1; *List 4:* 35

Abronia maritima, Acanthomintha obovata ssp. *cordata, Allium howellii* var. *clokeyi, Antirrhinum ovatum, Aphanisma blitoides, Astragalus brauntonii, A. pycnostachyus* var. *lanosissimus, Atriplex pacifica, A. serenana* var. *davidsonii, Baccharis plummerae* ssp. *plummerae, Boykinia rotundifolia, Calandrinia breweri, C. maritima, Calochortus catalinae, C. plummerae, C. weedii* var. *vestus, Cercocarpus betuloides* var. *blancheae, Chorizanthe procumbens, Cordylanthus maritimus* ssp. *maritimus, Delphinium inopinum, D. parryi* ssp. *blochmaniae, D. parryi* ssp. *purpureum, Dichondra occidentalis, Dudleya abramsii* ssp. *parva, D. blochmaniae* ssp. *blochmaniae, D. cymosa* ssp. *marcescens, D. cymosa* ssp. *ovatifolia, D. verityi, Eriogonum crocatum, E. kennedyi* var. *alpigenum, Eriophyllum jepsonii, Erysimum insulare* ssp. *suffrutescens, Fritillaria ojaiensis, Galium cliftonsmithii, Gilia latiflora* ssp. *cuyamensis, Hemizonia minthornii, Hordeum intercedens, Hulsea vestita* ssp. *gabrielensis, H. vestita* ssp. *parryi, Juglans californica* var. *californica, Juncus acutus* ssp. *leopoldii, Lasthenia glabrata* ssp. *coulteri, Layia heterotricha, Lepechinia fragrans*, *Lessingia tenuis*, *Lilium humboldtii* ssp. *ocellatum*, *Lupinus elatus*, *Malacothrix incana*, *Monardella linoides* ssp. *oblonga*, *Mucronea californica*, *Orcuttia californica*, *Oxytheca caryophylloides*, *O. parishii* var. *abramsii*, *Pentachaeta lyonii*, *Perideridia pringlei*, *Phacelia exilis*, *Polygala cornuta* var. *fishiae*, *Sagittaria sanfordii*, *Senecio aphanactis*, *Sidalcea neomexicana*, *Suaeda esteroa*, *S. taxifolia*, *Swertia neglecta*, and *Thermopsis californica* var. *argentata*.

Yolo County

***Total number of taxa:** 28*

***List 1A:** 0; **List 1B:** 15; **List 2:** 1; **List 3:** 1; **List 4:** 11*

Asclepias solanoana, *Astragalus breweri*, *A. clevelandii*, *A. rattanii* var. *jepsonianus*, *A. tener* var. *ferrisiae*, *A. tener* var. *tener*, *Atriplex depressa*, *A. joaquiniana*, *Collomia diversifolia*, *Cordylanthus palmatus*, *Cryptantha excavata*, *Eriogonum nervulosum*, *Fritillaria pluriflora*, *F. purdyi*, *Hesperolinon drymarioides*, *Hibiscus lasiocarpus*, *Juglans californica* var. *hindsii*, *Layia septentrionalis*, *Lepidium latipes* var. *heckardii*, *Lessingia hololeuca*, *Lomatium ciliolatum* var. *hooveri*, *Madia hallii*, *Malacothamnus helleri*, *Navarretia jepsonii*, *Neostapfia colusana*, *Psilocarphus brevissimus* var. *multiflorus*, *Senecio clevelandii* var. *clevelandii*, and *Tuctoria mucronata*.

Yuba County

***Total number of taxa:** 13*

***List 1A:** 0; **List 1B:** 4; **List 2:** 0; **List 3:** 0; **List 4:** 9*

Allium sanbornii var. *sanbornii*, *Arctostaphylos mewukka* ssp. *truei*, *Astragalus pauperculus*, *Clarkia mildrediae*, *Cypripedium fasciculatum*, *Darlingtonia californica*, *Erigeron petrophilus* var. *sierrensis*, *Fritillaria eastwoodiae*, *Lilium humboldtii* ssp. *humboldtii*, *Lupinus dalesiae*, *Navarretia heterandra*, *Pseudobahia bahiifolia*, and *Pyrrocoma lucida*.

Islands

Anacapa Island

***Total number of taxa:** 26*

***List 1A:** 0; **List 1B:** 10; **List 2:** 0; **List 3:** 1; **List 4:** 15*

Abronia maritima, *Achnatherum diegoense*, *Aphanisma blitoides*, *Arabis hoffmannii*, *Astragalus miguelensis*, *Atriplex coulteri*, *A. pacifica*, *Baccharis plummerae* ssp. *plummerae*, *Berberis pinnata* ssp. *insularis*, *Calandrinia maritima*, *Castilleja lanata* ssp. *hololeuca*, *Eriogonum grande* var. *grande*, *E. grande* var. *rubescens*, *Erysimum insulare* ssp. *insulare*, *Gilia nevinii*, *Hazardia detonsa*, *Hemizonia clementina*, *Heuchera maxima*, *Hordeum intercedens*, *Lavatera assurgentiflora* ssp. *assurgentiflora*, *Lotus dendroideus* var. *dendroideus*, *Malacothrix foliosa*, *M. squalida*, *Mimulus flemingii*, *Quercus tomentella*, and *Suaeda taxifolia*.

San Clemente Island

***Total number of taxa:** 56*

***List 1A:** 1; **List 1B:** 33; **List 2:** 4; **List 3:** 1; **List 4:** 17*

Abronia maritima, *Aphanisma blitoides*, *Astragalus miguelensis*, *A. nevinii*, *Atriplex coulteri*, *A. pacifica*, *Bergerocactus emoryi*, *Brodiaea kinkiensis*, *Calandrinia maritima*, *Calystegia macrostegia* ssp. *amplissima*, *Camissonia guadalupensis* ssp. *clementina*, *Castilleja grisea*, *Convolvulus simulans*, *Crossosoma californicum*, *Cryptantha traskiae*, *Delphinium variegatum* ssp. *kinkiense*, *D. variegatum* ssp. *thornei*, *Dendromecon harfordii* var. *rhamnoides*, *Dissanthelium californicum*, *Dudleya virens*, *Eriogonum giganteum* var. *formosum*, *E. grande* var. *grande*, *Eriophyllum nevinii*, *Eschscholzia ramosa*, *Euphorbia misera*, *Galium catalinense* ssp. *acrispum*, *Galvezia speciosa*, *Gilia nevinii*, *Hazardia cana*, *Hemizonia clementina*, *Hordeum intercedens*, *Isocoma menziesii* var. *decumbens*, *Jepsonia malvifolia*, *Lavatera assurgentiflora* ssp. *glabra*, *Linanthus pygmaeus* ssp. *pygmaeus*, *Lithophragma maximum*, *Lomatium insulare*, *Lotus argophyllus* var. *adsurgens*, *L. dendroideus* var. *traskiae*, *Lupinus guadalupensis*, *Lycium brevipes* var. *hassei*, *Lyonothamnus floribundus* ssp. *asplenifolius*, *Malacothamnus clementinus*, *Malacothrix foliosa*, *Microseris douglasii* var. *platycarpha*, *Mimulus flemingii*, *Muhlenbergia appressa*, *Nama stenocarpum*, *Phacelia floribunda*, *Quercus tomentella*, *Scrophularia villosa*, *Sibara filifolia*, *Stephanomeria blairii*, *Suaeda taxifolia*, *Trifolium gracilentum* var. *palmeri*, and *Triteleia clementina*.

San Miguel Island

***Total number of taxa:** 22*

***List 1A:** 0; **List 1B:** 12; **List 2:** 0; **List 3:** 0; **List 4:** 10*

Abronia maritima, *Achnatherum diegoense*, *Astragalus miguelensis*, *Atriplex coulteri*, *Castilleja lanata* ssp. *hololeuca*, *C. mollis*, *Dichondra occidentalis*, *Dithyrea maritima*, *Dudleya candelabrum*, *D. greenei*, *Eriogonum grande* var. *rubescens*, *Erysimum insulare* ssp. *insulare*, *Eschscholzia ramosa*, *Galium buxifolium*, *G. californicum* ssp. *miguelense*, *Helianthemum greenei*, *Lavatera assurgentiflora* ssp. *assurgentiflora*, *Lotus dendroideus* var. *veatchii*, *Malacothrix incana*, *M. indecora*, *Orobanche parishii* ssp. *brachyloba*, and *Phacelia insularis* var. *insularis*.

San Nicolas Island

***Total number of taxa:** 24*

***List 1A:** 2; **List 1B:** 10; **List 2:** 0; **List 3:** 1; **List 4:** 11*

Abronia maritima, *Achnatherum diegoense*, *Aphanisma blitoides*, *Astragalus traskiae*, *Atriplex pacifica*, *Calystegia macrostegia* ssp. *amplissima*, *Cryptantha traskiae*, *Dithyrea maritima*, *Dudleya virens*, *Eriogonum grande* var. *timorum*, *Eschscholzia ramosa*, *Gilia nevinii*, *Hemizonia clementina*, *Hordeum intercedens*, *Jepsonia malvifolia*, *Lavatera assurgentiflora* ssp. *assurgentiflora*, *Lomatium insulare*, *Lycium verrucosum*, *Malacothrix foliosa*, *M. incana*, *Orobanche parishii* ssp. *brachyloba*, *Phacelia cinerea*, *Suaeda taxifolia*, and *Trifolium gracilentum* var. *palmeri*.

Santa Barbara Island

Total number of taxa: *17*

List 1A: *0;* ***List 1B:*** *8;* ***List 2:*** *0;* ***List 3:*** *1;* ***List 4:*** *8*

Aphanisma blitoides, Astragalus traskiae, Calandrinia maritima, Calystegia macrostegia ssp. *amplissima, Dudleya traskiae, Eriogonum giganteum* var. *compactum, Eriophyllum nevinii, Eschscholzia ramosa, Galvezia speciosa, Gilia nevinii, Hemizonia clementina, Hordeum intercedens, Malacothrix foliosa, Phacelia floribunda, Platystemon californicus* var. *ciliatus, Suaeda taxifolia,* and *Trifolium gracilentum* var. *palmeri.*

Santa Catalina Island

Total number of taxa: *51*

List 1A: *2;* ***List 1B:*** *20;* ***List 2:*** *4;* ***List 3:*** *1;* ***List 4:*** *24*

Abronia maritima, Aphanisma blitoides, Arctostaphylos catalinae, Atriplex coulteri, A. pacifica, Bergerocactus emoryi, Calandrinia maritima, Calochortus catalinae, Cercocarpus betuloides var. *blancheae, C. traskiae, Convolvulus simulans, Crossosoma californicum, Dendromecon harfordii* var. *rhamnoides, Dichondra occidentalis, Dissanthelium californicum, Dithyrea maritima, Dudleya greenei, D. virens, Eriogonum grande* var. *grande, Eriophyllum nevinii, Eschscholzia ramosa, Euphorbia misera, Galium catalinense* ssp. *catalinense, G. nuttallii* ssp. *insulare, Galvezia speciosa, Gilia nevinii, Harpagonella palmeri, Helianthemum greenei, Hemizonia clementina, H. parryi* ssp. *australis, Hordeum intercedens, Isocoma menziesii* var. *decumbens, Jepsonia malvifolia, Lavatera assurgentiflora* ssp. *glabra, Lepechinia fragrans, Lotus dendroideus* var. *dendroideus, Lycium brevipes* var. *hassei, Lyonothamnus floribundus* ssp. *floribundus, Microseris douglasii* var. *platycarpha, Mimulus traskiae, Nemacaulis denudata* var. *denudata, Orobanche parishii* ssp. *brachyloba, Pentachaeta lyonii, Quercus engelmannii, Q. tomentella, Ribes viburnifolium, Scrophularia villosa, Sibara filifolia, Solanum wallacei, Suaeda taxifolia,* and *Trifolium gracilentum* var. *palmeri.*

Santa Cruz Island

Total number of taxa: *61*

List 1A: *1;* ***List 1B:*** *22;* ***List 2:*** *1;* ***List 3:*** *1;* ***List 4:*** *36*

Abronia maritima, Achnatherum diegoense, Aphanisma blitoides, Arabis hoffmannii, Arctostaphylos tomentosa ssp. *insulicola, A. tomentosa* ssp. *subcordata, A. viridissima, Astragalus miguelensis, Atriplex coulteri, A. pacifica, Baccharis plummerae* ssp. *plummerae, Berberis pinnata* ssp. *insularis, Calandrinia breweri, C. maritima, Calochortus catalinae, Castilleja lanata* ssp. *hololeuca, Ceanothus megacarpus* var. *insularis, Cercocarpus betuloides* var. *blancheae, Chorizanthe wheeleri, Convolvulus simulans, Dendromecon harfordii* var. *harfordii, Dichondra occidentalis, Dudleya candelabrum, D. greenei, D. nesiotica, Erigeron sanctarum, Eriogonum grande* var. *grande, E. grande* var. *rubescens, Erysimum insulare* ssp. *insulare, Eschscholzia ramosa, Galium buxifolium, G. nuttallii* ssp. *insulare, Gilia nevinii, Hazardia detonsa, Helianthemum greenei, Heuchera maxima, Hordeum intercedens, Jepsonia malvifolia, Lepechinia fragrans, Lepidium virginicum* var. *robinsonii, Lilium humboldtii* ssp. *ocellatum, Lotus argophyllus* var. *niveus, L. dendroideus* var. *dendroideus, Lyonothamnus floribundus* ssp. *asplenifolius, Malacothamnus fasciculatus* var. *nesioticus, Malacothrix incana, M. indecora, M. squalida, Mimulus brandegei, M. flemingii, Orobanche parishii* ssp. *brachyloba, Piperia michaelii, Quercus parvula* var. *parvula, Q. tomentella, Ribes thacherianum, Sanicula hoffmannii, Senecio aphanactis, Sibara filifolia, Solanum clokeyi, Suaeda taxifolia,* and *Thysanocarpus conchuliferus.*

Santa Rosa Island

Total number of taxa: *54*

List 1A: *0;* ***List 1B:*** *23;* ***List 2:*** *1;* ***List 3:*** *1;* ***List 4:*** *29*

Achnatherum diegoense, Aphanisma blitoides, Arctostaphylos confertiflora, A. tomentosa ssp. *insulicola, A. tomentosa* ssp. *subcordata, Astragalus miguelensis, Atriplex coulteri, A. pacifica, A. serenana* var. *davidsonii, Berberis pinnata* ssp. *insularis, Calandrinia maritima, Calochortus catalinae, Castilleja lanata* ssp. *hololeuca, C. mollis, Ceanothus megacarpus* var. *insularis, Cercocarpus betuloides* var. *blancheae, Chorizanthe wheeleri, Dendromecon harfordii* var. *harfordii, Dichondra occidentalis, Dudleya blochmaniae* ssp. *insularis, D. candelabrum, D. greenei, Erigeron sanctarum, Eriogonum grande* var. *rubescens, Erysimum ammophilum, E. insulare* ssp. *insulare, Eschscholzia ramosa, Galium buxifolium, G. californicum* ssp. *miguelense, G. nuttallii* ssp. *insulare, Gilia nevinii, G. tenuiflora* ssp. *hoffmannii, Hazardia detonsa, Helianthemum greenei, Heuchera maxima, Hordeum intercedens, Jepsonia malvifolia, Lasthenia glabrata* ssp. *coulteri, Lavatera assurgentiflora* ssp. *assurgentiflora, Lepechinia fragrans, Lilium humboldtii* ssp. *ocellatum, Lotus dendroideus* var. *dendroideus, Lyonothamnus floribundus* ssp. *asplenifolius, Malacothrix incana, Mimulus flemingii, Orobanche parishii* ssp. *brachyloba, Phacelia insularis* var. *insularis, Pinus torreyana* ssp. *insularis, Quercus tomentella, Salvia brandegei, Sanicula hoffmannii, Senecio aphanactis, Solanum clokeyi,* and *Suaeda taxifolia.*

Appendix IV: Plants by Family

The following list include all plants on CNPS Lists 1 through 4. Family arrangement matches that in *The Jepson Manual.*

Acanthaceae

Carlowrightia arizonica

Alismataceae

Sagittaria sanfordii

Anacardiaceae

Rhus trilobata var. *simplicifolia*

Apiaceae

Ammoselinum giganteum
Angelica callii
Cymopterus deserticola
C. gilmanii
C. ripleyi
Eryngium aristulatum var. *hooveri*
E. aristulatum var. *parishii*
E. constancei
E. pinnatisectum
E. racemosum
E. spinosepalum
Lilaeopsis masonii
Lomatium ciliolatum var. *hooveri*
L. congdonii
L. engelmannii
L. foeniculaceum ssp. *inyoense*
L. hendersonii
L. howellii
L. insulare
L. martindalei
L. parvifolium
L. peckianum
L. ravenii
L. repostum
L. rigidum
L. shevockii
L. stebbinsii
L. tracyi
Oreonana purpurascens
O. vestita
Perideridia bacigalupii
P. gairdneri ssp. *gairdneri*
P. leptocarpa
P. parishii ssp. *parishii*
P. pringlei
Podistera nevadensis
Sanicula hoffmannii
S. maritima
S. peckiana
S. saxatilis
S. tracyi
Tauschia glauca
T. howellii

Aristolochiaceae

Asarum marmoratum

Asclepiadaceae

Asclepias solanoana
Cynanchum utahense
Matelea parvifolia

Aspleniaceae

Asplenium septentrionale
A. trichomanes ssp. *trichomanes*
A. trichomanes-ramosum

Asteraceae

Ageratina shastensis
Ambrosia chenopodiifolia
A. pumila
Antennaria flagellaris
A. marginata
A. pulchella
A. suffrutescens
Arnica cernua
A. fulgens
A. sororia
A. spathulata
A. venosa
A. viscosa
Artemisia palmeri
Aster greatae
A. lentus
Baccharis plummerae ssp. *glabrata*
B. plummerae ssp. *plummerae*
B. vanessae
Balsamorhiza hookeri var. *lanata*
B. macrolepis var. *macrolepis*
B. sericea
Blennosperma bakeri
B. nanum var. *robustum*
Blepharizonia plumosa ssp. *plumosa*
Calycadenia hooveri
C. oppositifolia
C. villosa
Chaenactis carphoclinia var. *peirsonii*
C. douglasii var. *alpina*
C. parishii
C. suffrutescens
Chrysothamnus gramineus
Cirsium andrewsii
C. ciliolatum
C. crassicaule
C. fontinale var. *campylon*
C. fontinale var. *fontinale*
C. fontinale var. *obispoense*
C. hydrophilum var. *hydrophilum*
C. hydrophilum var. *vaseyi*
C. loncholepis
C. occidentale var. *compactum*
C. rhothophilum
Coreopsis hamiltonii
C. maritima
Corethrogyne filaginifolia var. *incana*
C. filaginifolia var. *linifolia*
C. leucophylla
Crepis runcinata ssp. *hallii*
Dimeresia howellii
Enceliopsis covillei
E. nudicaulis
Ericameria cuneata var. *macrocephala*
E. fasciculata
E. gilmanii
E. ophitidis
E. palmeri ssp. *palmeri*
Erigeron aequifolius
E. angustatus
E. biolettii
E. blochmaniae
E. bloomeri var. *nudatus*
E. breweri var. *bisanctus*
E. breweri var. *jacinteus*
E. calvus
E. cervinus
E. decumbens var. *robustior*
E. elegantulus
E. inornatus var. *calidipetris*
E. inornatus var. *keilii*
E. mariposanus
E. miser
E. multiceps
E. parishii
E. petrophilus var. *sierrensis*
E. petrophilus var. *viscidulus*
E. sanctarum
E. serpentinus
E. supplex
E. uncialis var. *uncialis*
Eriophyllum congdonii
E. jepsonii
E. lanatum var. *hallii*
E. lanatum var. *obovatum*
E. latilobum
E. mohavense
E. nevinii
E. nubigenum
Geraea viscida
Grindelia fraxino-pratensis
G. hirsutula var. *hallii*
G. hirsutula var. *maritima*
G. stricta var. *angustifolia*
Hazardia cana
H. detonsa
H. orcuttii
Helianthella castanea
Helianthus exilis
H. niveus ssp. *tephrodes*
H. nuttallii ssp. *parishii*
Hemizonia arida
H. clementina
H. congesta ssp. *calyculata*
H. congesta ssp. *leucocephala*
H. congesta ssp. *tracyi*
H. conjugens
H. floribunda
H. halliana
H. increscens ssp. *villosa*
H. minthornii
H. mohavensis
H. parryi ssp. *australis*
H. parryi ssp. *congdonii*

H. pungens ssp. *laevis*
Hesperevax sparsiflora var. *brevifolia*
Heterotheca villosa var. *shevockii*
Holocarpha macradenia
H. virgata ssp. *elongata*
Hulsea brevifolia
H. californica
H. mexicana
H. vestita ssp. *callicarpha*
H. vestita ssp. *gabrielensis*
H. vestita ssp. *inyoensis*
H. vestita ssp. *parryi*
Isocoma arguta
I. menziesii var. *decumbens*
I. menziesii var. *diabolica*
Iva hayesiana
Lasthenia burkei
L. conjugens
L. glabrata ssp. *coulteri*
L. leptalea
Layia carnosa
L. discoidea
L. heterotricha
L. jonesii
L. leucopappa
L. munzii
L. septentrionalis
Lembertia congdonii
Lessingia arachnoidea
L. germanorum
L. glandulifera var. *tomentosa*
L. hololeuca
L. micradenia var. *glabrata*
L. micradenia var. *micradenia*
L. occidentalis
L. tenuis
Machaeranthera asteroides var. *lagunensis*
M. canescens var. *ziegleri*
M. juncea
Madia doris-nilesiae
M. hallii
M. nutans
M. radiata
M. stebbinsii
M. yosemitana
Malacothrix foliosa
M. incana
M. indecora
M. saxatilis var. *arachnoidea*
M. squalida
Malperia tenuis
Micropus amphibolus
Microseris borealis
M. douglasii var. *platycarpha*
Palafoxia arida var. *gigantea*
Pentachaeta bellidiflora
P. exilis ssp. *aeolica*
P. lyonii
Perityle inyoensis
P. villosa
Pseudobahia bahiifolia
P. peirsonii
Psilocarphus brevissimus var. *multiflorus*
P. elatior
P. tenellus var. *globiferus*
Pyrrocoma lucida
P. racemosa var. *congesta*
P. uniflora var. *gossypina*
Raillardella pringlei
Raillardiopsis muirii
R. scabrida
Sanvitalia abertii
Saussurea americana
Senecio aphanactis
S. bernardinus
S. clevelandii var. *clevelandii*
S. clevelandii var. *heterophyllus*
S. eurycephalus var. *lewisrosei*
S. ganderi
S. hydrophiloides
S. ionophyllus
S. layneae
S. macounii
S. pattersonensis
Solidago gigantea
S. guiradonis
Stebbinsoseris decipiens
Stenotus lanuginosus
Stephanomeria blairii
Stylocline citroleum
S. masonii
S. sonorensis
Syntrichopappus lemmonii
Taraxacum californicum
Tetradymia argyraea
Tonestus eximius
T. lyallii
Townsendia parryi
Tracyina rostrata
Trichocoronis wrightii var. *wrightii*
Trimorpha acris var. *debilis*
Verbesina dissita
Viguiera laciniata
Wyethia elata
W. longicaulis
W. reticulata
Xylorhiza cognata
X. orcuttii

Azollaceae

Azolla mexicana

Berberidaceae

Berberis fremontii
B. nevinii
B. pinnata ssp. *insularis*
Vancouveria chrysantha

Boraginaceae

Amsinckia grandiflora
A. lunaris
A. vernicosa var. *furcata*
Cryptantha clevelandii var. *dissita*
C. clokeyi
C. costata
C. crinita
C. crymophila
C. excavata
C. ganderi
C. holoptera
C. hooveri
C. mariposae
C. rattanii
C. roosiorum
C. scoparia
C. traskiae
C. tumulosa
Hackelia amethystina
H. brevicula
H. cusickii
H. sharsmithii
Harpagonella palmeri
Mertensia bella
Plagiobothrys chorisianus var. *chorisianus*
P. diffusus
P. glaber
P. glomeratus
P. glyptocarpus var. *modestus*
P. hystriculus
P. lithocaryus
P. mollis var. *vestitus*
P. myosotoides
P. salsus
P. strictus
P. uncinatus

Brassicaceae

Arabis aculeolata
A. blepharophylla
A. bodiensis
A. breweri var. *pecuniaria*
A. cobrensis
A. constancei
A. dispar
A. fernaldiana var. *stylosa*
A. hoffmannii
A. johnstonii
A. koehleri var. *stipitata*
A. macdonaldiana
A. microphylla var. *microphylla*
A. modesta
A. oregana
A. parishii
A. pinzlae
A. pulchra var. *munciensis*
A. pygmaea
A. rigidissima var. *demota*
A. rigidissima var. *rigidissima*
A. serpentinicola
A. shockleyi
A. tiehmii
Cardamine nuttallii var. *gemmata*
C. pachystigma var. *dissectifolia*
Caulanthus amplexicaulis var. *barbarae*
C. californicus
C. simulans
Caulostramina jaegeri
Cochlearia officinalis var. *arctica*
Cusickiella quadricostata
Dithyrea maritima
Draba asterophora var. *asterophora*
D. asterophora var. *macrocarpa*
D. aureola
D. californica
D. cana
D. carnosula
D. cruciata
D. howellii
D. incrassata
D. monoensis
D. pterosperma
D. sharsmithii
D. sierrae
D. subumbellata
Erysimum ammophilum
E. capitatum ssp. *angustatum*
E. capitatum ssp. *lompocense*
E. franciscanum
E. insulare ssp. *insulare*
E. insulare ssp. *suffrutescens*
E. menziesii ssp. *eurekense*
E. menziesii ssp. *menziesii*

E. menziesii ssp. *yadonii*
E. teretifolium
Halimolobos virgata
Lepidium flavum var. *felipense*
L. jaredii ssp. *album*
L. jaredii ssp. *jaredii*
L. latipes var. *heckardii*
L. virginicum var. *robinsonii*
Lesquerella kingii ssp. *bernardina*
Lyrocarpa coulteri var. *palmeri*
Rorippa columbiae
R. gambellii
R. subumbellata
Sibara filifolia
Smelowskia ovalis var. *congesta*
Streptanthus albidus ssp. *albidus*
S. albidus ssp. *peramoenus*
S. batrachopus
S. bernardinus
S. brachiatus ssp. *brachiatus*
S. brachiatus ssp. *hoffmanii*
S. callistus
S. campestris
S. cordatus var. *piutensis*
S. drepanoides
S. farnsworthianus
S. fenestratus
S. glandulosus ssp. *pulchellus*
S. glandulosus var. *hoffmanii*
S. gracilis
S. hispidus
S. howellii
S. insignis ssp. *lyonii*
S. morrisonii ssp. *elatus*
S. morrisonii ssp. *hirtiflorus*
S. morrisonii ssp. *kruckebergii*
S. morrisonii ssp. *morrisonii*
S. niger
S. oliganthus
Thelypodium brachycarpum
T. stenopetalum
Thlaspi californicum
Thysanocarpus conchuliferus
Tropidocarpum capparideum

Burseraceae

Bursera microphylla

Cactaceae

Bergerocactus emoryi
Carnegiea gigantea
Echinocereus engelmannii var. *howei*
Escobaria vivipara var. *alversonii*
E. vivipara var. *rosea*
Ferocactus viridescens
Opuntia basilaris var. *brachyclada*
O. basilaris var. *treleasei*
O. curvospina
O. fragilis
O. munzii
O. parryi var. *serpentina*
O. pulchella
O. wigginsii
O. wolfii
Sclerocactus polyancistrus

Campanulaceae

Campanula californica
C. exigua
C. scabrella
C. sharsmithiae
C. shetleri
C. wilkinsiana
Downingia concolor var. *brevior*
D. pusilla
Githopsis diffusa ssp. *filicaulis*
Howellia aquatilis
Legenere limosa
Nemacladus gracilis
N. twisselmannii

Capparaceae

Wislizenia refracta ssp. *refracta*

Caryophyllaceae

Arenaria macradenia var. *kuschei*
A. paludicola
A. ursina
Loeflingia squarrosa var. *artemisiarum*
Minuartia decumbens
M. howellii
M. obtusiloba
M. rosei
M. stolonifera
Paronychia ahartii
Silene campanulata ssp. *campanulata*
S. invisa
S. marmorensis
S. occidentalis ssp. *longistipitata*
S. suksdorfii
S. verecunda ssp. *verecunda*
Stellaria longifolia
S. obtusa

Chenopodiaceae

Aphanisma blitoides
Atriplex cordulata
A. coronata var. *coronata*
A. coronata var. *notatior*
A. coulteri
A. depressa
A. joaquiniana
A. minuscula
A. pacifica
A. parishii
A. serenana var. *davidsonii*
A. tularensis
A. vallicola
Chenopodium simplex
Nitrophila mohavensis
Suaeda californica
S. esteroa
S. taxifolia

Cistaceae

Helianthemum greenei
H. suffrutescens

Convolvulaceae

Calystegia atriplicifolia ssp. *buttensis*
C. collina ssp. *oxyphylla*
C. collina ssp. *venusta*
C. macrostegia ssp. *amplissima*
C. malacophylla var. *berryi*
C. peirsonii
C. sepium ssp. *binghamiae*
C. stebbinsii
C. subacaulis ssp. *episcopalis*
Convolvulus simulans
Dichondra occidentalis

Crassulaceae

Dudleya abramsii ssp. *affinis*
D. abramsii ssp. *bettinae*
D. abramsii ssp. *murina*
D. abramsii ssp. *parva*
D. alainae
D. attenuata ssp. *orcuttii*
D. blochmaniae ssp. *blochmaniae*
D. blochmaniae ssp. *brevifolia*
D. blochmaniae ssp. *insularis*
D. calcicola
D. candelabrum
D. cymosa ssp. *costafolia*
D. cymosa ssp. *crebrifolia*
D. cymosa ssp. *marcescens*
D. cymosa ssp. *ovatifolia*
D. densiflora
D. greenei
D. multicaulis
D. nesiotica
D. saxosa ssp. *saxosa*
D. setchellii
D. stolonifera
D. traskiae
D. variegata
D. verityi
D. virens
D. viscida
Parvisedum leiocarpum
Sedum albomarginatum
S. divergens
S. eastwoodiae
S. laxum ssp. *flavidum*
S. laxum ssp. *heckneri*
S. niveum
S. oblanceolatum
S. paradisum

Crossosomataceae

Crossosoma californicum
Glossopetalon pungens

Cupressaceae

Cupressus abramsiana
C. arizonica ssp. *nevadensis*
C. bakeri
C. forbesii
C. goveniana ssp. *goveniana*
C. goveniana ssp. *pigmaea*
C. macrocarpa
C. nootkatensis
C. stephensonii

Cyperaceae

Carex albida
C. californica
C. comosa
C. congdonii
C. davyi
C. eleocharis
C. geyeri
C. gigas
C. halliana
C. hystricina
C. incurviformis var. *danaensis*
C. lasiocarpa
C. leptalea
C. limosa
C. livida
C. norvegica

C. obispoensis
C. parryana var. *hallii*
C. petasata
C. praticola
C. scoparia
C. sheldonii
C. tiogana
C. tompkinsii
C. vulpinoidea
Eleocharis parvula
E. quadrangulata
Fimbristylis thermalis
Kobresia bellardii
Rhynchospora alba
R. californica
R. globularis var. *globularis*
Scirpus clementis
S. heterochaetus
S. pumilus
S. subterminalis

Droseraceae

Drosera anglica

Dryopteridaceae

Dryopteris filix-mas
Polystichum kruckebergii
P. lonchitis
Woodsia plummerae

Empetraceae

Empetrum nigrum ssp. *hermaphroditum*

Equisetaceae

Equisetum palustre

Ericaceae

Arctostaphylos andersonii
A. auriculata
A. bakeri ssp. *bakeri*
A. bakeri ssp. *sublaevis*
A. canescens ssp. *sonomensis*
A. catalinae
A. confertiflora
A. cruzensis
A. densiflora
A. edmundsii
A. gabrielensis
A. glandulosa ssp. *crassifolia*
A. glutinosa
A. hispidula
A. hookeri ssp. *franciscana*
A. hookeri ssp. *hearstiorum*
A. hookeri ssp. *hookeri*
A. hookeri ssp. *montana*
A. hookeri ssp. *ravenii*
A. hooveri
A. imbricata
A. klamathensis
A. luciana
A. malloryi
A. manzanita ssp. *laevigata*
A. mendocinoensis
A. mewukka ssp. *truei*
A. montaraensis
A. montereyensis
A. morroensis
A. myrtifolia
A. nissenana
A. nortensis
A. obispoensis
A. osoensis
A. otayensis
A. pajaroensis
A. pallida
A. pechoensis
A. peninsularis ssp. *peninsularis*
A. pilosula
A. pumila
A. purissima
A. rainbowensis
A. refugioensis
A. regismontana
A. rudis
A. silvicola
A. stanfordiana ssp. *decumbens*
A. stanfordiana ssp. *raichei*
A. tomentosa ssp. *daciticola*
A. tomentosa ssp. *eastwoodiana*
A. tomentosa ssp. *insulicola*
A. tomentosa ssp. *subcordata*
A. virgata
A. viridissima
A. wellsii
Comarostaphylis diversifolia ssp. *diversifolia*
Monotropa uniflora
Ornithostaphylos oppositifolia
Pityopus californicus
Vaccinium coccineum
V. scoparium

Euphorbiaceae

Chamaesyce arizonica
C. hooveri
C. ocellata ssp. *rattanii*
C. platysperma
Croton wigginsii
Ditaxis californica
D. clariana
Euphorbia exstipulata var. *exstipulata*
E. misera
Tetracoccus dioicus
T. ilicifolius

Fabaceae

Astragalus agnicidus
A. albens
A. allochrous var. *playanus*
A. anxius
A. argophyllus var. *argophyllus*
A. atratus var. *mensanus*
A. bicristatus
A. brauntonii
A. breweri
A. cimae var. *cimae*
A. clarianus
A. clevelandii
A. crotalariae
A. deanei
A. douglasii var. *perstrictus*
A. ertterae
A. funereus
A. geyeri var. *geyeri*
A. gilmanii
A. insularis var. *harwoodii*
A. inversus
A. jaegerianus
A. johannis-howellii
A. kentrophyta var. *danaus*
A. kentrophyta var. *elatus*
A. lentiformis
A. lentiginosus var. *antonius*
A. lentiginosus var. *borreganus*
A. lentiginosus var. *coachellae*
A. lentiginosus var. *kernensis*
A. lentiginosus var. *micans*
A. lentiginosus var. *piscinensis*
A. lentiginosus var. *sesquimetralis*
A. lentiginosus var. *sierrae*
A. leucolobus
A. macrodon
A. magdalenae var. *peirsonii*
A. miguelensis
A. mojavensis var. *hemigyrus*
A. monoensis var. *monoensis*
A. monoensis var. *ravenii*
A. nevinii
A. nutans
A. oocarpus
A. oophorus var. *lavinii*
A. pachypus var. *jaegeri*
A. pauperculus
A. platytropis
A. preussii var. *laxiflorus*
A. preussii var. *preussii*
A. pseudiodanthus
A. pulsiferae var. *pulsiferae*
A. pulsiferae var. *suksdorfii*
A. pycnostachyus var. *lanosissimus*
A. rattanii var. *jepsonianus*
A. rattanii var. *rattanii*
A. serenoi var. *shockleyi*
A. shevockii
A. subvestitus
A. tener var. *ferrisiae*
A. tener var. *tener*
A. tener var. *titi*
A. traskiae
A. tricarinatus
A. umbraticus
A. webberi
Calliandra eriophylla
Dalea ornata
Lathyrus biflorus
L. delnorticus
L. glandulosus
L. jepsonii var. *jepsonii*
L. palustris
L. splendens
L. sulphureus var. *argillaceus*
Lotus argophyllus var. *adsurgens*
L. argophyllus var. *niveus*
L. argyraeus var. *multicaulis*
L. argyraeus var. *notitius*
L. crassifolius var. *otayensis*
L. dendroideus var. *dendroideus*
L. dendroideus var. *traskiae*
L. dendroideus var. *veatchii*
L. haydonii
L. nuttallianus
L. oblongifolius var. *cupreus*
L. rubriflorus
L. yollabolliensis
Lupinus albifrons var. *abramsii*
L. antoninus
L. cervinus
L. citrinus var. *citrinus*
L. citrinus var. *deflexus*
L. constancei
L. croceus var. *pilosellus*
L. dalesiae
L. duranii
L. elatus
L. excubitus var. *johnstonii*
L. excubitus var. *medius*
L. eximius
L. gracilentus

L. guadalupensis
L. holmgrenanus
L. lapidicola
L. lepidus var. *culbertsonii*
L. ludovicianus
L. magnificus var. *glarecola*
L. magnificus var. *hesperius*
L. magnificus var. *magnificus*
L. milo-bakeri
L. nipomensis
L. padre-crowleyi
L. peirsonii
L. sericatus
L. spectabilis
L. tidestromii
L. tracyi
Marina orcuttii var. *orcuttii*
Oxytropis deflexa var. *sericea*
Peteria thompsoniae
Phaseolus filiformis
Psorothamnus arborescens var. *arborescens*
Rupertia hallii
R. rigida
Senna covesii
Thermopsis californica var. *argentata*
T. californica var. *semota*
T. gracilis
T. macrophylla
T. robusta
Trifolium amoenum
T. bolanderi
T. buckwestiorum
T. gracilentum var. *palmeri*
T. howellii
T. lemmonii
T. macilentum var. *dedeckerae*
T. polyodon
T. trichocalyx

Fagaceae

Quercus dumosa
Q. engelmannii
Q. parvula var. *parvula*
Q. tomentella

Frankeniaceae

Frankenia palmeri

Gentianaceae

Gentiana affinis var. *parvidentata*
G. fremontii
G. plurisetosa
G. prostrata
G. setigera
Swertia fastigiata
S. neglecta

Grossulariaceae

Ribes amarum var. *hoffmannii*
R. canthariforme
R. divaricatum var. *parishii*
R. hudsonianum var. *petiolare*
R. laxiflorum
R. marshallii
R. menziesii var. *ixoderme*
R. sericeum
R. thacherianum
R. tularense
R. viburnifolium
R. victoris

Hydrangeaceae

Carpenteria californica
Fendlerella utahensis

Hydrophyllaceae

Eriodictyon altissimum
E. angustifolium
E. capitatum
Nama dichotomum var. *dichotomum*
N. stenocarpum
Nemophila parviflora var. *quercifolia*
Phacelia amabilis
P. anelsonii
P. argentea
P. ciliata var. *opaca*
P. cinerea
P. cookei
P. dalesiana
P. exilis
P. floribunda
P. greenei
P. insularis var. *continentis*
P. insularis var. *insularis*
P. inundata
P. inyoensis
P. leonis
P. mohavensis
P. monoensis
P. mustelina
P. nashiana
P. novenmillensis
P. orogenes
P. parishii
P. phacelioides
P. pulchella var. *gooddingii*
P. sericea var. *ciliosa*
P. stebbinsii
P. stellaris
P. suaveolens ssp. *keckii*
Pholistoma auritum var. *arizonicum*

Iridaceae

Iris bracteata
I. hartwegii ssp. *columbiana*
I. innominata
I. munzii
I. tenax ssp. *klamathensis*

Juglandaceae

Juglans californica var. *californica*
J. californica var. *hindsii*

Juncaceae

Juncus acutus ssp. *leopoldii*
J. dudleyi
J. duranii
J. hemiendytus var. *abjectus*
J. leiospermus var. *ahartii*
J. leiospermus var. *leiospermus*
J. marginatus var. *marginatus*
J. nodosus
J. regelii
J. supiniformis

Koeberliniaceae

Koeberlinia spinosa ssp. *tenuispina*

Lamiaceae

Acanthomintha duttonii
A. ilicifolia
A. lanceolata
A. obovata ssp. *cordata*
A. obovata ssp. *obovata*
Lepechinia cardiophylla
L. fragrans
L. ganderi
Lycopus uniflorus
Monardella antonina ssp. *antonina*
M. antonina ssp. *benitensis*
M. beneolens
M. candicans
M. cinerea
M. crispa
M. douglasii ssp. *venosa*
M. follettii
M. frutescens
M. hypoleuca ssp. *lanata*
M. leucocephala
M. linoides ssp. *oblonga*
M. linoides ssp. *viminea*
M. macrantha ssp. *hallii*
M. nana ssp. *leptosiphon*
M. palmeri
M. pringlei
M. robisonii
M. stebbinsii
M. undulata
M. villosa ssp. *globosa*
M. viridis ssp. *saxicola*
M. viridis ssp. *viridis*
Pogogyne abramsii
P. clareana
P. douglasii ssp. *parviflora*
P. floribunda
P. nudiuscula
Poliomintha incana
Salvia brandegei
S. dorrii var. *incana*
S. eremostachya
S. funerea
S. greatae
S. munzii
Satureja chandleri
Scutellaria bolanderi ssp. *austromontana*
S. galericulata
S. holmgreniorum
S. lateriflora
Trichostema austromontanum ssp. *compactum*
T. micranthum
T. ovatum
T. rubisepalum

Lennoaceae

Pholisma sonorae

Lentibulariaceae

Pinguicula vulgaris ssp. *macroceras*

Liliaceae

Agave shawii
A. utahensis
Allium atrorubens var. *atrorubens*
A. fimbriatum var. *purdyi*
A. hickmanii
A. hoffmanii
A. howellii var. *clokeyi*
A. jepsonii

A. munzii
A. nevadense
A. parishii
A. sanbornii var. *congdonii*
A. sanbornii var. *sanbornii*
A. sharsmithae
A. shevockii
A. siskiyouense
A. tribracteatum
A. tuolumnense
A. yosemitense
Androstephium breviflorum
Bloomeria humilis
Brodiaea coronaria ssp. *rosea*
B. filifolia
B. insignis
B. kinkiensis
B. orcuttii
B. pallida
Calochortus catalinae
C. clavatus ssp. *recurvifolius*
C. clavatus var. *avius*
C. clavatus var. *clavatus*
C. clavatus var. *gracilis*
C. dunnii
C. excavatus
C. greenei
C. longebarbatus var. *longebarbatus*
C. monanthus
C. obispoensis
C. palmeri var. *munzii*
C. palmeri var. *palmeri*
C. panamintensis
C. persistens
C. plummerae
C. pulchellus
C. raichei
C. simulans
C. striatus
C. tiburonensis
C. umbellatus
C. weedii var. *intermedius*
C. weedii var. *vestus*
C. westonii
Chlorogalum grandiflorum
C. pomeridianum var. *minus*
C. purpureum var. *purpureum*
C. purpureum var. *reductum*
Erythronium citrinum var. *citrinum*
E. citrinum var. *roderickii*
E. helenae
E. hendersonii
E. howellii
E. klamathense
E. pluriflorum
E. pusaterii
E. tuolumnense
Fritillaria agrestis
F. biflora var. *ineziana*
F. brandegei
F. eastwoodiae
F. falcata
F. lanceolata var. *tristulis*
F. liliacea
F. ojaiensis
F. pluriflora
F. purdyi
F. roderickii
F. striata
F. viridea
Lilium bolanderi
L. humboldtii ssp. *humboldtii*
L. humboldtii ssp. *ocellatum*
L. kelloggii
L. maritimum
L. occidentale
L. pardalinum ssp. *pitkinense*
L. pardalinum ssp. *vollmeri*
L. pardalinum ssp. *wigginsii*
L. parryi
L. rubescens
L. washingtonianum ssp. *purpurascens*
Muilla clevelandii
M. coronata
Nolina interrata
Trillium ovatum ssp. *oettingeri*
Triteleia clementina
T. crocea var. *crocea*
T. crocea var. *modesta*
T. hendersonii var. *hendersonii*
T. ixioides ssp. *cookii*
Veratrum fimbriatum
V. insolitum

Limnanthaceae

Limnanthes bakeri
L. douglasii ssp. *sulphurea*
L. floccosa ssp. *bellingeriana*
L. floccosa ssp. *californica*
L. floccosa ssp. *floccosa*
L. gracilis ssp. *parishii*
L. vinculans

Linaceae

Hesperolinon adenophyllum
H. bicarpellatum
H. breweri
H. congestum
H. didymocarpum
H. drymarioides
H. serpentinum
H. tehamense

Loasaceae

Eucnide rupestris
Mentzelia hirsutissima
Petalonyx thurberi ssp. *gilmanii*

Lycopodiaceae

Lycopodiella inundata
Lycopodium clavatum

Malvaceae

Eremalche kernensis
Herissantia crispa
Hibiscus lasiocarpus
Iliamna bakeri
I. latibracteata
Lavatera assurgentiflora ssp. *assurgentiflora*
L. assurgentiflora ssp. *glabra*
Malacothamnus abbottii
M. aboriginum
M. arcuatus
M. clementinus
M. davidsonii
M. fasciculatus var. *nesioticus*
M. gracilis
M. hallii
M. helleri
M. jonesii
M. mendocinensis
M. niveus
M. palmeri var. *involucratus*
M. palmeri var. *lucianus*
M. palmeri var. *palmeri*
M. parishii
Sidalcea calycosa ssp. *rhizomata*
S. covillei
S. hickmanii ssp. *anomala*
S. hickmanii ssp. *hickmanii*
S. hickmanii ssp. *parishii*
S. hickmanii ssp. *viridis*
S. keckii
S. malachroides
S. malvaeflora ssp. *patula*
S. neomexicana
S. oregana ssp. *eximia*
S. oregana ssp. *hydrophila*
S. oregana ssp. *valida*
S. pedata
S. robusta
S. stipularis
Sphaeralcea rusbyi var. *eremicola*

Marsileaceae

Marsilea oligospora

Martyniaceae

Proboscidea althaeifolia

Nyctaginaceae

Abronia alpina
A. maritima
A. nana ssp. *covillei*
A. umbellata ssp. *breviflora*
Acleisanthes longiflora
Mirabilis tenuiloba
Selinocarpus nevadensis

Onagraceae

Camissonia benitensis
C. boothii ssp. *alyssoides*
C. boothii ssp. *boothii*
C. guadalupensis ssp. *clementina*
C. hardhamiae
C. integrifolia
C. kernensis ssp. *kernensis*
C. lewisii
C. minor
C. sierrae ssp. *alticola*
C. tanacetifolia ssp. *quadriperforata*
Clarkia amoena ssp. *whitneyi*
C. australis
C. biloba ssp. *australis*
C. borealis ssp. *arida*
C. borealis ssp. *borealis*
C. breweri
C. concinna ssp. *automixa*
C. concinna ssp. *raichei*
C. delicata
C. exilis
C. franciscana
C. gracilis ssp. *albicaulis*
C. imbricata
C. jolonensis
C. lewisii
C. lingulata
C. mildrediae
C. mosquinii ssp. *mosquinii*
C. mosquinii ssp. *xerophila*
C. rostrata
C. speciosa ssp. *immaculata*
C. springvillensis
C. tembloriensis ssp. *calientensis*

C. virgata
C. xantiana ssp. *parviflora*
Epilobium howellii
E. luteum
E. nivium
E. oreganum
E. rigidum
E. septentrionale
E. siskiyouense
Oenothera caespitosa ssp. *crinita*
O. californica ssp. *eurekensis*
O. deltoides ssp. *howellii*
O. wolfii

Ophioglossaceae

Botrychium ascendens
B. crenulatum
B. lunaria
B. minganense
B. montanum
B. pinnatum
Ophioglossum californicum
O. pusillum

Orchidaceae

Corallorhiza trifida
Cypripedium californicum
C. fasciculatum
C. montanum
Listera cordata
Malaxis monophyllos ssp. *brachypoda*
Piperia candida
P. michaelii
P. yadonii

Orobanchaceae

Boschniakia hookeri
Orobanche parishii ssp. *brachyloba*
O. valida ssp. *howellii*
O. valida ssp. *valida*

Papaveraceae

Arctomecon merriamii
Canbya candida
Corydalis caseana ssp. *caseana*
Dendromecon harfordii var. *harfordii*
D. harfordii var. *rhamnoides*
Dicentra formosa ssp. *oregana*
D. nevadensis
Eschscholzia hypecoides
E. lemmonii ssp. *kernensis*
E. minutiflora ssp. *twisselmannii*
E. procera
E. ramosa
E. rhombipetala
Platystemon californicus var. *ciliatus*
Romneya coulteri

Philadelphaceae

Jamesia americana var. *rosea*

Pinaceae

Abies amabilis
A. bracteata
A. lasiocarpa var. *lasiocarpa*
Picea engelmannii
Pinus contorta ssp. *bolanderi*
P. edulis
P. longaeva
P. radiata
P. torreyana ssp. *insularis*
P. torreyana ssp. *torreyana*

Poaceae

Achnatherum aridum
A. diegoense
A. lemmonii var. *pubescens*
Agrostis blasdalei
A. hendersonii
A. hooveri
A. humilis
Alopecurus aequalis var. *sonomensis*
Blepharidachne kingii
Bouteloua trifida
Calamagrostis bolanderi
C. crassiglumis
C. foliosa
C. ophitidis
Deschampsia atropurpurea
Dichanthelium lanuginosum var. *thermale*
Dissanthelium californicum
Elymus californicus
E. scribneri
Enneapogon desvauxii
Erioneuron pilosum
Glyceria grandis
Hierochloe odorata
Hordeum intercedens
Lycurus phleoides var. *phleoides*
Melica spectabilis
Muhlenbergia appressa
M. arsenei
M. californica
M. fragilis
M. pauciflora
Munroa squarrosa
Neostapfia colusana
Orcuttia californica
O. inaequalis
O. pilosa
O. tenuis
O. viscida
Piptatherum micranthum
Pleuropogon davyi
P. hooverianus
P. refractus
Poa abbreviata ssp. *marshii*
P. atropurpurea
P. napensis
P. pattersonii
P. piperi
P. rhizomata
Puccinellia californica
P. howellii
P. parishii
P. pumila
Scleropogon brevifolius
Spartina gracilis
Sphenopholis obtusata
Swallenia alexandrae
Tuctoria greenei
T. mucronata

Polemoniaceae

Collomia diversifolia
C. larsenii
C. rawsoniana
C. tracyi
Eriastrum brandegeae
E. densifolium ssp. *sanctorum*
E. hooveri
E. luteum
E. virgatum
Gilia caruifolia
G. latiflora ssp. *cuyamensis*
G. maculata
G. nevinii
G. ripleyi
G. tenuiflora ssp. *arenaria*
G. tenuiflora ssp. *hoffmannii*
Ipomopsis effusa
I. tenuifolia
Leptodactylon californicum ssp. *tomentosum*
L. jaegeri
Linanthus acicularis
L. ambiguus
L. arenicola
L. bellus
L. concinnus
L. floribundus ssp. *hallii*
L. grandiflorus
L. killipii
L. nuttallii ssp. *howellii*
L. oblanceolatus
L. orcuttii
L. pygmaeus ssp. *pygmaeus*
L. rattanii
L. serrulatus
Navarretia eriocephala
N. fossalis
N. heterandra
N. jaredii
N. jepsonii
N. leucocephala ssp. *bakeri*
N. leucocephala ssp. *pauciflora*
N. leucocephala ssp. *plieantha*
N. myersii
N. nigelliformis ssp. *radians*
N. peninsularis
N. prolifera ssp. *lutea*
N. rosulata
N. setiloba
N. subuligera
Phlox dispersa
P. dolichantha
P. hirsuta
P. muscoides
Polemonium chartaceum

Polygalaceae

Polygala cornuta var. *fishiae*
P. heterorhyncha
P. subspinosa

Polygonaceae

Aristocapsa insignis
Chorizanthe biloba var. *immemora*
C. blakleyi
C. breweri
C. cuspidata var. *cuspidata*
C. cuspidata var. *villosa*
C. douglasii
C. howellii
C. leptotheca
C. orcuttiana
C. palmeri
C. parryi var. *fernandina*
C. parryi var. *parryi*
C. polygonoides var. *longispina*
C. procumbens
C. pungens var. *hartwegiana*
C. pungens var. *pungens*

C. rectispina
C. robusta var. *hartwegii*
C. robusta var. *robusta*
C. spinosa
C. valida
C. wheeleri
C. xanti var. *leucotheca*
Dedeckera eurekensis
Dodecahema leptoceras
Eriogonum alpinum
E. apricum var. *apricum*
E. apricum var. *prostratum*
E. argillosum
E. beatleyae
E. bifurcatum
E. breedlovei var. *breedlovei*
E. breedlovei var. *shevockii*
E. butterworthianum
E. congdonii
E. contiguum
E. crocatum
E. diclinum
E. eastwoodianum
E. eremicola
E. ericifolium var. *thornei*
E. foliosum
E. giganteum var. *compactum*
E. giganteum var. *formosum*
E. gilmanii
E. gossypinum
E. grande var. *grande*
E. grande var. *rubescens*
E. grande var. *timorum*
E. heermannii var. *floccosum*
E. heermannii var. *occidentale*
E. hirtellum
E. hoffmannii var. *hoffmannii*
E. hoffmannii var. *robustius*
E. intrafractum
E. kelloggii
E. kennedyi var. *alpigenum*
E. kennedyi var. *austromontanum*
E. kennedyi var. *pinicola*
E. libertini
E. luteolum var. *caninum*
E. microthecum var. *corymbosoides*
E. microthecum var. *johnstonii*
E. microthecum var. *lapidicola*
E. microthecum var. *panamintense*
E. nervulosum
E. nortonii
E. nudum var. *decurrens*
E. nudum var. *indictum*
E. nudum var. *murinum*
E. nudum var. *paralinum*
E. nudum var. *regivirum*
E. nutans
E. ochrocephalum var. *alexanderae*
E. ovalifolium var. *eximium*
E. ovalifolium var. *vineum*
E. pendulum
E. polypodum
E. prattenianum var. *avium*
E. prociduum
E. puberulum
E. shockleyi var. *shockleyi*
E. siskiyouense
E. strictum var. *greenei*
E. temblorense
E. ternatum
E. tripodum
E. truncatum
E. twisselmannii
E. umbellatum var. *glaberrimum*
E. umbellatum var. *humistratum*
E. umbellatum var. *juniporinum*
E. umbellatum var. *minus*
E. umbellatum var. *torreyanum*
E. vestitum
E. wrightii var. *olanchense*
Gilmania luteola
Goodmania luteola
Mucronea californica
Nemacaulis denudata var. *denudata*
N. denudata var. *gracilis*
Oxytheca caryophylloides
O. emarginata
O. parishii var. *abramsii*
O. parishii var. *cienegensis*
O. parishii var. *goodmaniana*
O. watsonii
Polygonum bidwelliae
P. marinense
P. polygaloides ssp. *esotericum*
Systenotheca vortriedei

Portulacaceae

Calandrinia breweri
C. maritima
Calyptridium parryi var. *hesseae*
C. pulchellum
C. quadripetalum
Claytonia lanceolata var. *peirsonii*
C. megarhiza
C. palustris
C. umbellata
Lewisia brachycalyx
L. cantelovii
L. congdonii
L. cotyledon var. *heckneri*
L. cotyledon var. *howellii*
L. disepala
L. longipetala
L. oppositifolia
L. serrata
L. stebbinsii
Montia howellii
Portulaca halimoides

Potamogetonaceae

Potamogeton epihydrus ssp. *nuttallii*
P. filiformis
P. foliosus var. *fibrillosus*
P. praelongus
P. robbinsii
P. zosteriformis

Primulaceae

Androsace elongata ssp. *acuta*
A. filiformis
Trientalis arctica

Pteridaceae

Argyrochosma limitanea var. *limitanea*
Aspidotis carlotta-halliae
Astrolepis cochisensis
Cheilanthes wootonii
Pellaea truncata

Rafflesiaceae

Pilostyles thurberi

Ranunculaceae

Delphinium bakeri
D. californicum ssp. *interius*
D. gypsophilum ssp. *gypsophilum*
D. gypsophilum ssp. *parviflorum*
D. hansenii ssp. *ewanianum*
D. hesperium ssp. *cuyamacae*
D. hutchinsoniae
D. inopinum
D. luteum
D. parishii ssp. *subglobosum*
D. parryi ssp. *blochmaniae*
D. parryi ssp. *purpureum*
D. purpusii
D. recurvatum
D. stachydeum
D. uliginosum
D. umbraculorum
D. variegatum ssp. *kinkiense*
D. variegatum ssp. *thornei*
Myosurus minimus ssp. *apus*
Ranunculus hydrocharoides
R. lobbii

Rhamnaceae

Adolphia californica
Ceanothus confusus
C. cuneatus var. *rigidus*
C. cyaneus
C. divergens
C. ferrisae
C. foliosus var. *vineatus*
C. fresnensis
C. gloriosus var. *gloriosus*
C. gloriosus var. *porrectus*
C. hearstiorum
C. maritimus
C. masonii
C. megacarpus var. *insularis*
C. ophiochilus
C. purpureus
C. roderickii
C. sonomensis
C. verrucosus
Colubrina californica
Condalia globosa var. *pubescens*

Rosaceae

Cercocarpus betuloides var. *blancheae*
C. traskiae
Chamaebatia australis
Geum aleppicum
Horkelia bolanderi
H. congesta ssp. *nemorosa*
H. cuneata ssp. *sericea*
H. hendersonii
H. hispidula
H. marinensis
H. parryi
H. sericata
H. tenuiloba
H. truncata
H. tularensis
H. wilderae
H. yadonii
Ivesia aperta var. *aperta*
I. aperta var. *canina*
I. argyrocoma
I. arizonica var. *arizonica*
I. baileyi var. *baileyi*

I. baileyi var. *beneolens*
I. callida
I. campestris
I. jaegeri
I. kingii var. *kingii*
I. longibracteata
I. paniculata
I. patellifera
I. pickeringii
I. sericoleuca
I. unguiculata
I. webberi
Lyonothamnus floribundus ssp. *asplenifolius*
L. floribundus ssp. *floribundus*
Neviusia cliftonii
Potentilla basaltica
P. concinna
P. cristae
P. glandulosa ssp. *ewanii*
P. hickmanii
P. morefieldii
P. multijuga
P. newberryi
P. rimicola
Prunus fasciculata var. *punctata*
Rosa minutifolia
Rubus glaucifolius var. *ganderi*
R. nivalis
Sanguisorba officinalis

Rubiaceae

Galium andrewsii ssp. *gatense*
G. angustifolium ssp. *borregoense*
G. angustifolium ssp. *gabrielense*
G. angustifolium ssp. *gracillimum*
G. angustifolium ssp. *jacinticum*
G. angustifolium ssp. *onycense*
G. buxifolium
G. californicum ssp. *luciense*
G. californicum ssp. *miguelense*
G. californicum ssp. *primum*
G. californicum ssp. *sierrae*
G. catalinense ssp. *acrispum*
G. catalinense ssp. *catalinense*
G. clementis
G. cliftonsmithii
G. glabrescens ssp. *modocense*
G. grande
G. hardhamiae
G. hilendiae ssp. *carneum*
G. hilendiae ssp. *kingstonense*
G. hypotrichium ssp. *tomentellum*
G. jepsonii
G. johnstonii
G. munzii
G. nuttallii ssp. *insulare*
G. oreganum
G. serpenticum ssp. *scotticum*
G. serpenticum ssp. *warnerense*
G. wrightii

Salicaceae

Populus angustifolia
Salix bebbiana
S. brachycarpa ssp. *brachycarpa*
S. delnortensis
S. reticulata ssp. *nivalis*

Sarraceniaceae

Darlingtonia californica

Saxifragaceae

Bensoniella oregona
Bolandra californica
Boykinia rotundifolia
Heuchera abramsii
H. brevistaminea
H. chlorantha
H. duranii
H. elegans
H. hirsutissima
H. maxima
H. parishii
H. rubescens var. *versicolor*
Jepsonia heterandra
J. malvifolia
Lithophragma maximum
Saxifraga cespitosa
S. howellii
S. nuttallii
S. rufidula
Tiarella trifoliata var. *trifoliata*

Scheuchzeriaceae

Scheuchzeria palustris var. *americana*

Scrophulariaceae

Antirrhinum cyathiferum
A. ovatum
A. subcordatum
A. virga
Castilleja affinis ssp. *neglecta*
C. ambigua ssp. *humboldtiensis*
C. campestris ssp. *succulenta*
C. cinerea
C. gleasonii
C. grisea
C. hispida ssp. *brevilobata*
C. lanata ssp. *hololeuca*
C. lasiorhyncha
C. latifolia
C. mendocinensis
C. miniata ssp. *elata*
C. mollis
C. montigena
C. plagiotoma
C. schizotricha
C. uliginosa
Collinsia corymbosa
C. multicolor
Cordylanthus capitatus
C. eremicus ssp. *eremicus*
C. eremicus ssp. *kernensis*
C. maritimus ssp. *maritimus*
C. maritimus ssp. *palustris*
C. mollis ssp. *hispidus*
C. mollis ssp. *mollis*
C. nidularius
C. orcuttianus
C. palmatus
C. parviflorus
C. rigidus ssp. *littoralis*
C. tecopensis
C. tenuis ssp. *barbatus*
C. tenuis ssp. *brunneus*
C. tenuis ssp. *capillaris*
C. tenuis ssp. *pallescens*
Galvezia speciosa
Gratiola heterosepala
Limosella subulata
Maurandya antirrhiniflora ssp. *antirrhiniflora*
M. petrophila
Mimulus acutidens
M. aridus
M. brachiatus
M. brandegei
M. clevelandii
M. diffusus
M. exiguus
M. filicaulis
M. flemingii
M. glabratus ssp. *utahensis*
M. glaucescens
M. gracilipes
M. grayi
M. inconspicuus
M. laciniatus
M. microphyllus
M. mohavensis
M. norrisii
M. nudatus
M. pictus
M. pulchellus
M. purpureus
M. pygmaeus
M. rattanii ssp. *decurtatus*
M. rupicola
M. shevockii
M. subsecundus
M. traskiae
M. whipplei
Orthocarpus cuspidatus ssp. *cuspidatus*
O. pachystachyus
Pedicularis bracteosa var. *flavida*
P. centranthera
P. contorta
P. crenulata
P. dudleyi
P. howellii
Penstemon albomarginatus
P. barnebyi
P. calcareus
P. californicus
P. cinereus
P. cinicola
P. clevelandii var. *connatus*
P. filiformis
P. fruticiformis var. *amargosae*
P. heterodoxus var. *shastensis*
P. newberryi var. *sonomensis*
P. papillatus
P. personatus
P. purpusii
P. rattanii var. *kleei*
P. stephensii
P. thurberi
P. tracyi
Scrophularia atrata
S. villosa
Triphysaria floribunda
Veronica copelandii
V. cusickii

Selaginellaceae

Selaginella asprella
S. cinerascens
S. densa var. *scopulorum*
S. eremophila
S. leucobryoides

Simaroubaceae

Castela emoryi

Smilacaceae
Smilax jamesii

Solanaceae
Lycium brevipes var. *hassei*
L. parishii
L. verrucosum
Oryctes nevadensis
Physalis lobata
Solanum clokeyi
S. wallacei

Sterculiaceae
Ayenia compacta
Fremontodendron decumbens
F. mexicanum

Thelypteridaceae
Thelypteris puberula var. *sonorensis*

Thymelaeaceae
Dirca occidentalis

Typhaceae
Sparganium natans

Verbenaceae
Verbena californica

Violaceae
Viola aurea
V. langsdorfii
V. palustris
V. pinetorum ssp. *grisea*
V. primulifolia ssp. *occidentalis*
V. tomentosa

Ceanothus ophiochilus

Appendix V: Plants New to This Edition

The following list includes all the plants that are new to this edition. A few have been considered but rejected or overlooked in past editions, but most are either newly described or more endangered now than previously.

Abronia maritima, A. nana ssp. *covillei, Acanthomintha obovata* ssp. *cordata, Acleisanthes longiflora, Allium atrorubens* var. *atrorubens, A. jepsonii, A. nevadense, A. sanbornii* var. *congdonii, A. sanbornii* var. *sanbornii, Androsace elongata* ssp. *acuta, Antennaria marginata, A. pulchella, Arabis dispar, A. rigidissima* var. *demota, Arctostaphylos andersonii, A. bakeri* ssp. *sublaevis, A. gabrielensis, A. glandulosa* ssp. *crassifolia, A. malloryi, A. mendocinoensis, A. nortensis, A. osoensis, A. rainbowensis, A. regismontana, A. tomentosa* ssp. *daciticola, Arnica fulgens, A. sororia, Asclepias solanoana, Aspidotis carlotta-halliae, Astragalus allochrous* var. *playanus, A. lentiginosus* var. *antonius, A. lentiginosus* var. *sierrae, A. pachypus* var. *jaegeri, A. platytropis, A. pulsiferae* var. *pulsiferae, A. pulsiferae* var. *suksdorfii, A. tener* var. *ferrisiae, A. tener* var. *tener, Atriplex coronata* var. *coronata, A. coulteri, A. depressa, A. minuscula, A. pacifica, A. serenana* var. *davidsonii, Azolla mexicana, Baccharis plummerae* ssp. *glabrata, Balsamorhiza hookeri* var. *lanata, Berberis fremontii, Blepharizonia plumosa* ssp. *plumosa, Boschniakia hookeri, Botrychium ascendens, B. lunaria, B. minganense, B. montanum, B. pinnatum, Calandrinia breweri, Calochortus clavatus* var. *gracilis, C. palmeri* var. *palmeri, C. panamintensis, C. plummerae, C. weedii* var. *intermedius, C. weedii* var. *vestus, Calycadenia villosa, Calystegia sepium* ssp. *binghamiae, Camissonia lewisii, Canbya candida, Carex comosa, C. hystricina, C. incurviformis* var. *danaensis, C. leptalea, C. limosa, C. norvegica, C. scoparia, C. tiogana, C. vulpinoidea, Carlowrightia arizonica, Ceanothus ophiochilus, Chaenactis carphoclinia* var. *peirsonii, C. douglasii* var. *alpina, Chlorogalum pomeridianum* var. *minus, Chorizanthe biloba* var. *immemora, C. cuspidata* var. *cuspidata, C. cuspidata* var. *villosa, C. leptotheca, C. palmeri, C. parryi* var. *parryi, C. polygonoides* var. *longispina, C. procumbens, C. pungens* var. *hartwegiana, C. robusta* var. *hartwegii, C. xanti* var. *leucotheca, Clarkia concinna* ssp. *automixa, C. concinna* ssp. *raichei, C. gracilis* ssp. *albicaulis, C. xantiana* ssp. *parviflora, Claytonia umbellata, Collinsia corymbosa, Condalia globosa* var. *pubescens, Convolvulus simulans, Corallorhiza trifida, Crepis runcinata* ssp. *hallii, Cryptantha clevelandii* var. *dissita, C. clokeyi, Delphinium hansenii* ssp. *ewanianum, Dicentra nevadensis, Draba californica, D. monoensis, D. subumbellata, Dudleya cymosa* ssp. *costafolia, Eleocharis quadrangulata, Epilobium howellii, Equisetum palustre, Ericameria cuneata* var. *macrocephala, Erigeron angustatus, E. biolettii, E. breweri* var. *bisanctus, E. breweri* var. *jacinteus, E. calvus, E. inornatus* var. *calidipetris, E. inornatus* var. *keilii, E. mariposanus, E. petrophilus* var. *sierrensis, E. petrophilus* var. *viscidulus, E. serpentinus, Eriogonum foliosum, E. heermannii* var. *occidentale, E. kennedyi* var. *alpigenum, E. microthecum* var. *corymbosoides, E. microthecum* var. *lapidicola, E. nudum* var. *decurrens, E. nudum* var. *indictum, E. nudum* var. *paralinum, E. nudum* var. *regivirum, E. ochrocephalum* var. *alexanderae, E. ovalifolium* var. *eximium, E. prattenianum* var. *avium, E. tripodum, E. umbellatum* var. *humistratum, Erysimum menziesii* ssp. *eurekense, E. menziesii* ssp. *yadonii, Erythronium citrinum* var. *roderickii, E. helenae, Erythronium pluriflorum, Eschscholzia lemmonii* ssp. *kernensis, E. minutiflora* ssp. *twisselmannii, Fritillaria biflora* var. *ineziana, F. lanceolata* var. *tristulis, Galium andrewsii* ssp. *gatense, G. angustifolium* ssp. *gracillimum, G. angustifolium* ssp. *jacinticum, Gentiana plurisetosa, Geum aleppicum, Goodmania luteola, Hemizonia parryi* ssp. *congdonii, Hesperevax sparsiflora* var. *brevifolia, Hesperolinon serpentinum, H. tehamense, Heterotheca villosa* var. *shevockii, Heuchera chlorantha, Hierochloe odorata, Holocarpha virgata* ssp. *elongata, Hordeum intercedens, Horkelia yadonii, Hulsea mexicana, H. vestita* ssp. *callicarpha, H. vestita* ssp. *gabrielensis, H. vestita* ssp. *parryi, Isocoma arguta, I. menziesii* var. *decumbens, I. menziesii* var. *diabolica, Ivesia aperta* var. *canina, I. baileyi* var. *beneolens, I. campestris, I. longibracteata, Jamesia americana* var. *rosea, Jepsonia heterandra, Juglans californica* var. *californica,*

Holocarpha macradenia

Juncus marginatus var. *marginatus*, *Lasthenia glabrata* ssp. *coulteri*, *Lathyrus palustris*, *L. sulphureus* var. *argillaceus*, *Layia heterotricha*, *L. septentrionalis*, *Lepidium jaredii* ssp. *album*, *L. latipes* var. *heckardii*, *L. virginicum* var. *robinsonii*, *Leptodactylon californicum* ssp. *tomentosum*, *Lessingia arachnoidea*, *L. hololeuca*, *L. micradenia* var. *glabrata*, *L. micradenia* var. *micradenia*, *Lilium humboldtii* ssp. *humboldtii*, *L. humboldtii* ssp. *ocellatum*, *L. kelloggii*, *Limosella subulata*, *Linanthus acicularis*, *L. ambiguus*, *L. grandiflorus*, *L. pygmaeus* ssp. *pygmaeus*, *Lomatium hendersonii*, *Lotus argyraeus* var. *multicaulis*, *L. argyraeus* var. *notitius*, *L. haydonii*, *Madia yosemitana*, *Marsilea oligospora*, *Mertensia bella*, *Microseris borealis*, *M. douglasii* var. *platycarpha*, *Minuartia obtusiloba*, *M. stolonifera*, *Monardella beneolens*, *M. candicans*, *M. follettii*, *M. stebbinsii*, *M. villosa* ssp. *globosa*, *Montia howellii*, *Mucronea californica*, *Muhlenbergia appressa*, *M. californica*, *M. fragilis*, *Munroa squarrosa*, *Nama dichotomum* var. *dichotomum*, *N. stenocarpum*, *Navarretia leucocephala* ssp. *bakeri*, *N. myersii*, *N. nigelliformis* ssp. *radians*, *N. peninsularis*, *Nemacaulis denudata* var. *denudata*, *N. denudata* var. *gracilis*, *Neviusia cliftonii*, *Opuntia curvospina*, *O. fragilis*, *O. wolfii*, *Oxytheca parishii* var. *abramsii*, *Pedicularis centranthera*, *Phacelia exilis*, *P. mohavensis*, *P. pulchella* var. *gooddingii*, *P. stellaris*, *Phlox dispersa*, *Piperia candida*, *P. yadonii*, *Plagiobothrys glomeratus*, *P. salsus*, *Poa abbreviata* ssp. *marshii*, *Pogogyne floribunda*, *Poliomintha incana*, *Polygala heterorhyncha*, *Potamogeton epihydrus* ssp. *nuttallii*, *P. filiformis*, *P. foliosus* var. *fibrillosus*, *P. praelongus*, *P. robbinsii*, *P. zosteriformis*, *Potentilla basaltica*, *P. cristae*, *P. glandulosa* ssp. *ewanii*, *P. morefieldii*, *P. newberryi*, *P. rimicola*, *Psilocarphus brevissimus* var. *multiflorus*, *P. elatior*, *P. tenellus* var. *globiferus*, *Puccinellia howellii*, *Quercus dumosa*, *Q. tomentella*, *Rhus trilobata* var. *simplicifolia*, *Rupertia hallii*, *Salix bebbiana*, *Sanguisorba officinalis*, *Scirpus heterochaetus*, *Scutellaria bolanderi* ssp. *austromontana*, *S. galericulata*, *S. lateriflora*, *Sedum divergens*, *Selaginella asprella*, *S. densa* var. *scopulorum*, *S. eremophila*, *Senecio aphanactis*, *S. clevelandii* var. *heterophyllus*, *Sidalcea calycosa* ssp. *rhizomata*, *S. malachroides*, *S. malvaeflora* ssp. *patula*, *S. neomexicana*, *S. oregana* ssp. *eximia*, *S. oregana* ssp. *hydrophila*, *Solidago gigantea*, *S. guiradonis*, *Stellaria longifolia*, *Stenotus lanuginosus*, *Streptanthus brachiatus* ssp. *hoffmanii*, *S. campestris*, *S. morrisonii* ssp. *elatus*, *S. morrisonii* ssp. *hirtiflorus*, *S. morrisonii* ssp. *kruckebergii*, *S. morrisonii* ssp. *morrisonii*, *Stylocline citroleum*, *S. masonii*, *S. sonorensis*, *Suaeda taxifolia*, *Thelypteris puberula* var. *sonorensis*, *Thermopsis gracilis*, *T. robusta*, *Tiarella trifoliata* var. *trifoliata*, *Trifolium buckwestiorum*, *Triteleia hendersonii* var. *hendersonii*, *Vaccinium scoparium*, *Viola aurea*, *V. langsdorfii*, *V. palustris*, *V. pinetorum* ssp. *grisea*, *V. tomentosa*, and *Wislizenia refracta* ssp. *refracta*.

Appendix VI: Plants Listed or Candidates Under State Law

Minus the six plants mentioned below, this list includes the 216 native plants that as of November 1993 have been designated as Endangered (CE, 128 plants), Threatened (CT, 19 plants), Rare (CR, 68 plants), or Candidate (CC, 1 plant) by the Fish and Game Commission pursuant to Section 1904 (Native Plant Protection Act of 1977) and Section 2074.2 and 2075.5 (California Endangered Species Act of 1984) of the Fish and Game Code. For additional information, contact the Endangered Plant Program, 1416 Ninth Street, Sacramento, CA 95814, or the nearest Department of Fish and Game office (see *Appendix X*).

The following State-listed taxa have been rejected from the fifth edition of the *Inventory* for taxonomic reasons, and do not appear in the list below: *Arctostaphylos edmundsii* var. *parvifolia* (CR), *Arctostaphylos pacifica* (CE), *Agrostis blasdalei* var. *marinensis* (CR), *Mahonia sonnei* (CE), *Caulanthus stenocarpus* (CR), and *Eriastrum tracyi* (CR). *Erysimum menziesii* (CE) has been split into three subspecies in the *Inventory*; all are protected by State law. Note also that the list below uses the nomenclature adopted in the *Inventory*, which in a few cases differs from that of the State of California. Entries in this *Inventory* give the name under which a taxon is State-listed if it differs from the name used here.

Acanthomintha duttonii (CE)
A. ilicifolia (CE)
Allium munzii (CT)
A. yosemitense (CR)
Amsinckia grandiflora (CE)
Arabis macdonaldiana (CE)
Arctostaphylos bakeri ssp. *bakeri* (CR)
A. densiflora (CE)
A. hookeri ssp. *hearstiorum* (CE)
A. hookeri ssp. *ravenii* (CE)
A. imbricata (CE)
A. pallida (CE)
Arenaria paludicola (CE)
Astragalus agnicidus (CE)
A. clarianus (CT)
A. johannis-howellii (CR)
A. lentiginosus var. *sesquimetralis* (CE)
A. magdalenae var. *peirsonii* (CE)
A. monoensis var. *monoensis* (CR)
A. tener var. *titi* (CE)
A. traskiae (CR)
Atriplex tularensis (CE)
Baccharis vanessae (CE)
Bensoniella oregona (CR)
Berberis nevinii (CE)
B. pinnata ssp. *insularis* (CE)
Blennosperma bakeri (CE)
B. nanum var. *robustum* (CR)
Bloomeria humilis (CR)
Brodiaea coronaria ssp. *rosea* (CE)
B. filifolia (CE)
B. insignis (CE)
B. pallida (CE)
Calamagrostis foliosa (CR)
Calochortus dunnii (CR)
C. persistens (CR)
C. tiburonensis (CT)
Calystegia stebbinsii (CE)
Carex albida (CE)
C. tompkinsii (CR)
Carpenteria californica (CT)
Castilleja affinis ssp. *neglecta* (CT)
C. campestris ssp. *succulenta* (CE)
C. gleasonii (CR)
C. grisea (CE)
C. uliginosa (CE)
Caulanthus californicus (CE)
Ceanothus hearstiorum (CR)
C. maritimus (CR)
C. masonii (CR)
C. ophiochilus (CE)
C. roderickii (CR)
Cercocarpus traskiae (CE)
Chlorogalum purpureum var. *reductum* (CR)
Chorizanthe howellii (CT)
C. orcuttiana (CE)
C. valida (CE)
Cirsium ciliolatum (CE)
C. fontinale var. *fontinale* (CE)
C. fontinale var. *obispoense* (CE)
C. loncholepis (CT)
C. rhothophilum (CT)
Clarkia franciscana (CE)
C. imbricata (CE)
C. lingulata (CE)
C. speciosa ssp. *immaculata* (CR)
C. springvillensis (CE)
Cordylanthus maritimus ssp. *maritimus* (CE)
C. mollis ssp. *mollis* (CR)
C. nidularius (CR)
C. palmatus (CE)
C. rigidus ssp. *littoralis* (CE)
C. tenuis ssp. *capillaris* (CR)
Croton wigginsii (CR)
Cryptantha roosiorum (CR)
Cupressus abramsiana (CE)
Dedeckera eurekensis (CR)
Delphinium bakeri (CR)
D. hesperium ssp. *cuyamacae* (CR)
D. luteum (CR)
D. variegatum ssp. *kinkiense* (CE)
Dichanthelium lanuginosum var. *thermale* (CE)
Dithyrea maritima (CT)
Dodecahema leptoceras (CE)
Downingia concolor var. *brevior* (CE)
Dudleya blochmaniae ssp. *brevifolia* (CE)
D. cymosa ssp. *marcescens* (CR)
D. nesiotica (CR)
D. stolonifera (CT)
D. traskiae (CE)
Eriastrum densifolium ssp. *sanctorum* (CE)
Eriodictyon altissimum (CE)
E. capitatum (CR)
Eriogonum alpinum (CE)
E. apricum var. *apricum* (CE)
E. apricum var. *prostratum* (CE)
E. butterworthianum (CR)
E. crocatum (CR)
E. ericifolium var. *thornei* (CE)
E. giganteum var. *compactum* (CR)
E. grande var. *timorum* (CE)
E. kelloggii (CE)
E. twisselmannii (CR)
Eriophyllum congdonii (CR)
E. latilobum (CE)
Eryngium aristulatum var. *parishii* (CE)
E. constancei (CE)
E. racemosum (CE)
Erysimum capitatum ssp. *angustatum* (CE)
E. menziesii ssp. *eurekense* (CE)
E. menziesii ssp. *menziesii* (CE)
E. menziesii ssp. *yadonii* (CE)
E. teretifolium (CE)
Fremontodendron decumbens (CR)
F. mexicanum (CR)
Fritillaria roderickii (CE)
F. striata (CT)
Galium angustifolium ssp. *borregoense* (CR)
G. buxifolium (CR)
G. californicum ssp. *sierrae* (CR)
G. catalinense ssp. *acrispum* (CE)
Gilia tenuiflora ssp. *arenaria* (CT)
Gratiola heterosepala (CE)
Helianthus niveus ssp. *tephrodes* (CE)

Hemizonia arida (CR)
H. conjugens (CE)
H. increscens ssp. *villosa* (CE)
H. minthornii (CR)
H. mohavensis (CE)
Hesperolinon congestum (CT)
H. didymocarpum (CE)
Holocarpha macradenia (CE)
Ivesia callida (CR)
Lasthenia burkei (CE)
Layia carnosa (CE)
Lessingia germanorum (CE)
Lewisia congdonii (CR)
Lilaeopsis masonii (CR)
Lilium occidentale (CE)
L. pardalinum ssp. *pitkinense* (CE)
Limnanthes bakeri (CR)
L. douglasii ssp. *sulphurea* (CE)
L. floccosa ssp. *californica* (CE)
L. gracilis ssp. *parishii* (CE)
L. vinculans (CE)
Lithophragma maximum (CE)
Lotus argophyllus var. *adsurgens* (CE)
L. argophyllus var. *niveus* (CE)
L. dendroideus var. *traskiae* (CE)
Lupinus citrinus var. *deflexus* (CT)
L. milo-bakeri (CT)
L. nipomensis (CE)
L. padre-crowleyi (CR)
L. tidestromii (CE)
Machaeranthera asteroides var. *lagunensis* (CR)
Malacothamnus clementinus (CE)
M. fasciculatus var. *nesioticus* (CE)
Maurandya petrophila (CR)
Monardella linoides ssp. *viminea* (CE)
Navarretia leucocephala ssp. *pauciflora* (CT)
N. leucocephala ssp. *plieantha* (CE)
Nemacladus twisselmannii (CR)
Neostapfia colusana (CE)
Nitrophila mohavensis (CE)
Nolina interrata (CE)
Oenothera californica ssp. *eurekensis* (CR)
O. deltoides ssp. *howellii* (CE)
Opuntia basilaris var. *treleasei* (CE)
Orcuttia californica (CE)
O. inaequalis (CE)
O. pilosa (CE)
O. tenuis (CE)
O. viscida (CE)
Parvisedum leiocarpum (CE)
Pedicularis dudleyi (CR)
Pentachaeta bellidiflora (CE)
P. lyonii (CE)
Phlox hirsuta (CE)
Plagiobothrys diffusus (CE)
P. strictus (CT)
Pleuropogon hooverianus (CR)
Poa napensis (CE)
Pogogyne abramsii (CE)
P. clareana (CE)
P. nudiuscula (CE)
Potentilla hickmanii (CE)
Pseudobahia bahiifolia (CE)
P. peirsonii (CE)
Rorippa gambellii (CT)
R. subumbellata (CE)
Rosa minutifolia (CE)
Sanicula maritima (CR)
S. saxatilis (CR)
Senecio ganderi (CR)
S. layneae (CR)
Sidalcea covillei (CE)
S. hickmanii ssp. *anomala* (CR)
S. hickmanii ssp. *parishii* (CR)
S. oregana ssp. *valida* (CE)
S. pedata (CE)
S. stipularis (CE)
Silene campanulata ssp. *campanulata* (CE)
Streptanthus niger (CE)
Swallenia alexandrae (CR)
Thelypodium stenopetalum (CE)
Thermopsis macrophylla (CR)
Trifolium polyodon (CR)
T. trichocalyx (CE)
Tuctoria greenei (CR)
T. mucronata (CE)
Verbena californica (CC)
Verbesina dissita (CT)

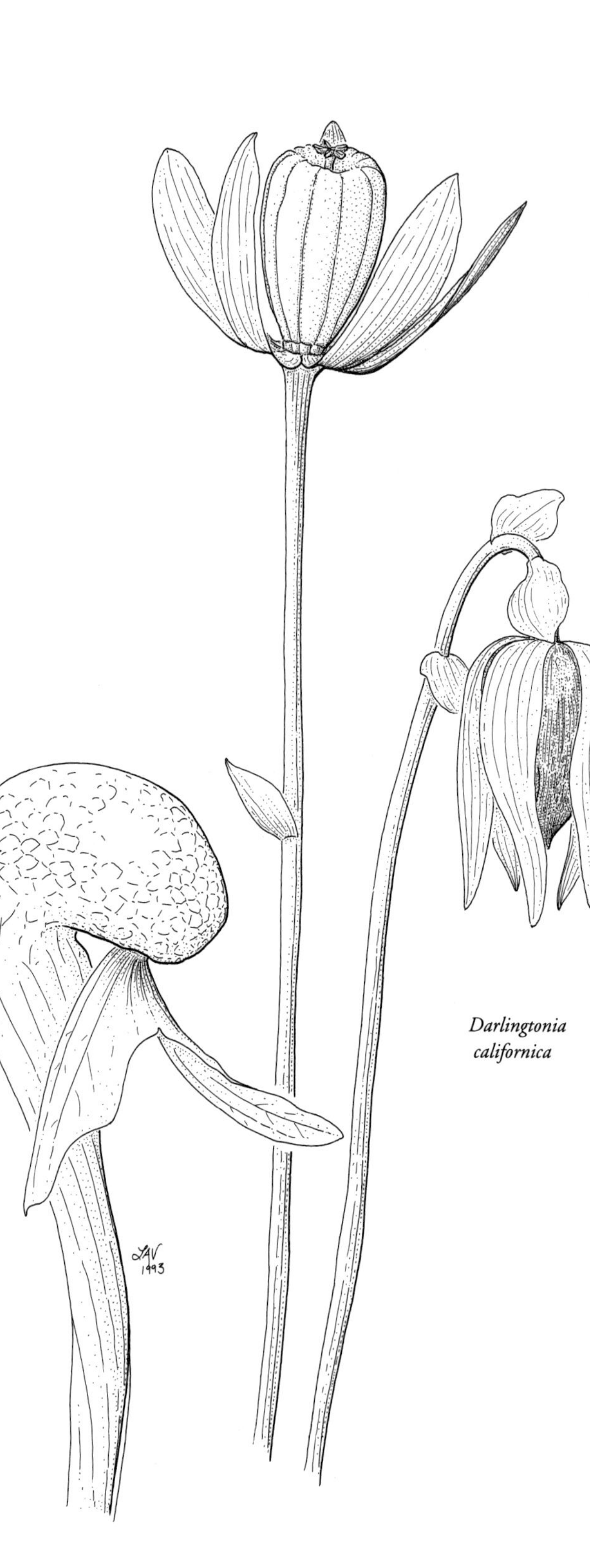

Darlingtonia californica

Appendix VII: Plants Listed or Candidates Under Federal Law

Minus the one plant mentioned below, this list includes the 48 California native plants that as of June 1993 have been designated as Endangered (FE, 45 plants) or Threatened (FT, 3 plants) by the U.S. Fish and Wildlife Service under Section 4 of the Endangered Species Act (1973). Also included are the 58 plants which have been proposed for listing under the Act as either Endangered (PE, 43 plants) or Threatened (PT, 15 plants). For additional information, contact the U.S. Fish and Wildlife Service, 2800 Cottage Way, Room E-1823, Sacramento, CA 95825.

Mahonia sonnei (FE) has been rejected from the fifth edition of the *Inventory* for taxonomic reasons, and does not appear on the list below. *Erysimum menziesii* (FE) has been split into three subspecies in the *Inventory*; all are protected by Federal law. Note also that this list uses the nomenclature adopted in the *Inventory*, which in a few cases differs from that of the U.S. Fish and Wildlife Service. Entries in this *Inventory* give the name under which a taxon is Federally-listed if it differs from the name used here.

Acanthomintha duttonii (FE)
Amsinckia grandiflora (FE)
Arabis macdonaldiana (FE)
Arctostaphylos glandulosa ssp. *crassifolia* (PE)
A. hookeri ssp. *ravenii* (FE)
A. morroensis (PE)
Arenaria paludicola (FE)
Astragalus albens (PE)
A. brauntonii (PE)
A. jaegerianus (PE)
A. lentiginosus var. *coachellae* (PE)
A. lentiginosus var. *micans* (PT)
A. lentiginosus var. *piscinensis* (PE)
A. lentiginosus var. *sesquimetralis* (PT)
A. magdalenae var. *peirsonii* (PE)
A. tricarinatus (PE)
Baccharis vanessae (PE)
Blennosperma bakeri (FE)
Calochortus tiburonensis (PT)
Camissonia benitensis (FT)
Castilleja affinis ssp. *neglecta* (PE)
C. campestris ssp. *succulenta* (PT)
C. grisea (FE)
Caulanthus californicus (FE)
Ceanothus ferrisae (PE)
Chamaesyce hooveri (PT)
Chorizanthe howellii (FE)
C. orcuttiana (PE)
C. pungens var. *hartwegiana* (PE)
C. pungens var. *pungens* (PE)
C. robusta var. *hartwegii* (PE)
C. robusta var. *robusta* (PE)
C. valida (FE)
Cirsium fontinale var. *fontinale* (PE)
C. fontinale var. *obispoense* (PE)
Clarkia franciscana (PE)
C. speciosa ssp. *immaculata* (PE)
Cordylanthus maritimus ssp. *maritimus* (FE)
C. palmatus (FE)
C. tenuis ssp. *capillaris* (PE)
Corethrogyne filaginifolia var. *linifolia* (PT)
Cupressus abramsiana (FE)
Delphinium variegatum ssp. *kinkiense* (FE)
Dodecahema leptoceras (FE)
D. abramsii ssp. *parva* (PT)
Dudleya blochmaniae ssp. *brevifolia* (PE)
D. cymosa ssp. *marcescens* (PT)
D. cymosa ssp. *ovatifolia* (PT)
D. setchellii (PE)
D. traskiae (FE)
D. verityi (PT)
Eremalche kernensis (FE)
Eriastrum densifolium ssp. *sanctorum* (FE)
E. hooveri (FT)
Erigeron parishii (PE)
Eriodictyon altissimum (PE)
Eriogonum ovalifolium var. *vineum* (PE)
Eriophyllum latilobum (PE)
Eryngium aristulatum var. *parishii* (FE)
E. constancei (FE)
Erysimum capitatum ssp. *angustatum* (FE)
E. menziesii ssp. *eurekense* (FE)
E. menziesii ssp. *menziesii* (FE)
E. menziesii ssp. *yadonii* (FE)
E. teretifolium (PE)
Gilia tenuiflora ssp. *arenaria* (FE)
Grindelia fraxino-pratensis (FT)
Hesperolinon congestum (PT)
Howellia aquatilis (PT)
Lasthenia burkei (FE)
Layia carnosa (FE)
Lembertia congdonii (FE)
Lesquerella kingii ssp. *bernardina* (PE)
Lilium occidentale (PE)
Limnanthes floccosa ssp. *californica* (FE)
L. vinculans (FE)
Lotus dendroideus var. *traskiae* (FE)
Lupinus tidestromii (FE)
Malacothamnus clementinus (FE)
Neostapfia colusana (PT)
Nitrophila mohavensis (FE)
Oenothera californica ssp. *eurekensis* (FE)
O. deltoides ssp. *howellii* (FE)
Opuntia basilaris var. *treleasei* (FE)
Orcuttia californica (FE)
O. inaequalis (PE)
O. pilosa (PE)
O. tenuis (PT)
O. viscida (PE)
Oxytheca parishii var. *goodmaniana* (PE)
Pentachaeta bellidiflora (PE)
P. lyonii (PE)
Pogogyne abramsii (FE)
P. nudiuscula (FE)
Pseudobahia bahiifolia (PE)
P. peirsonii (PE)
Rorippa gambellii (FE)
Sidalcea pedata (FE)
Streptanthus albidus ssp. *albidus* (PE)
S. niger (PE)
Suaeda californica (PE)
Swallenia alexandrae (FE)
Thelypodium stenopetalum (FE)
Tuctoria greenei (PE)
T. mucronata (FE)
Verbesina dissita (PT)

Appendix VIII: CNPS Policies Regarding Rare Plants

Policy Regarding Mitigation of Impacts to Rare and Endangered Plants

The policy of the California Native Plant Society is that all potential direct, indirect, and cumulative impacts to rare, threatened, or endangered plants and their habitats must be assessed and that appropriate measures be implemented to prevent such impacts resulting from projects. The policy of the Society is also that environmental documents and mitigation plans be based on complete, accurate, and current scientific information. Viability of rare, threatened, or endangered plants and their habitats takes precedence over economic or political expediency. Because of the tremendous diversity of rare plant habitats in California, and the dependence of rare plants on their local habitats, it is imperative that mitigation measures be developed on a site-specific basis. Local environmental conditions, species biology, land use patterns, and other factors must be incorporated into the design of mitigation plans.

The goals of this policy are to prevent the decline of rare plants and their habitats and to ensure that effective rare plant preservation measures are implemented.

Of the mitigation measures listed in the California Environmental Quality Act, the Society fully endorses only that of avoiding the impact. Measures to minimize, to rectify, or to reduce or eliminate the impact over time are recognized by the Society as partial mitigation. The Society does not recognize off-site compensation as mitigation.

Guidelines for project review and evaluation of mitigation proposals are available from the California Native Plant Society.

Adopted June 1987 by the CNPS Board of Directors

Policy on Appropriate Application of *Ex Situ* Conservation Techniques

The California Native Plant Society has always emphasized and will always emphasize *in situ* approaches to the conservation of rare and endangered plants. *In situ* conservation protects and enhances populations and species by protecting appropriate habitat within the historic range of the target species. *In situ* conservation is regarded by this society and by existing biological resource laws as the best available approach for conserving biological diversity at the species level.

In recent years, however, it has become clear that the rates of endangerment and extinction of entire species are accelerating. Habitat destruction remains the principle reason for accelerated endangerment in California, the nation, and the world. The destruction of biologically valuable habitat must be slowed, if not stopped, by using political, legislative, and economic strategies. Nevertheless, we recognize that for some species the risk of extinction is so high that only aggressive and extraordinary measures may affect their conservation. Such measures remove plants or propagules from their habitats into cold storage, gardens, or managed sites in the hope that the species can be re-established in the wild at some later date. Those measures fall under the realm of *ex situ* conservation.

Ex situ conservation involves a temporary, short-term set of germplasm preservation techniques that are usually applied as choices of last resort. The techniques include, but are not limited to: propagule collection from natural populations and cryogenic storage, garden propagation, tissue culture, transplantation, and the establishment of new populations in nature. Such techniques do not conserve all of the genetic variation, the metapopulation characteristics, the symbionts, the associated species, the community as a whole, the habitat, or the ecosystem of the endangered plant. Consequently, they do not conserve a species in its entirety and they do not conserve a species within its evolutionary and ecological contexts. For these reasons and also because we lack a solid knowledge of the effectiveness and limits of *ex situ* techniques at this time, we view the application of such techniques with scientific skepticism.

However, the members of CNPS recognize the gravity of the extinction problem and will not oppose the use of certain *ex situ* conservation techniques under certain circumstances. Those circumstances can be summarized as one of two general types:

1. During mandated recovery of endangered species, *ex situ* techniques may be essential for establishing new populations when all extant natural populations are fully protected *in situ.* This approach, if successful, can result in a gross (as opposed to net) increase in the number of extant populations of an endangered species. All recovery activities, including those which use *ex situ* techniques, must be designed and executed by qualified biologists with the approval of relevant state and federal government agencies (e.g., California Department of Fish and Game, U.S. Fish and Wildlife Service).

2. During the analysis of extinction probabilities for a given species (usually based on demographic or minimum viable population studies) it is determined that the remaining natural population(s) is (are) likely to go extinct in the near future due to stochastic, genetic, or natural ecological factors. This approach can minimize the effect of degenerative extinction processes that are effective when the

population(s) is (are) very small. Such analysis must be conducted by qualified biologists with the approval of relevant state and federal government agencies (e.g., California Department of Fish and Game, U.S. Fish and Wildlife Service).

We continue to strongly oppose the use of *ex situ* conservation techniques under the following circumstances:

1. When applied as mitigation for human-caused impacts to natural populations other than those impacts which operate on a global scale (e.g., global warming, acid rain). Losses of plant populations considered "significant" under the California Environmental Quality Act (CEQA) or the National Environmental Policy Act (NEPA) cannot be mitigated to less-than-significant levels using *ex situ* conservation techniques. Decreases in the number of individuals and populations of an endangered plant impacted by human activity cannot be compensated for in this manner.
2. When adequate ecological information does not exist for the remaining natural populations of a target species. Adequate studies of microhabitat selection, fecundity, essential pollinators, community relationships, and other important biological characteristics must be completed before seeds or other plant materials are removed from natural habitats for an *ex situ* conservation effort.
3. When reestablishment in the wild of plant material stored *ex situ* could result in genetic contamination of existing wild populations of the same or different taxa. Such reestablishment projects must be carefully reviewed to protect the target species from genetic contamination.
4. When the *ex situ* techniques are not based on scientific principles or facts, or when the methods and results of such projects are not fully documented.
5. When an *ex situ* conservation effort is not designed or conducted by qualified biologists or with the knowledge and consent of relevant governmental agencies (e.g., California Department of Fish and Game, U.S. Fish and Wildlife Service).

Glossary

cryogenic preservation - storage of rare materials at very low temperatures (usually below -25 C)

***ex situ* conservation** - conserving a germplasm, species, or natural community in the absence of its natural habitat or ecosystem, usually by removing propagules from the habitat and storing them temporarily "off site" (e.g., in a freezer, botanical garden, etc.)

germplasm - the contents of a species' genes, the actual genetic material (DNA contained in chromosomes)

***in situ* conservation** - conserving a germplasm, species, or community within its natural habitat or ecosystem

metapopulation - one of several interacting populations of a species

propagule - any living material that can be used to propagate a species; includes seeds, fruits, rhizomes, stolons, or buds

species - fundamental category in a taxonomic hierarchy, a "kind" of organism which shares similar appearance, ecology and behavior with others of its kind in time and space. For purposes of this document it includes subspecies, varieties, and other legally-recognized categories of organisms

stochastic - random, unpredictable

symbionts - different kinds of organisms that must live together in order to mutually prosper

---

Adopted September 1992 by the CNPS Board of Directors

Policy with Regard to Plant Collecting for Educational Purposes

The California Native Plant Society (CNPS) supports the use of plant and wildflower collections as a valid means of providing students at many educational levels with knowledge of and appreciation for the wonder, diversity, and beauty of plant life. However, to avoid breaking the law or damaging the viability of populations of plants, the instructor must make known several important points to the students who will be making the collections.

There are two levels of collection for educational and scientific purposes which are considered in this policy: (1) collection of plant specimens for herbaria, and (2) collections of plant and wildflower specimens for lower level science and biology classes. Most of the considerations discussed apply to both levels.

1. It is illegal to collect plants along a highway right of way, in National Parks, National Monuments or National Forests, State Parks, or most local parks without a collecting permit. Plants and wildflowers growing in such locations are part of a natural system designed for public enjoyment, and in most cases should be left to natural processes. Permits for collecting plants must be obtained from the appropriate supervising agency.
2. It is legal and permissible to collect wildflowers on private lands provided that permission of the landowner is obtained. Particularly appropriate sites for collection are lands slated for development.
3. It is the responsibility of the instructor to ensure that the students are made aware of rare plants endemic to the area in which the collecting is to take place, and to caution the students against collecting these plants. It is not appropriate (and there are substantial penalties) to allow collection of rare or

endangered plants, and areas known to contain rare plants should be avoided. The instructor may contact CNPS to find out about rare plants in the area in question.

4. To increase the environmental awareness associated with making the collection, the students should be requested to observe and describe the habitat in which the plant was growing.
5. Collecting (at the high-school or lower division college level) should be limited to the taking of as little of the plant as necessary to allow identification.
6. All collecting activities should be done inconspicuously, so that casual (uninformed) observers are not encouraged to do likewise.
7. CNPS encourages all botany and science instructors to use common, especially weedy or garden, species for demonstrating collecting techniques, plant structures, and diagnostic features.
8. The primary reason for collecting plants for herbaria is to increase knowledge of California's flora. Repeated collecting in well known areas may serve no useful purpose. While it is important to document the distribution of plants, including rare species, it is critical to first evaluate the impact of collecting on the plant population. Known and documented populations of rare plants should not be subject to additional collection.
9. A key to ensuring preservation of California's diverse flora and fauna is to develop a public informed about the value of these natural resources. For this reason, CNPS encourages limited and discriminating collection of plants as part of the educational process.

Adopted June 1993 by the CNPS Board of Directors

Lilium parryi

California Native Species Field Survey Form

Mail to:
Natural Diversity Data Base
California Dept. of Fish and Game
1416 Ninth Street, 12th Floor
Sacramento, CA 95814

Date of field work: ___-___-___ (mo day year)

For office use only
Source Code ______ Quad Code ______
Elm Code ______ Occ # ______
Copy to ______ Map Index # ______

Scientific Name *(no codes):*

Species Found? [ ] [ ] ______ (*yes no* / *If not, why?*)
Total # Individuals: ______ Subsequent visit? [ ]*yes* [ ]*no*
Compared to your last visit: [ ]*more* [ ]same [ ]fewer
Is this an existing NDDB occurrence? [______] [ ] [ ] (*Yes, Occ. # no unk.*)
Collection? If yes: ______ ______ (*number Museum/Herbarium*)

Reporter: ______
Address: ______
Phone: ( ) ______
Other knowledgeable individuals (name/address/phone):

Plant Information:

Phenology: ______ ______ ______ (*% vegetative % flowering % fruiting*)

Animal Information:

Age Structure: ______ ______ ______ (*# adults # juveniles # unknown*)
Site Function: [ ] *breeding* [ ] *foraging* [ ] *wintering* [ ] *roosting* [ ] *burrow site* [ ] *other*

Location: *(Please also attach or draw map on back.)*

County: ______ Landowner/Mgr: ______
Quad Name: ______ Elevation: ______ UTM: ______
T______R______ ______ 1/4 of ______1/4 Sec ______ T______R______ ______ 1/4 of ______1/4 Sec ______

Habitat Description: *(Plant communities, dominants, associates, substrate/soils, aspect/slope)*

Other rare spp.?

Site Information: Current/surrounding land use:

Visible disturbances, possible threats:

Overall site quality: [ ]*Excellent* [ ]*Good* [ ]*Fair* [ ]*Poor* Comments:

Determination: *(Check one or more, fill in the blanks)*
____ Keyed in a site reference: ______
____ Compared with specimen housed at: ______
____ Compared with photo/drawing in: ______
____ By another person (name): ______
____ Other: ______

Photographs: *(Check one or more)*

	Slide	*Print*
Plant/animal	___	___
Habitat	___	___
Diagnostic Feature	___	___
Other	___	___

May we obtain duplicates at our expense? [ ]yes [ ]no

California Natural Community Field Survey Form

Mail to:
Natural Diversity Data Base
California Dept. of Fish and Game
1416 Ninth Street
Sacramento, CA 95814
(916) 324-6857

For office use only
Source Code__________ Quad Code__________
Community Code __________ Occ #__________
Map Index #__________ Update Y _____ N _____

Please provide as much of the following information as you can. Please attach a map (if possible, based on the USGS 7.5 minute series) showing the site's location and boundaries. Use the back if needed.

Community name:__________

Reporter:__________ Phone __________
Affiliation and Address__________
Date of field work:__________ County:__________
Location:__________

Quad name: __________T_____R _____ _____ ¼ of____ ¼ sec_____ Meridian__________
UTM Zone _ _ Northing _ _ _ _ _ _ _ Easting _ _ _ _ _ _
Landowner/Manager: __________ Photographs: Slide ☐ Print ☐
Elevation: __________Aspect: __________ Slope (indicate % or °)__________ Drainage:__________
Site acreage: __________ Evidence of disturbance/threats: __________

Current land use: __________
Substrate/Soils: __________

General description of community: __________

Any Special Plants or Animals present: __________

Successional status/Evidence of regeneration of dominant taxa:__________

Overall site quality: Excellent____ Good ____Fair ____Poor ____Comments: __________

Basis for report: Remote image ____Binocular/Telescopic survey __________
Windshield survey ____ Brief walk-thru ____ Detailed survey ____ Other __________
Relevé: In the space below, indicate each species cover % within the following growth form categories:

Trees	Shrubs	Herbs/Graminoids

Continue on back if needed. Thank you for your contribution.

FG/NHD/1748 Rev. 7/92
survy.sr

Appendix X: California Department of Fish and Game Programs, Contacts, and Regions

Program	Role	Telephone
Natural Heritage Division		916/322-2493
Endangered Plant Program	Listing, management, research, recovery of State-listed plants	916/227-2318
Natural Diversity Data Base—Science	Inventory of distribution, ecology, and status information on rare, threatened, endangered, and sensitive plants, animals, and natural communities	916/322-2493
Natural Diversity Data Base—Information Services	NDDB products including *RareFind,* rare plant status reports, GIS text reports, map overlays, and lists of sensitive and listed species and natural communities	916/324-3812
Lands and Natural Areas Program	Coordination of land acquisition and management, identification of significant natural areas (Significant Natural Areas Program)	916/322-2493
Regional Offices		
Region 1 (Redding) Region 2 (Rancho Cordova) Region 3 North (Yountville) Region 3 South (Monterey) Region 4 (Fresno) Region 5 (Long Beach)	For local conservation issues and questions, contact regional biologists, environmental services staff, or plant ecologists	916/225-2300 916/355-0978 707/944-5500 408/649-2870 209/222-3761 310/590-5132

Department of Fish & Game Regions

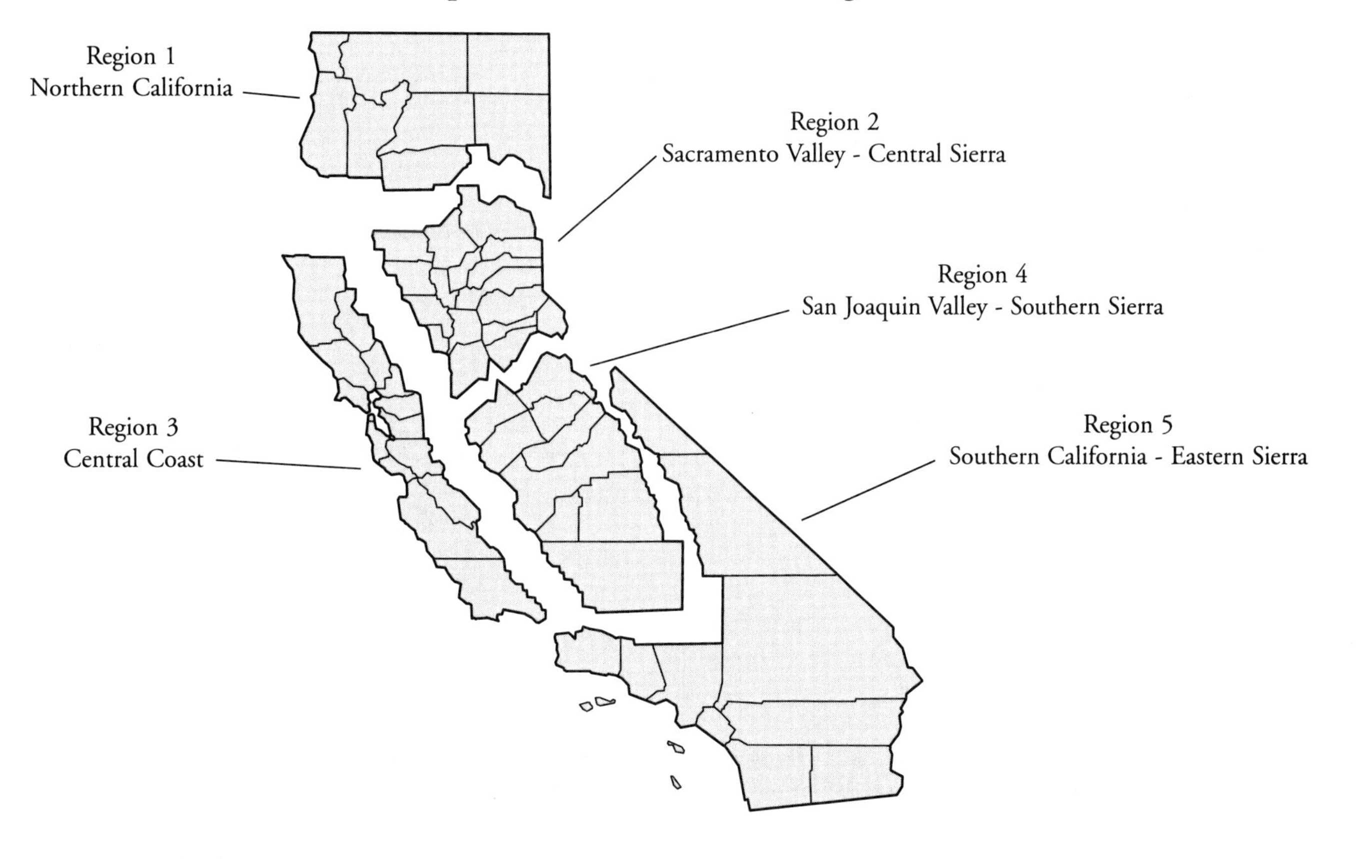

Appendix XI: Useful Addresses

Government Agencies

Federal

Fish and Wildlife Service

Sacramento Endangered Species Office
2800 Cottage Way, Room E-1823
Sacramento, CA 95825
Attn: Staff Botanist

Forest Service

Pacific Southwest Region
630 Sansome St.
San Francisco, CA 94111
Attn: Regional Botanist

Bureau of Land Management

California State Office
2800 Cottage Way, Room E-2841
Sacramento, CA 95825
Attn: Endangered Species Coordinator

State

California Department of Parks and Recreation

1416 9th St.
Sacramento, CA 95814

Non-Governmental Organizations

Audubon Society

Western Regional Office
555 Audubon Place
Sacramento, CA 95825

California Native Plant Society

1722 J St., Suite 17
Sacramento, CA 95814

Defenders of Wildlife

California Office
1228 N St., Suite 6
Sacramento, CA 95814

The Nature Conservancy

California Field Office
785 Market St.
San Francisco, CA 94103

Planning and Conservation League

926 J St., Suite 612
Sacramento, CA 95814

Sierra Club

State Legislative Office
923 12th St., Suite 200
Sacramento, CA 95814

Appendix XII: Publications of the California Native Plant Society

Inventory of Rare and Endangered Vascular Plants of California

Edited by Mark W. Skinner and Bruce M. Pavlik
Special Publication #1 (Fifth Edition)
1994 338 pp. $22.95

California's Changing Landscapes: Diversity and Conservation of California Vegetation

By Michael Barbour, Bruce Pavlik, Susan Lindstrom, and Frank Drysdale
1993 224 pp. $24.95

Conservation and Management of Rare and Endangered Plants

Edited by Thomas E. Elias
1987 640 pp. $24.95 softbound / $45.00 clothbound

Terrestrial Vegetation of California (New Expanded Edition)

Edited by Michael G. Barbour and Jack Major
1988 1,036 pp. $55.00

Flora of Mount Hamilton Range, California

By Helen K. Sharsmith
Special Publication #6
1982 100 pp. $7.50

Flora of Northern Mojave Desert, California

By Mary DeDecker
Special Publication #7
1984 184 pp. $8.95

A Flora of the San Bruno Mountains

By Elizabeth McClintock, Paul Reeberg, and Walter Knight
1990 223 pp. $9.50

Geographic and Edaphic Distribution of Vernal Pools in the Great Central Valley

By Robert Holland
1977 2 maps/text $7.50

Oak Action Kit

Edited by Joan Stewart and Pam Muick
Third edition, 3-ring binder format
1989 387 pp. $13.00

Plant Communities of Marin County

By W. David Shuford and Irene Timossi
1989 32 pp. $8.95

Bibliography for the Fifth Edition

James P. Smith, Jr.

The list of references that follows is meant to provide you with literature citations for some of the standard California floras, for statewide treatments of rare and endangered plants in California and selected western states, for theoretical and legal aspects of rarity and endangerment, for conservation biology and the management of rare plant populations, and for descriptions of plant communities in California.

Floras

Abrams, L.R. 1923-1960. *An Illustrated Flora of the Pacific States, Washington, Oregon and California.* Vol. 4 by R. Ferris. Stanford University Press. Stanford, CA. 4 vols.

Hickman, J.C., ed. 1993. *The Jepson Manual: Higher Plants of California.* University of California Press. Berkeley, CA. 1400 pp.

Jepson, W.L. 1925. *Manual of the Flowering Plants of California.* University of California Press. Berkeley and Los Angeles, CA. 1238 pp.

Jepson, W.L. 1907-1979. A *Flora of California.* Vol. 4, Part 2 by L.T. Dempster. Jepson Herbarium and Library, University of California at Berkeley. Berkeley, CA. Incomplete, in 4 vols.

Munz, P.A. 1959. *A California Flora.* In collaboration with D.D. Keck. University of California Press. Berkeley, CA. 1681 pp.

Munz, P.A. 1968. *Supplement to a California Flora.* University of California Press. Berkeley, CA. 224 pp.

Munz, P.A. 1974. *A Flora of Southern California.* University of California Press. Berkeley, CA. 1086 pp.

Rare Floras

Argus, G.W., and K.W. Argus. 1990. *Rare Vascular Plants in Canada: Our Natural Heritage.* Canadian Museum of Nature. Ottawa, Canada. 191 pp.

Ayensu, E.S., and R.A. DeFilipps. 1978. *Endangered and Threatened Plants of the United States.* Smithsonian Institution and the World Wildlife Fund. Washington, D.C. 403 pp.

California Department of Fish and Game. 1991. *Annual Report on the Status of California State Listed Threatened and Endangered Plants and Animals.* The Resources Agency, State of California. Sacramento, CA. 191 pp.

Colorado Native Plant Society. 1989. *Rare Plants of Colorado.* Rocky Mountain Nature Association. Rocky Mountain National Park. Estes Park, CO. 75 pp.

Eastman, D.C. 1990. *Rare and Endangered Plants of Oregon.* Beautiful America. Wilsonville, OR. 194 pp.

Kartesz, J.T., and R. Kartesz. 1977. *The Biota of North America — Part I. Vascular Plants.* Volume 1: Rare Plants. Biota of North America Committee. Pittsburgh, PA. 361 pp.

Lucas, G., and H. Synge, eds. 1978. *The IUCN Plant Red Data Book.* International Union for Conservation of Nature and Natural Resources. Morges, Switzerland. 540 pp.

Meinke, R.J. 1981. *Threatened and Endangered Vascular Plants of Oregon: An Illustrated Guide.* U.S. Fish and Wildlife Service. Office of Endangered Species. Portland, OR. 353 pp.

Morefield, J.D., and T.A. Knight, eds. 1992. *Endangered, Threatened, and Sensitive Vascular Plants of Nevada.* Nevada State Office of the Bureau of Land Management. Reno, NV. 46 pp.

Moseley, R., and C. Groves, eds. 1992. *Rare, Threatened and Endangered Plants and Animals of Idaho.* Nongame and Endangered Wildlife Program, Idaho Department of Fish and Game. Boise, ID. 38 pp.

Mozingo, H.N., and M. Williams. 1980. *Threatened and Endangered Plants of Nevada: An Illustrated Manual.* U.S. Fish & Wildlife Service and U.S. Bureau of Land Management Office. 268 pp.

Murray, D.F., and R. Lipkin. 1987. *Candidate Threatened and Endangered Plants of Alaska With Comments on Other Rare Plants.* University of Alaska Museum. Fairbanks, AK. 76 pp.

New Mexico Native Plant Protection Advisory Committee. 1984. *A Handbook of Rare and Endemic Plants of New Mexico.* University of New Mexico Press. Albuquerque, NM. 291 pp.

O'Kane, S.L., Jr. 1988. Colorado's rare flora. *Great Basin Naturalist* 48:434-484.

Oregon Natural Heritage Program. 1993. *Rare, Threatened and Endangered Plants and Animals of Oregon.* Oregon Natural Heritage Program. Portland, OR. 79 pp.

Powell, W.R., ed. 1974. *Inventory of Rare and Endangered Vascular Plants of California.* Special Publication No. 1 (first edition). California Native Plant Society. Berkeley, CA. iii + 56 pp.

Sheehan, M., and R. Schuller. 1981. *An Illustrated Guide to Endangered, Threatened and Sensitive Vascular Plants in Washington.* Washington Natural Heritage Program. Olympia, WA. 328 pp.

Sivinski, R., and K. Lightfoot, eds. 1992. *Inventory of Rare and Endangered Plants of Mexico.* New Mexico Forestry and Resources Conservation Division, Minerals and Natural Resources Department. Santa Fe, NM. 58 pp.

Smith, J.P., Jr., and J.O. Sawyer, Jr. 1988. Endemic vascular plants of northwestern California and southwestern Oregon. *Madroño* 35(1):54-69.

Smith, J.P., Jr., ed. 1981. *First Supplement, Inventory of Rare and Endangered Vascular Plants of California.* Special Publication No. 1 (second edition). California Native Plant Society. Berkeley, CA. 28 pp.

Smith, J.P., Jr., and K. Berg, eds. 1988. *Inventory of Rare and Endangered Vascular Plants of California.* Special Publication No. 1 (fourth edition). California Native Plant Society. Sacramento, CA. xviii + 168 pp.

Smith, J.P., Jr., and R. York, eds. 1982. *Second Supplement, Inventory of Rare and Endangered Vascular Plants of California.* Special Publication No. 1 (second edition). California Native Plant Society. Berkeley, CA. 28 pp.

Smith, J.P., Jr., and R. York, eds. 1984. *Inventory of Rare and Endangered Vascular Plants of California.* Special Publication No. 1 (third edition). California Native Plant Society. Berkeley, CA. xviii + 174 pp.

Smith, J.P., Jr., R.J. Cole, and J.O. Sawyer, Jr., eds. 1980. *Inventory of Rare and Endangered Vascular Plants of California.* Special Publication No. 1 (second edition). California Native Plant Society. Berkeley, CA. vii + 115 pp.

Steele, R., and F. Johnson. 1981. *Vascular Plant Species of Concern in Idaho.* Bull. No. 34. University of Idaho Forest, Wildlife and Range Experiment Station. Moscow, ID. 161 pp.

Welsh, S.L., and L.M. Chatterley. 1985. Utah's rare plants revisited. *Great Basin Naturalist* 45:173- 236.

Vovides, A.P. 1981. Lista preliminar de plantas mexicanas raras o en peligro de extinction. *Biotica* 6:219-228.

Washington Natural Heritage Program. 1990. *Endangered, Threatened & Sensitive Vascular Plants of Washington.* Department of Natural Resources. Olympia, WA. 52 pp.

Welsh, S.L., and K.H. Thorne. 1979. *Illustrated Manual of Proposed Endangered and Threatened Plants of Utah.* U.S. Fish and Wildlife Service, U.S. Bureau of Land Management, and the U.S. Forest Service. 318 pp.

Rarity, Threats, and Values

Cochrane, S. 1985. Why rare species? *Outdoor California* 46(5):20-23.

Davis, S., S. Droop, P. Gregerson, L. Henson, C. Leon, J. Lamelin Villa-Lobos, H. Synge, and J. Zantovska. 1986. *Plants in Danger: What Do We Know?* International Union for Conservation of Nature and Natural Resources. Gland, Switzerland. 461 pp.

Drury, W.H. 1974. Rare species. *Biological Conservation* 6(3):162-169.

Drury, W.H. 1980. Rare species of plants. *Rhodora* 82:3-48.

Duke, J.A. 1976. Economic appraisal of endangered plant species. *Phytologia* 34(1):21-27.

Durant, M., and M. Saito. 1985. The hazardous life of our rarest plants. *Audubon* 87(4):50-61.

Ehrlich, P., and A. Ehrlich. 1981. *Extinction: The Causes and Consequences of the Disappearance of Species.* Random House. New York, NY. 305 pp.

Fiedler, P.L. 1986. Concepts of rarity in vascular plant species, with special reference to the genus *Calochortus* Pursh (Liliaceae). *Taxon* 35:502-518.

Fiedler, P.L., and J.J. Ahouse. 1992. Hierarchies of cause: Toward an understanding of rarity in vascular plant species. Pages 23-47 *in*: P.L. Fiedler and S.K. Jain, eds. *Conservation Biology: The Theory and Practice of Nature Conservation, Preservation and Management.* Chapman and Hall. New York, NY.

Gunn, A.S. 1980. Why should we care about rare species? *Environmental Ethics* 2(1):17-37.

Huxley, A. 1985. *Green Inheritance: The World Wildlife Fund Book of Plants.* Anchor/Doubleday. Garden City, NY. 193 pp.

Jones and Stokes Associates. 1987. *Sliding Toward Extinction: The State of California's Natural Heritage, 1987.* The California Nature Conservancy. San Francisco, CA. 105 pp.

Kaye, T. 1989. Endemism and rarity in plants. *Bulletin of the Native Plant Society of Oregon* 22(3): 23-24.

Koopowitz, H., and H. Kaye. 1983. *Plant Extinction: A Global Crisis.* Stone Wall Press. Washington, D.C. 239 pp.

Kruckeberg, A.R., and D. Rabinowitz. 1986. Biological aspects of endemism in higher plants. *Annual Review of Ecology and Systematics* 16:447-479.

Lewis, H. 1972. The origin of endemics in the California flora. Pages 179-188 *in*: D.H. Valentine, ed. *Taxonomy, Phytogeography and Evolution.* Academic Press. New York, NY.

Mohlenbrock, R.H. 1987. Why should we save our plants? *The Nature Conservancy Magazine* 37(5):4-9.

Morse, L.E., and M.S. Henifin, eds. 1981. *Rare Plant Conservation: Geographical Data Organization.* New York Botanical Garden. Bronx, NY.

Myers, N. 1983. *A Wealth of Wild Species: Storehouse for Human Welfare.* Westview Press. Boulder, Co. 274 pp.

Prance, G.T., and T.S. Elias, eds. 1977. *Extinction is Forever: Threatened and Endangered Species of Plants in the Americas and their Significance in Ecosystems Today and in the Future.* New York Botanical Garden. Bronx, NY. 437 pp.

Raven, P.H., and D.I. Axelrod. 1978. Origin and relationships of the California flora. *University of California Publications in Botany* 72:1-134.

Reveal, J.L. 1981. The concepts of rarity and population threats in plant communities. Pages 41-47 *in*: L.E. Morse and M.S. Henifin, eds. *Rare Plant Conservation: Geographical Data Organization.* New York Botanical Garden. Bronx, NY.

Reveal, J.L., and C.R. Broome. 1979. Plant rarity — real and imagined. *The Nature Conservancy News* 29(2):4-8.

Simmons, J.B., R.I. Beyer, P.E. Brandham, G.L. Lucas, and V.T.H. Parry. 1976. *Conservation of Threatened Plants.* Plenum Press. New York, NY. 336 pp.

Smith, J.P., Jr. and J.O. Sawyer, Jr. 1988. Endemic vascular plants of northwestern California and southwestern Oregon. *Madroño* 35(1):54-69.

Stebbins, G.L. 1978a. Why are there so many rare plants in California? I. Environmental factors. *Fremontia* 5(4):6-10.

Stebbins, G.L. 1978b. Why are there so many rare plants in California? II. Youth and age of species. *Fremontia* 6(1):17-20.

Stebbins, G.L. 1980. Rarity of plant species: a synthetic viewpoint. *Rhodora* 82:77-86.

Stebbins, G.L. 1986. Rare plants in California's national forests: their scientific value and conservation. *Fremontia* 13(4):9-12.

Stebbins, G.L., and J. Major. 1965. Endemism and speciation in the California flora. *Ecological Monographs* 35:1-35.

Wilson, E.O., ed. 1988. *Biodiversity.* National Academy Press. Washington, D.C. 521 pp.

Legal Aspects

Council on Environmental Quality. 1981. *A summary of the legal authorities for conserving wild plants.* Council on Environmental Quality. Washington, D.C. 156 pp.

Cummings, E.W. 1987. Using the California Endangered Species Act consultation provisions for plant conservation. Pages 43-50 *in*: T.S. Elias, ed. *Conservation and Management of Rare and Endangered Plants.* Proceedings of a conference of the California Native Plant Society. California Native Plant Society. Sacramento, CA.

McMahan, L. 1980. Legal protection for rare plants. *American University Law Review* 29(3):515-569.

McMahan, L. 1981. *Plants protected by the convention on international trade in endangered species of fauna and flora.* International Convention Advisory Committee. Washington, D.C. 207 pp.

McMahan, L. 1987. Plant conservation laws: how effective are they? *The Nature Conservancy Magazine* 37(5):21-23.

McMahan, L. 1987. Rare plant conservation by state governments. Pages 23-31 *in*: T.S. Elias, ed. *Conservation and Management of Rare and Endangered Plants.* Proceedings of a conference of the California Native Plant Society. California Native Plant Society. Sacramento, CA.

Smith, E.L.V. 1980. Laws and information needs for listing plants. *Rhodora* 82:193-199.

United States Fish and Wildlife Service. 1981. *Placing Animals and Plants on the List of Endangered and Threatened Species.* U.S. Government Printing Office. Washington, D.C. 8 pp.

United States Fish and Wildlife Service. 1987. *Endangered & Threatened Wildlife and Plants.* U.S. Government Printing Office. Washington, D.C. 32 pp.

Conservation Biology and Management

Bratton, S.P., and P.S. White. 1980. Rare plant management — after preservation what? *Rhodora* 82:49-75.

Elias, T.S., ed. 1987. *Conservation and Management of Rare and Endangered Plants.* Proceedings from a conference of the California Native Plant Society. California Native Plant Society. Sacramento, CA. 630 pp.

Falk, D.A. 1987. Integrated conservation strategies for endangered plants. *Natural Areas Journal* 7(3):118-123.

Falk, D.A., and K.E. Holsinger, eds. 1991. *Genetics and Conservation of Rare Plants.* Oxford University Press. New York, NY. 304 pp.

Fiedler, P., and S. Jain, eds. 1992. *Conservation Biology: The Theory and Practice of Nature Conservation, Preservation, and Management.* Chapman and Hall. New York. 507 pp.

Jensen, D.B., M. Torn, and J. Harte. 1990. *In Our Own Hands: A Strategy for Conserving Biological Diversity in California.* California Policy Seminar, University of California. Berkeley, CA. 184 pp.

Kesseli, R.V. 1992. Population biology and conservation of rare plants. Pages 69-90 *in*: S.K. Jain and L.W. Botsford, eds. *Applied Population Biology.* Kluver Academic Publishers. Netherlands.

Menges, E.S. 1990. Population viability analysis for an endangered plant. *Conservation Biology* 4:52-62.

Miasek, M.A., and C.R. Long. 1985. *Endangered Plant Species of the World and Their Endangered Habitats: A Compilation of the Literature.* Revised and enlarged edition. Plant Bibliography No. 6. The Library of the New York Botanical Garden. Bronx, NY. 46 pp.

Mlot, C. 1989. Blueprint for conserving plant diversity. *BioScience* 39(6): 364-368.

Palmer, M.E. 1987. A critical look at rare plant monitoring in the United States. *Biological Conservation* 39:113-127.

Pavlik, B.M., and M.G. Barbour. 1988. Demographic monitoring of endemic sand dune plants, Eureka Valley, California. *Biological Conservation* 46: 217-242.

Pavlik, B.M., D.L. Nickrent, and A.M. Howald. 1993. The recovery of an endangered plant. I. Creating a new population of *Amsinckia grandiflora. Conservation Biology* 7: 510-526.

Pavlik, B.M. 1994. Demographic monitoring and the recovery of endangered plants. In press *in*: M.C. Bowles and C. Whelan, eds. *Recovery and Restoration of Endangered Species.* Cambridge University Press. Cambridge.

Primack, R.B. 1993. *Essentials of Conservation Biology.* Sinauer Associates. Sunderland, MA. 559 pp.

Schonewald-Cox, C.M., S.M. Chambers, B. MacBryde, and L. Thomas. 1983. *Genetics and Conservation: A Reference for Managing Wild Animal and Plant Populations.* Benjamin/Cummings Publishing Co. Menlo Park, CA. 722 pp.

Soule, M.E., ed. 1986. *Conservation Biology: The Science of Scarcity and Diversity.* Sinauer Associates. Sunderland, MA. 584 pp.

Synge, H., ed. 1981. *The Biological Aspects of Rare Plant Conservation.* John Wiley & Sons. New York, NY. 558 pp.

Plant Communities

Barbour, M., and J. Major, eds. 1988. *Terrestrial Vegetation of California.* New expanded edition. Special Publication No. 9. California Native Plant Society. Sacramento, CA. 1002 pp. + Supplement.

Barbour, M., B. Pavlik, F. Drysdale, and S. Lindstrom. 1993. *California's Changing Landscapes: Diversity and Conservation of California Vegetation.* California Native Plant Society. Sacramento, CA. 244 pp.

Cheatham, N.H., and J.R. Haller. 1975. *An Annotated List of California Habitat Types.* University of California Natural Land and Water Reserves System. Berkeley, CA. 77 pp.

Holland, R.F. 1986. *Preliminary Descriptions of the Terrestrial Natural Communities of California.* Nongame-Heritage Program. California Department of Fish and Game. Sacramento, CA. 156 pp.

Acknowledgements

This *Inventory* is based upon the contributions of many individuals — amateur and professional alike — who are its primary authors. Our task, as editors, has been to compile, evaluate, and organize. In the lengthy list below we specifically acknowledge the help of those who made a special effort on the current edition. They are, of course, adding to a foundation built by the hundreds of contributors to past editions. We and California's rare plants are pleased to acknowledge the contributions of Lowell Ahart, Bob Allen, Perry Allen, Mark Bagley, Susan Bainbridge, Bruce Baldwin, Katie Barrows, James Barry, Jim Bartel, Randall Bayer, Mitch Beauchamp, Don Behrens, Gary Bell, Jack Booth, Mark Borchert, Mona Bourell, Frederica Bowcutt, Jackie Bowland, Angelica Brinkman-Busi, Steven Broich, Pam Brown, Roy Buck, Don Burk, Ray Butler, Joe Callizo, Barb Castro, Kenton Chambers, Zoe Chandik, David Chipping, Curtis Clark, Glenn Clifton, Joanna Clines, Susan Cochrane, Ron Coleman, Lincoln Constance, Beth Corbin, Toni Corelli, Robin Cox, Virginia Dains, Charli Danielsen, Alva Day, Sally DeBecker, Bruce Delgado, Jim Dice, James Dillane, Mark Dodero, Wendie Duron, Steve Edwards, Kathy Erwin, Phyllis Faber, Stan Farwig, Mark Faull, Gary Fellers, Jean Ferreira, Wayne Ferren, Peggy Fiedler, Mike Foster, Al Franklin, Ken Fuller, Roman Gankin, Carl Geldin-Meyer, Vic Girard, James Griffin, Jack Guggolz, Jennie Haas, Linnea Hanson, Clare Hardham, Steve Hartman, Larry Heckard, Gordon Heebner, Mary Ann Henry, Diana Hickson, Steven Hill, Deborah Hillyard, Ken Himes, Jeri Hirshberg, Robert Holland, Doris A. Hoover, Lisa Hoover, Julie Horenstein, Marc Hoshovsky, Ann Howald, Alice Howard, Diane Ikeda, Dave Imper, Dave Isle, Lawrence Janeway, Steve Junak, Jimmy Kagan, Todd Keeler-Wolf, Dave Keil, Dean Kelch, Jan Knight, Teri Ann Knight, Walter Knight, Daryl Koutnik, Joyce Lacey, Dianne Lake, Meredith Lane, Melody Lardner, Larry LePre, Tom Lemieux, Russ Lewis, Rich Lis, Aaron Liston, John Little, Mickey Long, Lynn Lozier, David Magney, Sally Manning, Karlin Marsh, Craig Martz, Corky Matthews, Steve Matthews, Steve McCabe, Niall McCarten, Elizabeth McClintock, Hugh McDonald, Teri McGuire, Malcolm McLeod, Joe Medeiros, Mary Meyer, Orlando Mistretta, Diane Mitchell, Peggy Moore, James Morefield, Sandy Morey, Larry Morse, Marlyce Myers, Maile Neel, Julie Nelson, Guy Nesom, Gail Newton, Tim Nosal, Patti Novak, Bart O'Brien, Vernon Oswald, Laurie Parendes, Bruce Parfitt, Bob Patterson, Charlie Patterson, Andrea Pickart, Oren Pollak, Teresa Prendusi, Rob Preston, Roger Raiche, Rich Reiner, Craig Reiser, James Reveal, Mary Rhyne, Jim & Georgie Robinette, Blaine Rogers, Chris Rogers, Wayne Rolle, J. Hawkeye Rondeau, Cynthia Roye, Peter Rubtzoff, Christa Russell, Connie Rutherford, Suzanne Schettler, Rob Schlising, Gary Schoolcraft, Kevin Shaffer, Teresa Sholars, Dave Showers, Mary Ann Showers, Jake Sigg, Alan Smith, Susan Smith, Sydney Smith, George Snyder, Susan Sommers, Connie Spenger, Fred Sproul, Veva Stansell, G. Ledyard Stebbins, John Stebbins, Diane Steeck, Joan Stewart, Jon M. Stewart, Karen Stokkink, Doug Stone, Pat Stone, Dean Wm. Taylor, Terri Thomas, Tim Thomas, Heather Townsend, Billie Turner, Roy Van de Hoek, Kathy Van Zuuk, Julie Vanderweir, Sue Vrilakas, Phil Wells, Clare Wheeler, Howie Wier, Barbara Williams, Chuck Williams, Jim Wilson, Kirsten Winter, Carl Wishner, Vernal Yadon, and David Zippin. We sincerely apologize to anyone who was inadvertently omitted.

The Natural Diversity Data Base and the Endangered Plant Program of the California Department of Fish and Game, the United States Fish and Wildlife Service, The Nature Conservancy, the United States Bureau of Land Management, the National Park Service, the California Department of Parks and Recreation, the California Department of Transportation, the California Department of Food and Agriculture, the California Energy Commission, and the United States

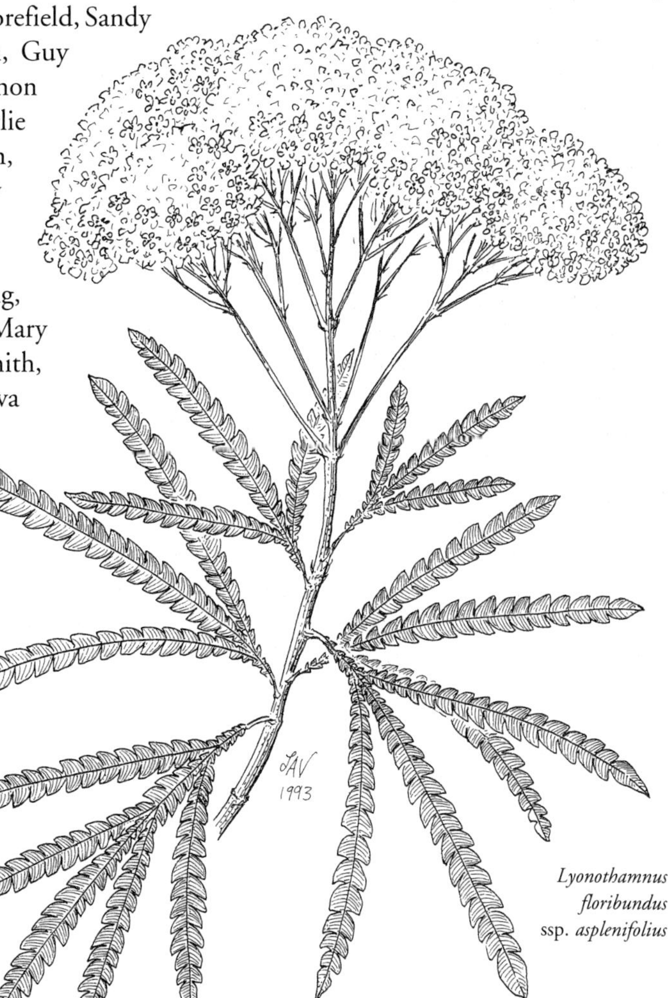

Lyonothamnus floribundus ssp. *asplenifolius*

Forest Service all deserve special recognition because of their many contributions, both of data and ideas. The Forest Service's many dedicated botanists are a nationwide model of agency cooperation and data sharing. The Department of Fish and Game's generosity in sharing office space and facilities since 1982 has made close collaboration with our most important partner a reality.

The curators and staffs of the herbaria at the University of California at Berkeley, Davis, Irvine, and Riverside, the California Academy of Sciences, Humboldt State University, and the Rancho Santa Ana Botanic Garden made their collections and libraries available to us at all times, for which we are most grateful. Barbara Ertter's efforts to provide us with rare plant records from the U.C. Berkeley herbaria are especially appreciated.

Before his untimely death, Editor Jim Hickman kept us alert to relevant botanical discoveries at *The Jepson Manual* Project, furnished us with crucial stacks of draft text, and worked with us to refine rarity codes for the *Manual*. Later, Project Manager Dieter Wilken smoothly continued this valuable exchange. Their cooperation allowed us to get the jump on thousands of changes to our information.

Tom Lupo and Mike Byrne drafted the terrific new quad map. John Ellison provided useful programming and computer assistance.

Mary DeDecker, Betty Guggolz, Jim Jokerst, Tim Messick, Brian Miller, Jim Nelson, and the other members of the Rare Plant Scientific Advisory Committee, along with important guests Ken Berg, Roxanne Bittman, Tim Krantz, John Sawyer, Jim Shevock, John Willoughby, and Rick York each made a major effort to review seemingly endless lists and contribute data and suggestions. Their help has been indispensable.

Dave Bramlet, Steve Boyd, Fred Roberts, Tom Oberbauer, Tim Ross, and Andy Sanders served as a vital southern California outpost of the Rare Plant Program, always willing to analyze lists, come to important meetings, or provide information from their home institutions. In particular, Andy contributed essential summaries of holdings from the U.C. Riverside herbarium for many of the newly recognized rare plants.

The CNPS Rare Plant Coordinators are uncommonly knowledgeable, and as a group provided much of the basic rare plant information for this *Inventory*. The contributions of R. Morgan, Virginia Norris, Brad Olson, Jake Rugyt, and Carol Witham deserve special notice.

The many Rare Plant Program assistants did much of the painstaking research involved in production of this volume, and helped lighten the load of the oft-beleaguered Botanist. Anne Gurnee, Tom Fraser, and Steve Henderson provided valuable general research support. Nina Bicknese and Amy Hiss did superior work while researching the new ecological information on plant blooming time and life form. Becky Yahr was an anchor in the program for two years, and performed essential comparisons with NDDB distributional information. Becky also carefully reviewed each record from the fourth edition for accuracy and made hundreds of necessary corrections. Staci Markos tackled the complex and pivotal tasks of library and herbarium research for the 313 plants new to this addition, and distilled much of this information into records. Staci also carefully prepared many of the provisional review lists. David Tibor distinguished himself as the Program's first official Assistant Botanist. David researched many of the new plants and improved all their records, checked the database for consistency in numerous creative ways, updated information on threats and natural communities from the NDDB, entered rarity information from other states, and entirely reworked the notes for the entries. We thank you all.

Roxanne Bittman, the Natural Diversity Data Base Botanist, merits special thanks for her close cooperation, encouragement, and advice during all phases of this project.

The past editors of the *Inventory*, Bob Powell, Jane Cole, John Sawyer, and especially James P. Smith, Jr., initially compiled much of the information presented here, and this book is as much theirs as ours. And finally, we wish to thank Rick York and Ken Berg for their fundamental contributions to this edition. As the past CNPS Botanists, both Ken and Rick co-edited various editions and paved the way for preparation of this one. These efforts, along with Ken's gracious assistance during the transition to a new CNPS Botanist, made our task much easier.